A DICTIONARY OF PLANT PATHOLOGY

Foot rot of black pepper in Sarawak caused by a species of Phytophthora. (Holliday & Mowat, Phytopath. Pap. 5, 1963.)

A DICTIONARY
OF PLANT PATHOLOGY

PAUL HOLLIDAY
formerly at CAB International Mycological Institute, Kew, UK

The right of the
University of Cambridge
to print and sell
all manner of books
was granted by
Henry VIII in 1534.
The University has printed
and published continuously
since 1584.

CAMBRIDGE UNIVERSITY PRESS

CAMBRIDGE

NEW YORK   PORT CHESTER   MELBOURNE   SYDNEY

Published by the Press Syndicate of the University of Cambridge
The Pitt Building, Trumpington Street, Cambridge CB2 1RP
40 West 20th Street, New York, NY 10011–4211, USA
10 Stamford Road, Oakleigh, Victoria 3166, Australia

First published 1989
Reprinted with corrections 1990
First paperback edition 1992

Printed in Great Britain at the University Press, Cambridge

British Library cataloguing in publication data
Holliday, Paul, 1923–
A dictionary of plant pathology
1. Plants. Pathology
I. Title
581.2

Library of Congress cataloguing in publication data
Holliday, Paul.
A dictionary of plant pathology/Paul Holliday.
 p. cm.
ISBN 0 521 33117 X
1. Plant diseases—Dictionaries.   I. title.
SB728.H65   1988
632′.3′0321—dc19   88–25636   CIP

ISBN 0 52133117 X hardback
ISBN 0 521 42475 5 paperback

FOR EUNICE and in memory of CLIFFORD.
Their art and history have balanced my science

Those who dwell in the clearer light of the next generation will build better than we have done, and will scarcely realise how slowly and painfully many of us have groped about for what seems to them so plain.
ERWIN FRINK SMITH, *Bacteria in relation to plant diseases*, 1905.

Those who believed in genes postulated the existence of an eternal quality R which they could take from wild plants, build into the genetical constitution of cultivated ones, and so make them disease resistant for ever. Those who thought, not in terms of mathematical abstractions but of the green flux of ever changing nature, saw little hope of such permanency, and no end to man's labours in defending the crops upon which he depended for life.
ERNEST CHARLES LARGE, *The advance of the fungi*, 1940.

It is a sobering thought that vast changes in social, educational and fiscal measures are made on less factual evidence than an applied biologist would think necessary before recommending how a farmer should manure his crop or whether he should use a pesticide.
FREDERICK CHARLES BAWDEN, *Annals of Applied biology* **58**:1, 1966.

CONTENTS

ACKNOWLEDGEMENTS

Throughout my preparation of this book I made constant use of the unique facilities at the CAB International Mycological Institute with the kind permission of the director, D. L. Hawksworth. I thank K. J. Hudson, the librarian at the institute, for his help. In the herbarium I was grateful for comments from E. Punithalingam and A. Sivanesan.

I acknowledge the help from other libraries: Bureau of Horticulture and Plantation Crops; CAB International Institutes of Entomology and Parasitology; Commonwealth Forestry Institute, Oxford; The Linnean Society of London; National Institute for Medical Research, London; National Vegetable Research Station, Wellesbourne; Rothamsted Experimental Station, Harpenden; Royal Botanic Gardens, Kew; Scottish Crop Research Institute, Invergowrie. I thank Tom Holliday for obtaining books through the British Public Libraries.

I am most grateful for all the helpful comments received from plant virologists: A. N. Adams (temperate fruit, hop); C. J. Asjes (gladiolus, hyacinth, iris, tulip); O. W. Barnett (clover); K. R. Bock (cassava); L. Bos (bean, broad bean, lucerne, onion, pea, spinach); A. A. Brunt (carnation, chrysanthemum, cucumber, pelargonium); J. I. Cooper (temperate trees and shrubs); P. R. Fridlund (stone fruit); R. E. Gingery (maize); R. M. Goodman (soybean); D. A. Govier (beet); S. A. Hill (potato); B. A. Jaffee (apricot, peach); A. T. Jones (*Rubus*); S. Kubo (tobacco); G. P. Martelli (globe artichoke, grapevine); F. A. van der Meer (lilac); R. J. Milne (tomato), W. P. Mowat (gladiolus, hyacinth, iris, lily, narcissus, tulip); A. F. Murant (carrot, parsley); Y. L. Nene (pigeon pea); R. T. Plumb (temperate cereals); D. C. Ramsdell (*Vaccinium*); D. V. R. Reddy (groundnut); C. N. Roistacher (citrus); E. Shikata (rice); B. J. Thomas (rose); G. Thottappilly (cowpea); J. A. Tomlinson (lettuce, turnip); L. G. Weathers (citrus).

I thank B. D. Harrison for suggesting most of those to whom I sent my draft lists of virus and mycoplasma diseases. M. R. Wilson, and his colleagues at the CAB International Institute of Entomology, were kind enough to check the names of vectors.

For data from time to time I acknowledge the help of: D. J. Allen, K. F. Boswell, J. F. Bradbury, C. M. Brasier, J. Bridge, R. H. Converse, H. Evans, J. N. Gibbs, C. S. Hodges, J. I. M. Irvine, J. A. G. Irwin, P. Jones, V. Lisa, J. W. Martens, I. R. McKinley, W. P. Mowat, C. Prior, J. Raine, N. F. Robertson, S. A. Rudgard, D. E. Shaw, E. G. Simmons, I. M. Smith, J. S. Soosai, H. Vermeulen, P. W. F. de Waard, J. M. Waller and G. A. Zentmyer.

Dilly Bradford undertook the task of producing the typescript with her usual efficiency, and deserves very many thanks. I thank the Cambridge University Press for its care in the final preparation and production. Last, that which should come first, I appreciated the tolerance of my wife Betty.

INTRODUCTION

I have for some time been convinced that a bibliographical list of authoritative names of plant pathogens is long overdue. This book attempts to fill a gap in the literature. These names make up most of the text. But I have put in entries which come under other categories: biography, crop names and pathology, disease names, disorders, fungicide names, specific epithets of bacteria and fungi, taxonomic groups, terminology, toxins and vectors with the pathogens that they transmit.

The fungi are the largest group of pathogens, although only c.8% of the 6000 genera contain them. Some selection was therefore inevitable. Fungus synonymy is so extensive, and some of it controversial, that virtually all synonyms have been omitted. But I have included specific epithets to assist in determining the approved name.

I have tried to give nearly all the proven viruses; they number only c.700. The etiology of many diseases, some caused by viruses and others by mollicutes, is inadequately known. Many of these are given; but many old, and seemingly little used, names are excluded. I have included some synonymy and some different diseases caused by the same virus. But no attempt has been made to list all the diseases caused by any one virus. Such a list would be very useful but is beyond the scope of the book.

In the Prokaryotae the nomenclature of the bacteria has been recently and fully covered. I have mostly selected those species and pathovars that cause the more important diseases. The Mollicutes have only begun to be characterised; but many diseases apparently caused by them are given.

It is perhaps presumptuous for one person to attempt a coverage of such a wide field. But it has its advantages. Those who have helped me with virus and mycoplasma names are in no way responsible for my interpretations of any comments that they may have made. I would be glad to have any notes of errors and omissions in the dictionary.

Richmond, Surrey, 1988.   P.H.

FOR THE USER

The field covered is the same as that of the *Review of Plant Pathology*, except that mycorrhizae are omitted and there is a little more data on nematodes. My main concern is with infectious diseases of the standing crop.

Where the binomial system is used the generic name is given in full at its first mention; but subsequently abbreviated to the initial capital. Where there are formae speciales, pathovars or subspecies, only the third element of the name is given in full after the first entry. Where the binomial system is not used (mycoplasmas and viruses) the first word of a name is usually a common plant name. After its first mention a dash represents it in subsequent entries. Sometimes the first word is the plant's generic term; in this case there is no capital, e.g. solanum. For bacteria and fungi the date after an author's name is not repeated if it is the same as an earlier entry.

In the fungi, mostly Ascomycotina, the information is given under their teleomorphs; but their anamorphs are placed in the order. For viruses and mycoplasmas where plants have 2 commonly used names both names must be referred to. The minimum requirement for entry of the name of a disease, possibly caused by an inadequately described virus, is graft transmission. Therefore this is not so stated. A virus that is adequately characterised has a lower case 'v.' as the last element of the name. This is omitted in an entry for a vector. Where the disease is probably caused by a Mollicute the abbreviation MLO is given; except in the very few cases where a binomial has been erected.

Fungus names, taxonomic groups and terms are based on the publications of, and usage at, the CAB International Mycological Institute. Hawksworth's, *Rev.Pl.Path.* **59**:473, 1980, abbreviations for authors were mostly adhered to. In compiling virus and MLO disease names I began with a shortened, unpublished list by Martyn. This was based on his publications of 1968 and 1971; it was supplemented by Smith's text of 1972 and the Association of Applied Biologists' descriptions. For viroids I first followed Diener 1979 and for bacteria Bradbury 1986.

Negative information, e.g. no vector known or not seedborne, is omitted. The geographical distribution of a pathogen is only mentioned where this appears to be limited; or, for MLO and virus, where the disease appears to have been first described. In most cases, where no distribution is given, the pathogen is probably widespread even if there are few verified records.

The references will lead into most of the main literature. For one that describes a new fungus species the page of the description is given. Therefore this will mostly differ from the first page given in an abstract. Journal titles, except those abbreviated differently, are shortened broadly in line with the *World list of scientific periodicals*. In some entries the final element is a reference(s) to an abstract(s) in the *Rev.appl.Mycol.* or *Rev.Pl.Path.* A volume number followed by a colon denotes a page, and by a comma an abstract number. Abstracts are not given where they contain only bibliographical details, only an outline of contents and for some taxonomy, e.g. new names of pathogens. Where a paper has not been seen, and for most annual reports, only author(s), date and abstract are shown. Fewer references have been noted for bacteria since these have been recently covered by Bradbury 1986, often as abstracts in the reviews cited.

Most characteristic names of diseases caused by bacteria and fungi are included in the order. But undistinctive ones are not. The latter category includes names where a crop name is followed by a common term for a pathogen group, e.g. powdery mildew, rust or smut; or a common disease term, e.g. anthracnose, dieback or leaf spot. For virus disease names see paragraph 3 of the introduction.

The literature for individual crops can be found under their common names. Their binomials are included in the order. Finally, for many pathogens, Koch's postulates have not been adequately satisfied. I must leave the reader to determine whether this is the case or not.

MAJOR TEXTS

These are the most frequently used books and are cited as author(s) or editor(s) and date(s) only. For other books the title is also given.

Agrios, G.N. 1978. *Plant pathology*, edn. 2.
Ainsworth, G.C.; Sussman, A.S. ed. 1965, 1966, 1968. *The fungi. An advanced treatise*, vol. I–III.
—— ; Sparrow, F.K.; Sussman, A.S. ed. 1973. *The fungi. An advanced treatise*, vol. IV, A and B.
—— . 1981. *Introduction to the history of plant pathology*.
Airy Shaw, H.K. 1966. *Willis' dictionary of the flowering plants and ferns*, edn. 7.
Allen, D.J. 1983. *The pathology of tropical food legumes*.
Anderson, H.W. 1956. *Diseases of fruit crops*.
Anon. 1976. *Pest control in rice, PANS Manual 3*, edn. 2.
—— . 1977. *Pest control in bananas, ibid 1*, edn. 3.
—— . 1978. *Pest control in tropical root crops, ibid 4*.
—— . 1981. *Pest control in tropical grain legumes*, Centre for Overseas Pest Research.
—— . 1983. *Pest control in tropical tomatoes*, as above.
Arthur, J.C. 1934. *Manual of the rusts in United States and Canada*.
Atkinson, J.D. 1971. *Diseases of tree fruits in New Zealand*.
Baker, J.J. 1972. *Report on diseases of cultivated plants in England and Wales for the years 1957–1968, Tech.Bull.Min.Agric.Fish.Fd.*, UK.
—— , K.F.; Snyder, W.C. ed. 1965. *Ecology of soilborne plant pathogens*.
Barr, M.E. 1978. *The Diaporthales in North America, Mycologia Memoir*, 7.
Blakeman, J.P. ed. 1981. *Microbial ecology of the phylloplane*.
Booth, C. 1971. *The genus* Fusarium.
Bos, L. 1978. *Symptoms of virus diseases in plants*, edn. 3.
—— . 1983. *Introduction to plant virology*.
Boswell, K.F.; Gibbs, A.J. 1983. *Viruses of legumes. Descriptions and keys from virus identification data exchange*, Australian Natn.Univ.
Bould, C.; Hewitt, E.J.; Needham, P. 1983. *Diagnosis of mineral disorders of plants*, vol. 1, *Principles*.
Boyce, J.S. 1961. *Forest pathology*, edn. 3.
Bradbury, J.F. 1986. *Guide to plant pathogenic bacteria*.

Brooks, F.T. 1953. *Plant diseases*, edn. 2.
Browne, F.G. 1968. *Pests and diseases of forest plantation trees*.
Bruehl, G.W. ed. 1975. *Biology and control of soilborne plant pathogens*.
Buczaki, S.T. ed. 1983. *Zoosporic plant pathogens, a modern perspective*.
Burnett, J.H. 1976. *Fundamentals of mycology*, edn. 2.
Butler, E.J.; Jones, E.G. 1949. *Plant pathology*.
Byrde, R.J.W.; Willets, H.J. 1977. *The brown rot fungi of fruit, their biology and control*.
Callow, J.A. ed. 1983. *Biochemical plant pathology*.
Cannon, P.F.; Hawksworth, D.L.; Sherwood-Pike, M.A. 1985. *The British Ascomycotina. An annotated checklist*.
Carmichael, J.W.; Bryce Kendrick, W.; Conners, I.L.; Sigler, L. 1980. *Genera of Hyphomycetes*.
Chattopadhyay, S.B. 1967. *Diseases of plants yielding drugs, dyes and spices*.
Coley-Smith, J.R.; Verhoeff, K.; Jarvis, W.R. 1980. *The biology of* Botrytis.
Cook, A.A. 1975. *Diseases of tropical and subtropical fruits and nuts*.
Cooper, J.I. 1979. *Virus diseases of trees and shrubs*.
Couch, H.B. 1962. *Diseases of turf grasses*.
Cowan, S.T. 1978. *A dictionary of microbial taxonomy*, L.R. Hill ed., a revised and enlarged edition of Cowan's *A dictionary of microbial taxonomic usage*, 1968.
Coyier, D.L.; Roane, M.K. ed. 1986. *Compendium of rhododendron and azalea diseases*.
Cummins, G.B. 1971. *The rust fungi of cereals, grasses and bamboos*.
—— . 1978. *Rust fungi on legumes and composites in North America*.
Daly, J.M.; Deverall, B.J. ed. 1983. *Toxins and plant pathogenesis*.
Daniels, M.J.; Markham, P.G. ed. 1982. *Plant and insect mycoplasma techniques*.
Day, P.R.; Jellis, G.J. ed. 1987. *Genetics and plant pathogenesis*.
Dennis, C. ed. 1983. *Post-harvest pathology of fruits and vegetables*.
—— , R.W.G. 1978. *British Ascomycetes*, edn. 3.
Dickinson, C.H.; Lucas, J.A. 1982. *Plant pathology and plant pathogens*.
Dickson, J.G. 1956. *Diseases of field crops*, edn. 2.
Diener, T.O. 1979. *Viroids and viroid diseases*.
Dixon, G.R. 1981. *Vegetable crop diseases*.

Drew-Smith, J. 1965. *Fungal diseases of turf grasses*, edn. 2 by N. Jackson & J. Drew-Smith.

Durbin, R.D. ed. 1981. *Toxins in plant disease.*

Ebbels, D.L.; King, J.E. ed. 1979. *Plant health. The scientific basis for administrative control of plant diseases and pests.*

Ellis, M.B. 1971. *Dematiaceous Hyphomycetes.*

——. 1976. *More dematiaceous Hyphomycetes.*

Fahy, P.C.; Persley, G.J. ed. 1983. *Plant bacterial diseases. A diagnostic guide.*

Fawcett, H.S. 1936. *Citrus diseases and their control*, edn. 2.

Feakin, S.D. ed. 1973. *Pest control in groundnuts, PANS Manual 2*, edn. 3.

Fischer, G.W. 1951. *The smut fungi. A guide to the literature with bibliography.*

——. 1953. *Manual of North American smut fungi.*

—— ; Holton, C.S. 1957. *Biology and control of the smut fungi.*

Fletcher, J.T. 1984. *Diseases of greenhouse plants.*

Francki, R.I.B. ed. 1985. *The plant viruses*, vol. 1, *Polyhedral virions with tripartite genomes.*

—— ; Milne, R.G.; Hatta, T. 1985. *Atlas of plant viruses*, 2 vol.

Frederiksen, R.A. ed. 1986. *Compendium of sorghum diseases.*

Gareth-Jones, D.; Clifford, B.C. 1983. *Cereal diseases, their pathology and control*, edn. 2.

Garnsey, S.M.; Timmer, L.W.; Dodds, J.A. ed. 1984. *Proc. 9th Conf.Int.Org.Citrus virologists.*

Garrett, S.D. 1970. *Pathogenic root-infecting fungi.*

Gäumann, E. 1950. *Principles of plant infection*, English edn. by W.B. Brierley.

Gibbs, A.; Harrison, B. 1976. *Plant virology. The principles.*

Gibson, I.A.S. 1975. *Diseases of forest trees widely planted as exotics in the tropics and southern hemisphere*, part 1, CAB International.

Gilmer, R.M.; Moore, J.D.; Nyland, G.; Welsh, M.F.; Pine, T.S. ed. 1976. *Virus diseases and noninfectious disorders of stone fruits in North America, USDA Agric.Handb. 437.*

Graham, J.H.; Frosheiser, F.I.; Stuteville, D.L.; Erwin, D.C. 1979. *Compendium of alfalfa diseases.*

Hagedorn, D.J. ed. 1984. *Compendium of pea diseases.*

Harris, K.F.; Maramorosch, K. ed. 1980. *Vectors of plant pathogens.*

—— ; —— . ed. 1982. *Pathogens, vectors and plant diseases: approaches to control.*

Hawksworth, D.L.; Sutton, B.C.; Ainsworth, G.C. 1983. *Ainsworth & Bisby's dictionary of the fungi*, edn. 7.

Heald, F.D. 1933. *Manual of plant diseases*, edn. 2.

Heitefuss, R.; Williams, P.H. ed. 1976. *Physiological plant pathology*, vol. 4, *Encyclopedia of plant physiology.*

Hepting, G.H. 1971. *Diseases of forest and shade trees of the United States. USDA Agric.Handb. 386.*

Hill, D.S.; Waller, J.M. 1982. *Pests and diseases of tropical crops*, vol. 1. *Principles and methods of control.*

Holliday, P. 1980. *Fungus diseases of tropical crops.*

Holton, C.S.; Fischer, G.W.; Fulton, R.W.; Hart, H.; McCallan, S.E.A. ed. 1959. *Plant pathology. Problems and progress 1908–1958.*

Hooker, W.J. ed. 1981. *Compendium of potato diseases.*

Horsfall, J.G.; Dimond, A.E. ed. 1959, 1960. *Plant pathology. An advanced treatise*, 3 vol.

—— ; Cowling, E.B. ed. 1977–80. *Plant disease. An advanced treatise*, 5 vol.

Horst, R.K. 1983. *Compendium of rose diseases.*

Howes, F.N. 1974. *A dictionary of useful and everyday plants and their common names.*

Hughes, C.J.; Abbott, E.V.; Wismer, C.A. ed. 1964. *Sugarcane diseases of the world*, vol. 2.

Jenkyn, J.F.; Plumb, R.T. ed. 1981. *Strategies for the control of cereal disease.*

Johnston, A.; Booth, C. ed. 1983. *Plant pathologist's pocket book*, edn. 2.

Knorr, L.C. 1973. *Citrus diseases and disorders.*

Kommedahl, T.; Williams, P.H. ed. 1983. *Challenging problems in plant health.*

Kranz, J.; Schmutterer, H.; Koch, W. ed. 1977. *Diseases, pests and weeds in tropical crops.*

Krieg, N.R. ed. 1984. *Bergey's manual of systematic bacteriology*, edn. 9, vol. 1.

Kurstak, E. ed. 1981. *Handbook of plant virus infections. Comparative diagnosis.*

Large, E.C. 1940. *The advance of the fungi.*

Lucas, G.B. 1975. *Diseases of tobacco*, edn. 3.

Maas, J.L. ed. 1984. *Compendium of strawberry diseases.*

Mace, M.E.; Bell, A.A.; Beckman, C.H. ed. 1981. *Fungal wilt diseases of plants.*

MacFarlane, H.M. compiler, 1968. *Review of Applied Mycology. Plant host-pathogen index to volumes 1–40 (1922–1961).*

Manners, J.G. 1982. *Principles of plant pathology.*

Maramorosch K. ed. 1969. *Viruses, vectors and vegetation.*

—— ; Harris, K.F. ed. 1981. *Plant diseases and vectors, ecology and epidemiology.*

—— ; Raychaudhuri, S.P. ed. 1981. *Mycoplasma diseases of trees and shrubs.*

—— ; McKelvey, J.J. ed. 1985. *Subviral pathogens of plants and animals.*

Martin, J.P.; Abbott, E.V.; Hughes, C.G. ed. 1961. *Sugarcane diseases of the world*, vol. 1.

Martyn, E.B. ed. 1968, 1971. *Plant virus names. An annotated list of names and synonyms of plant viruses and diseases. Phytopath.Pap.* 9 and suppl. 1.

Mathre, D.E. ed. 1982. *Compendium of barley diseases.*

Matthews, R.E.F. 1981. *Plant virology*, edn. 2.

——. 1982. *Classification and nomenclature of viruses.* 4th report of the international committee on taxonomy of viruses, *Intervirology* **17**(1–3).

Moore, W.C. 1959. *British parasitic fungi.*

——. 1979. *Diseases of bulbs, Min.Agric.Fish.Fd.UK Ref.bk.* HPD1, replaces Bull.117, 1939; revision by A.A. Brunt; D. Price; A.R. Rees; J.S.W. Dickens ed.

Mordue, J.E.M.; Ainsworth, G.C. 1984. *Ustilaginales of the British Isles, Mycol.Pap.* 154.

Mundkur, B.B.; Thirumalachar, M.J. 1952. *Ustilaginales of India.*

Neergaard, P. 1977. *Seed pathology*, 2 vol.

Nelson, P.E.; Toussoun, T.A.; Cook, R.J. ed. 1981. *Fusarium: diseases, biology and taxonomy.*

Nowell, W. 1923. *Diseases of crop plants in the lesser Antilles.*

Ogilvie, L. 1969. *Diseases of vegetables, Min.Agric.Fish.Fd.UK Bull.* 123, edn. 6.

O'Rourke, C.J. 1976. *Diseases of grasses and forage legumes in Ireland.*

Ou, S.H. 1985. *Rice diseases*, edn. 2.

Palti, J. 1981. *Cultural practices and infectious crop diseases.*

Peace, T.R. 1962. *Pathology of trees and shrubs.*

Phillips, D.H.; Burdekin, D.A. 1982. *Diseases of forest and ornamental trees.*

Plumb, R.T.; Thresh, J.M. ed. 1983. *Plant virus epidemiology.*

Porter, D.M.; Smith, D.H.; Rodriguez-Kabana, R. ed. 1984. *Compendium of peanut diseases.*

Purseglove, J.W. 1968, 1972. *Tropical crops. Dicotyledons*, 2 vol. *Monocotyledons*, 2 vol.

Ramakrishnan, T.S. 1963. *Diseases of millets.*

Reuther, W.; Calavan, E.C.; Carman, G.E. ed. 1978. *The citrus industry*, vol. 4, *Crop protection.*

Rich, A.E. 1983. *Potato diseases.*

Richardson, M.J. 1979. *An annotated list of seed-borne diseases, Phytopath.Pap.* 23, edn. 3, and suppl. 1, 1981; 2, 1983.

Roberts, D.A.; Boothroyd, C.W. 1984. *Fundamentals of plant pathology*, edn. 2.

Robinson, R.A. 1976. *Plant pathosystems.*

Safeeulla, K.M. 1976. *Biology and control of the downy mildews of pearl millet, sorghum and finger millet.*

Sampson, K.; Western, J.H. 1954. *Diseases of British grasses and herbage legumes*, edn. 2.

Scaif, A.; Turner, M. 1983. *Diagnosis of mineral disorders of plants*, vol. 2. *Vegetables.*

Scott, P.R.; Bainbridge, A. ed. 1978. *Plant disease epidemiology.*

Sherf, A.F.; MacNab, A.A. 1986. *Vegetable diseases and their control*, edn. 2 of Chupp & Sherf, same title.

Shurtleff, M.C. ed. 1980. *Compendium of corn diseases*, edn. 2.

Simmonds, N.W. ed. 1976. *Evolution of crop plants.*

Sinclair, J.B. ed. 1982. *Compendium of soybean diseases*, edn. 2.

Sivanesan, A. 1984. *The bitunicate Ascomycetes and their anamorphs.*

—— ; Waller, J.M. 1986. *Sugarcane diseases, Phytopath.Pap.* 29.

Smiley, R.W. 1983. *Compendium of turfgrass diseases.*

Smith, K.M. 1972. *A textbook of plant virus diseases*, edn. 3.

——, W.H. 1970. *Tree pathology: a short introduction.*

Spencer, D.M. ed. 1978. *The powdery mildews.*

——. ed. 1981. *The downy mildews.*

Sprague, R. 1950. *Diseases of cereals and grasses in North America (fungi, except smuts and rusts).*

Staples, R.C.; Toenniessen, G.H. ed. 1981. *Plant disease control: resistance and susceptibility.*

Starr, M.P.; Stolp, H.; Trüper, H.G.; Balows, A.; Schlegel, H.G. ed. 1981. *The prokaryotes. A handbook on habits, isolation and identification of bacteria*, 2 vol.

Stevens, R.B. 1974. *Plant disease.*

Stipes, R.J.; Campana, R.J. ed. 1981. *Compendium of elm diseases.*

Stover, R.H. 1972. *Banana, plantain and abaca diseases.*

Sutton, B.C. 1980. *The Coelomycetes.*

Talbot, P.H.B. 1971. *Principles of fungal taxonomy.*

Tarr, S.A.J. 1962. *Diseases of sorghum, Sudan grass and broomcorn.*

——. 1972. *Principles of plant pathology.*

Thorold, C.A. 1974. *Diseases of cocoa.*

Toussoun, T.A.; Bega, R.V.; Nelson, P.E. ed. 1970. *Root diseases and soil-borne pathogens.*

Turner, P.D.; Bull, R.A. 1967. *Diseases and disorders of the oil palm in Malaysia.*

——. 1981. *Oil palm diseases and disorders.*

Usher, G. 1974. *A dictionary of plants used by man.*

Van Regenmortel, M.H.V.; Fraenkel-Conrat, H. ed. 1986. *The plant viruses*, vol. 2, *The rod shaped viruses.*

Walker, J.C. 1952. *Diseases of vegetable crops.*
—— . 1969. *Plant pathology,* edn. 3.
Walkey, D.G.A. 1985. *Applied plant virology.*
Waterson, A.P.; Wilkinson, L. 1978. *An introduction to the history of virology.*
Watkins, G.M. ed. 1981. *Compendium of cotton diseases.*
Wehmeyer, L.E. 1975. *The pyrenomycetous fungi, Mycologia Memoir 6.*
Western, J.H. ed. 1971. *Diseases of crop plants.*
Wheeler, B.E.J. 1969. *An introduction to plant diseases.*
Whetzel, H.H. 1918. *An outline of the history of phytopathology.*
Whitcomb, R.F.; Tully, J.G. ed. 1979. *The mycoplasmas,* vol. 3, *Plant and insect mycoplasmas.*

Whitney, E.D.; Duffus, A.E. ed. 1986. *Compendium of beet diseases and insects.*
Wiese, M.V. 1987. *Compendium of wheat diseases,* edn. 2.
Wilson, M.; Henderson, D.M. 1966. *British rust fungi.*
Wolfe, M.S.; Caten, C.E. ed. 1987. *Populations of plant pathogens. Their dynamics and genetics.*
Wood, R.K.S. 1967. *Physiological plant pathology.*
—— ; Ballio, A.; Graniti, A. ed. 1972. *Phytotoxins in plant diseases.*
—— ; Jellis, G.J. ed. 1984. *Plant disease: infection, damage and loss.*
Wormald, H. 1955. *Diseases of fruits and hops,* edn. 3.
Ziller, W.G. 1974. *The tree rusts of western Canada.*

ABBREVIATIONS AND CONVENTIONS

AAB *Annals of Applied Biology*
abs. abstract
anam. anamorph(s)
A.R.Phytop. *Annual Review of Phytopathology*
asco. ascospore(s)
assoc. associated with
av. average
bean *Phaseolus vulgaris*
c. circa (about, approximately)
cf. compare
CJB *Canadian Journal of Botany*
Coelom. Coelomycete(s)
con. conidia
cv(s). cultivar(s)
Descr.B. ⎫ *Descriptions of pathogenic fungi and*
Descr.F. ⎭ *bacteria* issued by the CAB International
Mycological Institute
Descr.N. *Descriptions of plant parasitic nematodes*,
issued by the CAB International Institute of
Parasitology
Descr.V. *Descriptions of plant viruses*, edited by the
Association of Applied Biologists
diam. diameter
DNA deoxyribonucleic acid
ds double stranded
ed. editor(s)
edn. edition
e.g. for example
et al. and others
FBPP *A guide to the use of terms in plant
pathology*, *Phytopath.Pap.* 17, 1973, Federation of
British Plant Pathologists which, in 1981, became
the British Society for Plant Pathology
fide according to
filament an elongated plant virus particle which is
to some degree waved or flexuous, cf. rod
FRS Fellow of the Royal Society
f.sp.(ff.sp.) forma(e) specialis(es)
Hyphom. Hyphomycete(s)
ibid the same place
i.e. that is
loc.cit. place (already) cited
MLO an organism(s) like a mycoplasma(s)
Myz.pers. *Myzus persicae*
nm nanometre, 10^{-9} metre
PD *Plant Disease*

PDR *Plant Disease Reporter*
Phytop. *Phytopathology*
poss. possible
prob. probable
pv(s) pathovar(s) of a bacterium
q.v. which see
race physiologic race of a fungus
RNA ribonucleic acid
rod an elongated plant virus particle which is
straight, cf. filament
sp., spp., ssp. species (singular and plural),
subspecies
ss single stranded
str(s). strain(s)
TBMS *Transactions of the British Mycological
Society*
teleom. teleomorph(s)
temp. temperature
transm. transmission, transmits, transmitted by
transm. sap the mechanical transmission of an
agent causing a disease, mostly a plant virus, and
using plant sap
UK United Kingdom
μm micrometre, 10^{-6} metre
Univ. university(ies)
USA United States of America
USDA United States Department of Agriculture
USSR Union of Soviet Socialist Republics
v. virus, used only in conjunction with the name of
an adequately characterised plant virus; and only
where a viral etiology has been reasonably
established
var(s) variety(ies)
vol. volume(s) of a book
< less than
> more than
± more or less
= is a synonym of
( ) is caused by, e.g. silver leaf (*Chondrostereum
purpureum*); tulip Augusta (tobacco necrosis v.)
* see addenda

(The usual abbreviations for metric measurements of
length and symbols for chemical elements are used. All
temperatures are in degrees celsius or centigrade and C is
omitted.)

A

Asterisks indicate that further information relating to the entry exists in the Addenda, beginning on page 364.

abaca, *Musa textilis*, see banana.
— **bunchy top** = banana bunchy top.
— **mosaic**, a str. of sugarcane mosaic v.
Abacarus hystrix, transm. agropyron mosaic, ryegrass mosaic; see *Hirsutella*.
abelia latent v. Waterworth et al., *Phytop.* **65**:891, 1975; prob. a str. of eggplant mosaic v., USA (Maryland), *A. grandiflora* [**55**,758].
abeliophyllum mottle Schmelzer, *Zentbl.Bakt. ParasitKde.* 2, **129**:139, 1974; transm. sap, Germany [**54**,1183].
abelmoschi, *Pseudocercospora*.
Abelmoschus esculentus, lady's fingers, okra.
Abies, firs.
abieticola, *Grovesiella*.
abietina, *Gremmeniella*, *Verticicladiella*.
abietis, *Chrysomyxa*, *Cytospora*, *Valsa*, *Xenomeris*.
abietis-concoloris, *Lirula*.
abiotic, of a disease, one with a cause other than a living organism, i.e. a disorder q.v.
abnormal defoliation, excess Mn in Satsuma mandarin, Japan, Ishihara, *JARQ* **7**:38, 1973 [**53**,954].
— **leaf fall** (*Phytophthora botryosa*, *P. meadii*, *P. palmivora*) rubber.
— **sepal**, prob. genetic cause, pumpkin, Ragozzino, *Inftore. fitopatol.* **28**(9):13, 1978 [**58**,3528].
abrasive, fine particles of a material such as charcoal, carborundum or diatomaceous earth ('celite') in the inoculum, or dusted on leaves before inoculation, to aid the mechanical transm. of a plant virus, after FBPP.
abrus witches' broom Yang, Chen & Wang 1985; MLO assoc., Taiwan, *A. precatorius* [**65**,3202].
abundans, *Embellisia*.
*****abutilon mosaic**, see malvaceae infectious chlorosis and Cooper 1979.
Abyan root rot (*Thanatephorus cucumeris*) cotton.
acacia ringspot Marras, *Riv. Patol. Veg.* **2**:277, 1962 [**42**:495].
Acalymma, transm. wild cucumber mosaic; see *Fusarium oxysporum* f.sp. *melonis*.
acalypha little leaf van Velsen, *New Guin. agric. J.* **14**:128, 1961; New Britain, *A. wilkesiana* [**42**:127].
— **yellow mosaic** Chenulu & Phatak, *Curr. Sci.* **34**:321, 1965; transm. *Bemisia tabaci*, India, *A. indica* [**44**,3063].
acanthospermum little leaf Raju & Muniyappa,

Phytopath. Z. **102**:232, 1981; MLO assoc., India (Hebbal), *A. hispidum* [**61**,4568].
Acer, maples; *A. pseudoplatanus*, sycamore.
Aceratagallia calcaris, transm. beet yellow vein; *A. sanguinolenta*, transm. potato yellow dwarf.
Aceria cajani, transm. pigeon pea sterility mosaic; *A. ficus*, transm. fig. mosaic; *A. tritici*, transm. cereal spotting; *A tulipae*, transm. garlic mosaic, onion mosaic, wheat spot mosaic, wheat streak mosaic.
acer ribbon pattern Schmelzer, H.E. & H.B. Schmidt, *Arch. Forstw.* **15**:107, 1966; Europe [**46**,442].
acerina, *Melasmia*, *Mycocentrospora*; **acerinum**, *Rhytisma*; **aceris**, *Oidium*.
acer variegation Brierley, *PDR* suppl. **150**:414, 1944; Europe, Japan [**23**:438].
acervulus, an immersed conidioma, a flat layer of pseudoparenchyma on which con. are formed whilst still covered by host tissue.
aceti, *Acetobacter*.
Acetobacter Beijerinck 1898, Gram negative, obligately aerobic, when motile flagella peritrichous, causes postharvest damage, e.g. pineapple pink disease and brown rot of apple and pear; *A. aceti* (Pasteur) Beijerinck and *A. pasteurians* (Hansen) Beijerinck 1916; Bradbury 1986.
achyranthes mosaic Verma & Singh, *Phytopath. Z.* **73**:375, 1972; transm. aphid, India (N.); *A. aspersa* [**51**,4753].
acicola, *Lecanostica*, *Scirrhia*.
acid fast, of bacteria, resistance to decolourisation by mineral acids; normally applied to bacteria in which steaming carbol fuchsin is applied to stain organisms that are not readily stained with simple dyes.
— **rain**, and see air pollution, waldsterben; rain that contains as principal components the hydrolysed end products from oxidised N or S and halogen compounds; van den Burg, *Neth. Bosbouw Tijdschr.* **55**:371, 1983, forest decline in central Europe; Binns, *Res. Devel. Pap. For. Commission* UK 134, 1984; Evans, *A.R.Phytop.* **22**:397, 1984; *Bot. Rev.* **50**:449, 1984.
aconitum mosaic A.E. & E.P. Protsenko 1964 [**45**,1712].
Aconurella prolixa, transm. bermuda grass etched line.

acquired resistance, a non-inherited resistance response in a normally susceptible host following a predisposing treatment; acquired immunity should only be used if there is no degree of infection in the host.

acquisition, virus entry of, or its attachment to, a vector; —— **access time**, the length of time that a test vector is given access to a virus source in transm. tests; it is not implied that the vector feeds during all or any of this time; —— **feeding**, the feeding of a vector on a virus source in transm. tests; —— **feeding time**, the time that a vector feeds on a virus source in transm. tests; —— **threshold period**, the minimum time necessary for a vector to spend on a virus source in transm. tests to obtain an infective charge of the virus, after FBPP.

Acremonium Link 1809; Hyphom.; conidiomata hyphal, conidiogenesis enteroblastic, phialidic; con. aseptate, hyaline, in chains or heads, often in soil. Gams, Cephalosporium *artig Schimmelpilze*, 1971, keys; *TBMS* **64**:389, 1975; Samuels, *N.Z.J. Bot.* **14**:231, 1976, teleom.; Chesson et al., *TBMS* **70**:345, 1978, electrophoresis; Siegel et al., *PD* **69**:179, 1985, as endophytes in grasses causing animal toxicoses [**54**,4830; **56**,2396].

A. boreale Smith & Davidson, *CJB* **57**:2138, 1979; orange sclerotia, antagonist of snow mould complexes; Canada, Norway [**59**,3293].

A. crotocinigenum (Schol-Schwarz) W. Gams, Uchida & Aragaki, *PD* **66**:421, 1982; leaf spot of *Syngonium podophyllum* [**61**,6432].

A. kiliense Grütz 1925; Brady, *Descr.F.* 741, 1983; poss. causing maize black bundle.

*A. strictum W. Gams 1971; poss. causing disease as above; Natural et al., *PD* **66**:863, 1982, sorghum wilt; Mexico, USA (S.) [**62**,183].

A. typhinum, teleom. *Epichloë typhina.*

A. zonatum (Saw.) W. Gams 1971; Hawksworth, *Descr.F.* 502, 1976; zonate leaf spot, plurivorous.

Acrocalymma Alcorn & Irwin, *TBMS* **88**:163, 1987; Coelom.; conidiomata pycnidial, conidiogenous cells phialidic, con. becoming 1–3 septate. *A. medicaginis* Alcorn & Irwin loc.cit., con. 11–21 × 3.5–5 μm, with a mucilagenous appendage, globose to hemispherical, at each end, 2.5–3.5 × 2–4 μm; root and crown rot of lucerne in Australia; at first attributed to *Stagonospora meliloti* (Lasch) Petrak, see Irwin, *Australasian Pl.Path.Soc.Newsl.* 1:29, 1972 [**66**,3868].

Acroconidiella Lindq. & Alippi 1964; Hyphom.; conidiomata hyphal, conidiogenesis enteroblastic, tretic; con. multiseptate, pigmented, echinulate, solitary. *A. tropaeoli* (Bond) Lindq. & Alippi; Ellis, *Descr.F.* 161, 1968; nasturtium leaf spot, seedborne.

acromania, cotton, as for crazy top q.v.

acronym, letters, usually capitals, strung together and made up of the first letters of several words. They are widely used, usefully so, for plant viruses, i.e. sigla q.v. But since they are also used for plant viroids confusion may occur. Acronyms are sometimes used very undesirably, e.g. FOP or fop for *Fusarium oxysporum* f.sp. *phaseoli*. Their excessive use denotes at best laziness and at worst illiteracy.

Acrophialophora Edward, *Mycologia* **51**:784, 1959, published 1961; Hyphom.; conidiomata hyphal, conidiogenesis phialidic, enteroblastic; con hyaline, echinulate or verruculose; Samson & Mahmood, *Acta bot. neerl.* **19**:804, 1970, key. *A. fusispora* (Saksena) M.B. Ellis 1971, con. in long chains, sometimes pale brown, 6–11 × 3.5–5 μm; Purkayastha & Chakraborty, *PD* **65**:362, 1981, synergism with *Fusarium oxysporum* f.sp. *ciceris* [**61**,511].

Acrospermum Tode 1790; near Clavicipitaceae; Sherwood, *Mycotaxon* **5**:38, 1977. *A viticola* Ikata in Ikata & Hitomi, *Ann.Phytopath.Soc.Japan* **2**:39, 1931, causes zonate leaf spot of mulberry; anam. *Gonatophragmium mori* (Saw.) Deighton 1969; con. usually 3 septate, mostly 13–26 × 4.5 μm; Takahashi & Teramine, *ibid* **52**:404, 1986 [**66**,1543].

Acrothecium (Corda) Preuss 1851; Hyphom.; conidiomata hyphal, con. hyaline, 2 to many celled; not accepted by Carmichael et al. 1980. *A. carotae* Årsvoll 1965, decay of stored carrots in Norway [**45**,930].

Actinidia chinensis, Chinese gooseberry, Kiwi fruit.

actinidiae, *Diaporthe.*

Actinomyces Harz 1877, see *Streptomyces.*

active ingredient, a.i., the active component of a formulated product, FBPP.

aculeata, *Tunstallia*; **acuminatum**, *Fusarium*; **acutatum**, *Colletotrichum.*

acute dieback (*Cryphonectria cubensis*) clove. —— **phase**, or shock phase, applied to the initial severe symptoms caused by a virus infection. If the host survives this may be followed by a chronic phase where new growth shows less severe symptoms or be symptomless.

Acyrthosiphon pisum, transm. alfalfa latent, alfalfa Michigan, bean leaf roll, bean yellow vein banding, broad bean severe chlorosis, cherry latent, chicory blotch, milk vetch dwarf, narcissus latent, pea enation mosaic, pea seedborne mosaic, pea streak, red clover mild mosaic, soybean mosaic, subterranean clover red leaf.

A. pelargonii zerozalphum, transm. filaree red leaf.

adanensis, *Septoria.*

Adansonia digitata, Attafuah & Tinsley, *AAB*

46:20, 1958, transm. of poss. viruses to *A. digitata* and cacao by *Planococcoides njalensis*, 3 symptom patterns, Ghana [**37**:582].

adhatoda mosaic Verma, *Z. PflKrankh. PflPath. PflSchutz.* **81**:611, 1974; India, *A. vasica* [**54**,4071].

adhesorium, an organ formed from a resting zoospore of *Plasmodiophora* for attachment to, and penetration of, the host; Aist & Williams, *CJB* **49**:2023, 1971 [**51**,2959].

adiposa, *Ceratocystis*.

adjuvant, material added to improve some chemical or physical property, e.g. of a plant protectant, or a biological property, e.g. to improve antibody response to antigen, FBPP.

adlay, *Coix lachryma-jobi*.

adzamethica, *Ascochyta*.

adzuki bean, *Vigna angularis*, adzuki bean mosaic (cowpea aphid-borne mosaic v.).

Aecidium Pers. 1799, Uredinales; anam. genus for aecidial and pycnidial states; often used where there is little knowledge of life histories, and as an anam. name even where there is a named teleom.; Notoatmodo 1964 described *A. fragiforme* Ces. on *Agathis dammara*, an important timber in Indonesia. *A. mori* Barclay causes a destructive disease of mulberry in Taiwan; the aecidial state is repeated asexually on *Morus*, there are no pycnidia; Kaneko, *Trans. mycol. Soc. Japan* **14**:294, 1973; Wang, *Natn. Sci. Council Monthly* **8**:604, 1980 [**44**,2636; **47**,3228; **53**,3108; **60**,4588].

aeciospore, aecidiospore, stage 1 in the life cycle of a rust fungus, formed in aecia or aecidia, and resulting from dikaryotisation via pycnidiospores of opposite mating types. Aeciospores germinate to give a dikaryotic mycelium, typically they are catenate and verrucose; Sato et al. *Trans. mycol. Soc. Japan* **23**:51, 1982, surface structure; *TBMS* **85**:223,1985, morphology of aecia [**62**,587; **64**,5324].

aeglopsis vein clearing Nour Eldin, *Fruits* **18**:548, 1963.

aegopodium chlorotic mosaic Kochman & Stachyra 1957; Poland [**37**:340].

aeonium foliar variegation Reiter, *Protoplasma* **45**:509, 1956; Austria [**35**:747].

aerial blight (*Thanatephorus cucumeris*) soybean and other crops.

—— **photography**, this can reveal disease patterns not seen from the ground; it increases the scale of investigation, and poses questions that can only be answered by field work but which may never have been asked; Brenchley, *A.R.Phytop.* **6**:1, 1968; French & Meyer 1978; Toler, *PD* **65**:24, 1981 [**60**,2782].

aerobe, an organism needing free O_2 for growth.

aerosol, a dispersion, in a gas or gases, of droplets of diam. 0.1–5 μm, FBPP.

aesculi, *Fusicoccum, Guignardia*; **aesculicola**, *Leptodothiorella*.

Aesculus hippocastanum, horse chestnut.

aesculus line pattern, see horse chestnut yellow mosaic.

—— **necrosis** Smolák, *Biologia Pl.* **5**:59, 1963; Czechoslovakia [**42**:496].

aflaroot (*Aspergillus flavus*) groundnut.

afluidal variants, of bacteria, the loss of fluidity, i.e. no formation of copious extracellular polysaccharide, has been positively correlated with a loss of virulence, see *Pseudomonas solanacearum*, Woods.

African rice, *Oryza glaberrima*.

Agallia constricta, transm. potato yellow dwarf, wound tumour; *A. quadripunctata* transm. PYD.

Agalliopsis novella, transm. clover club leaf, wound tumour.

Agaricales, Hymenomycetes, the gill fungi, mushrooms; few plant pathogens, see *Armillaria, Crinipellis, Marasmiellus, Marasmius, Mycena*.

agarici, *Pseudomonas*.

Agaricus brunnescens, common cultivated mushroom.

Agave, sisal.

aggressin, see toxin.

aggressiveness, aggressivity, FBPP rejected these terms and considered them to be synonymous with pathogenicity and virulence, respectively.

Aglenchus costatus (de Man) Meyl 1961; Andrássy, *Descr.N.* 80, 1976; cosmopolitan, prob. causes little damage to plants.

Agonomycetales, mycelia sterilia, con. absent, chlamydospores in some genera, may be states of other fungi, see *Rhizoctonia, Sclerotium*.

agricultural development, effects on crop diseases, Waller, *Trop. Pest Management* **30**:86, 1984.

Agrobacterium Conn 1942; Bradbury 1986 fully discussed taxonomic characteristics and problems. Bacteria, aerobic, Gram negative, non-pigmented and non-sporing rods, motile, 1–6 peritrichous flagella. Three spp. are recognised on biochemical characteristics but without reference to any plant pathogenicity. The main diseases caused on many crops are crown gall and hairy root. The induction of disease is controlled by genes carried on plasmids which can be lost or exchanged between cells of different spp. Characteristics controlled in this way cannot be used to distinguish spp. Where a plasmid is absent the bacterium is saprophytic. The presence of a Ti plasmid causes pathogenicity shown by the crown gall symptom; likewise hairy root is mediated by an Ri plasmid. The genus has also been divided into biotypes or biovars. *A. rhizogenes* (Riker et al.) Conn (biovar 2) strs. mostly cause hairy root, but strs. with the crown gall Ti plasmid occur. *A. rubi* (Hildebrand) Starr & Weiss 1943 causes raspberry cane gall, not to be confused with *A. tumefaciens* (Smith & Townsend) Conn (biovar 1) on *Rubus*. This last sp. is saprophytic or causes crown gall; strs. causing hairy root are much fewer. *A. radiobacter* is a synonym of *A. tumefaciens* since the former differs only in that the tumour-inducing Ti plasmid is absent. See tumour, and: Cleene & de Ley, *Bot. Rev.* 47:147, 1981, hosts & hairy root; Merlo, *Adv. Pl. Path.* 1:139, 1982, crown gall; Daly & Knoche, *ibid*:83, crown gall; Drummond in Callow 1983, crown gall; Kerster & de Ley in Krieg 1984; Nester et al. *A. Rev. Pl. Physiol.* 35:387, 1984, crown gall; Cleene, *Phytopath. Z.* 113:81, 1985, *A. tumefaciens* on monocotyledons.

agrocin 84, a bacteriocin q.v., from the nonpathogenic str. 84 of *Agrobacterium tumefaciens* used in the control of crown gall by Kerr 1972; Moore & Warren, *A.R.Phytop.* 17:163, 1979.

agropyri, *Urocystis*.

agropyron mosaic v. McKinney 1937; Slykhuis, *Descr.V.* 118, 1973; Potyvirus, subgroup 3 transm. by mites, filament *c*.717×15 nm, transm. *Abacarus hystrix*, Europe, N. America, *A. repens*, Gramineae.

Agropyron repens, couch or twitch grass.

agrostis, *Cheilaria*.

—— **wilt** Roberts et al., *PD* 65:1014, 1981; bacterium assoc., USA (Illinois), *A. palustris*, creeping bent grass cv. Toronto [61,5045].

ailanthus leaf roll mosaic Svobodová 1963, in Bojňanský et al. 1963, Czechoslovakia [43,953].

*****air pollution**, a pollutant is any gas, liquid or solid air contaminant that causes undesirable effects on living organisms or materials; for terms, not given in the order here, see *Phytopath. News* 8(8), 1974, Am. Phytopath. Soc., and: acid rain, fluoride, sulphur dioxide, waldsterben. Darley & Middleton, *A.R.Phytop.* 4:103, 1966, problems in plant pathology; Heck, *ibid* 6:165, 1968; oxidant damage to plants; Heagle, *ibid* 11:365, 1973, effects on plant parasites; *Phytop.* 69:998, 1979; Heath, *A. Rev. Pl. Physiol.* 31:395, 1980, initial events in plant injury; Laurence, *Z. PflKrankh. PflSchutz.* 88:156, 1981, effects on plant & pathogen interaction; Laurence & Weinstein, *A.R.Phytop.* 19:257, 1981, effects on plant productivity; Smith, *Air pollution and forests*, 1981; Krupa et al., *PD* 66:429, 1982, effects on plant health; Unsworth & Ormrod, *Effects of gaseous air pollution in agriculture and horticulture*, 1982; Reinert, *A.R.Phytop.* 22:421, 1984, plant response; Koziol & Whatley ed. *Gaseous air*

pollutants and plant metabolism, 1984; Landolt & Keller, *Experientia* 41:310, 1985, evaluation & prediction of forest growth response; Drew-Smith, 1985, turf grasses [64,3869].

air spora, the population of airborne particles of biological origin; see: disease gradient, epidemiology, volumetric spore trap; Ingold, *Fungal spores. Their liberation and dispersal*, 1971; Gregory, *The microbiology of the atmosphere* edn. 2, 1973; *A.R.Phytop.* 15:1, 1977; Hirst, *TBMS* 61:205, 1973; Meredith, *A.R.Phytop.* 11:313, 1973, spore release & dispersal; Ingold, *TBMS* 85:575, 1985, water & spore discharge in ascomycetes & hymenomycetes.

Ajuga reptans, creeping bugle weed q.v.

akagare, nutrient imbalance, rice, Japan and elsewhere; **akiochi**, autumn decline, occurs in nutrient-deficient soil, Japan; Ou 1985.

alba, *Pezicula*.

albescens, *Monographella*.

albinism, complete absence of green colour in plants and due to chimeral growth, genetic factors or toxins formed by pathogens; Barmore et al. *PD* 68:43, 1984, for fungi & citrus seedlings where albinism can be a problem in USA (Florida) [63,2336].

albino fruit, strawberry; lower than normal translocation of sugar to the fruit during maturation; Maas 1984.

albiziae, *Camptomeris*.

albo-atrum, *Verticillium*.

Albuginaceae, Peronosporales; sporangiophores unbranched, clavate, bearing basipetal chains of deciduous sporangia in dense, subepidermal clusters, forming on the host white or cream sori which erupt to shed sporangia; oogonial periplasm persistent, conspicuous, haustoria like knobs; obligate plant parasites.

Albugo (Pers.) Roussel ex Gray 1821; Albuginaceae, the only genus in this family; white 'rusts' or white blisters; Biga, *Sydowia* 9:339, 1955, key; Holliday 1980; Sherf & MacNab 1986.

A. candida (Pers. ex Hooker) O. Kuntze 1891; Mukerji, *Descr.F.* 460, 1975; mostly on Cruciferae, frequently serious in W. Canada on turnip rape, staghead; Verma & Petrie, *Can. J. Pl. Sci.* 60: 267, 1980, oospore infestation & flower bud infection, q.v. for references; Verma et al., *Can. J. Pl. Path.* 5:154, 1983, race 7, effects of leaf age, detachment & temp. on development of the pathogen [59,5976; 63,2547].

A. ipomoeae-aquaticae Saw. 1922; Edie, *TBMS* 55:167, 1970; Edie & Bess *ibid*: 205; on water spinach, *Ipomoea aquatica*, E. Asia, causes an important disease in Hong Kong [50,540,1036].

A. ipomoeae-panduratae (Schwein.) Swingle 1891;

Mukerji & Critchett, *Descr.F.* 459, 1975; on Convolvulaceae, including sweet potato.

A. occidentalis G. Wilson 1907; on Chenopodiaceae, including spinach; Raabe & Pound, *Phytop.* **42**:448, 473, 1952, biology; Dainello & Jones, *PD* **68**:1069, 1984; **70**:240, 1986, chemical control [**32**:297, 419; **64**,2202; **65**,3550].

A. tragopogonis (Pers.) S.F. Gray; Mukerji, *Descr.F.* 458, 1975; on Compositae, economic hosts include salsify, *Tragopogon porrifolius*, sunflower; races prob. occur; Kajornchaiyakul & Brown, *TBMS* **66**:91, 1976, infection of sunflower; Whipps & Cooke, *ibid* **70**:285, 389, 1978, nomenclature; inoculation of Compositae with the race from *Senecio squalidus*; *ibid* **71**:121, 1978, zoosporangia & zoospore behaviour in infection of *S. squalidus*: Hartman & Watson, *Can. J. Pl. Path.* **2**:137, 173, 1980, f.sp. on ragweed, *Ambrosia artemisiifolia* [**55**,2833; **57**,4353; **58**,2176; **60**,4811–12].

Alebroides nigroscutellatus, transm. potato Indian purple top roll.

Aleurites, tung.

aleurites rough bark Large, *Phytop.* **39**:718, 1949, USA (Louisiana, Mississipi), *A. fordii* [**29**:183].

aleuritis, *Mycosphaerella, Pseudocercospora*.

alfalfa, and see lucerne.

— **cryptic v.**, and temperate v., see cryptic viruses.

— **dwarf** Weimer 1936, Goheen et al., *Phytop*, **63**:341, 1973; caused by a Gram negative, bacterium limited to the xylem; it is the same as, or similar to, the one causing Pierce's grapevine leaf scald, q.v. for vectors and wild hosts, causes decline and stunting, USA (California, S.) [**16**:103; **53**,635].

— **latent v.** Veerisetty & Brakke 1977; Veerisetty, *Descr.V.* 211, 1979; Carlavirus, rod or slightly flexuous *c*.635 nm long, transm. sap, *Acyrthosiphon pisum*, non-persistent; now considered as a str. of pea streak v.

— **Michigan v.** Thottappilly et al., *Phytop.* **67**:1451, 1977; prob. Luteovirus, isometric 23 nm diam., transm. *Acyrthosiphon pisum*, persistent, mostly symptomless in lucerne [**57**,4516].

— **mosaic v.** Weimer 1931, 1934; Jaspars & Bos, *Descr.V.* 229, 1980; forms a distinct group, bacilliform $25–58 \times 18$ nm, linear ssRNA, transm. sap, seed and 14 aphid spp., non-persistent, common, causes diseases in celeriac, celery, chick pea, clover, bean, cowpea, eggplant, lettuce, lucerne, lupin, mung bean, potato, pea, red pepper, tobacco, tomato; Regenmortel & Pinck in Kurstak 1981; Boswell & Gibbs 1983; Bailiss & Ollennu, *Pl. Path.* **35**:162, 1986, effect on lucerne yield in Britain [**65**,5579].

alfalfae, *Physoderma, Pleospora*.

algae, parasitic on plants, Joubert & Rijkenberg, *A.R.Phytop.* **9**:45, 1971, particularly *Cephaleuros virescens* Kunze 1832. Where this alga, mostly in warmer regions, is apparently causing significant disease there will be predisposing factors which need to be corrected; by itself it is of little or no economic importance; Holcomb, *PD* **70**:1080, 1986, hosts & literature.

Allard, Harry Ardell, 1880–1963; born in USA, Univ. N. Carolina, USDA; original and important work on tobacco mosaic v. *Phytop.* **54**:125, 1964.

allelopathy, first used by Molisch in 1937, the term was used for both harmful and beneficial effects. The effect of a higher plant, the donor, on the germination and growth of another, the receptor; it is exerted through the release of a chemical by the donor; the extension of the term to other organisms seems debatable; Putman & Duke, *A.R.Phytop.* **16**:431, 1978; Rice, *Allelopathy*, edn. 2, 1984.

allemanda witches' broom Ghosh, Srimathi & Raychaudhuri, *Sci. Cult.* **43**:495, 1977; India (Karnataka), *A. cathartica* [**57**:4500].

Allen, Ruth Florence, 1879–1963; born in USA, Univ. Wisconsin, USDA; noted for cytology and histology of plant infection by rusts. *Phytop.* **54**:885, 1964.

alliaria mosaic Rønde Kristensen 1960; Papa et al., *Phytopath. Z.* **78**:344, 1973; rod, modal length 745 nm, Denmark, Italy (S.), *A. officinalis* [**40**:580; **54**,87].

alligator pear, see avocado.

allii, *Botryotinia, Botrytis, Cladosporium, Embellisia, Pleospora, Puccinia*.

allii-cepae, *Ascochyta, Cladosporium, Mycosphaerella*.

Allium, *A. ampeloprasum*, leek; *A. cepa*, onion, shallot; *A. fistulosum*, Japanese bunching or Welsh onion; *A. sativum*, garlic; Walker & Larson, *Agric. Handb. USDA* 208, 1961; Bos, *Acta Hortic.* 127:11, 1982, viruses; Maude in Dennis 1983; Sherf & MacNab 1986.

allium yellows Ko & Chen, *Pl. Prot. Bull. Taiwan* **20**:83, 1978; MLO assoc., *A. fistulosum* [**57**,5794].

almond, *Prunus amygdalus*; Refati, *Tec. agricola* **26**:1159, 1974, MLO & viruses.

— **bud failure and calico** (prunus necrotic ringspot v.).

— **leaf scorch** Moller et al., 1974; Mircetich et al., *Phytop.* **66**:17, 1976; Davis et al., *ibid* **70**:472, 1980; caused by a Gram negative bacterium limited to the xylem, like the one causing alfalfa dwarf and Pierce's grapevine leaf scald q.v., USA (California) [**53**,2627; **55**,3226; **60**,2092].

— mosaic Quacquarelli & Martelli 1968; Lansac et al., *Acta phytopath. Acad. Sci. hung.* **15**(1–4):359, 1980; assoc. with apple chlorotic leaf spot v., prune dwarf v. or prunus necrotic ringspot v. [**49**,2522t; **61**,1833].
— stem pitting Quacquarelli & Savino 1981? Italy (Apulia) [**61**,3243].
— yellow bud mosaic (tomato ringspot v.).
alnea, *Phomopsis.*
Alocasia macrorrhiza, giant taro, see taro.
aloes, *Uromyces*; alopecuri, *Dilophospora.*
alphitoides, *Microsphaera.*
alsike clover, see *Trifolium.*
— — mosaic Zaumeyer & Wade, *J. agric. Res.* **51**:715, 1935; USA [**15**:274].
— — proliferation Chiykowski, *CJB* **43**:527, 1965; transm. *Macrosteles fascifrons*, Canada (Alberta); prob. the same as clover phyllody [**44**,2825].
— — vein mosaic v. Gerhardson & Lindsten, *Phytopath. Z.* **72**:76, 1971; Boswell & Gibbs 1983; isometric 30 nm diam., transm. sap, Europe (Scandinavia), experimental host range narrow [**51**, 1592].
alstroemeria latent v. Chiko & Godkin 1984; isometric *c*.29 nm diam., Canada (British Columbia), cv. Orange Beauty [**64**,3452].
— mosaic v. Brunt & Phillips 1981; poss. Potyvirus *c*.750 × 12 nm, England; see Kristensen, *Tidsskr. PlAvl.* **66**:545, 1962 [**42**:515; **60**,6227].
alternanthera stunt v. Hill & Zettler, abs. *Phytop.* **63**:443, 1973; filament assoc. av. 1717 nm long, USA (Florida), *A. philoxeroides*, alligator weed.
Alternaria Nees 1816; Hyphom.; conidiomata hyphal, conidiogenesis enteroblastic, tretic, polytretic, sympodial or sometimes monotretic, cicatrised, terminal becoming intercalary; con. pigmented, muriform, solitary or catenate, dry; cf. *Stemphylium*. Ellis 1971, 1976, described 44 spp. and arranged according to how abrupt is the transition from spore body to beak, and the ratio of the beak and body lengths. The spp. often cause economically important diseases, mostly as necrotic lesions on the upper parts of herbaceous crops; see zinniol and: Simmons, *Mycologia* **59**:67, 1967; Carmichael et al. 1980; Holliday 1980; Nishimura & Kohmoto, *A.R.Phytop.* **21**:87, 1983, toxins; Simmons, *Mycotaxon* **25**, 1986; Sherf & MacNab 1986.
A. alternata (Fr.) Keissler 1912; con. av. 37 μm long; a common saprophyte, as a pathogen it is prob. most assoc. with host damage. A f.sp. *lycopersici*, causing tomato stem canker, has been described, see Gilchrist & Grogan, *Phytop.* **66**:165, 1976. The fungus is often described as the cause of tobacco brown spot, this is prob. in

error, see *Alternaria longipes*; Prusky et al., *AAB* **98**:79, 1981, latent infection; *Phytop.* **71**:1124, 1981, on persimmon; Droby et al., *PD* **68**:160, 1984, potato tuber black pit; *Phytop.* **74**:537, 1984, on potato leaves [**55**,4281; **60**,6565; **61**,2982; **63**,2467, 4537].
A. brassicae (Berk.) Sacc. 1880; Ellis, *Descr.F.* 162, 1968; con. up to 350 μm long, brassica grey leaf spot, Cruciferae, seed infection is important, see Wiltshire, next entry, for taxonomy; Dixon 1981.
A. brassicicola (Schwein.) Wiltshire, *Mycol. Pap.* 20, 1947; Ellis, *Descr.F.* 163, 1968; con. up to 130 μm long, brassica black leaf spot, Cruciferae, seed infection is important; Knox-Davies, *TBMS* **73**:235, 1979; Maude & Humpherson, *AAB* **95**:311, 321, 1980, all seedborne phases, with *Alternaria brassicae*, & control in Britain; Bassey & Gabrielson, *Seed Sci. Technol.* **11**:403, 1983, on cabbage seedlings [**59**,3457; **60**,2257–8; **63**,1465].
A. burnsii Uppal, Patel & Kamat 1938; Anahosur, *Descr.F.* 581, 1978; con. 25–119 × 7–28 μm, beak length moderate, cummin blight, seedborne, India.
*A. carthami Chowdhuri 1944; Ellis & Holliday, *Descr.F.* 241, 1970; con. body mostly 60–110 × 15–26 μm, beak 25–160 μm long; safflower leaf spot, can be serious in Australia and USA, seedborne; Jackson et al., *Aust. J. Exp. Agric. Anim. Husb.* **22**:221, 1982, effects on yield & seed quality; Tietjen et al., *Physiol. Pl. Path.* **23**:387, 1983, toxins; McRae et al., *PD* **68**:408, 1984, effects of temp., dew period & inoculum density [**61**,7112; **63**,1904, 4518].
A. cassiae Jurair & Khan 1960; Simmons, *Mycotaxon* **14**:17, 1982; con. body 90 × 20 μm, beak 150–200 μm long; sickle pod seedling blight, *Cassia obtusifolia*, a weed, USA; Walker, *PD* **66**:426, 1982, hosts; Walker & Boyette, *ibid* **70**:962, 1986, dew & infection [**61**,6741; **66**,1728].
A. cheiranthi (Lib.) Bolle 1924; con. 20–100 × 13–32 μm, beak short or absent; wallflower leaf spot, Cruciferae, seedborne; Bambridge et al., *Crop Res.* **25**:27, 1985, chemical control [**65**,247].
A. cichorii Nattrass 1937; con. body mostly 60–90 × 15–20 μm, beak up to 280 μm long, chicory and endive leaf spot; Elia, *Phytopath. Mediterranea* **7**:7, 1968; Sarasola 1970 [**48**,1004; **52**,544].
A. cinerariae Hori & Enjoji 1931; con. body 15–140 × 15–50 μm, beak up to 80 μm long; cineraria leaf spot, *Senecio cruentus*; Cooper, *J. hort. Sci.* **31**:229, 1956 [**36**:188].
A. citri Ell. & Pierce 1902; Ellis & Holliday, *Descr.F.* 242, 1970; con, 8–60 × 6–24 μm, beak short when present, citrus fruit black rot, and in storage, leaf spot, Washington navel orange fruit drop, Emperor mandarin brown spot; different

physiological forms occur; Kohmoto et al., *Phytop.* **69**:667, 1979 toxins; Gardner et al., Kono et al. 1985 [**59**,3253; **65**,722–3].

A. crassa (Sacc.) Rands 1917; Ellis & Holliday, *Descr.F.* 243, 1970; con. 130–220 × 15–24 μm beak longer than body, *Datura stramonium* leaf spot, seedborne; Fornet, *Agrotec. Cuba* **13**:81, 1981 [**62**,923].

A. cucumerina (Ell. & Ev.) Elliot 1917; Ellis & Holliday, *Descr.F.* 244, 1970; con. 130–220 × 15–24 μm, beak pronounced, curcurbit leaf spot, storage decay.

A. dauci (Kühn) Groves & Skolko 1944; con. 100–450 × 16–25 μm, beak long, sometimes longer than the body, total length longer than any other *Alternaria* spp. on carrot; leaf blight, Umbelliferae, seedborne, can cause serious losses; Gillespie & Sutton, *Can. J. Pl. Path.* **1**:95, 1979; Soteros, *N.Z. J. agric. Res.* **22**:191, 1979; chemical control; Barash et al., *Physiol. Pl. Path.* **19**:7, 1981, toxin zinniol; Maude et al. 1985, biology & seed treatment in UK [**59**,987, 3465; **61**,911; **64**,4563].

A. dianthi Stevens & Hall 1909; con. 30–120 × 10–25 μm, beak short, more longitudinal septa than in *Alternaria dianthicola*; carnation leaf blight, seedborne.

A. dianthicola Neergaard 1945; con. 55–130 × 10–16 μm, beak moderate, few longitudinal septa; carnation petal blight or flower bud rot, seedborne, prob. more prevalent and severe than *Alternaria dianthi*.

A. eichhorniae Nag Raj & Ponnapa, *TBMS* **55**:124, 1970; con. body 21–69 × 9–19 μm, beak up to 145 μm long; water hyacinth blight, India [**50**,469].

A. euphorbiae (Barth.) Aragaki & Uchida in Yoshimura et al., *PD* **70**:75, 1986; con. body av. 55 × 15 μm, beak 89 μm long; poinsettia blight, USA (Hawaii) [**65**,4977].

A. godetiae Neergaard 1945; W.C. & F.J. Moore, *TBMS* **32**: 275, 1949; con. 27–66 μm long, beak 0–3 μm long; *Godetia* (= *Clarkia*) stem blight, can be damaging [**29**:318].

A. helianthi (Hansf.) Tubaki & Nisihara 1969; Anahosur, *Descr.F.* 582, 1978; con. 45–145 × 10–30 μm, beak absent; sunflower blight, causing premature defoliation and stem break, a serious disease in many parts of the world, seedborne; forms toxins, see radicin; Bhaskaram & Kandaswami, *E.Afr.Agric.For.J.* **43**:5, 1977, epidemiology; Allen et al., *Aust.J.Exp.Agric.Anim.Husb.* **21**:98, 1981, loss; *AAB* **102**:413, 1983, infection, sporulation & survival; *Phytop.* **73**:893,896, 1983, pathogen growth & infection; Jeffrey et al., *ibid* **74**:1107, 1984, virulence & residual inoculum; Carson, *ibid*

75:1151, 1985, epidemiology & loss; Simmons, *Mycotaxon* **25**:210, 1986 [**60**,1024,6044; **62**,3935; **63**,207–8; **64**,1210; **65**,2393].

A. linicola Groves & Skolko, *Can.J.Res.* C **22**:223, 1944; con. 150–300 × 17–24 μm, long slender beak; linseed seedling blight, seedborne; Moore, *TBMS* **29**:256, 1946; Loughnane, *Nature Lond.* **157**:266, 1946; for control in seed see Mercer et al., *Tests Agrochem. Cvs.* 6 *AAB* **106** suppl.: 56, 1985 [**25**:213; **26**:408].

A. longipes (Ell. & Ev.) Mason 1928; Ellis & Holliday, *Descr.F.* 245, 1970; con. 35–110 × 11–21 μm, beak ⅓ to ½ total length, tobacco brown spot, can be an important disease, prob. seedborne but this may be unimportant. This disease has frequently, and apparently incorrectly, been attributed to *Alternaria alternata*; Ellis 1971 gave the av. length of con. for this fungus as 37 μm and for *A. longipes* as 69 μm; Simmons, *Mycotaxon* **13**:16, 1981 [**60**,6334].

A. longissima Deighton & MacGarvie 1968; con. very variable up to 500 μm long, beak long; plurivorous mostly saprophytic; Yu et al., *TBMS* **78**:447, 1982, sesame foliage blight, stem necrosis, capsule spot & zonate leaf spot, seedborne, with *Alternaria sesami*, *A. sesamicola*, *A. alternata* [**61**,5893].

A. macrospora Zimm. 1904; Ellis & Holliday, *Descr.F.* 246, 1970; con. 90–180 × 15–22 μm, beak long; cotton leaf spot; Padaganur, *Madras agric. J.* **66**:325, 1979, seedborne; Bashi et al., *Phytop.* **73**:1145, 1983, external environment, leaf age & disease development; *Phytoparasitica* **11**:89, 1983, disease & yield; Crawley et al., *PD* **69**:977, 1985, on the weed *Anoda cristata* in USA cotton [**60**,1454; **63**,1286–7; **65**,4237].

A. padwickii (Ganguly) M.B. Ellis 1971; Ellis & Holliday, *Descr.F.* 345, 1972; con. 95–170 × 11–20 μm, beak long; rice stackburn, seedborne phase important.

A. panax Whetzel 1912; Simmons, *Mycotaxon* **14**:17, 1982; con. 150–160 × 12–20 μm, beak 80–90 μm long; on ginseng and other plants; Atilano, *PD* **67**:224, 1983; Yu et al., *Ann.phytopath.Soc. Japan* **50**:313, 1984; Uchida et al., *PD* **68**:447, 1984 [**62**,3074; **63**,4453; **64**,1213].

A. passiflorae Simmonds 1938; Ellis & Holliday, *Descr.F.* 247, 1970; con. 100–250 × 14–29 μm, beak long; passion fruit brown spot, can become severe in Australia.

A. porri (Ell.) Cif. 1930; Ellis & Holliday, *Descr.F.* 248, 1970; con. 100–300 × 15–20 μm, beak long; onion purple blotch, seedborne; Miller, *PD* **67**:284, 1983, leaf age & susceptibility; Black et al., *Trop.Pest Management* **31**:47, 1985, in

Thailand, chemical control & farming practice [**62**,3386; **65**,3639].

A. radicina Meir, Drechsler & Eddy 1922; Ellis & Holliday, *Descr.F.* 346, 1972; con. 27–57 × 9–27 μm, no beak; carrot black rot; celery, dill, parsley, parsnip infected; *Alternaria dauci* has longer con. and is more a disease of the leaves than *A. radicina* which attacks the tap root, seed infection is important, long soil survival, disease in storage; Espir et al., *Tests Agrochem. cvs.* 6 *AAB* 106 suppl. 50, 1985, chemical control on celery.

A. raphani Groves & Skolko 1944; con. 50–130 × 14–30 μm, beak usually short; radish black pod blotch, Cruciferae; *Alternaria brassicae* has a longer beak, *A. brassicicola* has no beak, *A. raphani* forms chlamydospores the other 2 spp. do not, internally seedborne.

A. ricini (Yoshii) Hansford 1943; Ellis & Holliday, *Descr.F.* 249, 1970; con. 70–170 × 13–27 μm, beak long; castor blight and capsule rot, internally seedborne.

A. scorzonerae (Aderhold) Loerakker, *Neth.J.Pl.Path.* **90**:37, 1984; con. 175–195 × 16–22 μm, beak up to 120 μm long, usually solitary; black salsify, *Scorzonera hispanica*, leaf spot, Europe (W.) [**63**,5254].

A. sesami (Kawamura) Mohanty & Behera 1958; Ellis & Holliday, *Descr.F.* 250, 1970; con. 90–260 × 14–33 μm, beak av. 127 μm long; sesame blight; see *Alternaria longissima*, Yu et al., for morphology, infection & on seed.

A. sesamicola Kawamura 1931; see above Yu et al.; con. type A 11–78 × 7–20 μm body, 11–437 μm long beak; type B 5–178 × 5–21 μm body, 5–178 μm long beak, the av. lengths of the beaks 80 μm and 55 μm for A and B, respectively, is less than in *A. sesami*, on sesame, seedborne.

A. solani Sorauer 1896; Ellis & Gibson, *Descr.F.* 475, 1975; con. 150–300 × 15–19 μm, usually solitary; potato early blight and tuber rot, tomato early blight and fruit rot. The pathogen has been intensively studied in the hot, arid regions of Israel; not to be confused with *Septoria lycopersici* or *Stemphylium solani*: seedborne, forms alternaric acid q.v.; Madden & Pennypacker, *Phytopath.Z.* **95**:364, 1979, epidemiology in USA; Handa et al., *Physiol.Pl.Path.* **21**:295, 1982, toxins; Pennypacker et al., *PD* **67**:287, 1983, forecasting [**59**,1885; **62**,2366,3244].

A. tagetica Shome & Mustafee, *Curr.Sci.* **35**:370, 1966; con. mostly 186–217 × 9–31 μm; marigold blight, *Tagetes* spp., poss. seedborne; Hotchkiss & Baxter, *PD* **67**:1288, 1983, pathogenicity; Cotty & Misaghi, *Phytop.* **75**:366, 1985, effects of light on the pathogen, lesion & zinniol q.v. formation [**45**,3563; **63**,3409; **64**,4970].

A. tenuissima (Kuntze:Pers.) Wiltshire 1933; con. 22–95 × 8–19 μm, beak short, plurivorous, secondary invader, forms tenuazonic acid q.v.

A. tomato (Cooke) L.R. Jones, *Bull.Torrey bot.Club* **23**:353, 1896; the fungus is distinct from *Alternaria solani*, also on tomato, E.G. Simmons, personal communication, 1984; their patterns of conidial development differ, con. body 55 μm long, beak 153 μm long, both dimensions are smaller than those of *A. solani*; tomato nailhead spot, both fungi cause similar symptoms on the foliage but those of *A. tomato* are distinctive on the fruit. The disease, USA, has long been insignificant since the introduction of resistant cvs., but Goode & Montgomery 1979 reported an outbreak in Arkansas; Rosenbaum, *Phytop.* **10**:9, 1920 [**59**,2334].

A. triticina Prasada & Prabhu 1963; Anahosur, *Descr.F.* 583, 1978; con. 20–90 × 9–30 μm, beak short; wheat blight, Africa (N.), Asia (W.), India, seedborne.

*****A. zinniae** M.B. Ellis 1972; con. body 55–105 × 19–28 μm, beak av. 108 μm long; leaf spot, Compositae, seedborne; McDonald & Martens, *Phytop.* **53**:93, 1963, in Canada; Gambogi et al., *Seed Sci.Technol.* **4**:333, 1976, in seed [**42**:554; **55**,5798].

alternaric acid, toxin from *Alternaria solani*, Brian et al., *J.gen.Microbiol.* **5**:619, 1951; *AAB* **39**:308, 1952 [**32**:140,517].

alternariol, and related toxins from *Alternaria* q.v.

alternata, *Alternaria*.

alternate host, either of the 2 hosts of a heteroecious pathogen, typically a rust, or of a heteroecious pest e.g. an aphid. When several plant spp. are hosts of one pathogen they may be described as alternative hosts, after FBPP.

Altson, Ralph Abbey, 1894–1963; born in England, Imperial College of Science and Technology, London; British Guiana, now Guyana, 1922–8, Federated Malay States, now Malaysia (W.) 1928–41, Rubber Research Institute of Malaysia 1946–52, CAB International Mycological Institute 1952–60; an authority on rubber diseases; instigated the work on South American leaf blight which is restricted to tropical America, *Microcyclus ulei* q.v.; *Commonw.phytopath.News* **7**(1):6, 1961.

aluminium tris (ethyl phosphonate), systemic fungicide, particularly against Oomycetes; can be translocated from shoots to roots.

amaranthus leaf mottle v. Lovisolo & Lisa 1976; *Phytopath. Mediterranea* **18**:89, 1979; prob. Potyvirus, filament, transm. sap, *Myz.pers.*, Italy (including Sardinia), Morocco, Spain; serologically related to bean yellow mosaic v. [**62**,4638].

— **mosaics** Phatak, *Curr.Sci.* **34**:645, 1965; *A. blitum, A. viridis*; Ramakrishnan et al., *Madras agric.J.* **58**:679, 1971; transm. sap, *A. gangeticus*; said to differ from Phatak's disease; Srifah & Sutabutra, *Kasetsart J.* **19**:26, 1985; prob. Potyvirus, filament 600–750 nm long, transm. sap to a few spp. in Amaryllidaceae and 2 other families, resembles hippeastrum mosaic v. [**45**,1610; **51**,2299; **65**,4436].

— **mottle v.** Qureshi & Mahmood, *Acta bot.ind.* **8**:31, 1980; isometric 25 nm diam., transm. sap, *Myz.pers.*, non-persistent, India [**60**,4162].

— **yellow mosaic** Rubio-Huertos & Vela-Cornejo, *Protoplasma* **62**:184, 1966; rod assoc., *A. lividus*.

amaryllis cryptic v., see cryptic viruses, Raizada et al., *Indian J.exp.Biol.* **21**:299, 1983 detection by Elisa [**63**,3939].

— **mosaic** = hippeastrum mosaic v.

amazonense, *Cladobotryum*.

Ambari hemp, see *Hibiscus*.

ambiguum, *Phaeocytostroma*.

American apple rust (*Gymnosporangium juniperi-virginiane*); — **dagger nematode**, *Xiphinema americanum*; — **gooseberry mildew** (*Sphaerotheca mors-uvae*); — **hawthorn rust**, also apple and pear (*G. globosum*); — **plum**, see *Prunus*.

americanum, *Pucciniastrum*.

Amomum, cardamom; *A. aromaticum*, Bengal cardamom.

amorphophallus mosaic Capoor & Rao, *Indian Phytopath.* **22**:438,1969; *A. campanulatus* [**50**,1035].

ampelicida, *Phyllosticta*.

ampelina, *Elsinoë*; **ampelinum**, *Sphaceloma*.

Ampelomyces Ces. ex Schlecht. 1852; Coelom., Sutton 1980; conidiomata pycnidial, conidiogenesis enteroblastic, phialidic; con. aseptate, pigmented. *A. quisqualis* Ces., con. 4–6.5 × 2.5 μm, hyperparasitic on Erysiphaceae; Jarvis & Slingsby, *PDR* **61**:728, 1977; Philipp et al., *Z. PflKrank.PflSchutz.* **86**:129, 1979; **91**:438, 1984; Sundheim, *Pl.Path.* **31**:209, 1982; parasitism, control of *Sphaerotheca fuligena* [**57**,1893; **58**,6125; **62**,447; **64**,1341].

ampelopsidis, *Physopella*.

Amphicytostroma quercinum, see *Fusicoccum quercus*.

amphigynous, having the antheridium through which the oogonial intercept grows, Pythiaceae; Ho, *Mycologia* **71**:1057, 1979; cf. paragynous [**59**,3639].

Amphiporthe Petrak 1971; Barr 1978; Valsaceae; asco. narrowly ellipsoid, fusoid, 1 septate; *A. castanea* (Tul.) Barr; asco. 11–16 × 3.5 μm with short appendages; formerly in *Cryptodiaporthe*, causing chestnut dieback and canker; Défago, *Phytopath.Z.* **10**:168, 1937; Fowler, *Phytop.*

28:693, 1938; *A. leiphaemia*, see *Fusicoccum quercus* [**16**:845; **18**:148].

Amphisphaeriaceae, Sphaeriales; ascomata mainly single, asci with apical ring, usually amyloid; asco. usually with germ pores; *Clethridium, Discostroma, Lepteutypa, Monographella, Pestalosphaeria, Physalospora*.

Amphorophora, transm. black raspberry necrosis, raspberry leaf mottle, raspberry leaf spot, rubus yellow net, thimble berry ringspot.

ampla, *Davisomycella*.

amygdali, *Fusicoccum, Pseudomonas*.

amyloid, of fungal structures, staining blue black with I.

Amylostereum Boidin 1958, Stereaceae; fungal symbionts of *Sirex* wood wasps which may damage or kill pine trees; only in Australia and New Zealand has *Sirex* been considered to be a major pest of *Pinus*; Talbot, *A.R.Phytop.* **15**:41, 1977.

amylovora, *Erwinia*.

amylovorin, toxin from *Erwinia amylovora*; Eden-Green & Knee, *J.gen.Microbiol.* **81**:509, 1974; Goodman et al., *Science N.Y.* **183**:1081, 1974; Beer et al., *Phytop.* **73**:1328, 1983, lack of host specificity [**53**,4032; **54**,916; **63**,1319].

anabe (*Ganoderma lucidum*) betel.

anacardii, *Oidium, Phomopsis, Pseudocercospora*.

Anacardium occidentale, cashew.

anaerobiosis in soil, Drew & Lynch, *A.R.Phytop.* **18**:37, 1980, micro-organisms & root function.

anamorph, see fungi and teleomorph.

ananas, *Erwinia*.

Ananas comosus, pineapple.

andersenii, *Drechslera*.

Anderson, Harry Warren, 1885–1971; born in USA, Wabash College, Indiana, Univ. Illinois, demonstrated the translocation of streptomycin in plants, wrote *Diseases of fruit crops* 1956. *Phytop.* **64**:907, 1974.

andersonii, *Inonotus*.

androgynous, antheridium and oogonium on one hypha cf. diclinous and monoclinous.

andropogonis, *Epicoccum, Pseudomonas*.

anemone alloiophylly Klebahn 1926, Smith 1972; transm. sap, Germany [**5**:753].

— **brown ring** Hollings, *Pl.Path.* **7**:95, 1958; transm. sap, England (Cheddar); or anemone brown ringspot [**38**:7].

— **necrosis** (tobacco ringspot v.).

anemones, *Peronospora, Urocystis*.

anethi, *Mycosphaerella*.

Anethum graveolens, dill.

angelonia little leaf Hegde, Siddaramaiah & Prasad, *PDR* **61**:588, 1977; poss. MLO, transm. *Lepidosaphes ulmi*, India (Dharwar) [**57**,625].

—— witches' broom Raju, Subrahmanyam & Rao, *Sci.Cult.* 41:165, 1975; poss. MLO, India [55,269].

*Angiosorus Thirum. & O'Brien, *Sydowia* 26:201, 1972, published 1974; Ustilaginales; sori as locules in host, bounded by sporiferous hyphae from which ustilospore spore balls are differentiated, 2–several spores, filling cavity, dark and powdery. *A. solani* Thirum. & O'Brien, for *Thecaphora solani* Barrus, *Phytop.* 34:714, 1944; ustilospores released through tuber disintegration, 7.5–20 × 18 μm, densely verrucose on one side; potato smut, sometimes called gangrene, tubers misshapen and have internal, brown, sporulating lesions, N. and W. of South America, can be severe; Barrus & Müller, *ibid* 33:1086, 1943; Bazán de Segura & del Carpio, *Informe Minist.Agricultura Peru* 35, 1974; Zachmann & Baumann, *PDR* 59:928, 1975 [23:118; 24:70; 54,3456,4117; 55,2864].

angiosperm parasitism, higher plants that parasitise other such plants are not considered; see Musselman, *A.R.Phytop.* 18:463, 1980, biology of *Orobanche*, *Striga* & other weeds parasitic on roots.

Anguina agrostis (Steinbuch) Filipjev 1936; Southey, *Descr.N.* 20, 1973; *Agrostis* flower gall, Gramineae; the galls formed become colonised by *Clavibacter rathayi* which forms animal toxins, Vogel et al., *Aust.J.Exp.Agric.Anim.Husb.* 24:617, 1984.

—— **graminis** (Hardy) Filipjev; Southey, *Descr.N.* 53, 1974; causes galls on grasses in Europe.

—— **tritici** (Steinbuch) Chitwood 1935; Southey *Descr.N.* 13, 1972; cereals, mostly wheat; galls, ear cockles, now rare due to seed cleaning; may be assoc. with *Dilophospora alopecuri*, vector of *Clavibacter tritici*.

angular black leaf spot (*Protomycopsis*) black gram, cowpea.

—— **leaf spot** (*Phaeoisariopsis griseola*) bean, Lima bean, (*Pseudomonas syringae* pv. *lachrymans*) cucurbits, (*P. syringae* pv. *tabaci*) tobacco, (*Xanthomonas campestris* pv. *malvacearum*) cotton, (*X. fragariae*) strawberry.

angulata, *Pyricularia*.

angustata, *Puccinia*, see *P. menthae*.

anilazine, protectant fungicide.

Anisogramma Theiss & Sydow 1917; Gnomoniaceae; ascomata in tree or shrub bark, clustered, often with a well developed stroma, ostioles erumpent, asco ± hyaline, 1 septate, septum eccentric. *A. anomala* (Peck) E. Müller in Müller & von Arx 1962; asco. 8–12 × 4–5 μm; filbert or hazel nut canker; Cameron, *PDR* 60:737, 1976, on *Corylus avellana* in USA;

Gottwald & Cameron, *Mycologia* 71:1107, 1979, morphology & life history, also on *C. americana*, infects through wounds [56,2692].

anjermozaic = carnation ringspot v.

annosum, *Heterobasidion*.

anomala, *Anisogramma*.

anomaly, bean; stunt, leaf crinkle, necrotic spotting, abnormal flowering, due to Mn toxicity; Menten et al., *Fitopat. Brasiliera* 6:179, 1981 [61,4463].

antagonism, a relationship between different organisms in which one, partly or completely, inhibits the growth of another or kills it; usually applied to the effects of toxic metabolites of one organism on another, but does not exclude mutual inhibitions between organisms, after FBPP.

anther mould (*Botrytis anthophila*) red clover, —— **rot** (*Ustilago violaceae*) carnation, —— **smut** (*U. vaillantii*) Scilla.

anthophila, *Botrytis*.

Anthoxanthum odoratum, vernal grass.

anthoxanthum latent blanching v. Catherall & Chamberlain 1981; poss. Hordeivirus, rod, mode 135 nm long, 22 nm diam., Wales [61,3214].

—— **mosaic v.** Catherall, *Pl.Path.* 19:125, 1970; filament *c.*740–760 × 13 nm, transm. sap to Gramineae, including barley, oat, wheat [50,1278].

anthracnose, usually used to describe a disease caused by *Colletotrichum*, i.e. a black, slightly sunken lesion with pink to white, mucilagenous masses of water dispersed con. *Gloeosporium* is a rejected name and should not be used for fungi with *Colletotrichum* states. The term has been used for diseases caused by *Sphaceloma*, anam. of *Elsinoë*, but scab is preferred, see Jenkins, *Phytop.* 23:289, 1933. Scab lesions differ from those of anthracnose in that they usually result in surface layer thicknening, a localised hyperplasia [12:595].

anthriscus yellows v. Murant & Goold, *AAB* 62:123, 1968; Murant & Roberts, *ibid* 85:403, 1977; isometric *c.* 22 nm diam., transm. *Cavariella aegopodii*, semi-persistent, Scotland; a helper virus, see parsnip yellow fleck v. [47,3650; 56,5240].

anthurium mosaic v. Herold, abs. *Phytop.* 57:8, 1967; rod *c.*750 nm long, transm. sap. *Bemisia tabaci*, Venezuela; see Verplancke 1930, Belgium [9:388, 46,1854c].

antibiotic, a substance formed by a micro-organism and which kills, or inhibits the growth of, other such organisms; Hawksworth et al. 1983; Lowe & Elander, *Mycologia* 75:361, 1983, antibiotic industry in USA.

antibody, see antigen and Cowan 1978.

antifungal substance, one which reduces the

germination of fungus spores or the growth rate of hyphae, see FBPP for an expansion.

antigen, a substance, usually a protein or a polysaccharide, which induces the formation of antibodies, proteins, when a living, warm blooded animal is injected with it. The antigen reacts specifically with the antibody; and the reaction can be observed in vitro, after FBPP.

anti-penetrants, disease control by non-fungitoxic concentrations of melanin biosynthesis inhibitors which block the polyketide pathway and prevent melanisation of appressorial walls. Such walls do not appear to have the proper architecture and rigidity to support penetration of the epidermis; Sisler, *Crop Protect.* **5**:306, 1986 [**66**,1759].

antirrhini, *Cercospora, Peronospora, Puccinia.*

antiserum, blood serum containing antibody, see antigen and FBPP for antiserum titre.

antisporulant, a material preventing or decreasing spore formation without killing vegetative growth of a fungus, after FBPP.

antiviral compounds, especially ribavirin; Simpkins et al., *AAB* **99**:161, 1981; Hensen and Mancino & Agrios, *PD* **68**:216,219, 1984 [**61**,2748; **63**,3447,3676].

— **factor**, Sela & Applebaum, *Virology* **17**:543, 1962, in plants infected by viruses; Sela, *Adv.Virus Res.* **26**:201, 1981, plant & virus interactions, & resistance [**42**:252].

aphanidermatum, *Pythium.*

Aphanomyces de Bary 1860; Saprolegniales; oogonia smooth or sculptured, depending on the subgenus, zoosporangia linear, zoospores in 1 row, aquatic; Scott, *Tech.Bull.Va.agric.Exp.Stn.* 151, 1961; Holliday 1980.

A. **brassicae** M.S. Pavgi & S.L. Singh; *Mycopathologia* **61**:171, 1977; oospores 6.5–21.5 μm diam., root rot of cabbage and cauliflower; Singh & Pavgi, *Phytopath. Mediterranea* **16**:129, 1977 [**57**,1890; **59**,1477].

A. **cochlioides** Drechsler 1928; *J.agric.Res.* **38**:309, 1929; oospores av. 19 μm diam., sugarbeet black root rot or stringy root; Papavizas & Ayers, *Tech.Bull.USDA* 1485, 1974, monograph with *Aphanomyces euteiches* [**8**:606].

*A. **euteiches** Drechsler 1925; Stamps, *Descr.F.* 600, 1978; oospores 18–25 μm diam.; pea root rot, legumes, serious losses can occur in USA, sometimes assoc. with infection by *Fusarium solani* f.sp. *pisi.* Sundheim, *Physiol Pl.Path.* **2**:301, 1972, delineated 5 races in Norway, but Manning & Menzies, *N.Z.J.agric.Res.* **27**:569, 1984 questioned the race concept; see Papavizas & Ayers, *Aphanomyces cochlioides*; Pfender et al., *Phytop.* **71**:1169, 1981; **72**:306,1200, 1982; **73**:1109, 1983,

soil inoculum density, on bean, ff.sp. *phaseoli* & *pisi*, disease spread; Greenhalgh et al., *Australasian Pl.Path.* **14**:34, 1985, on subterranean clover [**51**,4702; **61**,3140,6066; **62**,1275; **63**,1480; **64**,2797; **65**,770].

A. **iridis** Ichitani & Kodama in *Ann.phytopath.Soc.Japan* **52**:593, 1986, basal rot of *Iris hollandica*, pathogenicity restricted to iris [**66**,1504].

A. **raphani** J.B. Kendrick 1927; oospores 21.5–30 μm diam.; radish black root rot or constricted root; Imoto et al., *Bull. Hiroshima prefect.agric.Exp.Stn.* 41, 1979; Sherf & MacNab 1986 [**6**:455; **7**:4; **61**,449].

Aphasmatylenchus straturatus Germani 1970; *Descr.N.* 104, 1977; migratory root ectoparasite, damage to groundnut, pigeon pea.

aphelandra ring mosaic Paludan, *Statens Plantepatol.Forsøg* 460:80, 1971; Denmark [**52**,418].

Aphelenchoides arachidis Bos 1977; Bridge & Hunt, *Descr.N.* 116, 1985; groundnut testa nematode, root endoparasite, reduces seed weight, Nigeria.

A. **avenae** Bastian 1865; Hooper, *Descr.N.* 50, 1974; cosmopolitan, primarily mycophagous.

A. **besseyi** Christie 1942; Franklin & Siddiqi, *Descr.N.* 4, 1972; rice white tip, strawberry crimp or summer dwarf; Fortuner & Orton Williams, *Helminth.Abs.* **44**:1, 1975.

A. **bicaudatus** (Imamura) Filipjev & Schuurmans Stekhoven 1941; Siddiqi, *Descr.N.* 84, 1976; in soil around crops, mycophagous, little damage to the common, cultivated mushroom.

A. **blastophthorus** Franklin 1952; Hooper, *Descr.N.* 73, 1975; ecto- and endoparasite on leaves and buds of *Scabiosa*, causing blind bud; on, inter alia, *Callistephus, Caltha, Convallaria, Narcissus, Viola*; readily cultured.

A. **composticola** Franklin 1957; Hesling, *Descr.N.* 92, 1977; saprophytic, ubiquitous in temperate regions, mycophagous, can attack the common, cultivated mushroom.

A. **fragariae** (Ritzema Bos) Christie 1932; Siddiqi, *Descr.N.* 74, 1975; strawberry crimp or spring dwarf; obligately ecto- and endoparasitic on above-ground parts, other diseases caused include: lily dieback, peony bud decay, primula blackened flower head, violet dwarfing.

A. **ritzemabosi** (Schwartz) Steiner & Behrer 1932; Siddiqi, *Descr. N.* 32, 1974; chrysanthemum foliar nematode, on other ornamentals, infects buds, growing points and outer layers of stems and leaves.

aphids, see vector.

*Aphis citricola**, transm. arauja mosaic, passion fruit ringspot, zucchini squash yellow mosaic; *A.*

coreopsidis, sonchus yellow net; *A. craccivora*, many viruses; *A. cytisorum*, pumpkin mosaics; *A. fabae*, artichoke latent, beet yellows, cassia mosaic, chicory blotch, leek chlorotic streak, soybean mosaic; ssp. *solanella*, negro coffee mosaic; *A. glycines*, soybean Indonesian dwarf; *A. gossypii*, many viruses; *A. grossulariae*, black currant wildfire; *A. idaei*, raspberry leaf curl, raspberry vein chlorosis; *A. medicaginis*, cowpea witches' broom; *A. nerii*, aranja mosaic, cowpea rugose mosaic; *A. rubicola*, raspberry leaf curl; *A. rubifollii*, raspberry leaf curl; transm. for each *Aphis* sp.

Aphrodes bicinctus transm. clover phyllody, clover small leaf, clover yellow edge, potato stolbur, strawberry green petal.

Aphyllophorales, Holobasidiomycetidae; distinct, external, macroscopic basidiomata, development gymnocarpus, hymenium flattened or like clubs, or covering dentate processes, lining tubes or on lamellae, firmly united to basidioma, basidia not forked, sterigmata relatively small, usually 4, basidiospores not repetitive; poroid and lamellate forms are tough not fleshy like the Agaricales.

apical chlorosis (*Pseudomonas syringae* pv. *tagetis*) Jerusalem artichoke, sunflower, *Tagetes*.

apiculata, *Atropellis*.

apii, *Cercospora*, *Puccinia*; **apiicola**, *Septoria*.

Apiognomonia Höhnel 1917; Gnomoniaceae; asco. with a septum near the base, anam. *Discula*, the spp. are often found under *Gnomonia* which has asco. that are not always septate near the base; Barr 1978.

A. errabunda (Roberge ex Desm.) Höhnel 1918; anam. *Discula umbrinella* (Berk. & Br.) Sutton 1980; asco. 15–17 × 3.5–4 µm, con. 9.5–12 × 3.5–5 µm; Rathke 1978, *Platanus* leaf vein, Germany [58,2956].

A. erythrostoma (Pers.) Höhnel; asco. 17–20 × 6 µm, cherry leaf scorch, *Prunus*, Europe; Wormald 1955.

A. quercina (Kleb.) Höhnel 1921; asco. 13–15 × 3–4 µm, oak dieback and leaf spot, also called anthracnose q.v.; Europe, N. America; Neely & Himelick, *Phytop.* **57**:1230, 1967 [47,911].

*A. veneta (Sacc. & Speg.) Höhnel 1917; commonly known as *Gnomonia platani* Kleb. 1914; *Platanus* leaf scorch, bud and twig blight, canker; Europe, North America, New Zealand; Neely & Himelick, *Mycologia* **57**:834, 1965, nomenclature; Neely, *J.Arboricult.* **2**:153, 1976, in USA, severe disease on *P. occidentalis*; Cellerino & Anselme, *Infiore.fitopatol.* **28** (11–12):53, 1978 in Italy; Hitchcock & Cole, *N.Z.J.Sci.* **23**:69, 1980, climate; Gibbs & Refford, *Eur.J.For.Path.* **12**:395, 1982; Gibbs & Burdekin, *J.Arboricult.* **7**:227,

1983, on common or London plane in UK; Spiers et al., *NachrBl.dt. PflSchutzdienst.* **37**:17, 1985, in Germany [**45**, 614; **56**, 5212; **58**, 3482; **60**,1644; **62**,3271, 4467].

Apion africanum, *A. varipes* transm. red clover mottle; *A. vorax*, transm. broadbean stain, echtes ackerbohnenmosaik.

Apiosporina Höhnel 1910; Venturiaceae; Sivanesan 1984; a superficial subiculum present, ascomata unilocular, asco. septate at the lower end.

A. collinsii (Schwein.) Höhnel; Corlet, *Fungi Canadenses* 76, 1975; asco. 12–15 × 4.5–6 µm, witches' broom of *Amelanchier* spp.; Sprague & Heald, *Trans.Am.microsc.Soc.* **46**:219, 1927; Kennedy & Stewart, *CJB* **45**:1597, 1967 [**8**:144; **47**,764].

A. morbosa (Schwein.) v. Arx 1954; Sutton & Waterston, *Descr.F.* 224, 1970 as *Dibotryon morbosum* (Schwein.) Theiss & Sydow 1915; a *Fusicladium* anam. was described by Ellis 1976; asco. 14–18 × 4.5–6 µm; black knot of *Prunus* spp., N. America, physiologic forms; Rosenberger & Gerling, *PD* **68**:1060, 1984, chemical control; Wall, *Can.J.Pl.Path.* **8**:71, 1986, disease distribution & intensity on *P. pensylvanica* [**64**,2086; **65**,5743].

Apium graveolens, var. *dulce*, celery; var. *rapaceum*, celeriac.

aplerotic, oospores, of Pythiaceae, not filling the oogonium.

apoplexy, a name for the terminal syndrome in a complex of apricot diseases, fide Klement, *EPPO Bull.* **7**(1) 1977, descriptions of the pathogens [**57**,1291–1302].

Aposphaeria ulei, pycnidial state of *Microcyclus ulei*.

apothecium, a sessile or stipitate ascoma, shaped like a cup or saucer, in which the hymenium is exposed at maturity.

Appalachian blister rust (*Cronartium appalachianum*), *Pinus virginiana*.

appalachianum, *Cronartium*, *Peridermium*.

appendiculatus, *Uromyces*.

applanata, *Didymella*; **applanatum**, *Ganoderma*.

*apple, *Malus*; Porrott et al., *Publ.Agric.Canada* 1737, 1982, postharvest.

— **blister bark** Parish, *HortScience* **16**:52, 1981; key to complex, cv. Delicious, Australia [**60**,6002].

—— **canker** Blattný & Janečková 1980; Czechoslovakia, cv. Starkspur golden Delicious [**60**,5487].

— **brown line** = apple stem grooving v.

—— **ringspot** Novák, *Ochr.Rost.* **1**:75, 1965; Czechoslovakia [**44**,3398].

— **chat fruit** Beakbane et al., *J.gen.Microbiol.* **66**:55, 1971; England [**50**,3912].

—— **chlorotic leaf spot v.** Cadman 1963; Lister, *Descr.V.* 30, 1970; poss. Closterovirus, filament *c*.600 × 12 nm, transm. sap, natural spread very limited, in woody Rosaceae, no symptoms in apple cvs.

—— **dapple** Smith, Barrat & Rich, *PDR* **40**:765, 1956; USA (New Hampshire); Yamaguchi et al., *Bull.Fruit Tree Res.Stn.* C2: 73, 1975 [**36**:327; **55**,5815].

—— **dead spur** Parish, Williams & Misley 1981; Parish & Williams, abs. *Phytop.* **76**:1074, 1986; USA (Washington State) [**61**,2342].

—— **decline** Refatti, *Riv.Patol veg.* **6**:255, 1970; Kefalli et al. 1970; Italy; cv. Red Delicious; Welsh & Spangelo, *Phytoprotection* **52**:58, 1971; Canada (British Columbia), *Malus robusta* 5 stock [**50**,2355; **51**,2633a; **52**,181].

—— **depression** Palmiter, *PDR* **53**:329, 1969; USA (New York), cv. McIntosh [**48**,2452].

—— **flat apple** (cherry rasp leaf v.), Parish, *Acta Hortic.* 67: 199, 1976 [**57**,1773c].

—— —— **limb** Thomas, *Phytop.* **32**:435, 1942; USA (California); apple Gravenstein flat limb may be a form of apple rubbery wood.

—— **freckle scurf** Cheney, Lindner & Parish, *PDR* **54**:44, 1970; USA (Washington State) [**49**,2106].

—— **fruit wrinkle** Welsh & May, *PDR* **54**:493, 1970; Canada (British Columbia), cv. Newton.

*—— **green crinkle** Atkinson & Robbins, *N.Z.J.Sci.Technol.*A 33(3): 58, 1951; Chamberlain et al., *N.Z.J.agric.Res.* **17**:137, 1974; MLO assoc. [**31**:612; **53**,4034].

—— —— **pit** Giunchedi, *Inf.agrar.Verona* **26**:585, 1970, Italy [**50**,3921].

—— **H(ypertrophic) M(itochondrial)** Weintraub & Ragetli, *Phytop.* **61**:431, 1971; Canada, Virginia crab apple [**50**,3920].

—— **infectious variegation** = apple mosaic v.

—— **internal bark necrosis** Parish, *HortScience* **16**:52, 1981; Australia, cv. Delicious; Miller & Schubert, *Proc.W.Va.Acad.Sci.* **49**:97, 1977; soil pH & Mn. assoc. [**59**,4235; **60**,6002].

—— **Jonadel bark necrosis** Bovey & Brugger 1971; Switzerland [**51**,2633d].

—— **"kikei-ka"** Sawamura, *Bull.hort.Res.Stn.Morioka* C 3:25, 1965; Japan, affects fruit only, sunken areas on young apples which become severely malformed [**44**,3089].

—— **latent narrow leaf** Larsen, *Tidsskr.PlAvl.* **81**:359, 1977; Denmark [**57**,4029].

—— **leaf pucker** Welsh & Keane 1957; Welsh et al., *Phytop.* **63**:50, 1973; Canada (British Columbia), clover yellow mosaic v. from trees with apple leaf pucker [**37**:359; **52**,3746].

—— **mosaic v.** Bradford & Joley 1933; Fulton, *Descr.V.* 83, 1972; Ilarvirus, isometric *c*.26 nm

diam., transm. sap, difficult from apple and rose but readily from many herbaceous hosts; occurs naturally in birch and hop, wide range of experimental hosts.

—— **necrotic stem pitting** Welsh & Spangelo, *Phytoprotection* **52**:58, 1971; Canada (British Columbia), *Malus robusta* stock [**52**,181].

—— **platycarpa dwarf and scaly bark** Luckwill & Campbell, *J.hort.Sci.* **34**:248, 1959, England [**39**:592].

—— **proliferation** Mulder, *Tijdschr.PlZiekt.* **59**:72, 1953; MLO assoc., Europe; see Davies et al., *Pl.Path.* **35**:400, 1986, for references, they described a single occurrence of the disease in England [**33**:33; **66**,651].

—— **pustule canker** Cheney, Lindner & Parish, *PDR* **54**:44, 1970; USA (Pacific N.W.), cv. Red Delicious [**49**,2106].

—— **red ring** Coyier, Cheney & Parish, *PDR* **51**:281, 1967; USA (Oregon), cv. Red Delicious [**46**,2262].

—— **ring russetting** Kunze & Kennel 1973; van de Meer 1974; Germany, Netherlands, cvs. Belle de Boskoop, Golden Delicious [**54**,650, 2606c].

—— **ringspot** Atkinson, Chamberlain & Hunter, *N.Z.J.Sci.Technol.* A. **35**:478, 1954; also called Henderson spot, mainly cv. Granny Smith.

—— **rosette** Van Katwijk, *Tijdschr.PlZiekt* **59**:233, Netherlands, cv. Belle de Boskoop [**33**:487].

—— **rough bark** Posnette 1962; England [**42**:650].

—— —— **skin** Van Katwijk; Mulder, *Tijdschr.PlZiekt.* **61**:11 1955; Netherlands [**34**:374].

—— **rubbery wood** Cropley 1963, see apple chat fruit reference; Seemüller et al., *Z.PflKrankh.PflSchutz.* **91**:371, 1984, colonisation patterns with pear decline MLO; Clark & Davies 1984 [**63**,4810; **64**,1168].

—— **russet wart** Posnette & Cropley 1969; England [**49**,1074b].

——**"sabi-ka"** Sawamura, as for apple "kikei-ka"; fruit with corky scars and reduced in size.

—— **scar skin** Millikan 1957; Parker & Agrios, *Phytop.* **65**:707, 1975, histology; Koganezawa, *Ann.phytopath. Soc.Japan* **51**:176, 1985, poss. caused by a viroid [**36**:411; **55**,794; **64**,5439].

—— **spy 227 epinasty and decline** Gardner et al. 1946; see apple stem pitting; Smith 1972.

—— **star crack** Jenkins & Storey, *Pl.Path.* **4**:50, 1955; England (E. Anglia) [**35**:303].

—— **stem grooving v.** Lister, Bancroft & Nadakavukaren 1965; Lister, *Descr.V.* 31, 1970; filament *c*.600–700 × 12 nm, transm. sap, England, prob. widespread in apple but not causing symptoms in many clones, these occur in Virginia Crab; van der Meer, *Acta Hortic.* 67: 293, 1976, occurrence with apple stem pitting, seed transm. in *Chenopodium quinoa, Malus platycarpa*; Sawamura & Osada, *Bull.Fac.Agric Hirosaki Univ.* **33**:19, 1980, hosts [**57**,1775d; **60**,3834].

— — **pitting** Guengerich & Millikan, *PDR* **40**:934, 1956; USA (Missouri); Desvignes, *Acta phytopath.Acad.Sci.hung.* **15**:371, 1980, poss. same as apple spy decline, pyronia decline, quince sooty ringspot [**36**:328; **61**,1835].

— **top working disorder** (apple chlorotic leaf spot v.), Yanase et al., *Ann.phytopath.Soc.Japan* **45**:369, 1979 [**59**,3821].

— **Tulare mosaic v.** Yarwood 1965; Fulton, *Descr.V.* 42, 1971 and in Kurstak 1981; Ilarvirus, isometric *c*.33 nm diam., transm. sap, USA (California, Tulare country), can cause severe mosaic in apple, found naturally only once in apple and now only known in experimental material.

— **union incompatibility** Lana et al., *Phytopath.Z.* **106**:141, 1983; tobacco ringspot v. assoc., Canada [**62**,3103].

— — **necrosis and decline** (tomato ringspot v.).

— **Virginia Crab decline** Welsh & Nyland, *Can.J.Pl.Sci.* **45**:443, 1965; Waterworth, *Phytop.* **62**:695, 1972; Canada (British Columbia) and prob. outside N. America, conspicuous yellowing in Virginia Crab [**45**,487; **52**,444].

— — — **graft union breakage** Larsen *Tidsskr.PlAvl.* **78**:422, 1974; Denmark [**54**,4973].

— — — **stem pitting** Welsh & Nyland, see apple Virginia Crab decline.

appressorium, or infection structure; an organ formed from a hypha for attachment to a host before penetration; it may be a distinct cell, sometimes with a thickened wall, or it may be undifferentiated from the parent hypha. In the latter case it is distinguishable only by its adherence to host tissue; Emmett & Parbery, *A.R.Phytop.* **13**:147, 1975; Muirhead in Blakeman 1981, dormancy & latent infection.

apricot, *Prunus armeniaca.*

— **bare twig and fruit failure** Blattný & Janečkova, 1977; *Acta phytopath.Acad.Sci.hung.* **15**:383, 1980; Čech et al., *ibid* **15**:391, 1980; transm. sap, assoc. with cucumber green mottle mosaic v. and strawberry latent ringspot v., Czechoslovakia [**57**,5582; **61**,1838–9].

— **blotch v.** Ragozzino & Pugliano, *Riv.Ortoflorofruttic.ital.* **58**:136, 1974; filament 550–600 nm long, Italy (Campania) [**54**,2344].

— **chlorotic leaf mottle** Wood, *N.Z.J.agric.Res.* **18**:255, 1975; New Zealand (central Otago) [**55**,1352].

— — — **roll** Morvan 1957; Morvan et al.,

Phytopath.Z. **76**: 33, 1973; MLO assoc., France; see apoplexy, Klement, page 37 [**37**,292; **52**,3768].

— **moorpark mottle** Chamberlain, Atkinson & Hunter, *N.Z.J.Sci. Technol.* A **35**:471, 1954, New Zealand [**33**:734].

— **necrotic leaf roll**, and yellow mosaic, Wolfswinkel, *Zašt. Bilja* **16**:469, 1965; South Africa (Cape Province) [**45**,3363k].

— **pucker leaf** Wadley 1966, USA (Utah), Gilmer et al. 1976.

— **ring pox** Reeves, *Bull.Wash.Sta.Dep.Agric.* 1, 1943; same as apricot ringspot Bodine & Kreutzer, *Phytop.* **32**:179, 1942; USA (Colorado, Washington State); Hansen et al. in Gilmer et al. 1976 [**21**:339, **23**:391].

— **rosette** Morvan & Castelain, *Annls.Épiphyt.* Hors sér. **18**: 205, 1967; France [**48**,62r].

— **viruela** Pena & Ayuso, *Annls.Phytopath.* Hors sér.:85, 1971; Spain (Murcie) [**51**,2633g].

Arabica coffee, see coffee.

arabis mosaic v. Smith & Markham 1944; Murant, *Descr.V.* 16, 1970; Nepovirus, isometric *c*.30 nm, transm. sap readily, *Xiphinema coxi, X. diversicaudatum*, natural spread in Europe, prob. widespread, diseases caused include: celery necrosis and stunt, cucumber stunt mottle, lettuce chlorotic stunt, raspberry yellow dwarf, strawberry mosaic and yellow crinkle; hop nettlehead q.v.

arachidicola, *Cercospora, Didymosphaeria.*

arachidis, *Mycosphaerella, Oidium, Puccinia.*

Arachis hypogaea, groundnut, peanut.

arachnoidea, *Cercosporella.*

aralia dwarf Koike, *Proc.Assoc.Pl.Prot.Hokuriku* **32**:105, 1984; MLO assoc., *A. cordata*, Japanese udo salad.

— **leaf chlorosis** Dale, *PDR* **63**:472, 1979; MLO assoc., USA (Arkansas), *A. spinosa* [**59**,1897].

araujia mosaic v. Charudattan et al., *Phytop.* **70**:909, 1980; Potyvirus, filament mostly 740 nm long, transm. sap, *Aphis citricola, A. nerii, Myz. pers.*; Argentina, *Araujia, Morrenia* spp., including *M. odorata*, mikweed vine [**60**,6777].

arborescens, *Peronospora.*

archangelica chlorotic mosaic Kochman & Stachyra 1957; Poland, *A. officinalis*, angelica [**37**:341].

arctostaphyli, *Chrysomyxa.*

arcuata, *Rosellinia.*

arecae, *Phytophthora.*

areca nut, areca palm, betel nut, *Areca catechu*; Nayar, *PDR* **55**:170, 1971, stated that *A. cathecu* is the correct spelling; Reddy et al., *J.Plantation Crops* **6**(1):28, 1978; Bavappa et al. 1982, a text on all aspects of the crop, see *Hort.Abs.* **53**,5603.

— — **yellow leaf** Menon 1963; Nayar, as above; Nayar & Seliskar, *Eur.J.For.Path.* **8**:125, 1978;

MLO assoc., India (S.) [**43**,1109; **50**,2422; **57**,5635].

arenaria, *Stagonospora*; **areola**, *Mycosphaerella*.

areolate leaf spot (*Thanatephorus cucumeris*) citrus; —— **mildew** (*Mycosphaerella areola*) cotton.

areolatum, *Pucciniastrum*.

Aristastoma Tehon 1933; Coelom.; Sutton 1980; pycnidial, conidiogenesis holoblastic; con. hyaline, multiseptate, solitary; Sutton, *Mycol.Pap*. **97**, 1974, key. *A. camarographioides* Sutton; con. 4–6 septate, 33–45 × 7.5–9 μm; soybean zonate leaf spot, India, Maiti & Dhar, *TBMS* **79**:547, 1982. *A. oeconomicum* (Ell. & Tracy) Tehon, con. 2–4 septate, 19–32 × 3.5–5.5 μm; cowpea zonate leaf spot, USA (S.), Lefebvre & Stevenson, *Mycologia* **37**:42 1945 [**24**:353; **62**,2211].

arizonicus, *Inonotus*.

Arkansas cowpea mosaic v. (cowpea severe mosaic v.).

Arkoola J. Walker & Stovold, *TBMS* **87**:28,1986; Venturiaceae, large black, setose ascomata; the genus was erected for *A. nigra*, asco. 50–70 × 16–22 μm; causes a black leaf blight of soybean in Australia, a superficial black web forms on the host, seedborne [**66**,2604].

armeniacae, *Eutypa*.

*****Armillaria** (Fr.) Staude 1857; Tricholomataceae; basidioma with ring but no volva, gills decurrent, spore print white to cream, pathogens of woody hosts and saprophytic; Watling et al., *TBMS* **78**:271, 1982, described the morphology and taxonomy of the genus and its type sp. *A. mellea* (Vahl:Fr.) Kummer; Pegler & Gibson, *Descr.F.* 321, 1972 as *Armillariella mellea* (Vahl:Fr.) P. Karsten. The fungus (honey agaric, shoestring or bootlace) has been considered a complex with forms differing in pathogenicity, morphology, ecology and cultural characteristics; sometimes conspicuous black rhizomorphs occur, see Garrett 1970. These forms are now considered to be distinct spp. A morphological, or macrospecies, may consist of several biological, or microspecies. The latter are clonal genotypes isolated by intersterility; there is a bifactorial, heterothallic, multiple allelic, mating system. The fungi occur in natural forests and may become pathogenic in broad leaved and conifer plantings; they can be destructive in urban areas.

See major texts and: Chandra & Watling, *Kavaka* **10**:63, 1982, spp. in India; Kile & Old, *TBMS* **79**: 366, 1982, pseudoparenchyma zones, spp. in Australia; Morrison, *ibid* **78**:459, 1982, variation in British *A. mellea*: Rishbeth, *Pl.Path.* **31**:9, 1982, spp. in S. England; Gramss, *Eur.J.For.Path.* **13**:142, 1983; Grieg & Strouts, *Arboric.leaflet* UK, 2, 1983; Kile & Watling,

TBMS **81**:129, 1983, spp. in S.E. Australia; Bennell et al., *ibid* **84**:447, 1985; Roll-Hansen, *Eur.J.For.Path.* **15**:22, 1985, spp. in Europe; Siepmann *ibid* **15**: 71, 1985, spp. in Europe & clones; Guillaumin et al., *ibid* **15**:268, 1985, systematics & pathology; Rishbeth in D. Moore et al., *Developmental biology of higher fungi*, *Br.mycol.Soc Sympos*. 10, 1985; Wargo & Shaw, *PD* **69**,826, 1985, in USA; Rishbeth, *TBMS* **86**:213, 1986, English spp. in culture [**61**,5974, 5994; **62**,396; **63**,260, 934; **64**,2171, 3344, 3740, 4501; **65**,2211, 4552].

A. borealis Marxmüller & Korhonen 1982; Gregory & Watling, *TBMS* **84**:47, 1985; basidiomata small, slender; stipe long; pileus light in colour, yellowish to ochraceous, tinged pinkish; N. Europe, pathogenicity characteristics are not yet clear [**64**,1762].

A. bulbosa (Barla) Watling 1983; basidiomata with bulbous base; rich development of rhizomorphs, black, up to 5 mm diam.; tough, white to purple mycelial sheets under the bark, N. Europe; in S. England the sp. only attacks weakened trees.

A. hinnulea Kile & Watling, *TBMS* **81**:131, 1983; may occur with *Armillaria novae-zelandiae* but can be distinguished by the non-viscid pileus and predominantly pink brown colours; distinguished from *A. luteobubalina* by the lack of yellow, less scaly pileus and habit; Australia (S.E.), New Zealand, mixed wet forests, causes localised lesions on living eucalypts, root and butt rot of understory plants, assoc. with regrowth dieback q.v. of *Eucalyptus* in Tasmania; Kile, *Eur.J.For.Path.* **10**:278, 1980; *TBMS* **87**:312, 1986 [**60**,1643; **66**,1671].

A. luteobubalina Watling & Kile, *TBMS* **71**:79, 1978, who gave the diagnostic features of 5 *Armillaria* spp.; pileus with a strong yellow ground colour, no reddish and cinnamon brown hues, Australia, widely distributed in eucalypt communities, a primary pathogen causing decline and death; most infection occurs within 5 years of harvest, very long survival in stumps; 44 genotypes have been detected, only one type occurred in 71% of infected areas, maximum number in any one area was 3; Podger et al., *TBMS* **71**:77, 1978, spread & effects in *Eucalyptus regnans*; Kile & Watling, *ibid* **77**:75, 1981, morphology & distribution; Kile, *Aust.For.Res.* **11**:63, 1981, pathogenicity; *Aust.J.Bot.* **31**: 657, 1983, genotypes & clonal development; Kile et al., *Australasian Pl.Path.* **12**:18, 1983 [**58**,2410; **61**,1403, 1597; **63**,1402, 3045].

A. mellea, see *Armillaria*.

A. novae-zelandiae (Stev.) Boesewinkel 1977; Kile & Watling, *TBMS* **81**:129, 1983; distinguished by the viscid, thin, striate, honey yellow to isabelline pileus, and the semi-bulbous to bulbous stipe; a secondary pathogen in eucalypt dieback and decline diseases, mixed wet forest, Australia, New Zealand, S. America.

A. obscura (Pers.) Herink 1973; basidiomata brownish with a reddish stain, conspicuous, persistent scales are characteristic, ring thick and conspicuous; N. Europe; a primary pathogen prob. more so to conifers than to broad leaved trees; Durrieu et al., *Eur.J.For.Path.* **15**:350, 1985, pathogenicity in pine, France (E. Pyrenees) [65,2471].

Armoracia rusticana, horseradish.

Armstrong, George Miller, 1893–1986; born in USA, Univ. Wisconsin and Washington, Clemson College, S. Carolina, a specialist on ff.sp. of *Fusarium oxysporum. Phytop.* **76**:849, 1986.

aroids, edible, see *Colocasia* and taro.

aromatic compounds, see pathogenesis, Rohringer & Samborski.

aronia ringspot Bremer, *Annls.Agric.Fenniae* **23**:176, 1984, transm. sap, Finland, J.I. Cooper, personal communication, 1986.

arracacha A v. Jones & Kenten 1978; *Descr.V.* 216, 1980; Nepovirus, isometric 26 nm diam., transm. sap readily, Bolivia, Peru (Andes), occurs naturally only in *Arracacia xanthorhiza*, yellow mosaic, serologically unrelated to arracacha B v.

— **B v.** Kenten & Jones 1979; *Descr.V.* 270, 1983; isometric 26 nm diam., transm. sap and seed, Bolivia, Peru (Andes), occurs naturally in *Arracacia xanthorhiza, Oxalis tuberosa*, potato, see above.

arrhenatherum blue dwarf v. Schumann 1969; Milne et al., *Phytopath.Z.* **79**:315, 1974; Milne & Leseman, *Virology* **90**:299, 1978; Phytoreovirus subgroup 3, isometric 70 nm diam., a form of oat sterile dwarf v. [**49**, 2008b; **54**,1319; **58**,3172].

arrhenomanes, *Pythium*.

arrowroot mosaic Celino & Martinez, *Philipp.Agric.* **40**:285, 1956; Philippines, poss. str. of sugarcane mosaic v. [**36**:646].

arthric, thallic conidiogenesis q.v., characterised by the conversion of a pre-existing determinate hyphal element into a conidium, an arthroconidium, which is often cylindrical with truncate ends.

Arthrobacter Conn & Dimmick 1947; Gram positive rod, strictly aerobic, on complex media cells change from rods to cocci; *A. ilicis* Mandel et al.) Collins, Jones & Kroppenstedt 1981; leaf and twig blight of *Ilex opaca*, USA (Maryland); Bradbury 1986.

arthroconidium, see arthric.

arthropods, see vector

Arthur, Joseph Charles, 1850–1942; born in USA, Agricultural Experiment Station, Geneva, New York, Univ. Purdue, extensive and classic work on the morphology and taxonomy of rusts; *The plant rusts (Uredinales)*, 1929; *Manual of the rusts in United States and Canada*, 1934, illustrated by G.B. Cummins. *Mycologia* **34**:601, 1942; *Phytop.* **32**:833, 1942; *Česká Mykol.* **25**:185, 1971; *A.R.Phytop.* **16**:19, 1978.

*artichoke, globe, *Cynara scolymus*; Martelli et al., Congresso Internaz. Studi Carciofo, Bari, November 1979, published 1981, viruses; Marras et al., *Inftore.fitopatol.* **35**(9):19, 1985, mostly fungi [65,3644].

— **curly dwarf v.** Leach & Oswald 1950; Morton, *Phytop.* **51**:731, 1961; filament 582 × 15 nm, transm. sap, USA (California), Compositae [41;349].

— **Italian latent v.** Majorana & Rana 1970; Martelli & Rana, *Descr. V.* 176, 1977; Nepovirus, isometric c.30 nm diam., transm. sap, *Longidorus apulus, L. fasciatus*; Bulgaria, Greece, Italy; herbaceous and woody plants, causes: chicory chlorotic mottle, grapevine fan leaf and reduced growth, pelargonium leaf malformation and stunt; Savino et al., *Phytopath.Mediterranea* **16**:41, 1977, comparison of 4 isolates [59,202].

— **latent v.** Costa et al., *Phytop.* **49**:49, 1959; Rana et al., *AAB* **101**:279, 1982; Potyvirus, filament c.746 × 12 nm, transm. sap, *Aphis fabae, Brachicaudus cardui, Myz.pers.*, non-persistent; Israel, Turkey, Mediterranean (W.), USA (California); Rana et al. concluded that previously reported isolates from different geographical regions were indistinguishable from the Italian one studied by them [**38**:440; **62**; 1319].

— **mosaic** Gigante, *Boll.Staz.Patol.Veg.Roma* **7** (1949):177, 1951; transm. sap, Italy (Sicily) [**32**:299].

— **mottled crinkle** (str. of tomato bushy stunt v.).

— **vein banding v.** Gallitelli, Rana & Di Franco 1978; Gallitelli et al., *Descr.V.* 285, 1984; poss. Nepovirus, isometric c. 30 nm diam., transm. sap, natural infection only in artichoke, Italy (S.).

— **yellow band** Chagas, Flores & Caner 1969; Salamão 1976 (tobacco rattle v. Brazilian isolate) [**49**,1533; **57**,4742].

— — **ringspot v.** Kyriakopoulou & Bem 1973; Rana et al., *Descr.V.* 271, 1983; Nepovirus, isometric c.30 nm diam., transm. sap, seed and pollen; Greece (Peloponnesus), Italy (Sicily), wide experimental host range, causes: chlorotic blotches and rings in cardoon, ringspot and oak leaf patterns in tobacco cv. Black of Argos, stunt leaf yellowing and malformation in bean and broad bean.

artificial leaf coating, protection against infection by fungi; Osswald et al., *Z.PflKrankh.PflSchutz.* **91**:337, 1984 [**64**,1354].

arum lily mosaic v. Kolbasina & Protsenko 1973; filament mostly 750 nm long, USSR, *Zantedeschia aethiopica* [**53**,982].

arvalis, *Laetisaria.*

asclepias mosaic Kochman & Stachyra, *Roczn.Nauk roln* A **81**:287, 1960; Poland.

Ascocalyx Naumov 1925; Dermateaceae; Groves, *CJB* **46**:1273, 1968, key; see *Gremmeniella,* Müller & Dorworth. *A. laricina* (Ettl.) Schläpfer-Bernhard, *Sydowia* **22**:42, 1968, published 1969; anam. *Brunchorstia laricina* Ettl. 1945; asco. mainly 1 septate, 10–16 × 3–4 μm, con. 15–23 × 3–4 μm, N. hemisphere; Funk, *CJB* **47**:1509, 1969, described this sp. as *Encoeliopsis laricina* (Ettl.) Groves causing a shoot blight of *Larix occidentalis. A. pinicola* Kondo & Kobayashi, *J.Jap.For.Soc.* **66**:60, 1984, causes canker of *Pinus taeda* [**49**,868, **64**,1306].

ascochitine, toxin formed by *Ascochyta fabae* and *A. pisi*; Lepoivre, *Phytopath.Z.* **103**:25, 1982, for literature and pea cvs. [**61**,5326].

Ascochyta Lib. 1830; Coelom.; Sutton 1980; conidiomata pycnidia, conidiogenesis enteroblastic, phialidic, has been interpreted as annelidic; con. hyaline, mostly 1 septate; not to be confused with *Phoma* with aseptate con. or *Phyllosticta* with aseptate con. which have an apical appendage; Punithalingam, *Mycol.Pap.* 142, 1979, gave a full generic description and spp. on Gramineae; Holliday 1980; see vegetables.

A. adzamethica, teleom. *Didymosphaeria arachidicola.*

A. allii-cepae Punithalingam, Gladders & McKeown, *TBMS* **85**:559, 1985; con. 16–19 × 5–6 μm; assoc. with onion white leaf blotch, poss. a coloniser of scorched or senescent foliage, England [**65**,490].

A. avenae (Petrak) Sprague & Johnson 1948; Punithalingam, *Descr.F.* 731, 1982; con. 17–20 × 6–7 μm; oat leaf spot.

A. brachypodii (Sydow) Sprague & Johnson 1950; Punithalingam, *Mycol.Pap.* 142:60, 1979; con. 18–20 × 5–7 μm; Zeiders, *PD* **66**:502, 1982, teleom. in *Didymella,* on grasses in USA (New York, Pennsylvania) [**61**,6434].

A. caulicola Laub. 1903; con. 10–15 × 4–5.5 μm; sweet clover, *Melilotus,* grey stem canker, gooseneck; seedborne, Berkenkamp et al., *PDR* **53**:348, 1969 [**48**,2442].

A. chrysanthemi, teleom, *Didymella ligulicola.*

A. corticola McAlp. 1899; con. 7–9 × 2–3 μm; lemon bark blotch, Australia, New Zealand;

Brien, *N.Z.J.Agric.* **42**:341; **43**:421, 1931 [**10**:727; **11**:365].

A. cucumis, teleom. *Didymella bryoniae.*

A. desmazieresii Cav. 1893; Punithalingam, *Descr.F.* 661, 1980; con. 14–20 × 2–5 μm; glume and leaf spot of Italian and perennial ryegrasses.

*A. fabae Speg. 1898; Punithalingam & Holliday, *Descr.F.* 461, 1975; con. 16–24 × 3.5–6 μm; broad bean leaf, pod and stem spot, largely specialised on host, seedborne; Lockwood et al., *Pl.Path.* **34**:341, 1985, genotypic effects on incidence of infection [**65**,451].

A. gossypii Woronichin 1914; Holliday & Punithalingam, *Descr.F.* 271, 1970; con. 12–14 × 8 μm; cotton wet weather blight or ashen spot, seedborne.

A. heveae Petch 1917; Punithalingam, *Descr.F.* 631, 1979; con. 9–12 × 3–4.5 μm; rubber rim blight or marginal leaf scorch, Asia, Africa, prob. assoc. with poor growing conditions.

A. hordei Hara 1916; Punithalingam, *Mycol.Pap.* 142:88, 1979, described 3 vars., con. 17–20 × 3.5–4 μm; var. *americana* has broader conidia than var. *hordei* whose conidia are longer than those of var. *europea*; ring spot of barley, wheat; Kiewnick, *NachrBl.dt.Pflschutzdienst* **35**:145, 1983, in Germany [**63**,1693].

A. hyalospora (Cooke & Ell.) Boerema, Mathur & Neergaard, *Neth.J.Pl.Path.* **83**:156, 1977; con. 20–30 × 8–12 μm, in seeds of *Chenopodium quinoa,* infects leaves and stems.

*A. lentis Vassiljevsky 1940; see *Rev.appl.Mycol.* suppl.3:23, 1941; *ibid* **20**:232, 1941; con. 13.5–17 × 4.5–7 μm lentil blight, apparently specialised on host, seedborne; Morrall & Sheppard, *Can.Pl.Dis.Surv.* **61**:7, 1981; Gossen & Morrall, *Can.J.Pl.Path.* **5**:168, 1983; **6**:233, 1984; **8**:28, 1986, effects on seed quality & yield; spread from seed & plant residues; Kaiser & Hannan, *Phytop.* **76**:355, 1986, incidence in seed [**61**,2576; **63**,2600; **64**,1859; **65**,5300, 5807].

A. lethalis, teleom. *Didymella lethalis.*

A. majilis, teleom. *Mycosphaerella convallariae.*

A. oryzae, see *Phomopsis oryzae-sativae.*

A. paspali (H. Sydow) Punith. 1979; Punithalingam, *Descr.F.* 821, 1985; con. 17–21 × 4.5–5.5 μm; leaf blotch, on *Paspalum dilatum,* Dallis grass, seedborne, New Zealand, USA; can cause losses in pastures; Buchanan, *N.Z.J.Bot.* **22**:515, 1984; *N.Z.J.agric.Res.* **27**:451, 1984 [**64**,2063,3474].

A. phaseolorum Sacc. 1878; Sutton & Waterston, *Descr.F.* 81, 1966; con. 13–16.5 × 3.5–5 μm; bean leaf and pod spot, legumes, seedborne.

A. pinodes, teleom. *Mycosphaerella pinodes.*

A. pisi Lib. 1830; Punithalingam & Holliday,

Descr.F. 334, 1972; con. 10–16 × 3–4.5 μm; pea leaf, pod and stem spot. legumes, seedborne, forms ascochitine q.v., often found with *Ascochyta pinodes* which on oat agar forms a light buff to flesh exudate of conidia, that of *A. pisi* is carrot-red; Holliday 1980; Maude et al., *Tests Agrochem. Cvs.* 7, *AAB* **108** suppl.: 70, 1986, chemical treatment of seed; Darby et al., *Pl.Path.* **35**:214, 1986, 5 major pathotypes & differential hosts [**65**,5780].

A. pteridis, see *Pteridium aquilinum*.

A. rabiei (Pass.) Labrousse 1931; Punithalingam & Holliday, *Descr.F.* 337, 1972; con. 10–16 × 3.5 μm; chickpea blight, seedborne, review by Nene, *Trop.Pest Management* **28**:61, 1982; Reddy & Kabbabeh, *Phytopath.Mediterranea* **24**:265, 1985, 6 races delineated [**66**,1240].

A. sorghi Sacc. 1875; Punithalingam, *Descr.F.* 632, 1979; *Mycol. Pap.* 142:144, 1979, and for a discussion on *Ascochyta sorghina* Sacc. 1878, whose taxonomic status is in doubt; sorghum rough leaf spot.

A. tarda Stewart, *Mycologia* **49**:430, 1957; con. 9–14 × 2–3 μm; Firman, *TBMS* **48**:161,1965; coffee leaf blight & stem dieback [**36**:760; **44**,3055].

Ascodichaena Butin, *TBMS* **69**:249, 1977; Ascodichaenaceae. *A. rugosa* Butin; anam. *Polymorphum quercinum* (Pers.) Chev. fide Hawksworth, *Taxon* **32**:212, 1983; asco. aseptate, 18–24 × 13–16 μm; con. 22–30 × 12–16 μm; beech bark black scab. Europe; Butin, *Eur.J.For.Path.* **11**:299, 1981 [**61**,3085; **63**,262].

Ascodichaenaceae, Rhytismatales, ascomata carbonaceous, excipulum absent, opening by a slit, asci splitting open, asco. sheath absent; *Ascodichaena*.

ascoma, plural ascomata, any fungus structure which bears asci and asco.

Ascomycotina, Ascomycetes, Eumycota; mycelium septate, sexual reproduction by asco. endogenously in asci, the largest group in the true fungi, there is no general agreement on their classification. Past groupings were based on how the asci were arranged, e.g. plectomycetes and pyrenomycetes; these have given way to emphasising development of the ascoma and the presence or absence of the bitunicate (double-walled) ascus, and variations. There are *c.*37 orders, of which *c.*15 contain plant pathogens.

ascospore, spore formed by free cell formation in an ascus.

ascus, the typical cell of the Ascomycotina, shaped like a bag or hollow club and containing asco (generally 8) which are formed after karyogamy and meiosis in free cell formation. The separation

into 2–3 ascus types e.g. bitunicate (double-walled), prototunicate, unitunicate (one wall) is a simplification, see *Ainsworth & Bisby's Dictionary of the fungi* for the 9 main ascus types.

aseptate, of a fungus, having no cross walls.

asexual state, imperfect state, anam., see fungi.

ash, *Fraxinus*.

—— **chlorotic necrotic spot** Casalicchio, *Monti Boschi* **16**(6):39, 1965; transm. sap, Italy (Bologna) [**46**,750].

—— —— **vein banding** (arabis mosaic v.), Cooper, *Pl.Path.* **24**:114, 1975 [**55**,403].

—— **decline**, Castello et al., *PD* **69**:243, 1985; etiology unknown, USA (N.E.), *Fraxinus americana*; this diseased condition has been well known since the early 1950s, it may have several causes; a decline has also been reported in UK, see ash chlorotic vein banding [**64**,3354].

—— **mosaic** Schmelzer, H.E. & H.B. Schmidt, *Arch.Forstw.* **15**:107, 1966; Europe [**46**,442].

—— **ringspot** (poss. str. tobacco ringspot v.), Hibben & Bozarth, *Phytop.* **62**:1023, 1972; USA (New York State), decline [**52**,1676].

—— **yellows**, see Matteoni & Sinclair, *Phytop.* **75**:355, 1985; part of ash dieback and decline syndrome, USA (N.E.), MLO assoc. [**64**,5127].

Ashby, Sidney Francis, 1874–1954; born in England, Univ. Liverpool and Edinburgh; Jamaica, worked on Panama wilt of banana, chair at Imperial College of Tropical Agriculture, Trinidad 1921–6; Director, Imperial Mycological Institute 1935; student of *Phytophthora*. *Nature, Lond.* **173**:802, 1954.

ashbyi, *Eremothecium*.

ashy stem blight (*Macrophomina phaseolina*) plurivorous.

asiaticum, *Gymnosporangium*.

asparagi, *Cercospora*, *Puccinia*.

asparagus, *Asparagus officinalis*; Weissenfels, *Arch.Gartenbau* **21**: 235, 1973, viruses; —— **bean**, see *Vigna*.

—— **B** = asparagus 1 v.; —— **C** = asparagus 2 v.; —— **latent 2** = asparagus 2 v.

—— **1 v.** Hein 1960, 1969; Mink & Uyeda, *PDR* **61**:398, 1977 Fujisawa et al., *Ann.phytopath.Soc.Japan* **49**:299, 1983; Howell & Mink, *PD* **69**:1044, 1985; filament av. 740 nm long, transm. sap, *Aphis craccivora*, *Myz.pers.*, non-persistent [**39**:760; **57**,891; **63**,3153].

—— **2 v.** Hein 1963; Uyeda & Mink, *Descr.V.* 288, 1984; Ilarvirus, quasi-isometric *c.*26–36 nm diam., transm. sap and seed; Europe, Japan, N. America, occurs naturally only in asparagus, no obvious symptoms, vigour and yield are reduced, wide experimental host range; Phillips & Brunt,

Pl.Path. **34**:440, 1985, in cv. Limbergia with cucumber mosaic v. [**65**,494].
— **3 v.** Fujisawa, *Ann.phytopath.Soc.Japan* **52**:193, 1986; Potexvirus, filament 580 × 13 nm, transm. sap to 26 spp. in 8 families, no distinct symptoms in inoculated asparagus [**66**,1257].
— **bean rugose mosaic** (blackeye cowpea mosaic v. + cucumber mosaic v.), Chang, *Pl.Prot.Bull.Taiwan* **25**:177, 1983, *Vigna unguiculata* ssp. *sesquipedalis* [**63**,325].
— **stunt** (tobacco streak v.), Brunt & Paludan, *Phytopath.Z.* **69**: 277, 1970, [**50**,2059].
aspen, *Populus.*
— **necrotic leaf spot** Boyer, *CJB* **40**:1237, 1962; Boyer & Navratil, *ibid* **48**:1141, 1970, aggregates suggesting a virus, transm. seed, Canada (Quebec); *Populus tremuloides* [**42**:282; **51**,4396].
Aspergillus Micheli ex Link 1809; Hyphom.; conidiomata synnematal; conidiogenesis enteroblastic, phialidic; con. aseptate; common moulds, conidia borne densely in coloured heads, often green or yellowish; cause diseases in man and other animals, aspergillosis; not primarily plant pathogens, although a few spp. are often assoc. with diseased conditions, especially on seed or after harvest and storage; Smith & Pateman ed., *Genetics and physiology of* Aspergillus, *Br.mycol.Soc.Sympos.* 1, 1977; Holliday 1980; Onions et al., *Smith's introduction to industrial mycology,* edn. 7, 1981; *Ainsworth & Bisby's dictionary of the fungi,* edn. 7, 1983.
A. flavus Link:Fr. 1832; Onions, *Descr.F.* 91, 1966; warmer regions, saprophytic and causing disease in plants and animals, forming aflatoxins when growing on plant products and causing aflatoxicoses in animals, occurs on seed and causes diseases inter alia in cotton, boll rot; groundnut, aflaroot; maize, soybean.
A. niger v. Tieghem 1867; Onions, *Descr.F.* 94, 1966; saprophytic, black mould of fruits and vegetables, post- and pre-emergent rot and collar rot of groundnut; see Maude et al. 1984, for storage rot of onions [**63**,5686].
Asperisporium Maubl. 1913; Ellis 1971, 1973; Hyphom.; conidiomata sporodochial, conidiophore growth sympodial, conidiogenesis holoblastic; con. pigmented, 1 septate; *A. caricae* (Speg.) Maubl.; Ellis & Holliday, *Descr.F.* 347, 1972; papaya leaf black spot.
aspidistra ring mottle Khristova & Alexsandrova, *Ovoshtarstvo* **9**(12):30, 1962; Bulgaria [**42**:464].
assistor virus, see helper virus and dependent transm.
aster chlorosis Kitajima & Costa, *Fitopat. Brasileira* **4**:55, 1979; bacilliform particles assoc.,

200–300 × 60–80 nm, Brazil, *Callistephus chinensis* [**58**,5279].
— **ringspot** (tobacco rattle v.), Corbett, *Phytop.* **57**:198, 1967 [**46**,2208].
Asterinaceae, Dothideales; mycelium superficial, bearing the ascomata which are flattened, uniloculate, opening by breakdown of the shield by radiating clefts of a slit; *Aulographina.*
asteris, *Phialophora.*
Asteroma DC. 1815; Coelom.; Sutton 1980; conidiomata acervular, conidiogenesis enteroblastic, phialidic; con. hyaline, aseptate; *A. frondicola* poss. teleom. *Limospora ceuthocarpa*; *A. ulmeum* poss. teleom. *Stegophora ulmea.*
Asteromella Pass. & Thümen 1880; Coelom.; Sutton 1980; conidiomata pycnidial, conidiogenesis enteroblastic, phialidic; con. hyaline, aseptate; *A. brassicae,* teleom. *Mycosphaerella brassicicola.*
asterum, *Coleosporium.*
*****aster yellows** Smith 1902; Kunkel 1924, 1926; see Ploaie & Maramorosh under Mycoplasmataceae. The disease is of historical interest since it was the first one to be shown to be assoc. with MLO infection. In 1898 an organism of this type was found to cause a disease in cattle; transm. leafhoppers, crops infected include: carrot, celery, lettuce, onion, potato, spinach; see potato purple top wilt, Shiomi & Sugiura; Sinha & Benhamou, *Phytop.* **73**:1199, 1983, serological detection; Sinha, *Yale J.Biol. Med.* **56**:737, 1983; Lin & Chen, *Science N.Y.* **227**:1233, 1985, serologically differentiated from MLO assoc. with: ash yellows, elm phloem necrosis, loofah witches' broom, maize bushy stunt, paulownia witches' broom, sweet potato witches' broom [**63**,1290; **64**,2415].
Athelia rolfsii, see *Corticium rolfsii.*
atmospheric trajectory analysis, application to problems in epidemiology; *Peronospora tabacina,* Davis & Main.
atomise, to reduce a liquid to fine droplets by passing it under pressure through a suitable nozzle, or by applying drops to a spinning disc, FBPP.
atramentosa, *Myriogenospora.*
atropa mild mosaic = henbane mosaic v.
Atropellis Zeller & Goodding 1930; Helotiaceae; a distinct epithecium formed over the asci, black in section, encrustations on the paraphyses turning KOH blue green or in 1 sp. brown, ascus pore not amyloid; Reid & Funk, *Mycologia* **58**:417, 1966; key & morphology; pine cankers, N. America, greyish green to blue black discolouration of wood beneath [**45**,3231].
A. apiculata Lohman, Cash & Davidson 1942;

asco. 1–2, rarely 3, celled, 20–24 × 4.8–6.5 μm
[**22**:187].

A. pinicola Zeller & Goodding; asco. 1–6 celled, av. 40 × 2 μm.

A. piniphila (Weir) Lohman & Cash 1940; asco. 0–2, rarely up to 4 celled, 16–28 × 4–7 μm; Hopkins, *CJB* **39**:1521, 1961; **41**:1535, 1963, culture & etiology [**19**:629; **41**,262; **43**,1168].

A. tingens Lohman & Cash; asco. 1–4, rarely 5–6, celled, av. 30 × 3 μm; Diller, *J.For.* **41**:41, 1943 [**22**:282].

A. treleasei (Sacc.) Zeller & Goodding; asco. aseptate, 42–60 × 2–5.5 μm; transferred to *Discocainia* by Reid & Funk 1966; Takahashi & Saho 1975 [**55**,2404].

atropunctatum, *Hypoxylon*.

atrum, *Ulocladium*.

attachment organ, of a fungus, initiated from a single aerial hypha which attaches to a substratum and branches dichotomously; this is not, as in appressoria, a preliminary to host penetration; first described by de Istvanffi 1905, see Jarvis & Berry, *TBMS* **87**:543, 1986, in *Botryotinia porri*.

attenuation, lessening the capacity of a pathogen to cause disease, reduction of its virulence, FBPP.

attenuatum, *Sarocladium*.

aubergine, see eggplant.

aucuba ringspot v., bacilliform 180 × 30 nm, fide Francki et al., vol. 2, 1985.

aucupariae, *Monilinia*.

Aulacophora fevicollis, transm. cucumber green mottle mosaic.

Aulacorthum circumflexum, transm. subterranean clover red leaf; *A. solani*, transm. soybean dwarf, subterranean clover red leaf.

Aulographina v. Arx & E. Müller 1960; Asterinaceae; Wall & Keene, *TBMS* **82**:257, 1984, *A. eucalypti* (Cooke & Massee) v. Arx & Müller, eucalyptus defoliation, Australia [**63**,2516].

aurantiogriseum, *Penicillium*.

aurantius, *Hypomyces*.

Aureobasidium, see *Kabatiella*.

Auriculariales, Phragmobasidiomycetidae; metabasidia ± cylindrical, transversely i.e. horizontally septate, basidiomata hemiangiocarpus and sessile in Auriculariaceae; *Helicobasidium*.

australe, *Lophodermium*.

Australian scab (*Sphaceloma fawcettii* var. *scabiosa*) citrus.

australis, *Elsinoë*, *Sphaceloma*.

Austroagallia torrida, transm. rugose leaf curl.

austro-americana, *Thyronectria*; **austro-americanum**, *Gyrostroma*.

author's names, of pathogens; Hawksworth, *Rev.Pl.Path.* **59**:473, 1980, gave the abbreviations and conventions for authorities of fungi; for

bacteria see Bradbury 1986. Strict priorities for plant virus names, authors given in full and where a binomial system is not used, do not apply; no authors are given for insects but are, in most cases, given for nematodes.

autoecious, of a fungus, completing its life cycle on one kind of host or hosts, especially rusts, cf. heteroecious.

autofluorescence, occurring during interactions between plant cells and pathogens; Russo & Pappelis, *TBMS* **81**:47, 1983 [**63**,338].

autogenous necrosis, tomato, genetic abnormality, particularly cv. Syston Cross, fide Fletcher 1984.

autotrophic, using inorganic nutrient sources for growth.

auxotroph, auxotrophic, micro-organism variants having nutritional requirements not required by the wild type or prototroph; e.g. a biochemical mutant which will only grow on a minimal medium, on which the prototroph will grow, after adding one or more specific substances.

Avena, oat; **avena**, *Stagonospora*.

avenacea, *Gibberella*; **avenaceum**, *Fusarium*.

avenae, *Ascochyta*, *Drechslera*, *Pseudomonas*, *Pyrenophora*, *Septoria*, *Ustilago*.

avenaria, *Leptosphaeria*.

avesiculatum, *Cylindrocladium*.

avirulent, of a variant of a pathogen that does not cause severe disease; but non-virulent is the preferred synonym to avoid confusion in speech with 'a virulent'.

avocado, *Persea americana*; Zentmyer, *Trop.Pest Management* **30**:388, 1984.

—— **black streak** Jordan, Dodds & Ohr 1981; *Phytop.* **73**:1130, 1983; 3 ds RNAs detected, transm. seed; RNA 2 alone or with RNA 3 may cause the disease, USA (California) [**63**:1335].

—— **sunblotch viroid** Horne & Parker 1931; Dale et al., *Descr.V.* 254, 1982; circular ss RNA molecule of 247 nucleotides, transm. sap difficult, in most growing areas, cv. Hass seedlings show yellow, orange or white streaks or spots on stems and petioles after 2 months to 3 years, leaf variegation and distortion.

Awka, coconut, see palm lethal yellowing.

awl nematode, *Dolichodorus heterocephalus*.

axenic culture, a growing system for one organism in the absence of all others, i.e. a pure culture cf. gnotobiotic.

axeny, inhospitality, 'passive' as opposed to 'active' resistance of a plant to a pathogen, FBPP.

axonopus chlorotic streak van Velsen, *Papua New Guin.agric.J.* **18**:139, 1967; transm. leafhopper [**47**,845].

Azalea, see rhododendron.

azaleae, *Ovulinia*, *Ovulitis*.

B

babiana mosaic, Smith & Brierley, *Phytop.* **34**:593, 1944; transm. aphid, USA [23:488].

baccata, *Gibberella*.

baccatin, toxin from *Gibberella baccata*, Gäuman et al., *Phytopath.Z.* **36**:111, 1959 [39:246].

bacilliform, cylindrical, sometimes restricted to a body with a length:breadth ratio of 2–3; a virus particle of this shape, see Phytorhabdovirus.

bacillus, a rod-shaped bacterium; **Bacillus* Cohn 1872 forms spores, none are active plant pathogens, sometimes assoc. with plants as saprophytes; Bradbury 1986.

bacteria, Procaryotae, Firmicutes and Gracilicutes; cell wall rigid, an elastic cell membrane, often unicellular but may be divided by one or more septa; cells are spherical, cocci; cylindrical, rods or bacilli; spiral, vibrios; multiplication mostly by division i.e. fission; chains of cells may occur; not dependent on chlorophyll; some motile with flagella; anaerobic, aerobic; facultative or obligate; all require small concentrations of CO_2; the Gram reaction q.v. is the basis for the 2 main divisions of bacteria; see phloem and xylem limited bacteria.

A.R.Phytop.: Schuster & Coyne, **12**:199, 1974, survival; Lacy & Leary, **17**:181, 1979, genetics; Hildebrand et al., **20**:235, 1982, DNA homology & taxonomy; Hirano & Upper, **21**:243, 1983, foliar pathogens, ecology & epidemiology; Starr, **22**:169, 1984; Mills, **23**:297, 1985, transposon mutagenesis & virulence genes; Huang **24**:141, 1986, fine structure & plant penetration. Also: Bradbury, *Rev.Pl.Path.* **49**:213, 1970, preliminary study; Dye et al., *ibid* **59**:153, 1980, nomenclature; Strobel, *A.Rev.Microbiol.* **31**:205, 1977, toxins; Rhodes-Roberts & Skinner ed., *Bacteria and plants*, 1982; Fahy & Persley 1983; Boelema, *Neth.J.Pl.Path.* **91**:189, 1985, dose response relationships; Bradbury 1986; Chatterjee & Vidaver, *Adv.Pl.Path.* **4**, 1986, genetics.

bacterial wetwood, see wetwood.

bactericide, a substance that kills bacteria; **bacteriostat**, a substance that inhibits the multiplication of bacteria but does not kill them.

bacteriocin, a highly specific antibacterial protein formed by bacterial strs. and mainly active against other strs. of the same sp., after FBPP; Vidaver, *PD* **67**:471, 1983.

bacteriophage, or phage, a virus infecting procaryotes; phage particles multiply in the host cell, e.g. a bacterium, which bursts, thus releasing them and causing lysis. Phage typing is a method of distinguishing bacterial spp. or types by testing reactions to selected phages; hence phagotype, distinguished by sensitivity to a specific phage; Okabe & Goto, *A.R.Phytop.* **1**:397, 1963; Ackerman et al., *Adv.Virus Res.* **23**:1, 1978.

bacteriosis, infection by bacteria or a disease so caused, FBPP.

baekea yellows Hiruki, *Pl.Path.* **35**:396, 1986; MLO assoc., *B. virgata*, Australia [66,817].

bainieri, *Mortierella*.

bakanae (*Gibberella fujikuroi*), rice.

Baker, Richard Eric Defoe, 1908–54; born in England, Univ. Cambridge, Imperial College of Agriculture, Trinidad, 1932–54, head department mycology and plant pathology, later with botany, diseases of cacao and citrus, edited flora of Trinidad and Tobago, *Nature Lond.* **174**:1128, 1954; *Trop.Agric.Trin.* **32**:159, 1955.

Balansia Speg. 1885; Clavicipitaceae; ascomata deeply immersed in stroma, that is rarely stalked and arises from a pseudo-sclerotium of host cells and hyphae, asco. filiform; mostly on tropical grasses, Diehl, *USDA Agric.Monogr.* 4, 1950 [30:491].

B. claviceps Speg.; poss. anam. *Ephelis mexicana* Fr.:Berk. 1868; con. prob. shorter than those of *E. oryzae* fide Sprague 1950; Tarr 1962; Ullasa, *Mycologia* **61**:572, 1969, anam. in vitro.

B. cyperi Edg., *Mycologia* **11**:259, 1919; con. 15–30 × 1.8 μm; Clay, *PD* **70**:597, 1986, on the serious weed *Cyperus rotundus*, purple nutsedge [65,5836].

B. oryzae-sativae Hashioka 1971; Booth, *Descr.F.* 640, 1979; anam. *Ephelis oryzae* Sydow 1914; con. mostly 18–22 μm long; rice udabatta, panicle emerges as a mummified, erect, greyish white axis, initially covered with white mycelium which darkens, becoming stromatic; seedborne, systemic; Mohanty, *Riso* **26**:243, 1977; **28**:133, 1979 [57,1180; 59:1250].

Baldulus tripsaci, transm. maize rayado fino.

ballistospore, a forcibly discharged basidiospore, see Buller's drop; Webster et al., *TBMS* **82**:13, 1984.

balsamea, *Dermea*, *Thyronectria*; **balsamicola**, *Potebniamyces*; **balsamiferae**, *Pollaccia*.

bambarra groundnut, see *Vigna*.

bamboo mosaic v. Lin, *Phytop.* **67**:1439, 1977; Lin et al. 1979; Chen, *Pl.Prot.Bull.Taiwan* **27**:111, 1985; filament 480–500 μm long, transm. sap,

Brazil (São Paulo), Taiwan [**57**,4502; **64**,1159; **65**,236].

*banana, and plantain, *Musa*; Meredith, *Rev.Pl.Path.* **49**:539, 1970, major diseases; *Trop.Agric.Trin.* **48**:35, 1971, transport & storage diseases; Stover 1972; Wardlaw, *Banana diseases*, 1972; anon., *PANS manual* 1, 1977; Waller, *Pestic.Sci.* **9**:478, 1978, chemical control; Slabaugh & Grove, *PD* **66**:746, 1982, postharvest; Stover, *A.R.Phytop.* **24**:83, 1986, disease management.

— **bunchy top** Magee 1927; *Proc.R.Soc.N.S.W.* **87**:3, 1953; the cause of this old and important disease, assumed to be a virus, is still unknown; transm. *Pentalonia nigronervosa*, Asia (S.E., including India), Australia (E.), central Africa, Egypt, Oceania; Allen in Plumb & Thresh 1983, spread & control by roguing [**7**:253; **34**:381].

— **dotted line mosaic** Lassoudière *Fruits* **29**:349, 1974; Ivory Coast [**53**,4851].

— **streak** Lockhart, *Phytop.* **76**:995, 1986; bacilliform particles assoc. 119 × 27 nm, Morocco [**66**,2443].

band canker (*Botryosphaeria ribes*) almond.

banded chlorosis, sugarcane, due to a sudden drop of temp. after heavy rain, but may have other causes; Faris, *Phytop.* **16**:885, 1926; — **leaf blight** (*Thanatephorus cucumeris*), arrowroot [**6**:184].

barclayana, *Tilletia*.

bare patch (*Thanatephorus cucumeris*) temperate cereals; MacNish, *Pl.Path.* **34**:159, 165, 175, 1985; Rovira, *Phytop.* **76**:669, 1986 [**64**,4224–5, 4266; **66**,154].

bark black scab (*Ascodichaena rugosa*) beech; — **blotch** (*Ascochyta corticola*) lemon; — **canker** (*Phytophthora palmivora* and bark beetles) cacao, Prior, *AAB* **109**:535, 1986; (*Potebniamyces discolor*) apple; — **hypertrophy** (*Seimatosporium etheridgei* assoc.) *Populus tremuloides*; — **measles**, apple, olive caused by B deficiency; Scott et al., *Phytop.* **33**:933, 1943; Bould et al. 1983; olive partial paralysis, Nicolini & Traversi 1950, in Argentina may have the same cause; — **necrosis** (*Nectria fuckeliana*) spruce, (*Cryptodiaporthe*) poplar; — **rot** (*N.striispora*) *Erythrina*, immortelle; — **stem necrosis** (*N.coccinea*) beech [**23**:69; **30**:377; **66**,2281].

barley, *Hordeum*; Colhoun, Manners in Western 1971; Habgood & Clifford in Jenkyn & Plumb 1981; Mathre 1982.

— **dark green dwarf** (poss. maize rough dwarf v.), Shen & Chen, *Biochem.biophys.sin.* **15**:191, 1983 [**62**,3834].

— **false stripe** = barley stripe mosaic v.

— **mosaic v.** Dhanraj & Raychaudhuri, *PDR*

53:766, 1969; isometric 40 nm diam., transm. sap to barley, oat, wheat, and *Rhopalosiphum maidis*, seedborne in barley, India [**49**,446].

— **striate mosaic v.** Conti, *Phytopath.Z.* **66**:275, 1969; Conti & Appiano, *J.gen.Virol.* **21**:315, 1973; Phytorhabdovirus, bacilliform 260–330 × 45–60 nm, transm. *Laodelphax striatellus* to barley, oat, wheat; Europe, Morocco [**49**,1581; **53**,1777].

— **stripe mosaic v.** McKinney 1951; Atabekov & Movikov, *Descr.V.* 68, 1971; Jackson & Lane in Kurstak 1981; Hordeivirus, type, rod *c*.128 × 20 nm, transm. in seed and pollen, barley and wheat are natural hosts; Chiko, *Virology* **63**:115, 1975, multiple virion components [**54**,3849].

*— yellow dwarf v. Oswald & Houston 1951; Rochow, *Descr.V.* 32, 1970; Bruehl, *Monogr.Am.phytopath.Soc.* 1, 1961; Plumb in Plumb & Thresh 1983; Hill in Wolfe & Caten 1987; Luteovirus, type, isometric *c*.20–24 nm, transm. aphid spp., persistent, temperate Gramineae. BYDV causes the most severe disease of barley, and causes problems in oat and wheat, it occurs in many strs., the diseases caused differ in epidemiology often because the vectors differ; strs. MAV, RMV, RPV, SGV are transm. specifically by *Sitobion avenae*, *Rhopalosiphum maidis*, *R. padi* and *Schizaphis graminum*, respectively; PAV is transm., non-specifically, by *R. padi* and *S. avenae*; strs. differ serologically; Rochow, *PD* **66**:381, 1982; Paliwal, *Can.J.Pl.Path.* **4**:59, 1982; *CJB* **60**:179, 1982, identification & detection in aphids; Burnett & Cuéllar 1983; Waterhouse & Helms, *Australasian Pl.Path.* **14**:64, 1985, *Metopolophium dirhodum* as a vector; Clement et al., *Phytop.* **76**:86, 1986, ecology & epidemiology in USA (Indiana) [**61**,5593, 6299, 6300; **65**,1152, 4374, 4807].

— **— mosaic v.** Ikata & Kawai, 1940; Inouye & Saito, *Descr.V.* 143, 1975; Huth, *NachrBl.dt.PflSchutzdienst.* **36**:49, 1984; *Phytopath.Z.* **111**:37, 1984; Hill, as above; Potyvirus sub-group 2, fungus transm., filament, 2 particle types 270–290 and 550–600 nm long, transm. sap and *Polymyxa graminis*, soilborne, can cause severe loss, Japan, parts of W. Europe [**63**,3852; **64**,2001].

— **— striate mosaic v.** Conti 1969; Milne & Conti, *Descr.V.* 312, 1986; Phytorhabdovirus sub-group 1, bacilliform *c*.330 × 55 nm, transm. *Javesella pellucida*, *Laodelphax striatellus*, persistent; causes chlorotic stripes in winter wheat, maize; Australia, Italy, France, Morocco; little economic importance; Australian virus is the MS str. in maize.

barney patch disorder, see beet soilborne v.

barn mould (*Botryosporium longibrachiatum*); — rot (*Rhizopus oryzae*) tobacco.

barnyard millet, see millets.

basal canker (*Phytophthora cambivora*) Norway maple, — glume rot (*Pseudomonas syringae* pv. *atrofaciens*) barley, oat, wheat; — injury, pine needles assoc. with scale insects and pollutants, fide Rice et al., *CJB* **64**:632, 1986; — node rot (*Fusarium oxysporum*) rice, fide Prabhu & Bedendo, *PD* **67**:228, 1983; — rot (*F.oxysporum* ff. sp. *cepae* and *narcissi*) onion, narcissus, (*Ganoderma zonatum*) palms, (*Phytophthora citricola*) tomato; — stem break (*Marasmiellus cocophilus*) coconut; — — canker (*Thanatephorus cucumeris*) tomato; — — rot (*Ganoderma* spp.) oil palm, (*Gibellina cerealis*) wheat [**62**,3029; **65**,4574].

basauxic, of a conidiophore, elongating by a basal growing point.

basicola, *Thielaviopsis*.

basidioma, basidiocarp, fungus structure forming basidia.

Basidiomycotina, Basidiomycetes, Eumycota; mycelium septate, sexual reproduction by basidiospores borne on basidia, on germination the primary mycelium is haploid becoming dikaryotic by diploidisation, paired nuclei, and forming a secondary mycelium frequently with clamp connections, septa are dolipores; macroscopic basidiomata present: Hymenomycetes, Gasteromycetes; or absent: Urediniomycetes, Ustilaginomycetes. Khan & Kimbrough, *Mycotaxon* **15**:103, 1982, proposed 4 classes based mainly on septal pore fine structure: Hemibasidiomycetes, Holobasidiomycetes, Phragmobasidiomycetes, Teliomycetes, the 4th and 1st include the rusts and smuts, respectively.

basidiospore, spore formed on a basidium, see ballistospore.

basidium, cell or organ typical of the Basidiomycotina, from which, after karyogamy and meiosis, the basidiospores, generally 4, are formed externally on an extension of its wall, the sterigma; Ingold, *Bull.Br.mycol.Soc.* **17**:82, 1983.

Bassi, Agostino, 1773–1856; born in Italy, Univ. Pavia; he demonstrated in 1835 that the muscadine disease in silkworms in France and Italy was caused by a fungus, *Beauveria bassiana*; he may be regarded as a founder of microbial pathogenesis; see Ainsworth, *Introduction to the history of mycology*, 1976; wrote: *Del mal del segno calcinaccio o moscardino*, 1835, 1836; *Phytopath.Classics* 10, 1958, translated by P.J. Yarrow.

batatas, *Elsinoë*, *Sphaceloma*.

batschiana, *Ciboria*.

baudonii, *Pseudophaeolus*; baudysii, *Bremiella*.

Bawden, Frederick Charles, 1908–72; born in England, Univ. Cambridge, potato virus research unit, Rothamsted Experimental Station, head plant pathology 1936, director 1958; a world authority on plant virus diseases; with E.T.C. Spooner he unravelled the etiology of potato diseases caused by viruses; with N.W. Pirie he demonstrated that viruses are nucleo-proteins; FRS 1949, Knight 1967, research medal of the Royal Agricultural Society of England, Elvin C. Stakman award from Univ. Minnesota, wrote: *Plant viruses and virus diseases*, 1939, edn. 4, 1964; *Plant disease*, 1950. *AAB* **70**:107, 1972; *Nature Lond.* **236**:128, 1972; *Indian Phytopath.* **24**(4), 1971, published 1972; *Biogr.Mem.Fel.R.Soc.* **19**:19, 1973.

bayberry yellows Raychaudhuri 1952, *Phytop.* **43**:15, 1953; USA (New Jersey), *Myrica carolinensis* [**32**:663].

bayoud (*Fusarium oxysporum* f.sp. *albedinis*) date palm.

Bdellovibrio, Gram negative bacteria which parasitise other bacteria and cause lysis, saprophytic types occur; Stolp, *A.R.Phytop.* **11**:53, 1973.

bean, common, dwarf, French, haricot, kidney, *Phaseolus vulgaris*; see asparagus, Bengal, broad, cluster, Lima, mung, runner, sword, urd, velvet, winged; and legumes; Hubbeling, *Meded.Inst.plziektenk.Onderz.* 83, 1955; Zaumeyer & Thomas, *Tech.Bull.USDA* 868, 1957; Zaumeyer & Meiners, *A.R.Phytop.* **13**:313, 1975; Schuster & Coyne, *Hortic.Rev.* **3**:28, 1981; Webster et al., *PD* **67**:935, 1983; last 2 references on bacteria; Allen 1983; Sherf & MacNab 1986.

— angular mosaic (prob. cowpea mild mottle v.), Costa et al., *Fitopat.Brasiliera* **8**:325, 1983; Brazil [**63**:4153].

— atypical mosaic Nagaich & Vashisth, *Indian J.Microbiol.* **3**:113, 1963; transm. sap, seedborne in sunn hemp, India (Simla) [**45**,1955].

— common mosaic v. Stewart & Reddick 1917; Bos, *Descr.V.* 73, 1971; Boswell & Gibbs 1983; Potyvirus, filament *c*.750 × 15 nm, transm. sap, aphid spp., non-persistent, seed, pollen; Hagita & Tamada, *Bull.Hokkaido Prefect.agric.Exp.Stn.* 51:83, 1984, detection in bean seed by immune electron microscopy [**64**,3611].

— crumpling (euphorbia mosaic v.), Costa, *F.A.O. Pl.Prot.Bull.* **13**:121, 1965; Brazil (São Paulo) [**45**,2295].

— curly dwarf mosaic (str. of quail pea mosaic v.), Meiners et al., *Phytop.* **67**:163, 1977; El Salvador [**56**,5273].

— distortion dwarf v. Xi, Xu & Mang, *Acta*

microbiol.sin. **22**:293, 1982; poss. Geminivirus, poss. str. of bean golden mosaic v., isometric 9–11 nm diam., transm. sap and white fly, China [62,3691].

—— **golden mosaic v.** Costa 1965; Goodman & Bird, *Descr.V.* 192, 1978; Geminivirus, isometric *c*.19 nm diam., transm. sap, *Bemisia tabaci*, tropical and subtropical America, poss. elsewhere, Leguminosae; Goodman et al., *Virology* **106**:168, 1980, composition, ssDNA; Haber et al., *Nature Lond.* **289**:324, 1981, divided genome [60,3074, 4883].

—— **leaf curl** Singh, *F.A.O.Pl.Prot.Bull.* **24**:100, 1976; transm. sap, India (Pantnagar) [56,4271].

—— —— **roll v.** Quantz & Volk 1954, as leaf roll of *Vicia* beans and peas; Ashby, *Descr.V.* 286, 1984; Luteovirus, isometric *c*.27 nm diam., transm. 7 aphid spp., main vector *Acyrthosiphon pisum*, causes diseases in bean, broad bean, chickpea, cowpea, lentil, pea; infection of lucerne and white clover usually symptomless, see bean yellow vein banding v.

—— —— **wilt** Johnson, *J.agric.Res.* **64**:443, 1942; transm. sap to bean, from *Lathyrus pusillus* without symptoms, USA [21:388].

—— **lethal necrosis** (tobacco yellow dwarf v.).

—— **line pattern mosaic** Camargo, Kitajima & Costa, *Fitopat. Brasileira* **1**:207, 1976; isometric particles assoc. 17–23 nm diam., Brazil (São Paulo) [56,3309].

—— **mild mosaic v.** Waterworth et al., 1977; *Descr.V.* 231, 1981; isometric *c*.28 nm diam., contains RNA, transm. sap, beetles; Colombia, El Salvador; Leguminosae.

—— **pod mottle v.** Zaumeyer & Thomas 1948; Semancik, *Descr.V.* 108, 1972; Comovirus, isometric *c*.30 nm diam., transm. sap, beetles, USA, *Glycine* and *Phaseolus* spp.; Lin & Hill, *PD* **67**:230, 1983, seedborne in soybean; Hopkins & Mueller 1984, effect on yield in soybean [62,3341; 65,993].

—— **ringspot** (tomato black ring v.).

—— **rugose mosaic v.** Gámez 1972; *Descr.V.* 246, 1982; Comovirus, isometric 28 nm diam., transm. sap, beetles, Central America, strs., Leguminosae.

—— **severe bud blight** (tobacco ringspot v.), Tu, *Phytopath.Z.* **101**:153, 1981 [61,924].

—— **southern mosaic v.** Zaumeyer & Harter 1942, 1943; Tremaine & Hamilton, *Descr.V.* 274, 1983; Sobemovirus, type, isometric, *c*.30 nm diam., transm. sap, seed, leaf beetles, Chrysomelidae; only Leguminosae, narrow host range, causes diseases in bean, cowpea, soybean, urd bean; McGovern & Kuhn, *Phytop.* **74**:95, 1984, strs. [63,2617].

—— **stipple streak** (tobacco necrosis v.), van der Want, *Tijdschr. PlZiekt.* **54**:85, 1948 [28:44].

—— **summer death** = tobacco yellow dwarf v.

—— **top necrosis** (str. bean yellow mosaic v.).

—— **western mosaic** (str. bean common mosaic v.), Silbernagel, *Phytop.* **59**:1809, 1969 [49,1843].

—— **yellow mosaic v.** Pierce 1934; Bos, *Descr.V.* 40, 1970; Boswell & Gibbs 1983; Potyvirus, filament 750 × 15 nm; transm. sap, > 20 aphid spp., nonpersistent, and seed, mostly Leguminoseae, many strs.; Uyeda et al., *Mem.Fac.Agric.Hokkaido Univ.* **13**:69, 1982; strs. causing bean top necrosis; Hunst & Tolin, *PD* **66**:955, 1982, str. from *Gibasis geniculata*, Tahitan bridal veil; Stein et al., *AAB* **109**:147, 1986, in corms of *Gladiolus grandiflorus* [61,5333; 62,227; 65,6075].

—— —— **stipple** (str. cowpea chlorotic mottle v.), Zaumeyer & Thomas, *Phytop.* **40**:487, 1950; Fulton et al., *ibid* **65**:741, 1975; Gámez, *Turrialba* **26**:160 [30:257; 55,650; 56,2278].

—— —— **vein banding v.** Cockbain 1978; Cockbain et al., *AAB* **108**:59, 1986; no particle typical of a virus; transm. sap, *Acyrthosiphon pisum*, more readily, and *Myz. pers.*, persistent; transm. only from plants also infected with a helper virus, usually pea enation mosaic v. After separation from PEMV, by passage through *Phaseolus vulgaris*, it was no longer transm. by aphids; bean leaf roll v. is also a helper; BYVBV is found in broad bean mixed with PEMV in S. England [65,5789].

bearded iris mosaic = iris severe mosaic v., A.A. Brunt, personal communication, 1987; Barnett & Brunt, *Descr.V.* 147, 1975.

Beaumont period, meteorological criteria for conditions which are favourable for the development of potato blight, thereby warning of an outbreak which can be expected 10 days later; devised by Beaumont, *TBMS* **31**:45, 1947, primarily for S.W. England. The period is defined as one of 2 days when the minimum temp. is never < 10° and the relative humidity exceeds 75%; see blitecast, Large, *Pl.Path.* **2**:1, 1953 [27:88, 32:585].

beech, *Fagus*; Nienhaus et al., *Eur.J.For.Path.* **15**:402, 1985, reported virus particles, Germany [65,2447].

beet, a cv. of *Beta vulgaris*; Hull in Western 1971; Duffus in Kurstak 1971, viruses; Bugbee, *PD* **66**:871, 1982, storage rot; *Aspects appl.Biol.* 2, 1983, *Assoc.Appl.Biologists*, weeds; Byford & Gambogi, *TBMS* **84**:21, 1985, fungi on seed; Whitney & Duffus 1986 [64,1804].

*—— **cryptic v.**, see cryptic viruses and Kassanis et al., *Phytopath.Z.* **90**:350, 1977; **91**:76, 1978 [57,3662, 4174].

—— **curly leaf** (beet curly top v.).

—— —— **top v.** Ball 1909; Thomas & Mink, *Descr.V.* 210, 1979; Bennett, *Monogr.*

Am.phytopath.Soc. 7, 1971; Geminivirus, isometric 20 nm diam., single particles; transm. sap, but only by special procedures, *Neoaliturus opacipennis*, *N. tenellus*, persistent, restricted to phloem, wide host range, of great historical and economic importance in Canada (S.W.), Mediterranean (E.), Mexico, USA (W.), vectors confined to arid or semi-arid climates, BCTV has caused losses in bean, tomato and cucurbits; Duffus in Plumb & Thresh 1983.

— **cyst nematode**, *Heterodera schachtii*.

— **latent rosette** Bennett & Duffus 1957; Nienhaus & Schmutterer, *Z.Pflkrankh.PflSchutz.* **83**:641, 1976; Green & Nienhaus, *ibid* **87**:745, 1980; bacterium assoc., transm. *Piesma quadratum*, Germany (W.), USA [**37**:566; **56**,3750; **60**,5591].

— **leaf curl v.** Wille 1928, 1929; Proesler *Descr.V.* 268, 1983; Phytorhabdovirus, bacilliform *c.*225 × 80 nm, transm. *Piesma quadratum*, persistent, central Europe, sandy soils, no economic importance.

— **mild yellowing v.** Russell, *AAB* **46**:393, 1958; **48**:721, 1960; Jadot, *Parasitica* **30**:37, 45, 1974; **31**:62, 1975; Duffus & Russell, *Phytop.* **65**:811, 1975; Luteovirus, isometric 26 nm diam., transm. *Myz.pers.*, persistent, serologically close to beet western yellows v.; Govier, *AAB* **107**:439, 1985; purification, partial characterisation & serological detection [**38**:170; **40**:389; **54**,4218–9; **55**,1515–6; **65**,5758].

— — **yellows**, also beet family 41 yellows, Clinch & Loughnane, *Scient.Proc.R.Dubl.Soc.* **24**:307, 1948; prob. a str. of beet mild yellowing v. [**28**:264].

— **mosaic v.** Lind 1915; Russell, *Descr.V.* 53, 1971; Boswell & Gibbs 1983; Potyvirus, filament *c.*730 × 13 nm, transm. sap, aphid spp., non-persistent; Katis & Gibson, *Pl.Path.* **33**:425, 1984; transm. cereal aphids [**63**,5591].

* — **necrotic yellow vein v.** Tamada & Baba 1970; Tamada, *Descr.V.* 144, 1975; Al Musa & Mink, *Phytop.* **71**:773, 1981; Hecht, *Bayer landw.Jb.* **58**:600, 1981, review; poss. wheat soilborne mosaic group, rod of 4 lengths *c.*85, 100, 265, 390 nm, transm. sap, *Polymyxa betae* q.v.; causes the serious rhizomania of beet in Austria, France, Germany, Italy, Japan, USA (California) [**61**,2175].

— **pseudo-yellows** Duffus, *Phytop.* **55**:450, 1965; transm. *Trialeurodes vaporariorum*, USA (California) [**44**,2277].

— **ring mottle** Duffus & Costa, *Phytop.* **53**:1422, 1963; transm. sap, aphid spp., non-persistent, USA (California) [**43**,1468].

— **ringspot** (tomato black ring v.).

— **rosette** Bennet & Duffus, *PDR* **41**:1001, 1957; poss. MLO, USA (California) [**37**:566].

— **savoy** (prob. beet leaf curl v.), Coons et al., *PDR* **42**:502, 1958 [**37**:614].

— **soilborne v.** Ivanović & MacFarlane 1982; Henry et al., *Pl.Path.* **35**:585, 1986; rod, *c.*19 nm wide with 3 predominant lengths: 65, 150, 300 nm, England, Netherlands, assoc. with barney patch disorder.

— **temperate v.** Natsuaki et al. 1981; *Ann.phytopath.Soc.Japan* **49**:709, 1983; isometric *c.*30 nm diam., contains ds RNA, transm. seed; see cryptic viruses.

— **western yellows v.** Duffus 1960; *Descr.V.* 89, 1972; Boswell & Gibbs 1983; Whitney & Duffus 1986; Luteovirus, isometric 26 nm diam., transm. aphid spp., persistent, many dicotyledons, the wide host range makes BWYV very difficult to control; Ashby et al., *Neth.J.Pl.Path.* **85**:99, 1979, from lettuce, other vegetables, weeds; Smith & Hinckes, *AAB* **107**:473, 1985; from oilseed rape [**59**,1923; **65**,5757].

— **witches' broom** (beet latent rosette).

— **yellow blotch** (tobacco rattle v.).

— — **net** Sylvester, *Phytop.* **38**:429, 1948; *J.econ.Ent.* **51**:812, 1958; Roland, *Parasitica* **4**:152; transm. *Myz.pers.*, persistent, Belgium, England, USA (California) [**28**:105, 202; **38**:435].

— — — **mild yellows** Watson, *AAB* **50**:451, 1962; transm. aphid, poss. helper for beet yellow net, England [**42**:165].

— **yellows v.** Roland 1936; Russell, *Descr.V.* 13, 1970; Closterovirus, type, filament *c.*1250 × 12 nm, transm. sap difficult and at least 22 aphid spp., the most important *Aphis fabae* and *Myz.pers.*, semi-persistent, widespread with crop, mainly Chenopodiaceae; Rochow & Duffus in Kurstak 1981.

— **yellow stunt v.** Duffus 1972; *Descr.V.* 207, 1979; Closterovirus, filament *c.*1400 × 12 nm, transm. spp., semi-persistent, USA (California), not serologically related to beet yellows v., also: differs in host range in Compositae, BYSV infects lettuce and sowthistle, *Sonchus oleraceus*, both of which are immune to BYV.; the most efficient vector is *Nasonovia lactucae* which is common on sowthistle, it transm. BYSV to lettuce, causing stunt, but very inefficiently to beet.

— — **vein** Robbins 1921; Bennett, *PDR* **40**:611, 1956; Ruppel & Duffus, *Phytop.* **61**:1418, 1971; Staples et al., *J.econ.Ent.* **63**:460, 1970; transm. sap, *Aceratagallia calcaris*, USA [**36**:367; **50**,2034, **51**,2941].

— — **wilt** Bennett & Munck 1946; Bennett et al., *J.Am.Soc.Sug.Beet Technol.* **14**:480, 1967; Ehrenfeld 1970; Arentsen et al., *Cultivo Remolacha Azucarera Chile* 3, 1973; bacterium assoc., transm. *Paratanus exitiosus*, Argentina, Chile [**25**:532; **47**,2312; **52**,268; **54**,1917].

beetles, see *Ceratocystis*, vector.
begoniae, *Microsphaera*, *Oidium*
begonia yellow spot (tobacco ringspot v.), Lockhart & Betzold, *PDR* 63:1046, 1979, USA (Minnesota) [59,4634].
Beijerinck, Martinus Willem, 1851–1931; born in Netherlands, Polytechnic Delft, Univ. Leiden, Wageningen, fundamental work in microbiology; notable particularly for the concept of the filterable 'contagium vivum fluidum' in work with tobacco mosaic v. in 1898; he saw the causal agent of the disease in molecular terms, and which only replicated in the living host; Leeuwenhoek Medal 1905, Emil Christian Hansen Medal 1922, Foreign Member of the Royal Society 1926. *Proc.R.Soc.* B **109**:i, 1932; *Phytopath.Classics* 7, 1942, translated by J. Johnson; Waterson & Wilkinson 1978.
Beǐlin, Isaac Grigorevich, 1883–1965; diseases of ornamentals in European Russia and central Asia, see Parnes 1983 [63,1549].
bell pepper mottle v. Gracia et al. 1968; Feldman & Oremianer 1972; Tobamovirus, red pepper; Wetter et al., *J.Phytopath.* **119**:333, 1987 [48,63; 52,3499].
belladona mosaic = tobacco rattle v.
— mottle v. Bode & Marcus 1959; Paul, *Descr.V.* 52, 1971; Tymovirus, isometric *c.*27 nm diam., transm. sap, *Epitrix atropae*, Europe, Solanaceae; Moline & Fries, *Phytop.* 64:44, 1974, serologically related str. from *Physalis heterophylla*, particles *c.*29 nm diam., USA (Iowa); Lee et al., *ibid* 69:985, 1979, str. from *Capsicum*, USA (Kansas) [53,2488; 59,4084].
Belonolaimus longicaudatus Rau 1958; Orton Williams, *Descr.N.* 40, 1974; ectoparasite, mainly in USA (S.), causes stunted roots in several crops.
Bemisia tabaci, cotton or tobacco whitefly, transm. many viruses.
Bengal bean, *Mucuna aterrima*; — cardamom, *Amomum aromaticum*; — gram, *Cicer arietinum*.
— famine, this major disaster occurred in India, in 1943 when *c.*2 million people died of starvation; the main cause was an epidemic of rice brown spot (*Cochliobolus miyabeanus*) brought about by weather extremely favourable for infection; the 1942 rice crop yield was 60–90% less than the one a year earlier; Padmanabhan, *A.R.Phytop.* 11:11, 1973.
Beniowskia Racib. 1900; Hyphom.; Carmichael et al. 1980; conidiomata hyphal, conidiogenesis holoblastic; con. aseptate, pigmented, solitary. *B. sphaeroidea* (Kalchbr. & Cooke) Mason 1928; Gramineae; Brown & Hanlin, *PD* 66:1197, 1982, con. 9.8 μm diam., blight of *Setaria geniculata*, knotroot bristle grass in USA (Georgia) [62,1765].

benodanil, systemic fungicide against *Rhizoctonia* and rusts.
benomyl, protectant and systemic fungicide, broad spectrum, particularly against powdery mildews, pathogens may develop a tolerance after extensive use.
bent grass, *Agrostis*.
benzimidazole fungicides, Davidse, *A.R.Phytop.* 24:43, 1986.
bergamot, *Citrus bergamia* or a var. of *C. aurantium*, oil of bergamot, restricted to Italy (Reggio Calabria); Terranova et al. in Garnsey et al. 1984, virus diseases.
— bud knot, poss. genetic disorder, La Rosa et al. in Garnsey et al. 1984.
*Berkeley, Miles Joseph, 1803–89; born in England, Univ. Cambridge, took orders 1827, formed an important mycological herbarium, and was the first notable English plant pathologist; his evidence that potato blight was caused by a parasitic fungus was given at a time when the theory of spontaneous generation was widely held, FRS 1879. *Nature Lond.* **40**:371, 1889; *Grevillea* **18**:17, 1889; *Proc.R.Soc.* B **47**:ix, 1890; *Ann.Bot.* 3:451, 1890; **11**:ix, 1897; *Phytopath. Classics* 8, 1948; *Taxon* 23:324, 1974.
berkeleyi, *Mycosphaerella*, *Veluticeps*.
Bermuda grass, *Cynodon dactylon*.
— — etched line v. Lockhart et al., *Phytop.* 75:1258, 1985; isometric *c.*28 nm diam., transm. by the leafhopper *Aconurella prolixa*, Morocco, causes narrow chlorotic streaks on *Sorghum halepense*, infected maize, oat, wheat; serologically related to maize rayado fino v. and oat blue dwarf v. [65,2338].
— — mosaic v. Bhargava, Joshi & Rishi, *Indian Phytopath.* 24:119, 1971; poss. Carlavirus, rod 509–632 nm long, transm. sap, aphid spp. [51,2606].
— — stunt (*Clavibacter xyli* ssp. *cynodontis*) and see xylem limited bacteria.
— — white leaf Yang 1971, MLO assoc., Taiwan, causes a severe disease with the Bermuda grass stunt bacterium [51,3407].
— — witches' broom Chen et al. 1977; Raju & Chen, *Z. PflKrankh.PflSchutz.* 87,37, 1980; spiroplasma assoc. [59,5237].
— — yellow leaf Zelcer et al. 1972; Bar-Joseph et al., *Phytop.* 65:640, 1975; MLO assoc., Israel [55,776].
bermudianum, *Gymnosporangium*.
berry disease, see *Colletotrichum coffeanum*, coffee, frequently known as CBD; — rot (*Trachysphaera fructigena*) coffee.
berseem clover, see *Trifolium*.
— — enation Verma & Mishra,

Zentbl.Bakt.ParasitKde. 2 **130**:709, 1975; India (N.) [**55**,2773].

betae, *Erysiphe, Phoma, Polymyxa, Uromyces.* **Beta vulgaris**, beet.

betel, pan, *Piper betle*; *Areca catechu* may be called betel nut since the palm nut forms a part of the masticatory used in parts of S.E. Asia; Singh & Shankar, *Mycopath.Mycol.appl.* **43**:109, 1971; Maiti & Sen, *PANS* **25**:150, 1979; Jain et al., 1982 [**50**,3086; **59**,399; **62**,4961].

bethelii, *Gymnosporangium, Peridermium.* **Bethel's juniper rust** (*Gymnosporangium bethelii*).

beticola, *Cercospora, Ramularia.*

betle, *Pseudomonas.*

Betula, birch.

betulinum, *Melampsoridium*; **betulinus**, *Piptoporus.*

Bewley, William Fleming, 1891–1976; born in England, director Cheshunt Experimental Station 1921–55, pioneer in the glasshouse industry; CBE 1934; wrote: *Diseases of glasshouse plants*, 1923; *Commercial glasshouse crops*, 1950. *Ann.Rep. Glasshouse Crops Res.Stn.* for 1976:12, 1977; *AAB* **86**:135, 1977.

bibliography, see major texts, other general texts are given as appropriate; for a journals list see *Rev.Pl.Path.* **62**:45, 1983; this abs. journal scans ±1400 serial publications; other bibliographic journals: *Agrindex* 1975, FAO; *Bibliographie Pflanzenschutz-literatur* 1914, Berlin; *Bibliography Agriculture* 1942, *Washington DC*; *Biological Abstracts* 1926, Philadelphia. The *Rev.Pl.Path.* is compiled by the CAB International Mycological Institute q.v. It began in 1922 as the *Review of Applied Mycology*, with volume 48 the name changed; from 1973 all information was put on a computerised data base. This journal, with its long history of meticulous editing, cites the correct names for bacteria and fungi; virus names tend to be those used by authors; also given are back references and new geographical records for pathogens. MacFarlane 1968 is a useful indexing tool for the first 40 volumes; but there is, unfortunately, no cumulative index to 'the review'.

bicolor, *Bipolaris, Cochliobolus, Leptosphaeria, Prospodium.*

bidens mosaic v. Kitajima, Carvalho & Costa 1961; Kuhn et al., *Fitopat.Brasileira* **5**:39, 1980; **7**:185, 1982; filament, *c.*735 nm long, transm. sap, aphid, Brazil, infects inter alia lettuce and sunflower [**59**,5666; **62**,100].

—— **mottle v.** Christie, Edwardson & Zettler 1968; Purcifull & Zitter, *Descr.V.* 161, 1976; Boswell & Gibbs 1983; Potyvirus, filament *c.*720 nm long, transm. sap, *Myz.pers.*, non-persistent, USA, causes a mottle in endive and lettuce; Logan et al., *PD* **68**: 260, 1984, in ornamentals [**63**,3381].

—— **witches' broom** Vega, Almeida & Costa, *Fitopat.Brasiliera* **6**:29, 1981; MLO assoc., Brazil (Paraná) [**61**,132].

bidwellii, *Guignardia.*

Biffen, Roland Harry, 1874–1949; born in England, Univ. Cambridge, Professor of Agricultural Botany 1908, first director of the Plant Breeding Institute, breeding cereals for resistance to rusts; the first to show, in 1904, that resistance to pathogens in plants can be inherited as a Mendelian character; FRS 1914, Darwin Medal 1920, Knight 1925. *TBMS* **33**:166, 1950; *Obit.Not.Fel.R.Soc.* **7**:9, 1950.

Bifusella Höhnel 1917; Hypodermataceae; asco. aseptate, constricted in the middle, conifer needle cast. *B. linearis* (Peck) Höhnel; Minter & Millar, *Descr.F.* 782, 1984; asco. 40–60 × 4–7 μm, with mucous sheath, N. America, more disease in lower crown.

biguttatum, *Verticillium.*

binapacryl, contact and protectant fungicide against powdery mildews.

biodeterioration, an undesirable change in the properties of a material caused by an organism; biodegradation, the use by man of decay by organisms to render a waste material more useful or acceptable; Allsop & Seal, *Introduction to biodeterioration*, 1986.

***biological control**, sensu stricto: the use of one organism to eliminate or reduce the disease caused by another, e.g. the use of *Peniophora gigantea* against *Heterobasidion annosum*; there are few other practical examples for plant pathogens, but the method is widely used against insects. Sensu lato: reduction, firstly, of inoculum of the pathogen by cultural practices and/or antagonistic organisms; and secondly of infection and disease level by host resistance; Baker & Cook, *Biological control of plant pathogens*, 1974; Papavizas & Lumsden, *A.R.Phytop.* **18**:389, 1980, soilborne fungus propagules; Blakeman & Fokkema, *ibid* **20**:167, 1982, leaf diseases; Linderman et al., *PD* **67**:1058, 1983, characterisation systems for soilborne pathogens; Cook & Baker, *The nature and practice of biological control of plant pathogens*, 1983; Lupton, *AAB* **104**:1, 1984, plant breeding; Baker et al., *Phytop.* **74**:1019, 1984, experimental method; Cook, *ibid* **75**:25, 1985, theory to application.

biotroph, an organism entirely dependent upon another living organism as a source of nutrients, applicable to mycorrhiza and obligate parasites, after FBPP; Lewis, *Biol.Rev.* **48**:261, 1973, fungal nutrition & origin of biotrophy; Gay in Wood & Jellies 1984, mechanisms in fungal pathogens.

biotype, in microbiology a subdivision of a sp.,

ssp.; serology q.v., or a physiologic race q.v., or a pathovar in bacteriology. In higher plants a group of genetically identical individuals, a clone or sometimes a cultivar; biovar has been recommended to avoid the suffix — type; Cowan 1978.

*Bipolaris Shoem., *CJB* 37:882, 1959; Hyphom.; conidiomata hyphal, conidiogenesis enteroblastic tretic; con. multiseptate, pigmented. The genus was erected as a segregate from *Drechslera*; was rejected by Ellis 1971 but used by Luttrell 1977 and Alcorn 1981. In 1983, see *Drechslera*, the last worker upheld *Bipolaris* and Alcorn's nomenclature is used here. Also Sivanesan, *Mycol.Pap.* 158, 1987, considered that *Bipolaris* showed no morphological differences from *Curvularia* q.v. The taxonomic position of the former genus is therefore confused; Holliday 1980, as *Drechslera*.

— bicolor, teleom. *Cochliobolus bicolor*.

— cactivora (Petrak) Alcorn 1983; con. 30–65 × 9–12 μm; Chase *PD* 66:602, 1982, shattering of Easter cactus, *Rhipsalidopsis gaertneri* [61,7044].

— cynodontis, teleom. *Cochliobolus cynodontis*;

— hawaiiensis, teleom. *C. hawaiiensis*.

— incurvata (Ch. Bernard) Alcorn 1983; Ellis & Holliday, *Descr.F.* 342, 1972, as *Drechslera incurvata* (Ch. Bernard) M.B. Ellis; con. 100–150 × 19–22 μm; coconut leaf spot; Fagan, *AAB* 111:521, 1987.

— iridis (Oudem.) Dickinson 1966; Ellis & Waller, *Descr.F.* 434, 1974 as *Drechslera iridis* M.B. Ellis 1971; con. 45–90 × 16–29 μm; ink of iris, Canada, Europe; Moore 1979.

— maydis, teleom. *Cochliobolus heterostrophus*;

— nodulosa, teleom. *C. nodulosus*; — oryzae, teleom. *C. miyabeanus*.

— sacchari (Butler) Shoem. 1959; Ellis & Holliday, *Descr.F.* 305, 1971, as *Drechslera sacchari* (Butler) Subram. & Jain; con. 35–96 × 9–17 μm; sugarcane eyespot, leaf spots of lemon grass and Napier or elephant grass; forms a host specific toxin helminthosporoside; Holliday 1980; Macko et al., *Experientia* 37:923, 1981; 39:343, 1983; Beier et al., *ibid* 38:1312, 1982; Lesney et al., *Phytop.* 72:844, 1982; Duvick et al., *Pl.Physiol.* 74:117, 1984; activity, characterisation & structure of toxin [61,851, 7152; 62,743, 2318; 63, 1939].

— setariae, teleom. *Cochliobolus setariae*.

— sorghicola (Lefebvre & Sherwin) Alcorn 1983; Ellis & Holliday, *Descr.F.* 491, 1976, as *Drechslera sorghicola* (Lefebvre & Sherwin) Richardson & Fraser; con. mostly 50–90 × 14–17 μm; sorghum

target leaf spot; Lin, *Pl.Prot.Bull.Taiwan* 20:283, 1978; Borges, *PD* 67:996, 1983 [59,1237; 63,601].

— sorokiniana, teleom. *Cochliobolus sativus*; — spicifera, teleom. *C. spicifer*.

— stenospila (Drechsler) Shoem. 1959; Ellis & Holliday, *Descr.F.* 306, 1971 as *Drechslera stenospila* (Drechsler) Subram. & Jain; con. 70–135 × 14–22 μm, sugarcane brown stripe.

— victoriae, teleom. *Cochliobolus victoriae*.

bipolaroxin, toxin from *Cochliobolus cynodontis*; Sugawara et al. 1985 [65,4990].

birch, *Betula*.

— dieback Hansbrough & Stout, *PDR* 31:327, 1947; Berbee 1957; Clark & Barter, *Forest Sci.* 4:343, 1958; N. America (N.E.) [27:50; 36:795; 38:340].

— line pattern Callahan 1962; Gotlieb & Berbee, *Phytop.* 63:1470, 1973; N. America (prob. apple mosaic v.) [42:345; 53,3188].

— ringspot Schmelzer 1972; Cooper & Atkinson, *Forestry* 48:193, 1975; (prob. cherry leaf roll v.); *Descr.V.* 306, 1985 on CLRV calls this birch disease chlorotic ringspot and yellow vein netting.

— stem grooving, see Cooper 1979.

— — pitting Atanasoff, *Z.PflKrankh.PflPath.PflSchutz.* 74:205, 1967; Bulgaria [46,3574].

bird's eye rot (*Elsinoë ampelina*) grapevine.

— — spot (*Calonectria theae*) tea, (*Cercoseptoria ocellata* and *C. theae*) tea, (*Drechslera heveae*) rubber.

Bisby, Guy Richard, 1889–1958; born in USA, S. Dakota State College, Univ. Columbia, Minnesota, Professor plant pathology from 1920; Imperial Mycological Institute from 1937; Univ. Minnesota Gold Medal 1956; wrote: *The fungi of Manitoba*, 1929, with A.H.R. Buller and J. Dearness; *The fungi of Ceylon*, 1950 with T. Petch; *A dictionary of the fungi*, 1943, with G.C. Ainsworth, this classic text is now in its 7th edn. 1983. *Nature Lond.* 182:987, 1959; *Phytop.* 49:323, 1959; *Taxon* 8:2, 1959; *TBMS* 42:129, 1959.

bischofia witches' broom Jin 1983; MLO assoc., China (S.), *B. javanica*, Java bishopwood [64,5140].

Bitrimonospora indica, see *Monosporascus*.

*bitter gourd, *Momordica charantia*.

— pit, apple, Ca deficiency, Fiedler 1976, Delver 1978, reviews [55,5826; 58,4434].

— rot (*Greenaria uvicola*) grapevine, (*Glomerella cingulata*) apple, quince, (*Pezicula alba*, *P. malicorticis*) apple, pear.

bitunicate, see ascus.

bixae, *Oidium*.

björlingii, *Pleospora*.

black arm (*Xanthomonas campestris* pv.

malvacearum) cotton; —— **band** (*Botryodiplodia theobromae*) jute.
blackberry, see *Rubus*.
—— **dwarf** Zeller 1927; Smith 1972 [7:183].
—— **proliferation** Cazelles, *Revue suisse Vitic. Arboric.Hortic.* **8**:131, 1976; MLO assoc. [**56**,2152].
—— **variegation** Horn, *Phytop.* **38**:827, 1948; USA (Maryland) [**28**:179].
—— **yellow mosaic v.** Engelbrecht & van der Walt, *Phytophylactica* **6**:311, 1974; Engelbrecht, *Acta Hortic.* 66:79, 1976; filament, transm. sap, seedborne in *Chenopodium murale*; *Rubus rigidus*, wild blackberry [**54**,5008; **57**,690].
black blotch (*Cymadothea trifolii*) clover; —— **boll rot** (*Botryodiplodia theobromae*) cotton; —— **bundle** (*Acremonium kiliense*, *A. strictum*) maize; —— **canker** (*Itersonilia pastinacae*) parsnip; —— **chaff** (*Xanthomonas campestris* pv. *translucens*) barley; —— **choke** (*Balansia*, *B. orzyae-sativae*) Gramineae, rice; —— **cross** (*Phyllachora musicola*) banana; —— **crown rot** (*Mycocentrospora acerina*) celery, parsnip, —— **crust** (*Septoria paulliniae*) guarana, (*Mycoleptodiscus indicus*) vanilla.
blackcurrant, *Ribes nigrum* and other spp.
—— **infectious variegation** Posnette 1952; Ellenberger 1962 [**33**:97; **41**:729].
—— **reversion** Amos & Hatton 1927; Massee 1952; Thresh 1963; Jacob, *Z.PflKrankh.PflSchutz.* **83**:448, 1976; transm. *Cecidophyopsis ribis*; Silvere 1970, MLO assoc. [**7**:182; **33**:97; **42**:622; **50**,3m; **56**,1210].
—— **wildfire** Kalinitchenko & Gladkych, *Acta Hortic.* 66:85, 1976; transm. *Aphis grossulariae*, USSR (Siberia) [**57**,691].
—— **yellows** Posnette 1952, England [**33**:97].
black dead arm (*Botryosphaeria stevensii*) grapevine; —— **dot** (*Colletotrichum coccodes*) potato, tomato.
blackened flower head (*Aphelenchoides fragariae*) primula.
blackeye cowpea mosaic v. Anderson 1955; Purcifull & Gonsalves, *Descr.V.* 305, 1985; Potyvirus, filament *c.*750 nm long, transm. sap, *Aphis craccivora*, *Myz.pers.*, non-persistent, causes diseases in asparagus bean and cowpea, see this *Descr.V.* for differences from other cowpea viruses; Atiri et al., *AAB* **104**:339, 1984, effect of resistance in cowpea to *A. craccivora* infestation & BCMV transm. [**63**,4186].
blackfire (*Pseudomonas syringae* pv. *tabaci*) tobacco.
black fruit rot (*Didymella bryoniae*) cucurbits.
—— **gram**, see *Vigna*.
—— —— **aphid borne v.** Benigno, *Philipp. Agric.* **62**:328, 1979; filament [**60**,4112].

—— —— **mottle v.** Phatak 1974; Scott & Hoy, *Descr.V.* 237, 1981; isometric *c.*28 nm diam., contains ssRNA, transm. sap, beetles, seedborne, legumes; Asia, Australasia; Honda et al., *JARQ.* **16**:72, 1982, on mungbean, soybean [**62**,1692].
—— **head** (*Ceratocystis paradoxa*) banana.
*—— **heart**, celery, Ca deficiency, Cox & Dearman, *Exp.Hortic.* **30**:1, 1978; pineapple, prob. a disorder caused by chilling at temp. of 21° or less, Smith, *Trop.Agric.Trin.* **60**:31, 1983; Wills et al., *ibid* **62**:199, 1985; potato, O_2 deficiency in stored tubers; (*Gibberella fujikuroi*) banana [**57**,5174; **62**,2055; **64**,5037].
—— **hull** (*Thielaviopsis basicola*) groundnut; —— **kernel** (*Cochliobolus lunatus*) cereals; —— **knot** (*Apiosporina morbosa*) *Prunus* spp.; —— **leaf** (*Fusicladium pisicola*) pea; —— —— **blight** (*Arkoola nigra*) soybean; —— —— **spot** (*Alternaria brassicicola*) brassicas, (*Deightoniella torulosa*) banana, (*Septoria chrysanthemella*) chrysanthemum; —— —— **streak** (*Mycosphaerella fijiensis*) banana; —— **leathery rot** (*Phoma caricae*) papaya; —— **leg** (*Erwinia carotovora* ssp. *atroseptica*) potato, (*Leptosphaeria maculans*) brassicas, (*Pleospora betae*) sugarbeet, (*Sclerotium wakkerii*) tulip; —— **lesion root rot** (*Pezizella oenotherae*) strawberry.
—— **locust**, *Robinia pseudoacacia*; black locust true mosaic = robinia mosaic v.
—— **melanose** (*Mycosphaerella citri*) citrus; —— **mould** (*Aspergillus niger*) onion, (*Chalara thielavioides*) rose; —— —— **rot** (*Alternaria alternata*) tomato.
—— **mustard**, see *Brassica*.
—— **neck** (*Phytophthora cryptogea*) chrysanthemum.
—— **pepper**, *Piper nigrum*, the true pepper; Kueh, *Pests, diseases and disorders of black pepper*, Dept. Agric. Sarawak 1979; damaging viruses are suspected in this crop but only graft transm. has been described: stunt Holliday 1959; mottle vein clearing and mosaic Jamil 1966, Costa et al., 1970 [**39**:482; **46**,512b; **52**,207].
—— **pit** (*Alternaria alternata*) potato, (*Pseudomonas syringae*) citrus; —— **pod** (*Phytophthora megakarya*, *P. palmivora*) cacao, and see Kellam & Zentmyer, *Mycologia* **78**:351, 1986, *Phytop.* **76**:159, 1986, for other *Phytophthora* spp. assoc. with cacao; —— —— **blotch** (*Alternaria raphani*) Cruciferae, radish; —— **point** (*A. alternata*, *Cochliobolus sativus*) wheat, Huguelet & Kiesling, *Phytop.* **63**:1220, 1973; —— **pustule** (*Dothiora ribesia*) red- and blackcurrant; —— **root** (*Aphanomyces raphani*) radish; —— —— **rot** (*A. cochlioides*) beet, (*Calonectria crotalariae*) soybean, (*Phytophthora citricola*) hop, (*Rosellinia arcuata*, *R. bunodes*, *R. pepo*) many crops, (*Thielaviopsis basicola*) tobacco

and other crops, (*Xylaria mali*) apple; —— **rot**
(*Xanthomonas campestris* pv. *campestris*) crucifers;
—— —— **canker** (*Botryosphaeria obtusa*) pome fruit
[**53**,2961; **65**,3205; **66**,117].
—— **raspberry**, *Rubus occidentalis.*
—— —— **latent v.** Converse & Lister 1969, 1970,
Descr.V. 106, 1972; Ilarvirus, isometric *c*.26 nm
diam., transm. sap to a fairly wide range of
herbaceous plants, in pollen and seed of black
raspberry; Jones & Mayo, *AAB* **79**:297, 1975,
properties, serologically related to some strs. of
tobacco streak v., graft inoculation no symptoms
in black raspberry and 8 red raspberry cvs., but
symptoms of necrotic shock in *Rubus henryi, R.
phoenicolasius* and Himalayan blackberry
[**54**,4040].
—— —— **necrosis v.** Stace-Smith, *CJB* **33**:314, 1955;
Jones & Murant, *Pl.Path.* **21**:166, 1972; Jones &
Roberts *ABB* **86**:381, 1977; isometric 25–30 nm
diam., transm. sap, *Amphorophora* spp., prob.
semi-persistent; with rubus yellow net v. causes
raspberry vein banding mosaic; Jones, *Acta
Hortic.* 129:41, 1982, compared BRNV with
raspberry leaf mottle and raspberry leaf spot;
Jones & Mitchell, *AAB* **109**:323, 1986,
propagation & antiserum [**35**:201; **52**,1208;
57,697; **66**,2435].
—— —— **streak** Wilcox 1923, see Stace-Smith,
Rev.appl.Mycol. 47:104, 1968.
—— —— **witches' broom** Converse et al., *PD* **66**:949,
1982; poss. MLO, USA (Oregon) [**62**,277].
—— **rust** (*Puccinia chrysanthemi*) chrysanthemum;
—— **scab** (*Venturia carpophila*) almond, apricot,
peach, plum; —— **scorch** (*Ceratocystis paradoxa*)
date palm, Koltz & Fawcett, *J.agric.Res.* **44**:155,
1932, —— **scurf** (*Thanatephorus cucumeris*) potato
tubers, Hide & Cayley, *AAB* **100**:105, 1982; ——
seed (*Mycosphaerella fragariae*) strawberry; ——
shank (*Phytophthora nicotianae* var. *nicotianae*)
tobacco; —— **Sigatoka** (*M. fijiensis* and var.
difformis) banana; —— **slime** (*Sclerotinia
bulborum*), hyacinth; —— **snow mould**
(*Herpotrichia juniperi*) conifers; —— **spot**, potato
internal bruising, Rogers-Lewis, *AAB* **96**:345,
1980, (*Asperisporum caricae*) papaya,
(*Colletotrichum acutatum*) strawberry,
(*Diplocarpon rosae*) rose, (*Elsinoë ampelina*)
grapevine, (*Guignardia citricarpa*) citrus, (*Nectria
radicicola*) grapevine, (*Spilocaea eriobotryae*)
loquat, (*Stegophora ulmea*) elm; —— —— **and**
shothole (*Xanthomonas campestris* pv. *pruni*)
Prunus; —— **stain root** (*Verticicladiella*) conifers;
—— **stem** (*Dichotomophthora portulacae*) purslane,
(*Phoma medicaginis*) clover, lucerne, pea, (*Valsa
sordida*) poplar; —— —— **rot** (*Pythium
aphanidermatum*) tobacco; —— —— **rust** (*Puccinia*

graminis) wheat, temperate cereals and grasses ——
streak (*Leptotrochila porri*) leek; —— **stripe**
(*Cercospora atrofiliformis*) sugarcane,
(*Phytophthora botryosa, P. meadii*) rubber; —— **tip**
(*Deightoniella torryosa, Verticillium theobromae*)
banana fruit [**11**:509; **61**,5195].
Blackwell, Elizabeth Marianne, 1889–1973; born in
England, Univ. Liverpool, Imperial College of
Science; head, botany, Royal Holloway College,
Univ. London 1922–49; teacher, botanist,
naturalist; wrote important papers on
Phytophthora. TBMS **61**:611, 1973.
bladder (*Taphrina pruni*) plum.
*****blast** (*Pseudomonas syringae* pv. *syringae*) citrus,
(*Pyricularia curcumae*) turmeric, (*P. oryzae*) rice,
Gramineae, (*Pythium splendens*, primary invader)
oil palm; oat, withering of the basal spikelets, a
stress symptom arising at a critical time and
which could result from insect injury or adverse
weather, Empson, *Pl.Path.* **7**:85, 1958 [**38**:80].
blastcast, a simulator for rice blast, Ohta et al.,
Hashimoto et al., *Ann.Rep.Soc.Pl.Prot.N.Japan*
33:9, 12, 1982 [**62**,2463–4].
blastic, of conidiogenesis in fungi, one of the 2
basic sorts cf. thallic; characterised by a marked
enlargement, blown out, of a recognisable conidial
initial before the initial is delimited by a septum,
hence a blastospore; **holoblastic** is where both
walls of the conidiogenous cell contribute towards
forming the conidial wall; **enteroblastic** is where
only the inner wall contributes towards forming
the conidial wall; see phialidic.
Blastocladiales, Chytridiomycetes;
Physodermataceae, one of the 4 families, contains
pathogenic genera.
blastomany, abnormal tendency of a plant to
develop an unusual number of leaf buds.
Blastophaga psenes, fig wasp, see endosepsis.
bleeding, a plant exudate that is neither gummy nor
resinous; see slime flux, wetwood; (*Ceratocystis
paradoxa*) areca nut, coconut.
—— **canker** (*Phytophthora cactorum*) horse
chestnut.
blepharis, *Libertella.*
blight, a common term to describe many diseases;
it is best used to describe a rapid and extensive
necrosis of all or most of the above ground parts
of a plant, and caused by a largely airborne
pathogen; blight (*Phytophthora infestans*) of
potato is preferred to late blight, (*Xanthomonas
campestris* pv. *oryzae*) rice.
blind bud (*Aphelenchoides blastophthorus*) *Scabiosa
caucasia*; —— **seed** (*Gloeotinia granigena*) grasses;
Cole, *TBMS* **74**:199, 1980, noted a blind seed
condition in artichoke, sclerotia in seed resembled
those of *Sclerotinia sclerotiorum* [**59**,4401].

blindness, brassicas, cold shock; Scaife & Turner 1983.

blister (*Olpidium viciae*) broad bean; —— **blight** (*Exobasidium vexans*) tea; —— **canker** (*Nummulariella marginata*) apple, (*Physalospora persicae*) peach; —— **rusts** (*Cronartium*) pines; —— **smut** (*Entyloma dactylidis*) *Poa pratensis*, (*Ustilago maydis*) maize; —— **spot** (*Pseudomonas syringae* pv. *papulans*) apple.

blitecast, a computerised forecast of potato blight, and see Beaumont period; Krause et al., *PDR* **59**:95, 1975; MacKenzie, *F.A.O.Pl.Prot.Bull.* **32**:45, 1984 [**54**,4101; **64**,1713].

blossom blight (*Physalospora persea*) avocado, (*Pseudomonas syringae*) apple, pear, stone fruit; —— —— **& fruit rot** (*Peronophythora litchii*) litchi; —— **end rot** (*Botryotinia fuckeliana*) pear fruit in storage, Sommer et al., *PD* **69**:340, 1985; in tomato an induced Ca deficiency, Bould et al. 1983; *Capsicum* shows a similar symptom with which a Ca shortage is assoc.; —— **wilt & twig blight** (*Monilinia laxa*) stone fruit [**64**,4400].

blotch, a large area of discolouration of a leaf, fruit etc; on leaves its outline is unrelated to the distribution of main or minor veins; cf. mosaic and mottle, FBPP. —— (*Ascochyta avenae*) oat, (*Mycosphaerella macrospora*) iris, (*Septoria chrysanthemella*) chrysanthemum; —— **& char spot** (*Cheilaria agrostis*) Gramineae.

blotchy ripening, tomato, abnormal physiology, Hobson & Davies, *Ann.Rep.Glasshouse Crops Res.Inst.* UK for 1976; also cloud, greywall, waxy patch.

blue banana, prob. Mg deficiency, Moreira & Hiroce 1978 [**58**,5441].

blueberry, *Vaccinium*. —— **decline** Ramsdell, *PDR* **62**:1047, 1978; str. tobacco ringspot v. assoc., USA (Michigan) [**58**,4442]. —— **leaf mottle** v. Ramsdell & Stace-Smith 1979, *Descr.V.* 267, 1983; Nepovirus, isometric *c*.29 nm diam., transm. sap, seed, poss. pollen, USA (Michigan and New York strs.), occurs naturally in highbush blueberry, *Vaccinium corymbosum*, causing a severe dieback; and in grapevine; in *Vitis labrusca* cv. Concord BLMV causes delayed fruit break, irregular elongation of fruit, pale green foliage and straggly fruit clusters; Childress & Ramsdell, *Phytop.* **76**:1333, 1986, detection in pollen & seed; *ibid* **77**,176, 1987, spread in pollen through bees [**66**,3904, 4385]. —— **necrotic ringspot** (tobacco ringspot v.), Lister et al., *Phytop.* **53**:1031, 1963; Converse & Ramsdell, *PD* **66**:710, 1982 [**43**,517; **62**,276].

*——— **red ringspot** Hutchinson & Varney, *PDR* **38**:260, 1954; Kim et al., *Phytop.* **71**:673, 1981; isometric particles assoc. 42–46 nm diam., USA; prob. Caulimovirus fide Francki et al. 1985 [**33**:679; **61**,1289]. —— **sheep pen hill** Podleckis & Davis, abs. *Phytop.* **76**:1065, 1986; rod particle assoc. 750 × 12 nm, USA (New Jersey). —— **shoestring** v. Varney 1957; Ramsdell, *Descr.V.* 204,1979; prob. Sobemovirus, isometric *c*.27 nm diam., contains ssRNA, transm. sap, *Illinoia pepperi*; Canada, USA, only *Vaccinium angustifolium* and *V. corymbosum*, economically important; Ramsdell, *Phytop.* **69**:1087, 1979, physical & chemical properties; Morimoto et al. and Morimoto & Ramsdell, *ibid* **75**:709, 1217, 1985, aphid acquisition, transm., population & movement [**59**,4242; **65**,292, 2371]. —— **stunt** Wilcox 1942; Varney in Maramorosch & Raychaudhuri 1981; MLO assoc., transm. *Scaphytopius magdalensis*, *S. verecundus*; first reported in the mid 1920s in USA, occurs northwards from North Carolina to Canada (E.) and westwards to Michigan [**21**:496]. —— **witches' broom** Blattný & Stary 1940; Blattný & Váňa, *Biologia Pl.* **16**:476, 1974; Siller et al., *NachrBl. dt.PflSchutzdienst* **38**:1, 1986; MLO assoc.; Czechoslovakia, Germany (S.) [**54**,3408; **65**,5609].

blue mould (*Penicillium expansum*) apple, (*P. italicum*) citrus, (*Peronospora tabacina*) tobacco; —— **rot & wilt** (*Ceratocystis fimbriata*) cacao; —— **stain**, a deep seated discolouration of wood resulting from fungus attack, largely bluish but may have black, brown or grey tints, almost invariably in sapwood only, often assoc. with *Ceratocystis*.

blumea mosaic Verma, *Zentbl.Bakt.ParasitKde.* 2 **129**:533, 1974; India, *B. lacera* [**54**,1661]. —— **yellow vein mosaic** Wilson & Potty, *Agric.Res.J.Kerala* **10**:68, 1972; transm. *Bemisia tabaci*, India (Kerala), *B. neilgherrensis* [**53**,868].

Blumeriella v. Arx 1961; Dermataceae; ascomata remain immersed on rupture of overlying tissue. *B. jaapii* (Rehm) v. Arx, anam. *Phloesporella padi* (Lib.) v. Arx, asco. and con. tear shaped or filiform; cherry leaf spot or shot hole yellow leaf; formerly in *Coccomyces* and *Higginsia*; Jakobsen & Jørgensen 1986 [**66**,247].

boehmeriae, *Phytophthora*, *Pucciniastrum*.

Boehmeria nivea, ramie.

boerhavia mosaic Singh & Verma, *Phytopath.Z.* **73**:277, 1972; India, *B. diffusa* [**51**,4752].

boil smut (*Ustilago maydis*) maize.

bole rot (*Phytophthora nicotianae* var. *parasitica*) sisal.

bolleyi, *Microdochium*.

boll rot (*Aspergillus flavus*, *Eremothecium*) cotton.

boninense, *Ganoderma*.

bonnygate (*Nectria striispora*) banana.

bootlace (*Armillaria*) woody plants.

borbonol, toxic compound from *Persea* spp., inhibited vegetative growth of *Phytophthora cinnamomi*, not found in the highly susceptible *Persea indica*, Zaki et al., *Physiol.Pl.Path.* 16:205, 1980 [59,5280].

Bordeaux mixture, a fungicide discovered by Millardet q.v.; it can still be useful as a broad spectrum treatment at times when nothing else is available. It is an amorphous, flocculent, blue precipitate of copper hydroxide and calcium sulphate. It can be prepared on site by dissolving 4 lb. copper sulphate in 5 gal. water; slake 4 lb. quicklime (or 6 lb. hydrated lime) in 45 gal. water, strain through a fine sieve, add the copper sulphate to the lime stirring vigorously, use at once, this is the common 4-4-50 mixture.

borde blanco (*Marasmiellus*) maize.

boreale, *Acremonium*.

borealis, *Armillaria*, *Myriosclerotinia*.

bothrina, *Rosellinia*, see *R.arcuata*.

Botryodiplodia Sacc. 1884; Coelom., conidiomata pycnidial; Zambettakis, *Bull.Soc.mycol.Fr.* 70:219, 1955; synonymy in Sutton, *Mycol.Pap.* 141, 1977. Sutton 1980, following Zambettakis, stated that the correct generic name for *B. theobromae* is *Lasiodiplodia* Ell. & Ev. In this case it seems very doubtful whether plant pathologists will yet take up *L. theobromae* (Pat.) Griffon & Maubl. for this very common fungus of warmer regions and frequently assoc. with diseased conditions. *B. theobromae* is retained in the *Review of Plant Pathology*.

B. gallae (Schwein.) Petrak & Syd. 1926; con. 18–25 × 12–17 μm, 1 septate; oak canker, *Quercus* spp.; Schmidt & Fergus, *CJB* 43:731, 1965; Croghan & Robbins, *PD* 70:76, 1986, in USA (Pennsylvania and Michigan, respectively) [44,3171; 65,5180].

B. hypodermia (Sacc.) Petrak 1923; con. 24–32 × 16–22 μm, 1 septate; elm canker in USA; Riffle, *Phytop.* 68:1115, 1978; Krupinsky, *PD* 65:677, 1981; *Phytop.* 73:108, 1983 [58,3467; 61,1397; 62,2161].

B. theobromae Pat. 1892; Punithalingam, *Descr.F.* 519, 1976; con. at first aseptate, hyaline, when mature becoming 1 septate, brown, longitudinally striate, 20–30 × 10–15 μm, extruded from the pycnidia in clearly visible, black, powdery masses; a teleom. *Physalospora rhodina* Berk. & Curt., in Cooke 1889, *Botryosphaeria rhodina* (Cooke) v. Arx 1970, fide Sivanesan 1984, has been described but the asco. appear to have little or no role in infection. The fungus is ubiquitous in warmer

regions; it is saprophytic and a secondary invader, particularly in dieback complexes or as a component of postharvest diseases, e.g. banana finger rot, cacao pod rot and dieback, citrus stem end rot and dieback, cotton black boll rot, tuber decay in cassava, sweet potato and yam; Holliday 1980; Punithalingam, *Biblthca. Mycologia* 71,1980.

botryosa, *Phytophthora*.

botryose, of conidia, clustered like grapes.

Botryosphaeria Ces. & de Not. 1863; Botryosphaeriaceae; Sivanesan 1984; asco. usually hyaline, aseptate; distinguished from *Guignardia* by multilocular to unilocular, usually stromatic, ascomata; large asco. and different anam.; see *B. ribis*, Kobayashi & Oishi, for some comparisons of *Botryosphaeria* spp.

Botryosphaeriaceae, Dothidiales, *Botryosphaeria*, *Guignardia*.

Botryosphaeria corticis (Demaree & Wilcox) v. Arx & Müller 1954; asco. 24–37 × 10–16 μm; blueberry cane or stem canker, USA (E.), 7 races; Demaree & Wilcox, *Phytop.* 32: 1068, 1942; Milholland & Galleta, *ibid* 59:1540, 1969; Milholland, *ibid* 60:70, 1970; *PD* 68:522, 1984 [22:214; 49,1093, 2128; 63,4504].

B. dothidea (Moug.:Fr.) Ces. & de Not. 1863; may be the same as *Botryosphaeria ribis* q.v.; Pennycook & Samuels, *Mycotaxon* 24:445, 1985, described the fungus, anam. *Fusicoccum aesculi* Corda 1829, and *Botryosphaeria parva* new sp., anam. *F. parvum* new sp., assoc. with ripe fruit rot of kiwi fruit [65,1966].

B. obtusa (Schwein.) Shoem. 1964; Punithalingam & Waller, *Descr.F.* 394, 1973; asco. 25–33 × 7–12 μm; plurivorous, temperate regions, woody hosts, largely saprophytic, canker and dieback of pomaceous fruit and grapevine, on apple causes New York or black rot canker, frog eye leaf spot and fruit black rot; sometimes reported with *B. dothidea* and *B. ribis*; Britton & Hendrix, *PD* 70:134, 1986, peach gummosis canker with other *Botryosphaeria* spp. [65,2899].

B. pseudotsugae Funk, *CJB* 53:2300, 1975; asco. 50–67 × 21–30 μm; *Can.J.Pl.Path.* 7:355, 1985; assoc. with a canker of Douglas fir [65,3017].

B. rhodina, see *Botryodiplodia theobromae*.

B. ribis Grossenb. & Duggar 1911; Punithalingam & Holliday, *Descr.F.* 395, 1973; anam. in *Dothiorella*; asco. 17–23 × 7–10 μm, con. 17–25 × 5–7 μm; causes cankers, diebacks and fruit rots of woody plants; Holliday 1980; temperate crops: Weaver, *Phytop.* 64:1429, 1974, peach gummosis; English et al., *ibid* 65:114, 1975, almond band canker; Crist & Schoenewiss, *ibid* 65:369, 1975, blackberry canker; Kobayashi &

Oishi, *Trans.mycol.Soc.Japan* **20**:429, 1979, *Castanea crenata* black rot; Xiang et al., *Acta phytophylac.sin.* **11**(4):27, 1981, poplar blister canker; Brown & Hendrix, *Phytop.* **71**:375, 1981, on apple; Kohn & Hendrix, *ibid* **72**:313, 1982, on birch; Maas & Uecker, *PD* **68**:720, 1984, blackberry canker; Brown & Britton, *ibid* **70**:480, 1986, on apple & peach [**54**,2886, 4566, 5562; **60**,2207; **61**,307, 5255, 5829; **63**,5488].

B. stevensii Shoem. 1964; anam. *Diplodia mutila* (Fr.) Mont. 1834; asco. 30–40 × 12–16 μm, con. 20–27 × 10–16 μm; Lehoczky, *Acta phytopath.Acad.Sci.hung.* **9**:319, 329, 1974, grapevine black dead arm; Vajna, *Eur.J.For.Path.* **16**:223, 1986, branch canker & dieback in *Quercus petraea*, Hungary [**54**,5022–3; **66**,331].

B. xanthocephala (H. & P. Sydow & Butler) Theissen 1916; anam. *Fusicoccum cajani* (H. & P. Sydow & Butler) Samuels & Singh, *TBMS* **86**:295, 1986, who redescribed the fungus causing pigeon pea stem canker [**65**,4649].

B. zeae (Stout) v. Arx & Müller 1954; Sivanesan & Holliday, *Descr.F.* 774, 1983; anam. described as *Macrophoma zeae* Tehon & Daniels, *Mycologia* **19**:121, 1927, but belongs in *Dothiorella*; asco. 18–28 × 8 μm, con. 17–40 × 5–11 μm; maize grey ear rot, of little importance.

Botryosporium Corda 1831; Hyphom.; conidiomata hyphal, conidiogenesis holoblastic; con. aseptate, hyaline, botryose; Carmichael et al. 1980. *B. longibrachiatum* (Oudem.) Maire 1903; Anderson & Welacky, *PD* **67**:1158, 1983, tobacco barn mould in Canada [**63**,3553].

Botryotinia Whetzel, *Mycologia* **37**:628, 1945; Sclerotiniaceae; sclerotium ± definite, like a loaf or hemispherical, formed usually on or just beneath the cuticle or epidermis, medulla lacking, differs from *Sclerotinia* in the presence of a *Botrytis* anam.; some spp. cause important diseases of *Allium*: *B. allii*, *B. porri*, *B. squamosa* and *Botrytis allii*; others occur on temperate flower bulb crops, see Moore 1979 as *Sclerotinia*; Coley-Smith et al. ed., *The biology of* Botrytis, 1980 [**25**:235].

B. allii (Saw.) Yamamoto, *Sci.Rep.Hyogo Univ.Agric.Biol.* **2**:17, 1956; anam. *Botrytis byssoidea* Walker, *Phytop.* **15**:709, 1925; sclerotia rare or absent, con. mostly 10–14 × 6–9 μm, length:breadth ratio 1.5–1.65; one of the 4 *Botrytis* spp. that cause onion bulb neck rot; Owen et al., *ibid* **40**:749, 1950, not conspecific with *B. allii* [**5**:273; **30**:211; **37**:196].

B. convoluta (Drayton) Whetzel 1945; anam. *Botrytis convoluta* Whetzel & Drayton, *Mycologia* **24**:475, 1932; asco. 11.7–19.5 × 5.2–9.1 μm, con. mostly 10–12 × 8–10 μm, sclerotia convoluted; iris

rhizome root rot; MacWithey, *Phytop.* **57**:1145, 1967, infection & temp.; Maas & Powelsen, *Mycopath.Mycol.appl.* **41**:283, 1970 [**47**,839; **50**,1836].

B. draytonii (Buddin & Wakef.) Seaver 1951; anam. *Botrytis gladiorum* Timmerm. 1941; asco. 12–17 × 6–8 μm, con. 10–22 × 8–13 μm, sclerotia abundant, aggregated in large masses, gladiolus core rot, the most severe form of the disease are the flower spots, leaves and corms can be rotted.

B. fuckeliana (de Bary) Whetzel 1945; anam. *Botrytis cinerea* Pers. 1822; Ellis & Waller, *Descr.F.* 431, 1974 *as Sclerotinia fuckeliana* (de Bary) Fuckel; con. mostly 8–14 × 6–9 μm, length:breadth ratio 1.35–1.5; the anam. causes a ubiquitous mould, parasite and saprophyte, often called botrytis disease or grey mould, as a blight or rot of immature, fleshy or senescent tissue, mostly in cooler regions; under high relative humidity the fungus can cause great damage to perishable plant produce at harvest and in storage. Major diseases of soft fruit are caused, particularly of grapevine, see *EPPO Bull.* 12(2), 1982; and most vegetables are liable to be infected; for infection of broad bean see *Botrytis fabae*; Coley-Smith et al. under *Botryotinia*; Holliday 1980, as *S. fuckeliana*, and see *Botrytis* [**62**,299–333].

B. globosa Buchw., *Phytopath.Z.* **20**:250, 1953; anam. *Botrytis globosa* Raabe, *Hedwigia* **78**:71, 1938; con. mostly 12–18 μm diam., sclerotia linear, formed on leaf midrib, characteristic concertina collapse of conidiophores, like *B. squamosa*; garlic leaf spot; Webster & Jarvis, *TBMS* **34**:187, 1951, on *Allium ursinum* [**33**:463].

B. narcissicola (Gregory) Buchw. 1949; *Index Fungi* **1**:277, 1950; **2**:97, 1953; anam. *Botrytis narcissicola* Kleb. 1906; asco. 10–20 × 5–9 μm, con. 8–16 × 7.5–12 μm, sclerotia black, 1–1.5 mm diam.; *Narcissus* smoulder, a bulb and leaf rot; O'Neil et al., *Pl.Path.* **31**:65, 101, 1982, infection & epidemiology in Britain [**61**,5797, 7053].

B. polyblastis (Gregory) Buchw., as above; anam. *Botrytis polyblastis* Dowson, *TBMS* **13**:102, 1928; asco. 12–21 × 6–12 μm, con. 30–50 μm diam.; *Narcissus* fire; Chastagner, *PD* **67**:1384, 1983, epidemiology & control in USA (Washington State, W.) [**7**:581; **63**:1823].

B. porri (v. Beyma) Whetzel 1945; anam. *Botrytis porri* Buchw., *K.Vet.Højsk.Aarsskr.* **32**:137, 1949; con. mostly 11–14 × 7–10 μm, length:breadth ratio 1.35–1.5, sclerotia very large, often cerebriform; *Allium*, seedborne, an important pathogen on leek in storage, see Tahvonen, *J.Scient.Agric.Soc. Finland* **52**:331, 1980; **53**:27, 1981; Presly, see *Botrytis*, studied the spp. of this genus on leek and onion, and the implications of transfer between these crops.

B. ricini (Godfrey) Whetzel; anam. *Botrytis ricini* Buchw. 1949; Hennebert 1973 placed this sp. in the new genus *Amphobotrys*; con. mostly 7–10 μm diam., sclerotia usually 3–9 mm long; castor grey capsule mould; Godfrey, *J.agric.Res.* **23**:679, 1923 [**3**:377].

B. squamosa Viennot-Bourgin, *Annls.Épiphyt.* **4**:38, 1953; anam. *Botrytis squamosa* Walker 1925; con. 15–20 × 12–15 μm, length:breadth ratio 1.25–1.45; sclerotia flat, like scales, 0.5–2 mm diam., characteristic concertina collapse of conidiophores, like *B. globosa*; onion small sclerotial neck rot and a leaf blight, an important disease at moderate temp., seedborne; Presly & Maude, *AAB* **94**:197, 1980, chemical control; Alderman, Lacy et al., *Phytop.* **73**:670, 1020, 1983; **74**:1461, 1984; **75**:808, 1985; *CJB* **62**:2793, 1984, conidial release & weather, effects of moisture & temp. on infection, growth & sporulation, leaf position; Sutton *et al.*, *Can.J.Pl.Path.* **5**:256, 1983, weather & host in an epidemic [**59**,6053; **62**,4565; **63**,1039, 4198; **64**,1871, 4620; **65**,488].

*****Botrytis** Micheli: Pers. 1801; Hyphom.; Carmichael et al. 1980; conidiomata hyphal, conidiogenesis holoblastic; con. aseptate, hyaline, botryose; sometimes forms sclerotia, teleom. in *Botryotinia*; Ellis 1971 described 18 spp., 4 of them are important on *Allium* spp. especially onion; their conidia differ in length:breadth ratios; *B. porri* and *B. squamosa* have large and small sclerotia, respectively; in *B. allii* sclerotia are rare, and in *Botrytis allii*, teleom. unknown, they are small; Presly, *Pl.Path.* **34**:422, 1985, compared these fungi on leek & onion crops, & inoculum movement between them, in UK; Jarvis, *Monogr.Res.Branch Agric.Canada* 15, 1977; Moore 1979; Coley-Smith et al. ed., *The biology of* Botrytis, 1980; Holliday 1980; Mansfield et al., *Physiol.Pl.Path.* **17**:131, 1980; **19**:41, 1981, comparisons between pathogens & non-pathogens; Backhouse et al., *TBMS* **82**:625, 1984, taxonomy & electrophoresis; Harrison & Williamson, *ibid* **86**:171, 1986, red raspberry, survival & infection [**60**,4328; **61**,940; **63**,3776; **65**,492, 2372].

B. allii Munn 1917; Ellis & Waller, *Descr.F.* 433, 1974; con. mostly 7–11 × 5–6 μm, length:breadth ratio 1.7–2.5, narrower than those of the other *Botrytis* spp. on *Allium*, sclerotia 1–5 mm diam.; this is the most important of these spp. causing *Allium* neck rots, particularly onion; bulb infection is called grey mould neck rot, seedborne, controlled by chemical treatment of the seed and

high temp. drying for the harvested bulbs, Maude et al., *Pl.Path.* **31**:247, 1982; **33**:263, 1984, persistence in field soil, effects of direct harvesting & drying on disease incidence & control; Stewart & Mansfield, *ibid* **33**:401, 1984, development on onion & other *Botrytis* spp. [**62**,514; **63**,4670, 5693].

B. anthophila Bondartsev 1913; Ellis 1971; con. mostly 11–16 × 4–5 μm, sclerotia apparently absent; on red clover, this sp. has an unusual life history in that the fungus is systemic, no damage is evident until flowering when the conidiophores replace the anthers and pollen, internally seedborne; Silow, *TBMS* **18**:239, 1923; Noble, *ibid* **30**:84, 1948, comparison with fungi, as *Sclerotinia*, on clover [**28**:176].

B. byssoidea, teleom. *Botryotinia allii*; **Botrytis cinerea**, teleom. *Botryotinia fuckeliana*.

B. convallariae (Kleb.) Ondrej 1972, see *Index Fungi* **4**:403, 1977; lily of the valley, *Convallaria majalis*, stem rot; the fungus has been thought to be a form of *Botrytis cinerea*, and *B. paeoniae* may infect this plant.

B. convoluta, teleom. *Botryotinia convoluta*.

B. elliptica (Berk.) Cooke 1902; Ellis 1971; con. mostly 20–30 × 13–18 μm; lily leaf spot; van Beyma & van Hell, *Phytopath.Z.* **3**:619, 1931 [**11**:108].

B. fabae Sardiña 1929; Ellis & Waller, *Descr.F.* 432, 1974; con. mostly 16–25 × 13–16 μm, length:breadth ratio 1.25–1.45, sclerotia 1–1.7 mm diam.; broad bean chocolate spot; an important, seedborne pathogen which, in the aggressive phase at high humidities, causes spreading lesions, extensive blight and death. In Britain the disease is the most severe one on broad bean, less so in Scotland. The conidia have a diurnal periodicity with, on dry days, a forenoon peak; lesions develop at > 92% relative humidity but only rarely at < 86%, the sclerotia overwinter. *Botrytis cinerea* also causes chocolate spot, particularly early in the season but is less virulent; both fungi can be isolated from the same lesion; Harrison, *AAB* **95**:53, 63, 1980, effects of relative humidity & temp. on lesion growth & toxins; *TBMS* **72**:389, 1979; **80**:263, 1983; **82**:245, 1984; **83**:295, 631, 1984, overwintering, conidial survival, effect of relative humidity on infection & with light & temp. on sporulation, *B. cinerea* as a cause of disease; Fitt et al., *ibid* **85**:307, 1985, conidial dispersal; Bainbridge et al., *Pl.Path.* **34**:5, 1985, chemical control; Creighton et al., *Crop Protect.* **4**:235, 1985, conidial dispersal & chemical control; Fitt et al., *J.agric.Sci.* **106**:307, 1986, irrigation, benomyl & yield. [**59**, 1517; **60**,1154–5; **62**,4059; **63**,2602; **64**,372, 1366, 4048, 4591, 5553; **65**,4169].

B. galanthina (Berk. & Br.) Sacc. 1886; Ellis 1976; con. 9.5–20 × 8–11 μm, sclerotia 1–2 mm diam.; one of the few diseases, grey mould, of snowdrop, *Galanthus nivalis*; young shoots are killed and a rot spreads to the bulbs.

B. gladiorum, teleom. *Botryotinia draytonii*; **Botrytis globosa**, teleom. *Botryotinia globosa*.

B. hyacinthi Westerdijk & van Beyma, *Meded. phytopath.Lab.Willie Commelin Scholten* **12**:15, 1928; con. mostly 14–18 × 11–13 μm; hyacinth fire [8:41].

B. narcissicola, teleom. *Botryotinia narcissicola*.

B. paeoniae Oudem. 1898; Ellis 1971; con. mostly 12–18 × 8–10 μm, sclerotia 1 mm diam., peony blight.

B. polyblastis, teleom. *Botryotinia polyblastis*; **Botrytis porri**, teleom. *Botryotinia porri*; **Botrytis ricini**, teleom. *Botryotinia ricini*; **Botrytis squamosa** teleom. *Botryotinia squamosa*.

B. tulipae Lind 1913; Ellis 1971; con. mostly 16–20 × 10–13 μm, sclerotia commonly 1–2 mm diam.; tulip fire, has caused severe damage; Beaumont et al., *AAB* **23**:57, 1936, in England [15:508].

bottle gourd, or white flowered gourd, *Lagenaria siceraria*.

—— **mosaic** v. Vasudeva & Lal, *Indian J.agric.Sci.* **13**:182, 1943; Shankar et al., *Indian J.Microbiol.* **11**:43, 1971; prob. a str. of tobacco mosaic v. [**23**:125; **55**,5400].

*****bottom rot** (*Thanatephorus cucumeris*) cabbage, a complex in lettuce, see Kooistra 1983 [63,2633].

bougainvillea chlorotic blotch Bestagno, *Riv.Patol.veg.Pavia* **3**:213, 1963; Italy [43,2626].

bougainvilleae, *Cercosporidium*.

boussingaultia mosaic v. Beczner & Vassányi 1980; Potexvirus, Hungary [60,5769].

Boyce, John Shaw, 1887–1971; born in Belfast, Ireland, Univ. Nebraska, Lincoln, forest pathologist at Yale Univ. School of Forestry, wrote: *Forest pathology* 1938, edn. 3, 1961. *Phytop.* **62**:681, 1972.

Boyd, Andrew Edward Wilson, 1916–81; born in St Petersburg, now Leningrad, Univ. Glasgow, Schools of Agriculture, Univ. Nottingham 1943–6 and Univ. Edinburgh 1950–81, head of crop protection; a world authority on potato diseases, particularly the latent ones affecting seed tubers, blight and forecasting, see *Rev.Pl.Path.* **51**:297, 1972; etiology of sugarbeet strangles, Fellow of the Royal Society of Edinburgh 1970. *Bull.Br.mycol.Soc.* **15**:151, 1981; *Yrbk.R.Soc.Edinburgh* for 1982.

boysenberry, like a loganberry which is prob. a hybrid between a blackberry and a raspberry.

brachiaria white leaf Chen et al. 1972; MLO assoc., *B. distachya* [**52**,3356].

Brachycaudus cardui, transm. artichoke latent v.; *B. helichrysi*, transm. senecio mosaics.

brachypodii, *Ascochyta*.

brachypodium yellow streak v. Edwards et al., *Pl.Path.* **34**:95, 1985; isometric particle assoc. *c.*32 nm diam., single species RNA, not transm., appears distinct, *B. sylvaticum*, England [**64**,3875].

bracken, *Pteridium aquilinum* q.v.

bract necrosis, sunflower, prob. caused by moisture and heat stress; Yang & Berry, *Ann. phytopath.Soc.Japan* **49**:724, 1983 [**63**,4520].

branch blight (*Stigmina mori*) mulberry; —— **canker** (*Ceratocystis fimbriata*) coffee, (*Cryptodiaporthe salicella*) willow, (*Poria hypobrunnea*) tea; —— **wilt** (*Hendersonula toruloidea*) fig, grapevine, walnut.

brand canker (*Coniothyrium wernsdorffiae*) rose.

brasiliensis, *Elsinoë*.

Brassica campestris, turnip; *B. carinata*, Ethiopian mustard; *B. chinensis*, Chinese cabbage; *B. juncea*, brown mustard; *B. napus*, rape, swede; *B. nigra*, black mustard; *B. oleracea* vars., broccoli, Brussels sprout, cabbage, cauliflower, kale, kohlrabi; Dixon 1981; Geeson in Dennis 1983; Tait in *Adv.Appl.Biol.* **8**, 1983; Kolte, *Diseases of annual edible oilseed crops* vol. 2, 1985; Sherf & MacNab 1986.

brassicae, *Alternaria*, *Aphanomyces*, *Asteromella*, *Olpidium*, *Plasmodiophora*, *Pyrenopeziza*.

brassicicola, *Alternaria*, *Cercospora*, *Mycosphaerella*.

braziliensis, *Cylindrocladium*.

breakdown, applied to: rot and disintegration of tissue; failure of control by chemicals, of host resistance; an expansion in FBPP.

brebissonii, *Helicobasidium*.

Brefeld, Oscar, 1839–1925; born in Prussia, Westphalia; chairs at Forestry Academy, Eberswalde, Munster and Breslau; primarily a mycologist, he developed pure culture methods for fungi and 10 years before Koch was using gelatin to make solid media. *Nature Lond.* **116**:369, 1925.

brefeldin A, toxin from *Alternaria carthami*, Teitjen et al., *Physiol.Pl.Path.* **23**:387, 1983; **26**:241, 1985 [**63**,1904; **64**,4442].

Bremia, Regel 1843; Peronosporaceae; sporophores branch at an acute angle and dichotomously; branches with enlarged tips, like a disc, and bear 3 or 4 projections 'sterigmata' which are characteristic, spores with an apical papilla and usually germinating directly, i.e. not by zoospores, oogonial wall unornamented, oospore av. 25–26 μm diam.; only 2 spp. recognised: *B. graminicola* Naoumov on the grass *Athraxon* and *B. lactucae*.

B. lactucae Regel; Morgan, *Descr.F.* 682, 1981;

causes a severe disease of lettuce in which the fungus can become systemic; occurs on artichoke, chicory, endive, safflower and ornamental Compositae. The control of this downy mildew has been intensively investigated on lettuce in the parts of N.W. Europe and N. America where the crop is important, and where cool, moist conditions occur. Many genera in this family are hosts and *B. lactucae* is specific on one or a few spp. in one genus, i.e. ff.sp.; spread is by the con. but oospores may be important as a primary source of inoculum; heterothallism occurs, see Michelmore & Ingram, *TBMS* **75**:47, 1980; **77**:131, 1981; **78**:1, 1982; for viability of, and infection by, oospores see Morgan and Norwood & Crute, *ibid* **80**:403, **81**:144, 1983; pathogenicity is very variable, and many matching virulence and major resistance genes have been delineated; Crute, *Crop Protect.* **3**:223, 1984, described control through mixed fungicides on cvs. with novel resistance genes in a race non-specific background. The large literature since 1980 on breeding for resistance has been mostly omitted; the last reviews were by Crute & Dixon in Spencer 1981, Dixon 1981 loc.cit.; Crute in Wolfe & Caten 1987. Eenink et al., *Euphytica* **31**:73, 1982; **32**:139, 1983, partial resistance; Norwood et al., *ibid* **32**:161, 1983, inheritance of field resistance; Morgan, *Crop Protect.* **3**:349, 1984, environmental & fungicide control in the glasshouse; Yuen & Lorbeer, *Phytop.* **74**:149, 1984, field resistance [**60**,2903; **61**,2035, 4531, 6120; **62**,4079–80, 4555; **63**,333, 3147, 4659; **64**,877].

Bremiella G.W. Wilson 1914; Peronosporaceae; differs from *Bremia* in that the tips of the sporophore branches are blunt, only slightly enlarged, also the oogonial wall is thick and ornamented; Constantinescu, *TBMS* **72**:510, 1979, revision; *B. baudysii* (Scalický) O. Const. & Negrean on *Berula*, *B. megasperma* (Berl.) Wilson on *Viola* spp.

brevicaulis, *Scopulariopsis*.

Brevicoryne brassicae, transm. broccoli necrotic yellows.

Brevipalpus phoenicis, transm. citrus leprosis.

brevirama, *Laterispora*.

breynia bunch Dabek & Jackson, *Phytopath.Z.* **90**:132, 1977; MLO assoc., Solomon Islands (Malaita) [**57**,2670].

Brian, Percy Wragg, 1910–79; born in England, Univ. Cambridge, Long Aston Research Station, Imperial Chemical Industries, chairs in botany Univ. Glasgow 1962 and Cambridge 1968; fundamental work on antibiotics from fungi, particularly griseofulvin used against animal

mycoses; FRS 1958, CBE 1978. *Nature Lond.* **282**:541, 1979; *Physiol.Pl.Path.* **15**:231, 1979; *Biogr.Mem.Fel.R.Soc.* **27**:103, 1981; *TBMS* **76**:1, 1981.

Brierley, Philip, 1899–1968; born in USA, Univ. Minnesota and Cornell; work on suspected virus diseases of ornamentals and with F.F. Smith on vectors; in 1949 they showed that the chrysanthemum stunt agent, reported in 1947, could be sap transm.; 23 years later this disease was shown to be caused by a viroid; Gladiolus Society, New England, Gold Medal; *Phytop.* **59**:715, 1969.

Brierley, William Broadhurst, 1889–1963; born in England, Univ. Manchester, Royal Botanic Gardens, Kew; Rothamsted Experimental Station; chair agricultural botany, Univ. Reading 1932; editor of *AAB* for 24 years, a journal with still one of the highest editorial standards; prepared the English edn., 1950, of Gäumann's classic *Pflanzliche Infektionslehr* 1946. *AAB* **51**:509, 1963; *Nature Lond.* **198**:133, 1963.

bristle top, coconut, Guam, etiology uncertain but symptoms differ from those of tinangaja caused by the coconut cadang cadang viroid.

Briton-Jones, Harry Richard, 1893–1936; born in Wales, Imperial College of Science and Technology, chair at Imperial College of Tropical Agriculture from 1926, wrote: *The diseases and curing of cacao* 1934; *The diseases of the coconut palm* 1940, revised and completed by E.E. Cheesman. *Trop.Agric.Trin.* **13**:318, 1936.

brittle root (*Spiroplasma citri*) horseradish.

broad bean, *Vicia faba*; Bos in Hawtin & Webb ed., *Faba bean improvement*, 1982; Hebblethwaite ed., *The faba bean* (Vicia faba L.), 1983.

—— **Evesham stain** = broad bean stain v.

—— **fire blight** Kang & Zhou, *Microbiology* **8**:153, 1981; transm. sap, China (Yunnan) [**61**,2551].

—— **male sterility** Scalla et al., *Revue Cytol.Biol.Végét.Le Botaniste* **5**:71, 1982; isometric particle 73 nm diam. assoc. [**61**,5345].

—— **mild mosaic** Yu, *Phytop.* **29**:448, 1939; transm. sap, aphid, China (Chekiang, Kiangsu) [**18**:649].

—— **mosaic and leaf roll** v. Zagh & Férault, *Annls. Phytopath.* **12**:153, 1980; isometric 25 nm diam., transm. sap, Algeria [**60**,1718].

—— **mottle** v. Bawden, Chaudhuri & Kassanis 1951; Gibbs, *Descr.V.* 101, 1972; Bromovirus, isometric *c*.26 nm diam., transm. sap readily and beetles to many legumes; Africa, Europe.

—— **necrosis** v. Fugano & Yokoyama 1951; Inouye & Nakasone, *Descr.V.* 223, 1980; Tobamovirus, rod *c*.150 × 25 nm and 250 × 25 nm,

transm. sap and through soil, limited host range, Japan, disease in broad bean only.
— — — (bean leaf roll v.+pea enation mosaic v. or BLRV+pea early browning v.), Cockbain et al., *AAB* **102**:495, 1983; in England [**62**,4058].
— — **phyllody** Nour 1962; see Jones et al., *Pl.Path.* **33**:599, 1984; MLO assoc., Sudan [**64**,1843].
— — **severe chlorosis** v. Thottappilly, Harris & Bath, *Phytopath.Z.* **84**:343, 1975; Boswell & Gibbs 1983; rod or filament modal lengths 650–820 nm and 1360–1640 nm, transm. sap, *Acyrthosiphon pisum*, USA (Michigan) [**55**,3366].
— — **stain** v. Lloyd, Smith & Jones 1965, Gibbs & Smith, *Descr.V.* 29, 1970; Comovirus, isometric *c.*25 nm diam, transm. sap, seedborne, necrotic staining of testa, Africa (N.W.), Europe, legumes only; see Jones & Barker, echtes ackerbohnenmosaik v.; Cockbain et al., *AAB* **81**:331, 1975, transm. *Apion vorax, Sitona lineatus* [**55**,2033].
— — **tip dying** v. Lapchik, Kutznetsova & Mel'nichenko, *Mikrobiol.Zhurnal* **35**:188, 1973; isometric, USSR, transm. sap, legumes [**52**,3362].
— — **true mosaic** (echtes ackerbohnenmosaik v.).
— — **wilt** v. Stubbs 1947; Taylor & Stubbs, *Descr.V.* 81, 1972; Boswell & Gibbs 1983; poss. Comovirus, isometric *c.* 25 nm diam., transm. sap and aphid, non-persistent, wide host range, causes pea streak, spinach blight; Uyemoto & Provvidenti, *Phytop.* **64**:1547, 1974, serotypes [**54**,4394].
— — **yellow band** v. Russo et al., *AAB* **105**:223, 1983; poss. Tobravirus, rod, modal lengths S particles 77 nm, L particles 202 nm, 2 species ssRNA, transm. sap, Italy (Apulia), induces a very mild disease in pea, reported to be a str. of pea early browning v. [**64**,1361].
— — — **green mosaic** v. Zagh & Férault, *Annls Phytopath.* **12**:153, 1980; isometric *c.*25 nm diam., transm. sap, Algeria [**60**,1718].
— — — **ringspot** Natsuki, abs. *Ann. phytopath.Soc.Japan* **45**: 84, 1979; isometric 28 nm diam., Japan, fide Boswell & Gibbs 1983.
— — — **vein** Natsuki, abs. *Ann. phytopath. Soc.Japan* **47**:410, 1981; bacilliform 230–250 × 110–130 nm; Japan, fide Boswell & Gibbs 1983.
broccoli see *Brassica*.
— **mosaic** = cauliflower mosaic v.
— **necrotic yellows** v. Hills & Campbell 1968; Campbell & Lin, *Descr.V.* 85, 1972; Phytorhabdovirus, bacilliform 275 × 75 nm, transm. sap difficult, and *Brevicoryne brassicae*, *Brassica oleracea* only known naturally infected host, economic importance doubtful.

brome grass, *Bromus*.
— **mosaic** v. McKinney, Fellows & Johnston 1942; Lane, *Descr.V.* 180, 1977; Bromovirus, type, isometric *c.*26 nm diam., transm. sap readily; Edwards et al., *Pl.Path.* **32**:91, 1983, natural hosts in UK; Erasmus & von Wechmar and von Wechmar et al., *Phytopath.Z.* **108**:26, 1983; **109**:341, 1984, transm. uredospores of *Puccinia graminis* & *P. recondita*, & seed [**62**,2528; **63**,2244, 4351].
— **stem leaf mottle** (phleum mottle v.).
— **streak mosaic** v. Milličić et al., *Acta bot.croat.* **39**:27, 1980; **41**:7, 1982; filament 650–700–nm long, transm. sap, Yugoslavia, *Bromus mollis*, *Hordeum murinum*, infected barley, oat [**60**,6519; **62**,4335].
bromi, *Drechslera, Pyrenophora*; **bromivora**, *Spermospora*.
bromomethane, methyl bromide, general fumigant for dead plant products, closed buildings and to fumigate soil.
Bromoviruses, type brome mosaic v., other members: broad bean mottle v., cowpea chlorotic mottle v., melandrium yellow fleck v.; isometric *c.*29 nm diam., 4 segments linear ssRNA, the 3 largest are required for infection, sedimenting at *c.*85S, thermal inactivation 70–95°, longevity in sap days to weeks, transm. sap readily and beetles, persistent, limited host range; only this group, Cucumoviruses and Ilarviruses have a tripartite genome; Lane, *Adv.Virus Res.* **19**:151, 1974; *Descr.V.* 215, 1979; in Kurstak 1981; Francki in Francki et al. 1985.
bronopol, bactericide.
bronze leaf wilt, coconut, see *Phytomonas*.
bronzing, bean, ozone injury, Weaver & Jackson, *Can.J.Pl.Sci.* **48**:561, 1968; rice, Fe toxicity, Ou 1985; tung, Zn deficiency, Mowry & Camp, *Bull.Fla.agric.Exp.Stn.* 273, 1934 [**14**:481; **48**,1401].
— , willow, see *Drepanopeziza sphaerioides*.
Brooks, Frederick Tom, 1882–1952; born in England, Univ. Cambridge, Botany School, chair 1936, trained many students who became plant pathologists in all parts of the world, FRS 1930, CBE 1947, wrote: *Plant diseases* 1928, edn. 2, 1953. *Proc. Lin.Soc.* **163**:254, 1950–1; *AAB* **39**:617, 1952; *Obit.Not.Fel.R.Soc.* **8**:34, 1953; *TBMS* **36**:177, 1953.
Brook's fruit spot (*Mycosphaerella pomi*) apple, after Charles Brooks 1872–1962, *Phytop.* **54**:249, 1964.
broomcorn, see *Sorghum*.
broomrape, *Orobanche*, parasitic by their roots on roots of other plants; Pierterse, *Abs.Trop.Agric.* **5**(3):9, 1979.

broom rust (*Gymnosporangium nidus-avis*), juniper.
Brown, William, 1888–1975; born in Scotland,
Univ. Edinburgh, Imperial College of Science and
Technology, Univ. London, chair 1928, classical
studies on the physiology of fungi and the
interactions between host and pathogen, FRS
1938. *AAB* **80**: 255, 1975; *Biogr.Mem.Fel.R.Soc.*
21:155, 1975; *TBMS* **65**:343, 1975; *A.R.Phytop.*
23:13, 1985; and see Brown, *ibid* **3**:1, 1965.
*****brown bast,** rubber, a physiological response to
over exploitation; — **blight** (*Glomerella cingulata*)
coffee, tea, (*Pyrenophora lolii*) temperate grasses;
— **blotch** (*Colletotrichum capsici*) cowpea,
(*Pseudomonas tolaasi*) common cultivated
mushroom; — **bordered leaf spot** (*Cercospora
capsici*) red pepper; — **canker** (*Cryptosporella
umbrina*) rose, (*Mycosphaerella holci*) Gramineae;
— **checking,** celery, B deficiency, Yamaguchi
et al., 1956–8; — **core** (*Phytophthora primulae*)
primula; — **cubical rot** (*Lentinus lepideus*)
conifers; — **etch** (*Didymella bryoniae*) pumpkin;
— **eyespot** (*Cercospora coffeicola*) coffee; — **felt
blight** (*Herpotrichia coulteri*) pine; — **foot rot**
(*Monographella nivalis*) barley, oat, wheat; —
heart, pear, in storage, assoc. with high CO_2;
swede, turnip B deficiency; — **leaf spot**
(*Cercoseptoria sesami*) sesame, (*Deightoniella
torulosa*) abaca, (*Mycosphaerella henningsii*)
cassava, (*Sirosporium diffusum*) pecan; —
margin (*Bremia lactucae*) lettuce [**37**:63, 751].
— **mustard,** see *Brassica.*
browning root rot (*Lagena radicicola*) temperate
cereals.
brown needle blight (*Mycosphaerella gibsonii*) pine;
— **patch** (*Thanatephorus cucumeris*) St Augustine
grass; — **pocket rot** (*Lentinus kauffmanii*) sitka
spruce; — **pod** (*Botryodiplodia theobromae*)
cacao; — **root rot** (*Phellinus noxius*) oil palm,
rubber, tea, tropical tree crops.
— **rot,** rotting of wood in trees invaded by fungi
which destroy cellulose, leaving a brown lignin
residue, after FBPP; (*Acetobacter aceti, A.
pasteurianus*) apple, pear, (*Erwinia cypripedi*)
orchids, (*Monilinia fructicola, M. fructigena, M.
laxa*) pome and stone fruits, (*Phytophthora* spp.)
citrus fruit, (*Pseudomonas solanacearum*) potato;
— **rust** (*Phakopsora pachyrhizi*) soybean and
other legumes, (*Puccinia chrysanthemi*)
chrysanthemum, (*P. hordei*) barley, (*P. recondita*)
rye, wheat; — **sheath rot** (*Gaeumannomyces
graminis* var. *graminis*) rice, Gramineae; —
spongy rot (*Inonotus glomeratus*) *Acer, Fagus;* —
spot needle blight (*Scirrhia acicola*) pine; — **stem
rot** (*Phialophora gregata*) soybean; — **stringy rot**
(*Echinodontium tinctorium*); — **stripe** (*Bipolaris
stenospila*) Gramineae, (*Pseudomonas avenae*) rice,

Gramineae, (*Sclerophthora rayssiae*) maize; —
top rot (*Fomitopsis rosea*) conifers.
brugmansia little leaf, Hiruki,
Ann.phytopath.Soc.Japan **52**:675, 1986; MLO
assoc., Australia (Queensland), *B. candida*
[**66**,1944].
Brunchorstia Jakob Eriksson 1891; Coleom.;
Sutton 1980; conidiomata stromatic,
conidiogenesis enteroblastic phialidic; con.
multiseptate, hyaline; *B. pinea,* teleom.
Gremmeniella abietina.
brunnea, *Marssonina, Ramularia;* **brunnescens,**
Agaricus.
Brussels sprout, see *Brassica.*
bryoniae, *Didymella.*
bryonia v. Milne et al., *Phytopath. Mediterranea*
19:115, 1980; poss. 3 Potyviruses, Italy (N.), *B.
cretica,* white bryony [**63**,66].
buckeye (*Phytophthora erythroseptica, P. nicotianae*
var. *parasitica*) tomato fruit.
Buchwald, Niels Fabritius, 1899–1986, Univ.
Copenhagen, chair at Royal Veterinary and
agricultural Univ. 1944–68, taxonomy of the
Sclerotiniaceae, diseases and fungi in Denmark,
several text books in Danish. *Seed Path.News*
17:13, 1986.
buckwheat, *Fagopyrum,* Sidorova 1963 in USSR
[**44**, 2950d].
— **mosaic** Krotov 1969, USSR [**49**,2479].
bud and twig blight (*Botryotinia fuckeliana*) rose,
(*Gibberella baccata*) pine; — **blast** (*Pycnostysanus
azaleae*) Rhododendron; — **blight** (*G. baccata*)
mulberry; — — **and twig canker** (*Pseudomonas
syringae* pv. *eriobotryae*) loquat; — **decay**
(*Aphelenchoides fragariae*) tree peony; — **failure,**
almond, genetic disorder, Kester et al. 1975; —
proliferating gall (*Nocardia vaccinii*) blueberry;
— **rot** (*Fusarium tricinctum*) carnation,
(*Gibberella baccata*) apple, (*Phytophthora
palmivora*) coconut, other palms, (*Ramularia
belluensis*) pyrethrum; — **rust** (*Chrysomyxa
woroninii*) spruce; — **stunting** (*Elsinoë batatas*)
sweet potato [**55**,804–5].
— **union and rootstock disorder,** an incompatibility
in Eureka lemon on Troyer citrange stock,
Weathers et al., *Calif. Agric.* **9**(11):11, 1955;
PDR **39**:665, 1955 [**35**:179, 446].
— **wilt** (*Diplodia pinea*) pine, other conifers.
buffalo gourd, *Cucurbita digitata* or *C. foetidissima;*
Rosemeyer et al., *PD* **70**:405, 1986, 5 viruses from
greenhouse cuttings, USA (Arizona) [**65**,5234].
buismaniae, *Pythium,* see *P. megalacanthum.*
bulb canker (*Embellisia allii*) garlic.
bulborum, *Sclerotinia;* **bulbosa,** *Armillaria;*
bulbosum, *Phragmidium.*
bulbs, see ornamentals.

bulb skin necrotic patch (*Embellisia hyacinthi*) hyacinth.

bullata, *Ustilago*.

Buller, Arthur Henry Reginald, 1874–1944; born in England; Univ. London and Leipzig, chair at Univ. Manitoba 1904–36, a prodigious worker on fungi: behaviour patterns, mating, cytology, spore formation, liberation and dispersal; Fellow Royal Society of Canada 1909, FRS 1929, Flavelle Medal Royal Society of Canada 1929, Royal Medal Royal Society of London; wrote: *Researches on Fungi*, 7 vol., 1909, 1922, 1924, 1931, 1933, 1934, 1950; see *Mycologia* **50**:794, 1958 for the Buller memorial library. *Nature Lond.* **194**:173, 1944; *Science N.Y.* **100**:305, 1944; *Obit. Not.Fel.R.Soc.* **5**:51, 1945; *Phytop.* **35**:577, 1945; *A.R.Phytop.* **24**:17, 1986.

Buller phenomenon, Quintanilha's term, the diploidisation of a haploid mycelium by fusion with another mycelium, the formation of conjugate nuclei, hence to diploidise, reported by Buller in 1930, see *Researches on fungi*, vol. 4:187, 1931.

Buller's drop, a liquid drop which appears shortly before discharge of a ballistospore q.v.; Webster et al., *TBMS* **83**:524, 1984.

bull's eye rot (*Pezicula malicorticis*) pear.

bulrush millet, see millets.

bumpy fruit, papaya, B deficiency, Wang & Ko, *Phytop.* **65**:445, 1975; Chan & Raveendranathan, *MARDI Res.Bull.* 12:281, 1984 [**55**,343; **65**,3426].

bunchosiae, *Cercospora*.

bunch rot (*Marasmius*, prob. *Marasmiellus*, *palmivorus*) oil palm.

bunodes, *Rosellinia*.

bunt, see stinking smut.

bupirimate, systemic fungicide against powdery mildews.

bupleurum yellows Shiomi, Choi & Sugiura, *Ann.phytopath.Soc. Japan* **49**:228, 1983; MLO assoc., transm. *Macrosteles orientalis*, Japan, *B. falcatum* [**63**,343].

burdock, *Arctium lappa* and other spp.

—— **mosaic v.** Inouye & Mitsuhata, *Nogaku Kenkyu* **54**:1, 1971; rod 600 × 18–20 nm, longer particles now referred to burdock yellows v., Japan [**52**,1060].

—— **mottle v.** Inouye, *Ber.Ôhara Inst.landw. Biol.Okayama* **15**:207, 1973; rod *c.*250 × 17 nm, transm. sap, Japan [**53**,452].

—— **stunt** Chen et al. 1982; *J.gen.Virol.* **64**:409, 1983; poss. viroid, China (N.) [**62**,2070, 4406].

—— **yellows v.** Inouye & Mitsuhata, see burdock mosaic v.; Nakano & Inouye, *Ann.phytopath.Soc.Japan* **46**:7, 1980;

Closterovirus, filament 1700–1750 × 12 nm, transm. sap, *Uroleucon* (*Uromelan*) *gobonis*, Japan [**60**,674].

Burgundy mixture, similar to Bordeaux mixture q.v. except that sodium carbonate replaces the calcium hydroxide to give a blue precipitate of basic copper carbonate; it was much less widely used.

burn (*Leptosphaerulina trifolii*) clover, lucerne.

burnsii, *Alternaria*.

Burrill, Thomas Jonathan, 1839–1916; born in USA, Univ. State Normal Illinois, Univ. Illinois, an early investigator of plant diseases caused by bacteria, etiology of pear and apple fire blight, now *Erwinia amylovora*. *J.Bact.* **1**:269, 1916; *Bot.Gaz.* **62**:153; *Phytop.* **8**:1, 1918.

Bursaphelenchus xylophilus, this nematode has caused a devastating wilt of mature pines in Japan; that such an epidemic, where millions of trees were killed, was caused by a nematode is highly unusual; the few other pine wilts are caused by *Ceratocystis* spp.; Mamiya, *A.R.Phytop.* **21**:201, 1983, tabulated 18 and 19 *Pinus* spp. as resistant and susceptible, respectively; Malek & Appleby, *PD* **68**:180, 1984, described an epidemic in USA (Illinois).

Butler, Edwin John, 1874–1943; born in Ireland, county Clare, Queens College Cork, medically qualified but never practised, India 1901 as cryptogamic botanist, classic study on *Pythium*, disease investigations in tropical crops which had hardly begun except for sugarcane; founder director Imperial Bureau of Mycology, Kew, now the CAB International Mycological Institute q.v. and began the *Review of Applied Mycology*; later secretary Agricultural Research Council; CIE 1921, FRS 1926, CMG 1931, Knight 1939; wrote: *Fungi and disease in plants*, 1918; *The fungi of India*, 1932, with G.R. Bisby; *Plant pathology*, 1949, with S.G. Jones. *Obit.Not.Fel.R.Soc.* **4**:455, 1943; *AAB* **31**:168, 1944; *Phytop.* **34**:149, 1944; *Kavaka* **3**:9, 1975.

Butlerelfia Weresub & Illman, *CBJ* **58**:144, 1980; Corticiaceae; *B. eustacei* loc. cit., assoc. with fish eye rot in stored apples; Butler, *J.agric.Res.* **41**:269, 1930, as *Corticium centrifugum* (*Lév*) *Bres.* [**10**:38; **59**,5258].

butleri, *Pythium*.

butter bean, see Lima bean.

butterbur mosaic v. Tochihara & Tamura, *Ann.phytopath.Soc.Japan* **42**:533, 1976; filament 670 nm long, transm. sap and *Myz.pers.*, Japan, *Petasites japonicum* [**56**,4573].

buttercup mosaic v. Padma, Singh & Verma, *Z.PflKrankh.PflSchutz.* **79**:710, 1972; filament 800 nm long, transm. sap and *Myz.pers.*, India (N.), *Ranunculus asiaticus* [**52**,3724].

butt rot (*Ganoderma lucidum*) tree crops, (*Heterobasidion annosum*) conifers, (*Phaeolus schweinitzii*) conifers, (*Rigidoporus ulmarius*) elm, other temperate hardwoods, (*Serpula himantioides*) conifers.

Buxton, Eric William, 1926–64, born in England; Univ. Cambridge, Rothamsted Experimental Station, original work on parasexual recombination in *Fusarium*. *Nature Lond.* **204**:426, 1964.

byssoidea, *Botrytis*.

C

cabbage, see *Brassica*.
— **black ring**, and black ringspot (turnip mosaic v.).
— **Chinese vein necrosis** *Acta Microbiol.sin.*
16:136, 1976; China (Xi'an) [**56**, 2259].
— **flower greening** Cvjetković & Panjan 1978;
MLO assoc., Yugoslavia (Croatia) [**58**,5065].
— **internal necrosis** (turnip mosaic v.), Walkey &
Webb, *AAB* **89**:435, 1978; Britain, stored white
cabbage; cauliflower mosaic v. did not cause
internal necrosis in the cabbage cv. Decema as
had been reported. TuMV was not assoc. with
pepper spot or vein streaking [**58**,1998].
— **mosaic** (cauliflower mosaic v.).
CAB, Commonwealth Agricultural Bureaux.
— **International**, the largest organisation for the
dissemination of agricultural and biological
information, and for the identification of plant
pests and pathogens. It was set up in 1929 to
follow the successful Imperial Bureaux of
entomology and mycology begun in 1913 and
1920, respectively. In 1982 it was designated an
international organisation. It consists of 4
institutes and 10 bureaux and in 1973 all new
information began to be put on a computerised
data base; based in UK; Scrivenor, *CAB – The
first 50 years*, 1980; *Perspectives in world
agriculture*, 1980, compiled by CAB; Metcalf,
Span **28**:28, 1985.
— — **Mycological Institute**, formerly the
Commonwealth Mycological Institute; founded in
1920 as the Imperial Bureau of Mycology at Kew,
England, and where it has remained ever since. In
1930 it became a part of the Imperial, later
Commonwealth, Agricultural Bureaux. Its founder
director was E.J. Butler and its first mycologist
E.W. Mason. The institute is now the largest
centre for systematic mycology in the world, and
issues many publications in mycology and plant
pathology; see bibliography; Ainsworth, *CMI –
The first sixty years*, 1980; Hawksworth, *Biologist*
32(1):7, 1985.
cacabata, *Puccinia*.
cacao, *Theobroma cacao*; Brunt & Kenten,
Rev.Pl.Path. **50**:591, 1971, viruses; Thorold 1975;
Roivainen in Harris & Maramorosch 1980,
vectors of viruses; Wood & Lass, *Cocoa*, edn. 4,
1985.
— **abnormality** Oliveira, *Revta.Theobroma* **14**:253,
1984; Brazil (Bahia) [**65**,1808].
— **mosaics** Posnette & Palma, *Trop.Agric.Trin.*
21:130, 1944; Hutchins, *Communic.Turrialba* 66,

1959; Semangun 1961; Costa Rica, Indonesia
(Java), Venezuela (Paria peninsula) [**39**:223;
42:533].
— **mottle leaf v.** Posnette, *AAB* **34**:388, 1947;
Kenten & Legg, *J.gen.Virol.* **1**:465, 1967;
bacilliform 143 × 26 nm, transm. sap, *Planococcus*
spp., other mealy bugs, Ghana, Nigeria (W.),
Togo, common on Accra plains in *Adansonia
digitata*, prob. a str. of cacao swollen shoot v.
[**27**:178; **47**,480].
— **necrosis v.** Posnette 1948; Kenten, *Descr.V.*
173, 1977; Nepovirus, isometric 24–26 nm diam.,
transm. sap, Ghana, Nigeria, wide host range,
may occur in *Cola nitida*.
— **swollen shoot v.** Posnette 1947; Brunt, *Descr.V.*
10, 1970; Legg in Ebbels & King 1979; Owusu in
Plumb & Thresh 1983; bacilliform
129–135 × 26–28 nm, transm. sap, mealy bugs
including *Dysmicoccus brevipes, Ferrisia virgata,
Planococcoides njalensis, Planococcus citri,
P. kenyae*; Ghana, Ivory Coast, Nigeria (W.),
Sierra Leone, Togo; in *c*.40 spp. in Bombacaceae,
Malvaceae, Sterculiaceae, Tiliaceae; many strs.
named after a locality; the severe New Juaben str.
A kills cacao trees in 3–4 years. Cacao swollen
shoot is a classic example of the enormous
destruction that can be caused to a crop when a
virus spreads from wild plants; CSSV is unknown
in western Amazonia where cacao is indigenous.
The disease was first noted in the 1920s and by
1930 dying trees were seen; spread of the virus
was originally from wild hosts but cacao is now
the major source of outbreaks. CSSV has been
most damaging in Ghana, production fell from
118,000 tons in 1936 to 39,000 tons in 1955.
Control by rogueing has not been completely
successful for economic, political and social
reasons; neither host tolerance nor resistance has
had much impact; Thresh & Owusu, *Crop
Protect.* **5**:41, 1986, evaluation of eradication
procedures [**65**,2733].
— **Trinidad red mottle**, and vein clearing,
Posnette, *Trop.Agric.Trin.* **21**:105, 1944; Baker &
Dale, *AAB* **34**:60, 1947; Kirkpatrick, *Bull.ent.Res.*
41:99, 1950; transm. mealy bugs, strs. A and B,
reduces yield but is not lethal in Trinidad [**23**:379;
26:538; **30**:458].
— **watermark** Liu, *Tech.Bull.Dep.Agric. Sabah*
4:19, 1979 [**59**,2099].
— **yellow mosaic v.** Nixon & Gibbs 1963; Brunt,
Descr.V. 11, 1970; Tymovirus, isometric *c*.28 nm

diam., transm. sap, Sierra Leone, *Cola nitida,*
Culcasia scandans are naturally infected
—— vein banding Reddy 1968; Liu & Liew,
reference as for cacao watermark, page 11 [48,
1030b; 59, 2098].
Cacopaurus pestis Thorne 1943; Franklin, *Descr.N.*
44, 1974; sedentary parasite, described from
walnut roots.
cactivora, *Bipolaris*; cactorum, *Phytophthora.*
cactus opuntia v. Sammons & Chessin, *Nature*
Lond. 191:517, 1961; Milbrath et al., *Phytop.*
63:1133, 1973; rod *c.*320 × 18 nm, USA
[41:155; 53,2219].
—— saguaro v. Milbrath & Nelson, *Phytop.* 62:739,
1972; isometric 35 nm diam., transm. sap, USA
(Arizona), *Carnegia gigantea* [52,420].
—— witches' broom Uschdraweit, *Kakteen Stuttg.*
16:91, 1965; *Opuntia*; Casper et al., *PDR* 54:85,
1970; MLO assoc., *O. tuna* [44,2498; 50,1269].
*—— X v. Amelunxen 1958; Bercks, *Descr.V.* 58,
1971; Potexvirus, filament *c.*520 nm long, Europe,
USA.
cadang-cadang, see coconut.
Cadman, Colin Houghton, 1916–71; born in
Scotland, Univ. Liverpool, raspberry research
Univ. College Dundee 1943, Scottish
Horticultural, now Crop, Research Institute 1951,
director 1965; an authority on *Rubus* and
soilborne viruses, and nematode vectors; Fellow
of the Royal Society of Edinburgh 1950, Scottish
Horticultural Medal of the Royal Caledonian
Society of Scotland. *AAB* 69: 277, 1971.
Caeoma Link 1809; Uredinales; form genus for
spp. with aecia only, no peridium. *C. deformans*
(Berk. & Br.) Tubeuf, fide Hama 1982, witches'
broom of *Thujopsis dolobrata* in Japan [62,2668].
cajani, *Fusicoccum, Mycovellosiella, Uredo.*
Cajanus cajan, pigeon pea.
calcitrape, *Puccinia.*
calcium, for physiological disorders related to, see
Bangerth, *A.R.Phytop.* 17:97, 1979; in plant
senescence and fruit ripening, see Ferguson, *Plant*
Cell Environm. 7:477, 1984.
calendulae, *Entyloma.*
calendula mosaic Padma, Verma & Singh,
Gartenbauwissenschaft 40:133, 1975; transm. sap,
aphid spp., India, *C. officinalis*, common marigold
[54,4957].
—— yellows Rougier & Marchoux, *Annls.Phytopath.*
8:477, 1976; MLO assoc., France (S.E.)
[57,1226].
Caliciaceae, Caliciales; almost all form lichens,
Roesleria.
Caliciales, Ascomycotina; Caliciaceae.
Caliciopsis Peck 1880; Coryneliaceae; asco.
aseptate, smooth, *c.*5–6 × 3–4 μm; assoc. with
conifer cankers, North America, see Funk, *CJB*

41:503, 1963 for *C. orientalis* Funk, *C. pinea*
Peck, *C. pseudotsugae* Fitzpatrick.
Callbeck, Lorne Clayton, 1912–79; born in
Canada, Nova Scotia Agricultural College, Univ.
McGill; Canada Agriculture, Charlottetown,
authority on potato diseases particularly blight.
Can.J.Pl.Path. 2:102, 1980.
callosity, of fungi, wall thickenings assoc. with the
penetration by hyperparasites; lignituber q.v.;
Swart, *TBMS* 64:511, 1975.
callus, parenchymatous tissue of cambial origin
that forms in response to wounding or infections
by pathogens, FBPP.
Calonectria de Not. 1867; Hypocreaceae; ascomata
lacking a stroma, orange to scarlet or dark umber,
purplish with KOH, scaly to warty walls, asco. 1
to multiseptate, anam. *Cylindrocladium*, fide
Rossman, *Mycotaxon* 8,321, 485, 1979
[58,5735–6].
C. camelliae Shipton, *TBMS* 72:163, 1979; anam.
Cylindrocladium camelliae Venkataramani &
Venkata Ram 1961; Peerally, *Descr.F.* 428, 1974;
asco. 1, rarely 2, septate, 7–11 × 3–4 μm; con. 1
septate, 10–16 × 2 μm, on tea roots and other
crops [58,4277].
C. colhounii, Peerally 1973, *Descr.F.* 430, 1974;
anam. *Cylindrocladium colhounii* Peerally; asco. 3
septate, 34–84 × 4.5–8 μm; con. 3 septate,
38–42 × 3–6 μm, on tea foliage, secondary,
Mauritius.
C. kyotensis Terashita 1968; anam. *Cylindrocladium*
scoparium Morgan 1892, fide Rossman 1983;
Booth & Gibson, *Descr.F.* 362, 1973, anam. only;
Peerally, *Descr.F.* 421, 1974, gave the anam. as
C. floridanum Sobers & Seymour 1967; asco.
mostly 1 septate, 29–42 × 4–8 μm, con. 1 septate,
30–56 × 3–5 μm; mainly on woody plants, root
rots, including pine and eucalyptus; Barnard, *PD*
68:471, 1984, causing stem canker of eucalyptus
seedlings in USA (Florida) [63,4576].
C. pyrochroa (Desm.) Sacc. 1878; fide Rossman,
Mycol.Pap. 150, 1983, whose synonyms include:
Calonectria crotalariae (Loos) Bell & Sobers,
C. hederae Booth & Murray, *C. quinqueseptata*
Figueiredo & Namekata, *C. theae* Loos, all
described by Peerally, *Descr.F.* 423, 424, 426, 429,
1974; anam. *Cylindrocladium ilicicola* (Hawley)
Boedijin & Reitsma 1950; asco. 3–5 septate,
40–80 × 4–8.5 μm; con. mostly 3 septate,
40–120 × 8–9 μm, forms microsclerotia;
plurivorous, warmer regions, partly saprophytic;
causes a collar and root rot, particularly studied
in USA in black rot of groundnut and as
C. crotalariae; seedborne in groundnut; Holliday
1980; Phipps & Beute, *Phytop.* 69:240, 1979, soil
population dynamics; Diomande & Beute, *ibid*
71:491, 1981; *PD* 65:339, 1981, effects of

Macroposthonia ornata & *Meloidogyne hapla*;
Taylor et al., *Phytop.* **71**:1297, 1981; Tomimatsu
& Griffin, *ibid* **72**:511, 1972, inoculum density &
infection with microsclerotia; Pataky & Beute, *PD*
67:1379, 1983, survival of microsclerotia;
Johnson, *ibid* **69**: 434, 1985, on seed; Bolland et
al., *Eur.J.For.Path.* **15**:385, 1985, on eucalyptus
[**59**,2611; **61**,498, 1494, 4480, 6703; **63**,2083; **64**,
4604; **65**,2453].
Caloscypha Boud. 1885; Humariaceae; ascomata
large, deep cupulate, orange, staining greenish
when bruised, asco. aseptate, globose, hyaline.
C. fulgens (Pers.) Boud. 1907; anam.
Geniculodendron pyriforme Salt 1974, see Cannon
et al. 1985; asco. 5–6 μm diam., con. 7–5 × 5–3 μm
on *Picea sitchensis* seed, causing germination
failures in Canada and Britain, Paden et al., *CJB*
56:2375, 1978, anam. & teleom. connection;
Woods, *ibid* **60**:544, 1982, penetration of spruce
seed [**58**,2418; **61**,6626].
calostilbe, *Calostilbella*.
Calostilbella Höhnel 1919; Hyphom.; Carmichael
et al. 1980; conidiomata synnematal,
conidiogenesis holoblastic, con. solitary 2 to
several celled, pigmented; *C. calostilbe*, teleom.
Nectria striispora.
calotropis mosaic Mohan & Sharma, *Curr.Sci.*
56:274, 1987; transm. sap, *Aphis gossypii*,
Myz.pers., India; *C. procera*, a medicinal plant.
calyciformis, *Lachnellula*.
Calyptella Quelét 1886; Tricholomataceae; Singer,
Sydowia **30**:269, 1978; *C. campanula* W.B. Cooke,
Sydowia Beih. **4**:32, 1961; causes tomato root rot
in greenhouses in England, Clark et al., *Pl.Path.*
32:95, 1983 [**62**,2641].
calyx green end, apple cv. Golden Delicious, assoc.
with tissue fluoride, USA, Barritt & Kammereck
1983 [**62**,3565].
camarographioides, *Aristastoma*.
cambivora, *Phytophthora*.
Cambridge method, for investigating competitive
saprophytic colonisation q.v. of substrates by root
infecting fungi. The fungus is grown in a quartz
sand +3% maize meal mixture; the fully grown
cultures are then progressively diluted with
unsterilised soil to give a dilution series; standard
substrate units are then buried in each
soil+culture mixture; after an incubation period
the number of substrate units colonised is
determined, Butler, *AAB* **40**:284, 1953 [**33**:147].
camelliae, *Calonectria, Ciborinia, Cylindrocladium,
Exobasidium*.
Camellia, *C.sinensis*, tea.
camellia leaf and flower variegation Plakidas,
Phytop. **44**:14, 1954; **52**:77, 1962; USA
(Louisiana) [**33**:538; **41**:523].
—— **yellow mosaic** Sharma & Raychaudhuri,

Curr.Sci. **41**:267, 1972; Ahlawat & Sardar, *ibid*
42:181, 1973; transm. *Toxoptera aurantii*, by
graft to tea, India (W. Bengal) [**52**,3817; **54**,2434].
—— —— **mottle leaf** Milbrath & McWhorter, *Am.
Camellia Yrb.*: 51, 1946, USA [**30**:108].
campanula, *Calyptella*.
campestris, *Xanthomonas*.
Camptomeris Sydow 1927; Hyphom.; Ellis 1971;
conidiomata sporodochial, conidiogenesis
holoblastic, conidiophore growth sympodial, con.
multiseptate, pigmented. *C. albiziae* (Petch)
Mason, con. 30–70 × 8–12 μm, poss. pathogenic
on *Albizia*; Lenné, *PD* **64**:414, 1980, defoliation
of *Leucaena leucocephala* by *C. leucaenae* (Stev. &
Dalbey) Syd., con. 30–60 × 8–11 μm, Colombia
[**60**,371].
canaliculata, *Puccinia*; **canavaliae**, *Elsinoë*.
Canavalia ensiformis, *C. gladiata*, jack and sword
beans.
canavalia leaf pucker Doraiswamy & Jayarajan,
Madras agric.J. **71**:350, 1984; transm. sap, India
(Tamil Nadu), *C. ensiformis*, infected soybean
[**64**,4066].
—— **mosaic** Rodriguez et al. in Bird &
Maramorosch 1975, see legumes; transm. sap,
aphid spp., Puerto Rico, *C. maritima*, bay bean.
canberrianum, *Lophodermium*; **cancerogena**,
Erwinia.
cancroid spot, sweet orange, cytoplasmic or genetic
disorder, lesions on the fruit resemble those
caused by *Xanthomonas campestris* pv. *citri*;
cancrosis (*X. campestris* pv. *citri*) citrus.
candida, *Albugo*; **candidum**, *Cylindrocarpon,
Geotrichum*.
cane blight (*Didymella applanata, Leptosphaeria
coniothyrium*) blackberry, raspberry; —— ——
canker (*Botryosphaeria dothidea*) rose; —— **canker**
(*B. corticis*) blueberry; —— **dew** (*Peronosclerospora
sacchari*) sugarcane; —— **leaf rust** (*Kuehneola
uredinis*) Rubus; —— —— **spot** (*Phomopsis viticola*)
grapevine; —— **spot** (*Elsinoë veneta*) raspberry;
—— **tip blight** (*Diplodia natalensis*) grapevine.
canescens, *Cercospora*.
canker, and see hold over and perennial cankers; a
sunken, necrotic lesion of main root, stem or
branch arising from disintegration of tissues
outside the xylem cylinder, but sometimes limited
in extent by host reactions which can result in
more or less massive overgrowth of surrounding
tissues; concentric zonation may indicate
successive host responses to advancing infection,
FBPP.
—— **stain** (*Ceratocystis fimbriata*) London plane.
cannabis, *Septoria*; —— **phyllody** Phatak et al.,
Phytopath.Z. **83**: 281, 1975; MLO assoc., India
(N.) [**55**,1273].
Cannabis sativa, common hemp.

canna mosaic v. Fukushi 1932; Ocfemia et al. 1942, Japan, Philippines; Castillo et al., *PDR* **40**:169, 1956; rod 740 nm long, transm. sap, aphid spp. [**11**:797; **21**:491; **36**:32].

— **mottle v.** Gupta & Raychaudhuri, *Indian J.Hort.* **32**:106, 1975; *Acta bot.ind.* **5**:189, 1977; isometric 25–29 nm diam., transm. *Myz.pers.* [**55**,5776; **57**,3486].

— **yellow mottle v.** Yamashita et al., *Ann. phytopath.Soc.Japan* **51**:642, 1985; bacilliform *c*.120–130 × 28 nm [**68**,4439].

cannonballus, *Monosporascus.*

cantaloupe, see melon.

cape gooseberry, *Physalis peruviana.*

— — **mosaic R.** Singh, *Sci.Cult.* **34**:424, 1968; S.J. Singh et al., *Curr.Sci.* **44**:95, 1975; transm. sap, aphid, India (Gorakhpur) [**48**,1863; **54**,4591].

caper, *Capparis spinosa.*

— **latent v.** Gallitelli & Di Franco, *J. Phytopath.* **119**: 97, 1987; Carlavirus, filament *c*.660 nm long, poss. same as caper vein banding v., transm. sap, Italy (S.).

— **vein banding v.** Biga 1960; Majorana, *Phytopath.Mediterranea* **9**:106, 1970; poss. Carlavirus, rod 678 nm long, transm. sap, Italy (Apulia) [**40**:226; **50**,1925].

— — **yellowing** Franco & Gallitelli, *Phytopath.Mediterranea* **24**:234, 1985; bacilliform particles assoc., Italy [**65**,5676].

Cape St Paul wilt, coconut, see palm lethal yellowing.

Capparis spinosa, caper.

capsellae, *Pseudocercosporella.*

capsici, *Cercospora, Colletotrichum, Phomopsis, Phytophthora.*

capsicicola, *Phaeoramularia.*

Capsicum, red or chilli pepper, not the true pepper which is *Piper nigrum*; pepper q.v.

capsicum mosaic v. Pares, *AAB* **106**:469, 1985; Tobamovirus, rod *c*.297 × 18 nm, transm. sap, Australia; causes a decrease in height and weight of red pepper, with fruit malformation and mosaic; differs from tobacco mosaic v. and tomato mosaic v. in host reactions and serology [**65**,5335].

capsid, the protein coat or shell which surrounds the nucleic acid of a virus particle; also a trivial name for the family of plant bugs, Capsidae.

capsomeres, the sub-units on the surface of a virus particle and built up from protein.

capsule rot (*Alternaria ricini*) castor.

captafol, and captan; protectant fungicides, widely used for foliage, seed and soil.

caragana mosaic Kochman & Sikora, *Phytopath.Z.* **61**:147, 1968; transm. sap, Poland [**47**,1975].

caraway, *Carum carvi.*

— **yellow mottle** Wolf & Schmelzer, *Zentbl.Bact.ParasitKde* 2 **127**:665, 1972; Germany [**52**,3988].

carbendazim, systemic fungicide, broad spectrum; pathogen tolerance may occur after extensive use.

carbonacea, *Discella;* **carbonum,** *Cochliobolus, Drechslera.*

carboxin, systemic fungicide, against *Rhizoctonia* and smuts.

cardamom, *Elettaria cardamomum; Amomum* spp. are also called cardamom, see spice plants and Kumaresan & George 1980 [**61**,1545].

— **foorkey v.** Vasudeva 1956; Varma & Capoor, *Indian J.agric.Sci.* **34**:56, 1964; Alhwat et al., *Natn.Acad.Sci.Letters* **4**:165, 1981; isometric 37 nm diam., transm. *Pentalonia nigronervosa* to *Elettaria cardamomum*, in *Amomum subulatum* [**35**:661; **43**,3280; **62**, 1598].

— **Guatemala mosaic v.** Dimitman 1981; Gonsalves et al., *PD* **70**: 65, 1986; prob. Potyvirus, filament 700–720 nm long, transm. *Pentalonia nigronervosa* [**65**,5106].

— **Indian mosaic v.** Uppal, Verma & Capoor, *Curr.Sci.* **14**:208, 1945; Rao & Naidu, *J.Plantation Crops* suppl.:129, 1973; Randles & Hatta 1980; poss. Potyvirus, filament 650 × 15 nm, transm, sap, *Pentalonia nigronervosa*, other aphid spp., India (S.), called katte disease [**25**:10; **56**,969; **61**,32].

— — **streak** Raychaudhuri & Chatterjee 1958; Raychaudhuri & Ganguly, *Indian Phytopath.* **18**:373, 1965; *Phytopath.Z.* **62**: 61, 1968; transm. sap, *Brachycaudus helichrysi, Rhopalosiphum maidis*, India (N.), in *Amomum subulatum*, called chirke disease [**45**,2197; **48**,102].

cardamomi, *Sphaceloma;* **cardinale,** *Seiridium.*

cardoon, *Cynara cardunculus*, see artichoke, Martelli et al.

caricae, *Asperisporium, Mycosphaerella, Oidium.*

caricae-papayae, see *Oidium caricae, Phoma, Phomopsis.*

Carica papaya, papaya q.v., papaw, pawpaw.

caricapapayae, *Pseudomonas.*

caricicola, see *Oidium caricae;* **caricina,** *Puccinia.*

Carlaviruses, type carnation latent v., *c*.25 definitive members, filament slightly flexuous 600–700 × 12–15 nm, 1 segment linear ssRNA, sedimenting at 147–176 S, thermal inactivation 55–70°, longevity in sap a few days, transm. sap and aphid spp., usually non-persistent, narrow host ranges; Koenig, *Descr.V.* 259, 1982; Francki et al. vol. 2, 1985.

Carleton, Mark Alfred, 1866–1925; born in USA, a pioneer in agronomy and cereal rusts. *Phytop.* **19**:32, 1929.

carna 5 RNA Kaper, Tousignant & Lot 1976; a

satellite RNA of cucumber mosaic v. which
multiplies only in the presence of CMV RNAs 1,
2 and 3. It becomes encapsidated with this virus
genome, but is neither a part of it nor is it
essential for CMV replication. It regulates disease
expression and can cause a lethal necrosis of
tomato; it prob. caused the destructive disease
which occurred in France; Putz et al.,
Annls.Phytop. **6**:139, 1974; Kaper & Tousignant,
Virology **80**:186, 1977; Kaper & Waterworth,
Science N.Y. **169**:429, 1977; Waterworth et al.,
Phytop. **68**:561, 1978; *Science N.Y.* **204**:845,
1979; *Acta phytopath.Acad.Sci.hung.* **15**:123,
1980; Jacquemond & Lot, *Agronomie* **1**:927, 1981
[**54**,4146; **56**,4695, 5529; **58**,370; **61**,1794, 3635].

carnation, *Dianthus caryophyllus*; Oertel,
Arch.Gartenbau **25**:11, 1977; Lovisolo & Lisa,
Fitopat.Brasileira **3**:219, 1978, viruses; *Acta Hortic.*
141, 1983; Fletcher 1984 [**56**,4554; **63**,2349–61].

—— **bacilliform v.** Bennett & Milne, *Acta Hortic.*
59:61, 1976; particle *c.*55 × 26 nm, New Zealand.

—— **cryptic v.** Lisa, Luisoni & Milne, 1980; Lisa et
al., *Descr.V.* 315, 1986; isometric *c.*29 nm diam.,
contains 3 major size classes dsRNA, transm.
seed, no symptoms.

—— **D345 v.**, isometric *c.*30 nm diam., reference as
for carnation bacilliform v.

—— **etched ring v.** Hollings & Stone 1961; Lawson
& Civerolo, *Descr.V.* 182, 1977; Caulimovirus,
isometric *c.*45 nm diam., transm. sap, *Myz.pers.*,
naturally only in *Dianthus*.

—— **internode shortening** (arabis mosaic v.),
Hakkaart et al., *Neth.J.Pl.Path.* **78**:15, 1972
[**51**,2555].

—— **Italian ringspot v.** Hollings, Stone & Bouttell,
AAB **65**:299, 1970; Tombusvirus, isometric
*c.*29 nm diam., transm. sap; Britain, Italy, USA;
a str. of tomato bushy stunt v.[**49**,2881].

—— **latent v.** Kassanis 1954; Wetter, *Descr.V.* 61,
1971; Carlavirus, type, filament *c.*650 × 12 nm,
transm. sap, aphid, non-persistent; Australia,
Europe, Japan, New Zealand.

—— **mottle v.** Kassanis 1955; Hollings & Stone,
Descr.V. 7, 1970; Hull, *J.gen.Virol.* **36**:289, 1977;
Tombusvirus, isometric *c.*28 nm diam., transm.
sap, naturally only in Caryophyllaceae; Kummert,
Virology **105**:35, 1980, synthesis &
characterisation of complementary DNA [**56**,5504;
60,2625].

—— **necrotic fleck v.** Inouye & Mitsuhata 1973;
Inouye, *Descr.V.* 136, 1974; Closterovirus,
filament *c.*1400–1500 nm long, transm. sap,
difficult, *Myz.pers.*, semi-persistent; Smookler &
Loebenstein, *Phytop.* **64**:979, 1974; Poupet et al.,
Annls.Phytopath. **7**:277, 1975; Bar-Joseph &
Smookler, *Phytop.* **66**:835, 1976, properties,

purification & serology [**54**,2277; **55**,5772;
56,1143].

—— **Orange Triumph v.** Hakkaart 1971; prob =
carnation cryptic v., fide A.A. Brunt, personal
communication, 1985 [**51**,1039n].

—— **ringspot v.** Kassanis 1955; Tremaine & Dodds,
Descr.V. 308, 1985; Dianthovirus, type, isometric
34 nm diam., 2 species of RNA both needed for
infectivity, transm. sap, foliage and root contact,
poss. by nematodes; wide experimental host
range, occurs naturally in Caryophyllaceae, assoc.
with pear stony pit and decline of apple and sour
cherry.

—— **streak** (carnation mottle v.+carnation necrotic
fleck v.), fide A.A. Brunt, personal
communication, 1985.

—— **vein mottle v.** Kassanis 1955; Hollings & Stone,
Descr.V. 78, 1971; Potyvirus, filament *c.*790 nm
long, transm. sap, *Myz.pers.*, Europe, USA;
Hollings et al., *AAB* **85**:59, 1977, review for
Britain [**56**,3060].

—— **yellow fleck**, as for carnation necrotic fleck v.

—— —— **stripe v.** Galletti et al., *Phytopath.*
Mediterranea **18**:31, 1979; isometric *c.*30 nm
diam., contains RNA, transm. sap, Italy (Apulia),
related to the grapevine str. of tobacco necrosis
v. [**62**,4697].

Carneocephala fulgida, transm. xylem limited
bacteria q.v.

carotae, *Cercospora, Rhizoctonia*; **carotovora**,
Erwinia.

carpet bugle, creeping bugle weed q.v.

Carpinus betulus, hornbeam.

Carpobrotus, ornamental ice plants; MacDonald et
al., *PD* **68**:965, 1984; USA (California) [**64**,1143].

carpophila, *Stigmina, Venturia*; **carpophilum**,
Cladosporium.

carrier, (1) a material serving as diluent and vehicle
for the active ingredient; (2) an organism
harbouring a parasite without itself showing
disease; (3) a bacterial str. in which a population
equilibrium between phage resistant and phage
sensitive cells exists, the latter being constantly
infected by free phage particles, after FBPP.

*carrot, *Daucus carota*; Rader, *Bull.Cornell*
Univ.agric.Exp.Stn. 889, 1952; Årsvoll,
Meld.Norg.Lanbr.Høisk. **48**(2), 1969; Crête,
Publ.Agric.Canada 1615, 1977; Lewis & Garrod in
Dennis 1983; see *Conium maculatum*, Howell &
Mink.

—— **cyst nematode**, *Heterodera carotae*.

—— **latent v.** Ohki, Doi & Yora,
Ann.phytopath.Soc.Japan **44**:202, 1978;
bacilliform 220 × 70 nm, transm. *Semiaphis*
heraclei, Japan (Kanto) [**58**,1426].

—— **mosaic v.** Chod, *Ochr.Rost.* **3**:49, 1965;

Biologia Pl. **7**:463, 1965; **8**:53, 1966; filament, transm. sap, aphid, Czechoslovakia; Camargo et al., *Bragantia* **30**:31, 1971; poss. Potyvirus, Brazil (São Paulo); neither form is important [45,1262, 2672; **51**,3661].

—— **motley dwarf** (carrot mottle v. + carrot red leaf v.), Stubbs, *Aust.J.sci.Res.* B **1**:303, 1948; **5**:399, 1952; Watson et al., *AAB* **54**:153, 1964 [**28**:435; **32**:659; **44**,947].

—— **mottle v.** Watson, Serjeant & Lennon 1964; Murant, *Descr.V.* 137, 1974; isometric *c.*52 nm diam., contains RNA, transm. sap, *Cavariella aegopodii*, persistent, dependent on carrot red leaf v. for aphid transm.; Waterhouse & Murant, *AAB* **103**:455, 1983 [63,2017].

—— **proliferation** Giannotti et al., *C.r.hebd.Seanc.Acad.Sci.* D **287**:469, 1974; poss. MLO and bacterial mixed infection, see phloem limited bacteria.

—— **red leaf v.** Watson, Serjeant & Lennon 1964; Waterhouse & Murant, *Descr.V.* 249, 1982; Luteovirus, isometric *c.*25 nm, transm. *Cavariella aegopodii*, persistent, confined to phloem, Umbelliferae, frequently assoc. with carrot mottle v. which is dependent on it for aphid transm., both viruses together cause carrot motley dwarf; poss. variants in parsley.

—— **rusty root necrosis** Kemp & Barr, *Phytopath.Z.* **91**:203, 1978; tobacco necrosis v. and its vector, *Olpidium brassicae*, assoc., Canada [57,5740].

—— **temperate v.**, see cryptic viruses.

—— **thin leaf v.**, Howell & Mink 1976; *Descr.V.* 218, 1980; Potyvirus, filament *c.*736 × 11 nm, transm. sap, aphid, non-persistent, USA (N.W.) on wild carrot and *Conium maculatum*.

—— **yellow leaf v.** Yamashita et al., abs. *Ann.phytopath.Soc.Japan* **42**:382, 1976; Closterovirus, filament 1600 × 12 nm.

—— **yellows** Fedotina 1976; USSR, as for carrot proliferation [56,2924].

Carsner, Eubanks, 1891–1979; born in USA, Univ. Texas, USDA in California, successfully selected sugarbeet cvs. for resistance to beet curly top v. which had virtually destroyed the sugarbeet industry in much of western USA. *Phytop.* **70**:174, 1980.

carthagenesis, *Uromyces*, see *U. manihotis*.

carthami, *Alternaria, Cercospora, Puccinia, Septoria.*

Carthamus tinctorius, safflower.

Cartwright, Kenneth St George, 1891–1964; born in England, Univ. Oxford, Forest Products Research Laboratory 1927–48, a pioneer in the study of fungi that rot wood, wrote: *Decay of timber and its prevention*, 1946, with W.P.K. Findlay, *TBMS* **48**:151, 1965.

Carum carvi, caraway.

carunculoides, *Ciboria*; **caryae**, *Gnomonia.*

Carya illinoensis, *C. pecan*, pecan nut, hickory wood.

caryigena, *Mycosphaerella, Pseudocercosporella*; **caryigenum**, *Cladosporium.*

caryophyllacearum, *Melampsorella*; **caryophylli**, *Pseudomonas.*

caryopteris yellow spot Schmelzer 1963; Germany, *C. incana* [**42**: 614].

cashew, *Anarcadium occidentale*; Lim & Singh 1985, Malaysia [**65**,5187].

cassandrae, *Godronia.*

cassandra rust (*Chrysomyxa ledi* var. *cassandrae*) spruce.

cassava, *Manihot esculenta*; Ingram et al., *Trop.Sci.* **14**:131, 1972, in storage; Lozano et al., *PANS* **20**:30, 1974; *PANS Manual* 4, 1978, Centre Overseas Pest Res.

—— **African mosaic v.** Warburg 1894; Storey & Nichols 1938; Bock & Harrison, *Descr.V.* 297, 1985; Geminivirus, isometric, particle pair *c.*30 × 20 nm, each particle has a genome of a circular molecule of ssDNA, transm. sap, *Bemisia tabaci*, persistent, most natural hosts in Euphorbiaceae; Africa and nearby islands, India; great economic importance, a 60–80 % decrease in tuber yield can be caused, strs. occur; Fargette et al., *AAB* **106**:285, 1985, spread by white flies in the field [**65**,2126].

—— **antholysis** Jayasinghe, Pineda & Lozano, *Phytopath.Z.* **109**:295, 1984; MLO assoc., Colombia [63,4684].

—— **Brazilian common mosaic** = cassava common mosaic v.

—— —— **latent** Kitajima & Costa, *Fitopat.Brasileira* **4**:55, 1979; bacilliform particles assoc. [**58**,5279].

—— **brown streak** Storey 1936; Lister, *Nature Lond.* **183**:1588, 1959; Kitajima & Costa, *E.Afr.agric.For.J.* **30**:28, 1964; particle 600 nm long, transm. sap, E. Africa; Harrison et al., abs. *Phytop.* **76**:1075, 1986, found that cassava plants from Kenya, and showing brown streak, contained 2 viruses with filamentous particles, transm. sap to the same range of solanaceous plants.

—— **common mosaic v.** Costa 1940; Costa & Kitajima, *Descr.V.* 90: 1972; Potexvirus, filament *c.*495 × 15 nm, transm, sap readily, S. America; Zettler & Elliot, *Phytop.* **76**:632, 1986, a distinct str. in *Cnidoscolus aconitifolius*, chaya, a Mexican vegetable, USA (Florida) [**66**,431].

—— **frogskin**, a disease that occurs in Brazil, Colombia and Peru, the typical symptoms are confined to the roots which remain thin and fibrous, a yield loss of 90 % can occur and the root tubers cannot be processed for starch, the condition is spread through cuttings, B. Nolt, personal communication, 1986; Harrison et al., see cassava brown streak, found that affected plants contained a Potexvirus, filament 480 nm long, sap transm., narrow host range, not serologically related to cassava common mosaic v.; it was called cassava X v.

*—— green mottle v. Harrison et al., see cassava brown streak; isometric 26 nm diam., bipartite RNA genome, transm. sap and seed, Solomon Islands, wide host range

—— latent Bock & Guthrie 1976 = cassava African mosaic v.; Gabriel et al., Phytop. 77:92, 1987, detected a latent agent like a virus [66,3606].

—— vein mosaic v. Costa, J. Agronomia 3:239, 1940; Kitajima & Costa, Bragantia 25:211, 1966; poss. Caulimovirus, isometric 50–60 nm diam., Brazil [47,725].

—— witches' broom Silberschmidt & Campos, Arq.Inst.biol.S.Paulo 15:1, 1944; Brazil [24:440].

——X v., see cassava frogskin.

Cassia occidentalis, coffee senna, negro coffee.

cassia Australian yellow blotch v. Dale, Gibbs & Behncken, J.gen.Virol. 65:281, 1984; Bromovirus, isometric 25–27 nm diam., transm. sap, C. pleurocarpa, systemic infection in 8 legumes and Nicotiana clevelandii [63,2217].

—— Brazilian yellow blotch v. Paguio & Kitajima, Fitopat. Brasileira 6:187, 1981; Potyvirus, filament c.845 nm long, transm. sap, C. hoffmannseggii, systemic infection of some legumes and cucurbits [61,4302].

—— bright mosaic Singh et al., Indian J.Mycol.Pl.Path. 9:101, 1979; transm. sap, C. occidentalis [60,2481].

—— mild mosaic v. Lin et al., PDR 63:501, 1979; PD 64:587, 1980; poss. Carlavirus, filament 640 × 15 nm, transm. sap, Brazil, C. macranthera, C. sylvestris [59,1647; 60,2214].

—— mosaic van Velsen, Papua New Guin.agric.J. 14:124, 1961; transm. sap, C. occidentalis [42:139].

—— ringspot mosaic Mathur & Singh, Indian Phytopath. 25:314, 1972; transm. sap, India (Kanpur), C. occidentalis [53,1495].

cassiae, Alternaria; cassiicola, Corynespora.

cast, refers to falling needles, needle casts, from conifers with a disease or a disorder.

castagnei, Marssonina; castanea, Amphiporthe.

Castanea, chestnut; C. crenata, Japanese; C. dentata, American; C. mollisima, Chinese; C. sativa, European, Spanish or sweet; see horse chestnut.

castaneae, Diplodina; castaneicola, Coniella, Macrophoma.

castor, Ricinus communis.

casuarina witches' broom Wellman & Grant, PDR 35:498, 1951; Central America, C. equisetifolia; Zhang et al., Acta phytopath.sin. 13(4):37, 1983, reported an MLO and bacterium associated in the same plant in China [31:262; 63,4582].

Catacauma torrendiella, see Phyllachora.

catalpa chlorotic leaf spot Schmelzer 1970 (broad bean wilt v.).

—— flat limb Bojaňsky, Acta Biol.Iugoslavica B 10:37, 1973; transm. seed, Yugoslavia, C. bignonioides [54,575].

catenaria, Drechslera.

cat face, tomato, failure of fruit wall to develop normally, cracks and splits occur; the cause is apparently like those of frenching of tobacco and yellow strap leaf of chrysanthemum; frenching q.v., Woltz.

Catharanthus roseus, Madagascar periwinkle.

catharanthus little leaf Kar, Nanda & Kabi, Sci.Cult. 48:180, 1982; MLO assoc., India, C. roseus [62,2008].

—— wilt McCoy, PDR 62:1022, 1978; bacterium assoc., transm. Oncometopia nigricans, USA (Florida) [58,5400].

cat scratch, beet, B deficiency, Scaife & Turner 1983; —— tail (Epichloë typhina); —— millet, see millets.

cattleyae, Pseudomonas.

cattleya flower necrosis (cymbidium mosaic v.), Faccioli & Marani, Phytopath.Mediterranea 18:21, 1979 [62,4700].

—— mosaic Jensen, Phytop. 39:1056, 1949; transm. Myz.pers., USA (California) [29:309].

caulicola, Ascochyta.

cauliflower, see Brassica.

—— inflorescence phyllody Müller, Schmelzer & Kleinhempel, Arch.Phytopath.PflSchutz. 9:335, 1973; MLO assoc., Germany [53,3664].

*—— mosaic v. Tompkins 1937; Shepherd, Descr.V. 243, 1981; Caulimovirus, type, isometric 50 nm diam., transm. aphid spp., non- or semi-persistent, temperate regions, in crucifers, often in mixed infections with turnip mosaic v.; Schoelz et al., Phytop. 76:451, 1986, described an unusual str. causing an atypical veinal necrosis in brassica [65,5218].

Caulimoviruses, type cauliflower mosaic v., 10 definitive members, isometric 45–50 nm diam., circular dsDNA, sedimenting at 215–45 S, thermal inactivation 80–90°, longevity in sap up to 7 days, particles occur in characteristic cytoplasmic inclusion bodies, transm. sap, aphid spp., non- or semi-peristent, rather narrow host ranges; Hull, Descr.V. 295, 1984; Francki et al., vol.1, 1985; Hirth, Microbiological Sci. 3:260, 1986.

caulivora, *Kabatiella*.
causal organism, an organism causing disease, usually as the sole cause but sometimes as a member of a causal complex; see Koch's postulates.
Cavariella spp., transm. heracleum latent; *C. aegopodii*, transm. anthriscus yellows, carrot mottle, carrot red leaf, parsley 5, parsnip mosaic, parsnip yellow fleck; *C. theobaldi*, transm. parsnip mosaic.
cavity spot, carrot, Guba et al., *PDR* 45:102, 1961; the cause of this disease or disorder has for long not been clear; but White, *AAB* 108:265, 1986, showed that *Pythium* spp. caused cavity spot; *P. violae* Chesters & Hickman, relatively slow growing, gave the highest percentage of carrots with cavities. The poss. role of *Phytophthora* needs to be considered [40:504; 66,406].
Cecidophyopsis ribis, transm. blackcurrant reversion.
cedrela witches' broom Mukhopadhyay, Chowdhuri & Tarafder 1981; India (W. Bengal) [61,1620].
cedri, *Peridermium*.
Cedros wilt, coconut, see *Phytomonas*.
celastrus chlorotic mottle Schmelzer 1974; see Cooper 1979.
celery, *Apium graveolens*; Pemberton & Frost, *AAB* 108:319, 1986, viruses in England [66,374].
— **latent v.** Luisoni, *Atti Accad.Sci.Torino* 100:541, 1966; Verhoyen et al., *Parasitica* 32:158, 1976; Bos et al., *Neth.J.Pl.Path.* 84:61, 1978; filament *c.*860 nm long, transm. sap, seedborne; Belgium, Italy, Netherlands [46,1162; 56,5861; 57,5172].
— **strap leaf** (strawberry latent ringspot v.), Walkey & Mitchell, *Pl.Path.* 18:167, 1969; England [49,1858].
— **western mosaic v.** Severin & Freitag 1935, 1938; Shepard & Grogan, *Descr.V.* 50, 1971; Potyvirus, filament *c.*780 nm long, transm. sap, aphid, non-persistent; Europe, USA; only in Umbelliferae.
— **yellow mosaic** Kitajima & Costa 1968; Oliveira et al., *Fitopat.Brasileira* 6:35, 47, 1981, like celery western mosaic v. [51,1234; 61,1080–1].
— — **net** Hollings, *J.hort.Sci.* 39:130, 1964; Britain; see parsnip yellow fleck v. [44,2938].
— — **spot** Severin & Freitag, *Hilgardia* 16:375, 1945; transm. *Hyadaphis foeniculi*, USA, occurs in *Conium maculatum*; the causal agent may be the same as the one causing celery yellow spot in Britain described by Hollings 1964, and Pemberton & Frost, see celery, the latter sap transm. [25:287; 44,2938].
— — **vein** (tomato black ring v.).
— **036 v.** Pemberton & Frost, *Pl.Path.* 23:20,

1974, and see celery; filament *c.*780 nm long, transm. sap, *Myz.pers.*, in Umbelliferae, serologically related to celery western mosaic v., causes ringspot in some celery cvs. and loss of yield of up to 24% in England [54,600].
— **065 v.** Pemberton & Frost 1986, see celery; isometric 28–30 nm diam., transm. sap, England, wide host range, causes stunt and necrosis, loss of yield of up to 45%.
cell culture, see tissue culture.
cellulases, cellulolytic enzymes; Norkrans, *A.R.Phytop.* 1:325, 1963, degradation of cellulose; Mandels & Reese, *ibid* 3:85, 1965; Wood 1967; Garrett 1970.
cellulolysis adequacy index, a measure of the extent to which the rate of cellulose decomposition by a fungus is adequate to supply its needs for saprophytic survival. It is derived by dividing the value for the rate of cellulolysis by the mycelial growth rate on agar; Garrett, *TBMS* 49:57, 1966; Deacon, *ibid* 72:469, 1979 [45,2064; 59,2596].
cell wall degrading enzymes, Wood 1967; *Annls.Phytopath.* 10:127, 1978; Cooper in Wood & Jellis 1984 [58,3636].
Cenangium Fr. 1818; Helotiaceae; ascomata erumpent from bark or wood, relatively large, not gelatinous, outer face with a mealy or scruffy appearance, asco. aseptate, like *Encoelia*.
C. singulare (Rehm) Davidson & Cash, *Phytop.* 46:36, 1956; *Populus tremuloides*, aspen sooty bark canker, USA (Rocky mountains); Hinds & Ryan, *PD* 69:842, 1985, described the expansion of cankers with *Ceratocystis fimbriata* and referred the fungus to *Encoelia pruinosa* (Ell. & Ev.) Torkelsen & Eckblad, see *Index Fungi* 4:472, 1978 [35:560; 65,4113].
centrosema mosaic van Velsen & Crowley, *Nature Lond.* 189:858, 1961; *Aust.J.agric.Res.* 13:220, 1962; poss. Potexvirus, transm. aphid and *Nysius* spp., Papua New Guinea, mostly on *Crotalaria* spp. [40:524; 41:603].
cepacia, *Pseudomonas*.
Cephalenchus emarginatus (Cobb.) Geraert 1968; Hooper, *Descr.N.* 35, 1974; ectoparasitic, not damaging.
Cephaleuros, an alga on plants, see algae; inappropriately called red rust.
cephalotrichous, see flagella.
cepivorum, *Sclerotium*; **cepulae**, *Urocystis*.
cerasi, *Fusicladium, Venturia*; **cerasella**, *Mycosphaerella*.
Ceratobasideaceae, Tulasnellales; basidia with sterigmata, broad, ±straight and digitate, not inflated or deciduous; *Ceratobasidium, Koleroga, Oncobasidium, Thanatephorus*.
Ceratobasidium, Ceratobasideaceae; basidia

abruptly narrowed at the pedicels, hyphae with binucleate cells, sometimes forming web blights on tropical crops; Burpee et al., *Mycologia* **72**:689, 1980; *Phytop.* **70**:843, 1980, *C. cornigerum* (Bourd.) Rogers, pathogenicity [**60**,3057, 4339].

C. cereale Murray & Burpee, *TBMS* **82**:172, 1984; sclerotial state *Rhizoctonia cerealis* van der Hoeven, *Neth.J.Pl.Path.* **83**:191, 1977; wheat sharp eyespot, formerly attributed to *R.solani*; Burpee, *PD* **64**:1114, 1980, yellow patch of turf grasses; Lipps & Herr, *Phytop.* **72**: 1574, 1982, etiology on wheat; Clarkson & Cook, *Pl.Path.* **32**:421, 1983, yield loss in winter wheat in England and Wales [**60**,4511; **62**,1897; **63**,1218, 1639].

Ceratocystis Ell. & Halsted 1890; Ophiostomatales; ascomata small, dark, on wood, with long, slender beaks with an ostiole through which the small, hyaline, aseptate asco. are released in a mucilage; some include *Ophiostoma* in the genus but de Hoog & Scheffer, *Mycologia* **76**:292, 1984, retain *Ophiostoma* for spp. with an anam. other than in *Chalara*. Some spp. are general wound pathogens, often assoc. with blue stain of timber and bark beetles; young trees can be killed. Two spp. are important pathogens of several crops: *fimbriata* and *paradoxa*; 2 other spp., *fagacearum* and *ulmi*, are extremely destructive and have narrow host ranges; Basham, *Phytop.* **60**:750, 1970, wilt in *Pinus taeda*; Hepting 1971; de Hoog, *Stud.mycol.* 7, 1974; Holliday 1980; Upadhyay, *A monograph of* Ceratocystis *and* Ceratocystiopsis, 1981 [**49**,3497; **63**,3221].

C. adiposa (E. Butler) C. Moreau 1952; anam. *Chalara* sp.; Sivanesan & Waller 1986; asco. 6–9 × 3–5.5 μm, these spores have a different shape from those of *Ceratocystis paradoxa* which causes a similar but more serious disease of sugarcane setts called black rot; Sartoris, *J.agric.Res.* **35**:577, 1927, morphology; Byther, *PDR* **55**:7, 1971 [**7**:272; **50**,2481].

C. coerulescens (Münch) Bakshi 1950; *Mycol.Pap.* 35, 1971; anam. *Chalara ungeri* Sacc.; on pine and other trees, sap stain; Hepting, *Phytop.* **34**:1069, 1944, sugar maple sap streak, USA; Davidson, *Mycologia* **45**:584, 1953, new form, as *Endoconidiophora*, on Douglas fir [**24**:212].

C. fagacearum (Bretz) Hunt 1956; anam. *Chalara quercina* B.W. Henry; oak vascular wilt, an important disease; the fungus is restricted to USA (central, N.E., and recently reported from Texas) where it may be endemic. Its slow and erratic spread led to it being considered less important than had been originally thought. But the attention to oak wilt has intensified since the more recent epidemics of *Ceratocystis ulmi* in Europe.

Infection is through wounds; red oaks can be killed in a few weeks but white oaks are less susceptible; in the former group, particularly, spore mats are formed in the inner bark and adjacent xylem. No insects are intimately assoc. with the fungus but the mats attract nitidulid beetles which have been implicated in spread, as have bark beetles, these might not be important vectors, see Gibbs, *Eur.J.For.Path.* **10**:218, 1980; spread via natural root grafts is important; Gibbs, *Arboric.J.* **3**:351, 1978, salient features; in Ebbels & King 1979, measures against poss. European introduction; Gibbs & French, *For.Serv.Res.Pap.* NC 185; 1980, review of transm.; Jacobi & MacDonald, *Phytop.* **70**:618, 1980, colonisation of host; Bowen & Merrill, *PD* **66**:137, 1982, wilt foci & Pennsylvanian topography; Appel et al., *CJB* **63**:1325, 1985, in Texas & compatibility types [**60**,491, 2210; **61**,5254; **65**, 395].

C. fimbriata Ell. & Halsted 1890; Morgan-Jones, *Descr.F.* 141, 1967; anam. *Chalara* sp.; a saprophyte, restricted as a pathogen and prob. occurs as specialised strs., but the precise circumscription of such strs. is not clear; attacks woody hosts through wounds; the fungus can be distinguished from the also widespread *Ceratocystis paradoxa* by the unornamented ascomata and the hat shaped asco.; some of the more severe diseases caused are: cacao wilt, Iton in Thorold 1975; coffee canker, Pontis, *Phytop.* **41**: 178, 1951; London plane canker, Panconesi & Nembi, *Inftore fitopatol.* **28**(11–12):17, 1978; Gibbs, *EPPO Bull.* **11**:193, 1981; pimento canker, Leather, *TBMS* **49**:213, 1966; rubber mouldy rot, *Plrs'.Bull.Rubb.Res.Inst.Malaya* 118:3, 1972; sweet potato black rot, Holliday 1980 [**30**:366; **45**,2918; **51**,4303; **58**,3480; **61**,3657].

C. ips (Rumb.) C. Moreau 1952; pine blue stain; Himelick, *Arboric.J.* **8**:212, 1982, assoc. with a decline of *Pinus sylvestris* in USA (Illinois); Strobel & Sugawara, *CJB* **64**:113, 1986, on *P. contorta* as *Ceratocystis montia* (Rumb.) Hunt [**61**,7195; **65**,3529].

C. moniliformis (Hedgc.) C. Moreau 1952; anam. *Chalara* sp.; Morgan-Jones, *Descr.F.* 142, 1967; on pines and hardwoods; recorded on cacao and distinguished from *Ceratocystis fimbriata* and *C. paradoxa*, also on this crop, by the chlamydospore form and the short conical spines on the ascomata.

C. paradoxa (Dade) C. Moreau 1952; anam. *Chalara paradoxa* (de Seynes) Sacc. 1892; Morgan-Jones, *Descr.F.* 143, 1967 Sivanesan & Waller 1986; distinguished from the also widespread *Ceratocystis fimbriata* by the brown stellate appendages which ornament the ascomata;

saprophytic, the diseases caused include: areca and coconut palm bleeding, banana stem end rot and rhizome black head, cacao pod rot, oil palm dry basal rot, pineapple soft rot and water blister, sugarcane pineapple disease; the fungus gives off the characteristic odour of ethyl acetate; Holliday 1980.

*C. ulmi (Buisman) C. Moreau 1952; anam. *Pesotum ulmi* (Schwartz) Crane & Schoknecht 1973; Booth & Gibson, *Descr.F.* 361, 1973; the name *Ophiostoma ulmi* (Buisman) Nanf. is used by de Hoog 1974, *Ceratocystis* q.v.; Cannon et al. 1985 gave *O. ulmi* as a synonym; causes Dutch elm wilt, usually called Dutch elm disease; one of the most destructive of all tree diseases. An aggressive str. spread in elm logs from Canada to England in the mid 1960s where the resulting pandemic killed *c*.10.6 million elms or *c*.60% of all trees. The disease caused losses after the first report of its presence in Belgium, France and the Netherlands (1918) and USA (1920s). In Britain (1927) the disease was of little significance; these outbreaks were caused by a non-aggressive str. The N. American work had not detected the presence of an aggressive str., now the NAN subgroup. Further, the spread towards Europe from Asia, of another aggressive str., EAN subgroup, had been inadequately recognised as a threat, see Brasier, *Nature Lond.* **281**:78, 1979; *TBMS* **87**:1, 1986; Gibbs, *EPPO Bull.* **11**:193, 1981. *C. ulmi* is spread from dead or dying trees to green shoots of healthy ones by bark beetles, *Scolytus* spp., and in N. America by *Hylurgopinus rufipes*; there is an intimate assoc. between insect and fungus; spread also occurs through natural root grafts; the pathogen forms a toxin ceratoulmin q.v. American elms are more susceptible than Asiatic and European elms.

Gibbs, *A.R.Phytop.* **16**:287, 1978, epidemiology; Brasier & Gibbs in Scott & Bainbridge 1978, in Britain; Sinclair & Campana ed., *Search agric.Pl.Path.* 1, **8**(5), 1978, review; Holmes in Harris & Maramorosch 1980, bark beetles; Stipes & Campana 1981; Phillips & Burdekin 1982; Lea & Brasier, *TBMS* **80**:381, 1983; sporulation succession; Webber & Gibbs, *ibid* **82**:384, 1984, poss. control of bark beetles by *Phomopsis oblonga* (Desm.) Trav.; for 13 references to Brasier and others see Brasier 1986 loc.cit; Brasier in Wolfe & Caten 1987 [**52**,2352, 3905; **61**,3657; **62**,4456; **63**,2513; **66**,2519].

*C. wageneri Goheen & Cobb, *Phytop.* **68**:1193, 1978; anam. *Verticicladiella wagenerii* Kendrick, *CJB* **40**:793, 1962; causes root disease in conifers in USA (W.), pine black stain, including *Pinus ponderosa*; Goheen & Cobb, *Canadian*

Entomologist **112**:725, 1980, assoc. with bark beetles which infested diseased trees more; Cobb et al., *Phytop.* **72**:1359, 1982, spread; Harrington & Cobb, *ibid* **74**:286, 1984, variations in pathogenicity; Hessburg & Hansen, *ibid* **76**:627, 1986, effect of soil temp. on infection & colonisation of Douglas fir [**58**,3496; **61**,1966; **62**,1667; **63**,4102; **66**,353].

ceratosperma, *Valsa*.

cerato-ulmin, toxin from *Ceratocystis ulmi*; Takai, *Nature Lond.* **252**:124, 1974; Richards & Takai, *Can.J.Pl.Path.* **6**:291, 1984; Mitchell, *A.R.Phytop.* **22**:237, 1984 [**54**,1448; **64**,3203].

*Cercoseptoria Petrak 1925; Hyphom.; conidiomata sporodochial, conidiogenesis holoblastic, conidiophore growth sympodial; con. filiform, hyaline; like a *Cercospora* but the conidial scars are unthickened and inconspicuous; like *Pseudocercospora* and *Pseudocercosporella* but characterised by the narrow, acicular con.; Deighton, *Mycol.Pap.* 140, 1976; 151, 1983, who stated in 1983 that there is doubt whether *Cercoseptoria* can be clearly distinguished from *Pseudocercospora*; Holliday 1980.

C. ocellata Deighton *Mycol.Pap.* **151**:2, 1983; this is from *Cercospora theae* Breda de Haan 1900, con. 55–88 × 2–2.5 μm, 4–8 septate; tea bird's eye spot or eye spot. *Cercoseptoria theae* (Cav.) Curzi 1929, con. 55–75 × 2–2.5 μm, mostly 5 septate; also tea eye spot; Deighton loc.cit. refers to other Hyphom. on tea and newly described 2: *Mycocentrospora camelliae* and *Stigmina theae*; none of these fungi appear to be of any economic importance.

C. pini-densiflorae, teleom. *Mycosphaerella gibsonii*; **Cercoseptoria sesami**, teleom. *M. sesamicola*; both now in *Pseudocercospora*.

*Cercospora Fres. 1863; Hyphom.; conidiomata hyphal, conidiogenesis holoblastic, conidiophore growth sympodial; con. filiform, hyaline, septate; this large genus, ±2000 spp., is heterogeneous and the use of the host as a sole criterion for spp. distinction should be abandoned. About half the spp. are unlike the type sp. *C. apii*, most fall into one of 2 distinct taxonomic categories: those where the old conidial scars on the conidiophores are thickened, i.e. a true *Cercospora*, and those where the scars are not thickened. The hilum on the con. is thickened or unthickened corresponding to the scar on the conidiophore; Deighton has disposed many spp. to other genera; Ellis 1976, 107 spp. described; Carmichael et al. 1980, Holliday 1980.

C. antirrhini Muller & Chupp 1950; con. 35–125 μm long or longer; Jackson, *Phytop.* **50**:190, 1960 [**39**:584].

C. apii Fres. 1863; con. usually 60–200 μm long; celery early blight, an important disease in USA (Florida), seedborne; Berger, *Phytop.* **63**:535, 1973; infection rates; Sherf & MacNab 1986 [**53**,314].

C. arachidicola, teleom. *Mycosphaerella arachidis.*

C. asparagi Sacc. 1877; con. 80–130 × 4–5 μm, 7–16 septate; asparagus blight, seedborne, recently serious in USA (N. Carolina); Cooperman et al., *PD* **70**:392, 1986, dispersal & overwintering; *Phytop.* **76**:617, 1986, in vitro growth and inoculation; Conway et al., *PD* **71**:254, 1987, chemical control [**65**,5329; **66**,421].

*****C. beticola** Sacc. 1876; Kirk, *Descr.F.* 721, 1982; con. 40–160 × 3.5–4.5 μm; beet leaf spot, seedborne; Koch 1985, review of control [**65**,3029].

C. brassicicola Henn. 1905; Kirk, *Descr.F.* 722, 1982; con. 50–120 × 3–4.5 μm; *Brassica* dark border leaf spot.

C. bunchosiae Chupp & Muller 1942; con. 29–105 × 6–9 μm; leaf spot of Barbados cherry, *Malpighia acerola* or *M. glabra*; Holtzmann & Aragaki, *Phytop.* **56**:1114, 1966 [**46**,393].

C. canescens Ell. & Martin 1882; Mulder & Holliday, *Descr.F.* 462, 1975; con. 30–300 × 2.5–5 μm; leaf spot of bambarra groundnut, bean, cowpea; warmer regions; Dhingra & Asmus, *TBMS* **81**:425, 1983, detection in seed [**63**,2042].

C. capsici Heald & Wolf 1911; Kirk, *Descr.F.* 723, 1982; see Ellis 1976 for other names; con. 50–150 × 3–5 μm; red pepper leaf spot, seedborne.

C. carotae (Pass.) Solheim 1929; see Chupp, *A monograph of the fungus genus* Cercospora, 1954, for priority over *Cercospora carotae* (Pass.) Kaznowski & Siemaszko 1929; con. av. 95 × 2.2 μm; carrot leaf blight, can cause severe disease, seedborne; Thomas, *Phytop.* **33**:114, 1943; Hooker, *ibid.* **34**:606, 1944; Sherf & MacNab 1986 [**22**:285; **24**:45].

C. carthami Sundararaman & T.G. Ramakrishnan 1928; Waller & Sutton, *Descr.F.* 626, 1979; con. 80–220 × 2.5–4 μm; safflower leaf spot, seedborne.

*****C. circumscissa** Sacc. 1878; con. 30–80 × 3.5–4.5 μm, mostly 3–4 septate, often minutely verruculose; cherry leaf spot; Sztejnberg, *PD* **70**:349, 1986, described the disease and its chemical control. A similar disease has been attributed to *Mycosphaerella cerasella* Aderh. 1900, poss. anam. *Cercospora cerasella* Sacc., see Jenkins, *Phytop.* **20**:329, 1930 [**9**:661; **65**,5054].

C. cladosporioides Sacc. 1882; con. 25–65 × 4–6 μm, prob. a *Centrospora*; olive leaf spot; Hansen & Rawlins, *Phytop.* **34**:257, 1944; Pettinari, *Boll.Staz.Patol.veg.Roma* **18**:65, 1960 [**23**:349; **40**:700].

C. coffeicola Berk. & Cooke 1881; Mulder & Holliday *Descr.F.* 415, 1974; con. 40–150 × 2–4 μm; coffee brown eyespot, seed infection prob. unimportant; Siddiqi, *TBMS* **54**:415, 1970 [**50**,100].

C. corchori Saw. 1919; con. 80–150 × 3.5–4.5 μm; jute leaf spot; Chowdhury, *J. Indian bot.Soc.* **26**:227, 1948 [**27**:524].

C. elaeidis, see *Pseudospiropes.*

C. festucae Hardison, *Mycologia* **37**:492, 1945; con. 40–300 × 2–4 μm; Whitehead & Holt, *Phytop.* **40**:1023, 1950; on *Bromus inermis, Festuca elatior* [**29**:156; **30**:274].

C. fusimaculans Atk. 1892; con. 40–120 × 3–4.5 μm; Gramineae; Freeman, *Phytop.* **49**:160, 1959, on St Augustine grass, *Stenotaphrum secundatum*; Pauvert & Jacqua, *Annls.Phytopath.* **4**: 245, 1972; on guinea grass, *Panicum maximum* [**38**:524; **52**,2306].

C. gossypina, teleom. *Mycosphaerella gossypina.*

C. hayi Calpouzos 1955; con. 90–150 × 3–4 μm, banana fruit brown spot; Kaiser & Lukezic, *Phytop.* **55**:977, 1965; **56**:1290, 1966; morphology, symptoms, conidial dispersal & survival [**45**,518; **46**, 1052].

C. insulana Sacc. 1915; con. 55–135 μm long; Jackson, *Phytop.* **51**:129, 1961, statice leaf spot, *Limonium sinuatum*, USA (Florida) [**40**:611].

C. kikuchii (Matsumoto & Tomoyasu) M.W. Gardener 1927; Mulder & Holliday, *Descr.F.* 466, 1975; soybean purple seed stain, purple blotch or purple speck, internally seedborne, an important disease; Chen et al., *Mycologia* **71**:1158, 1979, infection; Yeh & Sinclair, *Mycotaxon* **10**:93, 1979; *Phytopath.Z.* **105**:265, 1982, con. ontogeny & morphology, effect on seed germination; Roy, *Can.J.Pl.Path.* **4**:226, 1982, comparison with other *Cercospora* spp.; Singh & Sinclair, *Seed Sci.Technol.* **14**:71, 1986, colonisation of seed [**59**,3013; **60**,1161; **62**,1290, 1717; **65**,4631].

C. longipes E. Butler 1906; Mulder & Holliday, *Descr.F.* 418, 1974; Sivanesan & Waller 1986; con. 20–200 × 3.5–8 μm, sugarcane brown spot, common, a minor disease.

C. longissima Cugini ex Traverso 1902; con. 20–100 × 3–5 μm; lettuce leaf spot; Szeto & Bau, *Agric.Hong Kong* **1**:278, 1975; Savary 1983, epidemiology in Ivory Coast [**56**,3788; **63**, 2103].

C. loti Hollós 1907; con. 22–224 μm long; *Lotus* leaf spot; Kreitlow & Yu, *PDR* **39**:236, 1955; *Phytop.* **46**:269, 1956, hosts, temp. & pathogenicity [**35**:22; **36**:34].

*****C. malayensis** F. Stevens & Solheim, *Mycologia* **23**:394, 1931; con. 100–270 × 4–5 μm; okra leaf spot, Reitsma & Sloof 1950 [**30**:445].

C. nicotianae Ell. & Ev. 1893; Mulder & Holliday,

Descr.F. 416, 1974; con. 35–300 × 3–4 μm; tobacco frogeye; during curing called barn spot, green spot, black barn spot, leaf burn or pole burn; Lucas 1975.

C. oryzae, teleom. *Sphaerulina oryzina*.

C. penniseti Chupp 1954; con. 30–240 μm long; Burton & Wells, *Phytop.* **71**:331, 1981, effect on yield of pearl millet in USA [**61**,1163].

C. piaropi Tharp, *Mycologia* **9**:113, 1917; con. 80–140 × 3 μm; Martyn, *J.Aquatic Pl.Managem.* **23**:29, 1985; disease in water hyacinth [**65**,2604].

C. platanicola, teleom. *Mycosphaerella platanifolia*.

C. rodmanii Conway, *CJB* **54**:1082, 1976; con. 66–374 × 3–5 μm; isolated from declining water hyacinth, USA (Florida), pathogenic, poss. of use in biological control; Conway, *Phytop.* **66**:914, 1976; Charudattan et al., *ibid* **75**:1263, 1985 [**55**,5649; **56**,1063; **65**,2606].

C. rosicola, teleom. *Mycosphaerella rosicola*.

C. sequoiae Ell. & Ev. 1887; Mulder & Gibson, *Descr.F.* 366, 1973; con. 25–80 × 4–7 μm, strongly constricted at the septa, prob. not a true *Cercospora*, poss. a *Heterosporium*; *Cryptomeria* needle blight and canker, other conifers, Japan, USA; Itô et al., *Bull.Gov.For.Exp.Stn.Tokyo* **268**:81, 1974, etiology & pathogenicity on *C. japonica*; Terashita, *J.Jap.For.Soc.* **57**:384, 1975, conidial formation [**54**,5591; **55**,4320].

C. sesami, teleom. *Mycosphaerella sesami*.

C. sojina Hara 1915; transferred to *Cercosporidium sojinum* q.v.

C. sorghi Ell. & Ev. 1887; Mulder & Holliday, *Descr.F.* 419, 1974; con. up to 300 × 3–4 μm; sorghum grey leaf spot.

C. taiwanensis, see *Leptosphaeria taiwanensis*.

C. traversiana Sacc. 1904; con. 45–130 × 3–3.5 μm; fenugreek leaf spot, seedborne; Leppik, *F.A.O.Pl.Prot.Bull.* **8**:19, 1959, distribution; Vörös & Nagy, *Acta phytopath.Acad.Sci.hung.* **7**:71, 1972, severe loss in Hungary; Singh & Pavgi, *Indian Phytopath.* **34**:414, 1981, infection [**39**:484; **52**,3022; **61**,6529].

C. vanderystii Henn. 1907; Skiles & Cardona-Alvarez, *Phytop.* **49**:133, 1959; con. up to 56 μm long, usually 1–3 septate; Chupp 1954 gave much longer measurements for con.; bean grey spot; distinguished from *Cercospora canescens*, on the same hosts, by the dense, grey, cushiony growth on the leaf lesions [**38**:437].

C. vicosae A. Muller & Chupp 1935; con. 25–100 × 4–6 μm; cassava diffuse leaf spot; Teri et al., *Trop.Agric.Trin.* **57**:239, 1980, effect on yield with *Mycosphaerella henningsii* [**60**,2928].

C. vitis (Lév.) Sacc. 1876; con. av. 63 × 7 μm; grapevine leaf spot; Harvey & Wenham,

N.Z.J.Bot. **10**:87, 1972, taxonomy, culture & infection [**52**,469].

C. zeae-maydis Tehon & Daniels, *Mycologia* **17**:248, 1925; con. 50–80 × 5–9 μm, but up to 180 μm long has been given; maize grey leaf spot was thought to be a minor disease until it recently became severe in USA, apparently because of the increased practice of minimum tillage, particularly where maize follows maize; severe losses can be caused when stem infection leads to lodging. Long periods of overcast weather and high relative humidities are required for epidemics to arise. Beckman & Payne, *Phytop.* **72**:810, 1982; **73**:286, 1983, infection, lesion development & culture in vitro; Rupe et al., *ibid* **72**:1587, 1982; effects of environment & plant maturity; Latterell & Rossi, *PD* **67**:842, 1983, review; Bair & Ayers, *Phytop.* **76**:129, 1986, variation in pathogenicity [**61**,7005; **62**,1918, 2999; **65**,3277].

****C. zebrina** Pass. 1877; con. 30–80 × 3–4.5 μm; on *Medicago, Melilotus, Trifolium*; Pratt, *Phytop.* **74**:1152, 1984, described a severe leaf spot of subterranean clover in USA (Mississippi), there was some host specificity in pathogenicity tests on 11 spp. of *Medicago* and *Trifolium*; see Pratt for references to *Cercospora* spp. on these plant genera; Barbetti, *Austr.J.Exp.Agric.* **25**:850, 1985, infection, black stem of subterranean clover [**64**,1164; **65**,5002].

C. zonata Winter 1883; con. 40–140 × 4.5–6.5 μm; broad bean zonate leaf spot; Woodward, *TBMS* **17**:195, 1932, described the disease as caused by *Cercospora fabae* Fautrey 1891 which was considered the same as *C. zonata*; the fungus may be little different from the type sp. *C. apii* [**12**:263].

Cercosporella Sacc. 1880; Hyphom.; Deighton, *Mycol.Pap.* 133, 1973; conidiomata synnematal, conidiogenesis holoblastic, conidiophore growth sympodial; con. filiform, hyaline; like *Cercospora* but with protuberant scars.

C. arachnoidea, teleom. *Mycosphaerella arachnoidea*.

C. rubi (Wint.) Plakidas, *J.agric.Res.* **54**:283, 1937; con. 13–96 × 2.7–4.7 μm; blackberry and dewberry rosette; Moore et al., *Arkansas Fm.Res.* **26**(2):10, 1977, chemical control, USA [**16**:475; **57**,706].

Cercosporidium Earle 1901; Hyphom.; Deighton, *Mycol.Pap.* 112, 1967; Ellis 1971; conidiomata synnematal, conidiogenesis holoblastic, conidiophore growth sympodial; con. multiseptate, hyaline; resembles *Cercospora* but differs in the fasciculate conidiophores with thickened bands, often incurved, and the broader con. with fewer septa.

C. bougainvilleae (Muntañola) Sobers & Seymour,

Proc.Fla.Sta.Hortic.Soc. 81:398, 1968, published 1969; con. 24–90 × 4.2–8.4 µm; bougainvillea leaf spot; Sobers & Martinez, *Phytop.* **56**:128, 1966 [**45**,1393; **49**, 330p].
C. henningsii, teleom. *Mycosphaerella henningsii*; **Cercosporidium magnoliae**, teleom. *M. milleri*; **C. personatum**, teleom. *M. berkeleyi*.
C. punctum (Lacroix) Deighton, *Mycol.Pap.* **112**:47, 1967, q.v. for a long synonymy; con. *c.*20–50 × 4–9 µm, mostly 1–2 septate; prob. teleom. *Mycosphaerella anethi* Petrak 1927; Deighton found no morphological differences between *Cercosporidium* from *Anethum*, *Foeniculum*, *Petroselinum*; on dill, fennel, parsley; prob. seedborne; Bougeard & Vegh, *Cryptogamie Mycologie* **1**:205, 1980, reported severe losses on fennel in France; Cirulli, *Inftore. fitopatol.* **31**(3):33, 1981, described infection of parsley in Italy [**60**,4609; **61**,343].
C. sojinum (Hara) Liu & Guo 1982, see *Index Fungi* **5**:191, 1983; usually called *Cercospora sojina* Hara 1915, see Lehman, *J.agric.Res.* **36**:811, 1928; con. 20–80 × 4–8 µm; soybean frogeye leaf spot; the pathogen, originally described from Japan, has been intensively studied in USA, forms races, seedborne, the seeds show grey and brown blotches; Phillips & Boerma, *Phytop.* **71**:334, 1981; **72**:764, 1982, races, resistance & its inheritance; Sinclair & Shurtleff 1982; Singh & Sinclair, *ibid* **75**:185, 1985, in seed; Bisht & Sinclair, *PD* **69**:436, 1985, effects on seed quality & yield [**7**:160; **61**,1487, 7244; **64**,3251, 4597].
cercosporin, toxin from *Cercospora* and like genera; Kuyama & Tamura, *J.Am.Chem.Soc.* **79**:5725, 1957, characterisation; Steinkamp et al., *Phytop.* **71**:1272, 1981, brief review, lesions caused on beet; Daub, *ibid* **72**:370, 1982, a photosensitising toxin [**37**:260; **61**,4410, 5534].
cereal African streak v. Harder & Bakker, *Phytop.* **63**:1407, 1973; Harder, *CJB* **53**:565, 1975; isometric 24 nm diam., transm. by a delphacid plant hopper, Kenya (Njoro) [**53**,2121; **55**,138].
— **chlorotic mottle** v. Greber 1977, 1979; *Descr.V.* 251, 1982; Phytorhabdovirus, bacilliform *c.*240 × 65 nm, transm. cicadellid leaf hoppers, persistent, Australia (E.), Gramineae, maize is the main economic host; Lockhart, *PD* **70**:912, 1986, in N. Africa; in Morocco caused severe necrotic and chlorotic leaf streaks in barley, durum wheat and oat; 5 maize cvs. were not infected [**66**,920].
— **cyst nematode**, *Heterodera avenae*.
cereale, *Ceratobasidium*; **cerealis**, *Gibellina*, *Hymenella*, *Rhizoctonia*.
cereal leaf spot Tsvetkov, *Rastit.Zasht.* **15**(4):29,

1967; transm. *Rhopalosiphum maidis*, *Schizaphis graminum*, Bulgaria [**46**,3041].
*— **northern mosaic** v. Ito & Fukushi, 1944; Toriyama, *Ann.phytopath.Soc.Japan* **42**:563, 1976; Hyung & Shikata, *Korean J.Pl.Prot.* **16**:87, 1977; Phytorhabdovirus, bacilliform 350 × 68 nm, transm. *Laodelphax striatellus* and other plant hoppers; Japan, Korea; temperate cereals, other Gramineae, resembles oat pseudo-rosette v.; Shirako & Ehara, *Phytop.* **75**:453, 1985; composition & detection by Elisa [**57**,2902; **64**,4846].
— **pale green dwarf** Onischenko et al., *Mikrobiol. Zhurnal.* **39**:621, 1977; MLO assoc., USSR [**57**,5393].
— **rhabdovirus** = barley striate mosaic v.
cereals, Johnson et al., *A.R.Phytop.* **5**:183, 1967, rusts; Slykhuis, *ibid* **14**:189, 1976, viruses; Catherall, *Welsh Pl.Breeding Rep.* 1978, Univ. Coll. Wales, viruses of Gramineae; Doodson in Ebbels & King 1979, resistant cvs.; Jenkins & Lescar, *PD* **64**:987, 1980, fungicides in Europe; Scott et al., *Appl.Biol.* **5**, 1980, host specificity in parasites & control; Fehrmann, *EPPO Bull.* **11**:259, 1981, chemical control; Jenkyn & Plumb 1981, control; Shaner, *A.R.Phytop.* **19**:273, 1981, environmental & fungal leaf blight; Mathre et al., *PD* **66**:526, 1982, seed treatment; Williams, *Adv.Pl.Path.* **2**, 1984, downy mildews of tropical cereals; Zadoks & Rijsdik, *Atlas of cereal diseases and pests in Europe*, 1984; Attwood ed., *Crop protection handbook – cereals*, Br.Crop Protect. Council, 1985.
cereal spotting Borodina et al., *Mikrobiol.Zhurnal* **44**(3):38, 1982; transm. sap, *Aceria tritici*, *Aculodes mckenziei*, USSR; barley, wheat, Gramineae [**61**,6975].
— **striate mosaic** = barley yellow striate v.
— **tillering** v. Lindsten, Gerhardson & Petterson, *Meddn. St. VäxtskAnst.* **15**:375, 1973; Phytoreovirus, isometric, *c.*65–70 nm diam., transm. *Dicranotropis hamata*, *Laodelphax striatellus*, Sweden [**53**,4769].
— **yellow dwarf** = barley yellow dwarf v.
cerebrum, *Peridermium*.
Cerotelium Arthur 1906, 1934; Uredinales; mostly heteroecious, teliospores ± adherent laterally, warmer regions; Sathe, *Indian Phytopath.* **25**:76, 1972, taxonomy.
***C. fici** (E. Butler) Arthur 1917; Laundon & Rainbow, *Descr.F.* 281, 1971; teliospores 14–22 × 10–13 µm, aecia unknown, fig rust, *Ficus*, *Morus* spp., often requires chemical control; distinguished from *Uredo ficina* on fig by the inconspicuous paraphyses, smaller uredospores with shorter spines.

Cerotoma ruficornis, transm. bean yellow stipple, cowpea chlorotic mottle; *C. trifurcata*, transm. cowpea chlorotic mottle.

cestrum v. Ragozzino, *Annali Fac.Sci.Agrar.Univ.Napoli Portici* 4 **8**:249, 1974; poss. Caulimovirus, isometric 35–40 nm diam., Italy, *C. parqui* [**55**,3491].

ceuthocarpa, *Linospora*.

Ceuthospora Grev. 1827; Coelom.; Sutton 1980; conidiomata stromatic, conidiogenesis enteroblastic phialidic; con. aseptate, hyaline; *C. lunata* Shear, a wound parasite, causes cranberry black rot, a storage disease in USA; Schwarz & Boone, *PD* **69**:225, 1985 [**64**,3497].

Chaetocnema pulicaria, transm. maize chlorotic mottle.

*Chaetosiphon, a subgenus of *Pentatricnopus* q.v., transm. strawberry mild yellow edge, strawberry stunt, strawberry vein banding.

*Chalara (Corda) Rabenh. 1844; Hyphom; conidiomata hyphal, conidiogenesis enteroblastic phialidic; con. 1 or 2 celled, hyaline; teleom. in *Ceratocystis*, Nag Raj & Kendrick, *A monograph of* Chalara *and allied genera*, 1975. *Chalara thielavioides* (Peyronel) Nag Raj & Kendrick; wound parasite, root rots, infections of grafts, elm, poinsettia, rose, walnut; storage rots, carrot. *C. paradoxa, C. quercina, C. ungeri* anam. of *Ceratocystis paradoxa, C. fagacearum, C. coerulescens*, respectively.

chamaecereus isometric v. Samyn & Welvaert, *Phytopath.Z.* **91**:276, 1978; *C. sylvestri*, in nurseries in Belgium (Ghent) on cacti imported from Brazil [**58**,253].

Chamberland, Charles, 1851–1908; born in France, bacteriologist, worked with Pasteur, noted for work on sterilisation of media for bacteria, he perfected the autoclave, and the bacterial filters which bear his name; see Bulloch, *The history of bacteriology*, 1938.

charcoal (*Hypoxylon mediterraneum*) oak, —— **base rot** (*Ustulina deusta*) oil palm, —— **rot** (*Macrophomina phaseolina*) plurivorous, —— **stump rot** (*U. deusta*) tea.

char spot (*Cheilaria agrostis*) Gramineae.

chartarum, *Leptosphaerulina, Pithomyces, Ulocladium*.

Cheilaria Lib. 1834; Coelom.; Sutton 1980; conidiomata stromatic, conidiogenesis enteroblastic phialidic, con. multiseptate, hyaline; *C. agrostis* Lib. 1837; Gibson & Sutton, *Descr.F.* 488, 1976; con. 22–25 × 3.5–4 μm; char spot of Gramineae.

cheiranthi, *Alternaria*.

chemodifferentiation, see thigmodifferentiation.

chemotaxis, a response of an organism, or a part of

it, to a chemical stimulus, see taxis; Adler, *A.Rev.Biochm.* **44**:341, 1975, bacteria; Zuckerman & Jansson, *A.R.Phytop.* **22**:95, 1984, nematodes.

*chemotherapy, or chemical control; the treatment of plants and pathogens by chemicals to prevent or cure disease. All chemicals, except the simplest, have a common name, a standard abbreviation; trade names can be confusing since they may become accepted, wrongly, as the common name. Chemicals are mostly applied to soil, seed or crop foliage; in the last case this is usually done by high, low or ultra-low volume spray using a liquid carrier; dusts may be used. Organic fungicides may move within the plant i.e. they are systemic; these are the newer therapeutants; there are now some that move downwards into the roots when the foliage is treated. Pathogens that attack only through the roots are the most difficult to control with chemicals. Only a few of the main chemicals, mostly fungicides, are given; the literature is very large and cannot be fully treated.

Martin & Worthing, *Insecticide and fungicide handbook*, edn. 5, 1976; Marsh, *Systemic fungicides*, edn. 2, 1977; Matthews, *Pesticide application methods*, 1979; Dekker & Georgopoulos, *Fungicide resistance in crop protection*, 1982; Hill & Waller 1982; Martin & Woodcock, *The scientific principles of crop protection*, edn. 7, 1983; Skylakakis, *A.R.Phytop.* **21**:117, 1983, theory & strategy of chemical control; Davidse & Waard, *Adv.Pl.Path.* **2**, 1984, systemic fungicides; Trinci & Ryley, *Mode of action of antifungal agents*, 1984; Vyas, *Systemic fungicides*, 1984; Worthing ed., *The pesticide manual; a world compendium*, edn. 8, 1987; *Tropical Pest Management pesticide index*, 1984.

chenopodium mosaic = sowbane mosaic v.

—— **necrosis** v. (str. of tobacco necrosis v.), Tomlinson et al. 1981; *AAB* **102**:135, 1983, England, rivers [**61**,1536; **62**,1851].

—— **star mottle** = sowbane mosaic v.

cherry, *Prunus avium*, sweet cherry; *P. cerasus*, sour cherry; *P. cerasifera*, cherry plum or myrobalan.

—— **albino** Zeller, Milbrath & Cordy 1944; Florance & Cameron, *Phytop.* **68**:75, 1978; MLO assoc., USA (Oregon) [**24**:156; **57**,5588].

—— **buckskin** (peach X MLO).

—— **chlorotic necrotic ringspot** v. Kegler, *Phytopath.Mediterranea* **2**:175, 1963; Smith 1972; isometric av. 19 nm diam., Germany [**43**,779r].

—— —— **ringspot** = prune dwarf v.; cherry chlorotic ring mottle Kegler, *Phytopath.Z.* **54**:305, 1965, is poss. as for CCR.

—— **decline** Crowley 1955; Heinis 1956; Kegler et al., *NachrBl dt. PflSchutzdienst. DDR* **27**:5, 1973;

Fos, *Revue Zool.Agric.Path. Vég.* **75**:134, 1976; Australia, France, Germany, USA (Oregon) [**35**:355,689; **53**,205; **56**,4608].

— **Eckelrade** (cherry leaf roll v. + raspberry ringspot v.).

— **Eola rasp leaf** (tomato ringspot v.).

— **European rasp leaf**, varying degrees of the rasp leaf symptom are induced by Nepoviruses: arabis mosaic v., raspberry ringspot v. or cherry leaf roll v. alone, usually mild symptoms; or with prunus necrotic ringspot v., moderate symptoms; or with prune dwarf v., severe symptoms. These diseases are different from the cherry rasp. leaf caused by one virus in N. America; A.N. Adams, personal communication, 1985.

— — **rusty mottle** Posnette & Cropley 1961; England [**40**:550].

— **flowering rough bark** Milbrath & Zeller, *Phytop.* **32**:428, 1942; USA (Oregon) [**21**:462].

— **fruit pitting** (tomato bushy stunt v.).

— **green ring mottle** Barksdale, *Phytop.* **49**:777, 1959; Canada, USA; Obata, *Res.Bull.Pl.Prot.Service Japan* **8**:1, 1970; Parker et al. in Gilmer et al. 1976 [**39**:330; **51**,1655].

— **leaf roll v.** Swingle et al. 1941, as elm mosaic; Posnette & Cropley 1955; Jones, *Descr.V.* 306, 1985; Nepovirus, isometric *c*.28 nm diam., transm. sap, seed, prob. pollen in some strs., not apparently by nematodes; common in many wild and cultivated woody plants, some diseases caused are: American elm chlorotic mosaic, ring pattern and dieback; birch chlorotic ringspot and yellow vein netting; cherry leaf roll and death; privet ringspot; walnut black line and leaf pattern; the many strs. are usually from different hosts.

— **line pattern** Gualaccini, *Boll.Staz.Patol.veg. Roma* 3 **17**:141, 1960; Paulechová 1968; Czechoslovakia, Italy [**40**:370; **49**, 2522p].

— **little cherry** Foster & Lott 1947; Raine et al., *Phytop.* **65**:1181, 1975; Ragetli et al., *CJB* **60**:1235, 1982; filamentous particle assoc.; Raine et al., *Can.J.Pl.Path.* **8**:6, 1986, transm. *Phenacoccus aceris*; it has not been confirmed that transm. is also by leafhoppers, J. Raine, personal communication, 1986. A serious disease in parts of N. America, sweet cherry bears small pale, angular fruit that do not ripen, see Slykhuis et al., *Can.Pl.Dis.Surv.* **60**:37, 1980, for the position in Canada [**26**:204; **55**,2307; **60**,3845].

— **Mahaleb leaf pattern** Pares, *Agric.Gaz.N.S.W.* **72**:635, 1961; Australia [**41**:470].

— **mid leaf necrosis**, and gummosis, on cv. Montmorency were described, respectively, by Milbrath, *Phytop.* **47**:637, 1957, USA (Oregon), and by Blodgett et al., *PDR* **48**:277, 1964, (Pacific N.W.) [**37**:293; **43**,2347].

— **Molières decline** Marenaud et al. 1979; Carles 1986; poss. MLO, transm. *Fieberiella florii*, France [**59**,3831; **65**,5039].

— **mottle leaf** Zeller 1934; Reeves, *J. agric.Res.* **62**:555, 1941; N. America (W.) [**20**:481].

— **Napoleon latent** Fridlund, *PDR* **50**:902, 1966; USA (Washington State) [**46**,1042].

— **necrotic ringspot** (prunus necrotic ringspot v.).

— — **rusty mottle** Posnette & Cropley, *Pl.Path.* **13**:20, 1964, England [**43**,2677].

— **Pfeffinger** (raspberry ringspot v.).

— **pinto leaf** Kienholz, *Phytop.* **37**:64, 1947; USA (Oregon) [**26**:247].

— **rasp leaf v.** Bodine & Newton 1942; Stace-Smith & Hansen, *Descr.V.* 159, 1976; poss. Nepovirus, isometric *c*.30 nm diam., transm. sap, *Xiphinema americanum*, seedborne, N. America (W.), wide host range including apple, cherry, peach; see cherry European rasp leaf; Jones et al., *AAB* **106**:101, 1985; properties of an isolate from a symptomless red raspberry selection imported to Scotland from Canada (Quebec) [**65**,2904].

— **rugose mosaic** (prunus necrotic ringspot v.).

— **rusty mottle** Reeves 1940; Wadley & Nyland in Gilmer et al. 1976; a group of diseases in N. America.

— — **spot** Wood, *N.Z.J.agric.Res.* **15**:155, 1972; New Zealand [**51**,2684].

— **small bitter cherry** Lott, *Sci.Agric.* **27**:260, 1947; Canada (British Columbia) [**27**:28].

— **sour fruit necrosis** Weber, *NachBl.dt.PflSchutzdienst.Stuttg.* **9**(12):179, 1957, Germany [**37**,361].

— — **ring pattern** Willison & Weintraub, *Phytop.* **43**:324, 1953; N. America [**33**:159].

— — **yellows** (prune dwarf v.).

— **spur** Blodgett, Jenkins & Aichele, *PDR* **49**:910, 1965; USA (Washington State), cv. Bing; Trifonov 1978, Bulgaria [**45**,1106; **59**,3832].

— **stem pitting** Mink & Howell; Al Musa et al., *PD* **64**:551, 1081, 1980; USA (Washington State), differs from similar diseases in USA (California and E.) [**60**,2093, 4573].

— **twisted leaf** Lott, *Sci.Agric.* **23**:439, 1943; Canada (British Columbia), USA (Pacific N.W.) [**22**:317].

— **veinal necrosis** (tomato bushy stunt v.).

— **xylem aberration**, Lott, *Can.Pl.Dis.Surv.* **47**:74, 1967; Canada (British Columbia), in stone fruits [**47**,851].

chestnut, *Castanea*, see horse chestnut.

— **leaf deformation**, see oak leaf deformation.

— **yellows** Shimada & Kouda 1954; Okuda et al., *Ann.phytopath.Soc.Japan* **40**:464, 1974; MLO assoc.

*chickpea, *Cicer arietinum*; Haware et al. 1986,

seedborne fungi; Grewal & Pal in Varma & Verma 1986, fungus diseases [65,5297, 5361].
— distortion mosaic Mali & Vyanjane, *Curr.Sci.* 49:599, 1980; transm. sap, *Aphis gossypii*, India (Parbhani) [60,3433].
— lethal wilt (lettuce necrotic yellows v.), Behncken, *Australasian Pl.Path.* 12:64, 1983 [64,863].
— stunt (str. of bean leaf roll v.), fide Boswell & Gibbs 1983.
chicory, *Cichorium intybus*.
— blotch v. Brčák 1960, 1961; *Phytopath.Z.* 45:335, 1962; Pozděna et al. 1980; poss. Carlavirus, rod 500–600 × 10–11 nm, transm. sap, *Acyrthosiphon pisum*, *Aphis fabae*, Czechoslovakia [41:498; 42:430; 60,5766].
— X v. Gallitelli & Franco, *Phytopath.Z.* 105:120, 1982; prob. Potexvirus, filament av. 553 nm long, transm. sap, Italy (S.) [62,1745].
— yellow mottle v. Volvas, Martelli & Quacquarelli 1971; *Descr.V.* 132, 1974; Piazzolla et al., *Phytopath.Mediterranea* 17:149, 1978; Nepovirus, isometric c.30 nm diam., transm. sap, seed; Italy (S.); occurs naturally in chicory and parsley; Piazzolla & Rubino, *Phytopath.Z.* 111:199, 1984, assoc. RNA may be a satellite; Piazzolla et al., *J.Phytopath.* 115:124, 1986, strs. [60,3082; 64,2397; 65,4223].
chilli, red pepper, see pepper.
chilling injury in plants, Lyons, *A.Rev.Pl.Physiol.* 24:445, 1973.
chimera, plant or plant part consisting of tissues of diverse genomes arising, as a result of grafting or somatic mutation, in such a way that the components occur as either separate layers in the apical meristem i.e. periclinal or mericlinal; or discrete segments involving all layers of the apical meristem i.e. sectorial, after FBPP.
Chinese cabbage, see *Brassica*; — gooseberry, see kiwi fruit; — shoeflower mosaic, see *Hibiscus*.
chitosan, in host and pathogen interactions, Hadwiger & Loschke, *Phytop.* 71:756, 1981.
chive, see *Allium*.
chlamydospore, of fungi, an asexual spore, primarily for perennation, a resting stage; arising from a hyphal segment and with an inner secondary wall; usually impregnated with hydrophobic material.
chloranil, a protectant fungicide for treatment of vegetable seed.
chloris striate mosaic v. Grylls 1963; Francki & Hatta, *Descr.V.* 221, 1980; Francki et al., *Virology* 101:233, 1980; Geminivirus, isometric, paired particles c.30 × 18 nm, transm. *Nesoclutha pallida*; Australia; Gramineae, including barley, maize, oat [60,164].

chloroneb, systemic fungicide for seed and soil application; effective against *Pythium* and *Rhizoctonia*, less so against *Fusarium*.
chloropicrin, soil fumigant.
Chloroscypha Seaver 1931; Helotiaceae; ascomata yellow green to blackish green; asco. mostly at least 15 × 6 μm; largely saprophytic but may cause conifer needle blights, e.g. Kobayashi 1965; Shoji 1978, 1981 [45,1923; 59,3440; 61,6007].
chlorosis, partial or complete absence of normal green colour; affected organs become yellow green, yellow or white; albinism q.v.
chlorothalonil, protectant fungicide and for sublimation in greenhouses, broad spectrum.
chlorotic banding, and white spot banding, in barley and wheat seedlings, caused by low and high temp. stress; Vanterpool, *Scient.Agric.* 29:334, 1949.
chlorquinox, fungicide with some eradicant and systemic activity, used against powdery mildews on cereals.
Choanephora Currey 1873; Choanephoraceae; Kirk, *Mycol.Pap.* 152, 1984; sporangiolum wall firmly attached to spore wall and not separable at maturity. *C. cucurbitarum* (Berk. & Rav.) Thaxter 1903, occurs in soil; it causes, under warm and moist conditions, a soft wet rot of the above ground parts of flowers and vegetables. It can be found in storage, plant wounding can increase infection. The conidial state has a silvery metallic sheen which distinguishes the fungus from the 2 other common fungal rots: *Botryotinia fuckeliana* and *Rhizopus stolonifer*; Holliday 1980.
Choanephoraceae, Mucorales; sporangia and sporangiola born on separate and distinct sporangiophores, zygospores striate and borne on suspensors like tongs; *Choanephora*.
chochamento das raizes, see *Phytomonas*.
chocolate browning, grapevine, B deficiency, Kuniyuki et al., *Bragantia* 44:187, 1985; — spot (*Botryotinia fuckeliana*, *Botrytis fabae*) broad bean [65,2930].
choke (*Epichloë typhina*) temperate grasses.
Chondrilla juncea, skeleton weed; chondrilla stunt Hasan, Giannotti & Vago, *Phytop.* 63:791, 1973; bacilliform particles assoc., Mediterranean [53,872].
chondrillina, *Puccinia*.
*Chondrostereum Pouzar 1959; Stereaceae; *C. purpureum* (Fr.) Pouzar; silver leaf of temperate fruit trees and bushes, particularly plum; Wall, *PD* 70:158, 1986, gave an example of pathogenicity to a non-fruit tree, *Betula alleghaniensis*; Butler & Jones 1949; Atkinson 1971; as *Stereum purpureum*; Corke, *AAB* 89:89, 1978; 1980, control with *Trichoderma viride*;

Spiers, *Eur.J.For.Path.* **15**:111, 1985, basidiospore release [**60**,4667; **64**,4502; **65**,2992].

Christensen, Jonas J., 1892–1964; born in USA, Univ. Minnesota, genetic variation in plant pathogenic fungi, monograph on *Ustilago maydis*. *Phytop.* **54**:1429, 1964.

chronic decline, citrus, a graft incompatibility, sweet orange budded on sour orange; Schneider, *Phytop.* **47**:279, 1957 [**36**:693].

chrysanthemella, *Septoria*; **chrysanthemi**, *Ascochyta*, *Erwinia*, *Mycosphaerella*, *Puccinia*; **chrysanthemicola**, *Phoma*.

***Chrysanthemum**, Scopes, *Pests, diseases and nutritional disorders of chrysanthemum*, Natn. Chrysanthemum Soc. Lond. edn. 2, 1975; *C. cinerariifolium*, pyrethrum.

chrysanthemum B v. Noordam 1952; Hollings & Stone, *Descr.V.* 110, 1972; Carlavirus, rod *c*.685 nm long, transm. sap, aphid, non-persistent, widespread with host, cvs. often without visible symptoms.

— C Noordam, *Tijdschr.PlZiekt.* **58**:121, 1952; Netherlands; — D Prentice, *Scient.Hort.* **13**:90, 1958; Hollings 1960; England, flower distortion and colour break; — E Hollings 1960; England, flower distortion [**32**:559; **38**:747; **40**:265].

— **chlorotic mottle viroid** Dimock & Geissinger 1969; Dimock et al., *Phytop.* **61**:415, 1971; Romaine & Horst, *Virology* **64**: 86, 1975; leaf mottle and chlorosis, dwarfing, delay in flowering, USA, occurs naturally in a few cvs. [**50**,3823; **55**,2753].

— **flower distortion** (tomato aspermy v.), TAV prob. causes other symptoms in chrysanthemum.

— **foliar necrosis** McGovern et al., abs. *Phytop.* **76**:655, 1986; MLO assoc. USA.

— — **nematode**, *Aphelenchoides ritzemabosi*.

— **green flower** Hollings, as for chrysanthemum D, MLO assoc.

— **latent** Hollings, *AAB* **45**:589, 1957; transm. sap, Britain [**37**:238].

— **mild mosaic** = chrysanthemum B v.

— **mottle v.** Tochihara, *Ann.phytopath.Soc.Japan* **36**:1, 1970; isometric *c*.25–30 nm diam., transm. sap, *Myz.pers.* [**49**,3331].

— **necrotic mottle v.** Hollings & Stone 1967; particle 700 nm long, England, found in chrysanthemum with a leaf mottling complex of several viruses [**47**,390j].

— **Q** Keller 1950 = chrysanthemum B v.

— **Rose Harrison v.** Hollings & Stone 1968 isometric 25–30 nm diam., England [**48**,1024h].

— **rosette** Brierley & Smith, *PDR* **35**:524, 1951; **42**:752, 1958; Kemp, *Can.Pl.Dis.Surv.* **41**:183, 1961; Ivory Seagull mosaic, transm. sap, Canada, USA [**31**:284; **38**:7; **41**:229].

— **stunt viroid** Dimock 1947; Diener & Lawson, *Virology* **51**:94, 1973; Hollings & Stone, *AAB* **73**:333, 1973; the electrophoretic mobility is distinct from that of the potato spindle tuber viroid; infected plants have paler and more upright foliage, can be < half as tall as normal ones, flower buds open prematurely; in some cvs. the red colour is bleached, leaves have conspicuous white spots *c*.3 mm diam., blossoms small; transm. in cultural operations; Canada, England, Netherlands, USA; occurs naturally in some chrysanthemum cvs.; Palukaitis & Symons, *J.gen.Virol.* **46**:477, 1980, purification, characterisation; Watermeyer & Thompson 1984 [**52**,3342; **53**,973; **59**,3799; **64**,2055].

— **temperate v.**, see cryptic viruses.

— **vein mottle** Hollings, *AAB* **45**:589, 1957; England [**37**:238].

— **virescence and witches' broom** Verhoyen et al. 1979; Pettersson & Tomenius 1979; Shiomi & Sugiura 1983; MLO assoc., Belgium, Japan, Sweden [**59**,2782, 5221; **63**,3159].

— **yellow band** Kitajima & Costa, *Fitopat.Brasileira* **4**:55,1979; bacilliform particles assoc., Brazil [**58**,5279].

Chrysomyxa Unger 1840; Uredinales; mostly heteroecious, telia on Ericaceae and related families, aecia on *Picea*, spruce rusts; the uredospores are formed in rows by basipetal abstriction and resemble aeciospores; the aecia occur only where the telial hosts are present, but the telia are not restricted to areas where spruces grow; Wilson & Henderson 1966; Ziller 1974.

C. abietis (Wallr.) Unger; Mordue & Gibson, *Descr.F.* 576, 1978; autoecious, only telia known, unusually these are on spruce, teliospores 14–42 × 9–16 μm, needle cushion rust, parts of Asia, Europe; Takahashi & Saho, *Trans.mycol.Soc.Japan* **26**:433, 1985, causing severe damage on *Picea engelmannii* and *P. rubens*, poss. a different form than the one in Europe [**65**,4572].

C. arctostaphyli Dietel 1894; telia on *Arctostaphylos uva-ursi*, teliospores 23–64 × 13–18 μm; anam. *Peridermium coloradense* (Dietel) Arthur & Kern; spruce yellow witches' broom, N. America; the rust is limited to where the hosts occur together, brooms are very conspicuous.

C. ledi de Bary 1879; Ziller 1974 described 4 vars.: var. *cassandrae* (Peck & G.W. Clinton) Savile 1950; telia on *Chamaedaphne calyculata*, teliospores 11–16 μm wide, cassandra rust, N. temperate. Var. *ledi*; telia on *Ledum*, teliospores 13–20 × 10–15 μm, spruce small spored rust, Labrador tea rust, N. temperate; in Canada spruce may be severely damaged where the

alternate ericaceous hosts are present. Var.
rhododendri (de Bary) Savile 1955, telia on
Rhododendron, teliospores 20–30 × 10–14 μm,
rhododendron short spored rust, has caused
damage to the telial host in several parts of the
world; Ziller stated that the rust is not known to
occur on spruce in N. America. Var. *vaccinii*
Ziller 1955; telia on *Vaccinium parvifolium*,
teliospores 12–22 × 7–12 μm, aecial state unknown,
red huckleberry rust, Canada (British Columbia).
C. ledicola Lagerh. 1893; telia on *Ledum*,
teliospores 13–18 × 10–14 μm, spruce large spored
rust, Labrador tea rust, Canada, Japan, USA (N.),
can become epidemic on Canadian spruce.
C. monesis Ziller, *CJB* **32**:435, 1954; telia on
Moneses uniflora, teliospores 12–22 × 6–12 μm,
coastal spruce cone rust, N. America (N.W.),
causes seed destruction and premature opening of
cones of *Picea sitchensis.*
C. piperiana Sacc. & Trott ex Cummins 1956;
anam. *Peridermium parksianum* Faull 1934; telia
on *Rhododendron macrophyllum*, teliospores
16–30 × 11–14 μm, long spored rhododendron
rust, occurs on spruce in USA (Oregon to
California) but not on the aecial host in Canada,
can cause defoliation of *Picea sitchensis.*
C. pirolata Winter 1882; telia on *Pyrola*, teliospores
12–19 × 7–9 μm; occurs on *Moneses uniflora* but
differs from *Chrysomyxa monesis* in the broad
polygonal warts of aeciospores and uredospores;
inland spruce cone rust, N. temperate; in many
areas the rust is known only on the telial hosts, it
attacks the cones of spruce not the needles, and
severe damage has been caused in Europe and N.
America to the seed crop; Pritam Singh,
Can.Pl.Dis.Surv. **61**:43, 1981, review; Nelson &
Krebill, *Great Baisin Naturalist* **42**:262, 1982,
epidemic on *Picea pungens* in 1969, USA (Utah);
Sutherland et al., *CJB* **62**:2441, 1984, spore
morphology & growth [**61**,3110; **63**, 283; **64**,2194].
C. weirii Jackson 1917; Sutton, *Descr.F.* 221, 1970;
only telia known, teliospores 17–40 × 5–8 μm and
these occur, unusually, on *Picea*, Weir's spruce
cushion rust; N. America, south central Asia;
needles shed prematurely when *c.*14 months old
but causes little damage; resembles *Chrysomyxa
abietis* but the teliospores are narrower.
C. woroninii Tranzschel 1903; telia on *Ledum
decumbens* and *L. groenlandicum*, causing witches'
brooms, teliospores 19–40 × 12–19 μm, spruce bud
or shoot rust, N. temperate, affects only young
emerging spruce shoots causing stunt and forming
aecial masses. The damage is relatively
insignificant; but McBeath, *Phytop.***74**:456, 1984,
who described the spores & symptoms, stated that
this rust has become increasingly important in

USA (Alaska) following a disease outbreak on
Picea glauca [**63**,4116].
chrysosperma, *Cytospora.*
Chupp, Charles David, 1886–1967; born in USA,
Univ. Cornell, wrote: *Manual of vegetable garden
diseases*, 1925; *A monograph of the fungus genus*
Cercospora, 1954; *Vegetable diseases and their
control*, 1960, with A.F. Sherf; an edn. 2 of this
last book was written by Sherf & MacNab in
1986. *A.R.Phytop.* **20**:41, 1982.
Chytridiales, Chytridiomycetes; mostly aquatic, on
algae, microscopic fauna and other fungi;
Synchytrium is terrestrial on higher plants; little
or no mycelium, zoospore germination
monopolar, resting spore functions as a
sporangium or prosporangium. There are 2
groups: sporangium inoperculate, opening by
deliquescence or rupture of one or more papillae,
Olpidiaceae, Synchytriaceae; sporangium
operculate, opening by one or more opercula.
Chytridiomycetes, Mastigomycotina; mostly
parasitic on fresh water algae and microscopic
fauna; thallus coenocytic, eucarpic or holocarpic;
monocentric, polycentric or mycelial, wall often
chitinous; zoospore with single, posteriorly
directed flagellum, whiplash, nuclear cap
conspicuous; zygote forms an encysted structure
or a diploid thallus.
Ciboria Fuckel 1870; Sclerotiniaceae; stroma
formed in the host tissues and showing little of
the aspect of a sclerotium; Schumacher,
Norwegian J.Bot. **25**:145, 1978 [**57**,5695].
C. batschiana (Zopf) Buchw. 1947; asco. 6–11 ×
4–6 μm; infects acorns of *Quercus*; Delatour,
Eur.J.For.Path. **8**:193, 1978 [**58**,2954].
C. carunculoides (Siegler & Jenkins) Wetzel in
Whetzel & Wolf, *Mycologia* **37**:484, 1945; asco.
6.4–9.6 × 2.4–4 μm, reniform; mulberry popcorn,
USA (E.). *Ciboria shiraina* (Henn.) Whetzel
loc.cit.; asco. 11–15 × 4.5–6 μm, ovoid to ellipsoid;
Lin & Lo, *Pl.Prot.Bull. Taiwan* **25**:157, 1983,
mulberry swollen fruit, E.Asia [**24**:458; **63**,198].
Ciborinia Whetzel 1945; Sclerotiniaceae; stroma
formed in host tissues and evident as black
sclerotia, in neither this genus nor in *Ciboria* is an
anam. known; Groves & Bowerman, *CJB* **33**:577,
1955, on *Populus*; Batra, *Am.J.Bot.* **47**:819, 1960,
key & on *Magnolia, Quercus, Salix* [**35**:404;
40:346].
C. camelliae Kohn, *Mycotaxon* **9**:399, 1979;
ascoma disc 5–18 mm diam., asco. 7.5–12.5 ×
4–5 μm, sclerotia 10–12 mm diam.; camellia flower
blight; Japan, USA; Hansen & Thomas, *Phytop.*
30:166, 1940, sclerotia 12 × 30 mm; Kohn &
Nagasawa, *Trans.mycol.Soc.Japan* **25**:149, 1984,
the smaller sclerotia [**19**:350; **64**,1142].

C. foliicola (Cash & Davidson) Whetzel; asco. 9–13 × 5.5 μm, sclerotia 10–40 mm long; black rib of willows, see Batra under the genus; Davidson & Cash, *Mycologia* **33**:266, 1933; Sutton & Lawrence, *PDR* **53**:101, 1969 [**48**,2015].

C. whetzelii (Seaver) Seaver 1951; asco. 12–15 × 6–9 μm, sclerotia 2–8 mm diam.; poplar inkspot, Canada; Baranyay & Hiratsuka, *CJB* **45**:189, 1967, comparison with *Ciborinia pseudobifrons* Whetzel 1955; Lavelée 1976 [**46**,2116; **55**,4880].

Cicadulina, transm. eleusine mosaic, streak and stunt, maize streak; *C. bimaculata* toxin, wallaby ear q.v.; *C. mbila* transm. maize vein enation, wheat eastern striate; *C. triangula* transm. maize mottle and chlorotic stunt.

Ciccarone, Antonio, 1909–82; born in Italy, Univ. Bologna, plant pathologist in Ethiopia, Kenya, Venezuela; Univ. Catania from 1952, Univ. Bari from 1957; mainly a mycologist but strengthened disease studies in the Mediterranean. *Phytop.* **74**:254, 1984.

Cicer arietinum, Bengal gram, chickpea.

ciceris-arietini, *Uromyces*.

cichoracearum, *Erysiphe*; **cichorii**, *Alternaria, Pseudomonas*.

Cifferi, Raffaele, 1897–1964; born in Italy, Univ. Bologna, Cuba and Dominican Republic 1925–32; Agricultural College Univ. Florence from 1936; Univ. Pavia from 1942; a versatile and prolific mycologist and plant pathologist; wrote *Trattato di patologia vegetale*, with E. Baldacci; founded *Mycopathologia* with P. Redaelli. *Atti Ist.Bot.Univ.Pavia* 21, suppl. 1964; *Mycologia* **57**:198, 1965.

cigar end (*Verticillium theobromae*) banana.

cincta, *Cytospora*; **cinctum**, *Leucostoma*.

cinerariae, *Alternaria*; **cinerascens**, *Diaporthe, Phomopsis*.

cinerea, *Botrytis, Monilia*; **cinerescens**, *Phialophora*.

cingulata, *Glomerella*.

cinnabarina, *Nectria*; **cinnamomi**, *Elsinoë, Phytophthora*.

Cionus, transm. scrophularia mottle.

circinans, *Colletotrichum*; **circinata**, *Periconia, Waitea*.

circulative viruses, see persistence.

circumnutation, the rotation of the tip of a growing stem so that it traces a helical curve in space; Kennedy et al., *Phytop.* **76**:712, 1986, effect of bacterial infection in bean [**66**,387].

circumscissa, *Cercospora*.

cirsium yellows Begtrup, *Tidsskr.Pl.Avl.* **79**:317, 1975; MLO assoc. Denmark, *C. arvense* [**55**,1159].

citrange, see *Citrus*.

citri, *Alternaria, Diaporthe, Mycosphaerella, Phomopsis, Septoria*.

citricarpa, *Guignardia, Phyllosticta*; **citricola**, *Phytophthora*; **citricolor**, *Mycena*.

citriformis, *Hirsutella*; **citri-grisea**, *Stenella*.

citron, see *Citrus*.

citrophthora, *Phytophthora*.

Citrullus lanatus, watermelon.

Citrus, *C. aurantifolia*, lime; *C. aurantium*, sour orange; *C. bergamia*, bergamot; *C. grandis*, pummelo; *C. limon*, lemon; *C. medica*, citron; *C. natsudaidai*, Japanese summer grapefruit; *C. paradisi*, grapefruit; *C. reticulata*, clementine, mandarin, satsuma, tangerine; *C. sinensis*, sweet orange. Hybrids include: citrange is sweet orange X *Poncirus*; orangelo is sweet orange X grapefruit; tangelo is mandarin X grapefruit; tangor is sweet orange X mandarin. Knorr 1973 lists citrus common names with the 2 binomial systems of Swingle and Tanaka; Hodgson proposed a compromise between them, see Simmonds 1976; Fawcett, *Citrus diseases and their control*, 1936; Knorr et al., *Bull.Univ.Fla.agric.Exp.Stns.* 587, 1957; Knorr 1973; *PANS* **19**:441, 1973, bibliography; Eckert, *Outl.Agric.* **9**:225, 1978, postharvest; Reuther et al. 1978; Raychaudhuri & Ahlawat, *Problems of citrus diseases in India* 1982; Salibe, *F.A.O.Pl.Prot.Bull.* **34**:49, 1986, viruses, Mediterranean.

citrus abnormal bud union McLean, *Proc. 6th Int.Conf.Organis.Citrus Virologists*:203, 1974; one of the bud union disorders which occur when some sweet orange cvs. are budded on rough lemon stock.

—— **Algerian navel orange v.** Garnsey, *PDR* **59**:689, 1975; rod 780 nm long, transm. sap to herbaceous plants [**55**,2719].

—— **blastomania** Mali, Choudhari & Rane, *Curr.Sci.* **44**:627, 1975; poss. MLO, India (Shirampur), Rangpur lime.

—— **blight**, the name usually used for a dieback and wilt in USA (Florida), its cause is unknown but its history goes back to the last century; losses of trees can still be high. Tucker et al., *PD* **68**:979, 1984, reported experimental transm.; Knorr 1973; Burnett et al., *Trop.Pest Management* **28**:416, 1982; Yokomi et al. in Garnsey et al. 1984 [**64**,1118].

—— **blind pocket** = citrus psorosis.

—— **blotchy mottle** McClean & Schwarz, *Phytophylactica* **2**:177, 1970; poss. a form of citrus greening [**51**,1486].

—— **bud union crease** Martinez 1970; similar to citrus abnormal bud union [**53**,557].

—— —— —— **disorder** Miyakawa, *Rev.Pl.Prot.Res.* **11**:1, 1978; prob. the same as citrus tatter leaf citrange stunt.

— **cachexia** = citrus xyloporosis.

— **concave gum** Roistacher & Calavan 1965; a form of citrus psorosis [45,1790e].

— **crinkly leaf** v. Fawcett 1936; Fraser 1961; Yot-Dauthy & Bové, *Proc.4th Int.Conf.Organis.Citrus Virologists* 1968; poss. Ilarvirus, isometric *c*.27 nm diam., transm. sap, similar to citrus infectious variegation v. and citrus psorosis A.

— **cristacortis** Vogel & Bové, *Fruits* 19:269, 1964; Vogel, *ibid* 29:417, 589, 645, 1974; no known natural spread, Mediterranean [43,2907; 54,2815].

— **decline**, or dieback; Garnsey et al. 1984 for citrus declines; for India, Naidu & Govindu in Maramorosch & Raychaudhuri 1981; citrus decline is preferably not used as a name for citrus blight q.v.

*— **exocortis viroid** Fawcett & Klotz 1948; Benton et al. 1950; Diener 1979; Semancik, *Descr.V.* 226, 1980; naked ssRNA, transm. sap readily, spread on contaminated tools which need sterilisation; bark scaly butt, shelling, splitting; susceptible cvs. include: some citranges, Eureka lemon, Rangpur and sweet limes, and trifoliate orange; Baksh et al. in Garnsey et al. 1984; see citrus variable viroid.

— **Florida scaly bark** = citrus psorosis A.

— **greening** van der Merwe & Andersen 1937; Wallace in Reuther et al. 1978; Garnier & Bové, *Phytop.* 73:1358, 1983; Bové, *F.A.O.Pl.Prot.Bull.* 34:7, 1986; phloem limited, Gram negative bacterium, transm. *Diaphorina citri*, *Trioza erytreae*, by dodder from sweet orange to periwinkle; Asian and South African forms. The disease may be the same as others in citrus: decline or dieback in India, likubin in Taiwan, yellow shoot and Wentan decline in China [65,6045].

— **gum pocket** Schwarz & McClean 1969; Schwarz 1971; gummy pitting is similar [48,2386; 51,3358]. .

— **hassaku dwarf** (citrus tristeza v.), a severe seedling yellows; Sasaki, *Rev.Pl.Prot.Res.* 12:80, 1979, on *Citrus hassaku*, protection by a mild str. [60,2032].

— **impietratura** Ruggieri 1955; Catara et al., *Proc.Int.Soc.Citriculture* 3:946, 1979, Mediterranean [59,5169].

— **infectious variegation** v. Fawcett & Klotz, *Phytop.* 29:911, 1939; Fulton in Kurstak 1981; Ilarvirus, isometric 28–33 nm diam., transm. sap, serologically like citrus leaf rugose v., and see citrus crinkly leaf [19:85].

— **inverse pitting** Reichert & Wincour, *Phytop.* 46:527, 1956 = citrus xyloporosis [36:317].

— **leaf mottle yellows** Salibe & Cortez,

F.A.O.Pl.Prot.Bull. 14:141, 1966; a form of citrus greening [46,2230].

— — **rugose** v. Garnsey 1975; Garnsey & Gonsalves, *Descr.V.* 164, 1976; Fulton in Kurstak 1981; Ilarvirus, isometric 25–32 nm diam., transm. sap readily, USA (Florida), tools contaminated; severe stunt in young Duncan grapefruit, leaf fleck in Eureka lemon, leaf rugosity in Mexican lime.

— — **yellow midvein** Sharma & Padney, *Int.J.Trop.Pl.Dis.* 1:149, 1983; poss. a form of citrus psorosis A, lime [64,3442].

— **leathery leaf** Ahlawat, Nariani & Sardar, *Indian Phytopath.* 32:198, 1979; transm. sap, *Aphis gossypii*, orange [60,2597].

— **lepra explosiva** = citrus leprosis.

— **leprosis**, see Knorr 1973; Chagas et al.; Chiavegato & Salibe in Garnsey et al. 1984; transm. *Brevipalpus phoenicis*; Kitajima et al., *Virology* 50:254, 1972, short rod assoc., poss. bacilliform particle. This is an old disease; for a long time it was assoc. with mite damage; it is severe in parts of S. America; large, rust brown, cracked spots are caused on the fruit, leaf spots, bark scaling which should not be confused with citrus porosis A bark scaling [52,1143].

— **likubin** = citrus greening.

— **mosaic** v. Ishigai & Kajitsu 1958; Yamada & Tanaka, *JARQ* 3:10, 1968; Tanaka & Imada 1976; isometric 27 nm diam., transm. sap, Japan; Ahlawat et al., *Curr.Sci.* 54:873, 1985, transm. aphid spp., India [48,2378; 56,5046; 65,1300].

— **multiple sprouting** Schwarz, *PDR* 54:1003, 1970; transm. sap; South Africa, Zimbabwe [51,2466].

— **nailhead rust** = citrus leprosis.

— **podagra**, see Knorr 1973; USA (Florida), Meiwa and Nagami kumquats budded on rough lemon, overgrowth and scaling of stock, stunting of top.

— **psorosis A** Swingle & Webber 1896; see Wallace in Reuther et al. 1978; spread through budwood, poss. some seed transm., all types show a characteristic vein banding chlorosis in immature leaves, fleck and oak leaf patterns; several disease types caused e.g. blind pocket, concave gum.

— **quick decline** = citrus tristeza v.

— **ringspot** Càtara & Grasso 1968; Wallace & Drake 1968; Garnsey, *PDR* 59:689, 1975; transm. sap, Algeria, Italy (Sicily), USA (California); Timmer, *Phytop.* 64:389, 1974; Timmer & Garnsey, *ibid* 69:200, 1979; necrotic str., relationship with citrus psorosis, hosts [48,1696, 2375f; 54,150; 55,2719; 59,2760].

— **root nematode**, *Tylenchulus semipenetrans*.

— **rubbery wood** Ahlawat & Chenulu, *Curr.Sci.*

54:580, 1985; India (Assam, Rajasthan), lemon, lime [**64**,5396].
— **sandhill decline** = citrus blight.
— **Satsuma dwarf v.** Yamada & Sawamura 1950; Usugi & Saito, *Descr.V.* 208, 1979; poss. Nepovirus, isometric *c*.26 nm diam., transm. sap, difficult, relatively wide host range; Japan, Turkey; serologically related to citrus natsudaidai dwarf, citrus navel orange infectious mottling and mosaic.
— **seedling yellows** (citrus tristeza v.), Roistacher et al., *Citrograph* **65**:167, 1979, natural spread; see citrus hassaku dwarf [**58**,5364].
— **stem pitting**, assoc. mostly with citrus tristeza v.; the early stages are detected when the bark is removed. Pits may be in the outer wood and the corresponding pegs or projections are on the inner face of the bark. Alternatively there are projections like teeth which form on the cambial face of the wood; these fit into minute holes in the corresponding face of the bark.
— **stubborn** (*Spiroplasma citri*).
— **tatter leaf citrange stunt complex** Wallace & Drake 1962; in Meyer lemon; a virus(es) etiology is apparently assumed, it is not known whether the pathogen(s) consists of 1 or 2 components; Wallace in Reuther et al. 1978; Roistacher, *Proc. Int.Soc.Citriculture* **1**:430, 1981, reviews [**41**:655].
— **tristeza v.** Fawcett & Wallace 1946; Meneghini 1946; Price, *Descr.V.* 33:1970; Closterovirus, filament *c*.2000 × 12 nm, transm. aphid spp., most efficiently by *Toxoptera citricidus*; but see Roistacher et al., *PD* **68**:494, 1984, for efficient transm. by *Aphis gossypii*; also spread through knife cuts. The disease caused, quick decline, is mainly one of citrus cvs. budded on sour orange, particularly of sweet orange, 3 forms: phloem necrosis at the bud union, seedling yellows q.v. and stem pitting q.v. The virus, prob. origin China, was spread worldwide in plants and budwood. The first major disease outbreak was in Argentina in 1930; *c*.30[6] trees died in S. America, *c*.3[6] in USA (California) and millions in Spain. Wallace in Reuther et al. 1978; Raccah et al. in Scott & Bainbridge 1978; Costa & Muller, *PD* **64**:538, 1980, control in Brazil by cross protection with mild strs.; Bar-Joseph et al., *Proc.Int.Soc.Citriculture* **1**:419, 1981; in Plumb & Thresh 1983; *Phytop.* **75**:195, 1985 [**63**,5433; **64**,3074].
— **variable viroid** Schlemmer, Roistacher & Semancik, *Phytop.* **75**:946, 1985; pathogenic RNA, poss. new viroid, poss. citrus exocortis viroid is not the only causal agent of exocortis [**65**,718].
— **variegation,** see citrus infectious variegation.

— **veinal chlorosis** Sawant, Choudhari & Desai, *Indian J.Mycol. Pl.Path* **13**:346, 1983, published 1985; transm. seed [**65**,2324].
— **vein enation** Wallace & Drake 1953; *PDR* **44**:580, 1960; McClean 1954; Fraser, *Proc.Linn.Soc.N.S.W.* **84**:332, 1959; Laird & Weathers, *PDR* **45**:877, 1961; transm. *Aphis gossypii, Toxoptera citricidus* [**32**:622; **39**:576; **40**:105; **41**:304].
— — **phloem degeneration** Tirtawidjaja et al. 1965; = citrus greening [**44**,3356].
— **Wentan decline** Huang & Chang 1980; = citrus greening [**60**,314].
— **witches' broom** J.M. Bové at British Society for Plant Pathology, December 1986, unpublished; MLO restricted to phloem assoc., a serious disease of lime in Oman.
— **woody gall** = citrus vein enation.
— **X** Franciosi, Cardich & Puiggros, *Proc.Int.Soc.Citriculture* **3**:897, 1979; Peru, navel orange, small chlorotic leaves, upright bushy growth, small crop of misshapen fruit [**59**,5156].
— **xyloporosis** Reichert & Perlberger, *Bull.Rehovot agric.Exp.Stn.* 12, 1934; Wallace in Reuther et al. 1978; Roistacher et al., *Citrograph* **68**:111, 1983, gave evidence for a viroid etiology [**14**:162; **63**,3366].
— **yellow mottle v.** Ushiyama, Usugi & Hibino in Garnsey et al. 1984; rod mostly 690–740 × 12–14 nm, transm. knife cut, Japan, Satsuma mandarin.
— — **shoot**, a form of citrus greening.
— — **vein** Weathers, *PDR* **41**:741, 1957; *Virology* **11**:753, 1960; *Nature Lond.* **200**:812, 1963; USA (California), naturally in limequat [**37**:234; **40**:223; **43**,1330].
— **young tree decline** = citrus blight.
Cladobotryum Nees 1816; Hyphom.; conidiomata hyphal, verticilliate branching; conidiogenesis enteroblastic phialidic; con. multiseptate, hyaline, in heads or irregular chains, chlamydospores or sclerotia; teleom. *Hypomyces*; Gams & Hoozemans, *Persoonia* **6**:95, 1970.
C. amazonense Bastos, Evans & Samson, *TBMS* **77**:274, 1981; con. 2–4 celled, smooth, 25–46 × 9–12 μm, mycoparasite of *Crinipellis perniciosa* which causes cacao witches' broom, on other Agaricaceae in Amazonian forests [**61**,2782].
C. dendroides, teleom. *Hypomyces rosellus*; **Cladobotryum variospermum**, teleom. *H. aurantius*.
cladosporioides, *Cercospora, Cladosporium*.
Cladosporium Link 1815; Hyphom.; Ellis 1971, 1976; conidiomata hyphal, conidiogenesis holoblastic, conidial development acropetal; con. mostly aseptate but can be > 3 septate, pigmented,

usually with a protruberant scar at each end;
teleom. include *Mycosphaerella* and *Venturia. C.
cladosporioides* (Fres.) de Vries 1952 and
C. herbarum, teleom. *M. tassiana*, are
cosmopolitan and frequently secondary invaders;
the spp. are often saprophytic and relatively
unimportant in causing leaf spots and fruit rots in
many crops; Holliday 1980.

C. allii (Ell. & G.W. Martin) P. Kirk & Crompton,
Pl.Path. 33:320, 1984; mostly 1–2 septate,
28–42 × 12–16 µm; leek leaf blotch, does not infect
onion; *Mycosphaerella allii-cepae* q.v., *Descr.F.*
842, 1986 [63,5692].

C. allii-cepae, teleom. *Mycosphaerella allii-cepae.*

C. carpophilum, teleom. *Venturia carpophila.*

C. caryigenum (Ell. & Lang.) Gottwald, *Mycologia*
74:388, 1982; con. 10–40 × 4.5–10 µm, stroma
present; pecan scab is very damaging in USA,
especially in the S.E.; the leaves are most
susceptible 7–21 days after bud break, con.
dispersal peaks at noon; Gottwald & Bertrand
and Gottwald, *Phytop.* 72:330, 1193, 1982;
75:190, 1985; epidemiology & infection; Latham,
ibid 72:1339, 1982 conidial dispersal & weather
[61,5980, 6613; 62,1225, 2166; 64,3209].

C. cucumerinum Ell. & Arthur 1889; Ellis &
Holliday, *Descr.F.* 348, 1972; con. mostly
aseptate, occasionally 2 celled, 4–9 × 3–5 µm;
cucurbit scab or gummosis; Dixon 1981; Sherf &
MacNab 1986.

C. echinulatum, teleom. *Mycosphaerella dianthi;*
Cladosporium herbarum, teleom. *M. tassiana.*

C. musae Mason in Martyn, *Mycol.Pap.* 13:2,
1945; con. 0–1 septate, 6–22 × 3–5 µm; banana
speckle on leaves; Stover, *TBMS* 65:328, 1975
[25:220; 55,1885].

*__C. phlei__ (C.T. Gregory) de Vries 1952; Ellis 1976;
con. 1–3 septate, 15–36 × 7–14 µm; *Phleum*
eyespot, seedborne, can be damaging; Sundheim
& Aarvold, *Meld.Norg.Landbrtløsk* 48(26):1,
1969, inoculations; Mino et al., *Ann.
phytopath.Soc.Japan* 45:463, 1979; Shimanuki &
Araki, *J.Japanese Soc.Grassland Sci.* 28:426, 1983,
toxin phleichrome [46,1406; 59,4661; 63,701].

C. vignae Gardner, *Phytop.* 15:457, 1925; con.
7–22 × 3–5 µm; cowpea scab, seedborne; Strider &
Toler, *PDR* 47:493, 1963, screening for resistance
[5:76; 43,305].

clamp connection, the clamp that is characteristic of
many basidiomycetes is a hyphal outgrowth
which, at cell division and septum formation,
makes a connection between the resulting 2 cells.

—— **rot** (*Botryotina fuckeliana*) beet.

clandestina, *Phytophthora, Podosphaera.*

classification, see taxonomy.

Clavariaceae, Aphyllophorales; erect basidiomata,

like clubs, often branched, surface smooth;
Corner, *A monograph of Clavaria and allied
genera,* 1950; *Typhula.*

clavariiforme, *Gymnosporangium;* **clavatum**,
Cylindrocladium.

Clavibacter Davis et al. 1984; Gram positive,
pleomorphic rod, not motile or acid fast, obligate
aerobes; Fahy & Persley 1983, as
Corynebacterium; Bradbury 1986 [63,4246].

C. iranicus (Carlson & Vidaver) Davis et al.; wheat
gumming, Iran.

C. michiganensis (Smith) Davis et al.; 5 ssp. but
these may be distinct enough to be separate spp.

C.m.ssp. insidiosus, Hayward & Waterston,
Descr.B. 13, 1964 as *Corynebacterium insidiosum*
(McCulloch) Jensen 1934; lucerne wilt and other
Medicago spp., Graham et al. 1979.

C.m.ssp. michiganensis, Hayward & Waterston,
Descr.B. 19, 1964 as *Corynebacterium
michiganense* (Smith) Jensen; tomato canker, also
causes a vascular wilt, leaf and fruit spots;
Latterrot et al., *Annals.Amél.Pl.* 28:579, 1978,
review on resistance, Jarvis, *Ann.Rep.Glasshouse
Crops Res.Inst.* for 1979:172, 1981 [58,5026;
60,6654].

C.m.ssp. nebraskensis, maize leaf freckles and wilt
or Goss's wilt and blight, USA (central); Shurtleff
1980; Smidt & Vidaver, *PD* 70:1031, 1986,
population dynamics [66,2347].

C.m.ssp. sepedonicus, Hayward & Waterston,
Descr.B. 14, 1964 as *Corynebacterium sepedonicum*
(Spieck. & Kotth.) Skaptason & Burkholder; ring
rot of potato, causing a vascular wilt and tuber
rot, one of the most serious diseases of the crop
especially of seed potatoes, Australia is still free
from the bacterium; Rich 1983; de Boer & Slack,
PD 68:841, 1984, detection & control in N.
America.

C.m.ssp. tessellarius, wheat leaf mosaic USA
(central); Carlson & Vidaver, *PD* 66:76, 1982
[61,4864].

C. rathayi (Smith) Davis et al.; Bradbury, *Descr.B.*
376, 1973 as *Corynebacterium rathayi* (Smith)
Dowson; cocksfoot yellow slime, upper parts of
plants are dwarfed and distorted; see *Anguina
agrostis.*

C. tritici (Carlson & Vidaver) Davis et al.;
Bradbury, *Descr.B.* 377, 1973 as *Corynebacterium
tritici* (Hutchinson) Burkholder; wheat gumming,
a limited distribution and absent from Europe and
N. America. *Anguina tritici* is the vector and
elimination of its galls from seed prevents the
disease.

C. xyli ssp. cynodontis Davis et al.; Bermuda grass
stunt, Taiwan, USA (Florida); Davis et al.,
Phytop 73:341, 1983, found that the disease

caused by this bacterium, which is limited to the xylem, became more severe when the MLO, which may cause Bermuda grass white leaf, was present; Davis & Augustin, *PD* **68**:1095, 1984, reported the presence of the bacterium in Florida [**62**,3083; **64**,2061].

C.x. ssp. xyli Davis et al.; sugarcane ratoon stunt, mostly with the crop. This disease, first described from Australia in 1950, was for many years thought to be due to virus. But *c*.25 years later a coryneform bacterium, which is present in the xylem only, was shown to cause the disease; Teakle in Fahy & Persley 1983; Damann & Benda, *PD* **67**:966, 1983, control by heat treatment; Sivanesan & Waller 1986 [**63**,850].

claviceps, *Balansia*.

Claviceps Tul. 1853; Clavicipitaceae; the stalked ascomata, with a globose head, are borne on an elongated blackish sclerotium; parasitic on the inflorescences of Cyperaceae, Gramineae and Juncaceae; the sclerotium, called an ergot, forms in the floral structures and contaminates the grain at harvest. It may contain alkaloids which are toxic to animals including man and cause the diseases called ergotism, see *C. purpurea*. The anam. *Sphacelia* forms a secretion, called honey dew, attracting insects which disperse the con., hence the term sugary disease in plants; Holliday 1980.

C. fusiformis Loveless, *TBMS* **50**:17, 1967; Thakur et al., *Phytop.* **74**:201, 1984; Chahal et al., *TBMS* **84**:325, 1985; asco. av. 128 × 4 μm, macroconidia 20.5–26.5 × 4.1–6.2 μm, microconidia 5.8–7.5 × 4–5.5μm; sclerotia 3.1–5.1 × 1.9–2.5 mm; causes a major disease of pearl millet, particularly in India, and see *Claviceps microcephala*; Thakur et al., *Phytop.* **70**:80, 1980; **72**:406, 1982; *AAB* **103**:31, 1983, pollination effects & control through resistance & pollen management; Prakash et al., *TBMS* **81**:65, 1983, spore germination & nuclear behaviour; Willingale & Mantle, *Ann.Bot.* **56**:109, 1985, stigma constriction & its effects on reproduction & disease [**59**,5123; **61**,5701; **62**,4893; **63**,135, 2861; **64**,2508, 4289].

C. gigantea Fuentes et al., *Phytop.* **54**:381, 1964; Ullstrup, *PANS* **19**:389, 1973; asco. 176–186 × 1.5 μm, macroconidia 8–27 × 4–6 μm; sclerotia variable in size, up to 8 × 5 cm; maize in Mexico [**43**,2599; **53**,1352].

C. maximensis Theis, *Mycologia* **44**:792, 1952; Ryley, *Australasian Pl.Path.* **10**:37, 1981; asco. 95–126 × 0.5 μm, sclerotia 2–9 × 1 mm; *Panicum maximum* [**32**:487; **61**,2854].

C. microcephala (Wallr.) Tul. 1853; this fungus has been described as an ergot of pearl millet in India, but its position on this host does not seem to be

clear, see Loveless, *Claviceps fusiformis*; Petch in 1937 regarded *C. microcephala* as a form of *C. purpurea*. But Thirumalachar in 1945, describing it from *Pennisetum hohenackeri*, gave the size of the con. as 17–24 × 3–7 μm, larger than those of *C. purpurea*; Siddiqui & Khan, *Trans.mycol.Soc.Japan* **14**:195, 1973, considered that *C. fusiformis* and *C. microcephala* were the same; Kulkarni, *Indian Phytopath.* **20**:139, 1967, viability of sclerotia; Siddiqui & Khan, *Trans. mycol. Soc.Japan* **14**:280, 1973, infection; Gupta et al., *Trop.Pest Management* **29**:321, 1983, weather & disease incidence; Thakur in Raychaudhuri & Verma 1984 [**25**:167; **47**,1863; **53**,1357, 3005; **63**,2303; **64**,1530].

C. oryzae-sativae Hashioka, said to be the teleom. of *Ustilaginoidea virens* but confused.

C. paspali F.Stev. & Hall 1910; Sprague 1950; asco. 101 × 0.5–1 μm, sclerotia *c*.3 mm diam.; *Paspalum* spp., particularly the forage *P. dilatum*, dallis grass; Cunfer et al., *Mycologia* **69**:1137, 1142, 1977, sclerotial germination & survival; Luttrell, *Phytop.* **67**:1461, 1977, life history [**57**,3795–6, 4514].

C. purpurea (Fr.) Tul. 1853; Sprague 1950; Sampson & Western 1954; asco. 50–76 × 0.6–0.7 μm, con. 4–6 × 2–3 μm, sclerotia 2–25 mm long. This fungus affords an early example of host specialisation due to the work of Stäger in 1903 and 1905. Loveless & Peach, *TBMS* **86**:603, 1986, examined asco. length, they found 3 groups and considered that this was morphological evidence that strs. of *Claviceps purpurea* are host restricted; on temperate cereals and grasses. The toxic ergot or sclerotium of rye, which contaminated bread in the Middle Ages, was from *C. purpurea*; it caused the disease in man known as ergotism, or St Anthony's fire in the fatal gangrenous form. Outbreaks of ergotism continued into the late 19th century in Finland, France, Germany and Russia; Fuller, *The day of St Anthony's fire*, 1969, described a French outbreak in 1951; for ergotism see also: Barger, *Ergot and ergotism*, 1931; Bové, *The story of ergot*, 1970; for the disease in plants: Butler & Jones 1949; Walker 1969; Mitchell & Cooke, *TBMS* **51**:721, 1968, effect of temp. on germination & longevity of sclerotia; Mantle & Shaw, *Pl.Path.* **26**:121, 1977, etiology & strs. [**48**,784; **57**,2873; **65**,6085].

C. sorghi P. Kulkarni, Seshadri & Hegde, *Mysore J.agric.Sci.* **10**:288, 1976; anam. *Sphacelia sorghi* McRae 1917; con. av. 16–17.5 × 4–6 μm, sclerotia 10–25 × 4–6 mm, sorghum ergot can be severe in India [**56**,2484].

Clavicipitaceae, Clavicipitales; asco. filiform,

multiseptate; *Balansia, Claviceps, Epichloë, Myriogenospora.*

Clavicipitales, Ascomycotina; ascomata fleshy, mostly brightly coloured, perithecioid, asci cylindrical, unitunicate, with a thick apical cap pierced by a narrow pore, asco. long filiform or fusiform, often multiseptate and fragmenting, some on Gramineae.

clavigignenti-juglandacearum, *Sirococcus.*

clavipes, *Gymnosporangium.*

clavus, *Kretzschmaria.*

claytonia mosaic v. Roland & Zenon-Roland, *Parasitica* 28:46, 1972; Potyvirus, rod av. 728 nm long, from *Stellaria media,* Belgium [52,3256].

cleistocarp, a fungus sporocarp with no characteristic opening e.g. an ascoma of the Erysiphaceae.

clematis ringspot and leaf distortion Thomson, *N.Z.J.Bot.* 16:167, 1978; transm. sap, New Zealand, *C. paniculata* [58,258].

—— **yellow mosaic** (tobacco streak v.), Bellardi et al., *Phytopath. Mediterranea* 24:255, 1985; *C. vitalba* [66,1023].

clementine, see *Citrus.*

clerodendron zonate ringspot Burnett & Youtsey, *PDR* 46:279, 1962; USA (Florida), *C. thomsoniae* [41:716].

Clethridium (Sacc.) Sacc. 1895; Amphisphaeriaceae; Holm, *Taxon* 24:477, 1975.

C. corticola (Fuckel) Shoem. & E. Müller, *CJB* 42:404, 1964; anam. *Seimatosporium lichenicola* (Corda) Shoem. & E. Müller; asco. 10–13 × 4–5 μm, aseptate; con. 13–15 × 5.5–6.5 μm, no appendages, see Zeller, *Mycologia* 17:33, 1925; 19:150, 1927; on woody stems of *Cornus, Prunus, Rosa, Rubus, Salix* and other genera; Ogawa et al., *Hilgardia* 28(10):239, 1959, almond leaf blight; Ferrata & D'Ambra, *Riv.Patol.veg.* 17:15, 1981, on raspberry [4:489; 38:492; 61,1291].

climate and weather, see disease warning, ecoclimate; Hepting, *A.R.Phytop.* 1:31, 1963; *Phytop.* 55:943, 1965; Bourke, *A.R.Phytop.* 8:345, 1970; Waller, *Rev.Pl.Path.* 55:185, 1976; Hatfield & Thomason ed., *Biometeorology in integrated pest management,* 1982; Pedgley, *Windborne pests and diseases. Meteorology of airborne organisms,* 1982; Sutton et al., *PD* 68:78, 1984.

clitoria mosaic B.N. & K.M. Srivastava & Singh, *Indian Phytopath.* 31:248, 1978; transm. sap, *C. ternatea* [59,1277].

—— **yellow vein v.** Bock, Guthrie & Meredith 1977; Bock & Guthrie, *Descr.V.* 171, 1977; Tymovirus, isometric *c.*28 mm diam., transm. sap, many spp. in Papilionaceae, Kenya, *C. ternatea.*

clone, of a micro-organism, isolates with the same genome; of a plant, genetically uniform

individuals originally derived from one individual by vegetative propagation, see FBPP.

Closteroviruses, type beet yellows v., *c.*14 definitive members, filament very flexuous, lengths in 3 subgroups *c.*730, 1250–1450, 1650–2000 nm, 1 segment linear ssRNA, sedimenting at *c.*96, 110 and 140S, thermal inactivation 40–55°, longevity in sap 12–48 hours, transm. sap difficult, mostly aphid transm., semi-persistent, typically narrow host ranges; Bar-Joseph et al., *Adv.Virus Res.* 25:93, 1979; Lister & Bar-Joseph in Kurstak 1981; Bar-Joseph & Murant, *Descr.V.* 260, 1982; Francki et al., vol. 2, 1985.

cloud, tomato, see blotchy ripening.

clove, *Syzygium aromaticum.*

—— **little leaf** Balasubramanian, *Indian J.agric.Sci.* 28:567, 1958; poss. MLO, India (Kerala, Tamil Nadu) [39:32].

*—— **sudden wilt,** this disease in Zanzibar and Pemba has been known for *c.*100 years; its etiology has still not been demonstrated. Dabek et al., *Phytopath.Z.* 114:180, 1985, reported an assoc. MLO; external symptoms are only seen in mature trees > 12–15 years old; there is a heavy green leaf fall and wilt. In 1949 *c.* half the clove trees in Zanzibar had been killed in the previous 10–12 years. Nutman et al., *E.Afr.agric.For.J.* 18:146, 1953, described the earlier attempts, which began in *c.*1914, to solve the problem [33:180; 65,1980].

*—— **Sumatra wilt** Hunt et al. 1979; Bennett et al. 1979; *Pl.Path.* 38:487, 1985; 36:45, 1987; Gram negative, xylem limited bacterium q.v. Deaths of clove trees in Sumatra have been observed over the past 60 years and recently there have been severe losses; leaf fall and dieback begin in the upper canopy, and death ensues in 6–8 months; trees > 10 years old become affected first, at 3–4 years an increasing proportion do not show symptoms; deaths are very rare in seedlings < 2 years old. The pattern of spread suggests that there is an aerial vector [65,1437; 66,3456].

clover, see *Trifolium*; viruses or suspected viruses see: alsike, berseem, crimson, red, subterranean, sweet, white clovers; some lists of viruses put these qualifying adjectives before the word clover, others put them after.

—— **big vein** = wound tumour v.

—— **blotch v.** Musil & Lešková, *Ochr.Rost.* 9:259, 1973; Musil et al., *Acta Virol.* 19:437, 1975; isometric *c.*28 nm diam., transm. aphid spp., non-persistent; Richter et al., *ibid* 23:489, 1979, a str. of peanut stunt v. [53,4014].

—— **Canadian yellows** Benhamou & Sinha, *Can.J.Pl.Path.* 3:191, 1981; bacterium assoc., Canada (Ontario) [61,6446].

— **club leaf** Black 1944; Windsor & Black, *Phytop.* **63**:1139, 1973; Markham et al., *AAB* **81**:91, 1975; phloem limited bacterium q.v. assoc., transm. *Agalliopsis novella*, England, USA [**23**:490; **53**,2241; **55**,298].

— **degradation** Benhamou, Giannotti & Louis, *Acta phytopath. Acad.Sci.hung.* **13**:107, 1978; bacterium assoc., France (S.) [**58**,3347].

— **Dutch enation** Bos & Grancini 1963; *Phytopath.Z.* **61**:253, 1968; Rubio-Huertos & Bos, *Neth.J.Pl.Path.* **75**:329, 1969; bacilliform particles assoc.; 2 other particles as rods, 475 and 665 nm long, assoc. with non-tumorous tissue [**47**,2515; **49**,781].

— **dwarf** Musil 1963; Valenta & Musil, *Phytopath.Z.* **47**:38, 1963; poss. MLO, Czechoslovakia [**42**:557, 750].

— **mosaic** = white clover mosaic v.

— **phyllody** Maillet, Gourret & Hamon 1968; Giannotti, *PDR* **53**: 173, 1969; Sinha & Paliwal, *Virology* **39**:759, 1969; **40**:665, 1970; Chiykowski, *Ann.Entomol.Soc.Am.* **68**:645, 1975; MLO assoc., transm. *Aphrodes bicinctus, Euscelidius variegatus, Macrosteles fascifrons* [**48**,1779; **49**,2388, 3728; **55**,5809].

— **primary leaf necrosis v.** Ragetli & Elder, *CJB* **55**:2122, 1977; prob. Dianthovirus, isometric av. 36 nm diam., transm. sap, Canada (British Columbia); Rao & Hiruki, *PD* **69**:959, 1985, a str. of red clover necrotic mosaic v. [**57**,1247; **65**,3967].

— **rugose leaf curl,** see rugose leaf curl.

— **small leaf** Chiykowski, *Can.Pl.Dis.Surv.* **49**:16, 1969; Canada, alsike and ladino clover; adventitious growth, witches' brooms, flowers reduced, transm. *Aphrodes bicinctus* [**49**,1408].

— **wound tumour** = wound tumour v.

— **yellow edge** Chiykowski 1969; *CJB* **54**:1171, 1976; *Can.J.Pl.Path.* **3**:139, 1981; MLO assoc., transm. *Aphrodes bicinctus*, Canada [**49**,1408; **55**,5807; **61**,2935].

— **yellow mosaic v.** Johnson 1942; Bos, *Descr.V.* 111, 1973; Potexvirus, filament *c.*540 nm long, transm. sap, seed; Canada, USA (W.); many spp. in Papilionaceae are susceptible, found in apple with apple leaf pucker q.v.; Rao et al., *Phytopath.Z.* **98**:260, 1980, vetch str. in Canada; Tollin et al., *J.gen.Virol.* **52**:205, 1981, diffraction patterns [**60**,3617, 5484].

— **yellows v.** Ohki, Doi & Yora, *Ann.phytopath.Soc.Japan* **42**:313, 1976; Closterovirus, filament 1700 × 12 nm, transm. *Aphis craccivora*, semi-persistent [**56**,1171].

— **yellow vein v.** Hollings & Nariani 1965; Hollings & Stone, *Descr.V.* 131, 1974; Potyvirus, filament *c.*760 nm long, transm. sap, aphid spp.,

non-persistent, causes vein clearing and mosaic in coriander, Australia, Europe, N. America, New Zealand; Lisa & Dellavale, *Phytopath.Mediterranea* **22**:49, 1983, bean leaf and stem necrosis, premature leaf and pod fall in Italy; Forster & Musgrave, *N.Z.J.agric.Res.* **28**:575, 1985, sweet pea leaf and stem necrosis, and flower break; Lawson et al., *Phytop.* **75**:899, 1985, *Limonium sinuatum* leaf strap and chlorotic mosaic [**63**,2037; **65**,749, 2711].

club root (*Plasmodiophora brassicae*) brassicas.

cluster bean, see guar.

Clypeoporthe Höhnel 1919; Gnomoniaceae; asco. narrowly fusoid to ellipsoid, 1 to several septate, on monocotyledons. *C. iliau* (Lyon) Barr 1978; Sivanesan & Holliday, *Descr.F.* 705, 1981; anam. *Phaeocytostroma iliau* (Lyon) Sivan. 1981; asco. 1 septate, hyaline, 16–21 × 3–4 μm; con. 13.5–23 × 5.8–7 μm; sugarcane iliau, symptoms near soil level, mycelium binds the leaf sheaths tightly together, the disease is now little reported.

clypeus, a fungal stromatic growth, like a shield, with or without host tissue and over an ascoma or a pycnidium.

coastal spruce cone rust (*Chrysomyxa monesis*).

cobnut, see *Corylus.*

***cobweb** (*Hypomyces rosellus*) common cultivated mushroom.

coccinia, *Nectria*; **coccodes,** *Colletotrichum.*

***Cochliobolus** Dreschsler 1934; Pleosporaceae or Pyrenophoraceae; asco. multiseptate, filiform, coiled like a helix in the ascus, minimum length mostly 80–200 μm, maximum mostly 120–420 μm, the ascoma has a well defined beak or cylindrical neck; anam. in *Bipolaris, Curvularia, Drechslera*; mostly on Gramineae; Luke & Gracen in Kadis & Ciegler ed., *Microbial toxins,* vol. 8, 1972; Holliday 1980; Alcorn, *Mycotaxon* **13**:349, 1981, ascus structure & function; Sivanesan 1984; *TBMS* **84**:548, 1985, key to spp. with *Curvularia* anam.

***C. bicolor** Paul & Parbery, *TBMS* **49**:386, 1966; anam. *Bipolaris bicolor* (Mitra) Shoem. 1959; con. mostly 40–80 × 14–18 μm; Muchovej, *PD* **64**:1035, 1980, on *Pennisetum ciliare* & *P. clandestinum* [**60**,4427].

C. carbonum Nelson, *Phytop.* **49**:809, 1949; anam. *Drechslera carbonum* (Ullstrup) Sivan. 1984; Ellis & Holliday, *Descr.F.* 349, 1972 as *D. zeicola* (Stout) Subram. & Jain 1966; con. mostly 60–80 × 7–8 μm; maize leaf spot, seedborne, forms a host specific toxin; Hamid et al., *Phytop.* **72**:1166, 1169, 1173, 1982, race 3 [**62**,1016–18].

C. cymbopogonis Hall & Sivan. 1972; Sivanesan & Holliday, *Descr.F.* 726, 1982; anam. *Curvularia cymbopogonis* (Dodge) Groves & Skolko 1945;

con. 28–50 × 12–20 μm; seedling blights of citronella and lemon grasses, cereals and other Gramineae; El Shafie & Webster, *TBMS* 75:141, 1980, asco. liberation [60,2421].

C. cynodontis Nelson, *Mycologia* 56:67, 1964; anam. *Bipolaris cynodontis* (Marignoni) Shoem. 1959; con. 30–75 × 10–16 μm; Bermuda grass leaf spot and on other crops; forms a toxin bipolaroxin q.v.; Tsuda & Ueyama, *Trans. mycol.Soc.Japan* 22:293, 1981 [61,5567].

C. geniculatus Nelson 1964; anam. *Curvularia geniculata* (Tracey & Earle) Boedijn 1933; Sivanesan & Holliday, *Descr.F.* 727, 1982; con. 26–48 × 8–13 μm; seedling blights, plurivorous, seedborne.

C. hawaiiensis Alcorn 1978; anam. *Bipolaris hawaiiensis* (Bugnic. & M.B. Ellis) Uchida & Aragaki, *Phytop.* 69:1115, 1979; Sivanesan & Holliday, *Descr.F.* 728, 1982; con. 12–37 × 5–11 μm; seedling blights, plurivorous.

C. heterostrophus (Drechsler) Drechsler, *Phytop.* 24:973, 1934; anam. *Bipolaris maydis* (Nisikado) Shoem. 1959; Ellis & Holliday, *Descr.F.* 301, 1971 as *Drechslera maydis* (Nisikado) Subram. & Jain 1966; con. 70–160 × 15–20 μm; maize southern leaf blight, seedborne. The disease aroused concern when race T caused extensive damage to the USA maize crop in 1970, Ullstrup, *A.R.Phytop.* 10:37, 1972. This race, first reported in the Philippines in 1961, attacked maize lines which contained Texas male sterile cytoplasm, 85% of hybrid seed was of this type. The epidemic led to a reassessment of the genetic vulnerability of major crops. T toxins q.v. from race T are specific to Tcms cytoplasm but those from race 0 are not. Shurtleff 1980; Tegtmeier et al., *Phytop.* 72:1492, 1982, T toxin formation by near isogenic isolates; Bekele & Sumner, *PD* 67:738, 1983, epidemiology in continuous maize culture [62,1486, 4888].

C. lunatus Nelson & Haasis 1964; anam. *Curvularia lunata* (Wakker) Boedijn 1933; Ellis & Gibson, *Descr.F.* 474, 1975; con. 20–30 × 9–15 μm; on angiosperms, particularly in the tropics, reduces seed germination and causes seedling blights, assoc. with discolouration of cereal grains; Muchovej, *AAB* 109:249, 1986, leaf health & infection in *Agrostis palustris* [66,2408].

C. miyabeanus (Ito & Kurib.) Drechsler ex Dastur 1942; see *Ann. Phytopath.Soc.Japan* 2:7, 1927 as *Ophiobolus*; anam. *Bipolaris oryzae* (Breda de Haan) Shoem. 1959; Ellis & Holliday, *Descr.F.* 302, 1971 as *Drechslera oryzae* (Breda de Haan) Subram. & Jain 1966; con. 63–153 × 14–22 μm; rice brown spot, seedborne. The disease was the main cause of the Bengal famine q.v.; under optimum weather conditions for infection brown spot can be extremely severe; most initial primary infection arises from infested seed and the disease often appears as a seedling blight; poorly drained and nutritionally unbalanced soils increase the likelihood of brown spot; Cholil & de Hoog, *TBMS* 79:491, 1982, morphological variability of the conidium; Ou 1985; Vidhyasekaran et al., *Phytop.* 76:261, 1986, host specific toxin [62,1949; 65,4924].

C. nodulosus Luttr. 1957; anam. *Bipolaris nodulosa* (Berk. & M.A. Curtis) Shoem. 1959; Ellis & Holliday, *Descr.F.* 341, 1972 as *Drechslera nodulosa* (Berk. & M.A. Curtis) Subram. & Jain 1966; con. mostly 50–65 × 16–19 μm; finger millet leaf blight, seedborne.

C. sativus (Ito & Kurib.) Drechsler ex Dastur 1942; anam. *Bipolaris sorokiniana* (Sacc.) Shoem. 1959; Sivanesan & Holliday, *Descr.F.* 701, 1981 as *Drechslera sorokiniana* (Sacc.) Subram. & Jain 1966; con. mostly 60–100 × 17–28 μm; foot and root rot, or spot blotch of temperate cereals, soil- and seedborne, con. air dispersed, the diseases are significant mainly in Australia, N. America and New Zealand; they are relatively unimportant in Europe, but see Kurppa 1985 for Finland; Harding 1980, 1981, bibliographies; Kidambi et al., *Can.J.Pl.Path.* 7:233, 238, 1985, epidemiology & effects on yield of spring barley in USA (Utah) [60,721; 61,3346; 64,4263–5; 65,1219–20].

C. setariae (Ito & Kurib.) Drechsler ex Dastur 1942; anam. *Bipolaris setariae* (Saw.) Shoem. 1959; Ellis & Gibson, *Descr.F.* 473, 1975 as *Drechslera setariae* (Saw.) Subram. & Jain 1966; con. mostly 50–70 × 10–15 μm; seedling blight and leaf spot of millets, mainly *Setaria italica*, seedborne, Shetty et al., *TBMS* 78:170, 1982; in seed of pearl millet [61,4101].

C. spicifer Nelson 1964; anam. *Bipolaris spicifera* (Bainier) Shoem. 1959; Sivanesan & Holliday, *Descr.F.* 702, 1981 as *Drechslera spicifera* (Bainier) v. Arx; con. mostly 30–36 × 9–14 μm; assoc. with Bermuda grass spring dead spot, seedborne.

C. victoriae Nelson 1960; anam. *Bipolaris victoriae* (Meehan & Murphy) Shoem. 1959; Sivanesan & Holliday, *Descr.F.* 703, 1981 as *Drechslera victoriae* (Meehan & Murphy) Subram. & Jain 1966; con. 40–120 × 12–19 μm, paler and narrower than those of *Cochliobolus sativus*; oat Victoria blight, Gramineae, seedborne, forms a host specific toxin victorin q.v.

cochlioides, *Aphanomyces*.

cock's eye (*Mycena citricolor*) coffee.

cocksfoot, *Dactylis glomerata*.

—— **mild mosaic** v. Huth & Paul, *Descr.V.* 107, 1972; now considered a str. of phleum mottle v.;

Torrance & Harrison, *AAB* **97**:285, 1981, Scottish isolates compared with others, RNA components [**60**,5983].

— **mosaic** = cocksfoot streak v.

— **mottle v.** Serjeant 1963; Catherall, *Descr.V.* 23, 1970; poss. Sobemovirus, isometric *c*.30 nm diam., transm. sap, *Oulema lichenis, O. melanopus*; England, New Zealand; Mohamed & Mossop, *J.gen.Virol.* **55**:63, 1981, comparison with lolium mottle v.; Toriyama, *Ann.phytopath.Soc.Japan* **48**:514, 1982; Rabenstein & Stanarius, *Arch.Phytopath.PflSchutz.* **20**:15, 1984, comparisons with other isometric viruses in Gramineae [**61**,2924; **62**,2023; **63**,5463].

— **necrosis and mosaic** Hariri & Lapierre, *Annls.Phytopath.* **9**:281, 1977; 1979; France [**58**,1098, 5760].

— **streak v.** Storey 1952; Catherall, *Descr.V.* 59, 1971; Potyvirus, filament *c*.752 × 13 nm, transm. sap, aphid, non-persistent, Europe, in a few grass spp.

cocoa, see cacao.

coconut, *Cocos nucifera*; Childs, *Coconuts*, edn. 2, 1974; for diseases prob. caused by MLO see palm lethal yellowing, and protozoa see *Phytomonas*.

— **cadang-cadang viroid** Ocfemia 1937; Randles determined the etiology in 1975; Randles & Imperial, *Descr.V.* 287, 1984; circular or linear ssRNA, basic size of 246 or 247 nucleotides as monomeric and dimeric forms, transm. mechanically using nucleic acid, very low seed transm.; in central Philippines, the infected zone is *c*.600 × 300 km, and Guam; in Palmae, lethal in coconut, *c*.30⁶ palms killed, the area of distribution has increased very little in the last 26 years but epidemics arise and decline within the infected zone. The eradication measures have largely failed prob. because early diagnosis is apparently not yet possible. The symptoms are a yellow brown in the lower leaf crown which becomes smaller and dies in 8–16 years after symptoms appear; palms < 10 years old are very rarely affected; disease incidence increases nearly linearly up to *c*.40 years old, thereafter remaining constant. Naturally infected oil palms show a general chlorosis of younger fronds and translucent orange spots, inflorescence formation ceases; other inoculated palms show a chlorotic or orange spotting. Tinangaja q.v. of coconut is a similar disease in Guam and the causal viroid is prob. a str. of coconut cadang-cadang viroid. This disease is the most economically important one of those caused by viroids; Randles in Maramorosch & McKelvey 1985; Mohamed et al., *Phytop.* **75**:79, 1985; Imperial et al., *Pl.Path.* **34**:391, 1985 [**64**,2679; **65**,329].

— **foliar decay**, see foliar decay.

cocophilus, *Marasmiellus*.

Cocos nucifera, coconut palm.

cocoyam, *Colocasia esculenta, Xanthosoma* spp., see taro.

Codinaea Maire 1937; Hyphom.; conidiomata synnematal, conidiogenesis enteroblastic phialidic; con. aseptate, hyaline; *C. fertilis* S. Hughes & Kendrick, *N.Z.J.Bot.* **6**:323, 1968, descriptions of genus & spp. inter alia; clover root rot in New Zealand and USA; Menzies, *N.Z.J.agric.Res.* **16**:239, 1973; Campbell, *PD* **64**:959, 1980; *Phytop.* **72**:501, 1982, hosts & in vitro growth; Campbell & Moyer, *PD* **67**:70, 1983, with clover yellow vein v., effects on growth [**52**,4120; **60**,3822; **61**,6449; **62**,2024].

Coelidia indica, transm. sandal spike; *C. olitoria*, transm. strawberry pallidosis.

Coelomycetes, Deuteromycotina; the so-called fungi imperfecti that, traditionally, are divided into the Melanconiales with acervular conidiomata and the Sphaeropsidales with pycnidial conidiomata. An alternative system where differences in conidiogenesis are also used has been described; Sutton 1980.

coerulescens, *Ceratocystis*.

Coffea, coffee; *C. arabica* is the main coffee of the trade; other coffees include *C. canephora*, robusta, and *C. liberica*, liberica.

coffeanum, *Colletotrichum*.

coffee, *Coffea* spp.; Waller in Clifford & Willson, *Coffee: botany, biochemistry and production of beans and beverage*, 1985.

— **blister spot** Wellman, *Turrialba* **7**:13, 1957; transm. *Toxoptera aurantii*, Costa Rica [**36**:760].

— **grease spot** Bitancourt, *Biológico* **24**:191, 1958; Brazil [**38**:320].

— **ringspot v.** Bitancourt 1939; Silberschmidt 1941; Reyes, *PDR* **45**:185, 1961; Valdez, *Philipp.Agric.* **50**:267, 1966; Chagas et al., *Phytopath.Z.* **102**:100, 1981; bacilliform up to 224 × 79 nm, transm. sap, seedborne; Brazil, Philippines [**18**:452; **20**:402; **40**:533; **47**,200; **61**,2885].

— **stem pitting** Riley, *Commonw.Phytopath.News* **3**:29, 1957; Tanzania [**36**:643].

— **sudden death** Ferrão & Serafim, *Rev.Café portug.* **7**:17, 1960; Angola [**41**:32].

coffeicola, *Cercospora, Hemileia*.

cohort senescence, a uniform loss of vigour of a tree canopy due to age and gradually increasing environmental stress; Mueller-Dombois et al., *Phytocoenologia* **11**:117, 1983, tree group death in Hawaiian & N. American forests [**62**,3999].

coicis, *Ustilago*.

coiled sprout, potato disorder; Lapwood et al.,

AAB **85**:203, 1977; see *Verticillium nubilum*
[**56**,5167].
coitre (*Coniella diplodiella*) grapevine.
Coix lachryma-jobi, adlay, job's tears.
coix mosaic Espeleta & Nuque, *Araneta J.Agric.*
6:45, 1959; Philippines, adlay [**39**:583].
Cola, see kola.
colchici, *Uromyces*.
cold pox, cucumber, squash; USA (Florida);
caused by a temp < 10°; Cox, *PDR* **39**:478, 1955;
55:350, 1971 [**35**:414; **51**,1003].
coleosporioides, *Cronartium*.
Coleosporium Lév. 1847; Uredinales; uredia and
telia on angiosperms, usually dicotyledons; aecia
on *Pinus*; uredospores are verrucose like the
aeciospores; damaging to pine seedlings, N.
Hemisphere, could be damaging if introduced
with pines to the tropics and S. Hemisphere;
Wilson & Henderson 1966; Ziller 1974; Laundon,
Mycotaxon **3**:154, 1975, typification; Kaneko,
Rep.Tottori mycol.Inst. 19, 1981, monograph
[**61**,3714].
C. asterum (Dietel) Sydow 1914; telia on *Aster* spp.
and other Compositae; anam. *Peridermium
montanum* Arthur & Kern 1906, pine defoliation,
N. America (W.).
C. ipomoeae (Schwein.) Burrill 1885; Laundon &
Rainbow, *Descr.F.* 282, 1971; telia on sweet
potato and other Convolvulaceae, warmer regions
of N. America, West Indies and prob. S. America;
in Africa the rust is found on sweet potato but
not on exotic, susceptible pines.
coleus mosaic Creager, *Phytop.* **35**:223, 1945, USA
(Illinois) [**24**:315].
colhounii, *Calonectria*, *Cylindrocladium*.
Colladonus clitellarius, *C. germinatus*, *C. montanus*
transm. peach X.
collapse (*Monosporascus*) melon.
collar branch canker (*Phomopsis theae*) tea, ——
canker (*Helicobasidium*) coffee, —— **root rot**
(*Ustulina deusta*) rubber; —— **rot**, rotting of the
stem at or about the level of the soil surface; the
name is given to many diseases in which this is a
typical symptom, ± synonymous with foot rot.
Colletogloeum nubilosum, teleom. *Mycosphaerella
cryptica*.
colletotin, toxin from *Colletotrichum*; Goodman,
Phytopath.Z. **37**:187, 1959; Lewis & Goodman,
Phytop. **52**:1273, 1962 [**39**:280; **42**:366].
colletotrichin, as above, Gohbara et al., *Agricultural
Biological Chem.* **42**:1037, 1978 [**60**, 2187].
*****Colletotrichum** Corda 1931–2; Coelom.; Sutton
1980; conidiomata acervular, conidiogenesis
enteroblastic phialidic; con. aseptate, hyaline;
either straight, cylindrical or elliptical; or curved,
falcate or lunate; appressoria brown, entire or

with crenate to irregular margins, characteristic of
the sp.; very large synonymies have been
proposed but much basic taxonomy is still needed.
The con. are water dispersed; sporulation is seen
as mucilagenous masses, often pinkish and like
dots, in typically sunken, irregularly outlined,
necrotic lesions. The diseases caused are usually
called anthracnose q.v.; fruit infections can at first
be latent; Sutton, *Mycol.Pap.* 141, 1977,
synonymy; Holliday 1980; Parberry in
Blakeman 1981, on leaf surfaces; Siddiqui et al.,
Seed Sci.Technol. **11**:353, 1983,
longevity & pathogenicity in seed at 5°;
Baxter et al., *Phytophylactica* **17**:15, 1985, review
[**63**,2003].
C. acutatum Simmonds 1965; Dyko & Mordue,
Descr.F. 630, 1979; con. straight, fusiform
8–16 × 2.5–4 μm; appressoria few, clavate to
obovate, smooth; plurivorous, causes: mango
anthracnose, pine terminal crook, strawberry
black spot; Fitzell, *PDR* **63**:1067, 1979, on mango
in Australia (New South Wales); Nair & Corbin,
Phytop. **71**:777, 1981; Nair et al., *TBMS* **81**:53,
1983, on pine, histology, survival in soil & host
debris; Van Zyl 1985, chemical control on
strawberry in South Africa [**59**,4697; **61**, 1961;
63,275; **65**,5062].
C. capsici (Sydow) E. Butler 1931; Mordue,
Descr.F. 317, 1971; con. falcate with acute apex
and narrow truncate base, 16–30 × 2.4–4 μm,
appressoria show exceptionally complex
development with assoc. hyphae in slide cultures;
close to *Colletotrichum dematium* but has wider
con.; plurivorous but mainly studied as a
pathogen of red pepper, mostly in warmer
regions, seedborne; Adikaram et al., *TBMS*
80:395, 1983, infection with *Glomerella cingulata*
[**62**,4569].
C. circinans (Berk.) Vogl. 1907; con. falcate,
fusiform apices, 19–21 × 3.5 μm, appressoria
clavate or circular, often becoming complex;
onion smudge, *Allium* spp., prob. seedborne,
persists in soil for years, red and yellow bulbs are
highly resistant.
C. coccodes (Wallr.) S. Hughes 1958; Mordue,
Descr.F. 131, 1967; con. straight, fusiform,
medianly constricted, 16–22 × 3–4 μm, appressoria
long clavate, sometimes almost crenate; sclerotia
setose, abundant, frequently on Solanaceae,
mainly as anthracnose of tomato fruit, black dot
of potato and tomato roots; eggplant and red
pepper are attacked; root inhabitant, high
inoculum build up in the soil is necessary for
serious root disease to appear; Tu, *CJB* **58**:631,
1980, ontogeny of sclerotia; Barksdale & Stoner,
PD **65**:71, 1981, on tomato fruit; Langerfield,

NachrBl.dt.PflSchutzdienst. **37**:49, 1985, review on potato [**59**,5647; **60**,6086; **64**,5092].

C. coffeanum Noack 1901; the fungus is prob. a form of *Colletotrichum gloeosporioides*, teleom. *Glomerella cingulata*; but *C. coffeanum* is usually used for the pathogen causing the important coffee berry disease in E. Africa. The nomenclatural situation is confused; con. av. 13.1 × 3.8 μm but up to 20 μm long, on aerial conidiophores not in acervuli, asco. unknown, coffee isolates differ in pathogenicity and cultural characteristics. A more suitable name is needed for the main disease caused on green berries by the virulent str. Other strs. cause fruit anthracnoses, brown blight, they do not attack green berries; stem dieback and leaf lesions occur. The virulent str. builds up during the wet season and causes epidemics especially at higher altitudes. Avirulent strs. can have higher optimum temps. for sporulation, thus at lower altitudes they have a competitive advantage over the virulent str.; this advantage disappears at high altitudes. In Kenya fungicide treatment can have complex effects on sporulation; and in recolonisation of the bark, disease may in fact be increased. Simultaneous control of *Hemileia vastatrix* is also needed. Fungicides can also lead to increased yields in the absence of obvious disease i.e. the tonic effect q.v.; Firman & Waller, *Phytopath.Pap.* 20, 1977, monograph.

C. corchori Ikata & Tanaka 1940; not *Colletotrichum corchori* Pavgi & U.P. Singh 1965, fide Sutton 1980; con. falcate, fusiform, 15–20 × 3.5–4 μm, appressoria clavate or circular, edge entire; jute anthracnose, seedborne; Purkayastha & Menon, *TBMS* **77**:183, 1981, appressorial formation [**61**,2303].

C. curvatum Briant & Martyn, *Trop.Agric.Trin.* **6**:258, 1929; con. falcate, apices acute, av. 19.8 × 3.6 μm, poss. a form of *Colletotrichum dematium*, sunn hemp stem break, seedborne [**9**:186].

C. dematium (Pers.:Fr.) Grove 1918; con. falcate, fusiform, apices acute, 20–24 × 2–3.5 μm, appressoria clavate to circular, edge usually entire, sclerotia black, abundant; plurivorous, mostly in temperate regions, a frequent coloniser of senescent tissue; Gourley, *Can.J.Pl.Sci.* **46**:531, 1966, pathogenicity to beet & other hosts; Chikuo & Sugimoto, *Ann.phytopath.Soc.Japan* **50**:249, 1984, on beet seed [**46**,186; **64**,829].

C. destructivum, teleom. *Glomerella glycines*; **Colletotrichum falcatum**, teleom. *G. tucumanensis*.

C. fuscum Laub. 1927; con. straight or slightly curved, 14–17 × 3.5–4 μm; appressoria absent, mycelium has dark, thick walls; on

Scrophulariaceae; Goto, *Ann.phytopath.Soc.Japan* **8**:1, 1938; Thomas, *Phytop.* **41**:997, 1951, *Digitalis* anthracnose [**17**:822; **31**:256].

C. gloeosporioides, teleom. *Glomerella cingulata.*

C. gossypii Southw. 1890; prob. a form of *Colletotrichum gloeosporioides*, anthracnose of cotton bolls. *C. indicum* Dastur, *Indian J.agric.Sci.* **4**:100, 1934, also on cotton, may be distinct. *Glomerella gossypii* (Southw.) Edg., which may be a form of *G. cingulata*, is assoc. with cotton seedling blight [**13**:507].

***C. graminicola** (Ces.) G. Wilson 1914; Mordue, *Descr.F.* 132, 1967; con. falcate, fusiform, apices acute, 24–29 × 3.5–5 μm, appressoria with a very irregular edge, lobed; Gramineae, sorghum red stalk rot, maize leaf infection, damages turf grasses, seedborne; Basu Chaudhary & Mathur, *Seed Sci.Technol.* **7**:87, 1979, on sorghum seed; Vizvary & Warren, *Phytop.* **72**:522, 1982; Lipps, *PD* **67**:102, 1983, survival in soil & maize residues, Danneberger et al., *Phytop.* **74**:448, 1984, forecasting on *Poa annua*; Nicholson et al., *ibid* **75**:654, 1985, effects on growth of maize with *Pratylenchus hexincisus* [**59**,755; **61**,6252; **62**,1917; **63**,3950; **65**,684].

C. higginsianum Sacc. in Higgins 1917; con. fusiform, 17–19 × 4 μm, details of appressoria apparently not known, prob. distinct from *Colletotrichum gloeosporioides* and always assoc. with brassicas, seedborne; Scheffer, *Tech.Bull.N.Carol.agric.Exp.Stn.* 92, 1950 [**30**:132].

C. lagenarium (Pass.) Ell. & Halsted 1893; this name is the one most commonly used for the fungus which causes the important cucurbit anthracnose; Sutton 1980 placed it as a synonym of *Colletotrichum orbiculare* q.v.; Thompson & Jenkins, *Phytop.* **75**:1422, 1985, yield loss & fungicides; Sherf & MacNab 1986 [**65**,2516].

C. lindemuthianum (Sacc. & Magnus) Brioso & Cavara 1889; Mordue, *Descr.F.* 316, 1971; con. straight cylindrical, apices obtuse, 10–24 × 4–5 μm, appressoria clavate or circular, not becoming complex, differs from *Colletotrichum gloeosporioides* sensu lato in the slow growth and dark pigmentation in culture, causes important diseases in legumes, seedborne, great variability in pathogenicity; Tu, *PD* **65**:477, 1981; **66**:781, 1982; **67**:402, 1983, epidemiology in Canada (Ontario); Dixon 1981; Allen 1983; Sherf & MacNab 1986 [**61**,1475; **62**,462, 3695].

C. linicolum Pethybr. & Lafferty 1918; con. cylindrical, straight or slightly curved, tapering at both ends. Sutton 1980 gave the flax pathogen as *Colletotrichum lini* (Westerd.) Tochinai 1926; con. falcate, apices obtuse, 16–19 × 3–4.5 μm, appressoria long clavate to irregular, often becoming complex. This name is from *Gloeosporium lini* Westerd. 1915 but Dickson 1956 considered that Westerdijk's fungus was not a *Colletotrichum*; seedborne, Zarzychka 1976 described races [55,5218].

C. malvarum (A. Braun & Casp.) Southw. 1891; poss. a form of *Colletotrichum gloeosporioides*; Kirkpatrick et al., *PD* **66**:323, 1982, poss. use in control of *Sida spinosa*, a weed in cotton [61,6147].

C. musae (Berk. & Curt.) v. Arx 1957; Sutton & Waterston, *Descr.F.* 222, 1970; con. cylindrical, straight, apices obtuse, 12–17 × 4.5–5.5 μm, appressoria irregular in shape, large lobes; causes an important disease of banana fruit after harvest, assoc. with other fungi in banana crown rot and main stalk rot; Stover 1972; Harper et al., *J.gen.Microbiol.* **121**:169, 1980; Graham & Harper, *ibid* **129**:1025, 1983; Brown & Swinburne, *TBMS* **77**:119, 1981, role of Fe in conidial germination & lesion formation; Swinburne & Brown, *ibid* **80**:176, 1983; Muirhead & Deverall, *Physiol.Pl.Path.* **19**:77, 1981, appressoria, latent & quiescent infections [60,3554; 61,1295; 62,3143, 4381].

C. orbiculare (Berk. & Mont.) v. Arx 1957; con. cylindrical, straight, apices obtuse, 14–15 × 4.5–6 μm, appressoria long clavate or slightly irregular, often becoming complex. The pathogen causing the important cucurbit anthracnose, most severe on cucumber, melon and watermelon, is usually called *Colletotrichum lagenarium* q.v.; seedborne, races occur; Amin & Ullasa, *Phytop.* **71**:20, 1981, control with thiophanate methyl. Thompson & Jenkins, *ibid* **75**:828, 1985, factors affecting lesion size & sporulation; *PD* **69**:833, 1985, disease assessment key [60,5616; 65,432, 4156].

C. sublineolum Henn. in Kabát & Bubák 1905; Sutton 1980 adopted this name for isolates of the genus on sorghum; con. falcate fusiform, apices acute, 19–28 × 3–4.5 μm, appressoria not as irregular as those of *Colletotrichum graminicola* q.v.

C. trifolii Bain in Bain & Essary 1906; see Tiffany & Gilman, *Mycologia* **46**:61, 1954; con. straight, 11–13 × 3–4 μm, poss. a form of *Colletotrichum gloeosporioides*; clovers, lucerne and other legumes; Welty et al., *PD* **64**:476, 1980; **66**:48, 653, 1982, effects of temp. & light on lucerne anthracnose, races 1 & 2; Elgin et al., *Crop.Sci.* **21**:457, 1981, resistance in lucerne increases yield; Welty & Rawlings, *Phytop.* **75**:593, 1985, effects of inoculum concentration & temp. on disease [60,2074; 61,2938, 5059; 62,250; 64,4990].

C. truncatum (Schwein.) Andrus & Moore 1935; often cited as *Colletotrichum dematium* f.sp. or var. *truncatum* (Schwein.) v. Arx 1957; con. falcate, apices obtuse, 16–24 × 3.5–4 μm, appressoria clavate or circular, often complex; soybean anthracnose, and other legumes, seedborne; Backman et al., *PD* **66**:1032, 1982, loss in soybean in USA; Hepperley et al., *Seed Sci.Technol.* **11**:371, 1983, seed assay; Sinclair 1982 [62, 810; 63,2055].

collinsii, *Apiosporina*.

colocasia bobone v. Kenten & Woods 1972; Shaw et al., *Papua New Guin.agric.J.* **30**:71, 1979; large and small bacilliform particles 300–335 × 50–55 nm, 125 × 28–29 nm, transm. *Pseudococcus longispinus*, *Tarophagus proserpina*, parts of Oceania and E. Indies [51,4593c; 60,4155].

colocasiae, *Phytophthora*.

Colocasia esculenta, cocoyam, dasheen, eddo, taro, see under last name; *Xanthosoma* spp. are also called cocoyam.

Colombia lance nematode, *Hoplolaimus colombus*.

colonisation, the growth of a pathogen, particularly a fungus, in the host; infection is the early stage of colonisation.

coloradense, *Peridermium*.

colour banding, or rugby stocking, wheat seedlings and related to the depth of sowing; Wilcox, *Pl.Path.* **8**:34, 1959.

— breakdown, anthurium, Ca deficiency; Higaki et al., *J.Am.Soc.hortic.Sci.* **105**:441, 1980 [60,1464].

— changes, in plants and caused by viruses, see Bos 1978.

Colpoma Wallr. 1833; Hypodermataceae; ascomata in bark or wood, erumpent. *C. quercinum* (Pers.) Wallr., anam. *Conostroma didymum* (Fautrey & Roum.) Moesz 1920–1; asco. mostly 40–60 μm long, con. 5–7 × 1.5 μm; oak dieback; Twyman, *TBMS* **29**:234, 1946; Butin, *Eur.J.For.Path.* **11**:33, 1981, fungi on the bark of young oaks [26:362; 61,2465].

columnea latent viroid Owens et al., *Virology* **89**:388, 1978, *C. erythrophae* [58,2370].

comandra blister rust (*Cronartium comandrae*) pine.

comandrae, *Cronartium*; **comari**, *Gnomonia*.

commelina diffusa mosaic v. Morales & Zettler, *Phytop.* **67**:839, 1977; Migliori & Lastra, *Annls.Phytopath.* **10**:467, 1978; **12**:145, 1980; Potyvirus, rod *c*.750 nm long, transm. sap, aphid; Guadeloupe, USA (Florida), no other hosts [57,1225; 59,701; 60,3806].

— jacobii mosaic Padmanabhan, Jaganathan & Kandaswamy, *Madras agric.J.* **59**:361, 1972; transm. sap, aphid; India (S.) [52,1830].

— mosaic v. Spire, Férault & Bertrandy,

Annls.Phytopath. **2**:268, 1970; filament 700 nm long, transm. aphid, W. Indies, infected abaca, *Canna indica*, maize.
— X v. Stone & Hollings 1979; Potexvirus [**59**,3546].

commensal, an organism that can colonise the surface of healthy tissue without attacking the tissue or itself being damaged by any tissue response. Commensalism is a system of mutual convenience without apparent benefit or ill effect to either organism, after Cowan 1978; cf. symbiosis.

common blight (*Xanthomonas campestris* pv. *phaseoli*) bean and other legumes, — **bunt** (*Tilletia laevis, T. tritici*) wheat, — **gall rust** (*Gymnosporangium tremelloides*) juniper, — **leaf spot** (*Pseudopeziza medicaginis*) lucerne, (*Ramulispora sorghicola*) sorghum, — **millet**, see millets, — **scab** (*Elsinoë fawcettii*) citrus, (*Streptomyces* spp.) potato.

Commonwealth Agricultural Bureaux, and Commonwealth Mycological Institute; see CAB International.

Comoviruses, type cowpea mosaic v., *c*.16 definitive members, isometric *c*.30 nm diam., 2 segments linear ssRNA, particles sediment as 3 components *c*.58, 98, 116S, thermal inactivation 60–75°, longevity in sap 7–35 days, transm. sap readily and leaf feeding beetles, narrow host ranges; Bruening, *Descr.V.* 199, 1978; Stace-Smith in Kurstak 1981; Francki et al., vol. 2, 1985.

compactum, *Helicobasidium*.

compartmentalisation, the responses of woody plants to injuries caused by pathogens, pests, external environment and management practices; Shigo, *A.R.Phytop.* **22**:189, 1984.

compatible, used in several senses, FBPP defines the first 2 but not the third. (1) A relation between host and pathogen in which disease can develop; where it does not the relation is incompatible. (2) A relation between a plant rootstock and the scion where a complete functional union is formed between the two. Where anomalies, causing defective growth or breakage, occur between the two the union is then incompatible. Signs of incompatibility may resemble the symptoms caused by virus infection. (3) Fungus strs. of a sp. are said to be compatible if, when growing together, the sexual or perfect state, i.e. the teleom., or a dikaryon, is formed. The genetically controlled mating system may be simple with 2 compatibility types + or 'A' strs. and — or 'a' strs. as in some oomycetes, many ascomycetes, rusts and smuts. In the hymenomycetes complex mating systems have evolved; there are multiple genetic

or mating factors which control distinct populations of strs. in the same sp.; Burnett 1976.

competitive saprophytic ability, the summation of physiological characteristics that make for success in competitive colonisation of dead organic substrates, first proposed by Garrett in 1950 and defined in 1956. In 1970 he stated: '...we are beginning to realise that the kinds of competitive saprophytic ability are almost as diverse as the types of substrate open to saprophytic colonisation.' A close analogy is to be found in the term 'pathogenicity'; pathogen q.v.

complanata, *Phoma*.

complement, a thermolabile substance in the blood serum of higher animals that combines with antigen and antibody complexes. The complement on combining with such complexes is removed from the serum. This removal, or complement fixation, is the basis of a sensitive test for detecting antigen and antibody interactions, after FBPP.

composite disease, in virology; one caused by 2 or more unrelated viruses which, together in a host, cause a disease with symptoms that are not predictable from knowledge of the symptoms caused by the components separately, after FBPP.

composts, in the control of soilborne pathogens, Hoitink & Fahy, *A.R.Phytop.* **24**:93, 1986.

comptoniae, *Cronartium, Peridermium*.

concentrica, *Plectophomella*; **concentricum**, *Cylindrosporium*.

concentric canker (*Ganoderma applanatum*) citrus.

concors, *Mycovellosiella*.

cone rust (*Chrysomyxa monesis, C. pirolata, Pucciniastrum areolatum*) spruce.

confluent orange spotting, oil palm, K deficiency, Turner 1981.

confusa, *Tympanis*; **confusum**, *Gymnosporangium*.

conidiogensisis, in a fungus, the process of conidium formation from the conidiogenous cell in a conidiophore, see blastic and thallic.

conidioma, plural conidiomata, any fungus structure which bears conidia, types: acervular, cupulate, hyphal, pycnidial, pycnothyrial, sporodochial, stromatic, synnematal.

conidiophore, a hypha or cell which bears a conidiogenous cell(s).

conidium, conidia, an asexual fungus spore formed and liberated from a conidiogenous cell; it is not motile, not developed by cytoplasmic cleavage like a sporangiospore or by free cell formation like an ascospore. When conidia are described in the text as aseptate they are not filiform, helical or stellate unless so stated; Sutton, *TBMS* **86**:1, 1986.

Coniella Höhnel 1918; Coelom.; Sutton 1980;

conidiomata pycnidial, conidiogenesis
enteroblastic phialidic; con. aseptate, pigmented.
C. castaneicola (Ell. & Ev.) B. Sutton 1980; con.
15–29 × 2.5–3.5 μm; Kaneko,
Ann.phytopath.Soc.Japan **47**:80, 1981, leaf blight
of chestnut and oak, does not usually cause
defoliation [**61**,1391].
C. fragariae (Oudem.) B. Sutton 1977; Sutton &
Waterston, *Descr.F.* 82, 1966 as *Coniella
diplodiella* (Speg.) Petrak & Sydow 1927; con.
7.5–11 × 5.5–5.7 μm; grapevine white rot or goitre,
mainly on fruit but also causing pedicel lameness
in fruit stalks and injury to shoots, stems and
leaves; on other plants; Triolo et al. 1984, effect
of infection on graft take [**64**,3158].
coniferarum, *Potebniamyces*.
*****conifers**, see trees; Millar and Mishra & Das
described needle infection and fungal succession,
respectively, in Blakeman 1981.
conigenum, *Cronartium, Lophodermium*.
Coniophoraceae, Aphyllophorales; a thelephoroid,
flattened form, *Serpula*.
coniothyrium, *Leptosphaeria*.
Coniothyrium Corda 1840; Coelom.; Sutton 1980;
conidiomata pycnidial, conidiogenesis holoblastic,
conidiophore growth percurrent; con. aseptate or
2 celled, pigmented, thick walled, verruculose;
requires revision.
C. fuckelii, teleom. *Leptosphaeria coniothyrium*.
C. minitans Campbell 1947; Punithalingam,
Descr.F. 732, 1982; con. 4–7 × 2.5–3.5 μm, dark
brown in mass; hyperparasite, on sclerotia of
Botrytis, Sclerotinia; promising for control of
Sclerotium cepivorum; for parasitism of
Sclerotinia sclerotiorum see: Trutmann et al., *Soil
Biol.Biochem.* **12**:461, 1980; *TBMS* **78**:521, 1982;
Phillips & Price, *Phytopath.Z.* **107**:193, 1983; Tu,
ibid **109**:261, 1984 [**60**,4334; **61**,6075; **63**,1088,
4229].
C. scirpi, teleom. *Paraphaeosphaeria michotii*.
C. wernsdorffiae Laub.; not cited by Sutton 1980;
con. 4.2–8.4 × 3.6–7.2 μm; rose brand canker,
Europe, N. America; Westcott, *Mem.Cornell
agric.Exp.Stn.* 153, 1934; Semina et al. 1982
[**13**:703; **63**,4466].
Conium maculatum, poison hemlock; Howell &
Mink, *PD* **65**:277, 1981, viruses, with wild carrot
[**61**,1525].
conjuncta, *Lophodermella*.
conk, a N. American term for the basidioma
of, usually, a polypore, Polyporaceae, on
wood.
Conoplea globosa, poss. teleom. *Urnula craterium*;
for *Conoplea* Pers. 1801; Hyphom., see Hughes,
CJB **38**:659, 1960.
Conostroma didymum, teleom. *Colpoma quercinum*;

for *Conostroma* Moesz 1921; Coelom., see Sutton
1980.
constricted root (*Aphanomyces raphani*) radish.
contact oleocellosis, rind oil spotting, citrus; caused
by fruit contact and compression on the tree; Lo
Giudice & Catàra, *Tec.agricola* **24**(3):5, 1972
[**52**,2944].
contaminant, a substance, micro-organism or virus
which accidentally appears in any culture, medium
or other substrate; any unwanted material in an
experimental system, hence to contaminate, which
should not be used instead of infect or inoculate.
context, of hymenomycetes, the hyphal mass
between the superior surface and the
subhymenium or the trama of a basidioma.
contraria, *Mycosphaerella*; *Pseudocercospora*.
control of plant diseases, the application and
effectiveness of control measures varies with the
differing degrees of agricultural development
throughout the world; as does that of any applied
science. Where development is high plant diseases
cause at most a temporary inconvenience. Where
it is low disease epidemics are common, crop
production inadequate and income at subsistence
levels. Garrett's phrase in 1955 still applies in
many countries: 'The first basic and most
important step in control must be a gradual
rationalisation of a husbandry still entirely
empirical.' This covers measures taken against
disease during crop growth. Breeding and
selection for plant resistance follow. Chemical
treatments, largely preventative, augment
resistance. Finally, since some important
pathogens are limited in their distribution,
measures are taken to prevent their spread.
Control measures require effective communication
between investigator, advisor and grower. Such
communication varies in efficiency as much as do
the crops and their culture. Wheeler 1969 and
Manners 1982 wrote introductions to the subject.
    Lester, *Pl.Path.* **35**:2, 1986; Curl, *Bot.Rev.* **29**:
413, 1963, crop rotation; Hardison, *A.R.Phytop.*
14:355, 1976, fire; Berger, *ibid* **15**:165, 1977,
epidemiology; Mulder ed., *Soil disinfestation*, vol.
6 of Developments in agricultural and managed
forest ecology, 1979; Harris & Maramorosch
1982; Fry, *Principles of plant disease management*,
1982; Sill, *Plant protection: an integrated
interdisciplinary approach*, 1982; Bos,
Adv.appl.Biol. **7**:105, 1983, viruses; Jacobsen,
A.R.Phytop. **21**:137, 1983, advisory; Scopes &
Ledieu, *Pest and disease control handbook*, edn. 2,
1983; White & Antoniw, *Crop Protect.* **2**:259,
1983, viruses; Conway ed., *Pest and pathogen
control: strategic, tactical and policy models*,
1984; Teng, *F.A.O.Pl.Prot.Bull.* **32**:51, 1984,

disease management; Rudd-Jones & Langton ed., *Healthy planting material: strategies and technologies, Monogr.Br.Crop Protect.Counc.* 33, 1986; McGrath et al., *USDA Agric.Handb.* 656, 1986; Palti & Ausher ed., *Advisory work in crop pest and disease management*, 1986.

controversa, *Tilletia*.

convallariae, *Botrytis*, *Mycosphaerella*.

Convallaria majalis, lily of the valley.

convoluta, *Botryotinia*, *Botrytis*.

Cook, Melville Thurston, 1869–1952; born in USA, Univ. De Pauw, Leland Stanford, Ohio State; New Jersey Agricultural Experiment Station 1911–23, worked in Cuba and Puerto Rico 1923–40, wrote: *The diseases of tropical plants*, 1913; *Enfermedades de las plantas economicas de las Antillas*, 1939. *Phytop.* **43**:591, 1953.

Cooke, Mordecai Cubitt, 1825–1914; born in England, naturalist, wrote books on British fungi including *Fungoid pests of cultivated plants*, 1906, left an important herbarium, Victorian Medal of Honour, Royal Horticultural Society 1902, Linnean Gold Medal 1903. *TBMS* **5**:169, 1915; *Phytop.* **6**:1, 1916; English, *Mordecai Cubitt Cooke – Victorian naturalist, mycologist, teacher and eccentric*, 1987.

copper spot (*Gloeocercospora sorghi*) turf grasses, — **web** (*Helicobasidium brebissonii*). plurivorous.

Coprinaceae, Agaricales, basidiospores fuliginous to black, *Coprinus*.

Coprinus Pers. 1797; largely saprophytic; *C.psychromorbidus* Redhead & Traquair, *Mycotaxon* **13**:382, 1981; a snow mould pathogen in Canada; Traquair et al., *Can.J.Pl.Path.* **4**:27, 106, 1982, sclerotial strs. & pathogenicity on lucerne; Gaudet & Kokko, *CJB* **63**:955, 1985, penetration & infection [**61**,582, 6443, 7066; **64**,4254].

coral spot (*Nectria cinnabarina*) temperate, woody crops.

corchori, *Cercospora*, *Colletotrichum*.

Corchorus capsularis, *C. olitorius*, jute.

Cordana Preuss 1851; Hyphom.; Ellis 1971, 1976; conidiomata hyphal, conidiogenesis holoblastic, conidiophere growth sympodial; con. 1 septate, pigmented; Hughes, *CJB* **33**:259, 1955. *C. musae* (Zimm.) Höhnel 1923; Ellis & Holliday, *Descr.F.* 350, 1972; con. 11–18 × 7–10 μm; banana leaf blotch, often a secondary invader of leaf lesions caused by other fungi; Meredith, *AAB* **50**:263, 1962; *Ann.Bot.* **26**:233, 1962, spore discharge & dispersal [**41**:666; **42**:37].

cordiae, *Puccinia*.

core flush, apple, early senescence assoc. with high CO_2 concentration in storage; — **rot** (*Botryotinia draytonii*) gladiolus.

coreopsis mosaic Verma & Mishra, *Trudy Biol. – Pochvenn.Inst.* **31**:115, 1975; — **mottle** Joshi & Suteri, *Sci.Cult.* **42**:616, 1976; transm. sap, aphid, India, *C. drummondii* and *C. grandiflora*, respectively [**56**,4073, 5662].

coriander, *Coriandrum sativum*; Taylor & Dudley, *Pl.Path.* **29**:177, 1980, described coriander flowerstand blight caused by a bacterium similar to *Pseudomonas syringae*, in Britain, Hungary; the inflorescence became infected and seed decayed, seedborne [**60**,5061].

— **feathery red vein** v. Misari & Sylvester, *Hilgardia* **51**(2), 1983; bacilliform $c.216 \times 75$ nm, transm. *Hyadaphis foeniculi*, persistent, USA (California), symptoms in celery and *Nicotiana* spp., several Umbelliferae, including carrot, fennel, parsley, parsnip, were symptomless hosts [**63**,2440].

— **interveinal yellowing** (heracleum latent v.).

— **mild yellows** v. Tomlinson & Carter 1969, 1970; isometric, transm. sap to *Nicotiana* spp. [**48**,3246d; **49**,3555k].

— **vein clearing and mosaic** (clover yellow vein v.).

coriandri, *Ramularia*.

Coriandrum sativum, coriander.

cork, see phellophagy.

corky pit, apple, see internal cork; — **root** (*Pyrenochaeta lycopersici*) tomato; lettuce, Amin & Sequeira, *Phytop.* **56**:1047, 1966; — **scab** (*Xanthomonas campestris* pv. *vesicatoria*) red pepper, tomato; — **stunt**, genetic disorder, tomato, Stevenson et al., *Phytop.* **66**:132, 1976 [**46**,500; **55**,4862].

corm dry rot (*Stromatinia gladioli*) gladiolus.

— **rot** (*Botryotinia draytonii*, *Fusarium oxysporum* f.sp. *gladioli*) gladiolus.

corn, see maize, wheat.

— **Ohio stunt** = maize chlorotic dwarf v.

— **stripe and yellow stripe** = maize mosaic v.

corni, *Elsinoë*.

Corniculariella P. Karsten 1884; Coelom.; Sutton 1980. *C. pseudotsugae*, teleom. *Durandiella pseudotsugae*.

cornigerum, *Ceratobasidium*; **cornui**, *Peridermium*.

cornus witches' broom Raju & Chen 1947; Raju et al., *PDR* **60**:462, 1976; MLO assoc., USA (New Jersey) [**56**,861].

coronaria, *Marssonina*; **coronata**, *Puccina*.

coronatine, toxin formed by some pvs. of *Pseudomonas syringae*, e.g. *atropurpurea*, *coronafaciens*; Mitchell, *A.R.Phytop.* **22**:221, 1984.

coronie wilt (see *Phytomonas*) coconut.

corrugata, *Pseudomonas*; **corticale**, *Cryptostroma*.

Corticiaceae, Aphyllophorales; basidioma thelephoroid, flattened; *Butlerelfia*, *Corticium*,

Laetisaria, Limonomyces, Peniophora, Veluticeps, Waitea.

corticis, *Botryosphaeria.*

Corticium Pers. 1794; Corticiaceae; many spp. have been placed in other genera, Talbot in Parmeter 1970; Holliday 1980.

C. anceps, see *Pteridium aquilinum.*

C. galactinum (Fr.) Burt 1926; fide White, *CJB* **29**:279, 1951; decays roots and stumps of trees, conifers and others; for eastern white root rot of apple: Cooley & Davidson, *Phytop.* **30**:139, 1940; Cooley, *ibid* **38**:110, 1948 [**19**:354; **27**:326; **31**:524].

*****C. rolfsii** Curzi 1932; sclerotial state *Sclerotium rolfsii* Sacc. 1911; Mordue, *Descr.F.* 410, 1974; *Corticium rolfsii* was transferred to *Athelia* by Tu & Kimbrough 1978, see *Rhizoctonia*; southern blight or stem rot, a plurivorous, unspecialised but important parasite; soil inhabitant with characteristic sclerotia, like mustard seed; more frequent in warm, moist conditions; the sclerotia have a long but variable survival time and germinate eruptively to infect plants directly; basidiospores may not be important in the epidemiology, but Punja & Grogan, *PD* **67**:875, 1983, found that they were infective; reviews: Aycock, *Tech.Bull.N.Carol.Exp.Stn.* 174, 1966; Punja, *A.R.Phytop.* **23**:97, 1985; Jenkins & Averre, *PD* **70**:614, 1986 [**63**,697].

C. salmonicolor Berk. & Broome 1873; anam. *Necator decretus* Massee 1898; Mordue & Gibson, *Descr.F.* 511, 1976; the disease which results from infection is called pink disease; pink crust, from Petch's description of the symptoms in 1911, is the preferred name which is derived from the flat, pinkish hymenial basidioma which covers branches; plurivorous on woody plants, mostly tropical and subtropical. This fungus has caused significant damage to black pepper, cacao, citrus, coffee, eucalyptus, rubber; it has been studied mostly on the last crop; Schneider-Christians et al. 1983, on cacao; *Z.PflKrankh.PflSchutz.* **93**:397, 1986, basidiospores [**64**,3001–2; **66**,118].

corticola, *Ascochyta, Clethridium, Cryptosporiopsis, Discostroma, Pezicula.*

corylea, *Phyllactinia*; **coryli**, *Nematospora.*

Corylus avellana, common hazel, cobnut, filbert, hazel nut.

corymbiferum, *Penicillium.*

Corynebacterium Lehmann & Neumann 1896; Gram positive; the genus has contained plant pathogens which are now in *Arthrobacter, Clavibacter, Curtobacterium, Rhodococcus*; Vidaver, *A.Rev.Microbiol.* **36**:495, 1982; Bradbury 1986; Davis, *A.R.Phytop.* **24**:115, 1986.

Coryneliaceae, Ascomycotina; Coryneliales, one family; *Caliciopsis.*

Corynespora Güssow 1906; Hyphom.; Ellis 1971; conidiomata hyphal, conidiogenesis enteroblastic tretic; con. multiseptate, tending to be distoseptate; i.e. the individual cells are each surrounded by a wall, like a sac and distinct from the outer wall.

C. cassiicola (Berk. & Curt.) Wei 1950; Ellis & Holliday, *Descr.F.* 303, 1971; con. 40–220 × 9–22 μm; target spot, plurivorous, poss. seedborne, diseases on: betel, cacao, cowpea, cucumber, jute, papaya, rubber, sesame, tomato; Holliday 1980; J.P. & J.B. Jones 1984, epidemiology & control on tomato; Kingsland & Sitterley, *Trop.Pest Management* **32**:31, 1986, chemical control on tomato in Seychelles [**65**,2978; **66**,716].

corynetoxins, animal toxins formed by *Clavibacter rathayi*, see *Anguina agrostis.*

cosmos yellow vein mosaic v. Srivastava, Aslam & Rao, *Curr.Sci.* **54**:1126, 1985; rod 640 × 13 nm, transm. *Bemisia tabaci*, India, *C. sulphureus* [**65**,2850].

cotton, *Gossypium*; Ebbels, *Rev.Pl.Path.* **55**:747, 1976, in Africa; *Outlook Agric.* **10**:176, 1980; Watkins 1981; Bird, *PD* **66**:172, 1982, resistance; Minton & Garber, *ibid* **67**:115, 1983, seedling disease control; Cauquil & Follin, *Coton Fibr.trop.* **38**:293, 1983, poss. MLO & viruses.

—— **anthocyanosis** Costa, *Phytopath.Z.* **28**:167, 1957; Mali, *Curr.Sci.* **47**:235, 1978; transm. *Aphis gossypii*; Brazil, India; poss. the same as cotton blue [**36**:405; **57**,4976].

—— **blue** Cauquil & Vaissayre, *Coton Fibr.trop.* **26**:463, 1971; Cauquil, *ibid* **32**:259, 1977; transm. *Aphis gossypii*, Africa, see above [**51**,2500; **57**,4481].

*—— **leaf crumple v.** Dickson, Johnson & Laird, *Phytop.* **44**:479, 1954; Brown & Nelson, *ibid* **74**:987, 1984; prob. Geminivirus, isometric, single 17–20 nm diam., double 30–32 × 17–20 nm diam.; transm. *Bemisia tabaci*, USA (W.), symptoms also on bean and *Malva parviflora* [**34**:299; **64**,201].

—— —— **curl** Jones & Mason, *Ann.Bot.* **40**:759, 1926; Tarr 1951; El Nur & Abu Salih, *PANS* **16**:121, 1970; transm. *Bemisia tabaci*; this important disease in Africa has been extensively investigated but its etiology remains undetermined [**6**:161; **31**:327; **49**,2485].

—— —— **wilt** Follin & Campagnac, *Coton Fibr.trop.* **36**:313, 1981, Argentina [**62**,211].

—— **mosaics**, Bink, *Cott.Grow.Rev.* **52**:233, 1975; Ebbels 1976, cotton q.v.; several have been described and some are transm. by whiteflies.

—— **phyllosis**, or phyllody, see cotton virescence.

—— **small leaf** Kottur & Patel, *Agric.J.India* **15**:64, 1920; Uppal et al., *Curr.Sci.* **13**:284, 1944; also called stenosis; Capoor et al., *Hindustan Antibiot.Bull.* **15**:40, 1972, found an assoc. MLO [**24**:100; **53**,560].

—— **veinal mosaic** Costa & Forster, *Revta Agric.S.Paulo* **13**:187, 1938; transm. *Aphis gossypii*, fide Ebbels 1980, cotton q.v., Brazil [**18**:520].

—— **virescence** Delattre 1965; *Coton Fibr.trop.* **23**:386, 1968; MLO assoc., transm. *Orosius cellulosus, Recilia trifasciata*, Africa (W.), see Ebbels 1976, cotton q.v. [**48**,477].

cottony blight (*Pythium aphanidermatum*) grasses, —— **leak** (*P. aphanidermatum*) cucurbits, —— **rot** (*P. butleri*) cucurbits, —— **soft rot** (*Sclerotinia sclerotiorum*) sunflower, vegetables.

couch grass streak mosaic = agropyron mosaic v.

coulteri, *Herpotrichia*.

cover, coverage, the proportion of the surface area of plants or plant parts on which an applied material is retained, following dusting or spraying; overall cover, application of a material to the whole crop rather than concentrated in rows or bands, after FBPP.

covered smut, a type where the masses of ustilospores are held together by the persistent grain membrane, cf. loose smut; diseases: (*Sporisorium sorghi*) sorghum, (*Tilletia holci*) *Holcus*, (*Ustilago segetum* var. *hordei*) barley, oat.

cowparsnip, *Heracleum sphondylium* q.v.

—— **mosaic v.** Wolf & Schmelzer, *Zentbl.Bakt.ParasitKde.* 2 **127**:665, 1972; Wolf, *Acta phytopath.Acad.Sci.hung.* **7**:353, 1972; Polák et al., *ibid* **12**:157, 1977; bacilliform *c*.265 × 90 nm, Europe [**52**,3988; **53**,1265; **57**,2862].

—— **ringspot** Cadman, *J.hort.Sci.* **31**:111, 1956; Scotland [**35**:619].

—— **yellow vein banding** Polák, *Biologia Pl.* **8**:73, 1966; transm. sap to coriander, dill, parsley, parsnip; Europe [**45**,2047].

cowpea, see *Vigna* and blackeye cowpea; Williams, *PANS* **21**:253, 1975, in Nigeria; *Trop.Agric.Trin.* **54**:53, 1977; Allen et al., *ibid* **58**:267, 1981, resistance; Allen 1983 [**55**,1561; **56**,3343; **61**,2566].

—— **aphid borne mosaic** = blackeye cowpea mosaic v.

—— **Arkansas mosaic** = cowpea severe mosaic v.

—— **banding mosaic v.** Sharma & Varma, *Indian Phytopath.* **28**:192, 1975; *Phytopath.Z.* **83**:144, 1975; *J.ent.Res.* **1**:29, 1977; isometric 25–30 nm diam., transm. *Aphis craccivora*, seedborne, serologically related to cucumber mosaic v. [**55**, 1562; **56**,1825; **59**,3023].

—— **bright yellow mosaic** Ahmad, *PDR* **62**:224, 1978; transm. *Bemisia tabaci*, Pakistan (Punjab),

like cowpea golden mosaic, yellow fleck and yellow mosaic [**57**,5227].

—— **chlorotic mottle v.** Kuhn 1964; Bancroft, *Descr.V.* 49, 1971; Bromovirus, isometric *c*.25 nm diam., transm. sap, *Cerotoma ruficornis, C. trifurcata, Diabotrica balteata, D. undecimpunctata*; Americas; cowpea, soybean; Adolf, *J.gen.Virol.* **28**:147, 1975, structural transition; Wyatt & Kuhn, *ibid* **49**:289, 1980, new str. from resistant cowpea; *Phytop.* **69**:125, 1979, replication & properties [**54**,4858; **59**,3021; **60**,1733].

—— —— **spot v.** Sharma & Varma, *Indian Phytopath.* **28**:192, 1975; prob. Tobamovirus, rod 300 × 17 nm, transm. sap, seedborne, prob. the same as sunnhemp mosaic v., G. Thottappilly, personal communication, 1984 [**56**,1825].

—— **common mosaic** (blackeye cowpea mosaic v.).

—— **cucumovirus** Díaz-Ruiz, *Microbiología esp.* **29**:9, 1976; isometric 20–22 nm diam., Spain, not infecting cucumber [**58**,455].

—— **golden mosaic**, *Ann.Rep.Int.Inst.Trop.Agric. Ibadan* for 1979; Vetten & Allen and Anno-Nyako et al., *AAB* **102**:219, 319, 1983; poss. Geminivirus, transm. *Bemisia tabaci*, Nigeria; like cowpea bright yellow mosaic, yellow fleck and yellow mosaic [**62**,3328, 3377].

—— **green vein banding v.** Lin, *Fitopat. Brasileira* **4**:203, 1979; Potyvirus, filament, transm. sap, *Myz.pers.*, Brazil (Goiás).

—— **little leaf v.** Benigno & Paje 1975; Talens, *Philipp. Phytopath.* **13**:43, 1977; Benigno & Favali-Hedayat, *F.A.O. Pl.Prot.Bull.* **25**:78, 1977; prob. isometric, transm. aphid, non-persistent, Philippines [**57**,467; **58**,4131].

—— **mild mottle v.** Brunt & Kenten 1973, *Descr.V.* 140, 1974; Carlavirus, filament *c*.650 nm long, transm. sap readily, seedborne in some legumes; Iwaki et al., *PD* **66**:365, 1982, from soybean in Thailand, transm. *Bemisia tabaci*, semi-persistent; Muniyappa & Reddy, *ibid* **67**:391, 1983, from groundnut in India, transm. *B. tabaci*, non-persistent; Brunt et al., *Intervirology* **20**:137, 1983; discussion of taxonomy [**61**,6688; **62**,3465; **63**,4802].

—— **mosaic v.** Chant 1959; van Kammen & Jager, *Descr.V.* 197, 1978; Comovirus, type, isometric *c*.24 nm diam., transm. sap readily, several genera of beetles, pigeon pea is susceptible; Fulton & Scott, *Phytop.* **69**:305, 1979, serogrouping for legume comoviruses [**59**,2414].

—— **mosaics**, many of these have been reported in the literature and they are almost all inadequately characterised; prob. most are of a Comovirus or a Potyvirus type, and typified by cowpea mosaic v. or blackeye cowpea mosaic v., respectively;

references to most of them and several synonyms are omitted.

— **mottle v.** Robertson 1966; Bozarth & Shoyinka, *Descr.V.* 212, 1979; isometric *c*.30 nm diam., contains ssRNA, transm. mainly by *Ootheca mutabilis*, Nigeria, many legumes are hosts.

— **necrosis v.**, reference as for cowpea chlorotic spot, isometric 25–30 nm diam., transm. aphid, India (N.W.), poss. = cucumber mosaic v.

— **ringspot v.** Phatak, Díaz-Ruis & Hull, *Phytopath.Z.* **87**:132, 1976; poss. Cucumovirus, isometric 26 nm diam., transm. sap, seedborne, Iran [56,3345].

— **rugose mosaic v.** dos Santos, Lin & Kitajima, abs. *Phytop.* **71**:890, 1981; poss. the same as cowpea green vein banding v.

— **severe mosaic v.** Smith 1924; Jager, *Descr.V.* 209, 1979; Comovirus, isometric *c*.25 nm diam., transm. sap readily, chrysomelid beetles, Americas; the severe str. of cowpea mosaic v. is now CSMV which is distinguished from CMV by: not infecting *Gomphrena globosa* systemically, a high middle to bottom component ratio and antigenic specificity; these properties and particle morphology distinguish CSMV from other viruses causing mosaics in *Vigna* spp.; Lin et al., *Phytop.* **74**:581, 1984, 2 new serotypes [63,4647].

— **stunt** (blackeye cowpea mosaic v. + cucumber mosaic v.), Pio-Ribeiro et al., *Phytop.* **68**:1260, 1978, synergistic interaction; Iwaki, *Rev.Pl.Prot.Res.* **12**:88, 1979; Indonesia, USA; these stunt diseases in the two countries are not the same; both BCMV and CMV are sap transm. and by aphids, non-persistent; the disease agent in Indonesia is not sap transm. and aphid transm. is persistent [58,3029].

— **Trinidad mosaic** = cowpea severe mosaic v.

— **yellow fleck** Sharma & Varma, *Indian Phytopath.* **29**:421, 1976; transm. *Bemisia tabaci*, India (N.); like cowpea bright yellow mosaic, golden and yellow mosaic [57,4718].

— — **mosaic** Ahmad, *PDR* **62**:224, 1978; transm. *Bemisia tabaci*, India, Pakistan; like cowpea bright yellow mosaic, golden mosaic and yellow fleck [57,5227].

——**witches' broom** Semangoen 1958; see cowpea stunt, Iwaki; transm. *Aphis medicaginis*, persistent, Indonesia; this is prob. the same as the form in this country and called cowpea stunt. Varma et al., see *Curr.Sci.* **47**:56, 1978, reported an assoc. MLO in India (N.) [57,4244].

cracked stem, celery, B deficiency; Purvis & Ruprecht, *Fla.agric.Exp.Stn.Bull.* 307, 1937 [16:792].

Craigie, John Hubert, 1887–, alive in Dec. 1986; born in Canada, Univ. Harvard, Minnesota, Manitoba; Dominion Rust Research Laboratory, Winnepeg 1925–45. In 1927 he discovered the 2 mating types of *Puccinia graminis*, i.e. the function of the pycnidia in dikaryotising the mycelium which leads to aeciospore formation. Hence the demonstration that inheritance in this rust followed Mendel's patterns and that new races could thereby arise. His numerous medals were donated to the Buller Memorial Library in Winnipeg shortly before he reached the age of 98. *A.R.Phytop.* **18**:19, 1980; *Can.J.Pl.Path.* **2**:1, 1980; *Tableau, Agric.Can.Res.Br.* Jan./Feb. 1986.

crameri, *Ustilago*.

cranberry, see *Vaccinium*.

— **false blossom** Dobroscky, *Science N.Y.* **70**:635, 1929; 1931; MLO assoc., transm. *Scleroracus striatulus*, USA (N.E.) [**9**:324; **10**:739].

— **ringspot** Boone, *PDR* **50**:543, 1966; USA (Wisconsin) [45,3582].

crassa, *Alternaria*.

Crataegus oxycantha, common hawthorn.

craterium, *Urnula*.

crater rot (*Rhizoctonia carotae*) carrot, celery, — spot (*Thanatephorus cucumeris*) celery, — stunt (*T. cucumeris* and prob. other fungi) wheat, Smith et al., *PD* **68**:582, 1984; Deacon & Scott, *TBMS* **85**:319, 1985 [63,5360; 64,5359].

crazy top (*Sclerophthora macrospora*) maize; in cotton related to soil compaction, Watkins 1981.

creeping bugle weed, *Ajuga reptans*; Shukla & Gough, *PD* **67**:221, 1983; viruses, Australia [62,3060].

— **soft grass**, *Holcus mollis*.

Criconema palmatum Siddiqi & Southey 1962; Orton Williams, *Descr.N.* 57, 1974; woody hosts.

Criconemoides morgensis (Hofmänner in Hofmänner & Menzel) Taylor 1936; Loof, *Descr.N.* 42, 1974.

Crimean yellows Valenta, *Phytopath.Z.* **33**:316, 1958; MLO assoc., flower proliferation in *Cuscuta campestris* [**38**:135].

crimp (*Aphelenchoides besseyi*, *A. fragariae*) strawberry.

crimson clover, see *Trifolium*.

— — **latent v.** Kenten, Cockbain & Woods, *AAB* **96**:79, 1980; isometric *c*.26 nm diam., contains ssRNA, transm. sap, seedborne, detected in seed from Europe, USA [60,2644].

Crinipellis Pat. 1889; Tricholomataceae; saprophytic except for one sp. which causes one of the most important diseases of cacao.

*C. **perniciosa** (Stahel) Singer 1942; Holliday, *Descr.F.* 223, 1970; prob. the only obligate pathogen in the Agaricales and causes the severe witches' broom of cacao; it occurs on wild

Theobroma spp. and cacao in the Amazon and Orinoco regions of S. America, on the Pacific side of the Andes in Colombia and Ecuador, in Trinidad, Tobago and Grenada. The fungus has been reported from Cuba but not the disease which is absent from the cacao crops of S.E. Brazil. The etiology was worked out by Stahel in 1915 and the life history by successive workers in Trinidad in 1943–55. Infection is caused only by the air dispersed basidiospores and initiated on any meristematic tissue; witches' brooms are formed in the canopy, the cusions and pods are destroyed. The infected tissue becomes necrotic and the sporophores then form from dikaryotic mycelium which, before necrosis sets in, is monokaryotic; the life cycle taking c.30 weeks. Infection is not systemic; each symptom pattern on the tree being the result of one infection. Crop destruction can be 80 % and the green canopy can become largely necrotic. Western and eastern forms of the fungus differ in pathogenicity, and others have been reported from other plants; the latter are not pathogenic to cacao. The heavy losses that have occurred recently, in W. Brazil, have been due to inadequate host resistance and a failure to apply the cultural control measures (broom removal) which were shown to be effective in Trinidad many years earlier; Baker & Holliday, *Phytopath.Pap.* 2, 1957, monograph; Holliday 1980; Evans, *Cocoa Growers Bull.* 32:5, 1981; Wheeler & Mepsted, *Proc.Int.Cocoa Res.Conf.*, Colombia 1981: 365, races; Rocha & Wheeler, *Pl.Path.* **34**:319, 1985; basidiomata formation, deposition & germination of basidiospores; Rudgard, *ibid* **35**:434, 1986, basidiomata formation on detached brooms in the field, Brazil (Rondonia) [**65**,629; **66**,1819].

crinkle, apple fruit, prob. due to a period of unusually hot weather; Carne & Martin, *J.Counc.scient.ind.Res.Aust.* 7:203, 1934 [**14**:242].

crinum potyvirus Pares & Bertus, *Phytopath.Z.* **91**:170, 1978 [**57**,4506].

Cristulariella Höhnel 1916; Hyphom.; emended by Redhead, *CJB* **53**:700, 1975; teleom. in *Grovesinia*; the hyphal conidiophore or propagulophore bears a multicellular propagule with internal branching, subglobose or cone shaped, hyaline, becoming brown with age, the branches are densely compacted and composed of clavate or subglobose cells which radiate from the conidiophore apex; the propagule functions as the conidium i.e. the infective unit; Holliday 1980 [**54**,4832].

C. depraedans (Cooke) Höhnel; propagule 100–150 μm diam.; Waterman & Marshall, *Mycologia* **39**:691, 1947; causes a leaf spot of

black olive, *Bucida buceras*; J. & H. Miller, *Proc.Fla.St.hort.Soc.* **88**:571, 1975, chemical control [**57**,2672].

C. moricola, teleom. *Grovesinia pyramidalis*; *Cristulariella pyramidalis* Waterman & Marshal 1947 is a synonym.

croci, *Uromyces*; **crocorum**, *Rhizoctonia*.

crocus breaking v. Rydén, *Phytopath.Z.* **80**:361, 1974; isometric. c.27 nm diam., transm. sap, Sweden [**54**,3962].

Cronartium Fr. 1815; Uredinales; telia erumpent, columnar or like hairs, teliospores in chains, aseptate, ± hyaline, wall smooth; aecia with a peridium, *Peridermium* spp., several cells thick with filaments, aeciospores verrucose, appearing tessellate in surface view; macrocyclic, telia on dicotyledons, aecia on pines which are the economically important hosts, the blister or pine stem rusts; Peterson & Jewell, *A.R.Phytop.* **6**:23, 1968, American stem rusts of pine; Peterson, *Rep.Tottori mycol.Inst.* 10:203, 1973; Ziller 1974; Hiratsuka & Powell, *Forestry Tech.Rep.* 83(4), 1976, Canadian stem rusts of pine [**53**,4342; **58**:402].

C. appalachianum Hepting, *Mycologia* **49**:898, 1957; anam. *Peridermium appalachianum* Hepting & Cummins 1952; telia on *Buckleya distichophylla*, aecia on *Pinus virginiana*, USA (E.), damage negligible, fide Boyce 1961.

C. coleosporioides Arthur 1907; anam. *Peridermium stalactiforme* Arthur & Kern 1906; Mordue & Gibson, *Descr.F.* 577, 1978; stalactiform blister rust, telia on *Castilleja* spp. and other dicotyledons, may damage *Pinus contorta* and *P. ponderosa* as young trees, N. America; Ziller, *CJB* **48**:1313, 1970 [**50**,1427a].

C. comandrae Peck 1879; anam. *Peridermium pyriforme* Peck 1875; Mordue & Gibson, *Descr.F.* 578, 1978; the obpyriform aeciospores distinguish it from other *Cronartium* spp., telia on *Comandra*, comandra blister rust, N. America, can cause moderate damage to hard pines.

C. comptoniae Arthur 1906; anam. *Peridermium comptoniae* Orton & Adams 1914; Mordue & Gibson, *Descr.F.* 579, 1978; telia on *Comptonia*, *Myrica gale*, sweet fern blister rust; N. America, can damage young stands of *Pinus banksiana*, *P. contorta*, *P. muricata*, *P. radiata*, *P. taeda*.

C. conigenum (Pat.) Hedgc. & Hunt, *Phytop.* 12:120, 1922, q.v. for differences from *Cronartium strobilinum*, telia on evergreen oaks, cone rust of *Pinus leiophylla*, Chihuahua pine; Mexico, USA (Arizona).

C. flaccidum (Alb. & Schwend.) Winter 1880; anam. *Peridermium cornui* Rostrup ex Kleb. 1890; Mordue & Gibson, *Descr.F.* 580, 1978; telia on many dicotyledons, *Vincetoxicum officinale* is an important host in Europe; Scots pine blister rust, resin canker or resin top, Eurasia; the germ tubes of aeciospores are long, growth is indeterminate and irregularly septate; those of *Endocronartium pini*, which also causes resin top of Scots pine, are short, growth is determinate and regularly septate; Ragazzi, *Phytopath.Z.* **108**:160, 1983, development on *V. officinale*; Ju et al. 1984, in China; Diamandis & Kam, *Eur.J.For.Path.* **16**:247, 1986; causing death of Scots pine in Greece (N.) [**63**,2530; **64**,2748; **66**,350].

C. fusiforme Hedgc. & Hunt ex Cummins 1956 = *Cronartium quercuum* f.sp. *fusiforme*.

C. himalayense Bagchee 1929, 1933, see Bakshi, *Forest pathology*, 1976; anam. *Peridermium himalayense* Bagchee which may be distinct from *P. pini*, see *Endocronartium* and *E. pini*; telia on *Swertia* spp., aecia on *Pinus canariensis*, *P. roxburghii*, Asia, pine felt rust.

C. occidentale Hedgc., Bethel & Hunt, *J.agric.Res.* **14**:413, 1918; anam. *Peridermium occidentale*, same authors; telia on *Grossularia*, *Ribes*; aecia on 1–2 needle pines, *Pinus edulis*, *P. monophylla*, not to be confused with *Cronartium ribicola* which infects only white, 5 needle pines, see Boyce 1961 for differences; piñon blister rust, USA.

*****C. quercuum** (Berk.) Miyabe ex Shirai 1899; anam. *Peridermium cerebrum* Peck 1873; telia on oaks; China, Japan, N. and Central America. Burdsall & Snow, *Mycologia* **69**:503, 1977, erected 4 ff.sp.: *banksianae*, primarily on jack pine, *Pinus banksiana*; *echinatae*, on short leaf pine, *P. echinata*; *fusiforme*, on loblolly and slash pines, *P. taeda* and *P. elliotii* var. *elliotii*, respectively; *virginianae* on Virginia pine, *P. virginiana*. The f.sp. *fusiforme*, fusiform rust, causes one of the most damaging diseases of forest trees in USA; incidence is highest in a corridor from central S. Carolina to S. Louisiana; the principal telial hosts are in the black oak group, most susceptible: *Quercus fakata* var. *pagodaefolia*, *Q. incana*, *Q. nigra*, *Q. phellos*; Powers et al., *A.R.Phytop.* **19**:353, 1981; Powers, *Eur.J.For.Path.* **14**:426, 1984, systemic fungicide for nurseries & host resistance [**64**,2753].

C. ribicola J.C. Fischer 1872; anam. *Peridermium strobi* Kleb. 1887; Laundon & Rainbow, *Descr.F.* 283, 1971; telia on *Grossularia*, *Ribes*, aecia on white, 5 needle pines; note that *Cronartium occidentale*, which has its telia on the same plant genera, only infects some 1–2 needle pines; Asia, Europe, N. America; white pine blister rust is a major disease; the fungus prob. originated in Asia. It appeared in USA, poss. in 1898, and was reported in the E. in 1906 and in W. Canada in

1910. The pathogen had been introduced on nursery stock of *Pinus strobus*, a native of N. America, from France and Germany; it caused enormous damage in N. American pines. The aeciospores spread hundreds of kilometres but the sporidia from currant to pines spread only 1–3 km. White pines in N. America are rather more susceptible than those in Eurasia. Yokota, *Eur.J.For.Path.* **13**:389, 1983, found that *P. monticola*, selected for resistance in USA, was susceptible to the Japan (Hokkaido) form and concluded that this form was a different race from the one in western N. America [**63**,3072].

C. strobilinum (Arthur) Hedgc. & Hahn, *Phytop.* **12**:113, 1922, q.v. for differences from *Cronartium conigenum*, telia on evergreen oaks; aecia on *Pinus elliotii*, slash pine, *P. palustris*, long leaf pine; southern cone rust, can destroy cones, USA.

crook rot (*Spongospora subterranea* f.sp. *nasturtii*) watercress.

crookwellense, *Fusarium*.

crop names, see plant names.

—— residue decomposition, effects on roots; Patrick et al., *A.R.Phytop.* **2**:267, 1964.

—— rotation, control of disease; Curl, *Bot.Rev.* **29**:413, 1963.

cross absorption test, a test of serological affinities of 2 antigens, e.g. viruses. This involves testing dilutions of an antiserum against the antigen used in its preparation after previously incubating the antiserum with an excess of dissimilar antigen, FBPP.

—— protection, see protection.

crotalariae, see *Calonectria pyrochroa*.

Crotalaria juncea, sunnhemp.

crotalaria mottle v. Bock, Guthrie & Meredith, 1977; filament 750 nm long, E. Africa, *C. incana*, *C. intermedia* [**57**,7e].

—— mucronata mosaic = sunnhemp mosaic v.

—— witches' broom Hadiwidjaja, *Tijdschr.PlZiekt.* **58**:1, 1952; Yang, *J.agric.Res.China* **28**:79, 1979; transm. *Nesophrosyne orientalis*; Indonesia (Java), Taiwan, sunnhemp, other *Crotalaria* spp. [**31**:557; **59**,2218].

—— yellow mosaic v. Igwegbe, *PD* **66**:74, 1982; Potexvirus, filament *c.*500 nm long, transm. sap, *Myz.pers.*, Nigeria, *C. spectabilis* [**61**,5437].

crotocinigenum, *Acremonium*.

croton yellow vein mosaic Nair & Wilson, *Agric.Res.J.Kerala* **7**:123, 1969; transm. *Bemisia tabaci*, India (Kerala), *C. sparsiflorus* [**50**,1136].

crown gall (*Agrobacterium*), **—— rot** (*Aspergillus niger*) groundnut, (*Deightoniella torulosa*) banana, (*Erwinia rhapontici*) rhubarb, (*Fusarium oxysporum* f.sp. *asparagi*) asparagus, (*Phytophthora cactorum*) strawberry,

(*P. cambivora, P. megasperma*) hollyhock, (*P. porri*) campanula, (*Thanatephorus cucumeris*) gerbera; —— **rust** (*Puccinia coronata*) oat, —— **sheath rot** (*Gaeumannomyces graminis* var. *graminis*) rice, grasses; —— **wart** (*Physoderma alfalfae*) lucerne.
cruciferarum, *Erysiphe.*
cruenta, *Mycosphaerella, Pseudocercospora, Sphacelotheca.*
crumbly fruit, raspberry, assoc. with virus infection, genetic disorder; Murant et al., *Hortic.Res.* **13**:49, 1973 [**53**,2629].
Crumenulopsis Groves, *CJB* **47**:48, 1969; Helotiaceae; asco. aseptate; on pines. *C. pinicola* (Rebent.) Groves; asco. 18–30 × 3–4 μm. *C. sororia* (P. Karsten) Groves, asco. 18–21 × 3–5 μm, anam. *Digitosporium piniphilum* Gremmen 1953; con. up to 60 μm long, 3–4 μm wide; the latter causes pine canker or resin bleeding; Batko & Pawsey, *TBMS* **47**:257, 1964, in Britain; Gremmen, *Bosbouw* **40**:176, 1968, review [**43**,3342; **47**,2874].
crus-galli, *Ustilago.*
*Cryphonectria (Sacc.) Sacc. 1905; Gnomoniaceae; Walker et al., *Mycotaxon* **23**:353, 1985, considered that the genus is distinct from, but very close to, *Endothia*; in *Cryphonectria* the asco. are 1 septate [**64**,5134].
C. cubensis (Bruner) Hodges, *Mycologia* **72**:547, 1980; asco. 6–8.5 × 2–3 μm; a wound pathogen which causes clove acute dieback and eucalyptus canker. *Cryphonectria gyrosa* (Berk. & Br.) Sacc. and *C. havanensis* (Bruner) Barr are also on eucalyptus; see Walker et al. above. The clove pathogen was transferred from *Cryptosporella* to *Endothia* by Reid & Booth, *CJB* **47**:1059, 1969, and to *Cryphonectria* by Hodges et al., *Mycologia* **78**:343, 1986; clove dieback became serious in Zanzibar because of the destructive harvesting methods which damaged the stems; Holliday 1980 as *Endothia*; clove: Nutman & Roberts, *PANS* **17**:147, 1971; Booth & Gibson, *Descr.F.* 363, 1973 as *E. eugeniae*; eucalyptus: Alfenas et al., *Eur.J.For.Path.* **13**:197, 1983, variations in virulence & resistance; *CJB* **62**:1756, 1984, isoenzyme & protein patterns of isolates differing in virulence; Sharma, *AAB* **106**:265, 1985, in India [**51**,1751; **63**,1414; **64**,320; **66**,327–8].
C. parasitica (Murrill) Barr 1978; Sivanesan & Holliday, *Descr.F.* 704, 1981; anam. *Endothiella*; asco. 7–12 × 3.5–3 μm, con. 3–5 × 1–1.5 μm; chestnut blight, causes stem girdling cankers, on *Acer, Castanea, Quercus*; shagbark hickory, *Carya ovata*; staghorn sumach, *Rhus typhina*; the American chestnut, *Castanea dentata*, is very susceptible but the European chestnut, *C. sativa*, is less so; widespread in the N. Hemisphere; the

fungus is saprophytic on the hosts other than chestnut. It appeared in the New York Zoological Park *c.*1904 and had prob. been introduced from E. Asia. The American chestnut was subsequently almost completely destroyed over its whole natural range, Appalachian mountains, *c.*25 years later; the pathogen later spread to Europe; it forms a toxin, diaporthin q.v. Avirulent or hypovirulent strs. are known; hypovirulence is under genetic control and is a condition that reduces pathogencity but not saprophytic vigour. Its presence has reduced blight in France and Italy; van Alfen, *A.R.Phytop.* **20**:349, 1982; Elliston, *Adv.Pl.Path.* **1**:1, 1982, *Phytop.* **75**:151, 170, 1405, 1985; Griffin et al., ibid **73**:1084, 1983, hypovirulence; see *Endothia*; Jaynes & De Palma, ibid **74**:296, 1984, seed infection [**63**,884; 4087; **64**,3201–2; **65**,2449].
cryptica, *Mycosphaerella.*
*cryptic viruses, also called temperate viruses; isometric *c.*30 nm diam., contain dsRNA, transm. pollen and seed only, cause no symptoms; Kassanis, *Phytopath.Z.* **111**:363, 1984; Francki et al., vol. 2, 1985; Kühne et al., *Arch.Phytopath.PflSchutz.* **22**:179, 1986; Natsuaki et al., *Intervirology* **25**:69, 1986 [**64**,2387; **65**,4775; **66**,85].
Cryptococcus fagisuga, see *Nectria coccinea.*
Cryptodiaporthe Petrak 1921; Valsaceae; Barr 1978; ascomata below a stromatic disc, conidial locules usually in separate stromata, asco. 1 septate; Butin, *Phytopath.Z.* **32**:399, 1958, on *Populus, Salix* [**38**:104].
C. lebiseyi (Desm.) Wehm. 1933; anam. *Phomopsis lebiseyi* (Sacc.) Died., fide Wehmeyer; asco. 9–11 × 2–2.5 μm; assoc. with sycamore diamond bark canker in England, see *Dichomera saubinetii.*
C. populea (Sacc.) Butin 1958; anam. *Discosporium populeum* (Sacc.) Sutton 1977, fide Sutton 1980; *Chondroplea populea* (Sacc. & Briard) Kleb., in Cannon et al. 1985; Booth et al., *Descr.F.* 364, 1973; asco. 16–23 × 6–9 μm, con. 9–11 × 7–8 μm; poplar bark necrosis, canker or trunk scab; the disease can be a major problem in young stands when assoc. with faulty cultivation, also can damage older stands; Gremmen, *Eur.J.For.Path.* **8**:362, 1978, review of work in Netherlands; Phillips & Burdekin 1982 [**58**,5035].
C. salicella (Fr.) Petrak 1921; anam. *Discella salicis* (Westend.) Boerema 1970, in Cannon et al. 1985; *Diplodina microsperma* (Johnston) Sutton 1977, fide Sutton 1980. *Cryptodiaporthe salicina* (Pers.) Wehm. 1933; poss. anam. formerly in *Discella carbonacea* (Fr.) Berk. & Br., see Sutton, *Mycol.Pap.* 141, 1977; asco. in both spp. are 15–22 μm long but they differ in width. The 2

teleom. have been confused, they are assoc. with willow dieback and canker but the disease(s) seem to be of little significance; Broekhuijsen, *Tijdschr.PlZiekt.* **40**:62, 1934; Bier, *CJB* **37**:229, 1959; **39**:139, 1961 [**13**:550; **38**:631; **40**:493]. **cryptogea**, *Phytophthora*.

cryptogram, a descriptive code or cipher which summarises the main properties of a virus or a virus group, Gibbs et al., *Nature Lond.* **209**:450, 1966. Each cryptogram consists of 4 pairs of symbols meaning: (1) nucleic acid and strandedness; (2) molecular weight of nucleic acid and the amount in infective particles; (3) outline or shape of particle and nucleocapsid; (4) kinds of host and vector; other symbols show where knowledge is uncertain or unknown. It appears that the device is not used as much as might have been expected. It may be that it takes longer to construct or decipher a cryptogram than it does to use words and figures. The situation is analogous to, if much more complex than, the use of acronyms q.v.; Gibbs & Harrison 1976.

Cryptosphaeria Grev. 1824; Valsaceae, Cannon et al. 1985; Hawksworth et al. 1983 gave Grev. 1822 and placed the genus in Diatrypaceae, see Diatrypales; ascomata not erumpent or well delimited, solitary or in small groups, asco. aseptate; *C. populina* (Pers.) Sacc. 1882, canker of *Populus tremuloides*, USA (Rocky mountains), Hinds, *Phytop.* **71**:1137, 1981 [**61**,3094].

Cryptosporella Sacc. 1877; Melanconidaceae; stroma may only be a discoloured area, ascomata embedded in bark, beaks erumpent. *C. umbrina* (Jenkins) Jenkins & Wehm., *Phytop.* **25**:886, 1935; asco. 8–11.2 × 3.2–4 μm, aseptate, rose brown canker, Jenkins, *J.agric.Res.* **15**:593, 1918. *C. viticola* Shear, *Phytop.* **1**:119, 1911, is a poss. teleom. of *Phomopsis viticola* [**15**:155].

Cryptosporiopsis Bubák & Kabát 1912; Coelom.; Sutton 1980; conidiomata acervular to stromatic, conidiogenesis enteroblastic phialidic; con. aseptate, hyaline, large, base abruptly tapered to a distinct truncate scar; **C. corticola**, teleom. *Pezicula corticola*; **C. malicorticis**, teleom. *P. malicorticis*. Taylor & Moore and Taylor, *Phytop.* **69**:236, 1979; *PD* **67**:984, 1983, described an unidentified sp. causing a canker of red maple, *Acer rubrum*, and assoc. with oviposition wounds caused by *Oecanthus niveus*, USA [**59**,1406; **63**,881].

Cryptostroma Gregory & Waller, *TBMS* **34**:593, 1951; Hyphom.; conidiomata stromatic, the stroma is very extensive with a base and roof separated by a cavity and connected by stromatic columns; conidiogenesis holoblastic, conidiophore growth percurrent; con. aseptate, pigmented.

C. corticale (Ell. & Ev.) Gregory & Waller; Sutton & Gibson, *Descr.F.* 539, 1977; con. 5–12 × 3.5–4 μm; sycamore sooty bark, *Acer pseudoplatanus* and other *Acer* spp.; saprophytic on wood and timber; Europe, N. America. The fungus was originally described from wood but in 1945 it caused an outbreak of the disease in London, England; further outbreaks occurred in the country in the 1970s. Although, as a saprophyte, it has long been widespread, disease prob. appears only in hot summers when trees are under water stress; growth is greater in sycamores at 25° than at 15° especially in stressed trees; Abbott et al., *TBMS* **69**:507, 1977, attraction for the grey squirrel, *Sciurus carolinensis*; Dickenson & Wheeler, *ibid* **76**:181, 1981, growth, temp. & water stress; Alcock & Wheeler, *Pl.Path.* **32**:173, 1983, variability in growth; Bevercombe & Rayner, *ibid* **33**:211, 1984, compatibility groups [**57**,3118; **61**,2453; **62**,4453; **63**,4572].

cryptotaenia witches' broom Okuda & Nishimura, *Ann.phytopath.Soc.Japan* **40**:439, 1974; MLO assoc., transm. *Macrosteles orientalis*, Japan (Kanto), *C. japonica* [**55**,1575].

cubensis, *Cryphonectria*, *Pseudoperonospora*.

cubical heart rot (*Laetiporus sulphureus*).

cucumber, *Cucumis sativus*, and see cucurbits; Bassett & Derbyshire, *Booklet Min.Agric.Fish.Fd* UK 2096, part 7, 1979; Sumner et al., *PD* **65**:401, 1981, **67**:1071, 1983, chemical control [**61**,2528; **63**,3642].

—— **aucuba mosaic** (cucumber green mottle mosaic v.).

—— **fruit streak** Gallitelli, Vovlas & Avgelis, *Phytopath.Z.* **106**: 149, 1983; a closely related str. of cucumber leaf spot v. [**62**,3318].

—— **green mottle mosaic v.** Ainsworth 1935; Hollings et al., *Descr.V.* 154, 1975; Tobamovirus, rod *c.*300 × 18 nm, transm. sap, by plant handling, in soil; strs. cause different symptom patterns; Rao & Varma, *Phytopath.Z.* **109**:325, 1984, prob. transm. *Aulacophora fevicollis*; Wang & Chen, *Pl.Prot.Bull.Taiwan* **27**:105, 1985; str. from *Lagenaria siceraria*, Okada in van Regenmortel & Fraenkel-Conrat 1986 [**63**,4610; **65**,955].

—— **leaf spot v.** Weber et al. 1982; Weber, *Descr.V.* 319, 1986; isometric *c.*28 nm diam., contains RNA, transm. sap, seed; occurs naturally only in cucumber, affected plants are severely stunted; Britain, Germany, Greece, Jordan.

—— **mosaic v.** Doolittle 1916; Francki et al., *Descr.V.* 213, 1979; Cucumovirus, isometric *c.*28 nm diam., transm. sap, > 60 spp. aphids, seed; a very wide host range, mosaics of many angiosperms, including woody plants, see Martyn 1968, 1971; many symptom variants and strs., satellite carna 5 RNA q.v.; Quiot et al. in Plumb & Thresh 1983, ecology in warmer regions; Alberts et al., *Aust.J.agric.Res.* **36**:267, 1985, epidemic on lupins [**64**,5436].

— **necrosis v.** McKeen 1959; Dias & McKeen, *Descr.V.* 82, 1972; poss. Necrovirus, isometric 31 nm diam., transm. sap, *Olpidium radicale*, soilborne, Canada, Cucurbitaceae, other plant families; Stobbs et al., *Can.J.Pl.Path.* **4**:134, 1982 [**61**,7217].

— **pale fruit viroid** van Dorst & Peters, *Neth.J.Pl.Path.* **80**:85, 1974; young leaves small, blue green, rugose; flowers stunted and crumpled, fruit pale green, smaller; internodes shorter; disease is more severe at 30°; transm. in pruning, cucumber the only natural host, infects other cucurbits and tomato, Netherlands, in greenhouses, Sano et al., *Ann.phytopath.Soc.Japan* **47**:599, 1981, molecular weights & pathogenicity of the viroids of hop stunt & cucumber pale fruit were alike but that of potato spindle tuber differed; Uyeda et al., *ibid* **50**:331, 1984, purification [**61**,5575; **64**,1339].

— **ringspot v.** McKeen 1957; Spire et al., *Annls. Épiphyt.* hors sér. **17**:121, 1966; isometric, transm. sap, *Myz.pers.*; Canada, France [**36**:371; **46**,1492m].

— **soilborne v.** Koenig et al. 1982; *Phytop.* **73**:515, 1983; isometric 31 nm diam., contains RNA, transm. sap, Lebanon [**62**,3685].

— **stunt mottle** (arabis mosaic v.).

— **systemic necrosis** (tobacco necrosis v.), van Koot & van Dorst, *Tijdschr.PlZiekt.* **61**:163, 1955; Thomas & Fry, *N.Z.J.agric.Res.* **15**:857, 1972 [**35**:266, **52**,1317].

— **toad's skin v.** Lecoq, *Revue hort.* 223:15, 1982; bacilliform, transm. sap, France (S.E.) [**61**,3131].

— **vein yellowing v.** Harpaz & Cohen, *Phytopath.Z.* **54**:240, 1965, first described as bottle gourd mosaic; Cohen & Nitzany, *Phytopath.Mediterranea* **1**:44, 1960; Sela et al., *Phytop.* **70**:226, 1980; rod 740–800 × 15–18 nm, contains dsDNA, transm. sap, *Myz.pers.*; Israel [**41**:196; **45**,2314; **60**,758].

— **yellows v.** Yamashita et al., *Ann.phytopath.Soc.Japan* **45**:484, 1979; filament c.1000 × 12 nm, transm. *Trialeurodes vaporariorum*, Japan (Kanto); cucumber, melon [**59**,4816].

cucumerina, *Alternaria*; **cucumerinum**, *Cladosporium*; **cucumeris**, *Micronectriella*, *Thanatephorus*.

cucumeropsis mosaic v. Igwegbe, *PD* **67**:315, 1983; filament av. 740 nm long, transm. sap, *Myz.pers.*, Nigeria, *C. edulis* [**62**,3317].

cucumis, *Ascochyta*.

Cucumis melo, cantaloup, melon, muskmelon.

Cucumoviruses, type cucumber mosaic v.; other members; peanut stunt, robinia mosaic, tomato aspermy; isometric 29 nm diam., 4 segments linear ss RNA, the 3 largest required for infection, sedimenting at 99S, thermal inactivation 70°, longevity in sap 4 days, transm. sap and aphids, non-persistent; only this group, the Bromoviruses and Ilarviruses have a tripartite genome; CMV has been isolated from 775 plant spp. in 174 genera and 52 families; Kaper & Waterworth in Kurstak 1981; Francki, vol. 1, 1985; Francki et al., vol. 2, 1985.

Cucurbita, *C. foetidissima* or *C. digitata*, buffalo gourd; *C. maxima*, pumpkin, winter squash; *C. moschata*, *C. mixta*, winter squash; *C. pepo*, marrow, pumpkin, summer squash.

cucurbitacearum, *Septoria*; **cucurbitae**, *Phomopsis*, *Ulocladium*; **cucurbitarum**, *Choanephora*.

cucurbit downy mildew (*Pseudoperonospora cubensis*).

— **phyllody**, Sastry & Singh, *Curr.Sci.* **50**:955, 1981; MLO assoc., India (S.) [**61**,4425].

— **ring mosaic** = squash mosaic v.

— **ringspot** (tobacco ringspot v.), Sinclair & Walker, *PDR* **40**:19, 1956; McLean & Meyer, *ibid* **45**:137, 1961 [**35**:810; **40**:505].

cucurbits, Cucurbitaceae; Sitterly, *A.R.Phytop.* **10**:471, 1972, breeding for resistance; Lovisolo, *Acta Hortic.* 88:33, 1980; Bhargava, *J.Indian bot.Soc.* **62**:1, 1983; Nameth et al., *PD* **70**:8, 1986, viruses.

cufraneb, protectant fungicide, a complex dithiocarbamate containing Cu, Fe, Mn, Zn.

culmicolous smut (*Ustilago scitaminea*) sugarcane.

culmigenus, *Limonomyces*; **culmorum**, *Fusarium*.

culpad viruses, of cultivated plants; used by Harrison, *AAB* **99**: 195, 1981; they tend to have narrow host ranges; some groups are: Potex, spread by contact with infected plants or contaminated objects; Ilar, mainly in woody perennials, spread in pollen and seed; Tobamo, survive in crop debris, seed coats and contaminated surfaces, also spread by contact; barley yellow mosaic, a long survival in resting spores of fungal vector in soil, high incidence under monoculture. **Wilpad viruses**; they tend to have wide host ranges, long persistence in vectors, survive in wild plants; some groups are: Gemini and Luteo, spread by whiteflies and aphids or leaf hoppers, respectively, with a persistence in weeks; Nepo and Tobra, spread by nematodes, with a persistence mostly in months. Wilpad forms are adapted for long distance spread.

culture, see methods, preservation of fungi, tissue culture. A general term for a growth of a micro-organism in vitro or in vivo; its use in the sense of isolate or strain should be avoided. Also to cultivate a micro-organism, usually in vitro, in a prepared medium and in pure culture; but the term can be applied to maintaining an obligate parasite on its host. A culture medium is a prepared substrate, variously defined, for growing a micro-organism or other living cells in vitro, after FBPP.

cumin, *Cuminum cyminum.*

Cumminsiella Arthur 1933; Uredinales; like *Puccinia*, 2 germ pores in each cell of the teliospore; macrocyclic, autoecious; *Berberis, Mahonia*; Baxter, *Mycologia* **49**:864, 1957.

C. mirabilissima (Peck) Nannf. 1947; Sivanesan, *Descr.F.* 261, 1970; Parmelee & Carteret, *Fungi Canadenses* 288, 1984; teliospores 28–36 × 18–25 μm.

Cunningham, Gordon Herriot, 1892–1962; born in New Zealand, Univ. Wellington, a founder of mycology and plant pathology in this country, Hutton Memorial Medal 1935, Hector Medal 1949, CBE 1949; FRS 1950; wrote: *Fungous diseases of fruit trees in New Zealand*, 1925; *The rust fungi of New Zealand*, 1931; *Plant protection by the aid of therapeutants*, 1935; *The Gasteromycetes of Australia and New Zealand*, 1944. *Nature Lond.* **197**:17, 1963; *Biogr.Mem.Fel.R.Soc.* **10**:15, 1964.

cunninghamianum, *Gymnosporangium.*

cupressi, *Lepteutypa, Seiridium.*

Curcuma domestica, turmeric.

curcumae, *Pyricularia.*

curl tip, see *Pteridium aquilinum.*

currants, *Ribes*, see black and red currant.

curtisii, *Stagonospora.*

Curtobacterium Yamada & Komagata 1972; Gram positive, short rods, some pleomorphic, bending type of cell division, older cultures with some cocci but no definite rod and coccus cycle, usually motile with lateral flagella, not acid fast, no spores, form acid slowly and weakly from various sugars, G+C content of DNA 66–73 mol %, Bradbury 1986.

*C. **flaccumfaciens** (Hedges) Collins & Jones 1983; Hayward & Waterston, *Descr.B.* 43, 1965, as *Corynebacterium flaccumfaciens* (Hedges) Dowson 1942.

C.f.pv.betae (Keyworth et al.) Collins & Jones; Bradbury, *Descr.B.* 374, 1973, as *Corynebacterium betae* Keyworth, Howell & Dowson 1956; silvering of red beet and mangold, rare on sugarbeet, a vascular infection, plant dies sometimes in a few days, seedborne; England, Ireland.

*C.f.pv.**flaccumfaciens** (Hedges) Collins & Jones; vascular wilt of legumes, including *Glycine,*

Lablab, Phaseolus, Vigna, seedborne; soybean tan spot, Dunleavy, *PD* **68**,774, 1984; **69**:1036, 1985, yield loss & field spread [**64**,860; **65**,4182].

C.f.pv.oortii (Saaltink & Maas Geesteranus) Collins & Jones; Bradbury, *Descr.B.* 375, 1973, as *Corynebacterium oortii* Saaltink & Maas Geesteranus 1969; a systemic infection of tulip, bulb yellow pustule and leaf hell fire; Denmark, England, Japan, Netherlands.

C.f.pv.poinsettiae (Starr & Pirone) Collins & Jones; poinsettia, watersoaked stem and petiole streaks, leaf spots, stem cracks; Britain, New Zealand (N. Island), USA, poss. Romania.

curvatum, *Colletotrichum.*

Curvularia Boedijn 1933; Hyphom.; Ellis 1971; conidiomata synnematal, conidiogenesis enteroblastic tretic; con. curved, usually 3–4 septate; 1 or 2 of the middle cells larger, pigmented; Alcorn, *Mycotaxon* **16**:353, 1983, compared the genus with *Bipolaris* and *Drechslera* q.v.; Sivanesan considered that *Bipolaris* q.v. was not morphologically different from *Curvularia*; saprophytes, weak pathogens, secondary invaders; often assoc. with seed; some with teleom. in *Cochliobolus*; Holliday 1980; Muchovej & Couch 1985, review [**65**,3367].

C. **cymbopogonis,** teleom. *Cochliobolus cymbopogonis.*

C. **eragrostidis** (Henn.) J. Meyer 1959; teleom. described as *Pseudocochliobolus eragrostidis* Tsuda & Ueyama, *Trans.mycol.Soc.Japan* **26**:322, 1985, this genus is a synonym of *Cochliobolus*, fide Sivanesan 1984; con. 22–23 × 10–18 μm; maize leaf spot, oil palm seedling blight; Nelson, *PDR* **40**:210, 1956; Johnston, *Malay agric.J.* **42**:14, 1959 [**36**:21; **38**:761].

C. **geniculata,** teleom. *Cochliobolus geniculatus.*

C. **ischaemi** McKenzie, *TBMS* **77**:446, 1981; con. 24–35 × 13–17.5 μm; on *Ischaemum indicum*, Batiki bluegrass, an important pasture grass in Fiji and Solomon Islands, eyespot, causes premature leaf death.

C. **lunata,** teleom. *Cochliobolus lunatus.*

C. **trifolii** (Kauffm.) Boedijn; con. 28–38 × 12–16 μm; *Trifolium*; a f.sp. *gladioli* Parmelee & Luttrell 1956, only infects gladiolus, Laundon, *Descr.F.* 307, 1971; large, dark lesions are caused on the corms, young plants are killed; Moore 1979.

cushion canker (*Phytophthora palmivora*) cacao.

cuticle, of a plant; Martin, *A.R.Phytop.* **2**:81, 1964, role in resistance; MacNamara & Dickinson in Blakeman 1981, microbial degradation; Kolattukudy, *A.R.Phytop.* **23**:223, 1985, enzymic penetration by fungus pathogens.

cutinases, cutinolytic enzymes; van Ende &

Linskens, *A.R.Phytop.* **12**:247, 1974;
Kolattukudy, *A.Rev.Pl.Physiol.* **32**:539, 1981;
Dickman & Patil, *Phytop.* **76**:473, 1986 [**65**,5910].
Cyamopsis tetragonolobus, cluster bean, guar.
cyanogena, *Gibberella*.
*****cycas necrotic stunt v.** Kusunoki et al.,
Ann.phytopath.Soc.Japan **52**:302, 1986;
Nepovirus, isometric 29 nm diam., transm. sap,
host range limited, seed transm. from
Chenopodium amaranticolor, *C. serotinum*; *Cycas
revoluta* [**66**,1024].
cyclamen, largely from *Cyclamen persicum*;
Garibaldi 1981; Reimherr 1985 [**61**,2911;
65,4959].
Cyclaneusma DiCosmo, Peredo & Minter,
Eur.J.For.Path. **13**:208, 1983; Rhytismataceae;
asco. filiform, hyaline, smooth, 2 septate, mucous
caps at each end, on conifer needles [**63**,1437].
C. minus (Butin) DiCosmo et al.; Millar & Minter,
Descr.F. 659, 1980, as *Naemacyclus minor* Butin;
asco. 65–100 × 2.5–3 μm; pine premature needle
cast; Gadgil, van der Pas et al., *N.Z.J.For.Sci.*
14:179, 197, 210, 215, 1984, biology & chemical
control [**65**,3008–11].
C. niveum (Pers.:Fr.) DiCosmo et al.; Minter &
Millar, *Descr.F.* 660, 1980, as *Naemacyclus niveus*
(Pers.:Fr.) Fuckel; asco. 75–120 × 2.5–3.5 μm;
weakly parasitic on pines.
Cydonia oblonga, quince.
Cylindrocarpon Wollenw. 1913; Hyphom.;
Carmichael et al. 1980; conidiomata sporodochial,
conidiogenesis enteroblastic phialidic; con.
multiseptate, hyaline; con. very like those of
Fusarium, differing only in that they have no heel
to the foot cell, i.e. both ends are rounded;
teleom. *Nectria*; mostly saprophytic, often in soil,
some are weak pathogens; Booth, *Mycol.Pap.*
104, 1966; Holliday 1980.
C. candidum, teleom. *Nectria coccinea*;
Cylindrocarpon cylindroides, teleom.
N. neomacrospora; **C.c.** var. **tenue**, teleom.
N. fuckeliana; **C. destructans**, teleom. *N. radicicola*.
C. didymum (Hartig) Wollenw. 1926; con. mostly 2
septate, 24–36 × 4–6 μm; Olexa & Freeman, *PDR*
62:283, 1978, gall on *Rhizophora mangle*, red
mangrove, USA (Florida) [**57**,5367].
C. faginatum, teleom. *Nectria coccinea* var.
faginata; **Cylindrocarpon heteronemum** or **C. mali**,
teleom. *N. galligena*.
C. musae Booth & Stover, *TBMS* **63**:506, 1974;
con. 44–62 × 5–6 μm; commonly assoc. with
banana rhizome and root lesions caused by
Radopholus similis [**54**,2905].
C. panacis Matuo & Miyazawa,
Trans.mycol.Soc.Japan **9**:111, 1969; con. 1
septate, mostly 14–26 × 3.5–5 μm; 3 septate,

mostly 36–46 × 5–6 μm; ginseng dry root rot
[**49**,229].
Cylindrocladiella Boesewinkel, *CJB* **60**:2289, 1982;
Cylindrocladium spp. with small con., differs also
in conidiophore branching [**62**,925].
Cylindrocladium Morgan 1892; Hyphom.;
Carmichael et al. 1980; conidiomata hyphal,
conidiogenesis enteroblastic phialidic; con.
cylindrical, hyaline, 1 to multiseptate; the main
axis, or sterile filament, of the penicillate
conidiophore mostly forming a long, unbranched
thread, which ends in a globose or clavate apex,
the vesicle, this is characteristic; some spp. with
teleom. in *Calonectria*; Alfenas et al.,
Fitopat.Brasileira **4**:445, 1979, 3 spp. on
eucalyptus leaf spots; Boesewinkel, *TBMS*
78:553, 1982, heterogeneity & teleom. [**59**,5937].
C. avesiculatum Gill, Alfieri & Sobers, *Phytop.*
61:60, 1971; conidiophore main axis without a
vesicle, con. 51–78 × 3.5–4.7 μm; leaf spot,
defoliation and twig dieback of *Ilex* spp., USA
[**50**,2992].
C. braziliensis (Bat. & Cif.) Peerally, *Descr.F.* 427,
1974; conidiophore main axis with a vesicle, con.
24–38 × 2–2.8 μm, 1 septate; eucalyptus leaf spot,
damping off, defoliation and dieback, Brazil.
C. camelliae, teleom. *Calonectria camelliae*.
C. clavatum Hodges & May 1972; Peerally,
Descr.F. 422, 1974; conidiophore main axis with a
vesicle, con. 36–60 × 3–5 μm, 1 septate;
microsclerotia and chlamydospores formed in
culture; assoc. with conifer and other root rots;
Bolkan et al., *PD* **65**:47, 1981, pathogenicity of
an isolate causing potato tuber rot; Lopes &
Reifschneider, *ibid* **66**:951, 1982, pea foot rot
[**60**,6630; **62**,455].
C. colhounii, teleom. *Calonectria colhounii*;
Cylindrocladium ilicicola, teleom. *C. pyrochroa*;
C. scoparium, teleom. *C. kyotensis*.
cylindroides, and var. **tenue**, *Cylindrocarpon*.
Cylindrosporella Höhnel 1916 = *Asteroma*, fide
Sutton, *Mycol.Pap.* 141, 1977; *C. ulmea* has been
reported as the anam. of *Stegophora ulmea*.
Cylindrosporium Grev. 1823; Coelom.; Sutton
1980, requires revision; conidiomata acervular,
conidiogenesis enteroblastic phialidic; con.
aseptate, hyaline; **C. concentricum**, teleom.
Pyrenopeziza brassicae; **C. juglandis**, teleom.
Mycosphaerella juglandis.
C. pomi Brooks 1908; poss. teleom. *Mycosphaerella
pomi* (Pass.) Lindau 1897; con. 2–3.5 × 1 μm;
Brook's fruit spot of apple; Brooks & Black,
Phytop. **2**:63, 1912; Thomas 1923; Walton &
Orton, *Science N.Y.* **63**:236, 1926; Tsuyama et al.,
Rep.Tottori mycol.Inst. **10**:465, 1973; Yoder, *PD*
66:564, 1982, chemical control; Penrose,

Australasian Pl.Path. **13**:23, 1984; description, in Australia (New South Wales) [3:43; 5:432; **53**,3088; **61**,7075; **64**,2081].

C. rubi, teleom. *Sphaerulina rubi.*

Cymadothea Wolf 1935; Dothidiaceae; the genus is monotypic.

*C. trifolii (Pers.:Fr.) Wolf 1935, fide Booth & O'Rourke, *Descr.F.* 393, 1973; *C. trifolii* (Killean) Wolf 1935, fide Sivanesan 1984; anam. *Polythrincium trifolii* Kunze 1917; dark stroma in which ascomata arise in locules, erumpent with conical apices; asco. hyaline, 1 septate, 22–27 × 6–7.5 μm, con. 1 septate, 24 × 13–24 μm, anam. described by Ellis 1971; *Trifolium* sooty or black blotch; affects herbage yields, causes stunt and partial defoliation; toxic to animals.

cymbalariae, *Eremothecium.*

cymbidium mosaic v. Jensen 1951; Francki, *Descr.V.* 27, 1970; Potexvirus, filament *c.*475 × 13 nm, transm. sap, orchids; Hamilton & Valentine, *Can.J.Pl.Path.* **6**:185, 1984, in pollen & odontoglossum ringspot v. [**64**,1618].

—— **rhabdovirus** Hakkaart & van Balen, *Vakblad Bloemistrerij* **35**(11):34, 1980; Netherlands [**60**,5002].

—— **ringspot** v. Hollings & Stone 1963; *Descr.V.* 178, 1977; Tombusvirus, isometric *c.*30 nm diam., transm. sap; Martelli & Russo, *J.Ultrastruct.Res.* **77**:93, 105, 1981, fine structure [**61**, 2176–7].

cymbopogonis, *Cochliobolus, Curvularia.*

Cynara, *C. cardunculus*, cardoon; *C. scolymus*, artichoke, globe.

cynara rhabdovirus Peña-Iglesias et al., *Anales Inst.Nac.Investig.Agrarias Protecc.Veget.* **2**:123, 1972; bacilliform *c.*260 × 70 nm, transm. sap; Russo et al., *Phytopath.Z.* **83**:223, 1975; Italy, Spain [**52**,3906; **55**,2057].

cynodon chlorotic streak v. Lockhart et al., *Phytop.* **75**:1094, 1985; Phytorhabdovirus, particle *c.*240 × 72 nm, leaf dip preparation; Mediterranean, Bermuda grass, causes stunt and chlorotic streak in this plant and in maize, transm. *Toya propinqua* [**65**,2230].

Cynodon dactylon, Bermuda or star grass.

cynodontis, *Bipolaris, Cochliobolus, Phyllachora, Puccinia, Ustilago.*

cynosorus mottle = lolium mottle v.

cyperi, *Balansia.*

Cyphomandra betacea, tamarillo, tree tomato.

cypripedii, *Erwinia.*

Cyrtopeltis nicotinae, transm. velvet tobacco mottle.

Cyrtosperma, see taro.

cyrtosperma bacilliform Jones, Shaw & Gowanlock, *Australasian Pl.Path.* **9**(3):5, 1980; particle 121–133 × 26–29 μm, Solomon Islands, *C. johnstonii* [**60**,3807].

cyst nematodes, *Heterodera*

cytisi, *Peronospora.*

cytokinins, growth regulating substances in plants; many micro-organisms which are parasitic or symbiotic on plants form cytokinins which are assoc. with some disease symptoms; Greene, *Bot.Rev.* **46**:25, 1980.

Cytospora Ehrenb.: Fr. 1823; Coelom., fide Sutton 1980; conidiomata stromatic, conidiogenesis enteroblastic phialidic; con. small, aseptate, hyaline, allantoid, extruded in coloured droplets or tendrils; teleom. in *Leucostoma* and *Valsa*; Holliday 1980.

C. abietis, teleom. *Valsa abietis*; **Cytospora chrysosperma**, teleom. *V. sordida*; **C. cincta**, teleom. *Leucostoma cinctum.*

C. eucalypticola Westhuizen, *S.Afr.For.J.* **54**:10, 1965; con. 3–4 × 0.7–1 μm; eucalyptus canker; the author of the sp. considered other spp. on *Eucalyptus* [**45**,620].

C. kunzei, teleom. *Valsa kunzei*; **Cytospora leucostoma**, teleom. *Leucostoma persoonii*; **C. nivea**, teleom. *L. niveum.*

C. personata Fr.; Kastirr & Ficke, *Arch.Phytopath.PflSchutz.* **20**:383, 1984, found this sp. to be the most frequent and aggressive of *Cytospora* from severe bark necroses on apple in Germany [**64**,2632].

C. pini, teleom. *Valsa pini*; **Cytospora rubescens**, teleom. *Eutypella sorbi.*

C. sacchari E. Butler 1906; Sivanesan, *Descr.F.* 777, 1983; con. 2.5–3.5 × 1.5 μm; sugarcane sheath rot, unimportant.

D

Dactuliophora Leakey, *TBMS* **47**:341, 1964; forms a sclerotium and sclerotiophore which appear to be functionally analagous to the conidium and conidiophore of a Hyphom. *D. glycines* is the sclerotial state of *Pyrenochaeta glycines*; Mukiibi, *TBMS* **52**:496, 1969, described the morphogenesis of *D. tarri* Leakey.

dactylidis, *Entyloma, Pyrenophora.*

Dactylis glomerata, cocksfoot, orchard grass.

dactylorhiza yellowing Marwitz & Petzold, *Phytopath.Z.* **75**:360, 1972; MLO assoc., Germany [**52**,3717].

Dade, Harry Arthur, 1895–1978; born in England, Royal College of Science, London; Ghana 1921–35; Imperial, now CAB International, Mycological Institute, 1935–60, where he laid the foundation of the culture collection; wrote: *Gold Coast plant diseases* 1925, with R. H. Bunting; *Classwork with fungi* 1966, with J. Gunnel. *Bull. Br. Mycol. Soc.* **13**:74, 1979.

daffodil, *Narcissus.*

dagger nematodes, *Xiphinema.*

dahliae, *Verticillium.*

dahlia mosaic v. Brandenburg 1928; Brunt, *Descr. V.* 51, 1971; Richins & Shepherd, *Virology* **124**:208, 1983; Caulimovirus, isometric *c.*50 nm diam., transm. sap, aphid [**62**,1850].

— **variegation** (tobacco streak v.), Bellardi et al., *Phytopath. Mediterranea* **24**:215, 1985; Italy [**65**,5558].

dak pora (*Ditylenchus angustus*) rice.

dalbergiae, *Phyllactinia.*

Dalbulus elimatus, transm. maize bushy stunt, *Spiroplasma kunkelii*; *D. maidis*, transm. maize bushy stunt, maize rayado fino, maize red leaf stripe, *S. kunkelii.*

damage threshold, see disease warning, loss of crop; the level at which damage to the crop begins to exceed the cost of control of any pathogen causing disease, i.e. where economic loss begins; Zadoks, A. R. *Phytop.* **23**:455, 1985, on crop loss assessment & the threshold theory.

damping off, collapse and death of seedlings which results from a lesion caused by a pathogen at *c.* soil level. The term can be divided into pre-emergent, i.e. death before seedlings have appeared above soil level, and post-emergent as at first defined. Damping off is usually found in wet, nursery conditions and is often caused by *Pythium* spp. Varying the density of young plants has a similar effect on the frequency of primary infection as that caused by a variation in the inoculum potential. There is a simple negative relationship between the mean distance separating adjacent plants and both the rate of disease advance and that of disease multiplication in a randomly inoculated seedling stand; Burdon & Chilvers, *Aust. J. Bot.* **23**:899, 1975 [**55**,4337].

dandelion, *Taraxacum officinale.*

— **carlavirus** Dijkstra, Clement & Lohius, *Neth.J.Pl.Path.* **91**:77, 1985; particle normally 668 nm long, transm. sap, Netherlands, infected 24 out of 52 plant spp. [**64**, 4093].

— **latent v.** Johns, *Phytop.* **72**:1239, 1982; Carlavirus, filament 640 × 12–13 nm, transm. sap, *Myz.pers.*, non-persistent, Canada (British Columbia) [**62**,943].

— **yellow mosaic v.** Kassanis 1944, *AAB* **34**:412, 1947; Phillips & Brunt 1983; Bos et al., *Neth.J.Pl.Path.* **89**:207, 1983; isometric *c.*30 nm diam., transm. sap, *Myz.pers.* non-persistent, Europe, causes a severe veinal necrosis of lettuce of potential economic importance [**23**:372; **27**:214; **62**,2179; **63**,3146].

dangeardii, *Penicillium.*

daphne chlorotic mosaic Y v. Forster & Milne 1975; *N.Z.J.agric. Res.* **19**:359, 1976; prob. Potyvirus, filament normally 733 nm long, transm. sap, *Myz.pers.*, non-persistent, New Zealand, *D. odora* [**56**,1146].

— **latent X v.** Forster & Milne 1975; *Descr. V.* 195, 1978; Potexvirus, filament *c.*500 × 12 nm, transm. sap. New Zealand, *D. odora, D. oneorum*; infects cucumber and *Gomphrena globosa* unlike daphne leaf distortion S v.

— **leaf distortion S v.** Harrison et al. 1971; Forster & Milne, *N.Z.J.agric Res.* **18**:391, 1975; **21**:131, 1978; Carlavirus, filament normally 704–716 nm long, transm. *Myz.pers.*, New Zealand, symptoms only in *D. odora*, the virus is latent in other *Daphne* spp. and cvs.; Denmark 1980 in *D. mezereum* [**55**,3203; **57**, 5531; **61**,990].

dark border leaf spot (*Cercospora brassicicola*) brassicas.

dasheen, *Colocasia esculenta*, see taro.

— **bacilliform v.,** see colocasia bobone v.

*— **mosaic v.** Zettler et al. 1970; *Descr. V.* 191, 1978; Potyvirus, filament *c.*721 nm long, transm. sap, aphid spp., non-persistent, Araceae including *Colocasia, Xanthosoma.*

Dastur, Jahangir Fardunji, 1886–1971; born in India, Univ. Bombay, Agricultural Research

Institute, Pusa; a pioneer plant pathologist, work on: betel, castor, citrus, cotton. *Indian Phytopath.* **24**(2), 1971.

Dasyscyphus Gray 1821; Hyaloscyphaceae; ascomata usually ± stalked, mostly saprophytic, Dennis 1978.

D. ellisiana (Rehm) Sacc., emended Hahn & Ayers, *Mycologia* **26**: 169, 1934; asco. mostly 17–21 × 2–2.5; assoc. with cankers of Douglas fir, conifers, N. America [**13**:553].

D. pini (Brunch.) Hahn & Ayers, reference as above page 487; asco. 15–20 × 5–6 μm; pine twig canker, N. temperate [**14**:266].

D. pseudotsugae Hahn, *Mycologia* **32**:138, 1940; asco. 3.8–7.2 × 1.8–3.6 μm; assoc. with cankers of Douglas fir, N. America [**19**:503].

date palm, *Phoenix dactylifera*; Carpenter & Elmer, *Agric. Handb. USDA* 527, 1978; Djerbi 1983 [**63**,2984].

datura blister mosaic Chowfla & Sharma, *Acta Bot. Indica* **8**:184, 1980; transm. sap, aphid, India, a few hosts in Solanaceae, *D. stramonium* [**60**,5522].

—— **Colombian v.** Kahn & Bartels, *Phytop.* **58**:587, 1968; Potyvirus, filament *c*.720 nm long, transm. sap, *Myz.pers.*, non-persistent, Colombia (Putumayo), *D. candida*, *D. sanguinia*, see datura mosaics [**47**,2647].

*—— **distortion mosaic** Capoor & Varma, *Indian J.agric.Sci.* **22**:303, 1952; transm. sap, aphid spp. [**32**:627].

—— **mosaics**, 2 mosaics and a necrosis, all assoc. with filamentous particles, have been described; datura Colombian and datura shoestring viruses also have filaments. Qureshi & Mahmood, *Phytopath Z.* **93**:113, 1978, mosaic, particle 712 × 11–13 nm, India; Peralta, Beczner & Dezsery, *Acta phytopath.Acad.Sci.hung.* **16**:85, 1981, mosaic, particle *c*.750 nm long, Hungary; Badami & Kassanis, *AAB* **47**:90, 1959, necrosis, particle a filament, India, from *Solanum jasminoides*; all these viruses are sap transm. [**38**:523; **58**,3729; **61**,4303].

—— **quercina** = tobacco streak v.

—— **rhabdovirus** Kitajima & Costa, *Fitopat.Brasileira* **4**:55, 1979; particles 200–300 × 60–80 nm, transm. sap, *D. stramonium* [**58**,5279].

—— **shoestring v.** Giri & Agrawal, *Phytopath.Z.* **70**:81, 1971; Weintraub et al., *CJB* **51**:855, 1973; Potyvirus, filament 720–750 × 16 nm, transm. sap, from *Solanum jasminoides*, see datura mosaics, Badami & Kassanis [**50**,2999; **52**,3605].

Datura stramonium, Jimson weed, thorn apple.

dauci, *Alternaria*.

Daucus carota, carrot.

Davisomycella Darker, *CJB* **45**:1423, 1967; Hypodermataceae; *D. ampla* (Davis) Darker;

Minter & Gibson, *Descr.F.* 561, 1978; asco. aseptate, 60–130 × 8 μm, jack pine needle blight, *Pinus banksiana*, Canada.

dazomet, soil fumigant.

dead arm (*Eutypa armeniacae*) grapevine; —— **spur**, apple, poss. transm., Williams & Misley, *J.Am.Soc.hort.Sci.* **106**:622, 1981, USA (Washington State) [**61**,2342].

dearnessii, *Mycosphaerella*.

death, mangrove, see *Phytophthora vesicula*.

de Bary, Heinrich Anton, 1831–88; born in Prussia, Univ. Berlin, medicine; Strasburg Univ. chair from 1872. Ainsworth 1981 called him the outstanding mycologist of the 19th century, a meticulous worker who had a great impact on plant pathology; experimental proof of the life cycle of *Puccinia graminis* on barberry and wheat; made the first inoculations with *Phytophthora infestans*; a teacher of great influence; wrote: *Die Brandpilze*, 1853; *Morphologie und physiologie der Pilze, Flechten, und Myxomyceten*, 1866; edn. 2 1884; English translation by H. E. F. Garnsey, revised by I. B. Balfour, 1887. *Phytop.* **1**:1, 1911; *Phytopath. Classics* 11, 1969, transl. by R. M. S. Heffner, D. C. Arny & J. D. Moore; Ainsworth, *Introduction to the history of mycology*, 1976; *Mycologia* **70**:222, 1978; *A.R.Phytop.* **20**:27, 1982.

debaryanum, *Pythium*.

decay, decomposition of woody tissue by micro-organisms, FBPP; see wood decay.

Deccan hemp, see *Hibiscus*.

decemcellulare, *Fusarium*.

decline (*Helicotylenchus multicinctus*) banana, (*Phytophthora palmivora*) betel, (*Radopholus similis*) avocado, tea, (*Phomopsis foeniculi*) fennel; —— **canning peach**, a disorder assoc. with trunk cankers, wounds and crotch angles, Australia (Western), Doepel et al. and Cripps et al., *Aust.J.agric.Res.* **30**: 1089, 1979; **34**:517, 1983; —— **& mortality**, in *Chamaecyparis nootkatensis*, Alaska cedar, etiology unknown, USA (S. E. Alaska), Shaw et al., *PD* **69**:13, 1985 [**60**,1529; **63**,2952; **64**,2196].

decoratus, *Uromyces*.

decoy crop, one which affects a pathogen in such a way that its infection of a newly planted, different crop is lessened; these effects include the spontaneous germination of resting propagules or the dissipation of rhizomorphs. A well known, practical example of the latter is the planting of a leguminous cover in young rubber to reduce the inoculum potential of the white root rot pathogen, *Rigidoporus lignosus*.

decretus, *Necator*.

deep bark canker (*Erwinia rubrifaciens*) European walnut.

deformans, *Caeoma, Elytroderma, Hapalosphaeria, Taphrina.*

deformed fruit, papaya, as for bumpy fruit q.v.

degeneration, a gradual and continuing decline in health, usually of a vegetatively propagated crop and frequently caused by virus(es) infection; (*Rhizoctonia fragariae*) strawberry.

Deightoniella S. Hughes 1952; Hyphom.; Ellis 1971; conidiomata hyphal, conidiogenesis holoblastic, con. 1 septate or multiseptate, pigmented; conidiophore growth percurrent.

D. papuana D. Shaw, *Papua New Guin.agric.J.* **11**:77, 1959; con. 15–20 × 15–18 μm; *Saccharum* veneer blotch, lesions can be 60 cm long [**38**:624].

D. torulosa (Sydow) M. B. Ellis 1957; Subramanian, *Descr. F.* 165, 1968; con. usually 3–5 septate, 35–70 × 15–25 μm, banana black leaf spot, fruit swamp spot and black tip; brown leaf spot of abaca, a common inhabitant of banana debris; Holliday 1980.

Delacroix, Edouard Georges, 1858–1907; born in France, succeeded E. E. Prillieux at the Institut National Agronomique, Paris, wrote books on plant pathology including the first full account of diseases of tropical crops, *Maladies des plantes cultivées dans les pays chaud*, 1911. *Bull.Soc. mycol.France* **24**:48, 1908; Whetzel 1918, Ainsworth 1981.

deliense, *Pythium.*

delphinii, *Diplodina.*

delphinium, Ahmed & Bailiss, *J.hortic.Soc.* **50**:47, 1975, viruses in Britain [**54**,4520].

— **virescence and phyllody** Marwitz & Petzold, *Phytopath.Z.* **87**: 1, 1976; MLO assoc., Germany [**56**, 2086].

— **yellows** Posnette & Ellenberger, *AAB* **51**:69, 1963; transm. *Macrosteles sexnotatus*, England [**42**:455].

Dematiaceae, Hyphom., pigmented, dark con., cf. Moniliaceae; Ellis 1971, 1976.

dematioidea, *Drechslera*; **dematium**, *Colletotrichum.*

Dematophora Hartig 1883; Hyphom.; Ellis 1971; Carmichael et al. 1980; conidiomata synnematal, con. aseptate; teleom. in *Rosellinia*; **D. necatrix**, teleom. *R. necatrix.*

deme terminology, see FBPP.

dendrobium bacilliform, Lesemann & Begtrup 1971 in *Phalaenopsis*, Petzold 1971 in *Dendrobium* are synonyms of orchid fleck v. q.v. but Doi et al. did not refer to the bacilliform particles of Begtrup, *Phytopath. Z.* **75**:268, 1972; Ali et al. and Lawson & Ali, *Bull.Am. Orchid Soc.* 43:529, 1974; *J. Ultrastruct.Res.* **53**:345, 1975 [**50**,2989; **51**,426; **52**,2963; **53**,4002; **55**,2266].

— **vein necrosis** Lesemann, *Phytopath.Z.* 89:330,

1977; particles 1865 × 10–20 nm, Germany, *D. phalaenopsis* [**57**,1233].

Dendroctonus terebrans, see *Leptographium.*

dendroides, *Cladobotryum, Mycosphaerella.*

Dendryphion Wallr. 1833, Hyphom.; Ellis 1971; conidiomata hyphal, conidiogenesis enteroblastic tretic; con. multiseptate, pigmented; *D. penicillatum*, teleom. *Pleospora papaveracea.*

deoxyradicinin, see radicin.

dependent transmission, transm. of a virus by a vector only in the presence of a second or helper virus. Smith, *Nature Lond.* **155**:174, 1945, discovered the phenomenon, and see tobacco mottle v. This virus and tobacco vein distorting v. together cause tobacco rosette. TobMV is sap transm. but is only transm. by *Myz.pers.* if the source plant also contains TobVDV which is therefore the helper q.v. or assistor; other plant viruses show similar phenomena; Rochow, *A.R.Phytop.* **10**:101, 1972 [**24**:208].

deposit, amount and pattern of spray or dust deposited per unit area of plant surface, FBPP.

depraedens, *Cristulariella.*

Dermateaceae, Helotiales; ascomata with an excipulum of subglobose, often brown cells, sclerotia absent; *Ascocalyx, Blumeriella, Dermea, Diplocarpon, Drepanopeziza, Durandiella, Leptotrochila, Nothophacidium, Pezicula, Potebniamyces, Pseudopeziza, Pyrenopeziza, Trochila.*

Dermea Fr. 1825; Dermateaceae; asco. 0–3 septate; Groves, *Mycologia* **38**:351, 1946; Funk, *CJB* **54**:2852, 1976; poss. weakly pathogenic [**26**:31; **56**,3260].

D. balsamea (Peck) Seaver 1932; asco. 20–30 × 6–8 μm; Dodge, *Mycologia* **24**:421, 1932; assoc. with a *Tsuga* blight [**12**:131].

D. pseudotsugae Funk, *CJB* **45**:1803, 1967; asco. 16–28 × 4–7 μm; assoc. with phloem necrosis of young Douglas fir [**47**,671].

desert habitats for crops, potato, tomato; Rotem, *PD* **65**:315, 1981, Israel (Negev).

desmazieresii, *Ascochyta, Meloderma, Rosellinia.*

desmodiae, *Synchytrium.*

desmodium mosaic v. Edwardson et al., *PDR* **54**:161, 1970; poss. Potyvirus, filament 775 nm long, transm. sap. aphid spp.; seedborne in *D. canum*, transm. to *Cyamopsis tetragonoloba*, USA, experimental host range narrow [**49**,2248].

— **mottle** Joshi, Suteri & Dubey, *Indian J.Mycol.Pl.Path.* **7**:189, 1977; transm. sap, aphid; seedborne in *D. triflorum*, a common weed [**58**,3171].

— **yellow mottle v.** Walters & Scott 1972; Scott, *Descr. V.* 168, 1976; Tymovirus, isometric *c.*30 nm diam., transm. sap. USA, *Desmodium* spp.

destructans, *Cylindrocarpon*; **destructiva**, *Phoma*;

destructivum, *Colletotrichum*.

destructor, *Peronospora*; **destruens**, *Plenodomus, Sphacelotheca*.

deusta, *Ustulina*.

Deuteromycotina, Eumycota; fungi imperfecti, a miscellaneous assemblage characterised by the absence of a teleom.; to be treated as a subdivision of the Eumycota is not strictly correct since the other subdivisions are separated by teleom. characteristics. More spp. will be found to have a teleom. connection; others are assoc. with teleom. groups, i.e. it is traditional to exclude certain anam. from the Deuteromycotina, e.g. in Erysiphales and Uredinales. This subdivision, however, is the second largest group of fungi. There are 2 classes: Coelomycetes, where the con. are formed in distinct conidiomata, an acervulus or a pycnidium; Hyphomycetes, where the con. are not borne in discrete conidiomata but are on separate hyphae or hyphal aggregations, the conidiophores. *Ainsworth & Bisby's Dictionary of the fungi*, edn. 7, 1983, gave the classification based on conidioma type, conidial shape and septation, con. and conidiophore development. From these characteristics a code is derived and the code for each genus is given. The code in this book is given in the appropriate terms which are placed and defined in the order. Ellis 1971, 1976; Carmichael et al. 1980, Hyphomycetes; Subramanian, *Hyphomycetes*, 1971; *Hyphomycetes, taxonomy and biology*, 1982; Sutton, *Mycol.Pap.* 141, 1977; 1980; *TBMS* **86**:1, 1986, Coelomycetes, on conidia; Sivanesan 1984, anam. of bitunicate Ascomycetes; Cannon et al. 1985, anam. of British Ascomycetes.

*Deuterophoma Petri 1929; Coelom.; monotypic. *D. tracheiphila* Petri, Punithalingam & Holliday, *Descr.F.* 399, 1973; the fungus, causing the important mal secco of citrus, was transferred to *Phoma* q.v. in 1948; con. 4.5–6.5 × 1.5 μm, often formed from mycelium, pycnidia pseudoparenchymatous, colonies felty, white to grey. The pathogen is obligate and occurs particularly on lemon; it is restricted to the Mediterranean region, USSR (Caucasus, Georgia), Iraq; countries near the area of distribution and apparently free from infection are: Egypt, Libya, Morocco, Portugal, Spain, Yugoslavia. Mal secco is a vascular wilt; if infection, though a wound, is at the trunk base or on a main root above ground collapse can be rapid. Wilt is less severe if only the middle or upper branches are invaded. Direct penetration of the leaves appears not to occur but it can take place at the point of fruit attachment to the penduncle; fruit becomes chlorotic, withers and

falls. Infection takes place in spring and winter; any tree injury, including that from severe weather, increases it. There are no races but citrus spp. and cvs. vary in their reaction to infection; citron, lemon, pummelo and sour orange are very susceptible; a resistant stock is used; Nachimas et al., *Physiol.Pl.Path.* **14**:135, 1979, toxin malseccin; Holliday 1980; Barash et al., *ibid* **19**:17, 1981, a further toxin; Pionnat, *Fruits* **37**:237, 1982, control [**58**,4843; **61**,729; **62**,3043].

Diabrotica spp., transm. cowpea chlorotic mottle, maize chlorotic mottle, melon necrotic spot, passion fruit yellow mosaic.

diagnosis, determination of the nature or cause of a diseased or disordered condition; Wallace, *A.R.Phytop.* **16**:379, 1978; Grogan, *ibid* **19**:333, 1981.

Dialeurodes kirkaldi, transm. jasmine yellow ring mosaic.

diamond bark canker (see *Dichomera saubinetii, Cryptodiaporthe lebiseyi*) sycamore; —— **leaf spot** (*Haplobasidion musae*) banana.

dianthi, *Alternaria, Mycosphaerella, Peronospora, Uromyces*.

dianthicola, *Alternaria, Peronospora*.

Dianthoviruses, type carnation ringspot v., other members: clover primary leaf necrosis v., red clover necrotic mosaic v., sweet clover necrotic mosaic v.; isometric *c*.34 nm diam., 2 segments linear ssRNA, both required for infection, sedimenting at 130–135 S, thermal inactivation 85–90°, longevity in sap 10 weeks at 20°, transm. sap and through soil; Francki et al., vol. 2, 1985.

Dianthus caryophyllus, carnation.

Diaphorina citri, transm. citrus greening.

Diaporthales, Ascomycotina; ascomata perithecioid, immersed or erumpent, often with a stroma, mainly beaked, beaks often united by the stroma, asci often free in the cavity, asci unitunicate, on bark or wood.

Diaporthe Nitschke 1870; Valsaceae; ascomata beneath a blackened zone on the surface of the substrate, delimited by a blackened marginal zone; the stromatic crusts appear as fine black lines when the substrate is cut at right angles to the surface, asco. 1 septate, anam. in *Phomopsis*; Wehmeyer, *The genus* Diaporthe *Nitschke and its segregates*, 1933; Barr 1978.

D. actinidiae Sommer & Beraha, *Mycologia* **67**:650, 1975; asco. 9–9.5 × 3 μm, B con. 20.5 × 1.5 μm; Beraha, *PDR* **54**:422, 1970; kiwi fruit stem end rot in New Zealand [**49**,2944; **55**,820].

D. citri Wolf 1926; Punithalingam & Holliday, *Descr.F.* 396, 1973, as anam. *Phomopsis citri* Fawcett 1912; asco. 12–14 × 3–4.5 μm, B con. 20–30 × 0.5–1 μm; citrus melanose and stem end

rot of the fruit, disease is important only where summers are wet; a primary pathogen and a secondary invader of wounds and scars, Holliday 1980.
D. cinerascens, poss. anam. *Phomopsis cinerascens* q.v.
D. helianthi Muntañola-Cvetković, Mihaljčević & Petrov, *Nova Hedwigia* **34**:433, 1981; anam. *Phomopsis helianthi* new sp.; asco. 12.5–14.5 × 3–4.5 μm, B con. 22–32 × 0.5–1 μm; sunflower stalk rot, Hungary, Yugoslavia; Yang et al., *PD* **68**:254, 1984, in USA; Mihaljčević et al., *Phytopath.Z.* **113**:334, 1985, infection source; Muntañola-Cvetković et al., *TBMS* **85**:477, 1985, in vitro comparisons of isolates [**61**,3006; **63**,3498; **65**,323, 848].
D. lokoyae Funk, *CJB* **46**:601, 1968; anam. *Phomopsis lokoyae* Hahn, *Mycologia* **25**:372, 1933; asco. 10–16 × 2.2–4.5, B. con. 11–12 × 1.5–2.5 μm; Douglas fir canker; Boyce, *J.For.* **31**:664, 1933; Thomas, *Can.J.Res.* C **28**:477, 1950 [**13**:199, 200; **30**:253; **47**,2571].
D. manihotis Punith. 1975; Punithalingam, *Descr.F.* 734, 1982; anam. *Phomopsis manihotis* Swarup, Chauhan & Tripathi 1966; asco. 8–12 × 3–4 μm, B con. 14–26 × 0.5–1 μm; cassava leaf spot, can be severe in India.
D. melonis Beraha & O'Brien, *Phytopath.Z.* **94**:205, 1979; asco. 7.2–11 × 2.2–4.7 μm, anam. resembles *Phomopsis cucurbitae* q.v., on melon, USA [**58**,6130].
D. phaseolorum (Cooke & Ell.) Sacc. 1882; anam. *Phomopsis phaseoli* (Desm.) Grove 1917; Punithalingam & Holliday, *Descr. F.* 336, 1972; asco. 9–11 × 2.5–3.5 μm, B con. 20–30 × 0.5–1 μm; 4 vars. have been distinguished on size of con., pathogenicity and hosts: *batatatis* (Harter & Field) Wehm., on sweet potato, prob. unimportant; *caulivora*, on soybean, anam. rare; *phaseolorum*, on Lima bean; *sojae* (Lehman) Wehm., on soybean, often called *Phomopsis sojae* Lehman; Kulik, *Mycologia* **76**:274, 1984, rejected vars. and proposed that the form causing soybean diseases, the most important, be named f.sp. *caulivora*; soybean pod and stem blight, stem canker; seedborne, most of the recent work has been done in USA; Holliday 1980; Hepperly et al., *Phytop.* **69**:846, 1979, seed infection & soybean mosaic v.; Spilker et al., *ibid* **71**:1027, 1981, factors affecting seed quality; Hill et al., *PD* **65**:132, 1981, mycelium in seed; Stuckey et al., *ibid* **66**:826, 1982, seed infection & bean pod mottle; Tekrony et al., *Phytop.* **73**:914, 1983, seed infection & weather; Rothrock, *ibid* **75**:1156, 1985, effects of tillage & cropping pattern; Backman et al., *PD* **69**: 641, 1985, an assessment

[**59**,3963; **61**, 2017, 2554; **62**,474; **63**,314, 3617; **65**,995].
D. tanakae Kobayashi & Sakuma, *Trans.mycol.Soc.Japan* **23**:37, 1982; anam. *Phomopsis tanakae* Kobayashi & Sakuma; asco. 12.5–17 × 3–5 μm, A con. 9–12.5 × 2.5–4 μm, B con. 12–20 × 0.8–1.5 μm; new spp. for European pear dieback, see Tanaka, *Hokkaido agric. Exp.Stn.* 31:85, 1934, as *Diaporthe ambigua* Nitschke = *D. eres* Nitschke [**13**:525].
D. vaccinii Shear 1931; anam. *Phomopsis vaccinii* Shear; asco. 6–13 × 2–4 μm, A con. 6–11 × 2–5 μm, B con. 15–25 long; blueberry twig blight and canker, USA; Wilcox, *Phytop.* **29**:136, 1939; **30**: 441, 1940; Parker & Ramsdell, *ibid* **67**:1481, 1977, a wound pathogen, spread & chemical control; Milholland, *PD* **66**:1034, 1982, twig blight phase, flower bud penetration; Chao & Glawe, *Mycotaxon* **23**: 371, 1985, concluded that *Diaporthe phaseolorum* was distinct [**18**:402; **19**:550; **57**,4552; **62**,714; **64**,4795].
D. woodii Punith. 1974; Punithalingam & Gibson, *Descr.F.* 476, 1975; anam. *Phomopsis leptostromiformis* (Kühn) Bubák; asco. 8–12 × 2.5–4 μm; A con. 6–8 × 2–3 μm, B con. unknown; distinguished from *Diaporthe lupini* Harkness by the smaller asco.; lupin stem blight, seedborne, infected lupins are dangerously toxic to sheep, lupinosis; Wood & Petterson, *Aust.J.exp.Agric.* **25**:164, 1985, seed infection in Western Australia [**65**,259].
diaporthin, toxin from *Cryphonectria parasitica*; Bazzigher, *Phytopath.Z.* **21**:105, 1953; Boller et al., *Helv.chim.Acta* **40**:875, 1957; McCarroll & Thor, *Physiol.Pl.Path.* **26**:357, 1985, found no support for the concept that this toxin, rugulosin & skyrin q.v. affected mycelial advance into chestnut inner bark [**33**:569; **37**:646; **65**,386].
Diaporthopsis Fabre 1883; Gnomoniaceae; Barr 1978; asco. aseptate; *D. metrosideri* Roane & Fosberg, *Mycologia* **75**:165, 1983; asco. 4–7 × 1–2 μm; *Phomopsis* state, A con. 3–4 × 1 μm, B. con. 7–14 × 1 μm; poss. pathogenic to *Metrosideros collina* in USA (Hawaii) [**62**,3278].
Diatrypales, Ascomycotina; 1 family Diatrypaceae; ascomata perithecioid, immersed or erumpent, usually in circular groups connected by a carbonaceous stroma, ostiolar beaks separate, often projecting, asci unitunicate, asco. often extruded in a cirrhus; on bark; *Cryptosphaeria*, or in Valsaceae, *Eutypa*, *Eutypella*.
dibromoethane, soil fumigant.
dicaryon, see dikaryon.
dicentra witches' broom Hiruki & Shukla, *Phytop.* **63**:88, 1973; MLO assoc., Canada (Alberta), *D. spectabilis*, bleeding heart [**52**,3712].

dichaeta, *Pestalotiopsis*.
dichlofluanid, protectant fungicide, particularly for powdery mildews on fruit trees.
dichlone, seed protectant and fruit tree foliage spray.
Dichomera Cooke 1878; Coelom.; Sutton 1980; conidiomata stromatic, conidiogenesis holoblastic determinate, some spp. indeterminate, annelidic; con. muriform, pigmented.
D. saubinetii (Mont.) Cooke, con. 11–13 × 7–10 μm; assoc. with sycamore, *Acer pseudoplatanus*, diamond bark canker in England; Bevercombe & Rayner, *TBMS* **71**:505, 1978; *New Phytol.* **86**:379, 1980, sometimes replaced by *Cryptodiaporthe lebiseyi*; Rayner et al., *ibid* **87**:383, 1981, fungal growth & explanation of canker shape [**58**,4034; **60**,4669–70].
Dichotomophthora Mehrlich & Fitzp. ex M.B. Ellis, *Dematiaceous Hyphomycetes*: 388, 1971; conidiomata hyphal, conidiogenesis enteroblastic tretic, con. multiseptate; *D. portula* authors as given; con. 0–6 septate, 15–90 × 6–15 μm; black stem of *Portulaca oleracea*, purslane, a weed, may cause death; Mehrlich & Fitzpatrick, *Mycologia* **27**:543, 1935; Klisiewicz et al., *PD* **67**: 1162, 1983 [**15**:106; **63**,3700].
Dickson, James Geere, 1891–1962; born in USA, Univ. Washington State, Wisconsin; USDA, an authority on diseases of cereals, forage legumes and grasses; wrote: *Diseases of field crops* 1947, edn. 2, 1956. *Phytop.* **52**:1093, 1962; *Mycologia* **55**:537, 1963; *A.R.Phytop.* **18**:11, 1980.
diclinous, antheridium and oogonium on different hyphae, cf. androgynous and monoclinous.
diclobutrazol, broad spectrum systemic fungicide, powdery mildews and rusts.
dicloran, fungicide for fruit rots caused by fungi.
Dicranotropis hamata, transm. cereal tillering v.
dictyochlamydospore, a multicelled chlamydospore where the outer wall is separable from the walls of the component cells which are rather easily separated from each other.
dictyoides, *Drechslera*, *Pyrenophora*.
Didymascella Maire & Sacc. 1903; Rhytismataceae; Dennis 1978; ascomata well developed, immersed or erumpent, asci 2–4 spored, asco. unequally 2 celled, pigmented, on conifers.
D. thujina (Durand) Maire 1927; ascomata fall away leaving an empty pit in the leaf, asco. 22–25 × 15–16 μm; causes a serious disease of western red cedar, *Thuja plicata*, *Thuja* needle blight, in nurseries and in younger trees; Europe, N. America; Pawsey, *Forestry* **33**:174, 1960; Burdekin & Phillips, *AAB* **67**:131, 1971, chemical control in Britain [**40**:440; **50**,2553].
Didymella Sacc. 1880; Dothideaceae or

Pleosporaceae; Sivanesan 1984; asco. 1 septate, hyaline; Holliday 1980; Corlett, *CJB* **59**:2016, 1981 [**61**,2743].
*D. applanata (Niessl) Sacc. 1882; Punithalingam, *Descr.F.* 735, 1982; anam. in *Phoma*; asco. 12–18 × 5–7 μm, con. 4–7 × 2–3.5 μm; cane or spur blight of red raspberry, losses can be heavy; Pepin et al., *AAB* **106**:335, 1985, effects of cv. & isolate on susceptibility; Williamson & Jennings, *ibid* **109**:581, 1986, resistance with *Botryotinia fuckeliana* [**65**,5060; **66**,2931].
D. bryoniae (Auersw.) Rehm 1881; Punithalingam & Holliday, *Descr. F.* 332, 1972; Corlett et al., *Fungi Canadenses* 303, 1986; anam. *Ascochyta cucumis* Fautrey & Roum. 1891; asco. 14–18 × 4–7 μm, con. 6–10 × 3–4 μm; gummy stem blight is an important disease of Cucurbitaceae, also called black fruit rot, stem canker and vine wilt, a gummy exudate on the stem is the characteristic symptom, seedborne; Bergstrom et al., *PD* **66**:683, 1982, synergism with other disease agents; van Steekelenburg, *Neth.J.Pl.Path.* **88**:47, 1982; **89**:75, 1983, cucumber external fruit rot & epidemiology; *ibid* **91**:225, 277, 1985, effects of transition from day to night temps & RH; *ibid* **92**:81, 1986, factors affecting internal fruit rot of cucumber; Lee et al., *Phytopath.Z.* **109**:301, 1984, in seed & transm. in seedlings [**61**,6047; **62**,446; **63**,301, 4613; **65**,2513, 3050, 4606].
D. exitialis (Morini) E. Müller 1952; Punithalingam, *Descr.F.* 633, 1979; anam. in *Ascochyta*; asco. 12–14 × 4–5.5 μm; barley and wheat scorch, Ahrens & Schöpfer 1983 called it a weak parasite or a secondary invader [**63**,2805].
D. lethalis (R. Stone) Sivanesan 1984; anam. *Ascochyta lethalis* Ell. & Barth. 1903; asco. 13–19 × 7–9 μm, con. 10–17 × 3–5 μm; one of a fungus complex causing spring black stem of clovers and lucerne.
D. ligulicola (Baker, Dimock & Davis) v. Arx 1962; anam. *Ascochyta chrysanthemi* F. Stev. 1907; asco. 12–16 × 4–6 μm, con. 6–10 × 3–5 μm. Punithalingam, *Descr.F.* 662, 1980, gave the name of the fungus causing chrysanthemum ray blight as *Didymella chrysanthemi* (Tassi) Garibaldi & Gullino 1971; Walker & Baker, *TBMS* **80**:31, 1983, considered that *D. ligulicola* is the correct name. *D. chrysanthemi*, based on *Sphaerella chrysanthemi* Tassi 1900, is a *Mycosphaerella*; they found no evidence that the pathogen had occurred in Italy before 1960. Ray blight is an important disease and Punithalingam loc.cit. gave a full account [**62**,3063].
*D. lycopersici Kleb. 1921; Holliday & Punithalingam, *Descr.F.* 272, 1970; anam. *Phoma lycopersici* Cooke 1885, fide Sutton 1980;

asco. 12–15 × 5 μm, con. 6–10 × 2–3 μm.
P. lycopersici should not be confused with
P. destructiva; the latter has smaller con. and there
are in vitro differences, see Laundon, *N.Z.J.Bot.*
9: 610, 1971 and Sutton 1980. Tomato stem and
fruit rot; prob. only causes serious loss in cooler
climates, spread from infested host material is
more important than from seed; Maider & Burge,
TBMS **72**:504, 1979, anam. in vitro; Boukema,
Euphytica **31**:981, 1982, inheritance of resistance
[**51**,3903; **59**,1396; **62**,2152].

D. macropodii Petrak 1929; anam. *Phoma
nigrificans* (P. Karsten) Boerema, Loerakker &
Wittern, *J.Phytopath.* **115**:270, 1986. Whilst
screening for resistance in winter rape to
Leptosphaeria maculans this *Phoma* was found. It
is distinguished from *P. lingam*, the anam. of
L. maculans, by the larger con. and lower opt.
temp. for growth. *P. nigrificans* caused lesions on
the hypocotyls of rape seedlings [**65**,4143].

D. pinodes, see *Mycosphaerella pinodes*.

Didymosphaeria Fuckel 1870; Didymosphaeriaceae;
Sivanesan 1984; ascomata without setae, asco.
pigmented, 1 septate.

D. arachidicola (Chockr.) Alcorn, Punith. &
McCarthy 1976; anam. *Ascochyta adzamethica*
Schoschi 1940; Punithalingam, *Descr.F.* 736,
1982; asco. 13–16 × 5–6.5 μm, con. 4–9 ×
2.5–4 μm, groundnut net or web blotch, severe
in southern Africa and Australia (Queensland),
occurs in USA; Phipps, *PD* **69**:1097, 1985
[**65**,4195].

D. donacina (Niessl) Sacc. 1882; asco. 14–22 ×
7–11 μm; cluster yam, *Dioscorea dumetorum*, leaf
spot; Emua & Fajola, *PD* **65**:443, 1981, in
Nigeria; for chemical control see *Mycosphaerella
contraria* [**61**,1524].

D. oregonensis Goodding, *Phytop.* **21**:916, 1931, no
Latin diagnosis; asco. 18–21 × 7–9 μm; assoc. with
alder canker, USA (N.W.) [**11**:139].

Didymosphaeriaceae, Dothideales; ascomata
applanate to distinctly sunken, often clypeate,
peridium usually thickened over the apex, asco. 1
or more transversely septate; *Didymosphaeria*.

didymum, *Conostroma*, *Cylindrocarpon*.

dieback, necrosis of a shoot beginning at the apex
and spreading towards the older tissue, stem
death may occur.

dieffenbachia stunt v. Hakkaart & Waterreus, *Acta
Hortic.* **59**:175, 1976; poss. Potyvirus, filament
807 nm long, transm. sap, Netherlands, tip
necrosis and necrotic edges to the leaves
[**56**, 2533].

differential host, a host that gives reactions which
distinguish between race specific isolates of a
pathogen, e.g. a rust; or a plant that gives

reactions distinctive for a specific virus, permitting
its differentiation from other viruses commonly
present; a type of indicator plant q.v., FBPP.

diffusa, *Microsphaera*; **diffusum**, *Sirosporium*.

digera yellow vein mosaic Ahmad, *Phytopath.Z.*
96:21, 1979; transm. *Bemisia tabaci*, Pakistan,
D. alternifolius [**59**,3041].

digitalis, *Septoria*.

—— **phyllody** Munro, *Aust.Pl.Path.Soc.Newsl.*
7(1):10, 1978; MLO assoc., Australia (Tasmania)
[**57**, 4504].

Digitaria decumbens, pangola grass.

digitaria mosaic Celino & Martinez, *Philipp.Agric.*
40:285, 1956; transm. aphid, Philippines,
D. sanguinalis [**36**:646].

—— **striate v.** Greber 1972; *Aust.J.agric.Res.* **30**:43,
1979; Phytorhabdovirus, bacilliform 288 × 55 nm,
transm. *Sogatella kolophon*, Australia
(Queensland), *D. ciliaris*, *D. decumbens*, infects
barley, oat and other Gramineae [**54**,1774;
58,5915].

digitatum, *Penicillium*.

Digitosporium Gremmen 1953; Coelom.; Sutton
1980; conidiomata stromatic, conidiogenesis
holoblastic; con. solitary, digitate, branches of
unequal length, 1–5 septate, apical cell long conic;
D. piniphilum, teleom. *Crumenulopsis sororia*.

dikaryon, of a fungus, a cell having 2 genetically
distinct haploid nuclei; hence dikaryotisation, i.e.
the conversion of a homokaryon into a dikaryon.

Dilophospora Desm. 1840; Coelom.; Sutton 1980;
conidiomata stromatic, conidiogenesis
enteroblastic phialidic; con. multiseptate, hyaline,
with several, simple, dichotomously or irregularly
branched, extracellular setulae. *D. alopecuri* (Fr.)
Fr. 1894; Gibson & Sutton, *Descr.F.* 490, 1976;
con. up to 13 × 1–1.5 μm, teleom. poss. *Lidophia
graminis* (Sacc.) J. Walker & B. Sutton, *TBMS*
62:232, 1974; twist of grasses, on temperate
cereals, flowering shoots may be partially
enveloped in a black stroma; disease outbreaks
have occurred in Europe, assoc. with nematodes,
Anguillulina spp.

diluent, an inert material added to a plant
protectant to reduce its concentration, FBPP.

dilution end point, the stage of a serial dilution of
cells or virus preparations at which growth or
infection from a standard sample of the
suspension no longer occurs, FBPP.

dimethirimol, systemic fungicide, powdery mildews
on ornamental and vegetable crops, in the latter
particularly Cucurbitaceae.

Dimond, Albert Eugene, 1914–72; born in USA,
Washington State College, Univ. Wisconsin,
Connecticut Agricultural Experiment Station; did
important basic work in the search for systemic

fungicides; Bronze Medal, 11th International Botanical Congress. *Phytop.* **63**:657, 1973.
dimorphotheca virescence Grimaldi & La Rosa, *Difesa Delle Piante* **10**:179, 1987, prokaryote assoc., *D. sinuata*.
dinebra fine speckled mosaic v. Greber, *Australasian Pl.Path.* **11**: 17, 1982; isometric *c.*28 nm, Australia (Queensland), *D. retroflexa* [**62**,248].
dinobuton, contact fungicide for powdery mildews.
dinocap, contact and persistent fungicide for powdery mildews.
Dioscorea, yam.
dioscorea green banding v. Ruppell, Delpín & Martin, *J.Agric. Univ.P.Rico* **50**:151, 1966; Hearon et al., *Phytop.* **68**:1137, 1978; filament 600–720 × 9–11 nm, transm. sap, aphid spp.; Puerto Rico, *D. floribunda* [**46**,1064; **58**,3607].
— **latent v.** Lawson et al. 1973; Waterworth et al., *J.Agric. Univ.P.Rico* **58**:351, 1974; Hearon et al., as above; Phillips et al., *AAB* **109**:137, 1986; Potexvirus, filament *c.*350–900 nm long, 2 modal lengths of 445 and 875 nm, transm. sap, from *D. composita, D. floribunda*, but not detected in any of 37 cvs. of *D. alata, D. bulbifera, D. esculenta, D. rotundata* from 8 countries, Puerto Rico [**54**,4718; **65**,6285].
Diospyros kaki, Japanese persimmon; *D. virginiana*, N. American ebony or persimmon.
Diplocarpon Wolf, *Bot.Gaz.* **54**:231, 1912; Dermataceae; asco. 1 septate.
D. earlianum (Ell. & Ev.) Wolf 1924; Sivanesan & Gibson, *Descr. F.* 486, 1976; anam. *Marssonina fragariae* (Lib.) Kleb. 1918; asco. 18–28 × 4–6 μm, con. 18–30 × 5–7 μm, strawberry leaf scorch, Maas 1984.
D. mali Y. Harada & K. Sawamura, *Ann.phytopath.Soc.Japan* **40**: 415, 1974; anam. *Marssonina coronaria* (Ell. & J. Davis) J. Davis, fide Harada et al., loc.cit.; asco. 22–33 × 5–6 μm, con. 20–24 × 6.5–8.5 μm; apple leaf blotch; Parmelee, *Can.Pl.Dis.Surv.* **51**:91, 1971 [**51**,461; **55**,318].
D. mespili (Sorauer) B. Sutton 1980; anam. *Entomosporium mespili* (DC.) Sacc. 1880; Sivanesan & Gibson, *Descr.F.* 481, 1976, as *Diplocarpon maculatum* (Atk.) Jørstad 1945; asco. 16–24 × 6–10 μm; con. 12–20 × 8–14 μm; pear and quince leaf blight or scald, Rosaceae, causes defoliation and reduces fruit quality; van der Zwet & Stroo, *Phytop.* **75**:94, 1985, loss of pathogenicity in culture; Cobb et al., *PD* **69**:684, 1985, chemical control; Baudoin *ibid* **70**:191, 519, 1986, infection of *Photina* [**64**,2638; **65**,751, 3359, 5568].
D. rosae Wolf 1912, see genus; anam. *Marssonina rosae* (Lib.) Died. 1915; Sivanesan & Gibson,

Descr.F. 485, 1976; asco. 20–25 × 5–6 μm, con. 18–25 × 5–6 μm; rose black spot, confined to *Rosa*, a serious disease, the fimbriate leaf lesions result in leaf fall, stem spots are not fimbriate; the fungus overwinters on bud scales and stems, races occur; Knight & Wheeler, *TBMS* **69**:385, 1977, perennation on leaves; *Phytopath.Z.* **91**:218, 346, 1978; resistance & conidial germination on cvs.; Horst 1983 [**57**,3494, 5539–40].
Diplodia Fr. 1834; Coelom.; Sutton 1980; conidiomata pycnidial, conidiogenesis holoblastic, con. 1 septate; the genus requires revision.
D. frumenti, poss. teleom. *Physalospora zeicola*.
D. macrospora, and *Diplodia maydis*, see *Stenocarpella*.
D. mutila, teleom. *Botryosphaeria stevensii*.
D. pinea (Desm.) Kickx 1867; Punithalingam & Waterston, *Descr.F.* 273, 1970; the fungus was transferred to *Sphaeropsis*: *S. sapinea* (Fr.) Dyko & B. Sutton, *The Coelomycetes*: 120, 1980; con. usually aseptate, later some 1 septate, becoming dark brown, wall rough, 30–45 × 10–16 μm; pine staghead, also: tip and twig blight, red top, bud wilt and seedling collar rot; on other conifers, assoc. with insects and wounds, losses in the S. Hemisphere can be serious, saprophytic on forest debris; Holliday 1980; Wingfield & Knox-Davies, *PD* **64**:221, 1980, in South Africa on stressed pine stands; Van Dam & Kam 1984, reported a damaging outbreak in the Netherlands in 1982–4; Palmer & Nicholls, *ibid* **69**: 739, 1985, on *Pinus resinosa* in USA (Wisconsin); Wang et al., *ibid* **69**:838, 1985, 2 morphological groups, con. either smooth or pitted, also differences in culture; Swart et al. 1985, in South Africa; Palmer et al., *ibid* **70**:194, 1986, chemical control [**59**,5949; **64**,1780; **65**,927, 3530, 4121, 5195].
D. tumefaciens (Shear) Zalasky, *CJB* **42**:1050; 1964; con. 28–40 × 9–15 μm; poplar branch gall, N. America; Hubert, *Phytop.* **5**: 182, 1915; Zalasky, *ibid* page 385 [**43**,2738; **44**,258].
Diplodina Westend. 1857; Coelom.; Sutton 1980; conidiomata acervular, conidiogenesis enteroblastic phialidic; con. 0–2 septate, usually medianly 1 septate, hyaline.
D. castaneae Prill. & Delacr. 1893; con. 9–12 × 2.5–3 μm; sweet chestnut dieback and canker; Day, *Q.J.For.* **24**:114, 1930.
D. delphinii Laskaris, *Phytop.* **40**:620, 1950; not *Diplodina delphinii* Golovin 1950; con. 9.9–19.8 × 3.3–4.1 μm; delphinium crown and root rot, black leaf spot, pod blight, USA [**30**:41].
D. microsperma, see *Cryptodiaporthe salicella*.
D. passerinii Allescher 1899; con. 7–12 × 2.5–3 μm;

assoc. with wilt of herbaceous ornamentals, England; Taylor, *AAB* **28**:91, 1941 [**20**:467].

dirty panicles (*Sarocladium attenuatum*) rice.

Discella Berk. & Br. 1850; Coelom.; genus doubtful, Sutton, *Mycol.Pap.* 141, 1977; *D. carbonacea*, *D. salicis*, see *Cryptodiaporthe salicella*.

Discocainia J. Reid & Funk, see *Atropellis treleasei*.

discomycetes, a traditional group of the Ascomycotina where the ascoma is frequently open and shaped like a cup or a saucer, called apothecia; the class is not now accepted.

Discosia Lib. 1837; Coelom.; Sutton 1980; conidiomata stromatic, conidiogenesis holoblastic, con. mostly 3 septate, an appendage at each end; mostly saprophytic on leaves and stems but *D. pini* Heald 1909 causes disease in conifer nurseries, snow scald; Georgescu & Petrescu 1956 in Romania, Suto 1976 in Japan [**37**:381; **57**,1869].

Discosporium Höhnel 1915; Coelom.; Sutton 1980; conidiomata stromatic, conidiogenesis holoblastic; con. aseptate, hyaline; conidiophore growth percurrent; *D. populeum*, teleom. *Cryptodiaporthe populea*.

Discostroma Clem. 1909; Amphisphaeriaceae; asco. 2 or more septate, hyaline; Brockman, *Nova Hedwigia* **28**:275, 1975; *D. corticola* (Fuckel) Brockman; asco. 12–20 × 5.5–8 μm; rose dieback and canker; Brooks & Alaily, *AAB* **26**:213, 1939 [**18**:597].

Discula Sacc. 1884; Coelom.; Sutton 1980; conidiomata stromatic, conidiogenesis enteroblastic phialidic; con. aseptate, hyaline; *D. umbrinella*, teleom. *Apiognomonia errabunda*.

disease, a condition where the normal functions are disturbed and harmed; may be divided into infectious, i.e. caused by a pathogen, and non-infectious, caused by other factors; FBPP recommends that the former be called diseases and the latter disorders. A plant infected by a pathogen may not show any symptoms, i.e. it is not obviously diseased; but since such a plant may be a source of inoculum it is probably best described as diseased. The word disease is often incorrectly used, i.e. as a synonym for a pathogen; disease means host + pathogen. Further, some host and parasite interactions are defined and described as diseases when Koch's postulates have not been adequately fulfilled; such a loose usage of the term disease is best avoided. Sometimes the word disease is used in the name of a disease, e.g. Dutch elm disease (*Ceratocystis ulmi*) and pink disease (*Corticium salmonicolor*). This has a tautological ring and is therefore undesirable; the preferred names Dutch elm wilt and pink crust are quite satisfactory.

—— **escape**, the pathogen is absent or, if present, either the host or the external environment inhibits infection; see klendusity; Agrios in Horsfall & Cowling, vol. 5, 1980.

—— **gradient**, the change in incidence of disease with distance from a source of infection; this can be expressed mathematically, Gregory, *A.R.Phytop.* **6**:189, 1968. Dispersal gradients are due to spatial variation in the amount of inoculum arriving; this is affected by eddy diffusion, splash ballistics, insect movement or pathogen death during transport. Gradients may be environmental due to variation in ecoclimate or soil. A gradient graph, in an ideal situation and from a local point source of inoculum, shows a hollow curve when plotted with linear scales of amount of disease, vertical axis, and distance from the source, horizontal axis; Ingold and Legg & Bainbridge in Scott & Bainbridge 1978; Thresh, *AAB* **82**:381, 1976, viruses; McCartney & Fitt in Wolfe & Caten 1987.

—— **lesion mimics**, in maize, Neuffer & Calvert, *J.Hered.* **66**:265, 1975, mutagen causing lesions that, in some cases, resemble those caused by *Cochliobolus carbonum* and *C. heterostrophus* [**55**, 2212].

—— **measurement**, see phytopathometry.

*—— **names**, the names of pathogens are mostly governed by international rules which need to be followed; but the names of diseases are only governed by usage and common sense. Those who compile lists of disease names would do well to digest the points made > 55 years ago, *TBMS* **17**:203, 1932. This paper was in reply to Cunningham's response, *ibid* **16**:145, 1931, to the first list of British plant disease names, *ibid* **14**:140, 1929. There are differences between, and debatable names in, 2 recent lists: an American one, *PD* **69**:649, 1985, and a British one, *Phytopath. Pap.* 28, 1984, edn. 5 of the one already given; the latter gives some American equivalents, and names used in some other countries in Europe. Four principles should guide compilers: (1) It is the disease not the pathogen that is being named. (2) A name should be one which the grower uses, or can use, readily; he is concerned with the disease not the pathogen. (3) A name should be taken from the most conspicuous abnormality caused. It can, or need not, be qualified by a crop name and/or a pathogen group name, e.g. downy mildew. (4) The introduction of a part of the pathogen's binomial, the generic term, is undesirable; it should be used sparingly. There is no objection to having 2 or more names for one crop and pathogen interaction, provided the syndromes are clearly

different and separate. It is principle (4) that has become increasingly ignored; particularly where the disease descriptor is undistinctive. The real solution is to have better descriptors. Further, a generic name may be changed; examples such as fomes root rot for a disease no longer caused by a fungus called *Fomes* are not particularly satisfactory. The use of generic terms also erodes principle (2); alternaria leaf spot and cercospora leaf spot mean something to a plant pathologist but not, necessarily, to a grower. Plant virologists, who have rejected binomials, are aware of the confusion which their nomenclature causes, but its study is still young; Merino-Rodriguez, *Lexicon of plant pests and diseases*, 1966; Millar & Pollard, *Multilingual compendium of plant disease names*, 1976, 1977; Gjærum et al.; *Nordic names of plant diseases and pathogens. Bacteria and fungi*, 1985.

— **warning**, or forecasting; predicting the probability that a disease outbreak will occur; see: Beaumont period, blitecast, climate and weather, damage threshold, epidem, epidemiology, epimay, epipre, loss of crop, Mills period. Predictions are largely based on weather known to be favourable for infection. One of the earliest attempts at warning that disease outbreaks are likely was by van Everdingen, *Tijdschr.Pl.Ziekt.* **32**:129, 1926, for potato blight (*Phytophthora infestans*); Krause & Massie, *A.R.Phytop.* **13**:31, 1975; Zadoks, *PD* **68**:352, 1984; Jones et al., *ibid* **68**:458, 1984.

disinfectant, an agent that frees from infection by destroying parasites in plants. The term is best avoided in plant pathology to avoid confusion with its popular meaning relating to destruction of bacteria etc. on objects or animals, not necessarily in situations where infection is present, FBPP.

disinfestant, an agent that kills organisms present on the surface of plants or plant parts, or in the immediate environment, e.g. in soil, FBPP.

disorder, a harmful deviation from normal functioning of physiological processes arising from causes other than pathogens, e.g. mineral deficiency or toxicity, genetic anomaly, low temp. injury, cf. disease, after FBPP; sometimes, less desirably, disorders are called abiotic diseases; Robinson, *Tech.Commun.Commonw.Bur.Hortic. Plantn.Crops* 34, 1974, annotated bibliography of colour illustrated mineral deficiency symptoms in tropical crops; Robinson ed. *Diagnosis of mineral disorders in plants*; Bould et al., vol. 1, *Principles* 1983; Scaife & Turner, vol. 2, *Vegetables* 1983; vol. 3, 4, 5 are on glasshouse crops, fruit and hops, cereals and forages, respectively; and see the disease compendium series published by the American Phytopathological Society.

dispersal of pathogens, see air spora, disease gradient, epidemic, epidemiology; Aylor & Wallace in Horsfall & Cowling, vol. 2, 1978.

dispora, *Gnomonia*.

dissemination, may be applied to the spread of fungus spores, see air spora, or other potentially infective units which may cause disease; it includes the phenomena which affect such spores or other units: release, dispersal and deposition; release of fungal spores into the air is through hydrostatic discharge, wind, rain tap and puff, splash.

dissolvens, *Erwinia*.

ditalimfos, protectant and curative fungicide, not systemic, for powdery mildews and apple scab.

dithianon, protectant fungicide, for foliage pathogens of fruit and vegetables.

dithiocarbamates, derivatives of dithiocarbamic acid, usually as a metallic salt; include some of the first organic fungicides and still widely used, broad spectrum; for their chemical names see Hill & Waller 1982.

ditissima, *Nectria*.

Ditylenchus angustus (Butler) Filipjev 1936; Seshadri & Dasgupta, *Descr.N.* 64, 1975; rice stem nematode, causing dak pora or ufra of this crop, obligate ectoparasite; infects seedlings and spreads, with plant growth, between folded leaves and sheaths, chlorosis at the young plant stage and panicles become distorted; Ou 1985.

D. destructor Thorne 1945; Hooper, *Descr.N.* 21, 1973; potato rot, infects tubers.

D. dipsaci (Kühn) Filipjev 1936; Hooper, *Descr.N.* 14, 1972; stem and bulb nematode endoparasite, attacks > 450 plant spp., can cause damage, mostly temperate regions, biological forms or races.

D. myceliophagus J. B. Goodey 1958; Hesling, *Descr.N.* 36, 1974; on the common cultivated mushroom and soil fungi.

dock, *Rumex*.

docking disorder (*Longidorus* and *Trichodorus* spp.) beet; Cooke, *AAB* **84**:451, 1976, economics of control.

dock mosaics, in England and New Zealand, transm. sap; Grainger & Cockerham 1930; Chamberlain & Matthews 1948 [**10**:199; **28**:507].

— **mottling mosaic** Juretić, Wrischer & Miličić 1976; poss. Potyvirus, Yugoslavia [**57**,2745f].

*dodder, *Cuscuta*; parasitic plants with slender, straggling stems, leafless, rootless, without chlorophyll and with haustoria. They can be used for transm. of plant viruses where other methods have failed; see stunting of dodder; Bennett, *Phytop.* **34**:905, 1944; Schmelzer, *Phytopath.Z.* **28**:1, 1957 [**24**:136; **36**:450].

— **latent mosaic**, Bennett, *Phytop.* **34**:77, 1944;
see above, Bennett; transm. sap, USA
(California), *Cuscuta californica* [**23**:247].
dodemorph, systemic fungicide, eradicant spray for
fruit crops and some ornamentals.
dodine, protectant fungicide with eradicant
properties, fruit trees.
dogwood ringspot (cherry leaf roll v.), Waterworth
& Lawson, *Phytop.* **63**:141, 1973; *Cornus florida*
[**52**,3843].
Doidge, Ethel Mary, 1887–1965; born in England;
Univ. Cape of Good Hope, public service in
South Africa; her book: *The South African fungi
and lichens, Bothalia* 5, 1950, was the foundation
for future mycological work in the region; *ibid* 9:
251, 1967.
dolichi, *Elsinoë.*
Dolichodorus heterocephalus Cobb 1914; Orton
Williams, *Descr.N.* 56, 1974; awl nematode,
ectoparasite, principally in USA, may cause crop
damage in wet sites in Florida.
dolicholi, *Uromyces.*
Dolichos lablab, or *Lablab purpureus*, hyacinth or
lablab bean; for *Dolichos* taxonomy see *Vigna*,
Verdcourt.
dolichos enation mosaic = sunnhemp mosaic v.
dolipore septum, a septum, characteristic of a
basidiomycete hypha, which flares out in the
middle part to form a structure like a barrel with
open ends; Moore & McAlear, *Am.J.Bot.* **49**:86,
1962.
dollar spot (*Sclerotinia homoeocarpa*) turf grasses.
donacina, *Didymosphaeria*; **donacis**, *Pseudoseptoria.*
dormancy, in fungal spores, Sussman in Ainsworth
& Sussman 1966.
dorotheanthus chlorotic spot v. Gupta & Singh,
Indian Phytopath. **32**:242; 1979; isometric 45–
50 nm diam., transm. sap, aphid, *D. bellidiformis*
[**60**,3203].
dose and dosage, (1) the quantity of fungicide etc.
applied per individual or per unit area, volume or
weight. (2) The quantity of a pathogen used to
inoculate living hosts particularly in quantitative
response studies; see median effective dose,
median lethal dose, FBPP.
dothidea, *Botryosphaeria.*
Dothideaceae, Dothideales; ascomata or
pseudothecia separate on or in a stroma, ostiolate
or not, asci numerous, aparaphysate, arising from
a basal or central or small cushion; *Cymadothea,
Didymella, Dothiora, Microcyclus, Mycosphaerella,
Scirrhia, Sphaerulina.*
Dothideales, Ascomycotina; the largest and most
varied group of this subdivision, the ascomata or
pseudothecia are immersed to erumpent; asci
bitunicate, a division of the loculascomycetes;

usually a stroma, in or on which the ascomata are
grouped; classification unsettled, at present there
are > 50 families; Sivanesan 1984.
Dothiora Fr. 1849; Dothideaceae or Dothioraceae;
ascomata multilocular, asco. usually with several
transverse septa and sometimes with vertical
septa, i.e. muriform. *D. ribesia* (Fr.) Barr 1972;
Sivanesan & Holliday, *Descr.F.* 707, 1981; asco.
16–26 × 4.5–7 μm; black pustule of currant and
gooseberry, a wound pathogen of little
importance.
Dothioraceae, Dothideales; Sivanesan 1984
included the family in Dothideaceae; *Dothiora,
Sydowia.*
Dothistroma Hulbary 1941; Coelom.; Sutton 1980;
conidiomata stromatic, conidiogenesis holoblastic;
con. multiseptate, filiform, hyaline; *D. septospora,*
teleom. *Scirrhia pini.*
dothistromin, toxin from *Scirrhia pini*; Basset et al.,
Chem.Ind. **52**:1659, 1970; Shain & Franich,
Physiol.Pl.Path. **19**:49, 1981 [**51**,1987; **61**,1424].
Douglas fir, *Pseudotsuga menziesii.*
downy mildew, see mildew; — **spot**
(*Mycosphaerella caryigena*) pecan.
Dowson, Walter John, 1887–1963; born in England,
Univ. Cambridge, an authority on plant
pathogenic bacteria, wrote: *Manual of bacterial
plant diseases*, 1949; *Plant diseases due to bacteria,*
1957. *Nature Lond.* **200**:630, 1963; *A.R.Phytop.*
19:29, 1981.
Draeculacephala minerva, transm. xylem limited
bacteria, q.v.; *D. portola*, transm. sugarcane
chlorotic streak.
draytonii, *Botryotinia.*
drazoxolon, used against foliage pathogens of fruit
trees, as a seed dressing and applied to the collar
region of trees to control root pathogens.
Drechslera Ito 1930; Hyphom.; Ellis 1971, 1976;
conidiomata hyphal, conidiogenesis enteroblastic
tretic; con. multiseptate, pigmented; many
graminicolous spp. were transferred from
Helminthosporium. Drechslera differs in that the
conidiogenous cells develop sympodially and have
scars where the con. were borne. The first con. is
formed at the apex of the conidiophore which
grows out laterally below the scar, pushing it to
one side and forming another con. at the newly
constituted apex, Ellis, *TBMS* **62**:228, 1974. The
genus and its 2 segregates, *Bipolaris* and
Exserohilum, were discussed by Alcorn,
Mycotaxon **17**:1, 1983, and the 3 genera
considered distinct; but see *Bipolaris*, Sivanesan.
Plant pathologists, incorrectly, still use
Helminthosporium for *Drechslera*. Alcorn loc.
cit.: con. commonly ± cylindrical, straight;
germinating from polar and intermediate cells, or

polar cells only; basal germ tube usually originating well clear of hilum, and growing at a wide angle to the longitudinal axis; hilum of the atrium type, and inserted within a rounded basal contour; first conidial septum delimiting the basal cell of the mature conidium, second septum ± median in the upper cell, the third septum distal; conidiogenous nodes smooth. Carmichael et al. 1980; Holliday 1980; Sivanesan 1984.

D. andersenii Lam, *TBMS* **85**:601, 1985; con. av. 94 μm long, av. 5 septate; on *Lolium* spp., see Lam & Chapman, *ibid* page 603 [**65**,1930–1].

D. avenae, teleom. *Pyrenophora avenae*; **Drechslera bromi**, teleom. *P. bromi*.

D. carbonum, teleom. *Cochliobolus carbonum*.

D. catenaria (Drechsler) Ito; con. av. 104 (40–170) μm long; leaf blight and crown rot of *Agrostis palustris*, see *Pyrenophora lolii*, Wilkins; Larsen et al. and Spilker & Larsen, *PD* **65**:79, 1981; **69**:331, 1985 [**60**,5981; **64**,4377].

D. dematioidea (Bubák & Wróbl.) Subram. & Jain 1966; con. av. 36 × 14.3 μm, mostly 3–4 septate; von Broembsen, *PD* **70**:33, 1986, pincushion, *Leucospermum*, blight in South Africa [**65**,4975].

D. dictyoides, teleom. *Pyrenophora dictyoides*.

D. gigantea (Heald & Wolf) Ito; Drechsler, *J.agric.Res.* **37**: 473, 1928; con. mostly 200–300 × 15–25 μm; zonate eyespot of grasses, on banana, coconut; Ahn and Kardin et al., *PD* **64**:878, 1980; **66**:737, 1982, on rice & wild rice, *Zizania aquatica* [**8**:384; **60**, 3171; **62**,249].

D. graminea, teleom. *Pyrenophora graminea*.

D. heveae (Petch) M.B. Ellis 1971; Ellis & Holliday, *Descr.F.* 343, 1972; con. mostly 90–130 × 15–21 μm; rubber bird's eyespot, only likely to be important in the nursery; Zainuddin & Lim, *Plrs′ Bull.* 158:3, 1979, chemical control in Malaysia [**59**,884].

*****D. nobleae** McKenzie & D. Matthews, *TBMS* **68**:309, 1977; con. 55–125 × 9–19 μm; on ryegrass, *Lolium*, leaves and seed in New Zealand; Morrison, *Mycologia* **74**:391, 1982, USA; Lam, *TBMS* **83**:339, 1984, England [**61**,6440; **64**,228].

D. poae (Baudys) Shoem., *CJB* **40**:827, 1962; con. mostly 60–100 × 17–23 μm; meadow grasses, *Poa*, melting out, a leaf spot and crown rot; Nutter et al., *PD* **66**:721, 1982, seasonal periodicity & role of mowing; Hagan & Larsen, *ibid* **69**:21, 1985, conidial source, dispersal peak at 1200–1600 hours, both references on Kentucky blue grass in USA [**62**,246; **64**,2064].

D. siccans teleom. *Pyrenophora lolii*; **Drechslera teres**, teleom. *P. trichostoma*, fide Cannon et al. 1985; **D. tritici-repentis**, teleom. *P. tritici-repentis*. **drechsleri**, *Phytophthora*.

Drepanopeziza (Kleb.) Höhnel 1917; Dermataceae; ascomata dark, soft, erumpent, not assoc. with a stroma in the host; asco. aseptate, hyaline; Rimpau, *Phytopath.Z.* **43**:257, 1962 [**41**:642].

D. populi-albae (Kleb.) Nannf. 1932; Pirozynski, *Fungi Canadenses* 14, 1974, as anam. *Marssonina castagnei* (Desm. & Mont.) Magnus 1906; asco. 14.5–18 × 7–9 μm, con. 17–21 × 6.5–7.5 μm; poplar leaves; Spiers, *Eur.J.For.Path.* **13**:218, 1983, host range & pathogenicity [**63**,1422].

D. populorum (Desm.) Höhnel; Pirozynski, *Fungi Canadenses* 15, 1974, as anam. *Marssonina populi* (Lib.) Magnus 1906; asco. 10–16 × 5–9 μm, con. 17–25 × 6–11 μm; poplar leaves; Boyer, *CJB* **39**:1409, 1961, anam. variability; Gremmen, *Ned.BoschbTijdschr.* **36**:149, 1964, asco. & con. inoculations [**41**:258; **43**,2424].

D. punctiformis Gremmen 1965; anam. *Marssonina brunnea* (Ell. & Ev.) Magnus 1906, fide Byrom & Burdekin, *TBMS* **54**:139, 1970; Pirozynski, *Fungi Canadenses* 13, 1974, as anam.; asco. 10–14 × 3–7 μm, con. 13–18 × 4.5–5.5 μm; for microconidia in vitro see *D. tremulae*, Spiers & Hopcroft; poplar leaves; de Kam, *Eur.J.For.Path.* **5**:304, 1975, asco. discharge & infection in Netherlands; Pinon & Poissonnier, *ibid* **5**:97, 1975, epidemiology in France; Spiers et al., *ibid* **13**:181, 305, 414, 1983; **14**:270, 1984, with *Drepanopeziza populi-albae* & *D. populorum*, seed transm.; *D. punctiformis* is a more serious pathogen in New Zealand than the other 2 spp.; effects of leaf age, surface & stomatal frequency [**49**,2668; **54**,4651; **55**,1965; **63**,902, 1423, 3057; **64**,1777].

D. ribis (Kleb.) Höhnel; anam. *Gloeosporidiella ribis* (Lib.) Petrak 1921; Booth & Waller, *Descr.F.* 638, 1979; asco. 10–15 × 6–7.5 μm, con. 14–21 × 4–7 μm; leaf spot of currant and gooseberry.

D. salicis (Tul. & C. Tul.) Höhnel; anam. *Monostichella salicis* (Westend.) v. Arx, see Cannon et al. 1985; asco. 10–17 × 5–8 μm, con. 10–17 × 5–8 μm; willow leaf spot, see Rimpau under the genus for the related fungi on willow and entry below; Anselmi, *Cellulosa Carta* **28**(6):19, 1977, biology in Italy; *Eur.J.For.Path.* **10**:438, 1980, ascomata [**60**,4685,5567].

D. sphaerioides (Pers.) Höhnel; anam. *Marssonina salicicola* (Bres.) Magnus 1906; see Cannon et al. 1985; asco. 13–22 × 6–8.5 μm; con. 14–17 × 5–8 μm; on willow; Vegh, *Annls.Phytopath.* **6**:309, 1974; Anselmi 1979, willow leaf bronzing in Italy; Rimfeldt 1979, willow canker in Norway; Vegh & Velastegui, *Cryptog.Mycologie* **4**:345, 1983, 3 *Marssonina* spp. on *Salix* [**55**, 411; **60**,3978; 4686; **63**,2522].

D. tremulae Rimpau 1962; poss. anam. *Marssonina*

tremulae (Lib.) Kleb. 1918; fide Rimpau under genus; asco. 8–9 × 1.8–2 μm, con. 13–17 × 4–5 μm; poplar leaves; Spiers & Hopcroft, *Eur.J.For.Path.* **16**:65, 1986, morphology & development of ascomata, temp, in New Zealand [**65**, 6204].

D. triandrae Rimpau 1962; anam. *Marssonina kriegeriana* (Bres.) Magnus, fide Rimpau under genus; asco. 10.5–12 × 5–5.6 μm, con. 13–17 × 3.5–6 μm; willow leaf spot; see *Drepanopeziza sphaerioides*, Vegh & Velastequi.

drippy gill (*Pseudomonas agarici*) common cultivated mushroom; — **nut** (*Erwinia quercina*) oak.

drop (*Sclerotinia minor, S. sclerotiorum*) lettuce.

— **spectrum**, the distribution, by number or volume of drops, of spray into intervals of drop size.

drummondii, *Pleospora, Stemphylium.*

dryadeus, *Inonotus.*

Dryad's saddle (*Polyporus squamosus*) temperate hardwoods.

dry basal rot (*Ceratocystis paradoxa*) oil palm; — **bubble** (*Verticillium fungicola*) common cultivated mushroom; — **bud rot** (see palm lethal yellowing) coconut; — **eye rot** (*Botryotinia fuckeliana*) apple.

— **rot**, (1) Rotting that proceeds at a rate which allows drying of the lesion to keep pace with lysis. (2) A term applied specifically to rotting of apparently dry timber by *Serpula lacrimans*, FBPP. Dry rot is used for several diseases which involve decays of corms, tubers and maize cobs caused by fungus pathogens.

— **top rot** (*Sorosphaera vasculorum*) sugarcane.

ducometi, *Peronospora.*

Duggar, Benjamin Minge, 1872–1956; born in USA, Univ. Alabama, Cornell, Harvard, Mississipi, Missouri, Wisconsin; early work on tobacco mosaic v., discovered aureomycin, wrote: *Fungous diseases of plants*, 1909; also a plant physiologist. *Nature Lond.* **178**:834, 1956; *Mycologia* **49**:434, 1957; *Phytop.* **47**:379, 1957; *A.R.Phytop.* **20**:33, 1982.

Duncan, Catherine Gross, 1908–68; born in USA; Univ. De Pauw, Wisconsin, pioneer in wood decay fungi. *Phytop.* **59**:1777, 1969.

Durandiella Seaver 1932; Dermataceae; Groves, *CJB* **32**:116, 1954; asco. filiform to fusoid, mostly 3–5 septate. *D. pseudotsugae* Funk, *CJB* **40**:332, 1962, asco. 0–3 septate, 45–95 × 2.5–3.5 μm; anam. *Corniculariella pseudotsugae* (W.L. White) Dicosmo 1978; Douglas fir canker; White, *Mycologia* **28**:433, 1936 [**16**:147; **41**:619].

durian, *Durio zibethinus*; Lim et al., *AAB* **111**:301, 1987, foliar blight (*Thanatephorus cucumeris*).

dust, finely divided particles, of a plant protectant etc., with or without a diluent, for application in the dry condition, FBPP.

Dutch elm wilt (*Ceratocystis ulmi*) elm; the preferred name, see disease, always called Dutch elm disease.

dwarf bean, see bean; — **bunt** (*Tilletia controversa*) wheat.

dwarfing, a decrease in overall size without alteration of the proportions between parts of the plant; synonymous with nanism, FBPP; — (*Aphelenchoides fragariae*) violet).

Dysmicoccus brevipes, transm. cacao swollen shoot.

dying arm, as for dead arm.

E

ear blight, rice, caused by several fungi; Ôhata, *Rev.Pl.Prot.Res.* **6**:101, 1973; — cockles (*Anguina tritici*) temperate cereals, mostly wheat.
earlianum, *Diplocarpon*.
early blight (*Alternaria solani*) tomato, (*Cercospora apii*) celery.
*— dying (mostly *Verticillium dahliae*) potato; Martin et al., *Phytop.* **72**:640, 1982; Rahimian & Mitchell and Kotcon et al., *ibid* **74**:327, 462, 1984; several authors in *Am.Pot.J.* **62**(4), 1985 [**61**,7138; **63**,4035, 4050; **65**,3454–9].
— leaf spot (*Mycosphaerella arachidis*) groundnut.
ear smut (*Ustilago bullata*) *Bromus*.
— stalk rot, maize, several fungi; Shurtleff 1980.
eastern gall rust (*Cronartium quercuum* f.sp. *virginianae*) *Pinus virginiana*; — white root rot (*Corticium galactinum*) apple.
ecballium mosaic Cohen & Nitzany, *Hassadeh* **42**:842, 1962; transm. sap, *Aphis gossypii*, non-persistent, Israel, infects cucumber, melon, squash; Ghosh & Mukhopadhyay, *Curr. Sci.* **48**: 788, 1979, in India, *E. elaterium*, squirting cucumber [**41**:757; **59**,2978].
*Echinochloa frumentacea, see millets.
Echinodontiaceae, Aphyllophorales; hydnoid form, basidiomata corky to woody, coloured; hymenophore toothed, basidiospores smooth, hyaline, walls thick, amyloid; Gross, *Mycopath.Mycol.appl.* **24**:1, 1964.
Echinodontium Ell. & Ev. 1900; Echinodontiaceae; *E. tinctorium* (Ell. & Ev.) Ell. & Ev.; brown stringy rot of conifer heart wood; Indian paint fungus, because of the brick red dye from the basidioma; Japan, N. America; Boyce 1961; Aho & Filip, *Can.J.For.Res.* **12**:705, 1982, wounding & infection of white fir, *Abies concolor*. Davidson & Chien, *Mycologia* **68**:1152, 1976, described *E. taxodii* (Lenz & McKay) Gross and a large white pocket rot of *Chaemocyparis formosensis* [**56**,4242; **63**,1430].
echinulatum, *Cladosporium*.
echtes ackerbohnenmosaik v. Quantz 1953; Gibbs & Paul, *Descr. V.* 20, 1970; Boswell & Gibbs 1983; Comovirus, isometric *c*.25 nm diam., transm. sap, seedborne; Africa (N.W.), Australia, Europe; causes mosaic in broad bean and other legumes; Jones & Barker, *AAB* **83**:231, 1976, comparison with broad bean stain v. q.v. Cockbain et al. for transm. by same insects [**55**,4941].
Eckelrade, see cherry.

eclipta phyllody Padma et al. 1973; Phatak et al. 1974; MLO assoc., India, *E. prostrata* [**54**,4865].
ecoclimate, or climate within the crop, see ecology; Waggoner, *A.R.Phytop.* **3**:103, 1965, microclimate & disease; Colhoun, *ibid* **11**:343, 1973, effects of environment; Burrage in Scott & Bainbridge 1978, monitoring the environment in relation to epidemiology.
ecofallow, a system where the weeds are controlled and soil moisture conserved in a crop rotation with a minimum of disturbance to the crop residue and soil, i.e. a reduced tillage; Doupnik & Boosalis, *PD* **64**:31, 1980, a 3-year cycle with sorghum and winter wheat in low rainfall areas of USA (Colorado, Kansas, Nebraska), sorghum stalk rot (*Gibberella fujikuroi*) was reduced and yield increased.
*ecology, study of the relations between organisms and their biological and physical surroundings; see: ecoclimate, epidemiology, epiphyte, host density, irrigation, salinity, stress, water in soil; Duffus, *A.R.Phytop.* **9**:319, 1971, weeds & virus diseases; *Proc.Am.Phytopath.Soc.* **1**:170, 1974, epidemics & ecosystems; Bruehl, *A.R.Phytop.* **14**:247, 1976, fungus colonists of cultivated soil; Cherrett & Sagar ed., *Origins of pest, parasite, disease and weed problems*, 1977; Burdon & Shattock, *Appl. Biol.* **5**, 1980, disease in plant communities; Perrin, *Protect.Ecol.* **2**:77, 1980, environment & crop protection; Harrison, *AAB* **99**:195, 1981, viruses; Thresh ed., *Pests, pathogens and vegetation*, 1981; Bos, *Adv.appl.Biol.* **7**:105, 1983, viruses; Dinoor & Eshed, *A.R.Phytop.* **22**:443, 1984, pathogens in natural plant communities.
ectotrophic infection, from fungi growing on the root surface or epidermal cells.
edaphosphere, the soil away from any effects by plant roots, i.e. beyond the rhizosphere, after FBPP.
eddo, *Colocasia esculenta*, see taro.
Edgerton, Claude Wilbur, 1880–1965; born in USA, Univ. Nebraska, Cornell, Louisiana State, original work on hybridisation in fungi, studies on *Glomerella*; wrote: *Sugarcane and its diseases*, 1955, edn. 2, 1958. *Mycologia* **60**:739, 1968.
edible aroids, see taro.
edifenphos, protectant fungicide with some eradicant action, used against rice pathogens.
eelworms, see nematodes.
effective inoculum dose, the amount of inoculum which is available for penetration of plant tissues.

This may be equivalent to the inoculum concentration used or only a proportion of it, e.g. in leaf scar inoculations the amount of inoculum which enters is dependent upon vascular tension in the plant and rate of evaporation etc., FBPP; see inoculum potential.

eggplant, *Solanum melongena.*

— **green mosaic** Ladipo, *PDR* **60**:1068, 1976; transm. sap, Nigeria [56,3371].

—**Indian mosaics** Sharma, *Phytopath.Z.* **65**:341, 1969; called crinkle mosaic; Naqvi & Mahmood, *Geobios* **2**:189, 1975; Mayee & Khatri, *Indian Phytopath.* **28**:238, 1975; transm. sap, may be seedborne and transm. by *Myz.pers.* [49,1243; 55,5633; 56,1841].

___ **leaf mottle v.** Khalil & Nelson 1977; Khalil et al., *Phytop.* **72**:1064, 1982; Carlavirus, filament 690 nm long, transm. sap; Sharja and Kalba, one of the United Arab Emirates, mostly Solanaceae [62,833].

— **little leaf** Thomas & Krishnaswami 1939; Varma et al., *Indian Phytopath.* **22**:289, 1969; Srinivasan & Chelliah, *Madras agric.J.* **64**:94, 1977; MLO assoc., transm. *Hishimonus phycitis*; Joshi & Bose, *Indian Phytopath.* **36**:604, 1983, yield loss; see potato purple top wilt, Harding & Teakle [19:61; 49,2026; 57,5804; 65,6283].

— **mild mosaic v.** Naqvi & Mahmood, *Indian Phytopath.* **29**:138, 1976; *Phytopath.Z.* **93**:86, 1978; filament 675–750 × 13 nm, transm. sap readily [57,2378; 58,3598].

— **mosaic v.** Ferguson 1951; Gibbs & Harrison, *Descr.V.* 124, 1973; Tymovirus, isometric *c.*30 nm diam.; transm. sap, *Epitrix* sp.; Brazil, S. America (N.W.), W. Indies; infects tomato; potato Andean latent v. q.v. is a str. of EMV.

—— **mottle v.** Rao, *Z.PflKrankh.PflSchutz.* **85**:313, 1978; rod 290 × 18 nm, transm. sap, India (N.), considered to be distinct from other eggplant mosaics [58,466].

— **mottled crinkle v.** Makkouk, Koenig & Lesemann, *Phytop.* **71**: 572, 1981; Tombusvirus, isometric 28–30 nm diam., transm. sap, Lebanon [61,1514].

—— **dwarf v.** Martelli 1969; Martelli & Russo, *Descr.V.* 115, 1973; Phytorhabdovirus, bacilliform 220 × 66 nm, transm. sap, Italy (S.); Martelli & Hamadi, *Pl.Path.* **35**:595, 1986, in Algeria [66, 2162].

— **ring and severe mosaic** Sharma, *Phytopath.Z.* **65**:341, 1969; transm. sap, India [49,1243].

Egyptian clover, see *Trifolium.*

ehrenbergii, *Sorosporium.*

Eichhornia crassipes, water hyacinth.

eichhorniae, *Alternaria*

elaeagni, *Phomopsis, Septoria.*

elaeidis, *Pseudospiropes,* transferred from *Cercospora* in 1985.

Elaeis guineensis, oil palm.

elettariae, *Phyllosticta.*

elder, elderberry, *Sambucus*; — **ringspot** (cherry leaf roll v.).

elderberry A = elderberry Carlavirus.

— **B.** = elderberry latent v.

— **carlavirus** van Lent, Wit & Dijkstra 1980; Dijkstra & van Lent, *Descr.V.* 263, 1983; filament *c.*680 × 12 nm; transm. sap, *Myz.pers.*, non-persistent, no symptoms; Netherlands, Scotland; *Sambucus* spp.

— **golden v.** Hansen & Stace-Smith, *Phytop.* **61**:1222, 1971; Jones & Murant, *AAB* **69**:11, 1971 = cherry leaf roll v. [51,2295, 2569].

— **latent v.** Jones 1972; *Descr.V.* 127, 1974; isometric *c.*30 nm diam., contains RNA, transm. sap, line patterns in American elder, *Sambucus canadensis*, no symptoms in *S. nigra..*

— **mosaics and rings,** Blattný 1930; Böning 1963; Rønde Kristensen 1964; Plavšic-Banjac & Miličić 1968; Czechoslovakia, Denmark, Germany, Yugoslavia [10:329; 43, 2491c; 44, 596c; 48, 3163].

— **vein mottle v.** Uyemoto, Gilmer & Williams, *PDR* **55**:913, 1971; filament 640 × 15 nm, transm. sap, USA (New York State) [51, 1682].

electron microscopy, M. A. & H. G. Ehrlich, *A.R.Phytop.* **9**:155, 1971, fine structure of host & parasite interface; Cooper in Staples & Toenniessen 1981, fine structural changes induced by the pathogen; Hayat, *Principles and techniques of electron microscopy,* edn. 2, 1981; Brown & White, *PD* **66**:282, 1982, applications of scanning electron microscopy; *Ainsworth & Bisby's dictionary of the fungi,* edn. 7, 1983, ultrastructure; Baker et al., *PD* **69**:85, 1985, applications to plant virology.

electrophysiology, applications; Tattar & Blanchard, *A.R.Phytop.* **14**:309, 1976.

Elettaria cardamomum, cardamom.

Eleusine coracana, see millets.

eleusine mosaic, streak and stunt, Khurana et al., *Indian Phytopath.* **26**:554, 1973; Nagaraju & Viswanath, *ibid* **34**:458, 1981; Maramorosch et al., *PDR* **61**:1029, 1977. These disease agents in India appear to fall into 2 groups, one transm. by aphid spp. the other by *Cicadulina* and *Sogatella,* the latter has an assoc. rhabdovirus; they occur in finger millet and other *Eleusine* spp. [54,1254; 57,3948; 61,6374].

eleusinis, *Melanopsichium.*

Elgon dieback (*Pseudomonas syringae* pv. *garcae*) coffee.

elicitor, a trigger for a host's response to infection by a pathogen. If the host has adequate resistance this response is shown by defence mechanisms, e.g. phytoalexin synthesis, deposition of material like lignin, accumulation of proteinase inhibitors and increases in activity of certain hydrolytic enzymes. The trigger can also be set off by mechanical damage, culture filtrates, other preparations of pathogens and by artificial elicitors. Keen, *Science N.Y.* **187**:74, 1975; Cramer et al., *ibid* **227**:1240, 1985; Wood in Varma & Verma 1986 [**55**,979; **64**,2804; **65**,5361].

*Elisa, acronym for enzyme linked immunosorbent assay; adapted from a medical immunodiagnostic procedure for the quantitative detection of plant viruses; Voller et al., *J.gen.Virol.* **33**:165, 1976; Clark & Adams, *ibid* **34**: 475, 1977; Clark, *A.R.Phytop.* **19**:83,1981; Katz, *Adv.Virus Res.* **29**: 169, 1984; Sutula et al., *PD* **70**:722, 1986, interpreting data [**56**, 606, 2905].

Elliot, Charlotte, 1883–1974; born in USA, Univ. Stanford and Wisconsin, USDA, bacterial diseases of cereals; wrote: *Manual of bacterial plant pathogens* 1930, edn. 2, 1951. *Phytop.* **66**: 237, 1976.

elliptica, *Botrytis*; ellisiana, *Dasyscyphus*.

elm, *Ulmus*; Stipes & Campana 1981.

— leaf scorch Wester & Jylkka, *PDR* **43**:519, 1959; Hearon et al., *CJB* **58**:1986, 1980; xylem limited bacterium q.v. assoc., Gram negative, USA (S.E.) [**38**:718; **60**,3967].

— mosaic (cherry leaf roll v.).

— mottle v. Schmelzer, H. E. & H. B. Schmidt 1966; Jones, *Descr. V.* 139, 1974; Ilarvirus, isometric 25–30 nm diam., transm. sap to many herbaceous plants, seedborne in elm, Europe, mosaic or ringspot in *Forsythia intermedia* and *Syringa vulgaris*; Cooper 1979.

— phloem necrosis Swingle 1938; Sinclair in Stipes & Campana 1981; MLO assoc., transm. *Scaphoides luteolus*, USA, epidemics caused in *Ulmus americana*, Asian and European elms escape infection in USA.

— witches' broom Pantanelli in Ciferri & Corte, *Atti Ist.bot. Univ.Pavia* **17**:122, 1960; Pisi et al., *Phytopath.Mediterranea* **20**:189, 1981; MLO assoc., parts of Europe [**39**:632; **63**,885].

— yellows = elm phloem necrosis.

— zonate canker Swingle & Bretz, *Phytop.* **40**:1018, 1950; USA [**30**:293].

elongata, *Phyllosticta*.

Elsinoë Racib. 1900; Myriangiaceae; Sivanesan 1984; asco. 2 or more septate, hyaline to pale yellow, anam. in *Sphaceloma*; on leaves, stems and fruits, the lesions are like scabs which is the term usually used to describe the diseases caused; obligate pathogens; a few cause significant diseases but many have scarcely been studied as pathogens. Since, however, they are potentially

damaging more spp. have been included than is prob. warranted from the literature; Holliday 1980.

E. ampelina Shear 1929; Sivanesan & Critchett, *Descr.F.* 439, 1974; anam. *Sphaceloma ampelinum* de Bary 1874; asco. 3 septate, 15–16 × 4–5 μm; con. 4–7.5 × 2–3.5 μm, grapevine bird's eye rot; Ozoe et al. *Bull.Shimane agric.Exp.Stn.* 10:120, 1972, ecology & control; Brook, *N.Z.J.agric.Res.* **16**:333, 1973, epidemiology; Suhag & Grover, *Indian Phytopath.* **30**:460, 1977, epidemiology; Cheema et al., *ibid* **31**;163, 1978, pathogenic & cultural variation; Mortensen, *J.Hered.* **72**:423, 1981, sources & inheritance of resistance [**53**,639; **55**,1893; **58**,5977; **59**,1337; **61**,4278].

E. australis Bitanc. & Jenkins 1936; Sivanesan & Critchett, *Descr. F.* 440, 1974; anam. *Sphaceloma australis*, same authors; asco. 1–3 septate, 12–20 × 4–9 μm; con. 4–6 × 2–4 μm; sweet orange scab, prob. mostly in S. America, control may be required, has a higher optimum temp. for growth than does *Elsinoë fawcettii*.

*E. batatas Viégas & Jenkins in Jenkins & Viégas, *J.Wash.Acad.Sci.* **33**:248, 1943; anam. *Sphaceloma batatas* Saw. 1931; asco. 7–8 × 3–4 μm, con. 6–7.5 × 2.5–3.5 μm; sweet potato scab, Sawada called the disease bud stunting, on leaves; S.E. Asia, S. America, Oceania; Goodbody, *Trop.Agric.Trin.* **60**:302, 1983, reduction in tuber yield in Papua New Guinea [**63**,2112].

E. brasiliensis Bitanc. & Jenkins 1942; anam. *Sphaceloma manihoticola* Bitanc. & Jenkins 1950, fide Zeigler & Lozano, *Phytop.* **73**:293, 1983; asco. 3 septate, 9.5–14 × 3–7 μm; con. 2.5–6.5 × 1–4.5 μm; cassava superelongation, forms a gibberellin which causes the characteristic elongation of the host's shoots; S. and Central America, W. Indies; Zeigler et al., *Phytop.* **70**:589, 1980, gibberellin; *Trop.Pest Management* **29**:148, 1983, resistance [**60**,2322; **62**,3398; **63**,1062].

E. canavaliae Racib. 1900; Sivanesan & Holliday, *Descr.F.* 313, 1971; asco. 2–3 septate, 9–12 × 2.9–3.5 μm; scab of jack and sword beans, E. Africa, S. E. Asia.

E. cinnamomi Pollack & Jenkins, *Mycologia* **38**:470, 1946; asco. 4–5 septate becoming muriform, 15–17 × 4–6 μm; con. 4–6 × 3.5 μm; on leaves and young stems of *Cinnamomum camphora*, USA [**26**:31].

E. corni Jenkins & Bitanc., *J.Wash. Acad.Sci.* **38**:362, 1948; asco. 3 septate, 12–15 × 5 μm; con. 4.5–6 × 2.5 μm, on *Cornus florida*; several authors in *PDR* 36:294, 1952 [**32**:81].

E. dolichi Jenkins, Bitanc. & Cheo, *J.Wash.Acad.Sci.* **31**:416, 1941; asco. 7–13 × 3–5 μm; hyacinth bean scab [**21**:45].

E. fawcettii Bitanc. & Jenkins 1936; Sivanesan &

Critchett, *Descr.F.* 438, 1974; anam. *Sphaceloma fawcettii* Jenkins 1925; asco. 1–3 septate, 10–12 × 5–6 μm; con. 5–10 × 2.5 μm; citrus common scab; this is the most widely studied of all the scabs caused by *Elsinoë* spp., Holliday 1980 gave a full account. Two other scabs occur on citrus: *Elsinoë australis* and *Sphaceloma fawcettii* var. *scabiosa*. *E. fawcettii* can cause a severe disease in the tropics and sub-tropics where the moisture and temp. conditions favour infection; this scab is unimportant in the drier, growing regions with the Mediterranean type of climate; Whiteside, *Proc.Fla.St.hort.Soc.* 94:5, 1981, chemical control [62,1971].

E. iwatae Kajiwara & Mukelar, *Contr.Cent. Res.Inst.Agric.Bogor* 23:8, 1976; anam. *Sphaceloma iwatae*, same authors; asco. 11–14 × 5–5.7 μm; mung bean scab, Indonesia; Mukelar et al., *ibid* 24, 1976, chemical control [58,2031, 3016].

E. leucospila Bitanc. & Jenkins, *Arq.Inst.biol. S.Paulo* 17:70, 1946 anam. *Sphaceloma theae* Kurosawa 1939; asco. 13–16 × 5–7 μm; tea white scab; Fukuda & Takaya 1977, see *Hort.Abs* 48, 5043 for factors affecting incidence in Japan [26:317].

E. mangiferae Bitanc. & Jenkins, *Arq.Inst.biol.S.Paulo* 17:218, 1946; anam. *Sphaceloma mangiferae*, same authors; asco. 3 septate, 10–13 × 4–6 μm; con. 6–29 × 2–4 μm; mango scab [26:552].

E. oleae Ciccarone & Graniti, *Arq.Inst.biol.S.Paulo* 26:20, 1959; anam. *Sphaceloma oleae*, same authors; asco. 12–15 × 3–4 μm, con. 2–3.5 × 3–6 μm, olive scab [40:119].

E. phaseoli Jenkins 1933; Sivanesan & Holliday, *Descr.F.* 314, 1971; asco. 3 septate, one of the central cells may become longitudinally septate; con. 10 × 4 μm; Lima bean scab, S. and Central America, W. Indies; seedborne, poss. races, can be damaging, Holliday 1980.

E. piri (Woronichin) Jenkins, *J.agric.Res.* 44:696, 1932; anam. *Sphaceloma pirinum* (Pegl.) Jenkins, *Phytop.* 36:460, 1946; asco. 3 septate, 10–15 × 4–6 μm; con. 4–4.5 × 2.5–4 μm; fruit and leaf spot, apple, pear; Jenkins, *Mycologia* 38:450, 1946 [11:723; 25:563; 26:65].

E. quercus-falcatae J. Miller, *Mycologia* 49:277, 1957; asco. 1–3 septate, 11–14 × 4–6 μm; on *Quercus falcata*, southern red oak, USA, an entire tree appeared yellow and there was some defoliation [36:623].

E. randii Jenkins & Bitanc., *Phytop.* 28:77, 1938; asco. 3 septate, 11–16 × 6–8 μm; con. 8–15 × 4–5 μm; pecan nursery blight, may require control; Brazil, USA [17:421].

E. rosarum Bitanc. & Jenkins, *Mycologia* 49:98, 1957; anam. *Sphaceloma rosarum* (Pass.) Jenkins 1932; asco. 10–14 × 5–7 μm; rose spot scab, usually called spot anthracnose.

E. theae Bitanc. & Jenkins, *Arq.Inst.biol.S.Paulo* 10:195, 1939; asco. 10–14 × 3–7 μm; tea mottle scab, Brazil, E. Africa, India, Sri Lanka [19:369].

E. tiliae Creelman, *Mycologia* 48:555, 1956; asco. 11–16 × 5–6 μm, con. av. 4.7 × 2 μm; *Tilia* scab, called spot anthracnose; Lindquist & Merlo 1964, in Argentina [36:217; 43, 3336].

E. veneta (Burkh.) Jenkins 1932; Sivanesan & Critchett, *Descr.F.* 484, 1976; anam. *Sphaceloma necator* (Ell. & Ev.) Jenkins & Shear 1946; asco. 3 septate, 18–21 × 6–8 μm; con. 5–7 × 2–3 μm, raspberry cane spot, on *Rubus*.

Elymana virescens, transm. wheat American striate mosaic.

Elytroderma Darker 1932; Hypodermataceae; asco. bacilliform, 1 septate, in a gelatinous sheath.

E. deformans (Weir) Darker 1932; Minter, *Descr.F.* 655, 1980; asco. 90–130 × 8–10 μm; pine needlecast and witches' broom, N. America; Lightle, *Phytop.* 44:557, 1954 [34:329].

E. torres-juanii Diamandis & Minter, *TBMS* 72:169, 1979; *Descr.F.* 654, 1980; asco. 130–165 × 9–12 μm; pine needle blight, parts of Europe (S.); on *Pinus brutia* in Greece, in assoc. with *Thaumetopoea pityocampa*, pine processionary caterpillar, destruction of the tree crown.

Embellisia Simmons, *Mycologia* 63:380, 1971; Hyphom.; conidiomata hyphal, conidiogenesis enteroblastic tretic; con. multiseptate, septa thick, dark, transverse and occasionally oblique or longitudinal; Ellis 1976; Simmons, *Mycotaxon* 17:216, 1983 [62,4628].

E. abundans Simmons 1983; con. mostly 3–6 transverse septa, occasionally 1–2 longitudinal septa, short 20–30 × 10–12 μm, long 40–50 × 8–12 μm; Lumyong et al., *Pl.Path.* 33:431, 1984, Brussels sprout leaf spot in Thailand (N.) [63,5606].

E. allii (Campanile) Simmons 1971; con. mostly 30–40 × 10–12 μm, chlamydospores up to 70 μm diam., often on the host but less frequent in culture, garlic bulb canker, on the membranous scales of onion; Campanile 1924; Moore, *TBMS* 26:22, 1943; Nakov et al. 1979 [4:325; 22:365; 59,4869].

E. hyacinthi de Hoog & Muller, *Neth.J.Pl.Path.* 79:85, 1973; con. 3–6 septate, mostly 32–38 × 9–12 μm, pigmented; hyacinth bulb skin necrotic patch, causes elliptical to lenticular leaf spots, also on *Freesia* and *Scilla* [53,193].

Emericella Berk. & Broome 1857; Eurotiales;
anam. *Aspergillus.*
Emericellopsis v. Beyma 1940; Eurotiales; anam.
Acremonium.
emilia yellow vein banding Loos, *Trop.Agriculturalist*
97:18, 1941; Sri Lanka, *E. scabra* [21:89].
—— —— —— mosaic Nair & Wilson,
Agric.Res.J.Kerala 7:123, 1969; transm. *Bemisia
tabaci*, India (Kerala) *E. sonchifolia* [50,1136].
Empoasca, transm. pigeon pea witches' broom;
E. payayae, papaya apical necrosis, papaya bunchy
top.
empty roots, see *Phytomonas*, cassava.
enation, an abnormal outgrowth from the surface
of an organ such as a stem or leaf.
encapsidation, the enclosure of the genome of a
virus in a protein coat.
*Encoelia pruinosa, see *Cenangium singulare.*
Encoeliposis laricina, see *Ascocalyx.*
endemic disease, a disease permanently established
in moderate or severe form in a defined area,
commonly a country or a part of a country,
FBPP.
endive, *Cichorium endivia.*
—— white chlorotic spot (turnip mosaic v.),
Verhoyen, *Meded.Fak.LandbWetensch.Gent*
48:859, 1983, Belgium, rapid spread [63,4199].
endobioticum, *Synchytrium.*
Endoconidium Prill. & Delacr. 1891; Hyphom;
Carmichael et al. 1980; conidiomata sporodochial,
conidiogenesis enteroblastic, phialidic; con.
aseptate, hyaline; *E. temulentum*, teleom.
Gloeotinia granigena.
Endocronartium Y. Hirats., *CJB* 47:1493, 1969;
Uredinales; the genus was erected to
accommodate 2 caulicolous, autoecious pine rusts
and previously referred to *Peridermium harknessii*
J. P. Moore and *P. pini* (Pers.) Lév. They occur
only on pine; the aecial state was considered to be
telioid with the aeciospores functioning as
teliospores and forming basidia on germination,
although the former resemble aeciospores;
Laundon in Ainsworth et al. 1973, vol. IVB,
stated that the genus is illegitimate because it
contains the type sp. of *Peridermium*; Ziller 1974
accepted the genus.
E. harknessii, author as above; telia with
teliospores resembling aeciospores, they form 1–4
septate germ tubes, function as basidia and infect
hard pines, aecia, uredia and basidiospores
absent, spermogonia rare or absent; western gall
rust; causes large, globose, pyriform galls on
branches, hence sometimes called globose gall
rust, N. America only; this rust is most damaging.
Ziller 1974 stated: 'Western gall rust is the most
common, most conspicuous and most destructive

stem rust of hard pines in Western Canada.' The
fungus is a danger to hard pines, especially Scots
pine, in regions outside N. America; Boyce 1961.
E. pini (Pers.) Y. Hirats., reference as for genus;
this rust is based on *Peridermium pini* (Pers.) Lév.
1826, emended by Kleb. 1890; Wilson &
Henderson 1966 referred to it, before Hiratsuka's
redisposition, as a non-alternating race of the
alternating rust *Cronartium flaccidum* q.v.
Endocronartium pini, on pine only, has aeciospores
whose germ tubes have a determinant growth and
are regularly septate; the aecial germ tubes of
C. flaccidum are indeterminate and irregularly
septate. Both rusts occur in Eurasia only and both
cause resin top of Scots pine. *E. pini* causes
girdling lesions which spread from the branches to
the main stem where large, blackened cankers are
formed; van der Kamp, *Forestry* 41:189, 1968,
review; *ibid* 42:185, 1969; 43:73, 1970, infection,
lesion development & anatomy; Murray et al.,
ibid 42: 165, 1969, disease incidence in S.E.
Scotland [48,961; 49,870–1; 3036].
endogenous inhibitor, a substance formed by an
organism which, at low concentration, will inhibit
the growth of itself or another organism, or some
physiological process assoc. with growth, in a
relatively specific manner. A characteristic of
higher plant growth inhibitors such as abscisic
acid is that their effects are often antagonised by
specific growth promoters, especially gibberellins
and cytokinins, FBPP; Gottlieb, *Phytop.* 63:1326,
1973, endogenous inhibitors of fungus spores; Lax
et al., *ibid* 75:386, 1985, endogenous inhibitor of
con. of *Glomerella cingulata* [64,4737].
Endogonales, Zygomycetes, 1 family
Endogonaceae; Hall, *TBMS* 68:341, 1977;
Glomus.
Endomyces geotrichum, anam. *Geotrichum candidum*
q.v.
Endomycetales, Ascomycotina; ascomata absent,
asci formed directly from free cells or hyphae
arising singly, clustered or catenate; protunicate,
globose to cylindrical, gametangia often
conspicuous, asco. globosed like a hat, or
fusiform to acicular, 0–1 septate.
*endophyte, an organism which completes its life
cycle in a plant which shows no external sign of
the infection. A group of fungi parasitic on
grasses are called endophytic and are toxic to
grazing animals; Latch & Christensen, *AAB*
107:17, 1985; Latch et al., *N.Z.J.agric.Res.*
28:129, 165, 1985; Siegel et al., *PD* 69:179, 1985
[64,3102, 4376, 4379; 66,640].
endosepsis, Smyrna fig internal fruit rot, caused by
fungi carried by the pollinating fig wasp
Blastophaga psenes; Caldis, *Hilgardia* 2:287, 1927;

Smith & Hansen, *Bull.Calif.agric.Exp.Stn.* 506, 1931; Obenauf et al., *PD* **66**:566, 1982 [**7**:104; **10**:678; **61**,7099].

***Endothia** Fr. 1849; Gnomoniaceae; Barr 1978; stroma of thin walled cells or compactly interwoven hyphae, brightly coloured externally and internally; asco. aseptate; close to *Cryphonectria* q.v., Walker et al., which has 1 septate asco.; Griffin et al., *Chestnut blight, other Endothia diseases and the genus* Endothia, 1986.

E. gyrosa (Schwein.:Fr.) Fr.; Snow et al., *Descr.F.* 449, 1975; asco. 6–10 × 2–2.5 μm; causes cankers at wounds on stems, branches and roots of hardwood, temperate trees. A serious disease is caused on *Liquidambar formosana* in USA; the tree is an Asiatic sp. and there is a need to prevent this virulent str. spreading to Asia. A blight of pin oak, *Quercus palustris*, also called hobnail canker, has been described from USA (Virginia), Roane et al., *Mycologia* **66**:1042, 1974; Appel & Stipes, *PD* **68**:851, 1984, found that only oaks under water stress were colonised. *Endothia gyrosa* is a weak pathogen on some trees in parts of Europe and S. Asia; in Portugal, Macara 1974, as orange 'rust' and the anam. *Endothiella*, reported on it on *Q. suber*, cork oak; Walker et al., see *Cryphonectria*, described it on eucalyptus and gave a full account of these 2 genera [**54**,4648–50; **64**,800].

Endothiella Sacc. 1906; Coelom.; teleom in *Endothia* fide Sutton 1980; conidiomata stromatic, conidiogenesis enteroblastic phialidic; con. aseptate, hyaline.

Endria inimica, transm. wheat American striate mosaic.

end rot (*Fusarium solani*) sweet potato.

enteroblastic, see blastic; **enterothallic**, see thallic.

entomogenous fungi, those that grow on insects, *Ainsworth & Bisby's dictionary of the fungi*, edn. 7, 1983.

Entomosporium Lév. 1856; Coelom.; Sutton 1980; conidiomata acervular, conidiogenesis enteroblastic phialidic; con. with larger basal and upper cells, and 2 or more smaller, lateral cells arising from the lower cell; apical and lateral cells each have a single, unbranched, flexuous appendage; Horie & Kobayashi, *Eur.J.For.Path.* **9**:366, 1979, on Rosaceae; *E. mespili*, teleom. *Diplocarpon mespili* [**59**,4088].

Entyloma de Bary 1874; Tilletiaceae; Mordue & Ainsworth 1984; sori tending to remain embedded in the host after maturity, ±agglutinated, forming distinct spots on leaves or other above ground parts, con. may be formed; no serious diseases caused.

E. calendulae (Oudem.) de Bary; Mordue, *Descr.F.* 801, 1984; ustilospores almost hyaline to pale yellow, smooth, 9–14 μm diam., con. commonly present, except in f.sp. *hieracii*, leaf spots on *Calendula* and other Compositae; 3 ff.sp., *bellidis* (Kreiger) Ainsworth & Sampson; *dahliae* (Sydow) Viégas on dahlia; Mordue, *Descr.F.* 802, 1984, *hieracii* Schröter.

E. dactylidis (Pass.) Cif. 1924; ustilospores 12–20 × 8–14 μm; Gramineae; Fushtey & Taylor, *Can.Pl.Dis.Surv.* **57**:29, 1977, blister smut of *Poa pratensis* in Canada [**57**,3502].

E. fuscum Schröter 1877; Mordue, *Descr.F.* 803, 1984; ustilospores, wall 2 layered, golden to pale brown, 11–17 μm diam., con. absent; leaf spot of *Papaver* spp.; Savile, *Can.J.Res.* **C24**:109, 1946, smuts on Papaveraceae.

E. oryzae H. Sydow & Sydow 1914; Mulder & Holliday, *Descr.F.* 296, 1971; ustilospores olivaceous brown to dark brown, firmly adhering in groups, 7.5–11.5 × 6–9 μm; rice leaf smut, restricted to the leaves, cf. *Tilletia barclayana*, the life history is incompletely known but there is little damage to rice; Ou 1985.

E. vignae Bat. et al. 1966; ustilospores 16–21 μm diam., on bean, cowpea, Central and S. America, W. Indies; Vakili, *F.A.O. Pl.Prot.Bull.* **26**:19, 1978; Prabhu & Albuquerque, *Pesqui. Agropec. Brasiliera* **17**:413, 1982 [**58**, 950; **62**,2219].

enveloped viruses, members of the Phytoreoviruses and Phytorhabdoviruses that have an outer lipid protein membrane surrounding the protein coat.

enzymes, see cell wall degrading enzymes.

Ephelis Fr. 1894; Hyphom.; Carmichael et al. 1980; conidiomata sporodochial, con. aseptate, hyaline, filiform; *E. mexicana*, poss. teleom. *Balansia claviceps*; **E. oryzae** teleom. *B. oryzae-sativae*.

Epichloë (Fr.) Tul. & C. Tul. 1865; Clavicipitaceae; ascomata form on or in a conspicuous, bright, crustose stromata which encircles grass stems; asco. multiseptate, filiform.

E. typhina (Pers.) Tul. & C. Tul.; anam. *Acremonium typhinum* Morgan-Jones & Gams, *Mycotaxon* **15**:315, 1982; Booth, *Descr. F.* 639, 1979; stromata golden yellow to orange, asco. almost 200 μm long, 1.5–2 μm diam., with septa at 8–12 μm intervals; choke or cat tail of temperate Gramineae, strs., seedborne in some genera; Hedley & Braithwaite, *Hortic.N.Z.* **9**:6, 1978; Siegel et al., *Phytop.* **74**:932, 937, 1984, as an endophyte, incidence, dissemination & control, on *Festuca arundinacea*, references to animal toxicity [**64**,231–2].

epicoccina, *Phoma*.

Epicoccum Link 1816; Hyphom.; Ellis 1971; Carmichael et al. 1980; conidiomata sporodochial, conidiogenesis holoblastic; con. muriform, pigmented, solitary, wall rough.

E. andropogonis (Ces.) Schol-Schwarz 1959; con. up to 30×27 μm, on Gramineae infected by *Claviceps* spp., substrate is the honeydew, can be confused with a smut.

E. purpurascens Ehrenb. ex Schlect 1824; pulvinate sporodochia appear as small black pustules, con. 15×25 μm, saprophytic, a very common invader of plants and other substrates, frequent in air spora, antagonistic to a few fungus pathogens. *Phoma epicoccina* Punith., Tulloch & Leach 1972 has an *Epicoccum* state which is indistinguishable from *E. purpurascens*, Punithalingam, *Descr.F.* 738, 1982, on *P.epicoccina*; Mulder & Pugh, *Int.Biodetn.Bull.* **7**:69, 1971, review.

epidem, a simulator of plant disease written for a computer; Waggoner & Horsfall, *Bull.Conn.agric.Exp.Stn.* 698, 1969 [**50**,3399].

epidemic, a widespread increase, usually limited in time, in the incidence of an infectious disease; usage favours this term in preference to epiphytotic, Wallace, *PD* **66**:761, 1982. An epidemic may occur over a single season, either in a plant growth or a weather sense; and over a relatively small area. It may extend over several seasons and years, over large areas, i.e. a pandemic. Requirements for an epidemic are: a virulent pathogen, a susceptible crop in monoculture, and favourable conditions for infection and spread. The last is mostly governed by weather and climate, but an animal vector may be important. Epidemics can occur in plants that are not strictly crops and are not being grown in monoculture. They may also be brought about by any breakdown in control measures and changes in agricultural methods, or unusual weather. Pandemics, other conditions being favourable, can arise in different ways; the dates are *c.* those of first detection of disease: (1) Pathogen and crop occur together; but the monoculture of a single, susceptible cv. is superimposed on this situation, *Fusarium oxysporum* f.sp. *cubense on* banana cv. Gros Michel, Central America 1910, Jamaica 1927. (2) The pathogen appears where it was before absent; *Phytophthora infestans* on potato, Ireland 1845; *Peronospora tabacina* on tobacco, Europe 1958. (3) A new str. of an already present pathogen appears; *Ceratocystis ulmi* on elm, England 1960s; *Cochliobolus heterostrophus* on maize, USA 1970. (4) A pathogen spreads from nearby wild hosts to the crop; *Crinipellis perniciosa* on cacao, Surinam 1895, Ecuador 1921; *Microcyclus ulei* on rubber, Surinam 1911, Brazil 1930. (5) A newly introduced crop is attacked by a potential pathogen which is already present;

Rigidoporus lignosus on rubber, Malaysia 1910–16; cacao swollen shoot virus on cacao, Ghana 1936. (6) Both pathogen and host crop are introduced and disease is detected later; prob. citrus tristeza virus on citrus, Argentina and Brazil 1937, USA (California) 1939. (7) Exceptionally favourable weather increases disease, *Cochliobolus miyabeanus* on rice, India, 1942; Holliday, *Rev.Pl.Path.* **50**:337, 1971; Klinkowski, *A.R.Phytop.* **8**:37, 1970; Thurston, *ibid* **11**:27, 1973; Kranz, *ibid* **12**:355, 1974.

epidemiology, the study of epidemics; Vanderplank, *Plant diseases: epidemics and control*, 1963, *Principles of plant infection*, 1975; Last, *A.R.Phytop.* **9**:341, 1971, nonfoliar pathogens & role of host; Young et al., *ibid* **16**:263, 1978, disease monitoring in prevention of epidemics; Scott & Bainbridge 1978; Zadoks & Schein *Epidemiology and plant disease management*, 1979; Zadoks, *EPPO Bull.* **9**:227, 1979, simulation of epidemics; Kranz & Hau, *A.R.Phytop.* **18**:67, 1980, systems analysis; Teng, *Z.PflKrank.PflSchutz.* **88**:49, 1981, computer models; Huisman, *A.R.Phytop.* **20**:303, 1982, root growth & invading fungi; Thresh, *ibid* **20**:193, 1982, cropping practices & virus spread; Kushalappa & Ludwig, *Phytop.* **72**:1372, 1982, infection rate, a method to correct for host growth; Østergaard, *ibid* **73**:166, 1983, cv. mixtures & prediction of epidemics; Jeger, *Pl.Path.* **32**:5, 1983, analysis in time & space; Plumb & Thresh 1983, viruses; Zadoks, *F.A.O.Pl.Prot. Bull.* **32**:38, 1984, method; Teng, *A.R.Phytop.* **23**:351, 1985, simulation modelling; Delp et al., *Phytop.* **76**:1299, 1986, field sampling for estimation of disease incidence.

Epilachna spp., transm, solanum nodiflorum mottle.

epilobii, *Pucciniastrum*.

epilobium yellows, and phyllody, as for cirsium yellows.

epimay, an epidemic model or simulator for southern leaf blight of maize; Waggoner et al., *Bull.Conn. agric. Exp.Stn.* 729, 1972 [**53**,135].

epinasty, downward bending of a petiole so that the angle between its base and the stem becomes obtuse. The orientation of the lamina may be vertical, with the apex hanging downwards; or it may continue the curve of the petiole so that its upper surface faces inwards towards the stem. The petiole and lamina remain turgid, and the condition must be distinguished from wilting, in which the tissues become flaccid, FBPP.

epiphyllum mosaic Blattný & Vukolov 1932; transm. sap and *Orthezia insignis*, Europe, *E. truncatum* [**12**:294].

epiphyte, a plant growing on another, not usually fed by it; a micro-organism living on plant surfaces in a non-parasitic relationship; Leben, *A.R.Phytop.* **3**:209, 1965; Preece & Dickinson ed., *Ecology of leaf surface micro-organisms*, 1971; Dickinson & Preece ed., *Microbiology of aerial plant surfaces*, 1976; Blakeman ed. 1981.

epiphytotic, see epidemic.

epipre, a disease and management system for winter wheat, Netherlands; Zadoks, *EPPO Bull.* **11**:365, 1981; Reinink, *Neth.J.Pl.Path.* **92**:3, 1986 [**61**,3396; **65**,3809].

epistar, a computer system for assessing pathogen, host and environment; Eisensmith et al., *PD* **64**:646, 1980.

epistasis, the process whereby the presence of one gene renders the phenotype insensitive to substitution or mutation in another gene; Sidhu, *Phytop.* **74**:382, 1984, parasitic epistasis.

epithet, in bacteriology and mycology the second word of a binomial, i.e. the specific epithet, Cowan 1978. In botany the second word is called the trivial name; Gardiner, *The Linnean* **1**(5):18, 1985, for a salutary comment.

Epitrix atropae, transm. belladonna mottle.

equiseti, *Fusarium*.

eradication (1) The elimination of a pathogen or pest from the host or from the host's environment, or both. (2) The complete removal of a host for the control of a disease or pest, e.g. of barberry to control wheat stem rust, *Puccinia graminis*, FBPP.

eragrostidis, *Curvularia*.

Eremothecium Borzi 1888; Metschnikowiaceae; only asci formed, no budding yeast cells, asco. acicular or semi-lunate, i.e. lunate with one end blunt; Batra, *Tech.Bull.USDA* 1469, 1973. *E. ashbyi* (Routien) Batra; Mukerji, *Descr.F.* 181, 1968, as *E. ashbyi* Guillierm.1935; asci intercalary, cylindrical, an ascus has 16–32 asco. which are semi-lunate, 16–25 µm long. *E. cymbalariae* Borzi; asci terminal, fusiform, an ascus with 30–60 asco. which are acicular, 19–25 µm long; cotton internal boll rot, stigmatomycosis, transm. mechanically by hemipterous insects; Marasas, *Bothalia* **10**:407, 1971 [**51**,390].

ergot, the sclerotia of *Claviceps*; ergotism, the disease caused in man and other animals by the toxic substances in the ergots which contaminate cereal grain, see *C. purpurea*.

erica root tumour Fliege, *Z.PflKrank.PflSchutz.* **81**:765, 1974; bacterium assoc., *E. gracilis* [**54**,4958].

Eriksson, Jakob, 1848–1931; born in Sweden, Univ. Lund, Academy of Agriculture, Stockholm; plant physiology with a main interest in the diseases of

crops; he noted in 1894 the existence of morphologically similar forms of *Puccinia graminis* on barley, oat, rye; from these forms the present day ff. sp. of many fungus genera originate; wrote a textbook in 1910, the edn. 2 was translated into English by W. Goodwin: *Fungous diseases of plants*, 1930. *Nature Lond.* **127**:945, 1931.

eriobotryae, *Spilocaea*.

Eriobotrya japonica, loquat.

Eriophyes insidiosus, transm. peach mosaic.

Erodium, see filaree.

errabunda, *Apiognomonia*.

eruptive germination, direct mycelial growth from a sclerotium; Punja & Grogan, *Phytop.* **71**:1092, 1981, for the germination of sclerotia of *Sclerotium rolfsii* [**61**,3334].

*****Erwinia** Winslow et al. 1920; Bradbury 1986; straight rods, mainly single, Gram negative, motile (except *E. stewartii*), flagella peritrichous; taxonomically heterogeneous; Dye, *N.Z.J. agric.Res.* **24**:223, 1981, numerical taxonomy; Pérombelon in Blakeman 1981, ecology on aerial plant surfaces; Starr, see Prokaryotae, in Starr et al. 1981; Fahy & Persley 1983.

*****E. amylovora** (Burrill) Winslow et al.; Hayward & Waterston, *Descr.B.* 44, 1965; fireblight of apple and pear, > 90 host spp. in the sub-family Pomoideae of the Rosaceae; cotoneaster, hawthorn and pyracantha are commonly attacked. This is one of the classic, and most important diseases caused by a bacterium, and a major one of the 2 primary hosts. It was also the first plant disease to be attributed to a bacterium, see Burrill. Fireblight, originally a disease of N. America, was first reported in the last quarter of the 18th century. In 1920 it was found in New Zealand and in 1956–7 in England; its additional distribution is; other parts of W. Europe but not Norway or Sweden, USSR (Crimea), absent from the Mediterranean except Cyprus and Egypt; in Asia: China, India, Korea, Turkey, Vietnam. The name of the disease well describes the characteristic syndrome, a blackening of flowers, leaves and twigs as if burnt by fire. The first symptoms are seen in early spring in warm, humid weather; blossoms become water soaked, shrivel and blacken; fruit at all ages can be attacked, causing severe necrosis, but mostly when it is young; cankers, slightly sunken, are caused on branches. Bacteria ooze from overwintered cankers and are disseminated by wind, rain, insects and spiders etc; secondary spread builds up from the first infected blossoms. *E. amylovora* forms amylovorin q.v., a capsular, extracellular polysaccharide; it is a non-specific toxin. But it induces wilt in susceptible hosts more rapidly than in resistant ones. Control is through resistance where commercially desirable, but where susceptible cvs. are grown, chemical and sanitation measures are taken; there is strict quarantine in several countries. Reviews: Eden-Green & Billing, *Rev.Pl.Path.* **53**:353, 1974; Schroth et al., *A.R.Phytop.* **12**:389, 1974; Aldwinkle & Beer, *Hortic.Rev.* **1**:423, 1979; Van der Zwet & Keil, *Agric.Handb. USDA* 510, 1979; *EPPO Bull.* **9**:3, 1979; Goodman in Callow 1983; *Acta Hortic.* 151, 1984; Byrde et al., *Pl.Path.* **35**:417, 1986.

E. ananas Serrano 1928; pineapple fruitlet rot, internal browning, infected fruits are dull and hard.

E. cancerogena Urošević 1966; poplar canker and dieback, Czechoslovakia, poss. Poland; Urošević 1968, spruce bark necrosis [**48**,2042].

E. carotovora (Jones) Bergey et al. 1923; see Bradbury 1986 for tabulated distinctions between the ssp. *atroseptica, betavasculorum, carotovora*; Pérombelon & Kelman, *A.R.Phytop.* **18**: 361, 1980, ecology of soft rot erwinias; McCarter-Zorner et al., *J.appl.Bact.* **59**:357, 1985, soft rot erwinias in the rhizosphere of weeds & crops [**65**,1485].

*****E.c.ssp. atroseptica** (van Hall) Dye 1969; Bradbury, *Descr.B.* 551, 1977; strongly pectolytic, causing a soft rot but prob. does not have a wide, natural host range; the main disease caused is potato blackleg; the stems are rotted and slimy, plants are stunted, wilt and die, tubers in storage are affected; unlike the ssp. *carotovora* it does not grow at 36–37°, black leg is a serious disease especially under wet conditions; Rich 1983; Langerfeld, *NachrBl.dt.PflSchutzdienst.* **36**:97, 1984, literature survey with ssp. *carotovora*.

E.c.ssp. betavasculorum Thomson, Hildebrand & Schroth, *Phytop.* **71**:1037, 1981; validated *Int.J.syst.Bact.* **34**:91, 1984; sugarbeet soft rot and vascular necrosis, a destructive disease in USA (Arizona, California, Idaho, Texas, Washington State), optimum temp. for growth of the bacterium is 26–28°; Whitney & Duffus 1986.

E.c.ssp. carotovora (Jones) Bergey et al.; Bradbury, *Descr.B.* 552, 1977; unlike ssp. *atroseptica* it grows at 36–37°, strongly pectolytic causing a rapidly progressing soft wet rot; very important as a disease in vegetables in transit; Sherf & MacNab 1986; causing tobacco hollow stalk see Katahira, *Bull. Morioka Tob.exp.Stn.* 18:1, 1984 [**64**,4486].

E. chrysanthemi Burkholder, McFadden & Dimock, *Phytop.* **43**:522, 1953; Bradbury, *Descr.B.* 553, 1977; *pv. chrysanthemi*; distinguished from *Erwinia carotovora* ssp. *atroseptica* and ssp.

carotovora by the 'fried egg' type colonies on potato dextrose agar, a blue pigment and by no acid from lactose, maltose or trehalose in 7 days, not growing in 5% NaCl and sensitivity to erythromycin; Bradbury 1986 tabulated the differences between the pvs.; causes vascular wilts, stunting, necrosis of the parenchyma and soft rots, many hosts; Dickey et al., *Phytop.* **74**:1388, 1984, serological relationships, strs. & hosts [**33**:355; **64**,1425].

E.c. pv. dianthicola (Hellmers) Dickey, *Phytop.* **69**:324, 1979; slow wilt, hosts include begonia, carnation, carrot, chrysanthemum, dahlia, potato, tomato; Saito, *Ann.phytopath.Soc. Japan* **51**:145, 1985 [**59**,2033; **64**,5411].

E.c. pv. dieffenbachiae (McFadden) Dye 1978; *Dieffenbachia* stem rot and spotting of young leaves, parts of Europe, Honduras, USA; McFadden, *Phytop.* **51**:663, 1961 [**41**:309].

E.c.pv. paradisiaca (Victoria & Barros) Dickey & Victoria, *Int.J.syst.Bact.* **30**:129, 1980; *Musa* and by inoculation other plants, banana rhizome rot or tip over, mainly a disease of the crop < 3 years old; Stover 1972 [**60**,997].

E.c.pv. parthenii (Starr) Dye 1978; stem and root rot of *Parthenium argentatum*, on celery and potato; Taiwan, USA; Starr, *Phytop.* **37**:291, 1947 [**26**:511].

E.c.pv. zeae (Sabet) Victoria, Arboleda & Muñoz 1975; maize stalk rot, many hosts; Thind & Payak, *Trop.Pest Management* **31**:311, 1985, review for maize in India; Tomlinson & Cox, *Pl. Path.* **36**:79, 1987, stem collapse of cardamom [**66**,3455].

E. cypripedii (Hori) Bergey et al. 1923; Bradbury, *Descr.B.* 554, 1977; orchid brown rot, Australia, Japan, South Africa, Taiwan, USA (California, Florida).

E. dissolvens (Rosen) Burkholder 1948; maize stalk rot, reported on tobacco, prob. confused with *Erwinia chrysanthemi* pv. *zeae*; Rosen, *Phytop.* **16**:241, 1926 [**5**:544].

E. herbicola (Löhnis) Dye 1964; Bradbury, *Descr.B.* 232, 1970; common on plant surfaces and in lesions as a secondary invader, usually considered non-pathogenic. But ff.sp. *gypsophilae* (Brown) Miller, Quinn & Graham, *Neth.J.Pl.Path.* **87**:167, 1981, and *millettiae* (Kawakami & Yoshida) Goto, Takahashi & Okajima, *Ann.phytopath. Soc.Japan* **46**:185, 1980, cause galls; they may just be strs. carrying a plasmid which induces tumours, fide Bradbury 1986 [**60**,3523; **61**,3503].

E. mallotivora Goto 1976; Bradbury, *Descr.B.* 691, 1981; leaf spot of *Mallotus japonica*, Japan.

E. nigrifluens Wilson, Starr & Berger, *Phytop.* **47**:669, 1957; Bradbury, *Descr.B.* 692, 1981;

shallow bark canker of European walnut, USA (California); see *Erwinia rubrifaciens* which causes more damage [37:319].

E. quercina Hildebrand & Schroth, *Phytop.* **57**:250, 1967; Bradbury, *Descr.B.* 693, 1981; oak drippy nut, copious oozing from acorns, USA (California) [**46**,2323].

E. rhapontici (Millard) Burkholder 1948; Bradbury, *Descr.B.* 555, 1977; rhubarb crown rot, wheat pink grain; causes browning in hyacinth bulbs and a soft rot in onion, a weak pathogen; Sellwood & Lelliott, *Pl.Path.* **27**:120, 1978, in hyacinth [**58**,4875].

E. rubrifaciens Wilson, Zeitoun & Fredricksen, *Phytop.* **57**:618, 1967; Bradbury, *Descr.B.* 694, 1981; deep bark canker of European walnut, also called phloem canker; dark, long necrotic streaks along the inner bark and cambium, the necrosis spreads outwards and a dark exudate emerges; spread principally through mechanical harvesters, the disease is more severe than the one caused by *Erwinia nigrifluens*, USA (California) [**46**,2845].

E. salicis (Day) Chester 1930; Bradbury, *Descr.B.* 122, 1967; willow watermark; England, Netherlands, Japan; the bacterium causes a very important disease of the cricket bat willow, *Salix alba* var. *caerulea*, which in UK only grows well in E. Anglia. Although other willows get the disease the spread is apparently very largely through this willow. The foliage wilts, leaves redden and brown; there is dieback, excessive leafy shoot proliferation below the dieback; trees are rarely completely killed; there is internal staining, red brown becoming black. Watermark has been known in E. England since 1920 but only in recent years has it been thoroughly investigated; Wong & Preece, *Pl.Path.* **22**:95, 1973, detection in the field by a specific antiserum; Wong et al., *ibid* **23**:25, 1974, in England (Essex); Wong & Preece, *Physiol.Pl.Path.* **12**, 321, 333, 349, 1978, histology, histochemistry, enzymic action in diseased wood, phenolic constituents; Preece et al. in Lovelock ed., *Plant pathogens, Tech.Ser.Soc.appl.Bact.* 12, 1979, diagnosis & characteristics; Preece & Wortley in Ebbels & King 1979, legislation as a means of control [**53**, 1967; **54**,1458; **58**,392–4].

E. stewartii (Smith) Dye 1963; Bradbury, *Descr.B.* 123, 1967; Stewart's wilt of maize; plants wilt, leaves show linear, pale green to yellow streaks with irregular margins, cavities form in the stalk pith near the soil line; infection is systemic and a yellowish exudate appears from the cut ends of vessels. The bacterium can overwinter in corn flea beetles, inoculum from them infects seedlings in

the spring; Pepper, *Monogr.Am.Phytopath.Soc.* 4, 1967; Shurtleff 1980.

E. tracheiphila (Smith) Bergey et al. 1923; Bradbury, *Descr.B.* 233, 1970; cucumber wilt, infection is systemic and exudate is usual at the cut ends of vessels; spread by cucumber beetles and prob. other insects, and mechanically; Sherf & MacNab 1986.

E. uredovora (Pon et al.) Dye 1963; very similar to *Erwinia herbicola*, on uredia of *Puccinia graminis*.

erysimum latent v. Shukla & Schmelzer 1972; Shukla & Gough, *Descr.V.* 222, 1980; Tymovirus, isometric *c*.27 nm diam., transm. sap, flea beetles, Germany, Cruciferae; Gough et al., *Aust.J.biol. Soc.* **35**:5, 1982, molecular weights [**61**,5603].

—— **mosaic** Ram, *Curr.Sci.* **44**:245, 1975; transm. sap, aphid spp., India (Simla), *E. hieraciifolium* [**55**,362].

Erysiphaceae, Erysiphales; Ascomycotina; 1 family; mycelium white, superficial, attached by haustoria, mostly on leaves; ascomata globose, no ostiole, called cleistothecia, often with appendages which are characteristic of the genus, asci bitunicate, asco. aseptate; anam. in *Oidium* and *Oidiopsis*; the powdery mildews, obligate parasites containing many very important pathogens; *Erysiphe, Leveillula, Microsphaera, Phyllactinia, Podosphaera, Sphaerotheca, Uncinula*; Spencer 1978; Boeswinkel, *Bot.Rev.* **46**:167, 1980; Holliday 1980; Braun, *Nova Hedwigia* **34**: 679, 1981; Zheng, *Mycotaxon* **22**:209, 1985; Hirata, *Host range and geographical distribution of the powdery mildew fungi*, 1986; Aust & Hoyningen-Huene, *A.R.Phytop.* **24**:491, 1986; Adams et al., *Phytop.* **76**:1239, 1986.

Erysiphe Hedw. f. ex DC. 1805; Erysiphaceae; anam. *Oidium*, distinguished from other genera in that the ascoma has unbranched appendages and several asci; Spencer 1978; Holliday 1980; see Cannon et al. 1985 for authorities.

E. betae (Vañha) Weltzien 1963; Kapoor, *Descr.F.* 151, 1967; ascomata 95–112 μm diam., appendages 100–250 μm long, asci 3–8, 50–70 × 25–40 μm, asco. 20–30 × 14–16 μm; anam. *Oidium erysiphoides* Fr., fide Kapoor, con. in short chains 30–50 × 13–20 μm; on beet; chemical control needed in high loss areas; Drandarevski in Spencer 1978; Whitney & Duffus 1986, as *Erysiphe polygoni* DC. 1805.

E. cichoracearum DC. 1805; Kapoor, *Descr.F.* 152, 1967; ascomata 90–135 μm diam., appendages much longer than diam. of ascoma, asci 2–3, 60–90 × 25–50 μm, con. in long chains 24–45 × 14–26 μm; anam. sometimes confused with that of *Sphaerotheca fuligena* whose con. have fibrosin bodies whilst those of *Erysiphe* *cichoracearum* do not; on Compositae, but also on other plants; important diseases on lettuce, safflower, sunflower, and some ornamentals; for tobacco see Cole in Spencer 1978; Lebeda & Buczkowski, *J.Phytopath.* **115**:21, 1986, ascomata on *Lactuca* spp. [**65**,3096].

E. cruciferarum Opiz ex Junell 1967; Purnell & Sivanesan, *Descr.F.* 251, 1970; ascomata 95–125 μm diam., appendages up to 3 times the diam. of ascoma, asci usually 6–8, 50–70 × 30–50 μm, con. single or in short chains 30–40 × 12–16 μm, Cruciferae, poss. races on *Brassica*; Brain & Whittington, *J.agric.Sci.* **93**:59, 1979; *AAB* **95**:137, 1980, chemical control, inheritance of resistance in swede; Dixon 1981; Sherf & MacNab 1986 [**59**,511; **60**,2254].

*****E. graminis** DC. 1815; Kapoor, *Descr.F.* 153, 1967; ascomata 135–250 μm diam., appendages rudimentary, asci 8–25, 70–109 × 25–40 μm, asco. 20–24 × 10–14 μm; anam. *Oidium monilioides* Link fide Kapoor; distinguished morphologically from other *Erysiphe* spp. by the large ascomata, poorly formed appendages and numerous asci; on temperate cereals and grasses, particularly barley. This sp. is the most studied of all powdery mildews, ff.sp. occur inter alia on barley, oat and wheat, and within each there are races; all aspects of the diseases and their control were discussed in Spencer 1978; for barley see Mathre 1982 and for wheat Wiese 1987; Wolfe & Schwarzbach, *A.R.Phytop.* **16**:159, 1978, race changes; Bennett, *Pl.Path.* **33**:274, 1984, review, resistance in wheat; Wolfe, *ibid* **33**:451, 1984; review, control in barley; Hau, *Acta Phytomedica* **9**, 1985, comparisons of models simulating epidemics; Wolfe in Wolfe & Caten 1987, f.sp. *hordei*.

E. heraclei Schleich. ex DC. 1815; Kapoor, *Descr.F.* 154, 1967; ascomata 85–120 μm diam., appendages 1–2 times the diam. of ascoma and once to many times irregularly branched, asci 3–8, asco. 20–28 × 10–15 μm, con. usually single 34–46 × 14–20 μm; distinguished from other *Erysiphe* spp. by the numerous branched appendages and the elongate cylindrical con.; Umbelliferae *c.*85 genera; crops attacked include carrot, cumin, fennel, parsley; seedborne, prob. races; see Dixon in Spencer 1978; Sherf & MacNab 1986.

E. pisi DC. 1805; Kapoor, *Descr.F.* 155, 1967; ascomata 85–126 μm diam., appendages up to 2–3 times the diam. of ascoma, asci 3–10, 50–60 × 30–40 μm; asco. 22–27 × 13–16 μm, con. usually single 31–28 × 17–21 μm; distinguished from *Erysiphe trifolii*, which is also found on Papilionaceae, by the shorter appendages; pea powdery mildew, on *Lens, Lupinus, Medicago,*

Phaseolus, Vicia; seedborne, poss. ff. sp. Dixon in Spencer 1978; Ayres, *TBMS* **81**: 269, 1983, visible light, conidial germination & germ tube growth; U. P. & H. B. Singh, *ibid* **81**:275, 1983, development on susceptible & resistant cvs.; Reeser et al. *Phytop.* **73**:1238, 1983, quantitative inoculation & infection efficiency; Hagedorn 1984 [**63**,1483, 2034–5].

E. polygoni DC. 1805; Sivanesan, *Descr.F.* 509, 1976; ascomata 90–150 μm diam., appendages 1–2 times the diam. of ascoma and forming a dense web around it, asci 3–12, asco. 20–30 × 10–12 μm, now restricted to *Polygonum* and *Rumex*; many forms of *Erysiphe polygoni* are now dispersed as distinct spp.; further analysis may show that these should be referred to ff.sp. only.

E. ranunculi Grev. 1824; Price & Linfield, *TBMS* **78**:378, 1982; ascomata 75–92 × 58–83 μm, asci 2–6, asco. 17–25 × 8–10 μm, on *Anemone, Ranunculus* [**61**,5788].

E. trifolii Grev. 1824; Kapoor, *Descr.F.* 156, 1967; ascomata 90–125 μm diam., appendages 2–12 times the diam. of ascoma, longer than those of *Erysiphe pisi*, asci 5–10, asco. 20–25 × 10–15 μm, con. single 28–40 × 16–22 μm; Leguminosae, common on *Lathyrus, Onobrychis, Trifolium*, races on clover, different forms from some plant genera; Stavely & Hanson, *Phytop.* **56**:309, 795, 940, 957, 1966; **57**:193, 1967, on *T. pratense* as *E. polygoni*, morphology, pathogenicity, races, effects of temp. & relative humidity on pathogen development, genetics of resistance [**45**,2412, 3568–9; **46**,2054].

erythrina mosaic Deighton & Tinsley 1958; Mulder, *Tea Q.* **33**:150, 1962, described a vein banding on *E. lithosperma*, dadap [**38**:190; **42**:417].

erythroseptica, *Phytophthora*; **erythrostoma,** *Apiognomonia*.

esculenta, *Ustilago*.

etheridgei, *Seimatosporium*.

Ethiopian mustard, see *Brassica*.

ethirimol, systemic fungicide, for cereal powdery mildews, absorbed by seedlings from seed dressing.

ethylene, one of the natural plant hormones which modify plant development. Its formation in plants often increases after physical or chemical stress, or invasion by pathogens. It is synthesised by many bacteria and fungi; Archer & Hislop, *AAB* **81**:121, 1975, host & pathogen relationship; Smith, *A.R.Phytop.* **14**:53, 1976, in soil; Primrose, *J.appl.Bact.* **46**:1, 1979, agriculture, role of micro-organisms; Knee et al., *AAB* **107**:581, 1985; use & removal in post-harvest handling of horticultural commodities [**58**,5621].

etiolation, internode extension and the lack of green

colour that result from growth of a plant in insufficient light or complete darkness, FBPP.

etiology, or aetiology; the study of the cause of disease or disorder; and the nature of the causal factor(s) or agent(s) and their relationships with the host; see Koch's postulates.

etridiazole, fungicide used for seed and soil application.

eucalypti, *Aulographina*; **eucalypticola**, *Cytospora*.

***eucalyptus little leaf** Sastry et al. 1971; Ghosh et al., *Phytop. Z.* **110**:207, 1984; MLO assoc., India, *E. citriodora*. Foddai & Marras 1963, described a mosaic on *E. rostrata* in Italy (Sardinia) [**44**:253; **51**,2888; **64**,319].

eucaryote, eukaryote, a cell or an organism where the DNA assoc. with chromosomes is enclosed by a nuclear membrane, i.e. animals, fungi and plants, cf. procaryote.

eucharis mottle (tobacco ringspot v.).

Eucheuma spinosum, and *E. striatum*, red algae cultivated in the Philippines for their content of carrageenan, an emulsifier used in the food and pharmaceutical industries, see *Penicillium waksmanii* and *Scopulariopsis*.

Eumycota, the true fungi, i.e. those typically mycelial.

euonymi-japonicae, *Oidium*, *Uncinula*.

euonymus chlorotic ringspot (tomato ringspot v.), Puffinberger & Corbett, *Phytop.* **75**:423, 1985 [**64**,4966].

— **clear vein** Chang, *Korean J.Pl.Path.* **2**:1, 1986; bacilliform particles assoc. 230–280 × 70–80 nm; Doi 1969 described a similar particle assoc. with a mosaic of *Euonymus*, fide Chang [**65**,4964].

— **fasciation** Codaccioni 1972; Jonsson, *Revue gén.Bot.* **81**:135, 1974; Codaccioni & Cossard, *C.r.hebd.Séanc.Acad.Sci.D.* **280**:1497, 1975; **284**:701, 1977; bacilliform particles assoc.; France, Yugoslavia; *E. japonica* [**54**,1769; **55**,266; **56**,5075].

— **mosaic** (tobacco necrosis v.), Mali, *Indian Phytopath.* **29**:262, 1976 [**57**,4158].

— **mottle v.** Larsen, Gergerich & Kim, abs. *Phytop.* **76**:1075, 1986; isometric 28 nm diam., 2 ss RNA's, transm. sap, seed; USA (Arkansas), *E. europaeus*.

— **Yugoslavian mosaic** Pleše & Wrischer, *Acta bot. croat.* **40**:31, 1981; rod *c*.650–680 × 18 nm assoc., *E. japonica* [**61**,3500].

eupatorium yellow vein v. Osaki & Inouye, *Ann.phytopath.Soc.Japan* **45**:111, 1979; prob. Geminivirus, isometric, pairs 25–30 × 15–20 nm, *E. japonica*; the strikingly regular symptoms were recorded in a poem by a Japanese empress in 752, the earliest known record of symptoms caused by a plant virus; see *ibid* **46**:49, 1980 [**60**,944].

euphorbiae, *Alternaria*.

euphorbia flexuous, 740–750 nm long, Denmark 1980, *E. loricata* [**61**,990].

— **mosaic v.** Costa & Bennett, *Phytop.* **40**:266, 1950; Kim & Flores, *ibid* **69**:980, 1979; Geminivirus, isometric, each half 18–20 nm diam., transm. sap, *Bemisia tabaci*; Brazil, *E. loricata*; Kim & Fulton, *Phytop.* **74**:236, 1984, fine structure in *Datura stramonium* inoculated with virus from wild *E. heterophylla*; Jaramillo & Lastra, *J.Phytopath.* **115**:193, 1986, purification & properties [**29**:493; **59**,4083; **63**,2783; **65**,3766].

Euphorbia pulcherrima, poinsettia.

euphorbia ringspot v. Bode & Lesemann, *Acta Hortic.* **59**:161, 1976; poss. Potyvirus, filament *c*.750 nm long, transm. *Myz.pers.*, non-persistent, and by knife cut, Germany [**56**,2534].

euphoria witches' broom v. Li, *Acta phytopath.sin.* **1**:211, 1955; So & Zee, *PANS* **18**:283, 1972; rod *c*.1000 × 12 nm, seedborne, China (S.E. including Hong Kong), *E. longana*, longan fruit [**37**:49; **52**,803].

eupyrena, *Phoma*.

European canker (*Nectria galligena*) apple, pear; — **cushion rust** (*Chrysomyxa abietis*) spruce; — **mildew** (*Microsphaera grossulariae*) gooseberry; — **plum**, see *Prunus*; — — **line** (apple mosaic v.); — **rusts** (*Gymnosporangium clavariiforme*) hawthorn, pear, quince, (*G. fuscum*) pear, (*G. tremelloides*) apple.

Eurotiales, Ascomycotina; ascomata, non-ostiolate, ±globose, no appendages, asci unitunicate, borne at all levels in the ascoma, asco. aseptate; anam. in *Acremonium*, *Aspergillus*, *Penicillium*; see *Talaromyces*.

Euscelidius variegatus, transm. clover phyllody, peach X, *Spiroplasma kunkelii*.

Euscelis plebeja, transm. strawberry green petal.

eustacei, *Butlerelfia*.

eustoma stunt Mayhew & Sorrell, abs. *Phytop.* **76**:1126, 1986; isometric particle 28 nm diam. assoc., transm. sap, USA, *E. russellianum*.

euteiches, *Aphanomyces*.

Eutypa Tul. & C. Tul. 1863; Diatrypaceae; stroma of fungus and host elements ±erumpent often pustulate, ascomata solitary or in small groups with ostioles not collectively erumpent; asco. aseptate, small, allantoid; Glawe & Rogers, *Mycotaxon* **14**:334, 1982, anam. & of *Eutypella* [**61**,4708].

E. armeniacae Hansf. & Carter 1957; Carter & Talbot, *Descr.F.* 436, 1974; Rappaz, *Mycotaxon* **20**:567, 1984, stated that the correct name is *Eutypa lata* (Pers:Fr.) Tul. 1863, given as *E. lata* (Pers.) Tul. & C. Tul. 1863, by Cannon et al. 1985; the anam. is in *Cytosporina* (Sacc.) Sacc.

1884, a synonym of *Dumortiera* Westend. 1857, fide Sutton 1977; asco. 7–11 × 1.5–2 μm, smooth; con. 18–25 × 1 μm, hyaline, extruded in a moist, shining, yellowish mass or as tendrils; the teleom. appears 2 or more years, on dead branches, after the anam.; a wound parasite with many hosts. It was the prob. cause of an apricot disease noted by *Hogg's fruit manual* in England in 1853. It was detected in Australia, causing apricot gummosis, in 1933; but not diagnosed in USA until 1962 and in Europe, assoc. with apoplexy q.v. of apricot, in 1964. The fungus is widely known on grapevine where it was recently shown to cause a canker and dieback called dead arm which had been attributed to *Phomopsis viticola* q.v. *E. armeniacae* occurs on *c*.60 plant spp., including almond, apple, apricot, pear, pistachio, plum, tamarisk, walnut; but some do not become diseased; strs. of differing pathogenicities have been detected from apricot; Carter & Moller, *EPPO Bull.* **7**:85, 1977; Carter et al., *Rev.Pl.Path.* **62**:251, 1983, reviews; *Aust.J.Bot.* **33**:361, 1985, strs. of high & low virulence to apricot; Rumbos, *J.Phytopath.* **116**:352, 1986, pistachio dieback in Greece [**65**,779, 6205].

Eutypella (Nitschke) Sacc. 1875; Diatrypaceae; like *Eutypa* but ascomata numerous, forming a dense cluster in a marked stromatic pustule, ostioles collectively erumpent, for anam. see *Eutypa*.

E. parasitica Davidson & Lorenz, *Phytop.* **28**:739, 1938; asco. 8–11 × 2–2.3 μm, assoc. with maple canker, mostly sugar maple, white to buff mycelial fans under bark at the margin of canker, N. America; Glawe, *Mycologia* **75**:742, 1983, anam. [**18**:147; **63**,449].

E. sorbi (Alb. & Schwein.) Sacc. 1882; Cannon et al. 1985 gave the anam. as *Cytospora rubescens* Fr., fide Grove 1937; con. 3.5–4 × 1 μm, released in deep red tendrils; *Sorbus* canker and dieback in Scotland; MacBrayne, *Eur.J.For.Path.* **11**:325, 1981, as anam. [**61**,2474].

eutypoides, *Monosporascus*.

exanthema, or summer dieback which is used for apple and pear; citrus, olive, prune; Cu deficiency; Bould et al. 1983.

excipulum, exciple, of an ascoma; tissue containing the hymenium in an apothecium or forming the walls of a perithecium.

excoriosis, as for dead arm.

exigua, *Phoma*; **exitialis**, *Didymella*.

Exobasidiaceae, Exobasidiales; Holobasidiomycetidae; 1 family; basidia emerge between epidermal cells, forming a ± continuous hymenium as a simple layer on the leaf surface, sometimes forming galls, *Exobasidium*.

Exobasidium Woronin 1867; on dicotyledons;

Savile *CJB* **37**:641, 1959; McNabb, *Trans.R.Soc.N.Z.* **1**:259, 1962; Nannfeldt, *Symbolae Bot. Upsalienses* **23**(2), 1981; Khan et al., *CJB* **59**:2450, 1981 [**39**:12; **61**,2156, 3956].

E. camelliae Shirai 1896; Akai, *Bot.Mag.Tokyo* **53**:118, 1939; *Ann.phytopath.Soc.Japan* **10**:105, 1940; Reid, *TBMS* **52**:19, 1969; basidiospores 14–22 × 4.8–8 μm; causes large fleshy galls, up to 7 × 4 cm, on *Camellia japonica*. On the same plant genus is *Exobasidium giganteum* S. Hirata, *Trans.mycol.Soc.Japan* **22**:395, 1981; basidiospores av. 21.4 × 10.2 μm; galls up to 63 cm in circumference [**18**:528; **20**:470; **62**,224].

E. japonicum Shirai 1896; Booth, *Descr.F.* 780, 1983; basidiospores 12–20 × 3–4.5 μm; azalea leaf gall; Coyier & Roane 1986.

E. perenne Nickerson, *Can.J.Pl.Path.* **6**:218, 1984; basidiospores 12–19 × 1.6–2.6 μm; cranberry red shoot, Canada (Newfoundland, Nova Scotia) [**64**,1658].

E. reticulatum Ito & Saw.; basidiospores 9–12 × 3–3.5 μm, tea blight; Japan, Taiwan; no blisters formed as in *Exobasidium vexans*; Ezuka, Study of tea; *Tea Div.Tokai-Kinki agri.Exp.Stn.* **15**:13, 1956; *ibid, Bull.* **6**:1, 1958; Chen, *NTU Phytopathologist Entomologist* **5**:57, 1977 [**38**:225; **58**,3998].

E. vaccinii (Fuckel) Woronin; Booth, *Descr.F.* 778, 1983; basidiospores 11–19 × 2–4 μm, *Vaccinium* red leaf, Ericaceae, N. temperate, poss. strs., see Coyier & Roane 1986 for azalea and rhododendron.

E. vexans Massee 1898; Booth, *Descr.F.* 779, 1983; basidiospores 13–27 × 4.3–6.5 μm; tea blister blight; the leaf blisters convex on the lower surface, dull at first grey becoming white, powdery, leaf distortion is caused by many blisters on one leaf; a major disease of tea; *Exobasidium vexans* has no other host and is confined to the tea growing areas of Asia; fungicidal control of infection by the airborne basidiospores is obligatory and has a long history; Holliday 1980; Venkata Ram & Mouli, *Crop.Protect.* **2**:27, 1983, interaction of dosage, spray interval & fungicide action in India [**62**,2626].

exopathogen, an organism that lives outside a susceptible plant or other organism and induces a disease by liberation of a toxin that can be absorbed by any part of the plant or organism and consequently results in symptoms of disease; after Woltz, *A.R.Phytop.* **16**:403, 1978.

expansum, *Penicillium*.

expose, to place a test plant where it can be expected to receive inoculum by natural means from a natural source, FBPP.

Exserohilum Leonard & Suggs, *Mycologia* **66**:289,

1974; Hyphom.; Carmichael et al. 1980; a
segregate from *Drechslera* and characterised by
the strongly protruberant hilum of the con.,
teleom. *Setosphaeria*. The genus was upheld by
Alcorn 1983 as being distinct from *Bipolaris* and
Drechslera q.v.; he summarised its characteristics:
con. fusoid to cylindrical or obclavate, straight or
curved; germinating from polar cells; basal germ
tube semiaxial from near the hilum, rarely lateral;
hilum protruding strongly, commonly with a
complex, double walled structure; first conidial
septum submedian, second septum in distal third
of maturing con., the third septum median;
conidiogenous nodes rough or smooth; Sivanesan,
TBMS **83**:319, 1984 [**54**,2095; **64**,85].

E. gedarefense (El Shafie) Alcorn 1983; see El
Shafie, *TBMS* **74**:437, 1980, as *Drechslera*, con.
mostly 50–90 × 15–18 μm, on sorghum grain,
Sudan [**60**,264].

E. rostratum, teleom. *Setosphaeria rostrata*;
Exserohilum turcicum, teleom. *S. turcica*.

extensa, *Macrohyporia*.

external ringspot, see potato corky ringspot.

exudate, material that has passed from within a
plant structure to the outer surface or into the
surrounding medium, e.g. by diffusion and not
usually through an aperture, as in leaf exudate,
root exudate etc., cf. ooze, FBPP; Schroth &
Hildebrand, *A.R.Phytop.* **2**:101, 1964, effects of
plant exudates on root infecting fungi; Rovira,
Bot.Rev. **35**:35, 1969, root exudates.

exude, to pass from within a plant structure to the
outer surface or into the surrounding medium,
e.g. by diffusion and not usually through an
aperture, cf. ooze. In ordinary English usage the
verbs 'to ooze' and 'to exude' are virtually
synonymous; but in plant pathology it is useful to
distinguish the movement of substances through
apertures of microscopic or macroscopic
dimensions as 'oozing', and that through
apertures of molecular dimensions over the whole
surface of a structure as 'exudation', FBPP.

eye spot (*Bipolaris sacchari*) sugarcane,
(*Cercoseptoria ocellata*, *C. theae*) tea, (*Curvularia
ischaemi*) Batiki bluegrass, (*Drechslera gigantea*)
banana, Bermuda grass, (*Kabatiella zeae*) maize;
see sharp and true eyespots.

F

faba bean, see broad bean.

fabae, *Ascochyta, Botrytis*.

Fabricus, Johann Christian, 1745–1808; born in Schleswig, N. Europe; chairs at Univ. Copenhagen, Kiel; wrote an early classification of diseases and a believer in the role of fungi as causes of disease; *Attempt at a dissertation on the diseases of plants* 1774, *Phytopath.Classics* 1, 1926, translated by M.K. Ravn; Whetzel 1918; Ainsworth 1981.

facultative, antonym of obligate; a facultative parasite can live as a saprophyte and grows readily in culture on laboratory media.

fagacearum, *Ceratocystis*.

Fagopyrum, buckwheat.

Fagus, beech.

fairy rings, of a fungus; centrifugal growth of some Gasteromycetes and Hymenomycetes through vegetation near the ground; free rings occur mostly in grassland and tethered rings in woodland. The former may be unsightly in ornamental and sports turf and warrant attempts at control. Free rings show as: (1) a zone of dying brown grass between inner and outer zones of stimulated, darker green grass, (2) a single ring of stimulated grass, (3) no obvious effect on grass growth. In the first group the causal fungi may be considered pathogenic, e.g. *Marasmius oreades* (Bolton: Fr.) Fr. 1838 q.v. The outer, dark green zone is due to a release of N by the fungus. The dying zone prob. has several causes including fungus attack, moisture lack, N toxicity and toxins. Rings may be very large in diam. and be very old, but most are 5–10 m diam.; basidiocarps may form on the rings; Gregory, *Bull.Br. mycol.Soc.* **16**:161, 1982; *Ainsworth & Bisby's dictionary of the fungi*, edn. 7, 1983; Smiley 1983.

—— ringspot (*Mycosphaerella dianthi*) carnation.

falcatum, *Colletotrichum*.

false acacia, see black locust; —— membrane, the sterile tissue of a smut which limits the sorus; —— mildew (*Plasmopara halstedii*) sunflower; —— root rot (*Nacobbus aberrans*); —— rust (*Synchytrium psophocarpi*) Goa or winged bean; —— smut (*Ustilaginoidea virens*) rice; —— tinder fungus (*Phellinus ignarius*) dicotyledonous trees.

fan mould (*Phialophora cinerescens*) carnation.

farinosa, *Peronospora*.

Farlow, William Gilson, 1844–1919; born in USA, Univ. Harvard, a pioneer plant pathologist but more notable as a mycologist, the Farlow Cryptogamic Herbarium and Library are famous. *Phytop.* **10**:1, 1920.

farlowii, *Inonotus, Melampsora*.

fascians, *Rhodococcus*.

fasciation (*Rhodococcus fascians*) sweet pea. The term fasciation means the proliferation of a shoot with incomplete separation of the elements, giving the appearance of a flattened, coalescent bundle of shoots; cf. witches' broom, in which shoots proliferate but are not coalescent, FBPP.

fasting, depriving a test vector of food for short periods. The term is inappropriate because fasting normally implies a voluntary abstinence from food by the subject, rather than an enforced deprivation; but its use in vector studies is now so common that a contra-recommendation is not justified. Fasting may be post-acquisition where it is done after acquisition feeding and before inoculation feeding; or pre-acquisition where it is done before acquisition feeding, synonymous with preliminary fasting, after FBPP.

faullii, *Isthmiella*.

fauna, associations of small animals with soilborne pathogens can increase the severity or incidence of plant disease; Beute & Benson, *A.R.Phytop.* **17**:485, 1979, considered microfauna > 100 μm and meiofauna > 1 cm but excluding nematodes.

Fawcett, Howard Samuel, 1877–1948; born in USA, Iowa State College, Florida and California Citrus Experiment Stations; wrote: *Citrus diseases and their control* 1926, edn. 1, with H.A. Lee; edn. 2, 1936; *Colour handbook of citrus diseases* 1941; edn. 2, 1948 with L.J. Klotz. *Phytop.* **39**:865, 1949.

fawcettii, *Elsinoë, Sphaceloma*; var. scabiosa, *Sphaceloma*.

feeding period, the period that a test vector actually feeds on a virus source or on a recipient plant, FBPP.

fenaminosulf, fungicide for *Phytophthora* and *Pythium*, on seed or in soil.

fenarimol, protectant and eradicant fungicide with a limited systemic activity.

fenuram, seed treatment for cereal smuts.

Fennel, Dorothy I, 1916–77; born in USA, Univ. Illinois, culture collections, an authority on *Aspergillus* and *Penicillium*; wrote: *The genus Aspergillus* 1965, with K.B. Raper. *Mycologia* **71**:889, 1979.

ferbam, protectant fungicide, mostly superseded.

fern leaf, potato, Zn deficiency, Boawn & Legget 1963, in Hooker 1981.

fern v. Hull, *Virology* **35**:333, 1968; rod 135 × 22 nm or 320 nm long, transm. sap, England, *Phyllitis scolopendrium*, a mottle. Nienhaus et al., *Z.PflKrankh.PflSchutz.* **81**:533, 1974; poss. Potyvirus, filament 785 × 12–14 nm, transm. *Myz.pers.*, Germany, *Dryopteris filix-mas*, *Polypodium vulgare* [**48**,193; **54**,3222].

Ferrisia virgata, transm. cacao swollen shoot v.

fertilis, *Codinaea*.

fescue cryptic v., see cryptic viruses.

festucae, *Cercospora*.

festuca leaf streak mosaic Lundsgaard & Albrechtsen, *Phytopath.Z.* **87**:12, 1976; **94**:112, 1979; bacilliform 87 × 57 nm and 286 × 61 nm, Denmark, *Festuca gigantea* [**56**,2106; **58**,4882].

— **mottle** (phleum mottle v.).

— **necrosis v.** Schmidt et al., *Phytopath.Z.* **47**:66, 1963; Bruehl et al., 1957; Gregor 1959; Closterovirus, filament 1725 × 18 nm, transm. sap, *Rhopalosiphum padi*; Germany, Scotland, USA; *Festuca pratensis*, *Lolium multiflorum*, transm. to oat [**37**:98; **38**:653; **43**,109].

fibre crops, Rao, *Sydowia* **30**:164, 1977, a list of fungi and references.

fici, *Cerotelium*; **ficina**, *Uredo*.

ficuserectae, *Pseudomonas*.

ficus ringspot Smolák & Brčák, *Biologia Pl.* **10**:81, 1968; 2 particle types, 370–640 × 12–13 nm, and those resembling tobacco rattle v., Czechoslovakia [**47**,1805].

Fieberiella florii, transm. cherry Molières decline.

field bean, see broad bean.

— **immune**, of plants, not becoming infected by a pathogen in the field although susceptible under experimental conditions, FBPP.

fig deformation v. Grbelja & Erić, *Acta bot.croat.* **42**:11, 1983; Potyvirus, filament 750–800 nm long, transm. sap to *Nicotiana* spp. and by *Myz.pers.*, nonpersistent, Yugoslavia [**63**,3476].

— **mosaic** Condit & Horne, *Phytop.* **23**:887, 1933; Flock & Wallace, *ibid* **45**:52, 1955; transm. *Aceria ficus* [**13**:252; **34**:465].

figwort mosaic v. Handley, Duffus & Shepherd 1982; Hull & Donson, *J.gen.Virol.* **60**:125, 1982; prob. Caulimovirus, transm. aphid spp., USA, *Scrophularia californica* [**61**,5578].

fijiensis, *Mycosphaerella*, *Paracercospora*.

Fijiviruses, see Phytoreoviruses.

filamentosum, *Peridermium*.

filaree red leaf Frazier, *Phytop.* **41**:221, 1951; Anderson, *ibid* **41**:699, 1951; **42**:110, 1952; transm. aphid, persistent, USA (California, Florida), *Erodium* spp.; Sylvester & Osler, *Environmental Entomol.* **6**:39, 1977, transm. *Acyrthosiphon pelargonii zerosalphum* [**30**:471; **31**:19,435].

filbert, see *Corylus*.

fimbriata, *Ceratocystis*.

Findlay, Walter Philip Kennedy, 1904–85; born in USA, Imperial College, London; mycologist, Forest Products Research Laboratory, Brewing Industry Research Foundation, England; books on decays of timber. *AAB* **109**:451, 1986; *TBMS* **87**:173, 1986.

fine structure, see electron microscopy.

finger millet, see millets; — **rot** (*Botryodiplodia theobromae*, *Trachysphaera fructigena*) banana; — **and toe**, see club root.

fir, *Abies*; Wachter, *Z.PflKrankh.PflSchutz.* **85**:361, 1978, German literature 1830–1978 on decline of silver fir, *A. alba*; — **broom rust** (*Melampsorella caryophyllacearum*).

fire (*Botryotinia polyblastis*) narcissus, (*Botrytis hyacinthi*) hyacinth, (*B. tulipae*) tulip, (*Stagonospora curtisii*) snowdrop.

fireblight (*Ceratocystis fimbriata*) pimento, (*Erwinia amylovora*) apple, hawthorn, pear, Pomoideae.

fireweed rust (*Pucciniastrum epilobii*) fir.

Firmicutes, Procaryotae.

fisheye rot (*Butlerelfia eustacei* assoc.) apple.

fitness, the ability of an organism to survive and reproduce, FBPP; estimating parasitic fitness: MacKenzie, *Phytop.* **68**:9, 1978; Groth & Barrett, *ibid* **70**:840; Skylakakis, *ibid* **70**:696, 1980; Fleming, *ibid* **71**:665, 1981; Barrett, *ibid* **73**:510, 1983; and see Nelson in Horsfall & Cowling, vol. 4, 1979.

five o'clock shadow, carrot, B deficiency in part; Scaife & Turner 1983.

flaccidum, *Cronartium*; **flaccumfaciens**, *Curtobacterium*.

flagella, singular flagellum; an appendage of a motile cell, sometimes of taxonomic significance. In fungi 2 types can be distinguished: whiplash, with a smooth continuous surface; tinsel, with the surface covered with processes like hairs. In bacteria the cell may have flagella arising from any part of the wall, i.e. it is peritrichous; or from one end, one or several, i.e. it is polar or cephalotrichous; see zoospore and Cowan 1978.

flag smut (*Urocystis agropyri*) barley, wheat, Gramineae.

flat apple (cherry rasp leaf v.).

flavofaciens, *Verrucalvus*; **flavovirens**, *Lachnellula*.

flavus, *Aspergillus*, *Talaromyces*.

flax, linseed, *Linum usitatissimum*; Johansen, *Tidsskr.PlAvl.* **48**:187, 1943; Muskett & Colhoun, *AAB* **33**:331, 1946; *Diseases of the flax plant*, 1947; Millikan, *Tech.Bull.Dep.Agric.Vict.* 9, 1951; Bedlan, *Pflanzenarzt* **37**:28, 1984.

— **leaf crinkle** (oat blue dwarf v.).

— **yellows** Rataj & Zapletalova 1959, Czechoslovakia [**40**:417].

fleck (*Ascochyta pteridis*) *Pteridium aquilinum*.

— **spot**, barley, poss. genetic disorder; McKinney & Menser, *PDR* **60**:1017, 1976 [**56**,2988].

flectens, *Pseudomonas*.

flesh browning, apple, low temp. disorder; Meheriuk et al., *J.Am.Soc.hort.Sci.* **109**:290, 1984 [**63**,5016].

Fletcher, James, 1852–1908; born in England, went to Canada in 1874 as a banker, a naturalist, became a botanist and entomologist, Experimental Farms 1887; a pioneer of Canadian plant pathology. *Can.J.Pl.Path.* **5**:120, 1983.

flies, see vector.

Florida gummosis, as for Rio Grande gummosis; — **velvet bean**, as for velvet bean.

floury leaf spot (*Mycovellosiella phaseoli*) bean.

flower break, petals become flecked, streaked or mottled as the amount of anthocyanin pigment is abnormally small or large; the cause is usually a virus, e.g. the famous break in tulips painted by the Dutch masters in the early 17th century and caused by tulip breaking v.; — **bud rot** (*Alternaria dianthicola*) carnation, (*Sclerotinia fuckeliana*) rose; — **gall** (*Anguina agrostis*) *Agrostis*; — **scorch** (*Itersonilia perplexans*) chrysanthemum.

flowers, see ornamentals.

flowerstand blight, (poss. *Pseudomonas syringae*) coriander q.v.

Fluiter, Hendrick Jacob de, 1907–70; born in the Netherlands, Univ. Leyden, Wageningen; in Java, diseases of coffee, rubber, tobacco; directorate Institute of Phytopathological Research, Wageningen 1954, strawberry viruses; Order of Oranje Nassau, Jozef van den Brande Prize, Belgium. *Neth.J.Pl.Path.* **76**:49, 1970.

fluorescens, *Pseudomonas*.

fluorescent antibody, one in which the gamma globulin proteins are conjugated with a fluorescent dye, such as fluorescein isothiocyanate, to facilitate microscopic observation on the distribution of antigens in cells, FBPP.

fluoride, causing injury to plants; Treshow, *A.R.Phytop.* **9**:21, 1971; Unwin, *ADAS Quart.Rev.* **39**:271, 1980; Granett, *HortScience* **17**:587, 1982; Lorenzini et al., *Inftore. Fitopat.* **37**(3):41, 1987.

fluotrimazole, protectant fungicide against powdery mildews.

flutolonil, systemic fungicide, Araki 1985 [**65**,5857].

fly speck (*Schizothyrium pomi*) apple, pear.

focus, a site of local concentration of diseased plants or disease lesions, either about a primary source of infection or coinciding with an area

originally favourable to establishment, and tending to influence the pattern of further transm. of the disease, FBPP.

foeniculi, *Phomopsis*, *Ramularia*.

foliage blight (*Pleospora herbarum*) tomato, (*Thanatephorus cucumeris*) beet, Naito, *Res.Bull.Hokkaido Natn.agric.Exp.Stn.* 139:145, 1984 [**64**,1326].

— **browning**, in *Picea sitchensis*, Britain, prob. cause climatic; Redfern, *Eur.J.For.Path.* **17**:166, 1987.

foliar blight (*Thanatephorus cucumeris*) cabbage, Abawi & Martin, *PD* **69**:158, 1985 [**64**,2772].

— **decay**, coconut, Vanuatu, induced by feeding of *Myndus taffini*, a cixiid plant hopper, also known as New Hebrides disease; Randles et al., *Phytop.* **76**:889, 1986, reported a positive correlation between the symptoms and ssDNA not typical of DNA from a plant virus group [**66**,1101].

— **disease**, damage and loss in yield as exemplified by barley pathogens: *Erysiphe graminis* f.sp. *hordei*, *Puccinia hordei*, *Rhynchosporium secalis*; and by *Botrytis fabae* on broad bean; Griffiths in Wood & Jellis 1984.

foliicola, *Ciborinia*, *Stagonospora*.

fomannosin, fomannoxin, toxins from *Heterobasidion annosum*; Bassett et al., *Phytop.* **57**:1046, 1967; Heslin et al., *Eur.J.For.Path.* **13**:11, 1983 [**47**,665; **62**,4028].

fomentarius, *Fomes*.

Fomes (Fr.) Fr. 1849; Polyporaceae; basidiocarp perennial, often very large with a crust and usually a stratified hymenophore, context light chestnut, basidiospores ellipsoid, brown; many are secondary and/or wound pathogens and are not given; many have been transferred to other genera, e.g. *Fomitopsis*, *Ganoderma*, *Heterobasidion*, *Phellinus*, *Rigidoporus*.

F. fomentarius (L.:Fr.) Kickx 1867; tinder fungus, mottled rot; causes a heart rot and attacks sapwood; beech, birch; MacDonald, *Trans.Proc. bot.Soc.Edinb.* **32**:396, 1938; Hilborn, *Bull.Maine agric.Exp.Stn.* 409:161, 1942 [**18**:214; **21**:475].

Fomitopsis P. Karsten 1881; Polyporaceae; basidiocarp applanate to ungulate, at times resupinate, with a crust, context white to wood coloured or pink.

F. rosea (Alb. & Schwein.:Fr.) P. Karsten; Pegler & Waterston, *Descr.F.* 191, 1968; context rose coloured, at least when young; brown top rot of conifers; Carranza-Morse & Gilbertson, *Mycotaxon* **25**:469, 1986, taxonomy of *F. rosea* complex, key, distribution [**65**,4099].

Fontana, Felice, 1730–1805; born in Italy, Univ. Padua and Bologna, chairs of philosophy at

Florence and Pisa; he gave a first description of wheat stem rust (*Puccinia graminis*) in 1767, and independently of the one given by Targioni-Tozzetti; *Phytopath.Classics* 2, 1932, English translation by P.P. Pirone; Ainsworth 1981.

food base, the energy and other nutrients necessary for supplying an inoculum potential q.v. are provided by the food base which, e.g. may be mycelium in infected, dead host tissue or a resting propagule, e.g. a sclerotium, sensu Garrett, *Biology of root infecting fungi*, 1956, and see his *Root disease fungi*, 1944. The significance of a food base came to be realised by those working on root diseases of tropical, perennial crops, which were frequently planted on land cleared from primary forest, in the 1920s and 1930s. This came about through the repeated failure of inoculation experiments with unsuitable inocula.

foot rot, mostly refers to a stem, cortical rot at or near soil level, ±synonymous with collar rot; it may extend to the proximal parts of roots, but the term is preferably not used for a rot of the distal parts; Shipton in Wood & Jellis 1984.

forage legumes, viruses, Edwardson & Christie 1986 [66,2410–1].

Ford, Henry, 1863–1947, see *Microcyclus ulei*.

forecasting diseases, see disease warning.

forest, see trees; — **decline**, see fir, spruce, waldsterben.

formaldehyde, a general sterilant.

forma specialis, f.sp., plural formae speciales, ff.sp.; an intra specific taxon of a fungus characterised only in physiological or biochemical terms, particularly in pathogenicity or host adaptation; most frequently used in *Fusarium*, especially in *F. oxysporum*, and the rusts of temperate cereals. The nomenclature of ff.sp. is not governed by the taxonomic code but a new f.sp. should not be erected without stringent tests of pathogenicity. The term is ± synonymous with pathotype q.v. and pathovar, the latter is used in plant bacteriology.

forsythia fasciation Codaccioni & Cossard 1975, France [55,266].

— **yellow net** (arabis mosaic v.), Cooper 1979.

foxtail millet, see millets.

— **mosaic v.** Paulsen & Sill 1969; Short, *Descr.V.* 264, 1983; Potexvirus, filament *c*.500 nm long, transm. sap, seedborne in *Briza maxima* and Clintland oat; Britain, USA; *Setaria italica*, *S. viridis*, Gramineae, including barley and causing a wheat mosaic, also in dicotyledons.

— **red leaf** Yu, Pei & Msu, *Acta phytopath.sin.* 3:1, 1957; transm. aphid spp., persistent, China (N.), *Setaria italica* [37:233].

Fragaria ananassa, strawberry.

fragariae, *Coniella*, *Marssonina*, *Mycosphaerella*, *Phytophthora*, *Rhizoctonia*, *Zythia*.

fragiforme, *Aecidium*.

frangipani, temple tree, *Plumeria rubra*.

— **mosaic v.** Francki, Zaitlin & Grivell 1971; Varma & Gibbs, *Descr.V.* 196, 1978; Tobamovirus, rod 300 × 18 nm, transm. sap; Australia, India, *Plumeria alba*, *P. acutifloria*; strs. Adelaide, Allahabad, Delhi, differ in symptoms caused on *Datura stramonium* and tobacco.

Frank, Albert Bernard, 1839–1900; born in Prussia, Univ. Leipzig, assistant professor of botany 1878; chair at the Agricultural College, Berlin, 1881. Wrote several text books including *Die Krankheiten der Pflanzen*, 1880; edn. 2, 3 vols., 1895–6; Whetzel 1918; Ainsworth 1981.

Franklin, Rosalind E., 1920–58; born in England, Univ. Cambridge, X-ray crystallographer at Birkbeck and Kings Colleges, Univ. London; fundamental work on the structure of DNA and tobacco mosaic v. *Nature, Lond.* **182**:154, 1958.

Frankliniella occidentalis, transm. tobacco streak; *F. schultzei*, transm, groundnut bud necrosis.

Fraser, William Pollock, 1867–1943; born in Nova Scotia, the year it became a part of Canada, Univ. Dalhousie, Cornell; 1916–25 in charge of the Dominion Laboratory, Univ. Saskatchewan, which was set up after a disastrous epidemic of wheat stem rust (*Puccinia graminis*) in W. Canada in 1916; later held a chair at this Univ.; important early work on the physiologic specialisation in the temperate cereal rusts. *Phytop.* **34**:707, 1944.

fraxinophila, *Perenniporia*.

Fraxinus, ash.

fraxinus tobravirus Cooper et al., *Pl.Path.* **32**:469, 1983; long and short particles, transm. sap, England, *F. mariesii* [63,1405].

freckle (*Guignardia musae*) banana, (*Pseudospiropes elaeidis*) oil palm, (*Venturia carpophila*) almond, apricot, peach, plum.

freesia leaf necrosis, see freesia severe leaf necrosis.

— **leaf yellowing and corm necrosis** (bean yellow mosaic v.), Derks et al., *Neth.J.Pl.Path.* **93**:159, 1987.

— **mosaic v.** Longford 1927; van Koot et al., *Tijdschr.PlZiekt.* **60**:157, 1954; rod, transm. sap, *Macrosiphum euphorbiae*; England, Netherlands [6:231; **34**:228].

— **severe leaf necrosis** (freesia mosaic v. + freesia leaf necrosis agent), van Dorst, *Neth.J.Pl.Path.* **79**:130, 1973; FLNA not transm. or identified [53,579].

— **streak** Brunt 1968; Casper & Brunt, *NachrBl.dtPflSchutzdienst. Stuttg.* **23**:89, 1971;

filament *c*.840 nm long, Europe [**48**,1024 n; **50**,3833].

freeze drying, a technique, used in the preservation of plant tissues and micro-organisms, whereby water is removed under vacuum while the tissue remains in the frozen state, synonymous with lyophilisation, FBPP.

freezing, and injury, in plants; Burke et al., *A.Rev.Pl.Physiol.* **27**:507, 1976.

French bean, see bean.

frenching, tobacco, a disorder where the lamina shows a green, net pattern of fine veins against a chlorotic background; internodes become shortened, abnormally small leaves form rosettes; the cause is thought to be a toxicity of biological origin; Woltz, *A.R.Phytop.* **16**:403, 1978.

friction discolouration, apple; Lougheed et al., *PD* **66**:1119, 1982, injury from abrasion or removal from storage and see the reference for pear [**62**,1567].

frog eye (*Botryosphaeria obtusa*) apple, (*Cercospora nicotianae*) tobacco, (*C. sojina*) soybean, (*Phytophthora nicotianae* var. *parasitica*) bitter orange, see Cutuli & Nicosia 1976 [**56**,1608].

frogskin, see cassava frogskin.

frondescence, ±synonymous with phyllody q.v., or can be restricted to the condition where only petals are transformed into organs like leaves.

frondicola, *Asteroma*.

frosty mildew (*Mycosphaerella pruni-persicae*) peach; —— **pod rot** (*Moniliophthora roreri*) cacao.

fructicola, *Monilinia*.

fructigena, *Monilia, Monilinia, Trachysphaera*.

fruit black rot (*Botryosphaeria obtusa*) apple, —— **stem end rot** (*Ceratocystis paradoxa*) banana. —— **bronzing**, tomato, shock reaction to infection by tomato mosaic v. —— **cracking**, apple cv. Golden Russet, prob. related to crop load and fluctuating water supply; Proctor & Lougheed, *Can.Pl.Dis.Survey* **60**:55, 1980 [**60**,3841].

fruitlet core rot (*Gibberella fujikuroi, Penicillium funiculosum*) pineapple; —— **rot** (*Erwinia ananas*) pineapple.

fruit pox, tomato, genetic disorder; Crill et al., *Phytop.* **63**:1285, 1973 [**53**,2322].

fruits, see major texts and: Crosse, *A.R.Phytop.* **4**:291, 1966, bacteria & temperate fruits; Lewis & Hickey, *ibid* **10**:399, 1972; Lewis & Fridlund, *PD* **64**:258, 826, 1980, fungicides & virus free trees in temperate fruits; Pathak, *Diseases of fruit crops*, 1980, India; 10th & 11th Int.Sympos.Fruit Tree Viruses, *Acta Hortic.* 67, 1976, *Acta phytopath.Acad.Sci.hung.* **15** (1–4), 1980; Dennis & Edney in Dennis 1983 [**57**,1767–75; **61**,1779–841].

frumenti, *Diplodia*.

fuberidazole, fungicide seed dressing against *Fusarium*.

fuchsiae, *Uredo*.

fuchsia latent Johns, Stace-Smith & Kadota, *Acta Hortic.* 110:195, 1980; assoc. particle a rod, Canada.

fuciformis, *Isaria, Laetisaria*.

fuckeliana, *Botryotinia, Nectria*; **fuckelii**, *Coniothyrium*.

fujikuroi, *Gibberella*.

fulgens, *Caloscypha*.

fuligena, *Pseudocercospora, Sphaerotheca*.

fulva, *Fulvia*.

Fulvia Cif. 1954; Hyphom.; Ellis 1971; Carmichael et al. 1980; conidiomata hyphal, conidiophore growth sympodial, conidiogenesis holoblastic; con. 0–3 septate, pigmented, catenate, chains often branched, conidiophores have unilateral, nodose swellings which may proliferate as short, lateral branches.

F. fulva (Cooke) Cif.; Holliday & Mulder, *Descr.F.* 487, 1976; frequently cited as *Cladosporium fulvum* Cooke; con. 12–47 × 4–10 μm, hilum thickened, conspicuous; tomato leaf mould; this airborne fungus causes an important disease in the cooler tropics and in temperate, greenhouse crops, seed becomes contaminated, many races; reviews in: Holliday 1980; Dixon 1981; Fletcher 1984; Sherf & MacNab 1986; Lazarovits et al., *Phytop.* **69**:1056, 1062, 1979, toxin; de Witt et al., *Physiol.Pl.Path.* **15**:257, 1979; **16**:391, 1980; **18**:143, 297, 1981; **21**:1, 1982, inhibitors & phytoalexins [**59**, 2925, 4765–6; **60**,1081, 6651; **61**,397; **62**,388].

fumigant, a chemical toxicant, used in volatile form against organisms; see control of plant diseases, Mulder; Munnecke & van Gundy, *A.R.Phytop.* **17**:405, 1979, movement in soil.

funerea, *Pestalotiopsis*.

*fungi, Eucaryotae; best considered as a kingdom separate from animals and plants; the key text is *Ainsworth & Bisby's dictionary of the fungi*, edn. 7, 1983. Fungi have an absorptive and heterotrophic nutrition; the true fungi have a thallus which is typically filamentous or hyphal, collectively known as a mycelium which forms an extremely wide array of microscopic and macroscopic structures; saprophytes, parasites, symbionts, pathogens; constituting the largest, and longest studied, group of plant pathogens; primary subdivision: Myxomycota, amoeboid or a plasmodium; Eumycota, the true fungi, mycelial, sometimes unicellular; the latter consist of: Mastigomycotina, Zygomycotina, Ascomycotina, Basidiomycotina, Deuteromycotina. Taxonomic literature is long lived unlike most literature in plant pathology. The valid publication of fungus names dates from 1753; Korf, *Mycologia* **74**:250, 1982; *Mycotaxon* **14**:476, 1982. Nomenclature is governed by the *International Code of Botanical Nomenclature*, last edn. 1978; important changes were made at the 13th International Botanical Congress, 1981. Some fungi have two valid names: a perfect state, the teleomorph, and an imperfect state, the anamorph; the former takes precedence, i.e. is the whole fungus or holomorph; plant pathologists should always cite the teleomorph even if it is only the anamorph that is occurring in the field and causing a disease.

See major texts and: Clements & Shear, *The genera of fungi*, 1931; Bessey, *Morphology and taxonomy of fungi*, 1950; Gäumann, *The fungi*, 1952, English translation by F.L. Wynd; Cochrane, *Physiology of fungi*, 1958; Hawksworth, *Mycologist's handbook*, 1974; Ainsworth, *Introduction to the history of mycology*, 1976; Alexopoulos & Mims, *Introductory mycology*, edn. 3, 1979; Webster, *Introduction to fungi*, edn. 2, 1980; Griffin, *Fungal physiology*, 1981; Turian & Hohled., *The fungal spore: morphogenetic controls*, 1981; Moore-Landecker, *Fundamentals of the fungi*, edn. 2, 1982; Garraway & Evans, *Fungal nutrition and physiology*, 1984.

fungicide, a substance that kills a fungus, spores and/or mycelium; described as: curative or eradicant when used for control after infection has occurred; protective when used to stop infection; systemic when it is absorbed by a plant surface and translocated from the site of application; fungicidal, the act of killing; a fungistat prevents fungus growth without killing; fungistatic the act of preventing growth; see chemotherapy.

fungicola, *Verticillium*.

fungicolous fungi, fungi growing on other fungi as parasites or saprophytes e.g. *Gliocladium* spp. and *Trichoderma* spp.

fungi imperfecti, Deuteromycotina.

fungistasis, or mycostasis; prevention of fungus spore germination and hyphal growth; if the inhibitor is removed or diluted growth is resumed. The term is applied to the widespread phenomenon in natural soils where spore germination is inhibited; the action is non-specific; first described by Dobbs & Hinson, *Nature Lond.* **172**:197, 1953; Lockwood, *A.R.Phytop.* **2**:341, 1964; Watson & Ford, *ibid* **10**:327, 1972.

funiculosum, *Penicillium*.

furalaxyl, protective fungicide with systemic properties, particularly against Oomycetes.

furcraea stunt v. Dabek & Castano, *Phytopath.Z.*

92:57, 1978; isometric *c*.27–31 nm diam., transm. sap, *Myz.pers.*, Colombia, *F. macana* [**58**,250].

*****Furoviruses**.

fusaric acid, toxin; a relatively common product from *Fusarium* spp.; Gäumann, *Phytop.* **47**:342, 1957; Wood 1967; Wood et al. 1972; Durbin 1981.

fusarioides, *Microdochium*.

*****Fusarium** Link 1809; Hyphom.; Booth 1971; Carmichael et al. 1980; conidiomata hyphal or sometimes sporodochial, conidiogenesis enteroblastic phialidic; macroconidia multiseptate, fusoid, with a foot cell bearing a characteristic heel, hyaline, sometimes coloured in mass; microconidia 1 or more septate, hyaline; chlamydospores globose, thick wall, intercalary, solitary or in chains or clumps, or terminal on short lateral branches, many form from cells of the macroconidia; teleom. sometimes in the Hypocreales. The genus is large, taxonomically extremely difficult, complex and notoriously variable; it is ubiquitous, saprophytic, generally or obligately pathogenic; frequently soilborne and occurs in disease complexes with other fungus pathogens; forms compounds toxic to both animals and plants.

Kern in Wood et al. 1972, toxins; Joffe, *Mycopathologia* **53**:101, 1974; Booth, Fusarium: *Laboratory guide to the identification of the major species*, 1977; Holliday 1980; Nelson et al. 1981; Fusarium *spp.: An illustrated manual for identification*, 1983; Gerlach & Nirenberg, *Mitt.biol.BundAnst.Ld-u.Forstw.* 209, 1982, a pictorial atlas; Burgess & Liddel, *Laboratory manual for* Fusarium *research*, 1983; Moss & Smith ed., *The applied mycology of* Fusarium, 1984; Marziano et al. 1984, on chlamydospores [**65**,1110–11].

F. acuminatum, teleom. *Gibberella acuminata*; **Fusarium avenaceum**, teleom. *G. avenacea*.

F. crookwellense Burgess, Nelson & Toussoun, *TBMS* **79**:498, 1982; several hosts, in soil debris, more abundant in temperate areas with high rainfall or irrigated, compared with 4 other *Fusarium* spp [**62**,1831].

F. culmorum (W.G.Sm.) Sacc. 1895; Booth & Waterston, *Descr.F.* 26, 1964; broad macroconidia, abundant microconidia, no chlamydospores; fast growth in culture, becoming reddish brown; coral coloured sporodochia on cereal spikelets, important as a cortical rot, foot rot and pre-emergence blight of temperate cereals, a soil inhabitant.

F. decemcellulare, teleom. *Nectria rigidiuscula*; **Fusarium equiseti**, teleom. *Gibberella intricans*; **F. graminearum**, teleom. *G. zeae*; **F. heterosporum**,

teleom. *G. gordonii*; **F. lateritium**, teleom.
G. baccata.

F. merismoides Corda 1838; fairly common in soil, polluted water and sludge; occasionally causing disease; Fletcher & Lord, *Pl.Path.* **34**:443, 1985, described a tomato stem rot [**65**,378].

F. moniliforme, teleom. *Giberella fujikuroi.*

F. nivale, preferred by some authorities for the anam. of *Monographella nivalis.*

F. oxysporum Schlecht.:Fr. emended Snyder & Hansen 1940; microconidia are borne on short, simple phialides, in *Fusarium solani* the phialides are long and the microconidiophores often elaborate, in *F. moniliforme* the microconidia are in chains. A soil inhabitant surviving almost indefinitely through chlamydospores and saprophytism. *F. oxysporum*, economically the most important sp., forms many variants differing in pathogenicity, i.e. ff.sp., see forma specialis, and which often have very narrow host ranges; each f.sp. may consist of several races. In nearly all cases the fungus colonises the vascular system causing discolouration and the plant wilts; epinasty is an early sign of wilt; the diseases are sometimes called yellows. A few ff.sp. do not cause the characteristic wilt but cause a foot rot; some would restrict the use of ff.sp. to the group causing true wilts, see *F. oxysporum* f.sp. *radicis-lycopersici*, Rowe; forms may occur on crops in which they cause no disease. The continuous cropping of a disease prone host leads to the development of a wilt sick soil; usually rotational cropping in the field exerts little or no control. G.M. & J.K. Armstrong, *Phytop.* **58**:1242, 1968, gave a list of ff.sp. which now number > 75; Booth 1971; Mace et al. 1981; Sherf & MacNab 1986; and under the genus see: Holliday, Nelson et al.

F.o.f.sp. albedinis (Killian & Maire) Gordon 1965; bayoud of date palm, serious in N. Africa; Louvet & Toutain in Nelson et al. 1981; Djerbi, *Date palm J.* **1**:153, 1982, reviews; Bellarbi-Halli & Mangenot, *CJB* **64**:1703, 1986, fine structure & root infection [**66**,274].

F.o.f.sp. aleuritidis Suelong 1981; tung wilt, China; *Aleurites fordii*, susceptible, grafted on a stock of *A. montana*, resistant, gives some control; Wu & Lin 1982 [**61**,3102; **65**,1542].

F.o.f.sp. allii Matuo, Tooyama & Isaka, *Ann.phytopath.Soc.Japan* **45**:305, 1979; bulb rot of *Allium bakeri*, Baker's garlic [**59**,3982].

F.o.f.sp. anethi Gordon 1965, dill root rot and wilt, *Anethum graveolens*; Janson, *Phytop.* **42**:152, 1952 [**31**:576].

*****F.o.f.sp. apii** (Nelson & Sherb.) Snyder & Hansen; celery yellows, a serious disease in parts of USA;

Hart & Endo, *Phytop.* **71**:77, 1981, infection; Orton, *Phytopath.Z.* **102**:320, 1981; *CJB* **60**:34, 1982, assoc. with seed; Puhalla, *CJB* **62**:546, 1984, 3 races; Opgenorth & Endo, *TBMS* **84**:740, 1985, factors affecting chlamydospore formation; Schneider, *Phytop.* **75**:40, 1985, disease suppression with K, chloride & nitrate; Correll et al., *ibid* **76**:396, 1986, race 2 [**60**,5605; **61**,4420–1; **63**,3100; **64**,2779, 5182; **65**,5232].

F.o.f.sp. asparagi Cohen 1946; asparagus decline in USA, stem and crown rot; Johnston et al., *Phytop.* **69**:778, 1979, assoc. with *Gibberella fujikuroi*; Inglis, *PD* **64**:74, 1980, seed contamination; Gilbertson et al., *Phytop.* **74**:1188, 1985, assoc. with *G. fujikuroi* and *Ophiomyia simplex*, asparagus miner [**59**,1982, 3983; **65**,2598].

F.o.f.sp. batatas (Wollenw.) Snyder & Hansen; Holliday, *Descr.F.* 212, 1970; sweet potato wilt, serious in USA but not in the tropics, causes a tobacco wilt, 2 races; Ogawa et al., *J.Central agric.Exp.Stn* **30**:97, 1979, spread & control [**59**,3988].

F.o.f.sp. benincasae Gerlach & Ester 1985; attacked *Benincasa cerifera* and watermelon, Netherlands [**65**,3561].

F.o.f.sp. carthami Klisiewicz & Houston 1963; safflower wilt, seedborne, 4 races; Klisiewicz. *Phytop.* **53**:1046, 1963; Klisiewicz & Thomas, *ibid* **60**:83, 1706, 1970; Klisiewicz, *PDR* **59**:712, 1975 [**43**,530; **49**,2151; **50**,1930; **55**,1901].

F.o.f.sp. cepae (Hanz.) Snyder & Hansen; onion basal rot, seedborne, a rot in storage, Naik & Burden, *Trop.PestManagement* **27**:455, 1981; Lacy & Roberts, *PD* **66**:1003, 1982, effect on yield with *Pyrenochaeta terrestris* in USA; Kodama, *Rep.Hokkaido Prefect.agric.Exp.Stns* 39, 1983; Everts et al., *PD* **69**:878, 1985, effects of wounds on disease incidence [**61**,4540; **62**,825, 4562; **65**,4219].

F.o.f.sp. chrysanthemi G.M. & J.K. Armstrong & Littrell 1970; Emberger & Nelson, *Phytop.* **71**:1043, 1981; chrysanthemum wilt; Locke et al. and Strider, *PD* **69**:167, 564, 1985, cv. susceptibility, biological control & fungicides in the greenhouse [**61**,2314; **64**,2575; **65**,2848].

F.o.f.sp. ciceris (Padw.) Matuo & Sato 1962; chickpea wilt, seedborne; Haware & Nene, *PD* **66**,809, 1982, 4 races; Kumar & Haware, *Phytop.* **72**:1035, 1982; Sindhu et al., *J.Hered.* **74**:68, 1983, inheritance of resistance; Allen 1983 [**62**,502, 816, 1732].

F.o.f.sp. citri Timmer et al. 1977; *Phytop.* **69**:730, 1979; Timmer, *ibid* **72**:698, 1982; wilt and dieback of Mexican lime seedlings in greenhouses in USA (Florida), poss. occurs in Brazil [**59**,4185; **61**,7029].

F.o.f.sp. conglutinans (Wollenw.) Snyder & Hansen; Subramanian, *Descr.F.* 213, 1970; yellows of crucifers; cabbage is attacked by race 1 only and radish by race 2 only, these races can therefore be distinguished, but both can cause disease in other brassicas; races 3 and 4 are separated on cvs. of *Matthiola incana* and stock, they are of little or no importance on the vegetables; Ramirez-Villupadua et al., *PD* **69**:612, 1985, race 5 infecting cabbage cvs. which are resistant to race 1; Sherf & MacNab 1986 [**65**,3038].

F.o.f.sp. coriandrii Narula & Joshii 1963; Srivastava, *Indian J.agric.Sci.* **42**:618, 1972; coriander wilt [**52**,3399].

F.o.f.sp. cubense (E.F. Smith) Snyder & Hansen; Subramanian, *Descr.F.* 214, 1970; Panama wilt of bananas. This is one of the classical plant diseases. Huge losses were caused in the universally grown triploid cv. Gros Michel, genotype AAA, in Central America and Jamaica in the first half of this century. The banana trade in this region was originally and entirely based on this cv. which was attacked and killed by race 1. Banana cvs. in other parts of the world based on another AAA group, Cavendish, were highly resistant to this race. In Central America the losses in Gros Michel were compensated for by a continual shift to new land, i.e. primary forest. Such a policy had disruptive social and political effects, see Carefoot & Sprott, *Famine on the wind*, 1967. The final solution was a shift to the Cavendish bananas. Panama wilt was therefore, in the early part of its history and unusually, a disease of a single cv. Where Gros Michel was not grown, Africa, Asia and Australasia, the disease never became a problem. The fungus showed little variation in pathogenicity until *c*.20 years ago. Race 2, less important, attacks the Bluggoe group, triploids ABB genotype; race 3 attacks wild *Heliconia* spp. In 1977 a prob. predictable event occurred with the delineation of race 4 in Taiwan; this race attacked the hitherto resistant Cavendish bananas, Sun et al., *Phytop.* **68**:1672, 1978. Race 4, like race 1, also attacks Gros Michel; and, like race 2, attacks Bluggoe cvs. The epidemic in Taiwan caused by race 4 began *c*.1970; infested fields rising from 25 hectares in 1969 to 1200 hectares in 1976. Race 4 has been reported from Australia, Canary Islands, Philippines, South Africa; Stover, *Phytopath.Pap.* 4, 1962; 1972; Sun & Su, *Trop.Agric.Trin.* **61**:7, 1984, detection of races 1 & 4; Su et al., *PD* **70**:814, 1986, position in Taiwan.

F.o.f.sp. cucumerinum Owen 1956; Holliday, *Descr.F.* 215, 1970; cucumber wilt; 3 races but see McMillan, *AAB* **109**:101, 1986, who found that

this f.sp. is prob. not distinct from *F.o.*f.sp. *niveum*, but only a different race; Takeuchi et al., *J.central agric.Exp.Stn.* 28:49, 1978, seed transm.; Iida et al., *Tech.Bull.Fac.Hort.Chiba Univ.* 30:35, 1982; soil population & chlamydospores; Jenkins & Wehner, *PD* **67**:1024, 1983, on seed [**58**,2014; **62**,63; **63**,970].

F.o.f.sp. cumini Patel et al. 1957; Chattopadhyay 1967; cumin wilt, India.

F.o.f.sp. dianthi (Prill. & Delacr.) Snyder & Hansen; Bickerton, *Bull.Cornell agric.Exp.Stn.* 788, 1942; wilt of carnation and pinks; Hood & Stewart, *Phytop.* **47**:173, 1957, 3 races; Scher & Baker, *ibid* **70**:412, 1980, suppressive soil; Baker, *PD* **64**:743, 1980, control; symposium in *Acta Hortic.* 141, 1983; Baayen & Elgersma, *Neth.J.Pl.Path.* **91**:119, 1985, colonisation & histology [**22**:387; **36**:589; **60**,3042, 3805; **63**,2349–60; **65**,238].

F.o.f.sp. elaeidis Toovey 1949; Holliday, *Descr.F.* 216, 1970; oil palm wilt, also called lemon frond, prob. restricted to Africa where only localised epidemics have apparently arisen; Turner 1981 described acute and chronic forms of the disease; Meunier et al. 1979 referred to heavy losses in the savannah of the Ivory Coast; Colhoun in Nelson et al. 1981; Ho et al., *Phytopath.Z.* **114**:193, 1985, pathogenicity; Oritsejafor, *TBMS* **87**:511, 1986, effects of moisture & pH on growth & survival [**59**,4283; **65**,2398; **66**,2462].

F.o.f.sp. fabae Yu & Fang, *Phytop.* **38**:587, 1948; broad bean wilt [**28**:107].

F.o.f.sp. fatshederae Triolo & Lorenzini, *AAB* **102**:245, 1983; wilt of *Fatshedera lizei*, *F. japonica* and *Hedera helix* [**62**,3068].

F.o.f.sp. fragariae Winks & Williams, *Qd.J.agric. Anim.Sci.* **22**:475, 1965; strawberry wilt [**46**,384].

F.o.f.sp. gladioli (Massey) Snyder & Hansen; gladiolus corm rot and yellows; Buxton, *TBMS* **38**:193, 202, 1955; Moore 1979; Hsieh, *Pl.Prot.Bull.Taiwan* **27**:247, 1985, ecology & control [**35**:296–7; **65**,1923].

F.o.f.sp. hebae Raabe, *PD* **69**:450, 1985; wilt of *Hebe* [**64**,4362].

F.o.f.sp. koae Gardner, *Phytop.* **70**:594, 1980; seedling wilt of *Acacia koa*, poss. seedborne [**60**,2203].

F.o.f.sp. lactucae Matuo & Motohashi, *Trans.mycol.Soc.Japan* **8**:13, 1967; lettuce root rot [**47**,2580].

F.o.f.sp. lagenariae Matuo & Yamamoto 1967; wilt of *Lagenaria siceraria* which is used as a stock for watermelon, therefore the watermelon crop wilts when grafted on bottle gourd which has been inoculated with this f.sp., seedborne; Kuniyasu and Kuniyasu & Takeuchi, *Bull.Veg.Ornam.Crops*

Res.Stn. A 5:177, 1979; 11:139, 1983; *JARQ*
14:157, 1980; *Ann.phytopath.Soc.Japan* **46**:607,
1980 [**59**,5459; **61**,917, 1459; **64**,5543].
F.o.f.sp. lentis (Vasudeva & Srinivasan) Gordon
1965, Khare 1980 [**60**,2899].
F.o.f.sp. lilii Imle 1942; lily scale rot, Moore
1979.
F.o.f.sp. lini (Bolley) Snyder & Hansen 1940; flax
wilt, races; Kommedahl et al., *Tech.Bull.Minn.
agric.Exp.Stn.* 273, 1970, review 1913–63.
F.o.f.sp. lupini Snyder & Hansen; lupin wilt, 3
races, seedborne; Richter, *Mitt.biol.Reichsanst.
Ld-u.Forstw.* **64**:50, 1941 [**23**:110].
F.o.f.sp. lycopersici (Sacc.) Snyder & Hansen;
Subramanian, *Descr.F.* 217, 1970; tomato wilt, a
major disease and one of the most studied of all
the fusarial vascular wilts, in particular the
mechanisms of pathogenicity and as forming
fusaric acid q.v.; seedborne, the pathogen occurs
frequently in synergistic complexes with
nematodes; 3 races, the second was reported in
1945 and the third in Australia (Queensland) in
1978, see Grattidge & O'Brien, *PD* **66**:165, 1982;
race 3 may be present in Tunisia, El Mahjoub
1974; Walker, *Monogr.Am.Phytopath.Soc.* 6,
1971; Holliday 1980; Jones & Woltz in Nelson
et al. 1981; Hirano, *Tech.Bull.Fac.Hortic.Chiba
Univ.* 32:129, 1983, synergism with root knot
nematodes; Abawi & Barker, *Phytop.* **74**:433,
1984, effects of cv., soil temp. & *Meloidogyne
incognita* populations [**56**,2661; **61**,5222; **63**,4560;
64,778].
F.o.f.sp. mathioli Baker 1948; prob. a race of f.sp.
conglutinans.
F.o.f.sp. medicaginis (Weimer) Snyder & Hansen;
lucerne wilt; Johnson et al., *Phytop.* **72**:517, 1982,
infection by this f.sp. reduced wilt caused by
Clavibacter michiganense ssp. *insidiosus*; Emberger
& Welty, *ibid* **73**:208, 1983, effects of soil
moisture; Hijano et al., *Crop Sci.* **23**:31, 1983,
inheritance of resistance; Salleh & Owen,
Phytopath.Z. **107**:70, 1983, resistance to 3 races,
seeds not colonised internally [**61**,6453; **62**, 3092,
3900, 4352].
F.o.f.sp. melonis Snyder & Hansen; Holliday,
Descr.F. 218, 1970; melon wilt, prob. more severe
in temperate rather than subtropical areas,
seedborne; Risser et al., *Phytop.* **66**:1105, 1976,
nomenclature for races; Mas et al. in Nelson et al.
1981; Marois et al., *Phytop.* **73**:680, 1983,
reinvasion of fumigated soil; Latin & Reed, *ibid*
75:209, 1985, wilt increased with root feeding by
larvae of *Acalymma vittatum*, striped cucumber
beetle [**56**,2729; **62**,4508; **64**,3232].
F.o.f.sp. momordicae Sun & Huang, *PD* **67**:226,
1983; bitter gourd wilt [**62**,3322].

F.o.f.sp. narcissi Snyder & Hansen; narcissus basal
rot, Moore 1979.
F.o.f.sp. niveum (E.F. Smith) Snyder & Hansen;
Holliday, *Descr.F.* 219, 1970; wilt of watermelon,
infects squash; f.sp. *cucumerinum* q.v. is prob. a
race of f.sp. *niveum*; Netzer & Weintall, *PD*
64:853, 1980, inheritance of resistance to race 1;
Caperton et al., *ibid* **70**:2207, 1986, effects of
fungus inoculum & *Meloidogyne* on resistance in
squash; Martyn, *ibid* **71**:233, 1987, race 2
[**60**,3405; **65**,4154].
F.o.f.sp. passiflorae Gordon in Purss, *Qd.J.agric.
Sci.* **11**:79, 1954; passion fruit wilt [**35**:203].
F.o.f.sp. pernicosum (Hept.) Toole 1941; *Phytop.*
42:694, 1952; *Albizia* wilt, races 1 and 2 attack
A. julibrissin and *A. procera*, respectively [**32**:597].
F.o.f.sp. phaseoli Kendr. & Snyder 1942; bean
yellows; Ribeiro & Hagedorn, *Phytop.* **69**:272, 859,
1979, 2 races & inheritance of resistance; Sherf &
MacNab 1986 [**59**,1507, 4379].
F.o.f.sp. pisi (van Hall) Snyder & Hansen; pea wilt
or near wilt, an important and widespread disease,
many interactions with other pathogens, both
fungi and nematodes: *Fusarium oxysporum* var.
redolens; *F. solani* f.sp. *pisi*, St John's wilt;
Pythium ultimum, *Pratylenchus penetrans* and
Rotylenchus robustus; seedborne, poss. 11 races,
see G.M. & J.K. Armstrong, *Phytop.* **71**:474, 1981,
for race classification; Holliday 1980; Hagedorn
1984 [**61**,1471].
F.o.f.sp. psidii Prasad et al. 1952; guava wilt;
Edward, *Indian Phytopath.* **13**:30, 168, 1960;
Pandey & Dwivedi, *Phytopath.Z.* **114**:243, 1985
[**31**:390; **41**:98; **42**:136; **65**,2375].
F.o.f.sp. pyracanthe G.M. & J.K. Armstrong 1981;
McRitchie, *PDR* **57**:389, 1973; *Pyracantha*
[**53**,977].
F.o.f.sp. radicis-lycopersici Jarvis & Shoem.,
Phytop. **68**:1679, 1978; tomato foot and root rot,
a necrosis of cortical and vascular tissue extends
upwards for not > 25 cm in the stem; the fungus
can kill seedlings but usually causes wilt by stem
girdling at fruiting; cool temp. *c.*18° are optimum;
the typical symptoms of a true vascular wilt are
absent; Rowe, *Phytop.* **70**:1143, 1980,
comparative pathogenicity with f.sp. *lycopersici*,
f.sp. *radicis-lycopersici* caused disease in certain
legumes, discussion of use of ff.sp. concept;
Marois et al., *ibid* **71**:1257, 1981, experimental
field control with fungus antagonists; Rowe &
Farley, *PD* **65**:107, 1981, control strategy in USA
(Ohio) greenhouses; Jarvis et al., *ibid* **67**:38, 1983,
cultural control in Canada (Ontario) greenhouses;
Sherf & MacNab 1986 [**59**,459; **60**,5122; **61**,2436,
4345; **62**,2147].
F.o.f.sp. ranunculi Garibaldi & Gullino,

Phytopath.Mediterranea **24**:213, 1985; wilt of *Ranunculus asiaticus* [**65**,5571].

F.o.f.sp. spinaciae (Sherb.) Snyder & Hansen; beet and spinach wilts, the beet pathogen was formerly f.sp. *betae* but is now race 2 of f.sp. *spinaciae*, see G.M. & J.K. Armstrong, *Phytop.* **66**:542, 1976; Reyes, *Can.J.Microbiol.* **25**:227, 1979; Naiki & Morita, *Ann.Phytopath.Soc.Japan* **49**:539, 573, 1983, populations in roots, rhizosphere & soil, inoculum on weeds; Sherf & MacNab 1986 [**55**,5974; **58**,4070; **63**,3089–90].

F.o.f.sp. tracheiphilum (E.F. Smith) Snyder & Hansen; Holliday, *Descr.F.* 220, 1970; cowpea and soybean wilts; f.sp. *glycines* G.M. & J.K. Armstrong 1965 also attacks soybean; there are 3 races of f.sp *tracheiphilum* fide Swanson & van Gundy, *PD* **69**:779, 1985, who described the effects of temp. & plant age on race differentiation in cowpea; see Sinclair 1982 for *Fusarium* spp. and ff.sp. on soybean; Allen 1983; Gbaja & Chant, *Trop.Agric.Trin.* **60**:272, 1983, synergism with sunnhemp mosaic v.; Sumner & Minton, *PD* **71**:20, 1987, race 1 on soybean, synergism with nematodes [**63**,1509; **65**,1012].

F.o.f.sp. trifolii sensu Pratt, *Phytop.* **72**:622, 1982; crimson clover wilt [**61**,7062].

F.o.f.sp. tuberosi Snyder & Hansen; potato wilt; Thanassoulopoulos & Kitsos, *Potato Res.* **28**:507, 515, 1985, natural occurrence of wilt & tuber infection, inoculations [**65**,5693–4].

F.o.f.sp. tulipae Apt 1958; tulip bulb rot, Moore 1979.

F.o.f.sp. vasinfectum (Atk.) Snyder & Hansen; Booth & Waterston, *Descr.F.* 28, 1964; Booth 1971; cotton wilt, a major disease which has been widely studied in East Africa, Egypt, India and USA; at least 6 races have been delineated: 1 and 2 in USA, 3 in Egypt, 4 in India, 5 in Sudan and 6 in Brazil. Breeding for resistance is also complicated by the extensive attacks on the root system by *Meloidogyne incognita* and other nematodes, these not only increase the severity of wilt but also break down the resistance to the fungus in some cvs.; Ebbels, *Cotton Grow.Rev.* **52**:295, 1975; Holliday 1980; Kappelman, *Crop Sci.* **20**:613, 1980, progress in breeding for resistance; Smith et al. in Nelson et al. 1981; Watkins 1981; Katan et al., *Phytop.* **73**:1215, 1983, control by solar heat in Israel; Hillocks, *Trop.Pest Management* **30**:234, 1984, resistant cvs. in Tanzania [**60**,4468; **63**,1810; **64**,3079].

F.o. var. redolens (Wollenw.) Gordon 1952; Booth & Waterston, *Descr.F.* 27, 1964 as *Fusarium redolens* Wollenw. 1913; temperate areas, can cause damping off, assoc. with *F.o.*f.sp. *pisi* in St John's wilt of pea.

F. pallidoroseum (Cooke) Sacc. 1886; the correct name for *Fusarium semitectum* Berk & Rav. whose type specimen is identical with *Colletotrichum musae* (Berk. & Curt.) v. Arx, fide Booth & Sutton, *TBMS* **83**:702, 1984; Booth, *Descr.F.* 573, 1978. It is prob. the only *Fusarium* sp. with a dry spored, holoblastic, form, and where airborne dispersal of conidia has been truly demonstrated; the fungus is common in tropical and subtropical areas; as a secondary invader it can cause storage rots and may be seedborne; Carter *PDR* **63**:1080, 1979, melon corky dry rot; Dhingra & Muchovej, *ibid* **63**:84, 1979, pod, seed & root rot of bean [**58**,6139; **59**,5457].

F. poae (Peck) Wollenw. 1913; Booth, *Descr.F.* 308, 1971; Neish, *Fungi Canadenses* 234, 1982; silver top or white head of temperate cereals and grasses, seedborne, a weak pathogen.

F. sambucinum, teleom. *Gibberella pulicaris.*

F. semitectum, see *Fusarium pallidoroseum.*

F. solani (Mart.) Sacc. 1881; Booth & Waterston, *Descr.F.* 29, 1964; teleom. *Nectria haematococca*; distinguished from *Fusarium oxysporum* by the long phialides and the branched, often elaborate, microconidiophores; *Hypomyces solani* for the teleom. of *F. solani* is untenable. The fungus is a widespread soil inhabitant which causes diseases on many plants but it has much fewer specialised pathogenic forms than *F. oxysporum*. It attacks the roots and collar that are not heavily cuticulerised or suberised, causing a general cortical rot and not a wilt or tracheomycosis; the stem further above soil level may be attacked and *F. solani* is sometimes assoc. with cankers in woody crops. It is frequently described in disease complexes with other pathogens; c.18 ff.sp. have been described but only a few appear to be of economic importance; Holliday 1980; Sherf & MacNab 1986.

F.s.f.sp. cucurbitae Snyder & Hansen 1941; foot rot of *Cucumis* and *Cucurbita*; 2 races, both seedborne, race 2 tends to be restricted and attacks the fruit only; Tousson & Snyder, *Phytop.* **51**:17, 1961; Sumner, *PDR* **60**:923, 1976 [**41**:77; **56**,3301].

F.s.f.sp. phaseoli (Burkh.) Snyder & Hansen; bean dry rot; causes a foot rot and a root rot; the latter can be much more severe and causes most damage if root growth is restricted; Kraft et al. in Nelson et al. 1981; Burke & Miller, *PD* **67**:1312, 1983, control through resistance & field culture; Dryden & van Alfen, *Phytop.* **74**:132, 1984, soil moisture, root density & infection under dry conditions [**63**,3117].

F.s.f.sp. piperis Albuquerque 1961; foot rot of black pepper in Brazil where a foot rot caused by

Phytophthora MF4 also occurs. The relative roles of these pathogens in the syndrome is not clear from the literature. It is also not clear whether this f.sp. can kill mature vines as can this sp. of *Phytophthora*. The disease caused by this *Fusarium* is reported to be severe, S.A. Rudgard and P.W.F. de Waard, personal communications, 1986; Freire & Bridge, *Fitopat Brasileira* **10**:559, 1985; **11**:131, 1986 [**65**,4491; **66**,1094].

F.s.f.sp. pisi (Jones) Snyder & Hansen 1941; pea foot rot; this pathogen is less serious than *Fusarium oxysporum* f.sp. *pisi* which may be assoc. with it; poor soil conditions favour the disease; Kraft et al. in Nelson et al. 1981; Rush & Kraft, *Phytop.* **76**:1325, 1986, inoculation [**66**,4551].

F.s. var. coeruleum (Sacc.) Booth, *The genus Fusarium*, 1971; commonly assoc. with plants but prob. only economically important in causing a storage rot, dry rot, of potato; other *Fusarium* spp. have been described as causes of the disease; Boyd, see potato; Hide & Cayley, *AAB* **107**:429, 1985, chemical control [**66**,292].

F. sporotrichioides Sherb. 1915; Vargo & Baumer, *PD* **70**:629, 1986, as a pathogen of wheat [**66**,541].

F. stilboides, teleom. *Gibberella stilboides*; **Fusarium sulphureum**, teleom. *G. cyanogena*.

F. tabacinum (v. Beyma) W. Gams 1968; the teleom. has been named as: *Plectosphaerella cucumeris* Kleb. 1930, *Micronectriella cucumeris* (Kleb.) Booth 1971, *Monographella cucumerina* (Lindfors) v. Arx, *TBMS* **83**:373, 1984, who gave the anam. as *Microdochium tabacinum* (v. Beyma) v. Arx 1984, loc. cit. Pascoe et al., *ibid* **82**:343, 1984, described the pathogenicity to tomato and other hosts, the con. are mostly 1 septate, 6–14 × 1.5–3 μm and unlike those of a *Fusarium*; Matta, *Riv.Patol.Veg.* **14**:119, 1978, pathogenicity to basil, *Ocimum basilicum*, & tomato; Mygind, *Tidsskr.Pl.Avl.* **90**:3, 1986, causing a wilt of *Campanula isophylla* in Denmark [**58**,3753; **63**,2493; **65**,5556].

F. trichothecioides Wollenw. 1912; potato powdery dry rot, storage disease, see *Fusarium solani* var. *coeruleum*.

F. udum, teleom. *Gibberella indica*. *Fusarium udum* f.sp. *crotalariae* (Kulkarni) Subram. in Booth 1971; wilt of *Crotalaria*, including sunnhemp, 3 races, poss. seedborne.

F. xylarioides, teleom. *Gibberella xylarioides*.

fuscous blight (*Xanthomonas campestris* pv. *phaseoli*) bean, Lima bean, scarlet runner bean.

fuscovaginae, *Pseudomonas*.

fuscum, *Colletotrichum, Entyloma, Gymnosporangium*.

Fusicladium Bonorden 1851; Hyphom.; Ellis 1971, 1976; Carmichael et al. 1980; Sivanesan 1984; conidiomata sporodochial, conidiogenesis holoblastic, conidiophore growth sympodial; con. 1 septate, pigmented.

F. cerasi, teleom. *Venturia cerasi*; **Fusicladium macrosporum**, teleom. *Microcyclus ulei*.

F. pisicola Linford, *Phytop.* **16**:554, 1926; con. 16–23 × 8–14 μm, pea black spot, an unimportant disease [**6**:70].

F. pyrorum, teleom. *Venturia pirina*.

fusicoccin, a non-specific toxin formed by *Fusicoccum amygdali*; Ballio et al., *Nature Lond.* **203**:297, 1964; Chain et al., *Physiol.Pl.Path.* **1**:495, 1971 [**44**,348; **51**,2183].

Fusicoccum Corda 1829; Coelom.; Sutton 1980; conidiomata stromatic, conidiogenesis holoblastic; con. aseptate, hyaline, solitary.

F. aesculi, teleom. *Botryosphaeria dothidea* q.v. fide Pennycook & Samuels.

F. amygdali Delacr., *Bull.Soc.Mycol.Fr.* **21**:184, 1905; con. 5–7 × 2–3 μm; almond and peach blight; forms a toxin fusicoccin q.v. and Ballio, *Annls.Phytopath.* **10**:145, 1978; Megino et al., *An.Inst.nac.Invest.Agrarias Prot.Veg.* **7**:107, 1977, epidemiology & control on almond in Spain; Jailloux & Froidefond, *Annls.Phytopath.* **10**:39, 1978, inoculum & leaf fall in peach in France; Sacchetti, *Inftore.fitopatol.* **31**(10):25, 1981, canker & blight of peach in Italy; Serizawa 1984, chemical control on peach in Japan [**58**,799, 2310, 3908; **61**,2354; **65**,5049].

F. cajani, teleom. *Botryosphaeria xanthocephala*; **Fusicoccum parvum**, teleom. *B. parva*, see *B. dothidea*.

F. putrefaciens Shear, *J.agric.Res.* **11**:36, 1917; con. mostly 10–12 × 2.5 μm; Creelman, *PDR* **42**:843, 1958, cranberry end rot, a storage disease, and highbush blueberry canker [**37**:728].

F. quercus Oudem. 1889; Butin, *Eur.J.For.Path.* **11**:33, 1981, described the fungus, which caused a bark dieback of young oak, and other fungi on oak; the con. are like those of *Phomopsis quercina* (Sacc.) Höhnel for which the new combination *Amphicytostroma quercinum* was proposed, teleom. *Amphiporthe leiphaemia*, see Butin, *Sydowia* **33**:18, 1980 [**60**,6667; **61**,2465].

fusiforme, *Cronartium*, see *C. quercuum*; **fusiformis**, *Claviceps*.

fusiform rust (*Cronartium quercuum* f.sp. *fusiforme*) loblolly and slash pines.

fusimaculans, *Cercospora*.

fusispora, *Acrophialophora*.

G

gaeumannii, *Phaeocryptopus.*

Gaeumannomyces v. Arx & Oliver, *TBMS* **35**:32, 1952; Gnomoniaceae; a superficial, hyphopodiate mycelium is formed, the single, sunken ascoma has a ± unilaterally inserted, short beak, asco. vermiculate, with pseudosepta when mature; on roots, crowns and lower parts of stems and leaf sheaths, on Cyperaceae, Gramineae; Walker, *Mycotaxon* **11**:1, 1980 [**60**,1314].

G. graminis (Sacc.) v. Arx & Oliver 1952; 3 vars.; Walker, *Descr.F.* 381, 382, 383, 1973; anam. like *Phialophora*; var. *avenae* (E. Turner) Dennis; hyphopodia simple, asco. av. 140 × 4 μm, oat, turf grasses; var. *graminis*; hyphopodia both simple and strongly lobed, asco. av. 120 × 4 μm, warm temperate and sub-tropical, rice crown and sheath rot, grasses; var. *tritici* Walker 1972; hyphopodia simple, asco. av. 110 × 4 μm, barley, wheat. Take all of temperate cereals, and grasses, a major soilborne disease, has been intensively investigated for many years. The emerged heads of infected plants appear bleached, whiteheads with empty spikelets; thick, brown, runner hyphae form on the roots and ascomata on the lower leaf bases; bare patches occur in a crop where seedlings have been killed. Asher & Shipton ed., *Biology and control of take all*, 1981, begin this monograph: 'More than 125 years after it was first recognised there is still no effective method of controlling the take all disease of cereals and grasses by chemical or genetic means. Frequent rotation with non-host crops continues to be the most widely advocated control measure. In spite of, and perhaps because of, this take all remains one of the major factors limiting cereal production and yields in many countries of the world, – one of the classic plant diseases.'

galanthina, *Botrytis.*

galinsoga mosaic v. Behncken 1970; Behncken et al., *Descr.V.* 252, 1982; Tombusvirus, isometric *c.*34 nm diam., transm. sap, usually restricted to roots and transm. through soil, Australia (Queensland), the weed *G. parviflora*; Hatta et al., *J.gen.Virol.* **64**:687, 1983, particle morphology & cytopathology.

gall, and see tumour; (*Nectria rigidiuscula*) cacao; eucalyptus, due to high relative humidity in a controlled environment and greenhouses, Warrington, *Aust.For.Res.* **10**:185, 1980 [**60**,3361].

gallae, *Botryodiplodia*; **galligena,** *Nectria.*

Galloway, Beverly Thomas, 1863–1938; born in USA, Univ. Missouri, USDA, played a notable part in the early development of plant pathology in America. *Science N.Y.* **88**:6, 1938.

gangrene (*Phoma exigua*) potato.

*****Ganoderma** P. Karsten 1881; Ganodermataceae; the broadly ellipsoid basidiospores have an apical thickening which collapses at maturity to give a truncate appearance; the spp. may be more serious pathogens in the tropics; several have been implicated in the etiology of the widespread oil palm stem rots, but their relative importance is not clear and it is unlikely that one sp. will be shown to be the sole cause of a disease in a particular area; Steyaert, *Bull.Jard.bot.natn.Belge* **37**:465, 1967; Holliday 1980; Turner 1981 [**47**,1945].

G. applanatum (Pers.) Pat. 1889; Steyaert, *Descr.F.* 443, 1975; white spongy heart rot and butt rot of trees, N. Hemisphere; Childs, *Phytop.* **43**:99, 1953, poss. cause of citrus concentric canker [**33**:25].

G. boninense Pat. 1889; Steyaert, *Descr.F.* 444, 1974; mainly on palms, basal stem rot with other *Ganoderma* spp.; S.E. Asia, Australasia, Japan.

*****G. lucidum** (Leyss:Fr.) P. Karsten; Steyaert, *Descr.F.* 445, 1975; root and butt rot of trees, N. hemisphere temperate zone, central Africa above 1500 m; a name that has been misapplied in the tropics; Adaskaveg & Gilbertson, *PD* **71**:251, 1987 on grapevine.

G. philippii (Bresad. & Henn.) Bresad. 1932; Steyaert, *Descr.F.* 446, 1975; red root rot of woody crops, S.E. Asia, mostly and intensively studied in Malaysia (W.) and parts of Indonesia on rubber with the 2 other polyporaceous fungi: *Phellinus noxius* and *Rigidoporus lignosus*. The roots attacked by *Ganoderma philippii* have a red rhizomorphic skin to which the soil adheres; less important economically than *R. lignosus* q.v.

G. tornatum (Pers.) Bresad. 1912; Steyaert, *Descr.F.* 447, 1975; butt and heart rot of trees, oil palm stem rot, perhaps the commonest sp. in the tropics.

G. zonatum Murrill 1902; Steyaert, *Descr.F.* 448, 1975; assoc. with basal rot of palms, widespread in Africa, prob. also in S. America.

Ganodermataceae, Aphyllophorales; hymenophore poroid, pileal surface with a crust; basidiospores have a complex wall structure, a hyaline epispore encloses a brown pigmented endospore which is ornamented with echinules covered by the epispore; lignicolous; *Ganoderma*.

gardeniae, *Phomopsis*.

garlic, see *Allium*; Peña-Iglesias & Ayuso, *Acta Hortic.* 127:183, 1982, viruses & their elimination by shoot apex culture.

—**chlorosis streak** Xie, Li & Pei, *Acta phytopath.sin.* **11**:57, 1981; China (Xinjiang).

— **latent** v. Lee et al., *Ann. phytopath.Soc.Japan* **45**:727, 1979; filament 650–700 nm long, comparison with garlic mosaic v. [**60**,603].

— **leaf deformation** v. Cadilhac et al., *Annls.Phytopath.* **8**:65, 1976; rod *c.*600 and 740 nm long, France, poss. a form of garlic mosaic v. [**56**,3359].

*— **mosaic** v. Lafon 1964; La, *Korean J.Pl.Prot.* **12**:93, 1973; Ahlawat, *Sci.Cult.* **40**:446, 1974; rod, transm. sap, aphid spp.; Abiko et al., *Bull.Veg.Ornamental Crops Res.Stn.* A 7:139, 1980, transm. *Neotoxoptera formosana*; Ahmed & Benigno, *Indian Phytopath.* **38**:121, 1983, published 1985, transm. *Aceria tulipae*, semi-persistent [**44**,2307; **54**,5175, 5177; **61**,5404; **66**,805].

— **stripe** Peterson, abs. *Phytop.* **71**:564, 1981; filament assoc., some particles > 1000 nm long, Canada (Quebec); Novák 1959 reported a garlic stripe in Czechoslovakia [**41**:192].

— **yellow streak** v. Mohamed & Young, *AAB* **97**:65, 1981; Potyvirus, filament 700–800 nm long, transm. sap, *Myz.pers.*, non-persistent, New Zealand, discussion on Potyviruses in *Allium* [**60**,4794].

Gäumann, Ernst Albert, 1893–1963; born in Switzerland, Berne Department of Agriculture; Buitenzorg, Java, 1919–22; Swiss Agricultural Experiment Station, Zurich-Oerlikon, 1922–7, chair special botany, Zurich from 1927. A prolific worker in fungus taxonomy and mycological plant pathology; one of the first to investigate the physiology of wilts caused by *Fusarium* spp. and a student of the rusts; wrote a monograph on *Peronospora* and: *Vergleichende morphologie der pilze*, 1926, translated by C.W. Dodge, 1928; *Biologie der pflanzenbewohnenden parasitischer pilze*, 1929, with E. Fischer; *Pflanzliche infektionslehre* 1946, edited for translation by W.B. Brierley 1950: *Principles of plant infection*, the first integrated account of the science; *Die pilze*, 1949; *Die pilze grundzüge ihrer entwicklungsgeschichte und morphologie*, 1952, translated by F.L. Wynd 1952; Die Rostpilze Mitteleuropas, *Beitr.Kryptog-fl.Schweiz* 12, 1959. *Verh.Schweiz Naturf.Ges.* 1963:194; *AAB* **53**:345, 1964; *TBMS* **47**:459, 1964; *Neth.J.Pl.Path.* **70**:99, 1964; *Mycologia* **57**:1, 1965.

gedarefense, *Exserohilum*.

*****Geminiviruses,** type maize streak v., *c.*14 definitive members, isometric, the particles occur in pairs, no other plant virus group has this characteristic, single particles 18–20 nm diam., pairs *c.*20–30 nm diam., each particle has one molecule of ssDNA, sedimenting at *c.*70 S, thermal inactivation 40–80°, longevity in sap 1–2 days, some transm. sap, some by *Bemisia tabaci*, persistent, or leafhoppers; Harrison, *A.R.Phytop.* **23**:55, 1985, grouped these viruses into 4 of which only the 4th is sap transm.: (1) In Gramineae, cause striate mosaics, leafhopper vectors; (2) in dicotyledons, cause stunting, yellowing, leaf curling and distortion, leafhopper vectors; (3) as for (2) but no yellowing and vectors are whiteflies, mostly *B. tabaci*; (4) in dicotyledons, cause yellow and green mosaics, vectors are whiteflies; Goodman, *J.gen.Virol.* **54**:9, 1981, and in Kurstak 1981; Bock, *PD* **66**:266, 1982; Francki et al., vol. 1, 1985; Stanley, *Adv.Virus Res.* **30**:139, 1985.

Gemmamyces Casagrande 1969; Pleosporaceae; Sivanesan 1984; asco. muriform, ascomata on a basal stroma.

G. piceae (Borthw.) Casagrande; anam. *Megaloseptoria mirabilis* Naumov 1925; asco. mostly 7 septate, usually with 1 vertical septum in all but the end cells, $35–50 \times 13–15 \ \mu m$; con. 7–21 septate, $150–200 \times 5–9 \ \mu m$; morphology, as *Cucurbitaria piceae* Borthw., in Shoemaker, *CJB* **45**:1243, 1967, con. 220–315 μm long; Norway spruce bud blight, *Picea* spp., Europe; Phillips & Burdekin 1982 as *C. piceae* [**47**,359].

*****gene for gene concept,** the concept that corresponding genes for resistance and virulence exist in host and pathogen, respectively, FBPP; see genetics, resistance to pathogens. The term originated in Flor's analysis 1946–7 of the inheritance of resistance in flax to *Melampsora lini*; Flor, *A.R.Phytop.* **9**:275, 1971; Loegering, *ibid* **16**:309, 1978; Ellingboe, *ibid* **19**:125, 1981; Keen, *Adv.Pl.Path.* **1**:35, 1982; Crute, *Pl.Path.* **35**:15, 1986; Damann, *Phytop.* **77**:55, 1987.

genestasis, inhibition of fungus sporulation; genestat, a substance inhibiting or reducing such sporulation without much affecting vegetative growth.

genetic engineering, or genetic manipulation; the deliberate alteration of the composition of a genome by man. R. Holliday, *New Scientist*, February:399, 1977, suggested the term heterogenetics, defined as: 'The synthesis and study of replicating DNA molecules containing nucleotide sequences from unrelated organisms.' Old & Primrose, *Principles of gene manipulation: an introduction to genetic engineering*, edn. 3,

1985; Hull & Davies, *Adv.Virus Res.* **28**:1, 1983; Comai & Stalker, *Crop Protect.* **3**:399, 1984; Mantell et al., *Principles of biotechnology. An introduction to genetic engineering in plants*, 1985.

*genetics, resistance to pathogens q.v. in plants is under genetic control; hence the importance of the genetics, both of host reaction and pathogenicity. Crops may have a genetic uniformity in genes for resistance which makes them vulnerable to attack by new genotypes of pathogens, Day, *A.R.Phytop.* **11**:293, 1973. One of the most recent examples of this was the destruction of much of maize crop in USA in 1970 by *Cochliobolus heterostrophus* q.v.
    Day, *Genetics of host-parasite interaction*, 1974; Webster, *A.R.Phytop.* **12**:331, 1974, genetics of plant pathogenic fungi; Person et al., *ibid* **14**:177, 1976, genetic change in host & parasite populations; Eenink, *Neth.J.Pl.Path.* **82**:133, 1976; Caten & Day, *A.R.Phytop.* **15**:295, 1977, diploidy in plant pathogenic fungi; Gracen, *ibid* **20**:219, 1982, genetics & etiology; Ellingboe, *Phytop.* **73**:941, 1983, host & soilborne pathogen interaction; *Adv.Pl.Path.* **2**, 1984; Tepper & Anderson, *Phytop.* **74**:1143, 1984, plant & pathogen interaction; Panopoulos & Peet, *A.R.Phytop.* **23**:381, 1985, bacteria; Yoder et al., *Phytop.* **76**:383, 1986, nomenclature & fungi.

genetic stripe, maize, Shurtleff 1980.

geniculata, *Curvularia*; **geniculatus**, *Cochliobolus*.

Geniculodendron Salt, *TBMS* **63**:339, 1974; Hyphom.; conidiomata hyphal, conidiogenesis holoblastic, conidiophore growth sympodial; con. aseptate, pyriform; *G. pyriforme*, teleom. *Caloscypha fulgens* [**54**,4205].

Geniculosporium serpens, teleom. *Hypoxylon serpens*.

genome, the minimum, complete set of genetic material; see multicomponent virus; —— **masking**, see heterologous encapsidation.

geographical distribution, of pathogens; one means of disease control is exclusion of the pathogens, many of which have a limited distribution. *The distribution maps of plant diseases*, CAB International Mycological Institute, is the most extensive publication of this type; the maps now number > 575. Country lists of pathogens have been compiled, see *Rev.Pl.Path.* **54**:963, 1975, Australasia & Oceania; **58**:305, 1979, Africa; **61**:519, 1982, Asia; **62**:121, 1983, America; **64**:363, 1985, Europe; also: Holliday, *ibid* **40**:337, 1971; Weltzien, *A.R.Phytop.* **10**:277, 1972; Johnston & Booth 1983; Brandburger, *Parasitische pilze und gefässpflanzen in Europe*, 1985; temperate pathogens have been commented on in lists by Boerema & Verhoeven,

Neth.J.Pl.Path. **78** suppl., 1972; **79**:165, 1973; **82**:193, 1976; **83**:165, 1977; **85**:151, 1979; **86**:199, 1980.

geotrichum, *Endomyces*.

Geotrichum Link 1809; Hyphom.; Carmichael et al. 1980; conidiomata synnematal, conidiogenesis holoarthric; con. aseptate, hyaline.

G. candidum Link; Carmichael, *Mycologia* **49**:820, 1957; teleom. *Endomyces geotrichum* E.E. Butler & Petersen, *ibid* **64**:367, 1972; ubiquitous, invading wounds, causes fruit and vegetable spoilage, e.g. Moline, *PD* **68**:46, 1984, on tomato; causes citrus sour rot, the form on citrus is sometimes considered to be physiologically distinct and called var. *citri-aurantii* (Ferraris) R. Cif. & F. Cif.; Morris and Baudoin & Eckert, *Phytop.* **72**:1336, 1592, 1982, synergism with *Penicillium digitatum* & factors affecting susceptibility in lemon [**51**,3823; **62**,1973, 1975; **63**,2494].

geranii, *Uromyces*.

geranium, the cultivated garden geraniums belong in *Pelargonium* q.v.

gerbera chlorotic mottle Aslam & Srivastava, *Indian J.Pl.Path.* **3**:38, 1985; transm. sap, *G. jamesonii* [**65**,2854].

—— **mosaic** Verma & Singh, *Gartenbauwissenschaft* **45**:61, 1980; India [**59**, 4642].

—— **phyllody** Belli, Amici & Osler 1972; MLO assoc. [**53**,9].

—— **ring and stripe** Schmelzer, *NachrBl.dt.PflSchutzdienst Berl.* **22**:71, 1968 [**48**,737].

—— **symptomless** Chang, Doi & Yora, abs. *Ann.phytopath.Soc.Japan* **42**:383, 1976; bacilliform particles assoc.

germination, Gottlieb, *The germination of fungus spores*, 1978.

ghost spot (*Botryotinia fuckeliana*) tomato.

giant granadilla, see *Passiflora*.

—— **hill**, potato, genetic disorder; plants are taller, stronger, more vigorous, profuse flowering, leaflets smaller and thicker; large coarse tubers; Hooker 1981.

—— **swamp taro**, *Cyrtosperma*; giant taro, *Alocasia macrorrhiza*, see taro.

Gibberella Sacc. 1877; Hypocreaceae; ascomata fuscous, blackish or violaceous; asco.±hyaline, 2 or more septate; some with anam. in *Fusarium*; Booth 1971; Holliday 1980.

G. avenacea R.J. Cook, *Phytop.* **57**:735, 1967; anam. *Fusarium avenaceum* (Corda) Sacc. 1886; Booth & Waterston, *Descr.F.* 25, 1964; asco. 1 septate, 13–19 × 4–5 μm, 2–3 septate, 17–25 × 5–6.5 μm, typical macroconidia phialidic, long narrow, no chlamydospores; prob. ubiquitous, mostly temperate, root damage to cereals, conifers, legumes; causes damping off, seedborne; f.sp. *fabae* (Yu) Yamamoto 1955, Ruan et al., *Acta phytopath.sin.* **12**(2):25, 1982, forms on broad bean; Booth & Spooner, *TBMS* **82**:178, 1984, teleom. on *Pteridium aquilinum* [**46**,3405; **61**,6685; **63**,1694].

G. baccata (Wallr.) Sacc. 1883; anam. *Fusarium lateritium* Nees 1817; Booth, *Descr.F.* 310, 1971; asco. 1–3 septate, 12–18 × 4.5–8 μm; macroconidia ± straight, beaked at apex; often on woody plants causing dieback, pine bud and twig blight; *F. lateritium* f.sp. *cerealis* Matuo & Sato 1962; f.sp. *mori* (Desm.) Matuo & Sato, mulberry bud blight; Afanide et al., *TBMS* **66**:505, 1976, leaf spot of *Celosia argentea*, vegetable, Nigeria. Pine pitch canker was formerly attributed to *F. lateritium* f.sp. *pini* Hepting, see Kuhlman et al., *Mycologia* **70**:1131, 1978, but the pathogen is now in *Gibberella fujikuroi*; Saito & Matuo, *Trans.mycol.Soc.Japan* **23**:73, 1982, heterothallism in ff.sp. [**56**,511; **58**,4531; **62**,716].

G. cyanogena (Desm.) Sacc. 1883; anam. *Fusarium sulphureum* Schlecht. 1824; Booth, *Descr.F.* 574, 1978; asco 3 septate, 20–25 × 5–7 μm; economically most important as causing a storage rot of potato, tuber dry rot, many other hosts and in soil; Boyd & Tickle, *Pl.Path.* **21**:195, 1972, on potato [**52**,2005].

G. fujikuroi (Saw.) Ito 1931; anam. *Fusarium moniliforme* Sheldon 1904, fide Booth 1971; Booth & Waterston, *Descr.F.* 22, 23, 1964; Kuhlman, *Mycologia* **74**:759, 1982, divided the fungus into 4 vars., based on ascoma and asco. size, phialide type, microconidial formation and mating group: *intermedia* Kuhlman, asco. mostly 1 septate, 14.6 × 4.8 μm; *fujikuroi* asco. mostly 1 septate, 12.5 × 4.7 μm; ascomata 140–300 μm diam., the smallest of the 4 vars.; *moniliformis* (Wineland) Kuhlman, asco. mostly 3 septate, 17.5 × 4.8 μm; *subglutinans* Edwards, mostly 3 septate, 22.4 × 5.6 μm. *Gibberella fujikuroi* is plurivorous, seed-, air- and soilborne; the main diseases caused are: banana black heart, cotton boll rot, maize and sorghum stalk rots; assoc. with mango malformation, see Dang & Daulta, *Pesticides* **16**(3):5, 1982; pineapple fruitlet core rot, see Rohrbach & Taniguchi, *Phytop.* **74**:995, 1984, infection, *Penicillium funiculosum* q.v.; pine pitch canker, see Kuhlman et al., *ibid* **72**:1212, 1982; Barrows-Broaddus & Dwinell, *ibid* **75**:1104, 1985; Dwinell et al., *PD* **69**:270, 1985; rice bakanae, Carolis et al., *Riso* **28**:141, 1979; Ou 1985; sugarcane pokkah boeng, sett rot and wilt, Sivanesan & Waller 1986 [**62**,589, 1243; **64**,265, 3571; **65**,2474].

G. gordonia Booth 1971; anam. *Fusarium heterosporum* Nees:Fr. 1832; Booth, *Descr.F.* 572, 1978; asco. 1–3 septate, 15.6–18.5 × 4–4.5 μm; cereal head blight, particularly in Africa.

G. imperatae Booth & Prior, *TBMS* **82**:181, 1984; anam. *Fusarium*; asco. 3 septate, 21–25 × 5–6 μm; growth habit resembles that of *Gibberella creberrima* Syd.; causing chlorosis and dieback of *Imperata cylindrica* in Australia (Queensland), Papua New Guinea; this grass, lalang, is one of the most serious weeds of S.E. Asia [**63**,1837].

G. indica Rai & Upadhyay, *Mycologia* **74**:343, 1982; anam. *Fusarium udum* E. Butler 1910; Booth, *Descr.F.* 575, 1978; asco. 10–17 × 4–6 μm; pigeon pea wilt, the severe form of this old disease may be restricted to India, the wilt syndrome is similar to those caused by *F. oxysporum* ff.sp. *F. udum* f.sp. *crotalariae* (Kulkarni) Subramanian, fide Booth 1971, causes a wilt of sunnhemp; Holliday 1980; Upadhyay & Rai, *TBMS* **78**:209, 1982, on pigeon pea, population dynamics in root region [**61**,6109, 6116].

G. intricans Wollenw. 1931; anam. *Fusarium equiseti* (Corda) Sacc. 1886; Booth, *Descr.F.* 571, 1978; asco. mostly 2–3 septate, 21–33 × 5.5 μm; plurivorous, very frequent in the tropics and subtropics, pathogenicity on avocado, cereal seedlings, cucurbits reported; Adams et al., *PD* **71**:370, 1987.

G. pulicaris (Fr.) Sacc. 1877; anam. *Fusarium sambucinum* Fuckel 1869; Booth, *Descr.F.* 385, 1973; asco. 3 septate, 20–28 × 6–9 μm; hop canker, potato storage rot, root rot of many crops.

G. stilboides Gordon ex Booth 1971; anam. *Fusarium stilboides* Wollenw. 1924; Booth & Waterston, *Descr.F.* 30, 1964, asco. often 1 septate, some 2–3 septate, 12–18 × 4–4.5 μm; Storey's bark of coffee in E. Africa, Siddiqi & Corbett, *TBMS* **51**:129, 1968; assoc. with citrus dieback in India, Banerjee & Mukherji, *Z.PflKrankh.PflPath.PflSchutz.* **74**:350, 1967 [**47**,818, 1891].

G. xylarioides Heim & Saccas 1950; anam. *Fusarium xylarioides* Steyaert 1948; Booth & Waterston, *Descr.F.* 24, 1964; asco. 1–3 septate, 15–20 × 5–6.5 μm; wilt or tracheomycosis of coffee in central Africa; van der Graaf and Pieters, *Neth.J.Pl.Path.* **84**:117, 1978; **86**:37, 1980, disease distribution pattern & resistance [**58**,705; **60**,323].

G. zeae (Schwein.) Petch 1936; anam. *Fusarium graminearum* Schwabe 1838; Booth, *Descr.F.* 384, 1973; the pathogen is frequently, and erroneously, called *Gibberella saubinetii*, see Booth 1971; asco. 3 septate, 19–24 μm long; common on temperate and tropical cereals, see the major texts on these crops; the diseases are: seedling blight, pre- and post-emergence blight, root and foot rot, culm decay, head or kernel blight, scab or ear scab and stalk rot; variations in pathogenicity but no races, seedborne; Holliday 1980; Sutton, *Can.J.Pl.Path.* **4**:195, 1982, review, epidemiology on maize & wheat; Duthie & Hall, *Pl.Path.* **36**:33, 1987, transm. seed to stem in wheat [**61**,6977; **66**,3302].

gibberellins, plant hormones significant in growth; their discovery arose from the investigations of Kurosawa 1926 on rice bakanae (*Gibberella fujikuroi*). The plant elongation symptoms characteristic of this disease are caused by gibberellin in some strs. of the fungus, where it is present in greater amounts than in plants; see growth regulators and Pegg in Heitefuss & Williams, 1976.

Gibellina Pass. 1886; Phyllachoraceae; ascomata deeply immersed in the host with long beaks, asco. 1 septate, light brown. *G. cerealis* Pass.; Booth, *Descr.F.* 534, 1977; asco. 28–34 × 7–10 μm; wheat basal stem rot, also white foot rot and white straw, parts of Europe, China (N.), USSR (Georgia); Glynne et al., *TBMS* **84**:653, 1985, pointed out the inappropriateness of the last 2 names since the infected tillers are black at harvest. They studied the disease in England at Rothamsted Experimental Station, to which it is confined, in a long term wheat and bare fallow experiment which receives no fertiliser. No spread has taken place in UK although airborne asco. can apparently spread infection. This is poss. because the fungus does not infect well fertilised wheat; it survives for several years in soil in the absence of this crop. Basal stem rot should not be confused with the 2 wheat diseases: true eyespot and sharp eyespot.

gibsonii, *Mycosphaerella.*

gigantea, *Claviceps, Drechslera, Peniophora*; **giganteum**, see *Exobasidium camelliae*; **giganteus**, *Leucopaxillus.*

gigantism, tobacco, genetic disorder; Wolf, *J.Elisha Mitchell scient.Soc.* **81**:144, 1965.

Gilbertella Hesselt. 1960; Mucoraceae; zygospores formed as in *Mucor* and with large projections on their walls; *G. persicaria* (Eddy) Hesselt. and var. *indica* M.D. & B.S. Mehrotra; Sarbhoy, *Descr. F.* 104, 105, 1966; the var. differs in not having branched sporangiophores; peach and mulberry rot.

Gilpatrick, John Daniel, 1924–82; born in USA, Univ. McGill, Alberta, MacDonald College, Quebec; Canada Department of Agriculture; USA including Univ. Cornell; an authority on diseases of temperate fruit, one of the first workers on pathogen tolerance of benzimidazole

fungicides; assoc. with the discovery of acquired resistance in plants to viruses. *Phytop.* **73**:635, 1983; **74**:45, 1984.

ginger, *Zingiber officinale*; Pegg et al., *Qd.agric.J.* **100**:611, 1974; Sharma & Jain, *PANS* **23**:474, 1977; Zakaullah & Badshah, *Pakistan J.For.* **29**:110, 1979.

— **blotch** (poss. *Pseudomonas fluorescens*) common cultivated mushroom.

— **chlorotic fleck v.** Thomas, *AAB* **108**:43, 1986; isometric *c*.30 nm diam., contains ssRNA, sap transm. to ginger only; found in Australia in ginger from several countries; prob. present in India, Malaysia, Mauritius, but not in Australian commercial ginger [**66**,272].

— **mosaic v.** So, *Korean J.Pl.Prot.* **19**:67, 1980; poss. Cucumovirus, isometric 28–32 nm diam., transm. sap; Nambiar & Sarma 1974 described a ginger mosaic, fide Thomas, see above.

ginseng, *Panax schinseng.*

gladioli, *Penicillium, Pseudomonas, Septoria, Stromatinia, Uromyces*; **gladiolicola**, *Urocystis.*

gladiolus flower malformation Mallozzi & Herbas 1968; Mallozzi & Barros, *Biológico* **40**:101, 1974; MLO assoc., Brazil [**47**,2181; **54**,1302].

— **grassy top** Smith & Brierley, *Phytop.* **38**:581, 1948; *PDR* **37**:547, 1953; van Slogteren, *Antonie van Leeuwenhoek* **40**:314, 1974; MLO, transm. *Macrosteles fascifrons*, *M. sexnotatus*; presumably the same as gladiolus grassiness [**28**:45; **33**:297; 53, 4477].

— **mosaic** (bean yellow mosaic v.), McWhorter et al., *Science N.Y.* **105**:177, 1947 [**26**:275].

— **notched leaf** (tobacco rattle v.), Cremer & Schenk, *Neth.J.Pl.Path.* **73**:33, 1967; Allam et al., *Annls.agric.Sci.Moshtohor* **2**:143, 1974 [**56**,2442; **55**,280].

— **virescence** Bertaccini & Marani, *Phytopath.Mediterranea* **19**:121, 1980; Ploaie et al., *An.Inst.Cerc.Prot.Pl.* **16**:35, 1980; MLO assoc. [**61**,744; **63**,167].

— **white leaf streak** (bean yellow mosaic v.); (cucumber mosaic v.), Brierley, *PDR* **36**:48, 1952; (tobacco ringspot v.), Bridgmon & Walker, *Phytop.* **42**:65, 1952 [**31**:385, 435].

— **witches' broom** Groen & van Slogteren, *Bloembollencultuur* **91**:327, 1980; MLO, transm. leafhoppers including *Macrosteles sexnotatus* [**61**,280].

gladiorum, *Botrytis.*

glassiness, glassy core, apple, see water core; lettuce, prob. due to an imbalance between water uptake and transpiration, in greenhouses, Netherlands; Maaswinkel & Welles 1986 [**65**,5823].

gleditschiae, *Linospora.*

gleditsia mosaic Atanasoff, *Phytopath.Z.* **8**:197, 1935; Bulgaria, *G. triaclanthos.*

Gliocladium Corda 1840; Hyphom.; Carmichael et al. 1980; conidiomata synnematal, conidiophores penicillate, conidiogenesis enteroblastic phialidic; con. aseptate, hyaline, mucilagenous heads, coloured in mass; sometimes antagonistic to, and parasitic on, other fungi, e.g. Barnett & Lilly, *Mycologia* **54**:72, 1962, *G. roseum* (Link) Bainier; for *G. virens* J.H. Miller, Giddens & Foster 1958, on *Sclerotinia sclerotiorum* & *Thanatephorus cucumeris*: Tu, *Phytop.* **70**:670, 1980; Tu & Vaataja, *CJB* **59**:22, 1981; Phillips, *J.Phytopath.* **116**:212, 1986; key in Morquer et al., *Bull.Soc.mycol.Fr.* **79**:137, 1963; Howell & Stipanovic, *Phytop.* **74**:1346, 1984, toxin viridiol; see *Trichoderma*, Papavizas [**42**:243; **43**,1844; **60**,1785, 6151; **64**,1131; **65**,5892].

globe artichoke, see artichoke.

Globodera pallida, *G. rostochiensis*; see *Heterodera pallida*, *H. rostochiensis.*

globosa, *Botryotinia*, *Botrytis*, *Conoplea*; globosum, *Gymnosporangium.*

globuliferum, *Stemphylium*, see *S. sarciniforme.*

Gloeocercospora Bain & Edgerton ex Deighton, *TBMS* **57**:358, 1971; Hyphom.; conidiomata hyphal, conidiogenesis holoblastic; con. solitary, pluriseptate, filiform, hyaline; Carmichael et al. 1980 considered the genus doubtful since the difference from *Ramulispora* is not clear.

G. sorghi Bain & Edgerton ex Deighton; Mulder & Holliday, *Descr.F.* 300, 1971; con. mostly 1–7 septate, 20–195 × 1.4–3.2 μm; sorghum zonate leaf spot, copper spot of grasses including Bermuda grass, sugarcane; the sclerotia are immersed in the necrotic tissue, whilst those of *Ramulispora sorghi* and *R. sorghicola* are superficial; Watanabe & Hashimoto, *Ann.phytopath.Soc.Japan* **44**:633, 1978, from sorghum seed; Holliday 1980 [**59**,265].

Gloeodes Colby 1920; Coelom.; conidiomata pycnothyrial, con. aseptate, hyaline; *G. pomigena* (Schwein.) Colby, apple sooty blotch; Anderson 1956; Brown & Sutton, *PD* **70**:281, 1986, chemical control [**65**,5028].

Gloeosporidiella Petrak 1921; Coelom.; Sutton 1980; conidiomata acervular, conidiogenesis enteroblastic phialidic; con. aseptate, hyaline; *G. ribis*, teleom. *Drepanopeziza ribis.*

gloeosporioides, *Colletotrichum.*

Gloeosporium Desm. & Mont. 1849; a rejected name = *Marssonina*; sometimes, incorrectly, used for *Colletotrichum.*

Gloeotinia M. Wilson, M. Noble & E. Gray, *TBMS* **37**:31, 1954; Sclerotiniaceae; ascomata form on mummified caryopses of grasses in assoc. with a pink slimy spored, conidial stage [**33**:759].

G. granigena (Quélet) T. Schumacher, *Mycotaxon* **8**:125, 1974; anam. *Endoconidium temulentum* Prill. & Delacr., Cannon et al. 1985; asco. aseptate, hyaline, 7.5–12 × 3–6 μm; blind seed of ryegrass, other temperate grasses; an important seedborne pathogen where early infection destroys the embryo and later infection causes germination failure; Calvert & Muskett, *AAB* **32**:329, 1945; Griffiths, *TBMS* **41**:461, 1958, mating types & variation; Hardison, *Mycologia* **54**:201, 1962, hosts; de Tempe, *Neth.J.Pl.Path.* **72**:299, 1966, epidemic; Hampton & Scott, *N.Z. J.agric.Res.* **23**:143, 149, 1980, disease & N; Matthews, *Seed Sci.Technol.* **8**:183, 1980, detection in seed [**38**:262; **42**:267; **46**,1285; **60**,955–6].

glomeratus, *Inonotus.*

Glomerella v. Schrenk & Spauld. 1903; Phyllachoraceae; ascomata single or in groups, not stromatic, mostly immersed in host tissue, asco. small, hyaline; anam. in *Colletotrichum*; v. Arx & Muller 1954 reduced many spp. to synonymy with *G. cingulata.*

*G. cingulata (Stonem.) Spauld. & v. Schrenk; anam. *Colletotrichum gloeosporioides* (Penz.) Penz. & Sacc. in Penz. 1884, fide Pennycook, *Mycotaxon.* **16**:507, 1983; Mordue, *Descr.F.* 315, 1971; con. straight cylindrical 9–24 × 3–4.5 μm; appressoria 6–20 × 4–12 μm; much of the literature was discussed by Holliday 1980. *Glomerella cingulata* is plurivorous and an important pathogen, and saprophytic, particularly in the tropics and subtropics. The anam. state is more commonly found with a diseased condition which is usually called an anthracnose. Latent infections are caused on tropical fruit, much studied on banana. The main diseases include: citrus, postbloom fruit drop, withertip; coffee, green berry disease, see *C. coffeanum*, brown blight; cotton, little broom, boll rot; legumes, southern anthracnose; muscadine grapevine, ripe rot; onion, seven curls; rubber, secondary leaf fall; tea, brown blight; other crops attacked include: apple, avocado, anthurium, areca palm, banana, betel, black pepper, cacao, cashew, cinnamon, cucurbits, kenaf, mango, papaya, red pepper, sisal, strawberry, tobacco, yam. Strs. differing in pathogenicity occur but these are often ill-defined.

G. glycines Lehman & Wolf 1926; anam. *Colletotrichum destructivum* O'Gara 1954, fide Manandhar et al., *Phytop.* **76**:282, 1986, descriptions of both states; asco. mostly 24.5–29 × 3.3–4 μm; con. mostly straight cylindrical, 14.5–17.3 × 4.1–4.8 μm; appressoria mostly 80–140 × 4.5–5.5 μm; soybean anthracnose

but less severe than *C. truncatum* on this crop and
which has curved con. [**65**,5273].

G. tucumanensis (Speg.) v. Arx & Müller 1954;
anam. *Colletotrichum falcatum* Went 1893;
Mordue, *Descr.F.* 133, 1967; con. falcate,
20–27 × 4–5 μm; appressoria edge entire,
12.5–14.5 × 9–12 μm; sugarcane red rot, one of the
oldest and most important fungus diseases of the
crop, investigated mostly in Hawaii, other parts of
USA, India. Red rot affects the setts, standing
cane and leaves; on splitting the cane the internal
tissue shows a reddening and there is a sour
odour; the pith dries, becomes necrotic, the rind
falls in and the cane may be readily broken. Other
Colletotrichum spp. occur on sugarcane but prob.
only *Glomerella tucumanensis* is host restricted to
spp. of *Erianthus, Saccharum, Sorghum*. Spread,
where cultivation is continuous, is through the
setts; con. are distributed in the vascular system.
The very large literature was reviewed by: Chona
et al., *Indian Phytopath.* **33**:191, 1980; Holliday
1980; Sivanesan & Waller 1986.

Glomus Tul. & C. Tul. 1845; Endogonales; tobacco
stunt, caused by one or poss. 2 spp. in USA
(Kentucky); Hendrix, *PD* **69**:445, 1985; Modjo &
Hendrix, *Phytop.* **76**:688, 1986 [**66**,310].

gloriosa fleck Araki et al., *Ann.phytopath.Soc.Japan*
51:632, 1985; Phytorhabdovirus assoc., *G.
rothschildiana* [**65**,4444].

— stripe mosaic v. Koenig & Lesemann,
Phytopath.Z. **80**:136, 1974; poss. Potyvirus,
filament 784 × 12 nm, transm. sap, Germany, *G.
rothschildiana* [**54**,2286].

glumae, *Pseudomonas*; **glumarum,** *Phoma*.

glume blight (*Phoma sorghina*) rice; **— blotch**
(*Leptosphaeria nodorum*) wheat, Gramineae,
(*Pithiomyces chartarum*) rice, sorghum; **— and
leaf spot** (*Ascochyta desmazieresii*) ryegrass.

Glycine max, soybean or soyabean.

glycine mosaic v. Bowyer, Dale & Behncken, *AAB*
95:385, 1980; Comovirus, isometric *c*.28 nm
diam., transm. sap, Australia, 2 strs., in *G.
clandestina, G. tabacina*, indigenous plants; not
found in soybean but infected 21 cvs. [**60**,1885].

— mottle v. Behncken & Dale, *Intervirology*
21:159, 1984; poss. Tombusvirus, isometric
c.30 nm diam., transm. sap, Australia
(Queensland), narrow experimental host range,
G. tomentella, systemic infection in *Glycine* spp.
but only local infection in soybean [**63**,4804].

glycines, *Dactuliophora, Glomerella, Pyrenochaeta,
Septoria, Sphaceloma*.

glycoproteins, from fungi, binding phenols as a
protective mechanism from toxic phenols formed
by plants; Nicholson et al., *Phytop.* **76**:1315, 1986
[**66**,4184].

Gnomonia Ces. & de Not. 1863; Gnomoniaceae;
Barr 1978; asco. 1 septate, hyaline, mostly
saprophytic; *Apiognomonia* spp. q.v. are often
found under *Gnomonia*, the former has asco. with
a basal septum, in the latter the septum may be
central or basal; for *G. caryae* and *G. caryae* var.
pecanae see *G. nervisida*.

Gnomoniaceae, Diaporthales; ascomata often small,
erect; beaks usually central, erumpent singly, asci
soon free, evanescent; *Anisogramma,
Apiognomonia, Clypeoporthe, Cryphonectria,
Diaporthopsis, Endothia, Gaeumannomyces,
Gnomonia, Ophiognomonia, Stegophora*; Monod,
Beiheft Sydowia 9, 1983.

Gnomonia comari P. Karsten 1873; Punithalingam,
Descr.F. 737, 1982; anam. *Zythia fragariae*
Laibach 1908; asco. 10–12 × 2–2.5 μm; strawberry
leaf blotch and stem end rot; Europe, N. America.

G. dispora Demaree & Cole, *Phytop.* **26**:1028, 1936
ex M.E. Barr 1978; asco. 23–39 × 4.5–9 μm; pecan
leaf spot [**16**:217].

G. leptostyla (Fr.) Ces. & de Not. 1863; anam.
Marssonina juglandis (Sacc.) Magnus 1906; asco.
17–21 × 2.5–4.5 μm, septum central; in an
emended description of the anam. Roquebert &
Fayret, *CJB* **60**:1320, 1982, referred it to
Marssoniella juglandis (Lib.) Höhnel; Sutton 1980
placed *Marssoniella* Höhnel 1916 as a synonym of
Marssonina; walnut anthracnose or leaf blotch,
not reported from Australasia. The fungus has
been studied recently in USA where it causes the
most destructive foliar disease of walnut in the
central areas; Black & Neely, *Phytop.* **68**:1054,
1978, free moisture, relative humidity, temp. &
disease; Matteoni & Neely, *Mycologia* **71**:1034,
1979, growth, sporulation & heterothallism; Cline
& Neely, *Phytop.* **73**:494, 1983; **74**:185, 1984, leaf
infection, juvenile leaf resistance & presence of
juglone, hydrojuglone glucoside; Kessler, *PD*
68:571, 1984, epidemics [**58**,2959; **59**,3911; **62**,83,
4019; **63**,3058, 5572].

G. nervisida Cole, *J.agric.Res.* **50**:92, 1935; anam.
Leptothyrium nervisidum Cole, ibid **46**:1083, 1933;
asco. 14–15 × 4–5 μm, con. 8–13 × 2–3 μm, pecan
vein spot. Barr 1978 considered *Gnomonia caryae*
var. *pecanae* Cole, ibid **47**:878, 1933, pecan liver
spot, to be a synonym and gave the asco. size for
G. nervisida as 16–25 × 2–2.5 μm. Wolf described
G. caryae, Annls. Mycol. **10**:491, 1912, from
Carya ovata, shag bark hickory. The anam. does
not appear to be a *Leptothyrium* since the type of
this genus has falcate con., Sutton. *Mycol.Pap.*
141, 1977.Vein spot occurs in USA and has only
recently been studied further by Sanderlin et al.,
PD **67**:1209, 1983 [**12**:798; **13**:409; **14**:537;
63,3589].

G. platani, see *Apiognomonia veneta*; *Gnomonia quercina*, see *A. quercina*.

G. rubi (Rehm) Winter in Rabenh. 1887, fide Barr 1978; = *Gnomonia rostellata* (Fr.) Bref. 1891, fide Monod, see Cannon et al. 1985; asco. 10–15.5 × 2.5–3.5 μm; rose dieback, on *Rubus*, Europe, N. America; Dowson, *Gdnrs.' Chron.* **76**:374, 1924 [**4**:287].

gnotobiotic, a culture or other growing system in which all the living components are known cf. axenic; Kreutzer & Baker in Bruehl 1975.

Goa bean, see winged bean.

godetiae, *Alternaria*.

Godronia Moug. & Lév. 1846; Helotiaceae; Groves, *CJB* **43**:1195, 1965; asco. several septate, hyaline, fusoid to subfiliform. *G. cassandrae* Peck 1887; asco. 3–7 septate, 50–70 × 2–3 μm; for anam. and form *vaccinii* see Groves; Shear & Bain, *Phytop.* **19**:1017, 1929, cranberry end rot; Weingartner & Klos, *ibid* **65**:105, 1975, blueberry canker and dieback; Melzer & Hoffmann, *Gartenbauwissenschaft* **45**:7, 89, 1980, biology in Germany [**9**:257; **45**,746; **54**,4575; **59**,3834, 4693].

goeppertianum, *Pucciniastrum*.

goitre (*Coniella fragariae*) grapevine.

golden elderberry = cherry leaf roll v.

—— **fleck**, tomato, genetic disorder, see fruit pox, Crill et al.

gomphrena bacilliform Kitajima & Costa, *Virology* **29**:523, 1966; Phytorhabdovirus assoc., 220–260 × 80–100 nm, Brazil, *G. globosa* [**46**,283].

—— **mosaic** Verma & Awasthi, *Phytopath.Z.* **95**:178, 1979; isometric particle assoc., 25 nm diam., India, *G. globosa* [**59**,322].

*****Gonatophragmium**.

gooseberry, *Ribes*.

——**Brodie**, Gray, *Gdnrs.'Chron.* **125**:198, 1949; young twigs short, spindly, clustered; leaves narrow; fruit few, small, Scotland [**28**:464].

—— **lethal**, 1972, filament assoc., poss. transm. sap, South Africa (Cape Province) [**53**,2b].

—— **mosaic** Basak & Maszkiewicz, *Acta Hortic.* **95**:49, 1980, Poland [**60**,3248].

—— **vein banding** Posnette 1952; Karl & Kleinhempel, *Acta phytopath.Acad.Sci.hung.* **4**:19, 1969; transm. aphid; England, Germany; Adams, *J.hort.Sci.* **54**:23, 1979, effect on growth & yield of gooseberry & red currant [**33**:97; **48**,3064; **58**,2863].

gooseneck (*Ascochyta caulicola*) *Melilotus*.

Gordon, William Laurence, 1901–63; born in Canada, Univ. McGill, Wisconsin, Dominion Rust Research Laboratory, Canada Department Agriculture Research Station, Winnepeg; a world authority on taxonomy of *Fusarium*; Fellow of

the Agricultural Institute of Canada, 1962. *Phytop.* **54**:1, 1964.

gordonia, *Gibberella*.

Goss's wilt and blight (*Clavibacter michiganensis* ssp. *nebraskensis*) maize.

gossypii, *Ascochyta*, *Colletotrichum*, *Nematospora*, *Phakopsora*, *Ramularia*.

gossypina, *Cercospora*, *Mycosphaerella*.

Gossypium, cotton.

Gottlieb, David, 1911–82; born in USA, City College of New York, Univ. Iowa State, Minnesota, Illinois; physiology of fungus plant pathogens, antibiotics, actinomycetes; wrote *The germination of fungus spores* 1978; Univ. Pavia medal, several American awards. *Phytop.* **73**:32, 1983.

Gracilicutes, Procaryotae.

graft canker (*Leptosphaeria coniothyrium*) rose.

grain rot (*Pseudomonas glumae*) rice.

Gram, Ernst, 1891–1964; born in Denmark, Univ. Copenhagen, Cornell; Director Statens Plantepatologiske Forsøg, Lyngby, from 1925; major role in the organisation of plant pathology in Denmark and Europe. Wrote: *The diseases of the potato*, 1940; *Plant diseases in orchard, nursery and garden crops*, with A. Webber, 1940, edn. 2, 1944, English translation by E. Ramsden, ed. R.W.G. Dennis 1952; *Diseases and pests of root crops*, with P. Bovien, 1942; *The life of plants*, 1951; *Diseases and pests of farm crops*, with Bovien & C. Stapel, 1956. *AAB* **55**:317, 1965; *Phytop.* **55**:373, 1965.

gram, see chickpea, *Vigna*.

graminea, *Drechslera*, *Pyrenophora*.

gramineae chlorosis Ploaie et al., *An.Inst.Cerc.Prot.Pl.* **11**:15, 1973, published 1975; MLO assoc., transm. *Psammotettix alienus*, Romania [**56**,1529].

graminearum, *Fusarium*.

Graminella nigrifons, transm. maize bushy stunt, maize chlorotic dwarf, maize rayado fino, *Spiroplasma kunkelii*; *G. sonora*, transm. maize chlorotic dwarf, sorghum stunt mosaic.

graminicola, *Colletotrichum*, *Mycosphaerella*, *Phialophora*, *Pythium*, *Sclerospora*.

graminin, toxin from *Acremonium*; Kobayashi & Ui, *Physiol.Pl.Path.* **14**:129, 1979 [**58**,4339].

graminis, *Erysiphe*, *Gaeumannomyces*, *Lidophia*, *Polymyxa*, *Puccinia*.

grammatophyllum ringspot Corbett, abs. *Proc.Am.phytopath.Soc.* **1**:149, 1974; bacilliform particle assoc., USA, *G. scriptum*.

Gram reaction, a method; devised by H.C.J. Gram, for staining bacteria in tissue sections. The section was stained using a violet dye, mordanted with I and decolourised in alcohol, not all bacteria were stained as some were decolourised. A counterstain, usually pink, is now used to show the decolourised cells; and the method is usually applied to smear of cultures. Bacteria that retain the violet stains are said to be Gram positive; those that are decolourised and take up the counterstain are Gram negative. The 2 groups have characteristic biochemical differences.

granadilla, see *Passiflora*.

granigena, *Gloeotinia*; **grantii**, *Grovesiella*; **granularis**, *Puccinia*.

granule, coarse particle of inert material impregnated or mixed with an active ingredient, i.e. plant protectant or fertiliser, for application to soil, after FBPP.

Granville wilt (*Pseudomonas solanacearum*) tobacco.

grapefruit, see *Citrus*.

grapevine, *Vitis*. The pathology of this crop is very much concerned with graft transm. disease agents, several of which have not been adequately characterised. Many diseases have been described; in some cases their etiology may be the same, in others it may be different. To some extent this uncertainty also occurs in woody fruit tree crops; Hewitt, *Rev.appl.Mycol.* **47**:433, 1968; *Riv.Patol.Veg.* **9**:217, 1973; Martelli in Scott & Bainbridge 1978; Bovey et al., *Virus and virus-like diseases of grapevines. Colour atlas of symptoms*, 1980; Baldacci, *Riv.Patol.Veg.* **17**:99, 1981; Hewitt, Bovey, Caudwell & Dalmasso, *Phytopath.Mediterranea* **24**:1, 8, 170, 1985; Martelli, *F.A.O. Pl.Prot.Bull.* **34**:25, 1986; for fungi: Bullit, *Bull.de l'O.I.V.* **53**(587):3, 1980; Egger et al. 1984, protection [65,829].

— **A and B**, virus particles, like Closteroviruses, were serologically distinct, Milne et al., *Phytopath. Z.* **110**:360, 1984; such particles are assoc. with corky bark, leaf roll, legno riccio and stem pitting. Rosciglione & Castellano, *Phytopath. Mediterranea* **24**:186, 1985, transm. a particle like A, using *Planococcus citri*, *P. ficus* and *Pseudococcus longispinus*, from grapevine to *Nicotiana clevelandii* [**64**:2101; **65**,5655].

— **ajinashika** Namba et al., *Ann.phytopath.Soc.Japan* **45**:70, 1979; prob. graft transm., isometric particle assoc., 26 nm diam. [59,844].

— **arricciamento** (grapevine fan leaf v.).

— **asteroid mosaic** Hewitt, *Bull.Calif.Dep.Agric.* **43**:62, 1954; Hewitt & Goheen, abs. *Phytop.* **49**:541, 1959; USA (California) [**34**:343; **39**:149].

*— **Australian yellows** Magarey & Wachtel, abs. *PD* **70**:694, 1986.

— **black wood** Caudwell, *Annls.Épiphyt.* **12**:241, 1961; poss. MLO, Europe [**41**:500].

— **Bulgarian latent v.** Martelli et al. 1976–7;

Descr.V. 186, 1978; Nepovirus, isometric *c.*30 nm diam., transm. sap, seedborne; Bulgaria, USA, Yugoslavia; Gallitelli et al., *Phytopath.Mediterranea* **22**:27, 1983, a distinct str. [63,1890].

— **chrome mosaic v.** Martelli, Lehoczky & Quacquarelli 1965; Martelli & Quacquarelli, *Descr.V.* 103, 1972; Nepovirus, isometric *c.*30 nm diam., transm. sap; England, Hungary.

— — **yellow discolouration** Savino, Boscia & Milkus, *Phytopath. Mediterranea* **24**:54, 1985 [65,5095].

— — — **leaf blotch** Savino, Boscia & Milkus, as above, from hybrid stocks grown in USSR (Ukraine).

— **corky bark** Hewitt et al., *Vitis* **3**:57, 1962; Goidanich & Canova, *Phytopath.Mediterranea* **2**:295, 1963; Moutous & Hevin, *Agronomie* **6**:387, 1986; transm. *Scaphoideus littoralis*, poss. the same as grapevine legno riccio [**42**:588; **43**,2485; 65,5082].

— **court noué** (grapevine fan leaf v.).

— **decline and degeneration**, names used for diseases in this crop caused by Nepoviruses.

— **enation** Gigante, *Boll.Sta.Patol.Veg.Roma* **17**:169, 1937; see Hewitt 1968, grapevine q.v. for references [**17**:221].

— **fan leaf v.** Hewitt 1950; Hewitt et al., *Descr.V.* 28, 1970; Nepovirus, isometric *c.*30 nm diam., transm. sap, *Xiphinema* spp., seedborne in *Chenopodium*, widespread, many variants, diseases caused in grapevine called degeneration in Europe; Raski et al., *PD* **67**:335, 1983, control in USA.

— **flat trunk** Hewitt 1971; *PDR* **59**:845, 1975; Hungary, Israel, Italy (S.), USA (California) [55,2820].

— **flavescence dorée** Gianotti et al., *C.r.hebd.Séanc.Acad.Sci.Paris* D **268**:845, 1969; Schvester et al., *Annls.Zool.Ecol.Anim.* **1**:445, 1969; Caudwell et al., ibid **10**:613, 1978; *Annls.Phytopath.* **3**:95, 107, 1971; poss. MLO, transm. *Scaphoideus littoralis*, *S. titanus*, these binomials may be synonymous; France, including Corsica, Italy [**51**,541, 1721–2; **59**,4703].

— **fleck** Hewitt et al., *Annls. Phytopath.* hors sér.:43, 1972; latent in most cvs., symptoms in *Vitis rupestris* St George.

— **graft incompatibility** Fallot et al. 1979; fide Bovey, 8*th Mtg.Int.Council Viruses, Virus diseases grapevine* Sept. 1984.

— **infectious degeneration**, or malformation (grapevine fan leaf v.).

— — **necrosis** Kvíčala & Pfeiferová 1955; Ulrychová et al., *Phytopath.Z.* **82**:254, 1975; bacterium assoc., Czechoslovakia [55,347].

—— **Joannes-Seyve** v. Dias, *Riv.Patol.Veg.* **9** suppl.:64, 1973; isometric *c*.26 nm diam., transm. sap, Canada (Ontario); Stobbs & van Schagen, *Can.J.Pl.Path.* **7**:37, 1985, considered that the virus is a str. of tomato black ring v. [**54**,942i; **64**,3911].

—— **Kerner** Gärtel 1981; *Phytopath.Mediterranea* **24**:152, 1985; bacteria assoc., Germany [**65**,5666].

—— **leaf roll** Scheu 1936; Hoeffert & Gifford, *Hilgardia* **38**:403, 1967; a major and widespread grapevine disease of unknown etiology. Both filamentous, some like Closteroviruses, and isometric particles are assoc. with this condition which has been extensively studied with other grapevine diseases whose etiology may or may not be the same, e.g. corky bark, legno riccio q.v., stem grooving, stem pitting q.v., and see grapevine A and B. The position is considered in several articles in *Phytopath.Mediterranea* **24**(1–2), 1985; Mossop et al., *N.Z.J.agric.Res.* **28**:419, 1985 [**65**,1969, 5636–8, 5641–2, 5650–2].

—— **legno riccio**, etiology not clear, poss. relationships with corky bark, leaf roll, stem pitting q.v.; assoc. particles are like those of a Closterovirus; Conti et al., *Phytop.* **70**:394, 1980; Boccardo & d'Aquilo, *J.gen.Virol.* **53**:179, 1981 [**60**,3271, 6569].

—— **little leaf** Singh, K.S.M. & K.S. Sastry, *Curr.Sci.* **44**:26, 1975; poss. MLO, India (S.) [**54**,4061].

—— **marbrure** = grapevine fleck.

—— **mosaic** Vuitenez, *Annls.Épiphyt.* **17**:hors sér.:67, 1966; France (S.E.), poss. the same as grapevine vein mosaic.

—— **new latent** Dimitrijević, *Zašt.Bilja* **31**:223, 1980 = grapevine Bulgarian latent v.

—— **Pierce's disease**, see Pierce's grapevine leaf scald and xylem limited bacteria.

—— **RNAs**, Sano et al., Flores et al., *J.gen.Virol.* **66**:333, 2095, 1985.

—— **rugose wood** = grapevine legno riccio.

—— **Shiraz**, see Engelbrecht & Kasdorf 1985 [**65**,5638].

—— **shoot necrosis** Martelli & Russo 1965; fide Hewitt 1968, see grapevine.

—— **small vein mosaic** = grapevine fleck.

—— **stem grooving**, prob. the same as grapevine legno riccio.

—— —— **pitting**, see grapevine legno riccio, Conti et al., who sap. transm. a particle like a Closterovirus from grapevine, with stem pitting, to *Nicotiana clevelandii* which showed vein clearing and stunt.

—— **summer mottle** Krake & Woodham, *Vitis* **17**:266, 1978; Australia; Woodham & Krake, *ibid*

22:247, 1983, considered the disease to be different from grapevine vein mosaic [**58**,5979].

—— **vein banding** (grapevine fan leaf v.). but see Prota et al., *Phytopath.Mediterranea* **24**:24, 1985; Krake & Woodham, *Vitis* **22**:40, 1983, on the etiology & grapevine yellow speckle [**65**,5089].

—— **vein mosaic** Legin & Vuittenez, *Riv.Patol.Veg.* **9** suppl.:57, 1973; indicator *Vitis riparia*, France [**54**,942h].

—— —— **necrosis**, as above; latent in many *Vitis* spp., indicator *V. rupestris* × *V. Berlandieri* 110R, France, Italy; Credi et al., *Phytopath.Mediterranea* **24**:17, 1985 [**65**,5088].

—— **wood pitting** Hewitt 1975, see grapevine flat trunk.

—— **yellow dwarf** H.L. Chen, Tzeng & M.J. Chen, *Natn.Sci.Counc.Monthly* **9**:584, 1981; assoc. particle isometric, Taiwan [**61**,2983].

—— —— **mosaic** (grapevine fan leaf v.).

—— **yellows** Küppers, Nienhaus & Schinzer, *PflKrankh.PflSchutz.* **82**:183, 1975; Germany [**55**,1378].

—— **yellow mottle**, Yang, Deng & Chen, *J.agric.Res.China* **35**:503, 1986, related to a str. of tomato ringspot v.

—— **yellow speckle** Taylor & Woodham, *Aust.J.agric.Res.* **23**:447, 1972; Australia, USA (California), origin poss. Europe; Mink & Parsons, *PDR* **59**:869, 1975, indexing; see grapevine vein banding [**52**,199; **55**,2821].

—— —— **vein** (tomato ringspot v.).

—— **Yugoslavian latent** = grapevine Bulgarian latent v.

*****grasses**, see cereals, major texts and: Catherall in Western 1971; Braverman & Oaks and Braverman, *Bot.Rev.* **38**:491, 1972; **52**:1, 1986, resistance; Wilkinson, *AAB* **98**:365, 1981, losses in conservation & use; Meyer, *PD* **66**:341, 1982, breeding for resistance; Drew Smith, *J. Sports Turf Res.Inst.* **61**:46, 1985.

grassiness, carnation, prob. genetic disorder; Fletcher 1984.

grass mosaic = sugarcane mosaic v.

greasy blotch (*Schizothyrium pomi*) carnation; —— **spot** (*Corynespora cassiicola*) papaya, (*Mycosphaerella citri, M. hori*, poss. *Septoria citri*) citrus, (*Pseudomonas syringae* pv. *passiflorae*) passion fruit.

green back, tomato, Mg deficiency.

—— **bridges**, living plants on which pathogens survive between cultivations of the same crop.

—— **death** (*Phytophthora dreschsleri*) cucurbits; —— **ear** (*Sclerospora graminicola*) bulrush, foxtail millets, sorghum.

Greeneria Scribner & Viala 1887; Coelom.; Sutton 1980; monotypic; conidiomata acervular, conidiogenesis enteroblastic phialidic, conidiophores branched, con. aseptate, pigmented. *G. uvicola* (Berk. & Curt.) Punith. 1974; Sutton & Gibson, *Descr.F.* 538, 1977; grapevine bitter fruit rot; M.S. & K.R.C. Reddy, *Indian Phytopath.* **36**:110, 1983 [63,2438].

green gram, see *Vigna*.

greenhouse plants, Fletcher 1984.

greenish spot, Japanese persimmon, prob. assoc. with excess Mn; Aoba & Konno 1983 [64,5458].

green islands, a localised area of tissue, usually of a leaf, which remains green around a pathogenic infection and against a background of general, spreading chlorosis. They are often seen in powdery mildews and rusts as they are prob. invariably assoc. with obligate pathogens. An unusual example is shown after infection of a very young cacao pod by *Crinipellis perniciosa*. If the pod subsequently goes necrotic, i.e. shows cherelle or physiological wilt, the infected part remains turgid and does not lose its colour. The term was first used by Cornu 1881, fide Allen, *Am.J.Bot.* **29**:432, 1942.

— **mould** (*Penicillium digitatum*) citrus; — **pitting**, red pepper, assoc. with high Ca; Hibberd, *Qd.J.agric.Anim.Sci.* **38**:47, 1981; — **point cushion gall** (*Nectria rigidiuscula*) cacao [61,3816].

gregata, *Phialophora*.

gregatins, toxins from *Phialophora gregata*; Kobayashi & Ui, *Physiol.Pl.Path.* **11**:55, 1977 [57,863].

Gregory, Philip Herries, 1907–86; born in England, Brighton Technical College, Imperial College, London; Manitoba Medical College, Winnipeg; chair, Imperial College 1954–8; Rothamsted Experimental Station 1940–67, except for 5 years. An epidemiologist who made fundamental contributions on spore dispersal in plant pathology, medical and veterinary mycology; wrote *The Microbiology of the atmosphere*, 1961, edn. 2, 1973; FRS 1962. *Indian Phytopath.* **39**(1):ii, 1986.

Gremmeniella Morelet 1969; Helotiaceae; Morelet, *Eur.J.For.Path.* **10**:268, 1980, in a taxonomic discussion, described the differences between the genus and *Ascocalyx* q.v. including their anam.; Müller & Dorworth, *Sydowia* **36**:193, 1983, placed *Gremmeniella* as a synonym of *Ascocalyx*, *G. abietina* became *A. abietina* (Lagerb.) Schläpfer-Bernhard; Cannon et al. 1985 retain *G. abietina* [60,1657].

G. abietina (Lagerb.) Morelet; anam. *Brunchorstia pinea* (P. Karsten) Höhnel 1905; Punithalingam & Gibson, *Descr.F.* 369, 1973; asco. 3 septate,

hyaline, 15–22 × 3–5–µm; con. mostly 3 septate, 25–40 × 3–3.5 µm; pine dieback; the fungus is virtually confined to pine and spruce, serious damage has been caused in Europe and it is apparently of growing importance in N. America; most damage is done to trees growing at the limit of their range, and therefore the selection of planting sites is important in control. Dorworth & Krywienczyk, *CJB* **53**:2506, 1975, delineated 3 races, Asian, European, N. American. Patton et al., *Eur.J.For.Path.* **14**:193, 1984, found that pine shoots are infected through stomata on the bracts, which subtend the short shoots, in late summer or early autumn, but only the following late January or early February does fungus colonisation begin; Yokoto et al., *ibid* **4**:155, 1974; **5**:7, 13, 1975, described the disease in Japan, particularly on Todo fir, *Abies sacchalinensis*; Dorworth, *PD* **65**:927, 1981, review & races [54,1900, 4658–9; 55,3768; 61,3105; 64,1784].

grey blight (*Pestalotiopsis theae*) tea; — **bulb rot** (*Rhizoctonia tuliparum*) herbaceous, temperate ornamentals, — **capsule rot** (*Botryotinia ricini*) castor; — **ear rot** (*Botryosphaeria zeae*) maize; — **leaf spot** (*Alternaria brassicae*) brassicas, (*Cercospora sorghi*) sorghum, (*C. zeae-maydis*) maize, (*Pestalotiopsis* spp.) guava, mango, palms, (*Pyricularia grisea*) Gramineae, (*Stemphylium solani*) potato, tomato; — **mildew** (*Mycosphaerella areola*) cotton; — **mould** (*Botryotinia fuckeliana*) many crops, (*Botrytis galanthina*) snowdrop; — — **neck rot** (*Botrytis allii*) onion; — **speck**, barley, oat, Mn deficiency; — **spot** (*Cercospora vanderystii*) bean; — **stem canker** (*Ascochyta caulicola*) sweet clover; — **sterile fungus** (see *Pyrenochaeta lycopersici*); — **tobacco**, nutrient imbalance, review by Arnold 1985 [65,4539].

greywall, tomato, see blotchy ripening.

grisea, *Magnaporthe, Pyricularia*.

griseofulvin, an antibiotic from *Penicillium griseofulvum*, has been used in crop protection; Brian, *TBMS* **43**:1, 1960.

griseola, *Phaeoisariopsis*.

grossulariae, *Microsphaera*.

groundnut, see peanut, *Arachis hypogaea*; McDonald, *Rev.appl.Mycol.* **48**:465, 1969; *PANS Manual 2* 1973, Centre Overseas Pest Res.; Smith & Littrell, *PD* **64**:356, 1980, fungicides; Allen 1983; Porter et al. 1984.

— **bud necrosis** (tomato spotted wilt v.), Ghanekar et al., *AAB* **93**:173, 1979; Reddy et al. in Plumb & Thresh 1983; transm. *Frankliniella schultzei, Scirtothrips dorsalis*; these groundnut diseases in India are omitted from the order as they have the

same etiology: bunchy top, chlorosis, ring mosaic, ring mottle, ringspot [59,3017].
— Chinese mild mottle Xu et al., *PD* 67:1029, 1983; a str. of peanut stripe v. [63,1012].
— — mosaic Shih & Hsu, *Acta phytopath.sin.* 9:131, 1979; poss. = peanut stripe v. [59,4391].
— chlorotic rosette (str. of groundnut rosette v.).
— crinkle v. Dubern & Dollett, *Phytopath.Z.* 95:279, 1979; 101:337, 1981; poss. Carlavirus, filament 650 × 13 nm, transm. sap, Ivory Coast, Leguminosae; Iizuka et al., *ibid* 109:245, 1984, described a str. of cowpea mild mottle v. occurring naturally in groundnut in India and which was serologically related to GCV from W. Africa [59,1532; 61,2561; 63,4635].
— eyespot v. Dubern & Dollet, *Oléagineux* 33:175, 1978; *AAB* 96:193, 1980; poss. Potyvirus, filament *c*.775 × 13 nm, transm. sap, *Aphis craccivora*, Ivory Coast [57,4710; 60,4086].
— green rosette (str. of groundnut rosette v.).
— leaf curl Morwood, *Qd.agric.J.* 61:266, 1945; Bolhuis, *Oléagineux* 10:157, 1955; Australia (Queensland), Indonesia [25:248; 34:340].
— — mottle Gourret & Triharso, *C.r.hebd.Séanc.Acad.Sci.Paris* D 284:179, 1977; Indonesia (Java) [56, 4792].
— mild mottle Storey & Ryland, *AAB* 45:318, 1957; transm. sap, *Aphis craccivora*, E. Africa, poss. = peanut mottle v. [36:747].
— root nematode, *Meloidogyne arenaria*.
— rosette v. Zimmerman 1907; Storey & Bottomley, *AAB* 15:26, 1928; Hull & Adams, *ibid* 62:139, 1968; Casper et al., *Phytopath.Z.* 108:12, 1983; Reddy et al., *AAB* 107:57, 65, 1985; 2 viruses cause this serious groundnut disease in Africa; 2 symptom patterns or strs., chlorotic rosette and green rosette; one virus induces symptoms, called GRV, and resembles carrot mottle v. and lettuce speckles v.; the other virus is latent and an assistor or helper q.v. called GRAV. GRV is sap transm. but depends on GRAV for transm. by *Aphis craccivora* and *A. gossypii*, persistent; the latter virus is prob. a Luteovirus, isometric *c*.25 nm diam.; CaMV and LeSV also depend on Luteoviruses for aphid transm. [7:486; 47,3640; 63,2613; 65,3597–8].
— testa nematode, *Aphelenchoides arachidis*.
— vein banding Klesser, *S.Afr.J.agric.Sci.* 10:515, 1967; transm. sap, *Aphis craccivora*, South Africa, Leguminosae [47,706].
— witches' broom Rutgers 1913 in Thung & Hadiwidjaja, *Tijdschr.PlZiekt.* 57:95, 1951; Indonesia (Java) [31:165].
ground rot (*Streptomyces ipomoeae*) sweet potato.
group dying (*Rhizina undulata*) conifers.
Grovesiella Morelet 1969; Helotiaceae; asco.

hyaline, filiform. *G. abieticola* (Zeller & Goodd.) Morelet; Gremmen & Morelet, *Eur.J.For.Path.* 1:80, 1971; asco. mostly 6–8 septate, 45–71 × 2.6–3.9 μm, assoc. with fir canker and dieback, Europe, N. America. *G. grantii* Funk, *CJB* 56:245, 1978; asco. 3 septate, 18–34 × 2–3 μm, *Abies grandis*, a secondary invader of canker and dieback, Canada (British Columbia) [51,2907].
Grovesinia M. Cline, Crane & S. Cline, *Mycologia* 75:989, 1983; Sclerotiniaceae; *G. pyramidalis* same authors, page 991; asco. 10–12 × 4–5 μm; anam. *Cristulariella moricola* (Hino) Redhead, *ibid* 71:1249, 1979; propagule 350–450 × 80–120 μm; zonate leaf spot, plurivorous; Holliday 1980 as *C. pyramidalis* Waterman & Marshall 1947 [63,1629].
growth regulators, in plant disease, Sequeira, *A.R.Phytop.* 1:5, 1963; Pegg in Wood & Jellis, 1984.
guar, *Cyamopsis tetragonolobus*.
— symptomless v. Hansen & Lesemann 1974; *Phytop.* 68:841, 1978; Potyvirus, filament, normally 760 nm long, transm. sap, prob. seedborne; India, Pakistan, USA, Zaire [58,2068].
Guatemala grass, *Tripsacum laxum*.
— — spikiness Mulder, *Tea Q.* 34:16, 1963; Sri Lanka [43,110].
guava, *Psidium guajava*, Williamson in Raychaudhuri et al. ed., *Advances in mycology and plant pathology*, 1975.
guazatine, seed dressing for cereals.
guepini, *Pestalotiopsis*.
Guignardia Viala & Ravaz 1892; Botryosphaeriaceae or Dothideaceae; Sivanesan 1984; anam. *Phyllosticta*; stromata not pulvinate or botryose, may be absent or greatly reduced, asco. aseptate; distinguished from *Botryosphaeria* by the unilocular ascomata, asco. not usually > 25 μm long, and by the anam.; Punithalingam, *Mycol.Pap.* 136, 1974; 149, 1981; Bissett, *Mycotaxon* 25:519, 1986.
*G. aesculi (Peck) Stewart 1916; anam. *Phyllosticta sphaeropsoidea* Ell. & Ev. 1883; asco. 12–18 × 7–9 μm, con. 10–20 × 9–13 μm; horse chestnut leaf blotch, can be damaging; Europe, N. America; Scaramuzzi 1954; Schneider, *Phytopath.Z.* 42:272, 1961; Neely & Himelick, *PDR* 47:170, 1963, disease ratings on *Aesculus* spp. [35:247; 41:337; 42:496].
G. bidwellii (Ell.) Viala & Ravaz; anam. *Phyllosticta ampelicida* (Engelman) van der Aa 1973; Sivanesan & Holliday, *Descr.F.* 710, 1981; Bisset & Darbyshire, *Fungi Canadenses* 273, 1984, anam.; asco. 12–17 × 6–7.5 μm, con. 5–12 × 4–7 μm; grapevine black rot, other hosts, widespread with main host, 3 strs. based on host specificity;

Ellis et al., *PD* **70**:938, 1986, chemical control [**66**,1089].

G. citricarpa Kiely 1948; anam. *Phyllosticta citricarpa* (McAlp.) van der Aa 1973; Sutton & Waterston, *Descr.F.* 85, 1966; asco. 8–17.5 × 3.5–8 µm; con. 6–13 × 5–9 µm; the fungus exists as 2 physiologic forms, one restricted to citrus and causing citrus fruit black spot; it is apparently absent from the Americas. The other is a saprophyte which can be found on citrus and many other plants. There are several symptom patterns on citrus, stored fruit may show the first symptoms; there is a long latent period; reviews: Holliday 1980; Kotzé, *PD* **65**:945, 1981.

G. laricina (Saw.) W. Yamamoto & Ito 1961; see *Index Fungi* **3**:355, 1966; larch shoot blight; Japan, Korea; Yokota 1962, 1966; Sato et al. 1971 [**42**:418; **45**,2979; **51**,740].

G. musae Racib. 1909; anam. *Phyllosticta musarum* (Cooke) van der Aa 1973; Punithalingam & Holliday, *Descr.F.* 467, 1975; asco. 13–16 × 4–8 µm, con. 8–15 × 4–8 µm; banana fruit freckle, may significantly damage the fruit; leaf infection is less important but leaves are sources of inoculum; Chuang, *TBMS* **77**:670, 1981; isolation & infection with con.; *Pl.Prot.Bull.Taiwan* **26**:335, 1984, epidemiology [**61**,3557; **64**,3138].

G. vaccinii Shear 1907; anam. *Phyllosticta elongata* Weidemann in *Mycologia* **74**:62, 1982; asco. 13–17 × 6–7 µm, con. 8–12 × 5–8 µm; latent infection of cranberry leaves and fruit, speckle; causes fruit rots in the field and in storage, N. America (N.E.); Weidemann & Boone, *PD* **67**:1090, 1983; *Phytop.* **74**:1041, 1984, pathogenicity & latent infection [**63**,3458; **64**,1184].

guinea grass mosaic v. Thouvenel, Givord & Pfeiffer 1976; *Descr.V.* 190, 1978; poss. Potyvirus, filament *c.*815 nm long, transm. sap, Ivory Coast, *Panicum maximum*, Gramineae; Lamy et al., *AAB* **93**:37, 1979, str. B from maize, transm. aphid, prob. *Rhopalosiphum maidis*; Kukla et al., *Phytopath.Z.* **109**:65, 1984, pearl millet str., transm. *Hysteroneura setariae*, *R. maidis* [**59**,3223; **63**,3887].

gumboil, rough bark, apricot, poss. genetic disorder; Thomas & Nyland, *PDR* **43**:106, 1959 [**38**:705].

gumming (*Clavibacter iranicus*, *C. tritici*) wheat, (*Xanthomonas campestris* pv. *vasculorum*) sugarcane.

gummosis, an exudate of gum from a plant, frequently external and from woody tissue; can be used for internal symptoms, e.g. filling of lysogenous cavities or xylem occlusion; (*Cladosporium cucumerinum*) cucurbits, (*Eutypa*

armeniacae) apricot (*Phytophthora citrophthora*, *P. nicotianae* var. *parasitica*, *P. palmivora*) citrus (*Xanthomonas axonopodis*) *Axonopus* spp.

gummy stem blight (*Didymella bryoniae*) cucurbits.

Güssow, Hans Theodor, 1879–1961; born in Germany, trained in Berlin and Leipzig; British Museum and Royal Agricultural Society, England; Dominion Botanist, Canada, 1911; then Chief, Botany and Plant Pathology, Canada Department of Agriculture; laid the foundations for work in these disciplines: seed potato certification, plant quarantine legislation, herbaria, seed exchange. *Phytop.* **51**:739, 1961.

guttata, *Phyllactinia*.

Gymnoconia Lagerh. 1894; Uredinales; Laundon, *Mycotaxon* **3**:139, 1975. *G. nitens* (Schwein.) Kern & Thurston 1929; Laundon & Rainbow, *Descr.F.* 201, 1969; autoecious, no uredospores, aecial infection results in telia formation, teliospores 1 septate, often ± distorted because of the prominent pores; *Rubus* orange rust; infected plants are stunted, do not bear fruit, prob. unimportant.

Gymnosporangium Hedw. ex DC. 1805; Uredinales; heteroecious, uredia and telia on Cupressaceae, aecia and spermogonia on Rosaceae; teliospores borne singly on pedicels, often 1 septate but can have several septa, characteristically gelatinising, perennial; causing witches' brooms, hypertrophies, galls on woody tissue; Boyce 1961; Kern, *A revised taxonomic account of* Gymnosporangium, 1973; Ziller 1974.

G. asiaticum Miyabe ex Yamada 1904; Laundon, *Descr.F.* 541, 1977; telia on *Juniperus chinensis*, *J. procumbens*, teliospores 32–47 × 15–25 µm; aecia on *Cydonia vulgaris*, *Pyrus* spp.; Japanese pear rust, Asia (N.E.)

G. bermudianum Earle in Seymour & Earle, *Economic Fungi* 249, 1893; autoecious and therefore unusual, teliospores 35–50 × 18–25, *Juniperus virginiana*, *J. lucayana*; Bermuda, USA (S.E.).

G. bethelii Kern 1907; telia on *Juniperus horizontalis*, *J. occidentalis*, *J. scopulorum*, teliospores 35–53 × 17–26 µm; aecia on *Crataegus*; disease severity depends on closeness of hawthorn to susceptible juniper, N. America.

G. clavariiforme (Wulf ex Pers.) DC. 1805; Laundon, *Descr.F.* 542, 1977; telia on *Juniperus communis*, other spp. of *oxycedrus* group, teliospores 45–100 × 12–20 µm; aecia principally on hawthorn, *Crataegus*; European hawthorn rust; can be severe on Rosaceae.

G. clavipes (Cooke & Peck) Cooke & Peck 1873; Laundon, *Descr.F.* 543, 1977; telia on *Juniperus* of the *oxycedrus* and *sabina* groups, teliospores 35–60 × 20–28 µm; aecia on apple, hawthorn, quince, many other Rosaceae; quince rust, can damage apple, ornamental junipers, quince; N. America.

G. confusum Plowr. 1889; Laundon, *Descr.F.* 544, 1977; telia on *Juniperus* as above, teliospores 35–60 × 19–30 μm; aecia on *Chaenomeles japonica, Cotoneaster, Crataegus, Cydonia vulgaris, Mespilus germanica, Sorbus*; medlar rust.

G. cunninghamianum Barclay 1890; telia on *Cupressus*, teliospores 56–72 × 25–30 μm; aecia on *Cotoneaster, Pyrus*; China (W.), India (N.).

G. fuscum DC. 1805; Laundon, *Descr.F.* 545, 1977; telia on *Juniperus* spp. of *sabina* group, teliospores 42–65 × 22–32 μm; aecia on pear, *Pyrus* spp., European pear rust or pear trellis rust; Africa (N.), Asia (W.), Canada (British Columbia), Europe, USA (California); Ormrod et al., *Can.J.Pl.Path.* 6:63, 1984, epidemiology, cv. susceptibility & chemical control [63,3983].

G. globosum (Farlow) Farlow 1886; Laundon, *Descr.F.* 546, 1977; telia on *Juniperus* as above, teliospores 35–40 × 17–24 μm; aecia mostly on *Crataegus*, also apple, *Malus* spp., pear, *Sorbus*; American hawthorn rust, Canada (E.), USA (Alaska, N.E.).

*G. juniperi-virginianae Schwein. 1822; Laundon, *Descr.F.* 547, 1977; telia on *Juniperus* of *sabina* group, teliospores 45–65 × 15–21 μm; aecia on apple and crab apple; American apple rust, the most serious disease caused by a *Gymnosporangium* in USA, also in Canada (E.).

G. libocedri (Henn.) Kern 1908; Laundon, *Descr.F.* 548, 1977; telia on *Calocedrus decurrens*, teliospores 20–95 × 18–30 μm; aecia on apple, *Malus* spp., pear, other Rosaceae; witches' brooms may be formed on telial hosts; may be severe on pear and quince if alternate host is close.

G. nelsonii Arthur 1901; fide Ziller 1974; telia on *Juniperus horizontalis, J. scopulorum*, teliospores 50–65 × 18–26 μm; aecia on *Amelanchier alnifolia, Malus diversifolia*, poss. other Rosaceae; Nelson's juniper rust, N. America; a common disease in Canada (British Columbia, S.) but control prob. not justifiable.

G. nidus-avis Thaxter 1891; fide Ziller 1974; telia on *Juniperus* of group *sabina*, teliospores 39–55 × 16–26 μm; aecia on *Amelanchier alnifolia, A. pumila, Cydonia oblonga, Sorbus* spp., juniper broom rust, causes witches' brooms on *J. horizontalis, J. scopulorum*, N. America, significant damage to junipers may occur.

G. tremelloides Hartig 1882; Laundon, *Descr.F.* 549, 1977; telia on *Juniperus communis*, spp. of *oxycedrus* group, teliospores 40–60 × 20–30 μm; aecia mostly on apple, *Sorbus*; European apple rust or common juniper gall rust; Africa (N.W.), N. America (W.), Europe but not Britain, can cause severe defoliation on apple in Europe.

G. yamadae Miyabe ex Yamada 1904; Laundon, *Descr.F.* 550, 1977; telia on *Juniperus chinensis*, teliospores 32–45 × 15–24 μm; aecia on apple, *Malus* spp.; Japanese apple rust, China, Japan, Korea; Sakuma, *Ann.phytopath.Soc.Japan* **51**:139, 1985, 2 races delineated by inoculation on apple [64,5440].

gynura latent v. Semancik & Weathers, *Virology* **36**:326, 1968; Weathers et al. 1974; rod 685 × 17 nm, USA, str. of chrysanthemum B v. [48,2240; **56**,5043].

gypsophila stunt Ulrychová et al., *Biologia Pl.* **25**:385, 1983; MLO assoc., Czechoslovakia [63,3398].

gyrosa, *Cryphonectria*, see *C. cubensis*; *Endothia*.

Gyrostroma austro-americanum, teleom. *Thyronectria austro-americana*.

H

hadacidin, from *Penicillium* spp.; Gray et al., *Pl.Physiol.* **39**: 204, 1964, induces dwarfing in plants without toxic effects.

hadromycosis, the synonymous term vascular wilt is preferred, see FBPP.

haematococca, *Nectria*.

Hainesia Ell. & Sacc. 1884; Coelom.; Sutton 1980; conidiomata cupulate, conidiogenesis enteroblastic phialidic; con. aseptate, hyaline, often curved fusiform; *H. lythri*, teleom. *Pezizella oenotherae*.

hair sprout, potato, see spindling sprout.

hairy root (*Agrobacterium*, particularly *A. rhizogenes*).

halacrinate, curative fungicide for cereals, not systemic.

Hales, Stephen, 1677–1761; an English gardener who recorded hop mildew (*Pseudoperonospora humuli*) in England in 1727; he referred to the transmission of canker diseases in budding; fide Whetzel 1918, Ainsworth 1981.

half leaf test, see local lesion assay.

—— life, used of plant virus inoculum for the time taken to lose half its infectivity on ageing, or in heat inactivation studies; also used to express the progressive decline in transmission of some viruses by their vectors as the period between acquisition and inoculation feeding increases, FBPP.

halo blight (*Pseudomonas syringae* pv. *coronofaciens*) oat, (*P.s.*pv. *garcae*) coffee, (*P.s.*pv. *phaseolicola*) legumes.

halstedii, *Plasmopara*.

Halticus tibialis, transm. sweet potato little leaf.

Halyomorpha, transm. paulownia witches' broom.

hamatum, *Trichoderma*.

Hansen, Hans Nicholas, 1891–1960; born in Denmark, Univ. California Berkeley, fungal variation especially in *Fusarium*, diseases of ornamentals and trees, worked on endosepsis q.v. *Phytop.* **52**:969, 1962.

Hapalosphaeria H. Sydow 1908; Coelom.; Sutton 1980; conidiomata pycnidial, conidiogenesis enteroblastic phialidic; con. aseptate, hyaline; monotypic. *H. deformans* (Syd.) Syd.; con. 3.5–4 μm diam.; stamen blight, on anthers of *Rubus*, N. America, parts of Europe, can be serious in S.E. Scotland on raspberry; Wilson, *TBMS* **7**:84, 1921; Zeller & Braun, *Phytop.* **33**:136, 1943, on blackberry; Dickens, *AAB* **60**:343, 1967; *TBMS* **51**:519, 1968, epidemiology & histology on raspberry [**22**:318; **47**,1209; **48**,226].

Haplobasidion Jakob Eriksson 1889; Hyphom.; Ellis, *Mycol.Pap.* 67, 1957; conidiomata hyphal, conidiogenesis holoblastic; con. aseptate, pigmented, catenate. *H. musae* M.B. Ellis 1957; Ellis & Holliday, *Descr. F.* 496, 1976; con. 4–6 μm diam., verruculose; diamond or Malayan leaf spot of banana, Asia (S.E.).

hardcore, sweet potato, tuber, a disorder induced by exposure to cold; Daines et al., *Phytop.* **64**:1459, 1974; **66**:582, 1976 [**54**,1988; **55**, 6053].

hard fruit, citrus, B deficiency, Morris, *J.Pomol.hort.Sci.* **16**: 167, 1938 [**17**:744].

—— rot (*Monilinia oxycocci*) cranberry, (*Septoria gladioli*) gladiolus.

haricot bean, see bean.

harknessii, *Endocronartium*.

Harris, Ralph Vernon, 1898–1980; born in England, Imperial College of Science, East Malling Research Station; a pioneer in the pathology of fruit and hop in UK, particularly of virus diseases of raspberry and strawberry. *AAB* **98**:167, 1981.

Hartig, Heinrich Julius Adolph Robert, 1839–1901; born in Prussia, Univ. Berlin, Marburg; chairs botany, Forest Academy, Eberswalde; Forest Experiment Station, Munich, classic texts on forest pathology: *Wichtige Krankheiten der Waldbäume*, 1874, in *Phytopath. Classics* 12, 1975, English translation by W. Merrill et al., *Important diseases of forest trees*; *Lehrbuch der Baumkrankheiten*, 1882, edn. 2, 1889, English translation by W. Somerville, revised by H.M. Ward, *Textbook of the diseases of trees*, 1894. *Phytop.* **5**:1, 1915; Whetzel 1918.

hartigii, *Hypodermina*.

harzianum, *Trichoderma*.

Hassall, Arthur Hill, 1817–94; born in England, doctor, naturalist, a founder of modern sanitary science. His biographer, E.A. Gray, *By candlelight: the life of Dr Arthur Hill Hassall 1817–94*, 1983, described how Hassall published observations, including inoculations, on potato blight in 1841–3. He considered that the observed fungus caused the disease and thus, apparently, anticipated Berkeley by 2–3 years. Hassall is not mentioned by the 3 historians of plant pathology: Whetzel 1918, Large 1940, Ainsworth 1981. Hassall wrote his autobiography: *The narrative of a busy life*, 1893. Desmond, *Dictionary of British and Irish botanists and horticulturalists*, 1977.

haustorium, of a fungus, a hyphal branch which

penetrates a host cell, does not cause lethal injury and presumably functions as an absorber of nutrients; Bushnell, *A.R.Phytop.* **10**:151, 1972; Harder & Chong in Bushnell & Roelfs ed. *The cereal rusts*, vol. 1, 1984.

havanensis, *Cryphonectria*, see *C. cubensis.*

hawaiiensis, *Bipolaris*, *Cochliobolus.*

hawthorn, common, *Crataegus oxycantha*; —— **ring pattern** Posnette 1957, see Cooper 1979.

hayi, *Cercospora.*

hazel, *Corylus*; Granata, *Inftore.fitopatol.* **35**(4) 19, 1985.

—— **dieback** Corte & Pesante, *Riv.Patol.veg.* **3**:177, 1963; transm. sap but not to hazel, Italy [**43**,1753].

—— **line pattern** Scaramuzzi & Ciferri 1957; Ragozzino et al., *Riv.Patol.veg.* **7**:83, 1971; MLO assoc., Italy [**37**:322; **50**, 3198].

—— **mosaic** Marénaud & Germain and Cardin & Marénaud, *Annls. Phytopath.* **7**:133, 159, 1975; Ragozzino, *Acta phytopath. Acad. Sci. hung.* **15**: 375, 1980; sap transm.; France, Italy [**55**,5366–7; **61**,1836].

—— **ring pattern** Cameron abs. *Proc.Am.phytopath.Soc.* **4**:202, 1977; isometric 35–57 nm diam., transm. sap, USA.

HC toxin, formed by *Cochliobolus carbonum*; Scheffer & Ullstrup, *Phytop.* **55**:1037, 1965; Walton & Earle, *Phytopath.Mediterranea* **24**:319, 1985 [**45**,438; **66**,975].

head atrophy, globe artichoke, caused by high temp. during apical bud transition and differentiation; Magnifico et al., *Inftore.fitopatol.* **35**(9):41, 1985 [**65**,3651].

—— **blight**, a general term for moulding or rotting of the cereal and grass spike; (*Alternaria helianthi*) sunflower, (*Gibberella gordonia*) cereals; —— **rot** (*Rhizopus oryzae*) sunflower; —— **smut** (*Sphacelotheca destruens*) common millet, (*Sporisorium reilianum*) maize, sorghum, (*Ustilago bullata*) temperate grasses, (*U. crameri*) foxtail millet.

Heald, Frederick DeForest, 1872–1954; born in USA, Univ. Wisconsin, Leipzig; worked at: Parsons College, Fairfield, Iowa; Univ. Nebraska, Texas and Washington State College; wrote the wide ranging and thorough text, *Manual of plant diseases* 1926, edn. 2, 1933; *Introduction to plant pathology* 1937, edn. 2, 1943. *Phytop.* **45**:409, 1955; *Science N.Y.* **121**:279, 1955; *A.R.Phytop.* **21**:13, 1983.

heart and butt rot (*Ganoderma tornatum*) oil palm.

heart rot, a decay of the heartwood, sometimes extended for a decay in the centre of a plant organ, e.g. a tuber; (*Inonotus hispidus*) ash, (*Phellinus pomaceous*) plum, Rosaceae,

(*Phytomonas*) coconut, (*Phytophthora nicotianae* var. *parasitica*) pineapple, (*Piptoporus betulinus*) birch; beet, B deficiency.

heartwood decay (*Inonotus glomeratus*) *Acer, Fagus.*

heat spot, strawberry; Smeets & Wassenaar, *Euphytica* **5**:51, 1956, at 23° virus indicator plants developed heat spot which was not caused by viruses, less distinct at 17° [**35**:619].

—— **therapy**, for control of viruses, see tissue culture; Nyland & Goheen, *A.R.Phytop.* **7**:331, 1969.

Hedera, ivy.

hederae, *Calonectria*, see *C. pyrochroa.*

Hedgcock, George Grant, 1863–1946; born in USA, Univ. Nebraska, Washington St Louis; USDA, important work on forest fungi, particularly conifer rusts. *Phytop.* **37**:603, 1947.

hedgcockii, *Leptostroma*, *Ploioderma.*

helenium green petal Begtrup, *Phytopath.Z.* **82**:356, 1975; MLO assoc., Denmark [**55**,769].

—— **S v.** Kuschki et al. 1978; Koenig & Lesemann, *Descr.V.* 265, 1983; Carlavirus, filament *c.*670–700 nm long, transm. sap, *Myz.pers.*, non-persistent.

—— **yellows** Begtrup, as for cirsium yellows.

—— **Y v.** Kuschki et al., *Phytop.* **68**:1407, 1978; Potyvirus, filament 720 nm long, transm. sap [**58**,3873].

helianthi, *Alternaria, Diaporthe, Phomopsis, Puccinia, Septoria.*

helianthicola, *Septoria*, see *S. helianthi.*

Helianthus annuus, sunflower; *H. tuberosus*, Jerusalem artichoke.

helicobasidin, toxin from *Helicobasidium mompa*; Takai, *Phytopath. Z.* **43**:175, 1961; *Bull.Govt.For.Exp.Stn.Tokyo* 195, 1966 [**41**,672].

Helicobasidium Pat. 1885; Auriculariales; basidiocarp resupinate of ± loosely interwoven hyphae, metabasidia emergent, not readily detached, typically circinnately coiled apically, mycelial mats often purplish; soil inhabitants, sometimes parasitic on roots of plants, diseases called violet root rot; Holliday 1980.

H. brebissonii (Desm.) Donk 1958; sclerotial state *Rhizoctonia crocorum* DC.: Fr.; Donk, *Persoonia* **4**:156, 1966; *H. purpureum* Pat. is a synonym fide Donk; Valder, *TBMS* **41**:283, 1958, biology; Hering, *ibid* **45**:488, 1962, hosts; *Leaflet Min.Agric.Fish.Fd. UK*, 346, 1980 [**38**:66; **42**:304; **60**,5165].

H. compactum (Boedijin) Boedijin 1930; mostly on tropical and subtropical trees; Reid, *TBMS* **64**:159, 1975, described the first specimen from Europe and discussed *Helicobasidium purpureum*, see above, and *H. mompa* [**54**,2705].

H. mompa Tanaka 1891; Itô, *Bull.Govt.For.Exp.Stn.Tokyo* **43**, 1949; Asia (E.) [**30**:337].

Helicotylenchus dihystera (Cobb) Sher 1961; Siddiqi, *Descr.N.* 9, 1972.

H. multicinctus (Cobb) Golden 1956; Siddiqi, *Descr.N.* 23, 1973; causes banana decline.

H. pseudorobustus (Steiner) Golden 1956; Fortuner, *Descr.N.* 109, 1985; semi-endoparasitic on roots, cosmopolitan, may require control.

helleborus ringspot Vukovits, *Pflanzenarzt* **11**:105, 1958; Austria [**38**:85].

hell fire (*Curtobacterium flaccumfaciens* pv. *oorti*) tulip.

helminthosporal, toxin from *Cochliobolus sativus*, Casinovi in Wood et al. 1972.

Helminthosporium Link 1809; Hyphom.; Ellis 1971, 1976; Carmichael et al. 1980; conidiomata hyphal, conidiogenesis tretic enteroblastic; con. multiseptate, pigmented; many spp. have been transferred to *Drechslera*, its 2 segregates *Bipolaris* and *Exserohilum*, and other genera. Many plant pathologists, unfortunately, still persist in using *Helminthosporium* for *Drechslera*. The former differs from the latter in that the conidiogenous cells are determinate, with the con. developing through pores beneath the septa whilst the conidiophore tip is actively growing; conidiophore growth ceases when the terminal conidium is formed; the conidiophores are never geniculate, i.e. bent like a knee, and never have dark scars, Ellis, *TBMS* **62**:228, 1974.

H. solani Durieu & Mont. 1849; Ellis, *Descr.F.* 166, 1968; Hughes, *Fungi Canadenses* 236, 1982; con. not catenate, av. 39 × 9.4 µm; potato silver scurf, most conspicuous after storage, a superficial blemish; Read & Hide, *Potato Res.* **27**:145, 1984, effects on seed potatoes [**65**,1479].

helminthosporoside, toxin from *Bipolaris sacchari*; Steiner & Byther, *Phytop.* **61**:691, 1971; **66**:423, 1976 [**51**,628; **56**,834].

Helochara communis, transm. Gram negative, xylem limited bacteria q.v.

Helotiaceae, Helotiales; excipulum usually of parallel hyphae, sclerotia lacking; *Atropellis*, *Cenangium*, *Chloroscypha*, *Crumenulopsis*, *Godronia*, *Gremmeniella*, *Grovesiella*, *Tympanis*.

Helotiales, Ascomycotina, inoperculate discomycetes, hymenium exposed before asci are mature, asci usually thickened at apex and with an apical pore through which the asco. are discharged.

helper virus, one which must be present for the transm. of a second, i.e. a dependent virus, by its vector, FBPP; see dependent transmission. The term is also used for a virus which is necessary

for the multiplication of a satellite virus q.v. in a plant cell.

Helvellaceae, Pezizales; ascomata large, stipitate, shaped like a cup or saddle, asco. ellipitical to fusiform with large oil drops, usually 1–2; *Rhizina*.

Hemibasidiomycetes, Teliomycetes, rusts and smuts.

Hemicriconemoides chitwoodii Esser 1960; Siddiqi, *Descr.N.* 41, 1974; ectoparasitic, on *Camellia japonica* roots.

H. mangiferae Siddiqi 1961; *Descr.N.* 99, 1977; ectoparasitic on roots, tropical and subtropical fruit crops, damages litchi in South Africa.

Hemicycliophora arenaria Raski 1958; Franklin & Stone, *Descr.N.* 43, 1974; sheath nematode, causes galls on root tips of rough lemon and other crops, USA.

Hemileia Berk. & Broome 1869; Uredinales; Gopalkrishnan, *Mycologia* **43**:271, 1951; uredia and telia only, autoecious, mostly on Rubiaceae, tropical Africa, Asia; characterised by discrete, hypophyllous uredosori formed separately from each leaf stoma, uredospores clustered on short pedicels, ± reniform, strongly warty or spiny on convex face; teliospores clustered on short pedicels, aseptate, ± spherical, smooth, no germ pores; Rajendren, *Mycopath.Mycol.appl.* **47**:81, 1972, development & parasitism in 17 spp. [**30**:345, 583; **52**,45].

H. coffeicola Maubl. & Roger 1934; Laundon & Waterston, *Descr.F.* 2, 1964; uredospores 34–40 × 20–28 µm, teliospores 20–26 µm diam., differs from *Hemileia vastatrix* in that: sori scattered over leaf surface instead of being confined to distinct spots; few feeder hyphae which swell to 20–30 µm diam. in passing from mesophyll to substomatal chamber; the uredospores have larger but fewer spines; coffee rust, Africa (W. and central), not a serious disease, if potentially so; requires wetter conditions than *H. vastatrix*; Saccas, *Bull.Inst.Fr.Café Cacao* **11**:5, 1972 [**52**,3700].

H. vastatrix Berk. & Broome 1869; Laundon & Waterston, *Descr.F.* 1, 1964; uredospores 28–36 × 18–28 µm, teliospores 20–28 µm diam., role of basidiospores unknown, differs from *Hemileia coffeicola* in that: sori confined to yellow orange spots on leaves; numerous, narrow feeder hyphae forming an interwoven mass in the substomatal chamber; uredospores have smaller and more spines; coffee rust, a major disease which causes extensive premature defoliation. The rust is found on wild coffee, the only host, in Ethiopia but the first report of its destructiveness came from Sri Lanka, then Ceylon, in 1867–8. The consequent epidemic reduced coffee yields in this island from c.4.5 cwt/acre to 2 cwt/acre by 1878; the industry never recovered and tea was substituted for coffee. The historically important papers by H. M. Ward 1881–2, see Large 1940, were among the first to show that a fungus could cause the destruction of a plant crop and an economic catastrophe. *H. vastatrix* subsequently spread rapidly in Africa and Asia; in 1970 it was discovered in Brazil and is now widespread in S. and central America; largely a disease of arabica coffee; there are many races, at least 30 with c.6 genes for resistance; the less frequently cultivated *Coffea canephora* has some resistance; at present control is through fungicides; reviews: Holliday 1980; Waller, *Crop Protect.* **1**:385, 1982; Bergam in Filho & Paiva, *Summa Phytopath.* **9**:155, 1983; Fulton ed., *Coffee rust in the Americas*, 1984; Schieber & Zentmyer, *PD* **68**:89, 1984, in Americas.

hemlock, *Conium maculatum* q.v.; *Tsuga*.

Hemmi, Takewo, 1889–1949; born in Japan, Univ. Tohoku, Hokkaido; chairs at Kyoto and Osaka; pioneer work on rice blast (*Pyricularia oryzae*). *Phytop.* **50**:687, 1960.

hemp, common, *Cannabis sativa*; Barloy & Pelhate, *Annls.Épiphyt.* **13**:117, 1962; see: banana for Manila hemp, *Hibiscus*, sunnhemp [42,386].

hen and chickens, see milleranderage.

henbane mosaic v. Hamilton 1932; Govier & Plumb, *Descr.V.* 95, 1972; Potyvirus, filament 800 or 900 × 12–13 nm, transm. sap, aphid spp., non-persistent; England, Germany, Italy, but may be widespread in weeds; *Datura* spp., *Hyoscyamus niger*, mainly in Solanaceae.

Henderson spot, see apple ringspot.

Hendersonula Speg. 1880; Coelom.; conidiomata stromatic, subcuticular, multilocular to convoluted but with a single ostiole, conidiophore growth percurrent, conidiogenesis holoblastic; con. 3 septate, pale brown. *H. toruloidea* Nattrass 1933; Punithalingam & Waterston, *Descr.F.* 274, 1970; con. formed from phialides and other differences; requires transfer, fide Sutton 1980; cankers, diebacks, woody hosts.

henningsii, *Cercosporidium, Mycosphaerella*.

heraclei, *Erysiphe*.

Heracleum sphondylium, cowparsnip, hogweed; Wolf, *Acta phytopath.Acad.Sci.hung.* **7**:353, 1972, viruses [53,1265].

heracleum latent v. Murant & Goold 1972; Bem & Murant, *Descr.V.* 228, 1980; Closterovirus, filament 730 × 12 nm, transm. sap to 39 spp. in 11 families of dicotyledons, and *Cavariella* spp., no symptoms in cowparsnip; infected carrot, celery, parsley, parsnip, causes interveinal yellowing in coriander.

—— **yellow vein banding** Rønde Kristensen 1963, Denmark, cowparsnip [43,1514].

—— **1–6 v.** Bem & Murant, *AAB* **92**:237, 1979; transm. sap and/or by *Cavariella* spp.; HV1 was parsnip yellow fleck v.; HV2 was heracleum latent v.; HV3 and HV5 transm. sap, aphid spp.; HV4 transm. sap; HV6 transm. aphid spp., poss. Closterovirus, particle up to 1400 nm long, distinct from HV2; all viruses could infect cultivated umbelliferous spp. but only PYFV is known to infect such crops under natural conditions; Scotland, cowparsnip [59,1146].

herbarum, *Cladosporium, Pleospora, Stemphylium*.

herbicides, effect on plant diseases; Grinstein et al., *Phytop.* **66**:517, 1976, effect of dinitroaniline on resistance to soilborne pathogens; Altman & Campbell, *A.R.Phytop.* **15**:361, 1977; Grinstein et al., *Physiol.Pl.Path.* **24**:347, 1984, induced resistance to wilt pathogens [56,122; 63,4819].

herbicola, *Erwinia*.

Herpotrichia Fuckel 1868; Pleosporaceae; Sivanesan 1984; ascomata on a subiculum of brown hyphae, asco. eventually pale to dark brown, 1–many septate, often with a gelatinous sheath.

H. coulteri (Peck) Bose 1961; Sivanesan & Gibson, *Descr.F.* 327, 1972; asco. 1 septate, 20–28 × 7–10 μm; pine brown felt blight, N. America, parts of Europe, behaves as a snow mould, develops in needles of lower branches and seedlings, survives through stromata.

H. juniperi (Duby) Petrak 1925; Sivanesan & Gibson *Descr.F.* 328, 1972; asco. becoming 3–4 septate, 25–34 × 8–12 μm; conifer black snow mould; thick, brownish, mycelial mats which block stomata; young trees and seedlings most susceptible; Bazzigher, *Eur.J.For.Path.* **6**:109, 1976, in Switzerland with *Herpotrichia coulteri*; Scharpf, *PD* **70**:798, 1986, colonising dwarf mistletoe, *Arceothobium abietinum* [55,5372; 66,730].

herpotrichioides, *Rosellinia*.

herpotrichoides, *Leptosphaeria, Pseudocercosporella*.

Heterobasidion Bref. 1888; Polyporaceae; *H. annosum* (Fr.) Bref.; anam. *Oedocephalum lineatum* Bakshi 1950, fide Pegler & Waterston, *Descr.F.* 192, 1968; formerly in *Fomes*, see Cannon, *Microbiological Sci.* **3**:286, 1986, for differences between these 2 genera. The fungus causes the important conifer butt and root rot, N. temperate; air dispersed basidiospores mostly infect stumps but can infect roots, spreads also through root contact, growth in roots is c.1–2 m/year; dying trees are seen 1–3 years after thinning, mature trees can be killed. The disease is less important on hardwoods but the pathogen may spread from these if a conifer crop follows. Stump infection is reduced by competing fungi and distance from a spore source. The standard control is to treat, i.e. inoculate, freshly cut stumps with oidia of the competitor *Peniophora gigantea*. This method of biological control sensu stricto is the oldest and the most well established example of control of a fungus pathogen by another fungus; it was developed by J. Rishbeth; see major texts and: Hodges, *A.R.Phytop.* **7**:247, 1969, infection & spread; McAree, *Irish Forestry* **32**:118, 1975, review; Rishbeth, *Eur.J.For.Path.* **9**:331, 1979, control & *Armillaria mellea*; Greig, *ibid* **14**:392, 1984, management of infected pine in England (E.) [**59**,4328; **64**,2752].

Heterococcus rehi, transm. rice chlorotic streak.

Heterodera, cyst nematodes.

H. avenae Wollenweber 1924; Williams & Siddiqi, *Descr.N.* 2, 1972; cereal cyst nematode Gramineae, causes molya of barley and wheat in India.

H. cacti Filipjev & Schuurmans Steckhoven 1941; Mulk, *Descr.N.* 96, 1977; likely to occur on roots of cactus where grown as ornamentals.

H. carotae Jones 1950; Mathews, *Descr.N.* 61, 1975; carrot cyst nematode, restricted to this crop, causing sickness, plant growth is poor and patchy, leaves are yellowish red.

H. cruciferae Franklin 1945; Stone & Rowe, *Descr.N.* 90, 1976; on brassicas, Europe.

H. glycines Ichinohe 1952; Burrows & Stone, *Descr.N.* 118, 1985; soybean yellow dwarf, a serious disease of this crop, causes severe losses, limited host range in Leguminosae; China, Egypt, Japan, Korea, Taiwan, USA; Sinclair 1982.

H. goettingiana Liebscher 1892; Stone & Course, *Descr.N.* 47, 1974; pea cyst nematode, temperate legumes.

H. humuli Filipjev 1934; Stone & Rowe, *Descr.N.* 105, 1977; hop cyst nematode; Moraceae, Urticaceae; weight reduction in hops shown experimentally.

H. oryzae Luc & Berdon 1961; Luc & Taylor, *Descr.N.* 91, 1977; rice cyst nematode, can reduce yields in this crop; Ou 1985.

H. pallida Stone 1973; *Descr.N.* 17, 1973; *Heterodera rostochiensis* Wollenweber 1923; Stone, *Descr.N.* 16, 1973; Williams & Bridge in Johnstone & Booth 1983, cite these potato cyst nematodes in *Globodera*; they cause poor growth, chlorosis, a decrease in tuber size, cysts occur on roots; Hooker 1981 tabulated the nematodes recorded on potato, on other Solanaceae; Rich 1983.

H. sacchari Luc & Merny 1963; Luc, *Descr.N.* 48, 1974; can cause stunting in sugarcane.

H. schachtii Schmidt 1871; Franklin, *Descr.N.* 1, 1972; beet cyst nematode, mainly on Chenopodiaceae, Cruciferae; Whitney & Duffus 1986.

H. trifolii Goffart 1932; Mulvey & Anderson, *Descr.N.* 46, 1974; clover cyst nematode, on many temperate legumes; see Maas & Heijbroek 1982 in Whitney & Duffus 1986 for the sugarbeet race.

heteroecious, of a fungus, having a life cycle, i.e. different parasitic stages, on 2 unlike hosts, as in the rusts, cf. autoecious.

heterogenetics, see genetic engineering.

heterokaryosis, forming a heterokaryon of a fungus, the condition of a hypha or cell having 2 or more genetically distinct haploid nuclei, cf. dikaryon a form of a heterokaryon; Buxton in Horsfall & Dimond 1960; Parmeter et al., *A.R.Phytop.* **1**:51, 1963; Davis in Ainsworth & Sussman, vol. 2, 1966.

heterologous encapsidation, or heterologous coating, where the nucleic acid of one virus becomes enclosed within a capsid or coat protein derived wholly or partly from another virus; such a particle is a complementation variant. Rochow, *A.R.Phytop.* **10**:117, 1972, illustrated 2 types: (1) trancapsidation or genome masking, where the genome of one virus becomes wholly coated with the capsid protein of another; (2) phenotypic mixing where each of 2 genomes is coated with protein from both viruses.

—— **reaction**, a serological reaction in which an antiserum is reacted against an antigen other than the one used in its preparation.

heteronemum, *Cylindrocarpon*; **heteropogoni**, *Peronosclerospora*; **heterosporum**, *Fusarium*; **heterostrophus**, *Cochliobolous*.

heterothallism and homothallism, in fungi; the terms were first used by Blakeslee in 1904. In the latter situation a single genome, a haploid nucleus in a single spore, can give rise to a mycelium which can reproduce sexually. In the former situation such reproduction only occurs between genetically different mycelia. In morphological heterothallism there are 2 distinct strs. one forming male, antheridia, organs and the other female, oogonia, organs. In physiological heterothallism each str., or mating type, bears both organs and is morphologically indistinguishable from the other; or the sexual organs are absent. The Ascomycotina usually show the first type of physiological heterothallism, as do some Mastigomycotina; the second type occurs in the Basidiomycotina. The simplest genetical control of heterothallism is called bipolar where the fungus has a pair of allelomorphs usually designated + and − or A and a; sexual reproduction only occurs when there is juxtaposition of an A and an a str. In the Hymenomycetes more complex mating types occur where > 2 allelomorphs operate to enable complex mating systems to arise; e.g. the genetically differing mycelia, forming distinct populations which are not morphologically distinct, as in *Armillaria* and *Thanatephorus* spp. Fungi also vary genetically through the phenomena of heterokaryosis and the parasexual cycle q.v.; Burnett 1976; Barrett, *Pl.Path.* **35**:158, 1986, statistical test for heterothallism & mating type in plant pathogenic fungi [**65**,5414].

heterotrophic, requires organic nutrient material for growth; fungi have a heterotrophic nutrition, cf. autotrophic.

Hevea brasiliensis, rubber, Para rubber.

heveae, *Ascochyta, Drechslera, Oidium, Phytophthora*.

hexachlorobenzene, seed dressing against wheat smuts, selective fungicide.

hibernalis, *Phytophthora*.

Hibiscus, *H. cannabinus*, Ambari and Deccan hemps, kenaf; *H. rosa-sinensis*, shoeflower; *H. sabdariffa*, roselle.

*hibiscus chlorotic ringspot v. Waterworth, Lawson & Monroe 1976; Waterworth, *Descr.V.* 227, 1980; isometric *c*.28 nm diam., contains RNA, transm. sap, USA, in many *H. rosa-sinensis* cvs., infected 12 spp. of Malvaceae including *H. cannabinus*.

—— latent ringspot v. Brunt et al. 1980; *Descr.V.* 233, 1981; Nepovirus, isometric *c*.28 nm diam., sediments as 3 components, 2 contain ssRNA, transm. sap, Nigeria (W.), *H. rosa-sinensis*, transm. to 22 spp. in 7 families.

—— leaf curl Mukherjee & Raychaudhuri, *Indian J.Hortic.* **21**:176, 1961; *Malvaviscus arboreus*; in 1951 a leaf curl was reported from *H. rosa-sinensis*, both India; Yassin & Nour, *AAB* **56**:207, 1965; Sudan, *Abelmoschus esculentus* [**31**:541; **45**,873].

—— ringspot Lana, *PDR* **58**:1040, 1974; *H. rosa-sinensis*, transm. sap to *H. cannabinus*, Nigeria [**54**,2846].

—— shoeflower mosaic Hendrix & Murakishi, *Bienn.Rep.Hawaii Univ.agric.Exp.Stn.* 1948–50:123, 1951; Verma, *Zentbl.Bakt.ParasitKde* 2 **131**:122, 1976, India (N.) [**55**,5232].

—— vein yellowing Plavšić, Erić & Miličić, *Phytopath. Mediterranea* **23**:52, 1984; bacilliform particle assoc., Canary Islands, Greece; *H. rosa-sinensis* [**64**,1615].

*—— witches' broom Vicente, Caner & July, *Arq.Inst.biol.S.Paulo* **41**:53, 1974; Caner et al., *Summa Phytopathologica* **3**:155, 1977; bacilliform

particle and MLO assoc., Brazil (São Paulo) [**54**,2288; **57**,3487].

—— yellow vein mosaic Capoor & Varma, *Indian J.agric.Sci.* **20**: 217, 1950; transm. *Bemisia tabaci*, India (Maharashtra), *Abelmoschus esculentus* [**31**:6].

Hickman, Clarence James, 1914–80; born in England, Univ. Birmingham, determined etiology of strawberry red core (*Phytophthora fragariae*), became an authority on this genus; chair, Univ. Western Ontario 1960. *TBMS* **79**:187, 1982.

hickory wood, *Carya*.

higginsianum, *Colletotrichum*.

highbush blueberry, *Vaccinium corymbosum*.

himalayense, *Cronartium, Peridermium*.

hinnulea, *Armillaria*.

hippeastrum latent v. Brölman-Hupkes, *Neth.J.Pl.Path.* **81:226, 1975; Jayasinghe & Dijkstra, *ibid* **85**:47, 1979; Brunt & Phillips 1979; rod *c*.600–650 × 13 nm, transm. sap, Netherlands, *H. hybridum* [**55**,3204; **58**,5402; **59**,3542].

—— mosaic v. Kunkel 1922; Brunt, *Descr.V.* 117, 1973; Potyvirus, filament 750 × 12 nm, transm. sap, aphid spp., occurs naturally in Amaryllidaceae.

—— streak van Velsen, *Papua New Guin.agric.J.* **19**:13, 1967; transm. sap, *H. vittatum* [**47**,2495].

Hirschmanniella oryzae (van Breda de Haan) Luc & Goodey 1964; Siddiqi, *Descr.N.* 26, 1973; rice root nematode, endoparasite, causes root necrosis and poor growth; Ou 1985.

H. spinicaudata (Schuurmans-Stekhoven) Luc & Goodey 1964; Luc & Fortuner, *Descr.N.* 68, 1975; mainly recorded parasitising flooded rice in Africa, also in USA (California), Venezuela.

Hirsutella Pat. 1892; Hyphom.; Carmichael et al. 1980; conidiomata hyphal to synnematal, conidiogenesis enteroblastic phialidic; con. aseptate, hyaline; on insects. *H. citriformis* Speare 1920; Brady, *Descr.F.* 607; poss. use for control of *Nilaparvata lugens*, vector of rice grassy stunt v. *H. thompsonii* Fisher 1950; Brady, *Descr.F.* 608, 1979, on mites; Minter et al., *TBMS* **81**:455, 1983, 3 spp. inter alia from *Abacarus hystrix*, vector of ryegrass mosaic v.

hirsutum, *Penicillium*; **hirta**, *Wojnowicia*.

Hishimonides chinensis, transm. ziziphus witches' broom; *H. sellatiformis*, transm. mulberry dwarf.

Hishimonus concavus, transm. loofah witches' broom; *H. phycitis*, transm, eggplant little leaf; *H. sellatus*, transm. mulberry curly little leaf, mulberry dwarf.

hispidus, *Inonotus*.

history, the known records of experimental plant pathology appear to begin in the middle of the 18th century. The early landmarks interlock, and the events interact, with the great explosion of discovery in biological science. In France first Tillet in 1755, and then Prévost in 1807, gave the first experimental proofs that fungi caused disease in plants, in man's crops. In *c*.1842 Darwin had written down the outlines of his theory of evolution. Three years later the Irish potato crop was destroyed by disease, the potato blight. There followed the greatest of the famines in Ireland; the consequent social disasters, economic and political effects. These events took place at a time when the germ theory of disease had not been proven and when the theory of spontaneous generation was widely held. The close association of a fungus, now called *Phytophthora infestans*, with potato blight led Berkeley to provide support for the first theory and to begin demolishing the second. De Bary gave proof that *P. infestans* caused potato blight. His book, *Untersuchungen über die Brandpilze*, 1853, finally confirmed that fungi could cause diseases in plants. Later, in the 1860s, Pasteur utterly disproved the theory of spontaneous generation and, with Koch on anthrax in 1876, showed that disease in animals could be caused by bacteria. During this period, in 1858, Darwin and Wallace gave their joint paper on the theory of evolution to the Linnean Society of London. Mendel in Austria demonstrated the particulate nature of inheritance; but his publications (1865, 1869), and their enormous significance, lay hidden until 1900. They were unearthed by de Vries and 2 others independently.

Between 1880 and 1898 there were many notable events. Marshall Ward showed clearly how a fungus, coffee rust (*Hemileia vastatrix*) in Ceylon, could destroy a crop and cause an economic crisis. The use of Bordeaux mixture by Millardet in France led to the first widespread application of a fungicide. Arthur in USA worked on apple fireblight (now *Erwinia amylovora*) and gave the first adequate proof of a bacterial etiology for a plant disease. In Russia Ivanovski was the first to show that the disease known as tobacco mosaic was caused by an entity that passed through a bacterial filter. The rediscovery of Mendel's work led to the demonstration by Biffen (1904) in England that resistance in cereals to the fungus rusts was inherited in a Mendelian manner. Meanwhile Eriksson's work foreshadowed the widespread studies done later on the formae speciales and races in fungi. The Swiss plant pathologist Stahel, whose work in Surinam has been neglected by historians, determined the etiology of 2 tropical plant diseases whose great economic importance remains with us today. His

2 monographs are classics; they are on witches' broom of cacao (now *Crinipellis perniciosa*) in 1915, and South American leaf blight of rubber (now *Microcyclus ulei*) in 1917. In the 1930s Stahel gave good evidence that protozoa could cause disease in plants. But his work on coffee phloem necrosis, now known to be basically correct, was neglected for nearly 36 years.

The nature of the agent that caused tobacco mosaic was not satisfactorily determined for many years, and the work is closely linked with progress in biochemistry. Allard (1915–18) demonstrated its particulate nature; Stanley (1935) detected a protein; and finally Bawden and his co-workers (1936) showed that the agent, early on called a virus, was a nucleoprotein. In 1956 Schram and Fraenkel-Conrat found that the infectivity of tobacco mosaic virus lay in its RNA. Diseases in plants called yellows were also thought to be caused by viruses. The etiologic agents had been transmitted by using insects. One of the oldest of these was peach yellows noted by E.F. Smith in N. America in 1888. In 1967 Japanese workers detected organisms, like mycoplasmas, that were associated with these diseases. In 1898 such a micro-organism had been found to cause a disease in cattle. The first plant mycoplasma sensu lato to be characterised was *Spiroplasma citri* in 1973 by Saglio and others; it causes citrus stubborn. Diener in 1971 showed that disease in plants could be caused by naked RNA molecules or viroids, i.e. without a protein coat like a virus, e.g. see coconut cadang cadang. Recently several old diseases, whose cause was unknown, were found to be caused by bacteria which differ fundamentally from other plant bacteria. Such xylem limited bacteria were first shown to cause disease in plants in 1978. Nine years later a devastating wilt of cloves in Indonesia was shown to be caused by a similar bacterium.

Whetzel 1918; Large 1940; Kiett in Horsfall & Dimond, vol. 1, 1959; Holton et al. ed., *Plant pathology: problems and progress*, 1959; Ainsworth, *Introduction to the history of mycology*, 1976; Fuchs in Heitefuss & Williams 1976; Waterson & Wilkinson 1978; Harrison, *AAB* **94**:321, 1980, 25 years of plant virus research; Ainsworth 1981; Whitcomb & Black in Daniels & Markham 1982, mycoplasmas; Kommedahl & Williams 1983; Starr, *A.R.Phytop.* **22**:169, 1984, bacteria.

hobnail canker (*Endothia gyrosa*) oak, other temperate hardwoods.

Hoggan, Ismé Aldyth, 1899–1936; Univ. Cambridge, Wisconsin; important work on aphid transm. of viruses. *Phytop.* **27**: 1029, 1937.

holci, *Mycosphaerella*, *Tilletia*.

Holcus lanatus, velvet grass, Yorkshire fog; *H. mollis*, creeping soft grass.

holcus rhabdovirus Amici et al., *Riv.Patol.veg.* 14:85, 1978 [58,3730].

— streak v. Catherall & Chamberlain, *Pl.Path.* 24:247, 1975; poss. Potyvirus, filament *c.*750 × 12 nm, transm. sap, Wales [55,2771].

— transitory mottle v. Catherall & Chamberlain, *Pl.Path.* 24: 217, 1975; isometric *c.*28 nm diam., transm. sap, Wales; a str. of phleum mottle v. [55,2765].

hold over canker, one in which the pathogen survives the winter and from which it may reinfect healthy tissues in the spring. A term particularly used of fireblight cankers caused by *Erwinia amylovora*, FBPP.

hollow heart, pea, induced by water stress, high temp., premature drying; Don et al., *Seed Sci.Technol.* 12:707, 1984; Halligan, *AAB* 109:619, 1986; in potato, due to rapid growth and the consequent formation of oversize tubers, Rich 1983; broad bean, Harrison, *Pl.Path.* 25:87, 1976 [56,1350; 64,2799; 66,2584].

— stalk (*Erwinia carotovora* ssp. *carotovora*) tobacco.

hollyhock yellow mosaic Singh & Misra, *Indian Phytopath.* 24:213, 1971 [25:252; 51,2574].

Holmes' ribgrass = ribgrass mosaic v.

holmii, *Setosphaeria*, see genus.

Holobasidiomycetidae, Hymenomycetes; metabasidium is not divided by a primary septum but may become adventitiously septate.

holoblastic, see blastic.

holodiscus witches' broom Zeller, *Phytop.* 21:923, 1931; USA (Oregon, Washington State), *H. discolor* [11:110].

holomorph, the whole fungus; Hennebert & Weresub, *Mycotaxon* 6: 207, 1977, anamorph + teleomorph; Cowan 1978, all the character states of an organism. The anamorph is the imperfect or asexual state, the teleomorph is the perfect or sexual state; in nomenclature the latter takes precedence over the former. Tribe, *Bull.Br.mycol.Soc.* 17:94, 1983, holomorphic map.

holonecrotic, see necrotic.

Holton, Charles Stewart, 1904–80, born in USA; Univ. Louisiana State, Minnesota, an authority on smuts of temperate cereals; wrote, with G. W. Fisher, *Biology and control of the smut fungi*, 1957. *Phytop.* 71:663, 1981.

Homalodisca coagulata, transm. xylem limited bacteria of peach phony, periwinkle wilt, ragweed stunt.

homoeocarpa, *Sclerotinia*.

homokaryon, of a fungus, having genetically identical nuclei, cf. heterokaryon.

homologous reaction, a serological reaction in which an antiserum is reacted against the antigen used in its preparation.

homothallism, see heterothallism.

honey agaric, *Armillaria mellea*.

— dew, see *Claviceps*.

honeysuckle latent v. van der Meer, Maat & Vink 1980; Brunt et al. 1980; Brunt & van der Meer, *Descr.V.* 289, 1984; Carlavirus, filament *c.*650 nm long, transm. sap, *Hyadaphis foeniculi*, non-persistent; commonly in *Lonicera japonica*, *L. periclymenum* in Europe, symptomless in many cvs.; serologically related to poplar mosaic v. and shallot latent v.

— yellow vein mosaic (tobacco leaf curl v.), Osaki et al., *Ann.phytopath.Soc.Japan* 45:62, 1979; *Lonicera japonica* [59,2228].

Hooghalen, oat, Mg deficiency; Smit & Mulder, *LandbHogesch, Wageningen* 46(3), 1942 [25:105].

hook leaf, oil palm, poss. assoc. with B deficiency.

hop, *Humulus lupulus*; Glassock in Western 1971; Neve in Ebbels & King 1979; Yu & Liu, *Pl.Path.* 36:38, 1987, 3 viruses in China; Munro, *Aust.J.agric.Res.* 38:83, 1987, viruses in Australia [66,3452].

— American latent v. Probasco & Skotland 1976; Barbara & Adams, *Descr.V.* 262, 1983; Carlavirus, filament or rod, *c.*680 × 15 nm, transm. sap, *Phorodon humuli*; England, Germany, USA; differs from hop latent v. and hop mosaic v. serologically, a wider host range and in causing obvious symptoms in *Chenopodium quinoa*, *Datura stramonium*.

— bare bine (a hop str. of arabis mosaic v.), hop nettlehead q.v.

*— chlorosis Salmon & Ware, *AAB* 17:241, 1930; Thresh & Adams, *Rep.E.Malling Res.Stn.* for 1983, suggesting an assoc. with arabis mosaic v. [9:742].

— crinkle Blattný 1930, Vanek et al., *Phytopath.Z.* 87:224, 1976; bacteria assoc., Europe [56,2614].

— cyst nematode, *Heterodera humuli*.

— infectious sterility Zielińska & Miciński 1983; filament 580 nm long assoc., Poland [64,697].

— latent v. Schmidt et al. 1966; Barbara & Adams, *Descr.V.* 261, 1983; Carlavirus, filament or rod *c.*675 × 14 nm, transm. sap, *Phorodon humuli*, non-persistent, narrow host range, Australia, China, Europe, USA (N.W.), see hop American latent v.

— mosaic v. Salmon 1923; Barbara & Adams, *Descr.V.* 241, 1981; Thresh & Adams, *Rep.E.Malling Res.Stn.* for 1982:173; Carlavirus, filament or rod $c.650 \times 14$ nm, transm. sap, aphid, non-persistent, Australia, China, Europe, N. America, no symptoms in tolerant cvs.

—— **necrotic crinkle mosaic** Schmidt, *Phytopath.Z.* **53**:216, 1965; Chod et al., *ibid* **80**:54, 1974; MLO assoc. [**45**,527; **54**,2921].

—— **nettlehead**, known in England as a disease of hop for over 400 years. Although serious losses are no longer caused to the British hop industry, the precise etiology of this virus disease remains uncertain. Thresh & Pitcher, in Scott & Bainbridge 1978, considered spread and stated: 'Nettlehead, the severe form of split leaf blotch, and bare bine, are 3 distinct but inter-related diseases. Some features of their etiology are still uncertain, but arabis mosaic v., hop str., is undoubtedly involved as it always occurs in affected plants of diverse origin and never in symptomless ones.' Davies & Clark, *AAB* **103**:439, 1983, detected a low molecular weight particle, like a satellite, and assoc. with AMV in hop. The virus +satellite caused unusually severe symptoms in *Chenopodium quinoa* and characteristic nettlehead symptoms in hop seedlings. The virus free from the satellite caused much milder symptoms in *C. quinoa* and no nettlehead symptoms in hop. These references list all pertinent, published work [**63**,1900].

—— **petiole crinkle** Kochman & Stachyra, *Roczn.Nauk roln* A **77**: 297, 1957; Poland [**37**:340].

—— **ring and band pattern mosaic** (apple mosaic v.), Sano et al., *AAB* **106**:305, 1985; Japan [**66**,265].

—— **ringspot** Schmidt, *Phytopath.Z.* **47**:192, 1963; transm. sap, Germany [**43**,526].

—— **split leaf blotch** (a hop str. of arabis mosaic v.), hop nettlehead q.v.

—— **stunt viroid** Yamamoto et al. 1970; hop plants show internode shortening and therefore a stunt, and leaf curl; transm. sap; infects cucurbits and tomato, the former show symptoms the latter do not; in Japan only; the sequence homology between this viroid and that of cucumber pale fruit viroid is high. Ohno et al., *Virology* **118**:54, 1982, purification & characterisation; Sano et al., *Ann.phytopath.Soc.Japan* **50**:339, 1984, comparison with cucumber pale fruit viroid; Yaguchi & Takahashi, *Phytopath. Z.* **109**:21, 32, 1984, survival in hop residues, in cucurbits; Shikata in Maramorosch & McKelvey 1985 [**50**,2407; **61**,5608; **63**,4024, 4139;**64**, 989].

—— **yellow leaf blotch** Hall, in Talboys, *Nature Lond.* **203**:1021, 1964; USA (California); resembles hop yellow mottling in New Zealand, Cunningham 1949 [**44**,196d].

—— —— **net** Legg & Ormerod 1961; England [**40**:552].

—— **246 v.** Macovei, *An.Inst.Cerc.Prot.Pl.* **15**:37, 1979; particle 32–34 nm diam. assoc., Romania, necrotic spots [**60**,1017].

Hoplolaimus columbus Sher 1963; Fassuliotis, *Descr.N.* 81, 1976; Columbia lance nematode, ecto- and endoparasitic on roots, can cause severe damage to cotton and soybean in USA (Georgia, S. Carolina), occurs on weeds, cover crops.

H. galeatus (Cobb) Thorne 1935; Orton Williams, *Descr.N.* 24, 1973; can damage cotton, pine in USA.

H. indicus Sher 1963; Khan & Chawla, *Descr.N.* 60, 1975; ecto-and endoparasitic, India, causes stunting in rice, sugarcane; can damage other crops.

H. seinhorsti Luc 1958; Van Den Berg, *Descr.N.* 76, 1976; root endoparasite, on tropical crops including cotton, cowpea, melon, okra, pigeon pea, sorghum.

hordei, *Ascochyta, Puccinia, Ustilago,* see *U. segetum.*

Hordeiviruses, type barley stripe mosaic v., other members: lychnis ringspot v., poa semi-latent v.; rod 100–150 × c.20 nm, linear ssRNA, divided genome, sedimenting at 192–202 S, thermal inactivation 65–70°, longevity in sap 14–21 days, transm. contact, sap readily, seed in BSMV, LRSV; narrow host ranges; Jackson & Lane in Kurstak 1981; Francki et al., vol. 2, 1985; Carroll and Atabetov & Dolja in van Regenmortel & Fraenkel-Conrat 1986.

Hordeum, barley.

hordeum mosaic v. Slykhuis & Bell, *CJB* **44**:1191, 1966; filament 700 × 15 nm, transm. sap, Canada (Alberta), temperate cereals [**46**,289].

horiana, *Puccinia*; **horii**, *Mycosphaerella.*

horizontal resistance, see resistance to pathogens.

hormones, sexual in fungi, see sterols; Sequeira, *A.Rev.Pl.Physiol.* **24**:353, 1973, hormone metabolism in diseased plants; Gooday, *A.Rev.Biochem.* **43**:35, 1974, in fungi; Trione in Staples & Toenniessen 1981, natural regulators of fungal development.

hornbeam, *Carpinus betulinus.*

—— **leaf mosaic** Ploaie & Macovei, *Revue roum.Biol.Bot.* **13**:269, 1968; Romania [**48**,2233].

—— **line pattern** Gualaccini in Seliskar 1964; Italy [**45**,3221j].

—— **yellow mosaic** Gualaccini, *Annali Ist.Patol.veg.* **1**:105, 1970; Italy [**51**,4387].

horse bean, see broad bean.

horse chestnut, *Aesculus hippocastanum.*

—— —— **mottled mosaic** Blattný, *Ochr.Rost.* **14**:86,

1938; —— —— necrosis, Smolák, *Biologia Pl.* **5**:59, 1963; Czechoslovakia [**17**:543; **42**: 496].

—— —— **spindle tumour** Buchwald, *Horticultura* **15**:102, 1961; Denmark, Germany [**41**:104].

—— —— **yellow mosaic** (apple mosaic v.), Sweet & Barbara, *AAB* **92**:335, 1979; prob. the same as aesculus line pattern [**59**,1890].

horsegram, *Macrotyloma uniflorum*.

*—— **yellow mosaic** Muniyappa et al., *Mysore J.agric.Sci.* **10**:605,611, 1976; transm. *Bemisia tabaci*, India; Muniyappa in Plumb & Thresh 1983; epidemiology in India (S.) [**58**,457–8].

horse-hair blights, see *Marasmius*.

horseradish, *Armoracia rusticana*; Kadow & Anderson, *Bull.Ill.agric.exp.Stn.* 469:531, 1940; Hickman & Varma, *Pl.Path.* **17**:26, 1968, viruses [**20**:78; **47**,1704].

—— **brittle root** (*Spiroplasma citri*).

—— **fern leaf** Novák & Vlk, *Ochr.Rost.* **23**:361, 1950; Czechoslovakia [**30**:553].

—— **geminivirus** Duffus, Milbrath & Perry, *PD* **66**:650, 1982; isometric, paired particles 35 × 20 nm, transm. *Neoaliturus tenellus*, from horseradish with brittle root and plants showing no symptoms of the disease, USA (Illinois), serologically related to beet curly top v., limited host range in Cruciferae [**62**,340].

—— **latent v.** Richins & Shepherd, *Phytop.* **76**:749, 1986; Caulimovirus, isometric 50 nm diam., transm. sap, *Myz.pers.*, causes mild chlorotic mottling in some brassicas, prob. confined to Cruciferae, Europe [**66**,368].

—— **mosaic** = turnip mosaic v.

—— **virescence** Eastman et al., *PD* **68**:968, 1984; MLO assoc., transm. *Neoaliturus tenellus*, USA (Illinois) [**64**,1330].

host, an organism harbouring a parasite or pathogen, see suscept and FBPP.

—— **density**, as a factor in plant disease ecology; Burdon & Chilvers, *A.R.Phytop.* **20**:143, 1982.

—— **predisposition**, see predisposition.

—— **response to infection**, Heitefuss & Williams 1976.

—— **specific toxin**, see toxin.

HS toxin, see helminthosporoside.

huberi, *Phyllachora*.

hull rot, almond; *Monilinia* and *Rhizopus* assoc., poss. toxins; Mirocha & Wilson and Mirocha et al., *Phytop.* **51**:843, 851, 1961 [**41**:397].

Humariaceae, Pezizales; ascomata small, often hairy, mostly on the ground, asci narrow not protruding beyond the hymenium, tip not blue with I; *Caloscypha*.

humuli, *Pseudoperonospora*, *Septoria*, *Sphaerotheca*.

Humulus lupulus, hop.

HV toxin, see victorin.

hyacinth, *Hyacinthus*; —— bean, see lablab.

hyacinthi, *Botrytis*, *Embellisia*.

hyacinth lisser (aster yellows MLO), van Slogteren, *Meded.Fak.LandbWetensch.Ghent* **37**:450, 1972; see gladiolus grassy top, van Slogteren [**53**,10].

—— **mosaic v.** Wakker 1885; Rønde Kristensen 1955; Asjes, *Bloembollencultuur* **90**:1396, 1980; Derks & van den Abeele, *Acta Hort.* 109:495, 1980; prob. Potyvirus, filament *c*.750 × 12 nm, transm. aphid, serologically related to, but different from, ornithogalum mosaic [**34**:577; **60**,5461].

—— **rattle** (tobacco rattle v.), van Slogteren 1958 [**38**:479].

Hyadaphis foeniculi, transm. celery yellow spot, coriander feathery red vein, honeysuckle latent, poison hemlock ringspot.

Hyaloscyphaceae, Helotiales; ascomata hairy, excipulum usually of soft prismatic cells, seldom on the ground; *Dasyscyphus*, *Lachnellula*.

hyalospora, *Ascochyta*.

Hyalothyridium Tassi 1900; Coelom.; placed as a queried synonym of *Camarosporium* Schultz 1870 by Sutton 1980, but used by the authors of *H. maydis* Latterell & Rossi, *Mycologia* **76**:506, 1984; con. muriform, cells often splitting apart, each germinating, 29–42 × 16–26 µm; assoc. with a severe maize leaf spot; Central America, Colombia [**63**,4403].

Hydnaceae, Aphyllophorales; hydnoid form, hymenophore on structures like teeth. Some *Hydnum* spp. cause a heartwood rot in living trees; Petersen, *Taxon* **26**:144, 1977, typification; Sankaran & Sharma, *TBMS* **87**:401, 1986, described *H. subvinosum* Berk. & Broome causing a serious stem canker in India (Kerala); it may also cause a tea root rot.

Hydrangea, Chiko & Godkin, *PD* **70**:541, 1986, found alfalfa mosaic v., hydrangea ringspot v., tobacco ringspot v.; Canada (British Columbia) [**65**,5562].

hydrangea mosaic v. Thomas, Barton & Tuszynski, *AAB* **103**:261, 1983; Ilarvirus, quasi-isometric *c*.30–38 × 28–32 nm, transm. sap, in *Chenopodium quinoa* seed, England [**63**,1299].

—— **ringspot v.** Brierley 1954; Koenig, *Descr.V.* 114, 1973; Potexvirus, filament *c*.490 nm long, transm. sap, leaf contact, pruning.

—— **virescence** Muth 1933; Welvaert et al., *Phytopath.Z.* **83**:152, 1975; Hearon et al., *Phytop.* **66**:608, 1976; MLO assoc., similar diseases called phyllody, proliferation or witches' broom; Cousin & Sharma, *J. Phytopath.* **115**:274, 1986, electron microscopy [**55**,1289, 5781; **65**,3953].

hydroponics, see soilless culture.

hydrosis, plant tissues not flaccid but with excessive intercellular water.

hydroxyisoxazole, soil and seed application against many soilborne pathogens.

Hylurgopinus rufipes, see *Ceratocystis ulmi*.

Hymenella Fr. 1823; Hyphom.; Tulloch, *Mycol.Pap*. 130, 1972; con. aseptate, hyaline; *Hymenula* Fr. 1828 is an incorrect name, reassessment of the generic position needed; not recognised by Carmichael et al. 1980. *H. cerealis* Ell. & Ev. 1894, Hawksworth & Waller, *Descr.F.* 501, 1976, as *Hymenula*; conidiogenous cells phialidic; con. in dense, slimy heads, 4–7 × 1–1.5 μm; leaf stripe of cereals, grasses; Bockus et al., *PD* **67**:1323, 1983, effects of 5 wheat management practices on disease incidence [63,1724].

hymenium, the spore bearing layer in a basidioma of a macro-fungus.

Hymenochaetaceae, Aphyllophorales; poroid form, basidiomata resupinate to pileate, or clavarioid, lignicolous, causing white rots; Pegler in Ainsworth et al. 1973, vol. IV B; Fiasson & David, *CJB* **61**:442, 1983; *Inonotus, Phaeolus, Phellinus, Vararia* [62,2887].

Hymenomycetes, Basidiomycotina; basidiomata macroscopic, typically gymnocarpus or semiangiocarpus, basidia in a ± well defined hymenium, basidiospores are ballistospores q.v.; Merrill, *A.R.Phytop*. **8**:281, 1970, spore germination & host penetration by heart rot spp.

hyoscyami, *Peronospora*.

Hyperomyzus lactucae, transm. lettuce necrotic yellows, sowthistle yellow vein.

hyperparasite, an organism parasitic on a parasite, FBPP; see biological control; Boosalis, *A.R.Phytop*. **2**:363, 1964; Kranz in Blakeman 1981.

hyperplasia, abnormal enlargement of plant tissue resulting from infection by a pathogen, causing galls, warts, witches' brooms, cf. hypoplasia.

hypersensitivity, Stakman, *J.agric.Res*. **4**:193, 1915; increased sensitivity by the host at the site of infection and shown by rapid cell death which prevents further growth by the pathogen and spread of infection; also see FBPP; Müller in Horsfall & Dimond 1959; Klement & Goodman, *A.R.Phytop*. **5**:17, 1967; Klement, see procaryotae, in Mount & Lacey.

hypertrophic mitochondrial A v. Weintraub & Ragetli, *Phytop*. **61**:431, 1971; rod 613 nm long, transm. sap, Virginia crab apple [50,3920].

hypertrophy, the state of enlargement of host tissue.

hypha, one of the filaments of a mycelium q.v.; Robertson, *TBMS* **48**:1, 1965; *A.R.Phytop*. **6**:115, 1968, growth.

hyphal tissue, textura; the fungus tissue formed in ascomycetes and coelomycetes, and superficially resembling plant parenchyma, has been divided into 7 textura types which are illustrated in *Ainsworth & Bisby's dictionary of the fungi* 1983: 378.

Hyphomycetes, Deuteromycotina q.v.

hypobrunnea, *Poria*.

hypochoeris mosaic v. Brunt & Stace-Smith 1978; *Descr.V*. 273, 1983; poss. Tobamovirus, fragile rod *c*.22 nm diam., predominant lengths 120–140 and 220–240 nm, transm. sap, Canada, *H. radicata*.

hypocotyl rot (*Thanatephorus cucumeris*) bean, other crops; Campbell & Pennypacker, *Phytop*. **70**:521, 1980 [60,2283].

Hypocreaceae, Hypocreales; ascomata fleshy, usually bright coloured, mostly ostiolate, often in a stroma or on a subiculum; asco. very variably septate; *Calonectria, Gibberella, Nectria, Nectriella, Neocosmospora, Sphaerostilbe, Thyronectria*.

Hypodermataceae, Rhytismatales; ascomata usually carbonaceous, determinate, often with lip cells; paraphyses often circinate, forming an epithecium; ascus with a pointed apex, usually with a pore; *Bifusella, Colpoma, Davisomycella, Didymascella, Elytroderma, Isthmiella, Lirula, Lophodermella, Lophodermium, Meloderma, Ploioderma, Rhabdocline*.

hypodermia, *Botryodiplodia*.

Hypodermina Höhnel 1916; Coelom; conidiomata forming multilocular structures, 1–several ostioles; conidiogenesis enteroblastic, proliferating percurrently; con. hyaline, 1.5–2 × 1–1.5 μm; *H. hartigii*, teleom. *Lirula macrospora*.

hypodytes, *Ustilago*; hypolateritia, *Poria*.

Hypomyces (Fr.) Tul. 1860; Hypomycetaceae; ascomata grouped on a hyphal web, ostioles shor* cylindrical or papillate, asco. hyaline, 1 septate, pointed at each end, cells of *c*. equal size; Tubaki, *Rep.Tottori mycol.Inst*. 12:161, 1975, spp. in Japan; Samuels, *Mem.N.Y.bot.Gdn*. **26**(3) 1976, the fungi formerly classified as *Nectria* subgenus *Hyphonectria*.

H. rosellus (Alb. & Schwein.:Fr.) Tul.; anam. *Cladobotryum dendroides* (Bull.:Fr.) W. Gams & Hoozemans 1970; mushroom cobweb, Fletcher 1984. *Hypomyces aurantius* (Pers.: Fr.) Tul., anam. *C. variospermum* (Link) S. Hughes, also occurs on mushroom; both spp. and others are found on higher fungi.

Hypomycetaceae, Clavicipitales; ascomata usually in a fleshy, bright coloured stroma, ascus with thick cap and slender pore, mainly fungicolous; *Hypomyces*.

hypoplasia, abnormal reduction of plant tissue
resulting from infection by a pathogen or due to
another factor, cf. hyperplasia.

hypovirulence, literally subnormal virulence;
includes all abnormal states in which pathogenic
fitness, i.e. virulence q.v., is reduced, fide Elliston
1982, see Cryphonectria parasitica.

Hypoxylon Bull.:Fr. 1825; Xylariaceae; stroma
superficial, sessile, not zoned; ascomata immersed
just beneath stromatic crust or protruding
through it, asco. aseptate, brownish, seldom
> 20 μm long; anam. hyphomycetous, con.
aseptate, ± hyaline, very small; inhabitants of
wood, mostly minor pathogens on dicotyledons,
host ranges generally not limited; Miller, A
monograph of the world species of Hypoxylon,
1961; Holliday 1980.

H. atropunctatum (Schwein.:Fr.) Cooke 1883; asco.
24–33 × 11–16 μm, N. America; Tainter et al., PD
67:195, 1983, early coloniser of oak showing
decline & death, USA (S. Carolina); Bassett &
Fenn, ibid 68:317, 1984, latent colonisation &
pathogenicity, oak [62,3268; 63,3584].

H. mammatum (Wahlenb.) Miller 1961;
Hawksworth, Descr.F. 356, 1972; stroma white
pruinose at first, asco. 15–36 × 5–14 μm; conidial
pillars, coremia; perennial canker of Populus alba,
P. grandidentata, P. tremula, P. tremuloides;
saprophytic on temperate trees; Europe, N.
America; in the latter region, where the fungus
has been most investigated, the disease is
geographically restricted. It is prevalent in the

N.E., Great Lakes and N. W. prairies, incidence is
low in central Rocky Mountains, but further N.
and in Alaska the disease is apparently absent.
Infection prob. occurs through wounds including
those made by insects, trees can be killed; the
fungus forms toxic metabolites, see mammatoxin;
Manion & Griffin, PD 70:803, 1986, review;
Griffin & Manion, Phytop. 76:1289, 1986,
multivariate method for estimating relative
resistance [66,4486].

H. mediterraneum (de Not.) Ces. & de Not. 1863;
Hawksworth, Descr.F. 359, 1972; anam. like
Nodulisporum; asco. 12.5–23 × 5–10 μm; charcoal
of Quercus suber, and Shorea robusta in India,
other hardwoods.

H. nummularium Bull. 1791; fide Cannon et al.
1985; asco. 11–15 × 6–9 μm; Miller 1961, see
genus, gave 6 vars.; reported to cause tarry root
rot of tea in India (N.E.); Agnihothrudu 1964
[44,3142].

*H. serpens (Pers.:Fr.) Kickx 1835; anam.
Geniculosporium serpens Chesters & Greenhalgh
1964; Hawksworth, Descr.F. 358, 1972; asco.
10–17 × 5–8 μm; common in Europe especially on
Fagus sylvatica, on other hardwoods, mostly
unimportant; causes a serious wood rot of tea in
India (S.); Venkata Ram 1967–9; Holliday 1980;
Jensen, CJB 59:40, 1981, developmental
morphology in culture [47,621b; 48, 587a;
49,2170c; 60,5754].

Hysteroneura setariae, transm. guinea grass mosaic
v.

I

iatrogenic, term derived from medicine; in plant pathology prob. first used by Horsfall 1972; a disease which results from, or is increased by, the use of a specific crop protection chemical; Griffiths & Berrie, *AAB* **89**:122, 1978; Griffiths, *A.R.Phytop.* **19**:69, 1981.

ice-ice (*Penicillium waksmanii* and *Scopulariopsis brevicaulis*), *Eucheuma spinosum, E. striatum*.

ice nucleation, some epiphytic bacteria cause damage to plants that are sensitive to frost by causing ice to form; Lindow, *PD* **67**:327, 1983; *A.R.Phytop.* **21**:363, 1983.

ice plants, see *Carpobrotus*.

idahoensis, *Typhula*.

identification, the practical application of the arts of classification and nomenclature, the third part of taxonomy; Cowan 1978.

ignarius, *Phellinus*.

Ilarviruses, type prunus necrotic ringspot v., *c*.15 definitive members, isometric, but Francki et al., vol. 2, 1985 refer to particles as quasi-spherical or slightly pleomorphic, 20–32 nm diam., some may be up to 73 nm long; 4 segments linear ssRNA, all required for infection except that the coat protein can be substituted for the smallest RNA, sedimenting into 3 or 4 components at 78–90, 89–98 and 100–114 S, thermal inactivation 45–65°, longevity in sap 9–12 hours for some and a week or more at 22–24° for others, transm. sap, seed; experimental host ranges wide; only this group, the Bromoviruses and the Cucumoviruses have a tripartite genome; Fulton in Kurstak 1981, *Descr.V.* 275, 1983; Francki, vol. 1, 1985; Francki et al., vol. 2, 1985.

iliau, *Clypeoporthe, Phaeocytostroma*; (*C. iliau*) sugarcane.

ilicicola, *Cylindrocladium*; ilicis, *Arthrobacter, Phytophthora*.

Illinoia pepperi, transm. blueberry shoestring v.

imazalil, protectant fungicide, broad spectrum, spray and seed dressing.

immune, exempt from infection; plant cannot be infected; freedom from attack by a potential pathogenic agent, hence immunity. It implies total exclusion of such an agent and is not the same as freedom from disease; see FBPP. This list recommended that the term 'immune reaction' should not be considered as a synonym of 'hypersensitive reaction' since the latter implies a transient infection; Cooper & Jones, *Phytop.*

73:127, 1983, suggested infectible as the antonym of immune.

impatiens latent v. Lockhart & Betzold, *Acta Hortic.* 110:81, 1980; particle 520 nm long, transm. sap.

—— leaf curl Verma & Singh, *Hort.Res.* **13**:55, 1973; transm. *Bemisia tabaci*, India (N.), *I. balsamina*; in USA (tobacco streak v.), Lockhart & Betzold, *PD* **64**:289, 1980, *I. holstii* [**53**,1843; 60,1475].

—— mosaic (tobacco ringspot v.), Lockhart & Pfleger, *PDR* **63**, 258, 1979; USA [**59**,1282].

imperatae, *Gibberella*.

imperfect state, see fungi and holomorph.

inaequalis, *Venturia*.

incarnata, *Typhula*.

incense (*Balansia*) Gramineae.

incite, incitant of a disease, often used instead of cause and causal pathogen or agent, respectively, but why is far from clear. Both words should be dropped; Luttrell, *PDR* **38**:321, 1954. FBPP omits incite but stated that, whilst incitant to describe the causal agent is incorrect, the term should refer to some other factor, which promotes the pathogenic action of the causal agent.

inclusion body, virus induced structures in the plant cell; apart from virus particles other characteristic structures may be seen, e.g. pinwheels; they may be used in diagnosis and classification; Martelli & Russo, *Adv.Virus Res.* **21**:175, 1977; Edwardson & Christie, *A.R.Phytop.* **16**:31, 1978.

incompatible, not compatible q.v.

incubation period, the time between infection or inoculation and the appearance of disease symptoms; Ercolani & Vannella, *AAB* **108**:275, 1986, response times in bacterial infection.

incurvata, *Bipolaris*.

indexing, any procedure for demonstrating the presence of known viruses in suspect plants, FBPP.

Indian paint fungus, *Echinodontium tinctorium*.

indica, *Bitrimonospora*, see *Monosporascus, Gibberella, Tilletia*.

indicator plant, one which reacts to a certain pathogen or environmental factor with obvious symptoms, and is used to assist in the detection of the pathogen or the environmental factor; synonymous with indicator host when applied to pathogens. Differential host q.v. is a type of indicator plant, FBPP.

indicum, *Colletotrichum*, see *C. gossypii*; *Oidium*, see *O. caricae*; **indicus**, *Mycoleptodiscus*.

indigenous pathogen, native to an area, not introduced, e.g. *Crinipellis perniciosa*, causing cacao witches' broom, and *Microcyclus ulei*, causing rubber South American leaf blight. Both fungi occur on wild plants, from which these crops arose, in the Amazon region.

indigoferae, *Uromyces*.

inducer, see elicitor.

indurated, made hard.

infectible, the antonym of immune q.v., fide Cooper & Jones; see invasion.

infection, penetration of a plant, now the host, by an organism and the establishment of a parasitic or pathogenic relationship; for viruses see invasion; Wood, *Pl.Path.* 33:3, 1984, establishment of infection; — **court**, the site of infection, see latent infection, Kerr & Flentje, *Nature Lond.* 179: 204, 1957; Flentje, *TBMS* 40:322, 1957; Heitefuss & Williams 1976; — **cushion**, of ectotrophic root parasites, formed from hyphae that become attached and grow along the junction lines between epidermal cells of young roots, the hyphae branch to form a cellular mass or cushion beneath which infection pegs grow to penetrate the plant; — **peg**, a structure formed by deposition of lignin, callose, cellulose, suberin etc. around a hypha which is penetrating the wall of a living host cell, FBPP; see callosity, lignituber; — **structures**, Staples, *Microbiological Sciences* 2:193, 1985.

infective, (1) of an organism or agent that can attack a plant and cause infection. (2) Of a vector carrying or containing a pathogen which can be transferred to a plant and then causes infection, after FBPP.

infectivity titration, the measurement of the relative effectiveness of bacteria to infect a plant and establish a parasitic or pathogenic relationship which has a measurable effect; Ercolani, *A.R.Phytop.* 22:35, 1984.

infest, to run over the surface of a plant, or to be dispersed through soil or other substrate. Used especially of insect and other animal pests. When used with reference to micro-organisms or virus particles on plant surfaces there is no implication that infection has occurred. The infesting organism may or may not be a contaminant; infest and contaminate are not synonymous, FBPP.

infestans, *Phacidium*, *Phytophthora*.

infiltration, the introduction of liquid or inoculum under pressure, e.g. into stomata, by spray or by applying a vacuum and then releasing it, FBPP.

inflata, *Phytophthora*.

inflorescence infection, reproductive disease, particularly in Gramineae; Shaw in Wood & Jellis 1984.

— **and pod rot** (*Alternaria ricini*) castor.

infuscans, *Monilochaetes*.

ingrown sprouts, potato sprouts from eyes grow through the tuber tissue; Davis, *Am.Pot.J.* 38:411, 1961; Wien & Smith, *ibid* 46:29, 1969.

injection, the introduction of liquid or inoculum with a hypodermic needle etc.; should not be used in the same sense as infiltration, after FBPP.

ink (*Bipolaris iridis*) iris, (*Phytophthora cambivora*, *P. cinnamomi*) sweet chestnut; — **spot** (*Ciborinia whetzelii*) poplar.

inland cone rust (*Chrysomyxa pirolata*) spruce.

inoculate, to apply inoculum to a plant or a culture medium; not to be used in the sense of contaminate or infest, after FBPP; Waterston, *Rev.appl.Mycol.* 47:217, 1968.

inoculation access time, the length of time a vector spends on a test host in transm. tests; it is not implied that the vector feeds throughout this time; — **feeding time**, the length of time a vector feeds on the test plant in transm. tests; — **threshold period**, the minium time necessary for a vector to feed on a plant so that transm. can occur, after FBPP.

inoculative, of a vector carrying or containing a pathogen, and that can place it into a plant; inoculation does not necessarily result in infection; the term is not strictly synonymous with infective, after FBPP.

inoculum, material containing micro-organisms or virus particles used to inoculate a plant or a culture medium; can refer to potentially infective material available in soil, air or water and which by chance results in the natural inoculation of a plant, after FBPP.

— **potential**, the energy of growth of a parasite, or fungus, available for infection of a plant or host, or colonisation of a substrate, at the surface of the plant organ to be infected, or substrate to be colonised. FBPP concluded that this definition, sensu S.D. Garrett 1956, is valuable as a philosophical concept in plant infection, even though it cannot be directly quantified. Inoculum potential, particularly in the complexities of a soil environment, continues to be studied practically and theoretically. As R.A. Fox pointed out in 1965, one of the first demonstrations of this phenomenon was by K. Bancroft in 1912 working on white root (*Rigidoporus lignosus*) of rubber, i.e. a certain level of inoculum was necessary before disease appeared; Baker and Dimond & Horsfall in Baker & Snyder 1965; Garrett 1970; Baker in

Horsfall & Cowling, vol. 2, 1978; Baker & Drury, *Phytop.* **71**:363, 1981.

inoderma, *Marasmiellus*.

Inonotus P. Karsten 1879; Hymenochaetaceae; basidiomata with context which is rusty brown, fibrous, soft, spongy; Pegler, *TBMS* **47**:175, 1964, key & descriptions; Gilbertson, *Mem.N.Y.bot.Gdn.* **28**:67, 1976, key, spp. in USA (Arizona).

I. andersonii (Ell. & Ev.) Černý 1963; basidiospores pigmented, 5.5–8 × 4–5 μm; white rot of heartwood of living oak; Campbell & Davidson, *Mycologia* **31**:161, 1939 [**18**:487].

I. arizonicus Gilbertson 1969; basidiospores pigmented, 4–6 × 3–4 μm; white heart rot, *Platanus*, USA, Goldstein & Gilbertson, *Mycologia* **73**:167, 1981, cultural morphology & sexuality [**60**,6322].

I. dryadeus (Pers.:Fr.) Murrill 1908; basidiospores hyaline, 7–8.5 × 6.5–8 μm, a smaller spored form, 4.4–5.6 × 3.6–5.2 μm, var. *brevisporus* (Thind & Chatrath) Pegler 1964, in India (N.W.), heartwood decay, weeping conk, N. temperate, often on oak, hardwoods mostly; Long, *J.agric.Res.* **1**:239, 1913; *Phytop.* **20**: 758, 1930 [**10**:141].

I. farlowii (Lloyd) Gilbertson 1976; basidiospores 6.8 × 4.5–6 μm, white heartwood rot of living willow; Long, *Lloydia* **8**:231, 1945 [**25**:143].

I. glomeratus (Peck) Murrill 1920; basidiospores pigmented, 4.5–5.8 μm diam.; brown spongy rot, heartwood decay and canker, particularly of *Acer*, *Fagus*: Campbell & Davidson, *Mycologia* **31**:606, 1939; Good & Nelson, *CJB* **29**:215, 1951 [**19**:125; **31**:524].

I. hispidus (Bull.:Fr.) P. Karsten 1879; Pegler & Waterston, *Descr.F.* 193, 1968; basidiospores pigmented, 8.3–11 × 7–9 μm, heart rot of *Fraxinus*, spongy rot of hardwoods, causes cankers; Toole, *Phytop.* **45**:177, 1955, on *Quercus* in USA; McCracken & Toole, *ibid* **64**:265, 1974, reducing basidioma formation [**34**:683; **53**,4133].

I. obliquus (Pers.:Fr.) Pilát 1942; Reid, *TBMS* **67**:329, 1976, synonymy, morphology, distribution in Britain; basidiospores becoming faintly coloured, 8–9.2 × 5.2–5.8 μm; white and heart rot, mostly on *Betula*, basidiomata often found to be sterile; Zabel, *For.Sci.* **22**:431, 1976, basidioma development & its inhibition; Ritter, *Gleditschia* **8**:183, 1980, in Germany; Blanchette, *Phytop.* **72**:1272, 1982, host discolouration & decay [**56**,4705; **61**,3086; **62**,1655].

I. pseudohispidus Kravts. 1950; basidiospores pigmented, 8.5–10.5 × 5.5–8 μm; white rot of *Populus nigra* in Egypt; Salem & Michail, *TBMS* **74**:107, 1980 [**59**,4336].

I. tomentosus (Fr.) Gilbertson 1974; basidiospores

hyaline, 5–6 × 3–4 μm; white pocket rot of conifers; Whitney, *CJB* **40**:1631, 1962, major cause of stand opening of *Picea glauca* in Canada (Saskatchewan) [**42**:419].

I. weirei (Murr.) Kotl. & Pouz. 1970; Pegler & Gibson, *Descr.F.* 323, 1972; basidiospores 3.6–4.5 × 2.7–3.5 μm; yellow ring rot of conifers, also laminated butt rot, yellow laminated rot; N. America, Japan; Hansen, *CJB* **57**:1573, 1579, 1979, incompatibility, cytology & heterokaryosis; Bloomberg & Hall 1986; effects of fungus on relationship between stem growth & roots in Douglas fir [**59**,1626–7; **65**,5753].

inoperculate, of an ascus or a sporangium, opening by an irregular split to discharge spores, as in the asci of the Helotiales, cf. operculate.

insects, by far the largest class of Arthropoda, see vector. The binomials are given in the order, without authorities, in nearly all cases as transm. mollicutes and viruses; Leach, *Insect transmission of plant diseases*, 1940; Browne 1968; Carter, *Insects in relation to plant disease*, edn. 2, 1973; Leftwich, *A dictionary of entomology*, 1976; Ordish, *The constant pest*, 1976; Kranz et al. 1977; Agrios in Harris & Maramorosch 1980; D'Arcy & Nault, *PD* **66**:99, 1982, vectors; Gilbert & Hamilton, *Entomology. A guide to information sources*, 1983; Hill, *Agricultural insect pests of the tropics and their control*, edn. 2, 1983; Johnston & Booth 1983.

insulana, *Cercospora*.

interferons, Isaacs & Linderman, *Proc.R.Soc.Lond.B* **147**:258, 1957; proteins induced in cells, particularly by viruses; but they can be induced by micro-organisms and some substances. They reduce susceptibility to infection in animal cells, and are not normally present in the non-induced cell. Chessin, *Bot.Rev.* **49**:1, 1983, considered whether interferons occur in plants & discussed a poss. mechanism of acquired resistance to viruses in plants.

inter fruitlet corking (*Penicillium funiculosum*) pineapple.

internal bark necrosis, see measles.

—— **black rot**, Japanese horseradish or wasabi, *Eutrema wasabi*, assoc. with soft rot bacteria and *Phoma wasabiae* Yokogi; Goto & Matsumoto, *Ann.phytopath.Soc.Japan* **52**:59, 69, 1986 [**66**, 685–6].

—— —— **spot**, potato, caused by bruising which injures tissue beneath the skin but does not break it; Jacob, *Mem.Cornell Univ.agric.Exp.Stn.* 368, 1959; Sawyer & Collin, *Am.Pot.J.* **37**: 115, 1960 [**39**:617–18].

—— **breakdown**, apple in storage; Porritt, *Can.J.Pl.Sci.* **55**:743, 1975; pear cv. Spadona in

storage; Ben-Arie & Guelfat-Reich, *ibid* **55**:593, 1975 [**54**,5485; **55**,319].
— **brown fleck**, potato, assoc. with high temp. and water conditions; Novak et al., *Aust.J.Exp.Agric.* **26**:129, 1986 [**65**,5147].
— **browning**, brassicas, Ca deficiency, Scaife & Turner 1983.
— **brown spot**, potato, prob. the most used of the many names for internal browning of the tuber, no external symptoms; the cause may be due to factors concerned in tuber growth; Iritani et al., *Am.Pot.J.* **61**:335, 1984 [**64**,1239].
— **cork**, apple, B deficiency, Atkinson 1971.
— **crown breakdown**, clover; assoc. with plant morphology, low density planting winter injury and weakly pathogenic fungi; Pratt & Knight, *Phytop.* **73**:980, 1983 [**63**,709].
— **necrosis** (turnip mosaic v.) cabbage.
— **rust spot**, potato, assoc. with low Ca in tubers; Collier et al., *J.agric.Sci.* **91**:241, 1978; **94**:407, 1980 [**57**,5667; **60**,464].
international organisations, Chiarappa, *A.R.Phytop.* **8**:419, 1970; *Rev. Pl.Path.* **58**:391, 1979; *PD* **64**:362, 1980; Thurston, *A.R.Phytop.* **15**:223, 1977.
intricans, *Gibberella*.
invasion, the penetration and colonisation of a plant by an organism, after FBPP. Cooper & Jones, see immune, discuss, for viruses, the distinction between infection q.v. and invasion; the first term is the act of plant cell entry, followed by nucleic acid replication; the second to denote the events following spread to other cells which may or may not lead to visible disease symptoms.
inventa, *Nectria*.
Ipomoea aquatica, water spinach; *I. batatas*, sweet potato.
ipomoeae, *Coleosporium*, *Streptomyces*; **ipomoeae-aquaticae**, *Albugo*; **ipomoeae-batatas**, *Phomopsis*; **ipomoeae-panduratae**, *Albugo*.
— **mosaic** Bird & Sánchez, *J.Agric.Univ.P.Rico* **55**:461, 1971; transm. *Bemisia tabaci*, Puerto Rico, *I. quinquefolia*, causes a tobacco leaf curl [**51**,4745].
iprodione, contact fungicide used for fruit diseases caused by fungi and as a seed dressing.
ips, *Ceratocystis*.
iranica, *Phytophthora*; **iridis**, *Bipolaris*.
iris, *Iris*; the name is also used to describe plants in other genera, see bearded iris.
— **bont** = narcissus latent v.
— **fulva mosaic v.** Travis 1957; Barnett & Alper 1977; Barnett, *Descr.V.* 310, 1986; Potyvirus, filament *c*.770 nm long, transm. sap, aphid spp., non-persistent, serologically distinct from iris mild

mosaic v. and iris severe mosaic v., occurs naturally only in *Iris fulva* hybrids, USA.
— **germanica rhabdovirus** Rubio-Huertos, *Phytopath.Z.* **92**:294, 1978; Spain, [**58**,3324].
— **grijs** (iris severe mosaic v.).
— **mild mosaic v.** van Slogteren 1958, 1960; Brunt & Phillips 1980; Brunt, *Descr.V.* 324, 1986; Potyvirus, filament *c*.750 × 12 nm, transm. sap, aphid spp., non-persistent, serologically distinct from iris fulva mosaic v. and iris severe mosaic v. Alone IMMV usually causes a mild leaf chlorosis; but it frequently occurs with ISMV, bean yellow mosaic v. and/or narcissus latent v., and then disease is more severe. IMMV is now so prevalent that stocks of commercially important cvs. free from it are prob. unobtainable by selection and roguing; meristem tip culture needs to be used.
— — **yellow mosaic** = narcissus latent v., fide Hammond et al., *Acta Hortic.* 164: 195, 1985.
— **severe mosaic v.** Brierley & McWhorter, *J.agric.Res.* **53**:621, 1936; this virus is the same as bearded iris mosaic v. q.v. and ISMV is apparently the preferred name. It is a Potyvirus and is serologically distinct from the other 2 Potyviruses, iris fulva mosaic v. and iris mild mosaic v.; Inouye et al., *Ann.phytopath.Soc.Japan* **47**:182, 1981 [**16**:254; **61**,2913].
— **tectorum mosaic v.** Yonaha 1978; filament 760 nm long, transm. sap, *Myz.pers.*, non-persistent, Okinawa [**59**,815].
— **yellow leaf stripe**, as for iris germanica rhabdovirus.
— — **mosaic** (iris severe mosaic v.), C. J. Asjes, personal communication, 1987.
Irish famines. The work of Tillet in 1755 and Prévost in 1807 had instilled doubt in the then widely held belief in Europe in the theory of spontaneous generation; see history. In 1845, and later, the epidemics of potato blight (now *Phytophthora infestans*) in Ireland became the first, meticulously documented, demonstrations of the economic destructiveness of a plant disease. This fungus played a significant part in American, English and Irish history; it led Berkeley to support the germ theory of disease. He therefore anticipated Koch by *c*.30 years; although it was de Bary who, in 1861 and 1863, put the etiology of potato blight beyond doubt. The potato had come to Europe from S. America in the mid 16th century. By late in the next century *Solanum tuberosum* was becoming the Irish potato, so close was it by then associated with Ireland in the public mind. The agricultural consequences of a terrible socio-political control had left almost the entire population existing at a bare subsistence level. The potato, for many reasons, became the

people's lifeline. The first major destruction of the potato crop by the fungus was in 1845, a second took place in 1846 and a third in 1848 which was the fourth year of famine. Salaman likened the consequent upheaval to the black death of 1348. It led to: the repeal of the corn laws in England by Peel and to the era of free trade; to the first well recorded attempts at food aid on a national scale; to emigrations, not the first in Irish history; and to castastrophic suffering and death. The population of Ireland fell from over 8 million in 1841 to just over 4 million in 1900. In no other European country did any such fall in population occur; Large 1940; Salaman, see potato; Woodham-Smith, *The great hunger, Ireland 1845–9*, 1962.

iron chelates, see siderophores.

ironuke and kubiore, tulip, B deficiency, disappearance of anthocyanin from flowers and transverse stem break, Japan; Ikarashi 1980 [**60**, 3215].

irregulare, *Pythium*.

irrigation, effect on disease, Rotem & Palti, *A.R.Phytop.* **7**:267, 1969; Palti 1981; Palti & Shoham, *PD* **67**:703, 1983; Johnson et al., *PD* **70**:998, 1986, application of crop protection chemicals.

Isaria Hill:Fr. 1832; Hyphom.; conidiomata synnematal, con. aseptate, hyaline; Carmichael et al. 1980 considered the name to be dubious; *I. fuciformis*, teleom. *Laetisaria fuciformis*.

ischaemi, *Curvularia*.

ishikariensis, *Typhula*, see *T. idahoensis*.

isodam, a line along which the mean annual damage caused by a pathogen or a pest is estimated to be the same at any point, used by Zadoks & Rijsdijk, *Atlas of cereal diseases and pests in Europe*, 1984.

isoelectric point, the pH at which a virus particle has a zero net charge.

isolation, essentially separating a micro-organism from a substratum and establishing the isolate as a pure culture, i.e. axenically.

isometric particle, of a virus, a basic type which is icosahedral in structure, it appears ± round; Gibbs & Harrison 1976.

isoprothiolane, systemic fungicide, used against rice blast.

Isthmiella Darker 1967; Hypodermataceae; asco. constricted in the middle. *I. faullii* (Darker) Darker; Cannon & Minter, *Descr.F.* 793, 1984; asco. 45–55 × 3.5–6 μm, with a conspicuous gelatinous sheath; needle blight of *Abies balsamea*, N. America.

Italian ryegrass, see ryegrass.

italicum, *Penicillium*.

Itersonilia Derx 1948; Hyphom.; Carmichael et al. 1980; conidiomata hyphal, conidiogenesis holoblastic; con. aseptate, hyaline, solitary, some are forcibly discharged, aerial ballistospores; there is a dikaryophase with clamp connections, and see *Tilletiopsis* which may be a synonym; Sowell & Korf, *Mycologia* **52**:934, 1960; Ingold, *TBMS* **86**:501, 1986, chlamydospore ontogeny [**41**:444].

I. pastinacae Channon, *AAB* **51**:13, 1963; ballistospores dikaryotic, 11.4–20 × 7.2–11.4 μm, chlamydospores thick walled, spherical 8.9–11.7 μm diam.; parsnip black canker. The pathogen was originally assigned to *Itersonilia perplexans* Derx which had been reported to cause petal blight or flower scorch of chrysanthemum by Dosdall, *Phytop.* **46**:231, 1956. *I. pastinacae* is not pathogenic to chrysanthemum and the other sp. does not attack parsnip; Webster et al. and Ingold, *TBMS* **80**:365, 1983; **82**:13, **83**:166, 1984, life cycle, structure & ballistospore discharge in *I. perplexans*; there are dikaryotic and monokaryotic forms with several spore types [**35**:826; **42**:508].

Ivanovski, Dmitri Iosifovich, 1864–1920; born in Russia, Univ. Petersburg, Warsaw; later director, Botanical Garden, Univ. Dorpat, Estonia; in 1892 he demonstrated that the agent that caused tobacco mosaic would pass though a Chamberland bacterial filter. Beijerinck did similar experiments 6–7 years later and his theory got closer than did Ivanovski's to the nature of the cause, a virus. *St.Petersb.Acad.Sci.Bull.* 35, 1892; *Phytopath. Classics* 7, 1942, translation by J. Johnson; *Bact.Rev.* **36**:135, 1972.

ivy, *Hedera*.

—— **fasciation** Witz et al., *C.r.hebd.Séanc.Acad.Sci.Paris* D **275**: 2437, 1972; rod assoc. 650 × 17 nm, France; also abnormal inflorescence, Plantefol, *ibid* **280**:507, 1991, 2607, 1975 [**52**, 1573; **55**,282–4].

—— **ringspot** Ploaie & Macovei, *Revue roum.Biol.Bot.* **13**:269, 1968; Romania [**48**,2233].

—— **vein clearing v.** Russo, Castellano & Martelli, *Phytopath.Z.* **96**:122, 1979; Castellano & Rana, *Phytopath.Mediterranea* **20**: 199, 1981; Phytorhabdovirus, *c*.325 × 55 nm, transm. sap to herbaceous hosts; Italy (S.), Yugoslavia [**59**,3277; **63**,685].

iwatae, *Elsinoë*, *Sphaceloma*.

ixia mosaic Smith & Brierley, *Phytop.* **34**:593, 1944; USA [**23**:488].

ixora witches' broom Chen, *Mem.Coll.Agric.Taiwan Univ.* 12:67, 1971; Taiwan, *I. chinesis* [**53**,185].

J

jaapii, *Blumeriella*.
jack bean, *Canavalia ensiformis*.
jacquemontia golden yellow mosaic Bird & Sánchez, *J.Agric.Univ.P.Rico* 55:461, 1971; transm. *Bemisia tabaci*, Puerto Rico [51,4745].
Jaczewski, Arthur Louis de, 1863–1932; born in Russia, Univ. Berne, Switzerland, St Petersburg Botanical Garden, first director Bureau of Mycology and Phytopathology, Russian Ministry Agriculture, later became the Jaczewski Institute, one of the founders of plant pathology in USSR.
jamaicensis, *Zygophiala*.
Japanese apple rust (*Gymnosporangium yamadae*); —— blight (*Exobasidium*) tea; —— bunching onion, see *Allium*; —— millet, see millets; —— pear rust (*G. asiaticum*); —— plum, see *Prunus*; —— summer grapefruit, see *Citrus*.
japonicum, *Exobasidium*.
jargon, see terminology.
jarrah dieback (*Phytophthora cinnamomi*) *Eucalyptus marginata*.
jasmine cholorotic ringspot Wilson, *Indian Phytopath.* 25:157, 1972; transm. *Bemisia tabaci*, India (Kerala) [52,1574].
—— infectious chlorosis Lawrence 1713, published 1715; fide Orton *Phytop.* 14:198, 1924, England; —— —— mosaic McLean 1960; fide Waterworth, see jasmine latent 1, USA.
—— interveinal chlorosis (tobacco ringspot v. assoc.), Morton 1977, fide Cooper 1979, *Jasminum officinale*, USA (S. Carolina).
—— latent 1 and mild mosaic Waterworth, *Phytop.* 61:228, 1971; jasmine latent, transm. sap to *Chenopodium*, *Jasminum odoratissimum*; jasmine mild mosaic, transm. *Myz.pers.* to *Chenopodium*, *J. multiflorum*, USA [50,2993].
—— phyllody Kandaswamy et al., *South Indian Hort.* 21:35, 1973; poss. MLO [54,1305].
—— yellow blotch Cooper & Sweet, *Forestry* 49:73, 1976; arabis mosaic v. isolated from jasmine, UK [55,5086].
—— —— ring mosaic Mariappan & Ramanujam, *South Indian Hort.* 23:77, 1975; transm. *Dialeuroides kikaldi* [56,2090].
—— —— ringspot mosaic v. Benigno, Favali-Hedayat & Retuerma, *Philipp.Phytopath.* 11:91, 1975; rod 700–750 nm long, transm. *Bemisia tabaci* [57,1230].
jatrophae, *Uromyces*, see *U. manihotis*.
jatropha mosaic v. Bird 1957; Bird & Sánchez, *J.Agric.Univ.P.Rico* 55:461, 1971; Kim et al.,

Phytop. 76:80, 1986; particles like a Geminivirus, transm. *Bemisia tabaci*, West Indies, *J. gossypifolia* [37:421; 51,4745; 65,5345].
Java black rot (*Botryodiplodia theobromae*) sweet potato; —— downy mildew (*Peronsclerospora maydis*) maize.
javart (*Diplodinia castaneae*) sweet chestnut.
Javesella, transm. barley yellow striate mosaic, oat sterile dwarf, wheat European striate mosaic.
Jensen, Jens Ludwig, 1836–1904; born in Denmark, began a company for selling tested seed, notable as the originator of the hot water treatment for control of smuts of temperate cereals; Gold Medal National Agricultural Society of France 1886. *Phytop.* 7:1, 1917.
Jerusalem artichoke, *Helianthus tuberosus*; Johnson, *J.agric.Res.* 43:337, 1931, storage rots; McCarter & Kays, *PD* 68:299, 1984, USA (Georgia) [63,3690].
jimson weed, or thorn apple, *Datura stramonium*.
job's tears, *Coix lachryma-jobi*.
Johnson, James, 1886–1952; born in USA, Univ. Wisconsin, tobacco diseases, early work on soil treatment by steam, a pioneer in the study of viruses, particularly tobacco mosaic v.; used chambers with controlled humidity and temp. *Phytop.* 44:335, 1954; *A.R.Phytop.* 22:27, 1984.
——, Thorvaldur, 1897–1979; born in Canada, Univ. Minnesota, Agriculture Canada, director rust program, Winnepeg; an authority on physiologic specialisation in temperate cereal rusts, genetics of *Puccinia graminis*; Fellow Royal Society of Canada 1950, Elvin Stakman award 1958, Gold Medal Professional Institute of the Public Service of Canada 1962. *Can.J.Pl.Path.* 2:3, 1980 *Phytop.* 70:173, 1980.
johnsonii, *Monilinia*.
Johnston fruit spot (*Pyricularia grisea*) banana.
Jones, Lewis Ralph, 1864–1945; born in USA, Univ. Michigan, Vermont Agricultural Experiment Station, Univ. Wisconsin; did early work on pectolytic enzymes, notable for studies on the effects of environment on disease using the Wisconsin soil temp. tank, breeding and screening for resistance. *Phytop.* 36: 1, 1946; *A.R.Phytop.* 17:13, 1979.
jonquil mild mosaic (narcissus late season yellows v.).
Jørstad, Ivar, 1887–1967; born in Norway, Univ. Oslo, government mycologist in 1919, a post held for 37 years. In 1940 he was still the only plant pathologist in the Norwegian civil service; fungus

systematics, particularly Uredinales; monographs on the Erysiphaceae and Ustilaginales of his own country. *Friesia* **8**:113, 1967; *Mycologia* **63**:697, 1971.

journals, and other serial publications, see bibliography.

juglandis, *Cylindrosporium, Marssonina, Mycosphaerella*.

Juglans, walnut.

jujube, *Ziziphus*.

juniper broom rust (*Gymnosporangium nidus-avis*), *Juniperus horizontalis, J. scopulorum*.

juniperi, *Herpotrichia, Kabatina*.

juniperi-virginianae, *Gymnosporangium*.

juniperivora, *Phomopsis*.

jute, *Corchorus*.

—— **mosaics**, several reports from India and Pakistan; Mitra et al., *Sci.Cult.* **50**:127, 1984; isometric particle assoc.; transm. by *Bemisia tabaci* was described, see: Bisht & Mathur, *Curr.Sci.* **33**:434, 1964; Varma, Rao & Capoor, *Sci.Cult.* **32**:466, 1966; Ahmad, *F.A.O.Pl.Prot.Bull.* **26**:169, 1978 [**44**,145; **46**,1599; **59**,805; **64**,1137].

K

Kabatiella Bubák 1907; Hyphom.; Carmichael et al. 1980; conidiomata pycnidial or acervular; con. aseptate, hyaline; Hermanides-Nijhof, *Stud.Mycol.* 15, 1977, gave *Kabatiella* as a synonym of *Aureobasidium* Viala & Boyer 1891 which was described; when parasitic on plants the spp. usually form acervuli with slimy masses of conidia which vary in shape and size; Holliday 1980.

K. caulivora (Kirchner) Karak 1923 or *Aureobasidium caulivorum* (Kirchner) W.B. Cooke 1962, fide Hermanides-Nijhof, see above; con. sickle shaped, 12–22 × 3.5–5 μm; clover scorch or northern anthracnose, infects lucerne, seedborne; Beale & Thurling, *Aust.J.agric.Res.* **31**:927, 935, 1980, genetic control of resistance in *Trifolium subterraneum*; Anderson et al., *Aust.J.exp.Agric.Anim.Husb.* **22**:182, 1982, sheep, weather & disease [**60**,4517–18; **62**,251].

K. zeae Nirita & Y. Hiratsuka 1959; con. av. 32.5 × 2.6 μm; Dingley, *N.Z.J.agric.Res.* **16**:325, 1973, gave the morphology and a proposal for transfer to *Aureobasidium*; maize eyespot, seedborne; Reifschneider & Arny, *TBMS* **75**:239, 1980; *Phytop.* **73**: 607, 1983, cultural & morphological variability, yield loss [**60**,4417; **62**,3849].

Kabatina R. Schneider & v. Arx 1966; Coelom.; Sutton 1980; conidiomata acervular, conidiogenesis enteroblastic phialidic; con. aseptate, hyaline. *K. thujae* R. Schneider & v. Arx; Gibson & Sutton, *Descr.F.* 489, 1976; *K. juniperi*, same authors, may be the same; both fungi cause diebacks; *Chamaecyparis, Cupressus, Juniperus*; Perry & Peterson, *PD* **66**:1189, 1982, on juniper in USA (New Jersey) [**62**,1676].

Kaincopé, coconut; etiology poss. the same as, or similar to, palm lethal yellowing q.v.

kalanchoë latent v. Hearon 1981; *Phytop.* **72**:838, 1982; **74**:670, 1984; Carlavirus, filament 600–620 nm long, transm. sap. USA, *K. blossfeldiana*; Paludan, *Tidsskr.PlAvl.* **89**:191, 1985, in Denmark, inactivation by heat & meristem tip culture [**61**,7052; **63**,5458; **65**,748].

— **mosaic** Burnett & Long, *PDR* **46**:692, 1962; USA (Florida), *K. flammea* [**42**:127].

— **top spotting v.** Hearon & Locke 1982; *PD* **68**:347, 1984; bacilliform particles assoc. 50–100 × 20–25 nm, transm. pollen, seed, USA, *K. blossfeldiana* [**63**,3399].

kale, see *Brassica*.

— **latent** Kitajima, Camargo & Costa, *Bragantia* **29**:181, 1970; poss. Carlavirus, particle 650 × 15 nm, Brazil [**51**,2022].

kalkoffii, *Rhizosphaera*.

kangaroo apple, *Solanum aviculare*; disease caused by *Phytophthora infestans* in USSR; Drozdovskaya 1983 [**65**,327].

karnal bunt (*Tilletia indica*) wheat.

kartoffel, and kartoffel rollmosaik = potato M v.

katsurae, *Phytophthora*.

Katte (cardamom Indian mosaic v.).

kauffmanii, *Lentinus*.

kawakamii, *Phellinus*.

Keitt, George Wannamaker, 1889–1969; born in USA, Univ. Wisconsin; notable for his work on apple scab (*Venturia inaequalis*) epidemiology to improve chemical control and the genetics of pathogenicity, an early example of the use of *Neurospora* genetical techniques in plant pathology; wrote a history of plant pathology in Horsfall & Dimond 1959. *Phytop.* **60**:1155, 1970; *A.R.Phytop.* **19**:35, 1981.

kenaf, see *Hibiscus*.

kennedya Y v. Dale, Gardiner & Gibbs, *Aust.Pl.Path.Soc.Newsl.* **4**:13, 1975; poss. Potyvirus, filament 710 nm long, transm. *Aphis craccivora*, non-persistent, Australia, *K. rubicunda* [**55**,109].

— **yellow mosaic v.** Dale, Gardiner & Gibbs 1975; Gibbs, *Descr. V.* 193, 1978; Tymovirus, isometric 28 nm diam., transm. sap, readily, Australia (eastern seaboard), *K. rubicunda*, occurs naturally in Papilionaceae.

*****Kentucky blue grass**, *Poa pratensis*.

Kerala wilt, coconut; see palm lethal yellowing.

Kerdill hampa = rice ragged stunt v.

Kern, Frank Dunn, 1883–1973; born in USA, Univ. Iowa, Purdue, Penn. State; an authority on rust taxonomy; wrote: *A revised taxonomic account of* Gymnosporangium, 1973, a previous edn. 1911 had a different title. *Phytop.* **64**:766, 1974; *Mycologia* **66**:739, 1974.

kernel smut (*Tilletia barclayana*) rice.

Keyworth, William Graham, 1915–75; born in England, Royal College of Science, London; East Malling Research Station, Kent; National Vegetable Research Station, Warwick; deputy director 1967; original work on *Fusarium* and *Verticillium* wilts, particularly of bean, hop, pea, tomato. *AAB* **81**:445, 1975; *Ann.Rep.Natn.Veg.Res.Stn.* for 1975:20.

khaira, or hadda, rice; Zn deficiency; Nene, *Indian J.agric.Sci.* **42**:87, 1972.

Khuskia Hudson 1963; ? Sphaeriales; *K. oryzae* Hudson; Sivanesan & Holliday, *Descr.F.* 311, 1971; anam. *Nigrospora oryzae* (Berk. & Broome) Petch 1924; asco. hyaline, curved, tapering to base, rounded ends, at first aseptate but on germination may form an off-centre, transverse septum, 16–21 × 5–7 μm; con. mostly 12–14 μm diam., largely saprophytic, assoc. with debris of monocotyledons, especially maize, rice, sorghum; one of the less important fungi in the maize stalk rot syndrome; Holliday 1980.

kidney bean, see bean.

kikuchii, *Cercospora*.

kikuyu grass yellows (an undescribed phycomycete near *Achlya*), serious on *Pennisetum clandestinum*, Australia (New South Wales); Wong, *PDR* **59**:800, 1975, etiology; Wong & Wilson, *Australasian Pl.Path.* **12**:47, 1983, reactions of cvs. & breeding lines [**55**,1780; **63**,3329].

kiliense, *Acremonium*.

kiwi fruit, *Actinidia chinensis*; Hawthorne & Reid, *N.Z.J.exp.Agric.* **10**:333, 1982, fungicides against storage rots; Opgenorth, *PD* **67**:382, 1983, storage rots; Harvey et al., *Crop Protect.* **5**:277, 1986, protective storage & transit environments [**62**,1132, 3596].

klendusity, a susceptible host does not become infected in the presence of a pathogen because of qualities preventing or hindering the operation of a vector or other inoculating agent, e.g. a var. that is resistant to an aphid will not become infected with viruses for which the aphid is a vector. Klendusity is a form of disease escape but the term should not be applied to other forms of disease escape q.v., FBPP.

Klotz, Leo Joseph, 1895–1984; born in USA, Michigan State College, Univ. Washington, New Hampshire, California Riverside; authority on citrus diseases particularly those caused by bacteria and fungi; wrote, *Colour handbook of citrus diseases*, edn. 4, 1974. *Phytop.* **75**:253, 1985.

Knight, Thomas Andrew, 1759–1838; born in England, Univ. Oxford, used flowers of sulphur against pear scab (*Venturia pirina*) and peach leaf curl (*Taphrina deformans*). In 1799 he recorded the differential response of cereal vars. to rusts; experimented with pea hybridisation and narrowly missed making the discoveries which Mendel did 60 years later; FRS 1805. *Dictionary of Scientific Biography*, vol. 7:408, 1973.

knobbiness, potato; mostly due to an irregular moisture supply during growth; Rich 1983.

knobby tuber, potato; genetic disorder; Katahdin

potato; Plaisted & Peterson, *Am.Pot.J.* **49**:285, 1972 [**52**,1246].

knot (*Pseudomonas syringae* pv. *savastanoi*) olive, (*Sphaeropsis tumefaciens*) citrus, some woody ornamentals.

koae, *Uromyces*.

Koch, Robert, 1843–1910; born in Klausthal, Hanover, Prussia, German bacteriologist, medicine at Univ. Gottingen, chair at Berlin Univ. 1885; the greatest of all bacteriologists. Koch and Pasteur laid the foundations of the subject. His etiological work on anthrax, published in 1876, gave the final refutation of the doctrine of spontaneous generation and showed that bacteria caused animal diseases. He set down the famous criteria or postulates which needed to be fulfilled to demonstrate pathogenicity. But this was nearly 70 years after Prévost had found that a fungus could cause a disease in plants. Foreign member of the Royal Society 1897, Nobel Prize 1905. *Proc.R.Soc.Lond.*B **83**:xviii, 1911; Bulloch, *A history of bacteriology*, 1938; Grainger, *A guide to the history of bacteriology*, 1958.

Koch's postulates, the essential procedural steps, with some modification, can be observed in nearly all cases. They lay down the criteria for proving the pathogenicity of an organism. There are 2 questions: is it a pathogen and does it cause the disease? The basis for them was laid by Robert Koch in 1884 and 1890; Rivers, *J.Bact.* **33**:1, 1937; FBPP stated them: (1) the suspected causal organism must be constantly assoc. with the disease; (2) it must be isolated and grown in pure culture; (3) when a healthy plant is inoculated with it the original disease must be reproduced; (4) added later by E.F. Smith, the same organism must be reisolated from the experimentally infected plant; Evans, *Yale J.Biol.Medic.* **49**:175, 1976, assessments of the postulates; Bos, *Neth.J.Pl.Path.* **87**:91, 1981, the postulates with particular reference to viruses.

kodo millet, see millet.

koepkei, *Mycovellosiella*.

kohlrabi, see *Brassica*.

kola, *Cola* spp., Oludemokun, *PANS* **25**:265, 1979 [**59**,1423].

Koleroga Donk 1958; Ceratobasidiaceae; characterised by the repent basidiiferous hyphae on the lower surface of the leaf. The genus was erected for the fungus causing coffee black rot in S. India; the local name is kole roga, meaning slimy disease. The unsettled taxonomic position of *K. noxia* was discussed by Muthappa, *TBMS* **73**:159, 1979, who gave a description of the sp. on coffee; the fungus may be largely tropical but see Kouyeas 1979. There are 2 stages: the pellicular, where the repent, basidia bearing hyphae form on green leaves, twigs and berries; and the sclerotial, where the dying leaves hang from the branches, bearing sclerotia 100–300 μm diam. Muthappa did not accept the transfer of *K. noxia* to *Ceratobasidium* [**59**,1741; **60**, 1463].

kole roga (*Koleroga noxia*) coffee, (*Phytophthora arecae*) areca palm.

korrae, *Leptosphaeria*.

kresek (*Xanthomonas campestris* pv. *oryzae*) rice.

Kretzschmaria Fr. 1849; Xylariaceae; Dennis, *Kew Bull.* 1957:308, described *K. clavus* (Fr.) Sacc. 1883 as a typical member of the genus, asco. 26–40 × 7–12 μm. Ko et al., *Phytop.* **67**:18, 1977; **72**:1357, 1982; *Pl.Path.* **35**:254, 1986, *K. clavus* causing a root decay of *Macadamia integrifolia* in USA (Hawaii) and infecting natural hosts in Hawaiian forests [**56**,3741; **62**,1663; **65**,5747].

Kribi, as for Kaincopé.

kriegeriana, *Marssonina*.

kromnek (tomato spotted wilt v.).

Kuehneola Magnus 1898; Uredinales; Wilson & Henderson 1966; teliospores in chains or with several cells which separate. *K. uredinis* (Link) Arthur 1906; Laundon & Rainbow, *Descr.F.* 202, 1969; Parmelee & Carteret, *Fungi Canadenses* 307, 1986; autoecious, *Rubus*, aecia uredinoid, uredospores present, teliospores in chains of 4–7 cells which fragment; the uredia can be distinguished from those of *Phragmidium* by not

having the distinctive paraphyses; this rust can be damaging; Fischer & Johnson, *Phytop.* **40**:199, 1950 [**29**:469].

kuehnii, *Puccinia*.

Kühn, Julius Gotthelf, 1825–1910; born in Pulsnitz, Saxony, Prussia; an agriculturist who became interested in plant diseases, Univ. Leipzig, agricultural chair at Univ. Halle; a founder of plant pathology whose studies on diseases caused by fungi demonstrated comprehensively how important these organisms were as pathogens. Kühn wrote the first, full text on plant diseases: *Die Krankheiten der Kulturgewächse, ihre Ursachen und ihre Verhütung*, 1858. Whetzel 1918; *A.R.Phytop.* **16**:343, 1978.

kumquat, *Fortunella*.

Kunkel, Louis Otto, 1884–1960; born in Mexico, Univ. Missouri, Washington, Colombia; Hawaiian Sugar Planters Association; Boyce Thompson Institute 1923–32; director, plant pathology, Rockefeller Institute; early, original work on insect transmission of the agents causing yellow diseases of plants, then thought to be caused by viruses; developed heat treatment methods to free plants from infections of the yellows type. *Phytop.* **50**:777, 1960.

kunzei, *Cytospora*, *Valsa*.

kusanoi, *Monilia*, *Monilinia*.

Kutilakesa, see *Nectriella pironii*.

kyotensis, *Calonectria*.

L

lablab, hyacinth bean, *Dolichos lablab* or *Lablab purpureus*; Kanapathipillai, *TBMS* **78**:503, 1982, fungi from seed [**61**, 6051].
— ringspot mottle Cheo & Tsai, *Acta phytopath.sin.* **5**:7, 1959; transm. sap, aphid, China [**39**:138].
— yellow mosaic Capoor & Varma, *Curr.Sci.* **19**:248, 1950; transm. *Bemisia tabaci*, India, *L. niger* [**30**:210].
laburnum, *Laburnum anagyroides*.
— mosaic Masters 1877; Bos, *Neth.J.Pl.Path.* **70**:168, 1964; Europe, Cooper 1979 [**44**,1452].
— yellow vein Schultz & Harrap, *AAB* **79**:247, 1975; Phytorhabdovirus assoc., particle *c*.250 × 95 nm, England [**54**,5574].
Lachnellula P. Karsten 1884; Hyaloscyphaceae; ascomata orange, usually sessile, on conifers; Dharne, *Phytopath.Z.* **53**:101, 1965, morphology; Oguchi, *Trans.mycol.Soc.Japan* **21**:435, 1980; **22**:165, 377, 1981, inoculations with 12 spp., on *Abies, Larix*; Baral 1984, 21 spp., Alps & Germany [**61**,890, 4389; **62**,412; **65**,6207].
L. calyciformis (Batsch) Dharne 1965; asco. 4–7.5 × 2.4–3.5 μm; canker on *Abies*, N. temperate; Yokota & Matsuzaki, *Bull. Govt.For.Exp.Stn.* 238:119, 1971, Japan, on *A. sachalinensis* causing cankers following damage, including freezing weather [**51**,1983].
L. flavovirens (Bresad.) Dennis 1962; asco. 10–22.5 × 3.5–11 μm; Frajo-Apor, *Eur.J.For.Path.* **6**:360, 1976, infection of *Larix, Pinus* through wounds; see *Lachnellula pini*, Kurkela & Norokorpi [**56**,2697].
L. occidentalis, see *Lachnellula willkommii*.
L. pini (Brunch.) Dennis; asco. 10.5–22 × 4–10.5 μm; Kurkela & Norokorpi, *Eur.J.For.Path.* **9**:65, 1979, on pine in Finland with *Lachnellula flavovirens*; Oguchi, *J.Jap.For.Soc.* **61**:215, 1979, canker on *Pinus pumila, P. strobus*, Japan, spreads in regions of heavy snowfall [**59**, 957, 959].
L. subtilissima (Cooke) Dennis; asco. 7–12 × 1.5–2.5 μm; conifer canker, Himalayas; Sharma et al., *Bangladesh J.Bot.* **9**:77, 1980, spp. in India [**60**,3592].
L. willkommii (Hartig) Dennis; Buczacki, *Descr.F.* 450, 1975 as *Trichoscyphella willkommii* (Hartig) Nannfeldt 1932; *Dasyscyphus willkommii* (Hartig) Rehm is also a synonym, fide Cannon et al. 1985; ascomata white with an orange or buff hymenial disc, usually up to 4 mm diam., asco. hyaline, 12–15 × 5–7 μm; larch canker, damaging, can be

confused with *Lachnellula occidentalis* (Hahn & Ayers) Dharne which follows it on cankers and grows on large debris, its ascomata and asco. are smaller. Historically *L. willkommii* has only caused serious harm on *Larix decidua* but severe disease has occurred on *L. leptolepis*, Japanese larch; some larches are effectively resistant; much work has been done on predisposing factors in outbreaks of canker but no clear cut picture has emerged; Asia, Europe; Manners, *TBMS* **36**:362, 1953; Ito et al., *Bull.Govt.For.Exp.Stn.Meguro* 155:23, 1963; Yde-Anderson, *Eur.J.For.Path.* **9**:211, 220, 347, 1979, reviews; Phillips & Burdekin 1982 [**34**:6; **43**,850].
lacquer scab (*Pseudomonas gladioli* pv. *gladioli*) gladiolus.
lactucae, *Bremia, Septoria*.
Lactuca sativa, lettuce; var. *capitata*, crisphead or iceberg.
lady's fingers, see okra.
laelia etch Jensen, *Calif.Agric.* **6**(2):7, 1952; transm. sap, USA (California), *L. anceps* [**32**:129].
— red leaf spot Peters, *J.Ultrastruct.Res.* **58**:166, 1977; bacilliform particles assoc., 190–220 × 80 nm, Germany, *Laelia* orchids [**56**,3592].
Laetiporus Murrill 1904; Polyporaceae; basidiomata forming large, irregular, imbricate pilei 10–40 cm diam., pore surface sulphurine.
Laetiporus sulphureus (Bull.:Fr.) Bond & Singer 1941; van der Westhuizen, *Descr.F.* 441, 1975, cited also as *L. sulphureus* (Bull.:Fr.) Murrill; cubical heart rot red brown; tree trunks, stumps, logs.
Laetisaria Burdsall, *TBMS* **72**:420, 1979; Corticiaceae; basidiomata effused, pellicular, smooth; basidia with 4 sterigmata [**59**,1130].
L. arvalis Burdsall in Burdsall et al., *Mycologia* **72**:729, 1980; has characters intermediate between *Laetisaria* and *Phanerochaete*; Martin et al., *Phytop.* **73**:1445, 1983; **74**:1092, 1984, suppression of *Pythium* in soil & effect on infection by *Pleospora betae*: Larsen et al., *PD* **69**:347, 1985, effects on *Thanatephorus cucumeris* in the field [**63**,2006; **64**,1325, 4544].
L. fuciformis (McAlp.) Burdsall 1979; anam. *Isaria fuciformis* Berk. 1872; red thread of turf grasses, cool and humid areas of Australasia, Europe, N. America where the disease may be called pink patch; Stalpers & Loerakker, *CJB* **60**:529, 1982, morphology, culture & symptoms; Cahill et al., *PD* **67**:1080, 1983, fertility & disease; Hims et al.,

Pl.Path. **33**:513, 1984, chemical control [**61**,6435; **63**,3412; **64**,1625].

laevis, *Tilletia*.

La France, a disease of *Agaricus brunnescens*, the common cultivated mushroom, assoc. with the presence of virus particles; Hicks & Haughton, *TBMS* **86**:579, 1986 [**65**,6278].

Lagena Vanterpool & Ledingham 1930; Lagenidiales; thallus of 1 cell, attached to an inner, host cell wall by a thick collar of callus, parasitic in the roots of temperate cereals; *L. radicicola*, same authors, browning root rot; Macfarlane, *TBMS* **55**:113, 1970 [**50**,594].

lagenariae, *Synchytrium*.

lagenaria witches' broom Chou, Yang & Huang, *PDR* **60**:378, 1976; MLO assoc., Taiwan, *L. siceraria*, bottle gourd [**56**,460].

lagenarium, *Colletotrichum*.

Lagenidiales, Oomycetes; zoospores not dimorphic, flagella laterally attached, holocarpic, endobiotic; parasites of algae, other Phycomycetes, microfauna and cereal roots; *Lagena*.

lamii, *Peronospora*.

laminaria coiling stunt Wang et al., *Acta microbiol.sin.* **23**:73, 1983; Shi et al., *ibid* **23**:108, 1983; MLO assoc., sea tangle [**63**,82–3].

laminated butt rot (*Inonotus weirei*) conifers.

lamium mild mosaic v. Lovisolo 1953; Lisa et al., *AAB* **100**:467, 1982; resembles broad bean wilt v. in particle morphology, containing RNA, in vitro properties and aphid transm., non-persistent, but considered to be distinct from BBWV [**61**,5615].

— **yellow mosaic** Hein, *Phytopath.Z.* **29**:79, 1957 [**36**:749].

lantana mosaic v. Arnott & Smith, *Virology* **34**:25, 1968; rod 1400 × 14–15 nm, transm. sap, USA (Texas), *L. horrida* [**47**,2736].

Laodelphax striatellus, transm. barley striate mosaic, cereal northern mosaic, cereal tillering, maize rough dwarf, oat pseudo-rosette, rice black streaked dwarf, rice stripe.

larch, *Larix*.

— **witches' broom** Nienhaus, Brüssel & Schinzer, *Z.PflKrankh. PflSchutz.* **83**:309, 1976; bacterium assoc., Germany (W.) [**56**,868].

Large, Ernest Charles, 1902–76; born in England, Univ. London, chemical engineer 1928–36; wrote 3 novels and the classic, *The advance of the fungi*, 1940, a history of plant pathology; in 1946 joined the Ministry of Agriculture, UK, and began work on quantitative disease assessment, and its relation to yield, see phytopathometry; these studies formed the basis for much of the epidemiological work which was to follow, Gregory, *Pl.Path.* **31**:7, 1982; wrote with A.E. Cox, *Potato blight*

epidemics throughout the world, *USDA agric.Handb.* 147, 1960. *TBMS* **69**:167, 1977.

large spored rust (*Chrysomyxa ledicola*) spruce.

laricina, *Ascocalyx*, *Guignardia*, *Mycosphaerella*; **laricinum**, *Triphragmiopsis*.

larici-populina, *Melampsora*.

laricis, *Meria*.

laricis-leptolepidis, *Mycosphaerella*, see *M. laricina*.

larici-tremulae, *Melampsora*, see *M. populnea*.

Larix, larch.

Lasiodiplodia theobromae, see *Botryodiplodia theobromae*.

lata, *Eutypa*, see *E. armeniacae*.

late blight (*Phytophthora infestans*) potato, tomato; American usage; blight is British usage; (*Septoria apiicola*) celery; — **leaf rust** or — **yellow rust** (*Pucciniastrum americanum*) *Rubus*; — **leaf spot** (*Mycosphaerella berkeleyi*) groundnut.

latent infection, an inapparent infection that is chronic and in which a certain host and parasite relationship is established, FBPP; viruses that occur in plants without causing macroscopic symptoms are said to be latent; Hayward, *A.R.Phytop.* **12**:87, 1974, bacteria; Verhoeff, *ibid* **12**:99, 1974, fungi; Swinburne in Dennis 1983.

— **period**, (1) elapsed time between infection and the appearance of disease symptoms; (2) elapsed time between phage infection or induction and lysis of bacterial cells; (3) period after acquisition of virus by a vector before it becomes infective, FBPP.

lateral bud necrosis, blackcurrant; poss. physiological disorder; England, Scotland; Gill, *Pl.Path.* **34**:297, 1985 [**64**,4413].

lateralis, *Phytophthora*.

Laterispora Uecker, Ayers & Adams, *Mycotaxon* **14**:492, 1982; Hyphom.; conidiomata hyphal, conidiogenesis holoblastic; con. pigmented, filiform. *L. brevirama*, same authors, con, 7–12 septate, 70–140 × 6–7 μm; on sclerotia of *Sclerotinia*; Parfitt et al., *Pl.Path.* **33**:441, 1984 [**63**,5313].

lateritium, *Fusarium*.

lathyrus chlorotic leaf spot Kulkarni et al. 1970; transm. sap, Kenya, *L. latifolius* [**50**,2637f].

Lathyrus odoratus, sweet pea.

launaea mosaic v. Padma, Verma & Singh, *Z.PflKrankh.PflSchutz.* **80**:68, 1973; Verma et al., *Zentbl.Bakt.ParasiKde* 2 **130**: 121, 1975; rod 750–930 × 6–18 nm, transm. sap, *Myz.pers.*, India (N.), *L. nudicaulis*; Naqvi & Mahmood, *Curr.Sci.* **45**:152, 1976, *L. asplenifolia* [**53**,647; **55**,1400, 3487].

— **phyllody** Phatak et al., *Zentbl.Bakt.ParasitKde* 2 **131**:205, 1976; MLO assoc., India, *L. nudicaulis* [**56**,328].

lavandula mosaic (poss. alfalfa mosaic v.), Marchoux & Rougier, *Annls.Phytopath.* **6**:191, 1974 [**54**,3780]. — **yellow decline** Cousin et al., *Annls.Phytopath.* **2**:227, 239, 1970; **3**:243, 1971; Moreau et al. 1974; MLO assoc., France [**50**,706; **51**,2578; **56**,5388].

lavender, *Lavandula*; — **spot** (*Cercospora kikuchii*) soybean.

laxa, *Monilinia*.

LC50, **LD50**, median lethal concentration, — lethal dose q.v.

leaf binding (*Myriogenospora*) sugarcane; — **blast** (*Paraphaeosphaeria michotii*) sugarcane, (*Botrytis allii*) onion; — **blight**, — **burn**, see separate entries; — **blight and dieback** (*Sydowia polyspora*) pines; — **blossom, fruit and stem spot** (*Septoria passifloricola*) passion fruit; — **casting mottle**, prune, prob. genetic disorder; — **circular yellow patch** (*Mycovellosiella nattrassii*) eggplant; — **curl** (*Taphrina*) several crops; — **distortion**, tomato, a variant of silvering q.v.; — **freckles** (*Clavibacter michiganensis* ssp. *nebraskensis*) maize; — **gall**, see separate entry; — **mosaic** (*C. m.* ssp. *tessellarius*) wheat; — **mottle** (*Verticillium dahliae*) sunflower; — **mould** (*Fulvia fulva*) tomato, (*Pseudocercospora abelmoschi*) hibiscus, okra; — **pod and stem spot** (*Ascochyta fabae*) broad bean, (*A. pisi*) pea; — **scald**, — **scorch**, see separate entries; — **scorch and surface canker** (*Sphaerulina rehmiana*) rose; — **and sheath blight** (*Alternaria triticina*) temperate cereals; — **sheath brown rot** (*Pseudomonas fuscovaginae*) rice; — **smut** (*Entyloma dahliae*) dahlia, (*E. oryzae*) rice; — **speckle** (*Cladosporium musae, Mycosphaerella musae, Veronaea musae*) banana; — **splitting** (*Peronosclerospora miscanthi*) sugarcane; — **and stalk spot** (*Ramularia rhei*) rhubarb; — **and stem blight** (*Pestalotiopsis funerea*) conifers; — **streak** (*Xanthomonas campestris* pv. *holcicola*) sorghum, (*X.c.* pv. *oryzicola*) rice; — **stripe**, see separate entry; — **vein** (*Apiognomonia errabunda*) plane.

leaf blight (*Alternaria dauci*) carrot, (*A. triticina*) wheat, (*Botrytis allii*) *Allium*, onion, (*Cochliobolus nodulosus*) finger millet, fowl foot grass, (*Leptosphaeria taiwanensis*) sugarcane, (*Phomopsis obscurans*) strawberry, (*Pleospora papaveracea*) opium poppy, (*Stemphylium lycopersici*) tomato, (*Xanthomonas campestris* pv. *oryzae*) rice.

— **burn**, cotton, accumulation of chloride and Na; Bhatt et al., *Cott.Grow.Rev.* **52**:228, 1975; sugarcane, high temp., Myatt, *Cane Grow.q.Bull.* **37**:115, 1974 [**54**,547; **55**,260].

— **gall** (*Exobasidium camelliae, E. giganteum*) camellia, (*E. japonicum*) azalea, (*Physoderma*

leproides) beet, (*Rhodococcus fascians*) tobacco, many other plants, sometimes leafy gall.

leafhoppers, see vector.

leaf lesions, Pascoe & Sutton, *Australasian Pl.Path.* **15**:78, 1986, proposed terms for the disposition of disease lesions and fungal structures on laminar leaves and phyllodes; epigenous: lesions visible only on leaf upper surface, extending partly through the lamina thickness; hypogenous: lesions visible only on lower leaf surface, extending partly through the lamina thickness; amphigenous: lesions on either leaf surface, each extending partly through the lamina; hologenous: lesions extending through the entire leaf thickness, equally visible on both surfaces.

leaflet shatter, oil palm; assoc. with B deficiency, Turner 1981.

leaf scald (*Monographella albescens*) rice, (*Rhynchosporium secalis*) temperate cereals, grasses (*Xanthomonas albilineans*) sugarcane; and see Pierce's grapevine leaf scald, xylem limited bacteria.

— **scorch** (*Apiognomonia veneta*) plane, (*Didymella exitalis*) barley, wheat, (*Diplocarpon earlianum*) strawberry, (*Gnomonia erythrostoma*) cherry, (*Leptosphaeria bicolor, Stagonospora sacchari*) sugarcane, (*S. curtisii*) narcissus; and see almond, elm, mulberry, sycamore leaf scorch, xylem limited bacteria.

— **stripe** (*Hymenella cerealis*) Gramineae, (*Peronosclerospora sacchari*) sugarcane, (*Pseudomonas andropogonis*) sorghum, Gramineae, Leguminosae (*Pyrenophora avenae*) oat, (*P. graminea*) barley.

leak (*Rhizopus stolonifer*) several crops, fruits, vegetables.

Leandria Rangel in Maublanc et al., *Bol.Agric.S.Paulo* **16**:310, 1915; Hyphom.; conidiomata hyphal, conidiogenesis holoblastic; monotypic. *L. momordicae* has globose, muriform con., at first hyaline, becoming dark, 27–50 μm diam., with 7–18 globose cells each 6–20 μm diam., originally described from *Momordica charantia*, balsam pear. This fungus, causing net blotch of cucumber leaves, was described by Blazquez, *PD* **67**: 534, 1983, in USA (Florida). It had been described earlier by Osner, *J.agric.Res.* **13**:299, 1918, as *Stemphylium cucurbitacearum* new sp. from USA (Indiana, Ohio); he was apparently unaware of Rangel's name. Osner's fungus is not a *Stemphylium* and should be reduced to synonymy with *L. momordicae* Rangel; E.G. Simmons, personal communication, 1984 [**62**,3686].

leathery pocket (*Penicillium funiculosum*) pineapple; — **rot** (*Phytophthora cactorum*) strawberry.

lebiseyi, *Cryptodiaporthe, Phomopsis.*

lecanii, *Verticillium.*

Lecanostica H. Sydow 1922; Coelom.; Sutton 1980; monotypic, conidiomata acervular, conidiogenesis holoblastic, conidiophore growth percurrent; con. multiseptate, pigmented; *L. acicola,* teleom. *Scirrhia acicola.*

ledi, ledicola, *Chrysomyxa.*

leek, see *Allium.*

— **chlorotic streak v.** Verhoyen & Horvat and Verhoyen, *Parasitica* **29**:16, 35, 1973; rod 820–840 nm long, transm. sap, *Aphis fabae,* non-persistent; garlic, onion, shallot [**53**,3265–6].

— **yellow chlorotic streak** (leek yellow stripe v. + shallot latent v.), Paludan, *Tidsskr. PlAvl.* **84**:371, 1980 [**60**,2316].

— — **dwarf** Graichen, *Arch.Phytopath.PflSchutz.* **14**:1, 1978; Germany, apparently different from onion yellow dwarf v. [**57**,5795].

— — **stripe v.** Bremer 1937; Bos, *Descr.V.* 240, 1981; Potyvirus, filament *c.*820 nm long, transm. sap, aphid, non-persistent, onion infected with difficulty and symptomless.

legislation, see plant quarantine.

legume little leaf Hutton & Grylls, *Aust.J.agric.Res.* **7**:85, 1956; Bowyer, *ibid* **25**:449, 1974; see tomato big bud, MLO assoc., pasture legumes [**36**:190; **54**,100].

*****legumes,** Carr in Western 1971; Bird & Maramorosch, *Tropical diseases of legumes* 1975; O'Rourke 1976; Meiners, *A.R.Phytop.* **19**:189, 1981, genetics of resistance; *Pest control in tropical grain legumes,* Centre Overseas Pest Res., London, 1981; Allen 1983; Boswell & Gibbs 1983, viruses.

legume yellows = bean leaf roll v.

leiphaemia, *Amphiporthe.*

lemon, see *Citrus.*

— **frond** (*Fusarium oxysporum* f. sp. *elaeidis*) oil palm.

lentil, *Lens culinaris*; Khare in Webb & Hawtin ed. *Lentils,* 1981.

— **yellows** (pea enation mosaic v.), Vovlas & Rana, *Phytopath. Mediterranea* **11**:97, 1972 [**52**,2055].

Lentinus Fr. 1825; Polyporaceae; basidiomata stipitate, hymenophore lamellate, triangular in section with a thin lamella edge, decurrent; some spp. are important in timber decay; Pegler, *Kew Bull.* additional series, 10, 1983, world monograph.

L. kauffmanii A.H. Smith in Bier & Nobles, *Can.J.Res.* C **24**:118, 1946; basidiospores 4.5–6.7 × 2.5–3.5 μm; brown pocket rot of *Picea sitchensis,* sitka spruce, on other conifers, no external symptoms of decay in standing trees,

basidiomata formed on felled and exposed timber; Japan, N. America [**26**:223].

L. lepideus (Fr.: Fr.) Fr. 1825; basidiospores 8.5–12.5 × 4.7 μm; brown cubical rot of conifer wood, mostly saprophytic, prob. the best known of all spp.; Wagener, *Phytop.* **19**:705, 1929 [**9**:147].

L. polychrous Lév. 1844; basidiospores 6–9 × 2.7–3.3 μm; mostly saprophytic, white stringy rot of *Shorea robusta.*

L. squarrosulus Mont. 1842; basidiospores 5.5–7.5 × 1.7–2.5 μm; common in tropics of Africa, Asia, Australasia; important as causing a white rot of *Shorea robusta* heartwood in India.

lentis, *Ascochyta.*

leontodon chlorotic mosaic v. Singh & McDonald, *Can.Pl.Dis.Surv.* **60**:47, 1980; rod 140 and 160 nm long, transm. sap, Canada, *L. autumnalis,* a common weed in potato in the east, serologically related to hypochoeris mosaic v. [**60**,4161].

lepideus, *Lentinus.*

Lepidosaphes ulmi, transm. angelonia little leaf.

leproides, *Physoderma.*

Lepteutypa Petrak 1923; Amphisphaeriaceae; ascomata not united in valsoid groups, mostly immersed, often covered by a structure like a clypeus; asco. 2 or more septate, pigmented.

L. cupressi (Nattrass et al.) Swart, *TBMS* **61**:79, 1973; anam. *Seiridium cupressi* (Guba) Boeswinkel 1983; Booth & Gibson, *Descr.F.* 325, 1972, as *Rhynchosphaeria cupressi* Nattrass, C. Booth & Sutton 1963; asco. 3 septate, 16–19 × 6–7.5, becoming brown; con. 27–32.5 × 7.5–10 μm, appendages present; assoc. with *Cupressus* canker, see *Seiridium* [**53**,711].

Leptodothiorella Höhnel 1923; Coelom.; Sutton 1980; conidiomata pycnothyrial, conidiogenesis enteroblastic, phialidic; con. aseptate, hyaline; spermatial states of *Botryosphaeria, Guignardia.*

Leptographium Lagerb. & Melin 1928; Hyphom.; fide Ellis 1971; Carmichael et al. 1980; conidiomata hyphal, conidiogenesis holoblastic; conidiophore growth percurrent, penicillate branching towards the apex; con. aseptate, hyaline; assoc. with *Verticicladiella* q.v., Harrington & Cobb, Wingfield, in staining of conifers. **L. terebrantis** Barras & Perry 1971; Highley & Tattar, *PD* **69**:528, 1985, assoc. with *Dendroctonus terebrans,* black turpentine beetle, blue stain and death of pine; Wingfield, *Eur.J.For.Path.* **16**:299, 1986, pathogenicity of 3 spp. to *Pinus strobus* [**64**,5525; **66**,1634].

Leptomelanconium Petrak 1923; Coelom.; Sutton 1980; conidiomata stromatic, conidiogenesis enteroblastic, phialidic; con. 0–1 septate, brown, thick wall, verruculose, base truncate. *L. pinicola*

(Berk. & M.A. Curtis) R.S. Hunt, *CJB* **63**:1157, 1985; con. mostly 3 septate, 20–25 × 6–12 μm; see Morgan-Jones, *CJB* **49**:1012, 1971, and Hunt for *Gloeocoryneum* Weindlmayre 1964 = *Leptomelanconium*, but see Sutton 1980; the fungus is assoc. with a needle blight of pine in N. America (W.) [**65**,407].

Leptosphaeria Ces. & de Not. 1863; Phaeosphaeriaceae or Pleosporaceae; Sivanesan 1984; ascomata solitary, small, scattered; asco. 2 or more septate, yellowish to yellow brown, smooth to finely verruculose; Lucas & Webster, *TBMS* **50**:85, 1967, anam. of British spp.; Holliday 1980; Shoemaker, *CJB* **62**:2688, 1984, 60 spp. & keys to allied genera [**46**,2192; **64**,4156].

L. avenaria Weber 1922; anam. *Septoria avenae* Frank 1895; Sivanesan, *Descr.F.* 312, 1971; the anam. has been referred to *Stagonospora avenae* (Frank) Bisset, *Fungi Canadenses* 239, 1982; asco. 3 septate, 23–28 × 4.5–6 μm; con. 3 septate, 24–45 × 3–4 μm; oat speckled blotch, f.sp. *avenaria*, seedborne; also on wheat f. sp. *triticea*.

L. bicolor D. Hawksw., W. Kaiser & Ndimande in Kaiser et al. *Mycologia* **71**:483, 1979; Punithalingam, *Descr.F.* 771, 1983; asco. mostly 3 septate, wider in the middle, 28–42 × 8–12 μm, anam. a *Stagonospora* which differs from *S. saccharis* Lo & Ling q.v., con. mostly 3 septate, 24–28 × 7–8.5 μm; *S. sacchari* Sawada is illegitimate; sugarcane leaf scorch in Kenya, weakly pathogenic to *Pennisetum purpureum* [**59**,895].

L. coniothyrium (Fuckel) Sacc. 1875; anam. *Coniothyrium fuckelii* Sacc. 1878; Punithalingam, *Descr.F.* 663, 1980; asco. 3 septate, 12–15 × 3.5–5; con. aseptate, 2.5–5 × 1.5–2.5 μm; cane or stem blight of *Rubus*, rose graft canker, other hosts; Jennings, *AAB* **93**:319, 1979, resistance in *Rubus*; Williamson & Hargreaves, *ibid* **97**:165, 1981, chemical control in *Rubus*; Horst 1983, on rose; Williamson et al., *ibid* **108**:33, 1986, development of cane blight in Germany, Scotland & USA (Washington State) [**59**,3323; **60**,5038; **66**,251].

L. herpotrichoides de Not. 1863; asco. 2–9 septate, 21–41 × 3.3–7.6 μm; causing leaf spots on barley, rye, wheat; Hosford, *Phytop.* **68**:591, 1978 [**58**,144].

L. korrae J. Walker & A.M. Smith, *TBMS* **58**:461, 1972, q.v. for some discussion of *Leptosphaeria*; asco. mostly 7 septate, 140–170 × 4.5 μm, spring dead spot of grasses, on cereals, symptoms like those of take all; Australia, USA; Endo et al., *PD* **69**:235, 1985, severe in California on Bermuda grass [**64**,3469].

L. lindquistii Frezzi 1968; anam. *Phoma macdonaldii* Boerema 1970; asco. 1–3 septate, 12.5–25 × 3.5–8.5 μm; con. 4.3–7.2 × 1.4–2.9 μm; see *Plenodomus*, Boerema et al. 1981, for a description of the anam.; sunflower black stem, and premature death q.v.; seedborne; McDonald, *Phytop.* **54**:492, 1964, in Canada (Manitoba); Marić et al., *Zašt.Bilja* **32**:329, 1981, teleom. in Yugoslavia; El Sayed & Marić, *ibid* **32**:13, 1981, seed loss in Yugoslavia, variable pathogenicity [**43**,2700; **61**,810, 3007].

*****L. maculans** (Desm.) Ces. & de Not. 1863; anam. *Phoma lingam* (Tode:Fr.) Desm. 1849; Punithalingam & Holliday, *Descr.F.* 331, 1972; asco. 5 septate, 35–70 × 5–8 μm; pycnidia globose or sclerotioid, con. 3.5 × 1.5–2 μm; black leg, canker, dry rot of brassicas and other crucifers; mostly temperate or subtropical, a major disease; seedborne, longevity in seed > 3 years, air dispersed asco. are important in spread; long persistence in soil, organic, including stubble, debris; Hammond et al., *Pl. Path.* **34**:557, 1985, described systemic infection; isolates differ from avirulent to highly virulent; Holliday 1980; Bonman et al., *PD* **65**:865, 1981; Humpherson-Jones, *AAB* **103**:37, 1983, differing virulences; Nathaniels & Taylor, *Pl.Path.* **32**:23, 1983, latent infection; Newman, *ibid* **33**:205, 1984; Hammond et al. *ibid* **34**:557, 1985, systemic infection; Petrie & Lewis, *Can. J.Pl.Path.* **7**:253, 1985, heterothallism; Humpherson-Jones, *Pl. Path.* **35**:224, 1986, virulent pathotypes in brassica seed crops in England; Hammond & Lewis, *ibid* **36**:53, 1987, differences in virulence [**61**,3125; **62**,2684, 5015; **64**,2771; **65**,1576, 1581,5765; **66**,3530].

L. narmari, authors as for *Leptosphaeria korrae*, page 459; asco. 3–7 septate, 45–62 × 4–5 μm; spring dead spot of grasses, Australia.

L. nodorum E. Müller 1952; anam. *Septoria nodorum* (Berk.) Berk. 1850; Bissett, *Fungi Canadenses* 240, 1982, gives the anam. as *Stagonospora nodorum* (Berk.) Castell. & Germaño 1975–6, this name is treated as a synonym by Sivanesan 1984; Sutton & Waterston, *Descr.F.* 86, 1966; asco. 3 septate, 19–23 × 4 μm; con. mostly 3 septate, 22–30 × 2.5–3 μm, those of *Septoria tritici* are up to 70 μm long; wheat glume blotch, infects barley, many temperate grass genera, seedborne; this very important disease has been reviewed several times, often discussed with *Mycosphaerella graminicola*: Shipton et al., *Bot.Rev.* **37**:231, 1971; Baker, *EPPO Bull.* **8**:9, 1978; Berggren, *Pl.Prot.Reps.Agric.* 19, Swedish Univ. agric. Sci., 1981; Eyal, *PD* **65**:763, 1981; Brönnimann, *Neth.J.agric.Sci.* **30**:47, 1982; King et al., *AAB* **103**:345, 1983. Also: Brennan et al.,

Phytopath.Z. **112**: 281, 291, 1985, con. dispersal by simulated rain & wind; Scharen et al., *Phytop.* **75**:1463, 1985, virulence genes; Scott et al., *Pl.Path.* **34**:578, 1985, effects of canopy & microclimate on infection of tall & short wheats; Luke, et al., *PD* **70**:252, 1986, disease development from infested & treated seed [64,4245–6; **65**,1198, 2276, 3246].

L. oryzicola Hara, anam. *Septoria oryzae* Catt., fide Teranaka et al., *Ann.phytopath.Soc.Japan* **48**:19, 1982, as *oryzaecola*; asco. 3 septate, 20–22 × 5 μm; con. mostly 2–3 septate, 20–33 × 3–3.5 μm; see Ou 1985 for other *Leptosphaeria* spp. and *Septoria* spp., latter under speckled blotch; this sp. causes a rice speckled blotch [**61**,5743].

L. pratensis Sacc. & Briard. 1885; asco. 3 septate, 19–21 × 5–6 μm; con. of both *Phoma* and *Stagonospora*, fide Sivanesan 1984; leaf spot and root rot of lucerne, on other legumes; Jones & Weimer, *J.agric.Res.* **57**:791, 1938; *Leptosphaeria* q.v. Lucas & Webster; Erwin et al., *PD* **71**:181, 1987 [**18**:320; **66**,3870].

L. sacchari van Breda de Haan 1892; Morgan-Jones, *Descr.F.* 145, 1967; asco. 3 septate, 18–25 × 3.5–6 μ; con. aseptate, anam. not characterised; sugarcane ringspot, etiology appears uncertain; Sivanesan & Waller 1986.

L. taiwanensis Yen & Chi 1952; Sivanesan, *Descr.F.* 506, 1976; anam. *Stagonospora tainanensis* Hsieh, *Mycologia* **71**:893, 1979; asco. 3 septate, 40–46 × 6.5–12.5 μm; con. mostly 3 septate, 28–40 × 11.5–16 μm; also 4.5–13 × 2–5 μm like *Phoma*. sugarcane leaf blight. *Cercospora taiwanensis* was thought to be the anam., see *Pseudocercospora taiwanensis* to which it was transferred by Yen.

Leptosphaerulina McAlp. 1902; ? Dothideaceae; ascomata small, immersed, erumpent at apex, asco. extremely variable, taxonomic status uncertain; Graham & Luttrell, *Phytop.* **51**:680, 1961, on legumes in USA; Irwin & Davis, *Aust.J.Bot.* **33**:233, 1985, on legumes in Australia (E.) [**41**:310; **64**,4829].

L. chartarum, anam. *Pithomyces chartarum* q.v.

L. trifolii (Rostrup) Petrak 1959; Booth & Pirozynski, *Descr.F.* 146, 1967, gave a long synonymy which included *Leptosphaerulina briosiana* (Poll.) Graham & Luttrell 1961, a name often used; asco. 3–4 transversely and 0–2 vertically septate, hyaline, slightly pigmented later, 25–50 × 10–20 μm; plurivorous, pepper spot or burn, the disease is a leaf spot and has been mostly studied on lucerne; Holliday 1980; Thal & Campbell, *Phytop.* **76**:190, 1986, spatial pattern analysis; see *Stemphylium vesicarium*, Lowe et al. [**65**,3381].

Leptostroma Fr. 1815; Coelom.; Sutton 1980; conidiomata stromatic, conidiogenesis holoblastic, conidiophore growth sympodial; con. aseptate, hyaline, straight; **L. hedgcockii, L. pinastri, L. strobicola**, anam. of *Ploioderma hedgcockii, Lophodermium pinastri, Meloderma desmazierii*, respectively; Minter, *CJB* **58**:906, 1980 [**60**,1106].

leptostromiformis, *Phomopsis*.

leptostyla, *Gnomonia*.

Leptothyrium Kunze 1923; Coelom.; conidiomata pycnothyrial, conidiogenesis enteroblastic phialidic; con. aseptate, hyaline, falcate; *L. nervisidum*, teleom. *Gnomonia nerviseda*; Sutton, *Mycol.Pap.* 141, 1977; 1980.

Leptotrochila P. Karsten 1871; Dermateaceae; Schüepp, *Phytopath.Z.* **36**:213, 1959, key [**39**:285].

L. medicaginis (Fuckel) Schüepp; asco. 7.5–10 × 3.5 μm; lucerne yellow leaf blotch; Berkenkamp & Meeres, *Can.J.Pl.Sci.* **59**:873, 1979 [**59**,1299].

L. porri v. Arx & Boerema, *Phytopath.Z.* **48**:289, 1963; asco. 16–21 × 6–9 μm; leek black streak, Netherlands; Kortleve, *Neth.J.Pl.Path.* **71**:63, 1965; Samuels et al., *N.Z.J.Bot.* **19**:131, 1981, in New Zealand, restricted to leek [**43**,1491; **44**,2022 g; **61**,3182].

lesion, and see leaf lesions; a localised area of diseased or disordered tissue, FBPP; local lesion is mostly used in bacteriology and virology to describe the first, macroscopic, host reactions which begin at the site·of inoculation; may be called primary lesions.

—— **nematodes**, *Pratylenchus*.

—— **test**, one to determine whether an organism causes a necrotic lesion on plant material usually from a high inoculum dose; the test is not indicative of field infection, after FBPP.

lespedezae, *Phyllachora*.

lethal bole rot (*Marasmiellus cocophilus*) coconut.

lethale, *Ploioderma*; **lethalis**, *Ascochyta, Didymella*.

lethal yellowing, coconut; see palm lethal yellowing.

***lettuce**, *Lactuca sativa*, Horvath 1980, viruses; Ceponis et al., *PD* **69**:1016, 1985, diseases & disorders in crisphead shipments to New York; Patterson et al., *ibid* **70**:982, 1986; Sherf & MacNab 1986 [**60**,4131].

—— **big vein** v. Jagger & Chandler 1934; Kuwata et al., *Ann. Phytopath.Soc.Japan* **49**:246, 1983, rod assoc., *c.*320–360 × 18 nm; Mirkov & Dodds, *Phytop.* **75**:631, 1985, dsRNA assoc.; transm. *Olpidium brassicae*; Tomlinson & Garrett, *AAB* **54**:45, 1964; Campbell et al., *Phytop.* **70**:741, 1980, described infection & control; Campbell, *CJB* **63**:2288, 1985, found the longevity of *O. brassicae* to be 21–23 years and the persistence end point of LBVV to be 19–21 years [**14**:283; **44**,308; **60**,4129; **63**,332; **65**,481, 3093].

— **blotchy interveinal yellows** van Dorst, Huijberts & Bos, *Neth.J.Pl.Path.* **86**:311, 1980; transm. *Trialeurodes vaporariorum*, Netherlands, poss. the same as lettuce pseudo yellows [**60**,6190].

— **infectious yellows v.** Duffus et al., 1982; *Phytop.* **76**:97, 1986; filament 1800–2000 × 13–14 nm; transm. *Bemisia tabaci*, semi-persistent, wide host range, serious losses caused in beet, cucurbits, lettuce in USA (S.W.); see watermelon curly mottle v. [**65**,5209].

— **internal rib necrosis** (lettuce mosaic v.), Coakley et al., *Phytop*, **63**:1191, 1973 [**53**,2384].

— **interveinal yellows** Lot, Onillon & Lecoq, *Revue hort*, 209: 31, 1980; transm. *Trialeurodes vaporariorum*, France [**60**,1174].

— **mosaic v.** Jagger 1921; Tomlinson, *Descr.V.* 9, 1970; Potyvirus, filament *c.*750 nm long, transm. sap, aphid spp., non-persistent, pollen, seedborne; Kuida et al., *Nogaku Kenkyu* **56**:33, 1977, in pea [**59**,528].

— **necrotic yellows v.** Stubbs & Grogan 1963; Francki & Randles, *Descr.V.* 26, 1970; Phytorhabdovirus, bacilliform *c.*227 × 66 nm, transm. sap, aphid, persistent; Australia, New Zealand; narrow host range. The main vector is *Hyperomyzus lactucae*; *Sonchus oleraceus*, common sowthistle, is the principle source of both virus and vector which does not colonise or breed on lettuce, and the crop is infected almost incidentally by migrant aphids from sowthistle; Francki & Randles, *Virology* **54**:359, 1973, RNA properties & in vitro transcription; Boakye & Randles, *Aust.J.agric.Res.* **25**:791, 1974, transm. & vector feeding; Martin in Plumb & Thresh 1983, main weed host & vector feeding [**53**,3399; **54**,3075].

— **phyllody** Müller & Kleinhempel, *Zentbl.Bakt.ParasitKde* **127**:637, 1972; MLO assoc. [**52**,3893].

— **pseudo yellows** Bos, van Dorst & Huijberts, *Gewasbescherming* **11**:107, 1980; van Dorst et al., *Groenten Fruit* **36**:166, 1980; transm. *Trialeurodes vaporariorum*, Netherlands, infects cucumber, poss. the same as lettuce blotchy interveinal yellows; *Hort. Abstr.* **51**,4561; *Rev.appl.Ent.*A. **69**,2045.

— **ringspot** (tomato black ring v.).

— **rusty brown discolouration**, as for lettuce internal rib necrosis.

— **speckles mottle v.** Falk, Duffus & Morris, *Phytop.* **69**:612, 1979; causes lettuce speckles when beet western yellows v. is also present; the coat protein of LSMV is unstable and its genome is masked by BWYV, both viruses are transm. by *Myz.pers.*, persistent, but LSMV is only so transm. when in mixed infection with BWYV;

LSMV is sap transm. and when separated from BWYV its aphid transm. is lost; aphid transm. from mixed infections extends the host range of LSMV compared with sap transm. BWYV causes disease in lettuce but stunting is more severe when both viruses are present; Falk et al., *Virology* **96**:239, 1979, unstable infectivity & sedimentable dsRNA [**59**,1166, 3442].

— **stunt** (beet yellow stunt v.).

— **veinal necrosis** (dandelion yellow mosaic v.).

— **yellow interveinal spotting** (sonchus yellow net v.), Falk et al., PD **70**:591, 1986; USA (Florida) [**65**,5818].

— — **mosaic** Vasudeva, Raychaudhuri & Pathanian, *Curr.Sci.* **17**:244, 1948; transm. sap., poss. seedborne, India (N.) [**28**:47].

— **yellows** (beet western yellows v.).

— **yellow stunt** (cucumber mosaic v.), Tomlinson et al., *AAB* **66**: 11, 1970, Britain [**50**,1021].

leucaenae, *Camptomeris*; **leucanthemi**, *Septoria*.

Leucopaxillus Boursier 1925; Tricholomataceae; *L. giganteus* (Fr.) Singer 1939; large, cream pileus, up to 30 cm diam., infundibuliform; forms fairy rings in pastures; Peace, *Forestry* **10**:74, 1936, destruction in young Scots pine in Britain [**15**:759].

leucospila, Elsinoë; **leucostoma**, *Cytospora*.

Leucostoma (Nitschke) Höhnel 1917; Valsaceae; like *Valsa* but has a better developed, distinct pseudostroma which is delineated by dark lines; asco. aseptate; Kern, *Pap.Mich.Acad.Sci.* **40**:9, 1955; *Phytopath.Z.* **30**:149, 1957; **40**:303, 1961 [**34**:820; **37**:215; **40**:522].

L. cinctum (Fr.) Höhnel 1928; anam. *Cytospora cincta* Sacc.; asco. 14–28 × 4–7 μm, con. 5.5–10 × 1–2 μm, fide Willison, *Can.J.Res.*C **14**:27, 1936; **15**:324, 1937, who compared *Leucostoma cinctum* and *L. persoonii*; apricot apoplexy, canker and dieback in Europe, Rozsnyay, *EPPO Bull.* **7**:69, 1977; peach perennial canker, on other stone fruits, also caused by *L. persoonii*. These fungi are fairly widespread but the disease on peach has been most fully studied in Canada (Ontario) and adjacent areas of USA where it is damaging. Infection is through leaf scars, pruning cuts, wounded or dead tissue. One sp. or the other has been reported to be the major pathogen, but Dhanvantari, *Can.J.Pl.Path.* **4**:221, 1982, showed that differences in the reports of relative pathogenicity can be rationalised in terms of the availability of canker sites, host susceptibility and temp. differences. *L. persoonii* is more virulent at or above 15° and in late summer; whilst *L. cinctum* is more virulent at 10–15°, and in late autumn and spring. There is also a differential response in cvs. to infection by both spp.; the seasonal availability of infection sites is also a factor affecting disease incidence. Hildebrand, *Mem.Cornell agric.Exp.Stn.* 276, 1947; Weaver, *Can.J.Pl.Sci.* 43:365, 1963; Wensley, *CJB* 42:841, 1964; Tekauz & Patrick, *Phytop.* 64:683, 1974; Helton, *Phytoparasitica* 3:39, 1975; Randall & Helton, *Phytop.* 66:206, 1976; Schulz & Schmidle, *Agnew.Bot.* 57:99, 1983 [15:447; 16:821; 27:569; 43,3265; 53,4516; 55,4774; 57,1295; 62, 1110, 4937].

L. niveum (Hoffm.) Höhnel; anam. *Cytospora nivea* Sacc., asco. 12–16 × 3 µm; assoc. with cankers on hosts given under *Valsa sordida* but prob. less common and less severe.

*****L. persoonii** Höhnel; anam. *Cytospora leucostoma* Sacc.; asco. 10–17 × 4.5 µ, con. 5–10 × 2–4.5 µm; peach perennial canker, and other stone fruit with *Leucostoma cinctum* q.v., Willison; Bertrand & English, *Phytop.* 66:987, 1976, release & dispersal of spores; *PDR* 60:106, 1976, non-vigorous trees less resistant; Wisniewski et al. and Biggs, *CJB* 62:2804, 2814, 1984, histology of canker development, response to infection & wounds in bark [55,4775; 56,1662; 64,4405–6].

leucostomoides, *Valsa*; **leucotricha**, *Podosphaera*.
leveillei, *Puccinia*.
Leveillula Arnaud 1921; Erysiphaceae; monotypic, differs from other genera in the family in its endoparasitic habit, and in the conidiophores which emerge through the stomata and bear single con., anam. *Oidiopsis* Scalia; ascomata and asci like those of *Erysiphe*; Braun, *Nova Hedwigia* 32:565, 1980 [60,3063].

L. taurica (Lév.) Arnaud; Mukerji, *Descr.F.* 182, 1968; asco. 25–40 × 12–22 µm, con. mostly 50–80 × 14–20 µm; an important powdery mildew of eggplant, globe artichoke, red pepper, tomato and other crops; the host range is wide. The disease tends to be of most significance in the more arid areas of the Mediterranean basin, central and S. Asia; and here there are more hosts. The xerophytic characteristics of the pathogen make it of particular importance on crops under irrigation, especially by the furrow method; Thomson & Jones, *PD* 65:518, 1981, described an epidemic on tomato in USA (Utah) where yield losses of up to 90 % were caused over 100 hectares; crops tend to become more susceptible as they mature; Palti in Kranz et al. 1977; Dixon in Spencer 1978; Holliday 1980; Homma et al., *Ann.phytopath.Soc.Japan* 46: 140, 1980; 47:143, 1981, conidiophore emergence, con. formation, germination, hyphal growth & penetration; Durrien & Rostam, *Cryptogamie Mycologie* 5:279, 1984, host specialisation [60,4143; 61,2437, 3201; 64,3341].

Libertella Desm. 1830; Coelom.; Sutton 1980; conidiomata stromatic, conidiogenesis holoblastic, conidiophore growth sympodial; con. aseptate, filiform, hyaline, curved, often markedly.

*****L. blepharis** A. L. Smith 1900; typified by Messner & Sutton, *Mycotaxon* 14:325, 1982; con. mean 39 × 1–1.3 µm; dying off of apple cv. McIntosh, brown rot of apple fruit, poss. widespread on Rosaceae; Europe, infection through wounds; Messner & Jähnl, *Z.PflKrankh. PflSchutz.* 88:18, 1981; Vajna, *Acta phytopath. Acad.Sci.hung.* 17:249, 1982; Jähnl 1985, infection & chemical control [60,6015; 63,718; 65,1383].

libocedri, *Gymnosporangium*.
lichenicola, *Seimatosporium*.
lichens, a biological, not a systematic, group; each is a stable assoc. of a fungus and an alga; Skye, *A.R.Phytop.* 17:325, 1979, as indicators of air pollution; *Ainsworth & Bisby's dictionary of the fungi*, edn. 7, 1983.
Lidophia graminis, poss. anam. *Dilophospora alopecuri* q.v.
life history, Andrews, *Adv.Pl.Path.* 2, 1984, strategies of plant parasites.
lightening strike, damaging crops, some references for cacao, coconut, coffee are given by Shaw, *Res.Bull.Dept.Primary Indust.Papua New Guin.* 33, 1984.
light leaf spot (*Pyrenopeziza brassicae*) brassicas.
Ligniera Maire & Tison 1911; Plasmodiophoromycetes; cysts united in loose or compact cytosori which are indefinite in shape, size and structure; sporangia usually numerous, small, united in loose sporangiosori; on temperate cereals, grasses; prob. of little or no pathogenic importance; Barr, *Can.J.Pl.Path.* 1:85, 1979 [59,3158].
lignin, Kirk, *A.R.Phytop.* 9:185, 1971, degradation by bacteria & fungi; Vance et al., *ibid* 18:259, 1980, lignification as a mechanism of plant resistance.
ligniperda, *Torula*.
lignituber, a proliferation, like a peg, of the host cell wall; induced by, and ensheathing, a penetrating hypha and apparently composed of lignin, after FBPP. Aist, *A.R.Phytop.* 14:145, 1976, prefers the term papillae; callosity q.v.
lignosus, *Rigidoporus*; **ligulicola**, *Didymella*.
ligusticum mosaic Smith & Markham, *Phytop.* 34:335, 1944; transm. sap, England, *L. scoticum* [23:346].
Ligustrum, *L. vulgare*, common or European privet; *L. ovalifolium*, evergreen privet; var. *aureum*, golden privet; Cooper 1979.
lilac, *Syringa vulgaris*, common lilac.
—— **chlorotic leaf spot v.** Brunt 1978; *Descr.V.* 202,

1979; Closterovirus, filament c1540 × 12 nm, transm. sap, Europe.
— **leaf roll** Schmelzer & Schmidt, *Arch.Gartenb.* **14**:303, 1966 [**47**,1587].
— **mosaics** Atanasoff, *Phytopath.Z.* **8**:197, 1935; Europe. E.P. & A.E. Protsenko, *Bull.bot.Gdn.Moscow* **5**:46, 1950; USSR [**14**:462; **32**:434].
— **mottle v.** Waterworth, *PDR* **56**:923, 1972; Carlavirus, filament 575–610 nm long, transm. sap, *Myz.pers.*, USA (Maryland), serologically related to carnation latent v. [**52**,2644].
— **ring mottle v.** van der Meer, Huttinga & Maat 1976; van der Meer & Huttinga, *Descr.V.* 201, 1979; Ilarvirus, isometric c.27 nm diam., transm. sap, seedborne, Netherlands, causes a severe stunt in *Chenopodium quinoa*.
— **ringspot and chlorotic ringspot**, the former (arabis mosaic v.), the latter (AMV + cherry leaf roll v.), Czechoslovakia; *Biologia Pl.* **17**:226, 1975.
— **streak mosaic** Smolák & Novák, *Ochr.Rost.* **23**:285, 1950; Czechoslovakia [**30**:520].
— **white mosaic** = elm mottle v.
— **witches' broom** Brierley 1951; Hibben et al., *PD* **70**:342, 1986; MLO assoc., USA [**65**,4974].
— **yellow ring** Novák 1969; Czechoslovakia, Poland [**49**,2008 g].
lily, *Lilium*, the true lilies; lily is used for many other plant genera; Mowat & Štefanac, *AAB* **76**:281, 1974; aphid transm. of viruses in Britain; Allen, *Lily Yrbk.N.Am.lily Soc.* 33:41, 1980, diagnosis of virus diseases; Bald et al., *PD* **67**:1167, 1983, control of bulb & root diseases in *Lilium longiflorum* [**53**,3519].
— **bulb brown ring** (lily symptomless v. + tulip breaking v.), Asjes et al., *Neth.J.Pl.Path.* **79**:23, 1973 [**52**,3346].
— **curl stripe** McWhorter & Allen, *Nature Lond.* **204**:604, 1964; (lily symptomless v.), Allen & McWhorter, abs. *Phytop.* **56**: 369, 1966 [**45**,814; **46**,12v.].
— **leaf spot and curl** (tobacco rattle v.), Derks, *Neth.J.Pl.Path.* **81**:78, 1975 [**54**,3970].
— **mottle** = tulip breaking v.
— **necrotic fleck** (cucumber mosaic v. + lily symptomless v.), in *Lilium longiflorum*; Brierley & Smith, *Phytop.* **34**:529, 1944; Civerolo et al., *ibid* **58**:1481, 1968; Allen & Lyons, *ibid* **59**:1318, 1969; Lyons & Allen, *J.Ultrastruct.Res.* **27**:198, 1969; see lily, Mowat & Štefanac [**24**:18; **48**,827; **49**,498, 1053].
— — **leaf mosaic** (arabis mosaic v.), Asjes & Segers, *Phytopath. Z.* **106**:115, 1983; *Lilium tigrinum* [**62**,3072].
— **Phytoreovirus** Varma, Gibbs & Ikin,

Australasian Pl.Path. **8**(3): 38, 1979; isometric 52 or 84 nm diam., Australia, in material from Japan [**59**,2788].
— **ringspot** Smith, *J.R.hort.Soc.* **75**:350, 1950; (poss. cucumber mosaic v.), Brierley, *PDR* **46**:625, 1962 [**30**:40; **42**,126].
— **rosette** Ogilvie, *AAB* **15**:540, 1928; transm. *Aphis gossypii, Macrosiphum lilii*; McWhorter & Brierley, *Phytop.* **41**:66, 1951; Moore 1979 [**8**:383, **30**:322].
— **streak mottle**, as for lily bulb brown ring.
— **symptomless v.** Brierley & Smith 1944; Allen, *Descr.V.* 96, 1972; Carlavirus, filament c.640 × 18 nm, prob. transm. *Myz.pers.*, non-persistent; Simmonds & Cumming, *Phytop.* **69**:1212, 1979, detection by immunodiffusion [**59**,4645].
— **yellowing v.** Inouye, Maeda & Mitsuhata, *Ann.phytopath.Soc.Japan* **45**:712, 1979; filament 650 × 12 nm, transm. sap to 33 spp. in 11 plant families, including lily, and in seeds of *Chenopodium quinoa*, lily, *Lilium longiflorum*; the virus is serologically related to citrus tatter leaf v.; the etiology of citrus tatter leaf citrange stunt complex q.v. is not entirely clear [**60**,344].
— **X v.** Stone, *Acta Hortic.* 110:59, 1980.
lily of the valley mosaic Blattný, *Ochr.Rost.* **9**:19, 1929; Czechoslovakia, *Convallaria majalis* [**8**:589].
Lima bean, *Phaseolus lunatus*.
— — **double yellow mosaic** Capoor & Varma, *Curr.Sci.* **17**:152, 1984; transm. *Bemisia tabaci*, India [**27**:549].
— — **green mottle**, rod c.800 nm long assoc., transm. sap, Nigeria 1976 [**58**,2089].
*— — **mosaic** Sawant & Capoor, *Indian Phytopath.* **36**:659, 1983; transm. seed [**65**,6250].
— — **mottle** Gay, abs. *Phytop.* **62**:803, 1972; rod c.680 × 25 nm, transm. sap, USA (Georgia) [**52**,2463].
— — **necrotic leaf** Zaumeyer & Goth, abs. *Phytop.* **59**:1562, 1969; filament av. 430 nm long, transm. sap, USA (New Jersey) [**49**,1247j].
lime, *Tilia* and see *Citrus*.
— **blotch**, Tahiti lime and lemon; prob. genetic disorder, like wood pocket q.v.; Knorr 1973.
limequat, *Citrus aurantifolia* × *Fortunella*.
lime sulphur, or S; useful fungicide against powdery mildews.
limonium line pattern (tobacco rattle v.), Dijkstra & van Dijke, *Neth.J.Pl.Path.* **87**:35, 1981; *L. latifolium* [**60**,5967].
— **yellows** Baker et al., *PD* **67**:699, 1983; MLO assoc., transm. *Macrosteles fascifrons*, USA (Michigan), *L. sinuatum* [**62**,4328].
Limonomyces Stalpers & Loerakker, *CJB* **60**:533, 1982; Corticiaceae; basidiomata resupinate, effused, ceraceous, clamp connections present;

basidia with 2–4 sterigmata, basidiospores hyaline, thin walled, not amyloid; compared with 4 related genera; causes pink growths on temperate grasses; *L. culmigenus* (J. Webster & D. Reid) Stalpers & Loerakker; *L. roseipellis* Stalpers & Loerakker [61,6435].

lindemuthianum, *Colletotrichum*; **lindquistii**, *Leptosphaeria*.

line, see FBPP for 4 definitions and references; this authority stated that in plant pathology the term should only be used, as applying to host plants, in the sense of an assemblage of sexually reproducing individuals of uniform appearance propagated by seeds or spores; its stability is maintained by selection to a standard.

linearis, *Bifusella*; **lineatum**, *Oedocephalum*.

line pattern, a leaf chlorosis centred on the venation and usually caused by a virus infection.

lingam, *Phoma*.

linhartiana, *Monilinia*.

lini, *Melampsora*, *Oidium*; **linicola**, *Alternaria*, *Mycosphaerella*, *Septoria*; **linicolum**, *Colletotrichum*.

Linospora Fuckel 1870; Valsaceae; Barr 1978; ascomata with a lateral beak, asco. filiform, hyaline.

L. ceuthocarpa (Fr.) Lind 1913; poss. anam. *Asteroma frondicola* (Fr.:Ficinus & Schubert) Morelet 1978; see Sutton 1980 for synonymy of anam.; asco. 2–5 septate, 100–130 × 1.9–2.7 μm; poplar leaf spot, Europe; Pinon & Morelet, *Eur.J.For.Path.* **5**:367, 1975; Kojwang & Kurkelia, *Karstenia* **24**:33, 1984 [**55**,2386; **64**,3983].

L. gleditschiae J.H. Miller & F.A. Wolf, *Mycologia* **28**:177, 1936; asco. aseptate, 70–90 × 3 μm, on *Gleditsia triacanthos*. honey locust, USA [**15**:620].

L. tetraspora Thompson, *Can.J.Res.* C **17**:236, 1939; asco. 6–8 septate, 175–255 × 2.5–3 μm; leaf blight of *Populus tacamahaca*, Canada [**18**:767].

Linum usitatissimum, flax q.v., linseed.

Liothrips pistaciae, transm. pistachio rosette.

Lipaphis erysimi, transm. radish stunt, shellflower mosaic.

liquorice rot (*Mycocentrospora acerina*) carrot.

liriodendri, *Phaeoisariopsis*, *Venturia*.

Lirula Darker *CJB* **45**:1420, 1967; Hypodermataceae or Rhytismataceae; ascomata simple, mono-ascocarpous, mostly linear, basal stromatic tissue weakly developed, asco. aseptate, on *Abies*, *Picea*.

L. abietis-concoloris (Mayr ex Dearn.) Darker; Dearness, *Mycologia* **16**:150, 1924; Scharpf, *PD* **70**:13, 1986, outbreak of a needle cast in *Abies concolor*, white fir, USA (California) [**65**,5192].

L. macrospora (Hartig) Darker; Cannon & Minter, *Descr.F.* 794, 1984; anam. *Hypodermina hartigii* Hilitzer 1929; asco. 56–77 × 5–3.5 μm; on spruce needles, N. temperate.

L. nervisequia (DC.:Fr.) Darker; Minter & Millar, *Descr.F.* 783, 1984; asco. 75–90 × × 3–4 μm; needle cast of fir, Europe.

lisianthus flower colour break (bean yellow mosaic v.), Lisa & Dellavalle, *Pl.Path.* **36**:214, 1987; Italy (Liguria), *L. russellianus*.

—— **necrosis v.** Iwaki et al., *Phytop.* **77**:867, 1987; isometric *c*.30 nm diam., contains ssRNA, transm. sap to 21 spp. in 10 plant families, soilborne, Japan, *Eustoma russellianum*.

litchii, *Peronophythora*.

lithospermum ringspot Follman & Bercks, *Naturwissenschaften* **46**:403, 1959; Germany [**40**:57].

Littauer, Franz Shimon, 1893–1978; born in Berlin, Univ. Leipzig, Breslau; worked in Palestine, now partly Israel, from 1923; an originator of much work on plant diseases in this region, especially postharvest fruit pathology and physiology. *Phytoparasitica* **8**:77, 1980.

little broom (*Glomerella cingulata*) cotton.

—— **leaf**, apple; Zn deficiency, see rosette; (*Rhadinaphelenchus* assoc.) coconut, oil palm; van Hoof & Seinhorst, *Tijdschr. PlZiekt.* **68**:251, 1962, Guyana, Surinam [**42**:335].

liver spot (*Gnomonia caryae* var. *pecanae*) pecan, see *G. nerviseda*.

livistona mosaic A.E. & E.P. Protsenko 1964; USSR [**45**,1712].

local lesion assay, a quantitative estimation of the infectivity of a preparation of a pathogen, e.g. a suspension of bacteria or viruses, from the number of lesions formed in inoculated leaves. Half leaves are the common experimental unit, and the leaves of an opposite pair differ less than leaves from different nodes; hence the terms half leaf test and opposite leaf test, FBPP.

—— —— **host**, a host which develops local lesions on inoculation with a virus, FBPP.

Loculoascomycetes, proposed by Luttrell in 1955 as a group of the Ascomycetes with bitunicate asci; Sivanesan 1984 divided the group into 7 orders; it is not accepted by *Ainsworth & Bisby's dictionary of the fungi*, edn. 7, 1983.

loganberry, *Rubus* cross, discovered by Logan in USA (California) *c*. 1881.

—— **degeneration** = raspberry bushy dwarf v.

—— **yellow blotch mosaic** Wilhelm, Thomas & Koch, *Calif.Agric.* **5** (1):11, 1951; USA (California) [**30**:573].

lokoyae, *Diaporthe*, *Phomopsis*.

lolii, *Pyrenophora*, *Spermospora*, *Tilletia*.

Lolium multiflorum, *L. perenne*, see ryegrass.

lolium enation Huth, *NachrBl.dt.PflSchutzdienst* 27(4):49, 1975; Phytoreovirus, isometric particle, Germany [55,777].

— latent Huth 1973; isometric 30 nm diam., Germany, Northern Ireland [54,646; 56,4310 c].

— mottle v. O'Rourke 1968; A'Brook 1972; Catherall et al., *AAB* 87:233, 1977; Huth & Paul, *Annls.Phytopath.* 9:293, 353, 1977; Mohamed, *J.gen.Virol.* 40:379, 1978; *N.Z.J.agric.Res.* 21:709, 1978; isometric, transm. sap, *Rhopalosiphum padi*; temperate cereals, Gramineae; causes a lethal ashy grey mottle on wheat, sometimes called cynosurus mottle v. [52,1580; 57,1757, 5353; 58,1100, 1111, 3877].

longevity end point, the storage time after which a virus in a crude sap preparation loses its infectivity, usually determined at 0° or 20°.

— in fungi, Sussman in Ainsworth & Sussman, vol. 3, 1968.

longibrachiatum, *Botryosporium*.

Longidorus, needle nematodes, root ectoparasites.

L. apulus, transm. artichoke Italian latent.

L. attenuatus Hooper 1961; Brown & Boag, *Descr.N.* 101, 1977; galling of root tips, damages arable crops, assoc. beet docking disorder, transm. tomato black ring, English str.

L. caespiticola Hooper; Boag & Brown, *Descr.N.* 63, 1975; assoc. beet docking disorder, Valdez, *AAB* 71:229, 1972, reported it transm. raspberry ringspot, English str. [52,642].

L. diadecturus, transm. peach rosette mosaic.

L. elongatus (de Man) Thorne & Swanger 1936; Hooper, *Descr.N.* 30, 1973; assoc. beet docking disorder, cooler regions, transm. raspberry ringspot, English, Scottish strs.; tomato black ring, Scottish str.

L. fasciatus, transm. artichoke Italian latent.

L. laevicapitatus Williams 1959; Hooper, *Descr.N.* 117, 1985; tropical and sub-tropical, particularly assoc. with sugarcane.

L. leptocephalus Hooper 1961; Boag & Brown, *Descr.N.* 88, 1976; assoc. beet docking disorder, transm. raspberry ringspot, English str.

L. macrosoma Hooper 1961; Brown & Boag, *Descr.N.* 67, 1975; causes root tip galls, hosts in 23 plant families, Europe, transm. raspberry ringspot, cherry and English strs.

L. martini, transm. mulberry ringspot.

longipes and longissima, *Alternaria, Cercospora*.

long smut (*Sorosporium ehrenbergii*) sorghum; — spored rust (*Chrysomyxa piperiana*) rhododendron.

Lonicera, see honeysuckle.

lonicera variegation Woods & DuBuy, *Phytop.* 33:637, 1943; USA [23:163].

— witches' broom Yakutkina 1979; MLO assoc., USSR (Leningrad) [60,5464].

— yellow vein mosaic (tobacco leaf curl v.), Osaki et al., *Ann.phytopath.Soc.Japan* 45:62, 1979 [59,2228].

loofah witches' broom Yang et al., *Pl.Prot.Bull.Taiwan* 16:162, 1974; Chung et al. and Chou et al., *ibid* 17:329, 384, 1975, MLO assoc., transm. *Hishimonus concavus, Luffa cylindrica* [55,958, 4341, 5992].

loose smut, a type where the entire grain is transformed into a mass of loose, unprotected ustilospores, cf. covered smut;

— — (*Sphacelotheca cruenta*) sorghum, (*Ustilago avenae*) oat, (*U. tritici*) barley, wheat.

Lophodermella Höhnel 1917; Hypodermataceae or Rhytismataceae; see *Lirula*, Darker.

L. conjuncta (Darker) Darker 1967; Millar & Minter, *Descr.F.* 658, 1980; asco. aseptate, 75–90 × 3–3.5 μm; pine needle blight, parts of central and N. Europe.

L. sulcigena (Rostrup) Höhnel 1917; Millar & Minter, *Descr.F.* 562, 1978; asco. aseptate, 27–40 × 4–5 μm; Swedish pine cast, Europe; Jalkanen, *Folia Forestalia* 476, 1981, review; *Karstenia* 25:53, 1985, on Scots pine in Finland [65,3013].

Lophodermium Chev. 1826; Hypodermataceae or Rhytismataceae; asco. long, filiform, often helically arranged, aseptate; prevalent on *Pinus*, needle casts, besides the spp. given others have been described by Cannon, Millar and Minter; *Descr.F.* 784–90, 795–8, 1984; Minter, *Mycol.Pap.* 147, 1981; Cannon & Minter, *Taxon* 32:572, 1983, nomenclatural history & typification [60,6681; 63,1634].

L. australe Dearness 1926; Minter & Millar, *Descr.F.* 563, 1978; asco. up to > 100 μm long; fairly widespread but not in Europe, commonest sp. in the tropics.

L. canberrianum Stahl ex Minter & Millar, *Descr.F.* 564, 1978; asco. 70–120 μm long, 2–3 needle pines of *ponderosa* group, Australia, can cause severe defoliation.

L. conigenum (Brunard) Hilitzer 1929; fide Cannon et al. 1985; Minter & Millar, *Descr.F.* 565, 1978, as *L. conigenum* Hilitzer 1929; asco. 90–130 μm long; mainly 2–3 needle pines, recorded on 5 needle pines, inhabits green needles causing no symptoms, fructifies saprophytically when needles are killed by some other cause; widespread in Europe, USA (E, uncommon), New Zealand.

L. nitens Darker 1932; Minter & Millar, *Descr.F.* 566, 1978; asco. 80–120 μm, on 5 needle pines, prob. unimportant; Japan, poss. Europe, N. America.

L. pinastri (Schrader) Chev. 1826; Minter & Millar, *Descr.F.* 567, 1978; anam. *Leptostroma pinastri* Desm. 1843; asco. 70–110 μm long, con. 4.5–6.3 μm long; on 2–3 needle pines. This sp. had been regarded as very variable with highly pathogenic strs., but the pathogenic form is now known as *Lophodermium seditiosum*, the complex was disentangled by Minter et al., *TBMS* **71**:295, 1978; *L. pinastri* causes no significant damage [58,2967].

L. seditiosum Minter, Staley & Millar, *TBMS* **71**:300, 1978; Minter & Millar, *Descr.F.* 568, 1978; asco. 90–120 μm long, wider than those of *Lophodermium pinastri*; this sp. is the severely pathogenic sp., previously placed in *L. pinastri*, seedlings and young trees can be killed, widespread in Europe, parts of USA; Phillips & Burdekin 1982.

loquat, *Eriobotrya japonica*; Weltzien, Atti primo Congresso dell-Unione Fitopat.Mediterranea: 394, 1969 [48,2706 w].

loss of crop, Madden, *Phytop.* **73**:1591, 1983, stated that there are few reliable estimates of crop losses due to plant diseases; nevertheless the destruction caused to crops has been well known from the beginnings of plant pathology; in *A.R.Phytop.*: Ordish & Dufour, **7**:31, 1969; James, **12**:27, 1974; Carlson & Main, **14**:381, 1976; Carlson, **17**:149, 1979; Grainger, **17**:223, 1979; Edens & Haynes, **20**:363, 1982; Weise, **20**:419, 1982; symposium in *Phytop.* **73**:1575, 1983; Brown, *AAB* **84**:448, 1976, nematodes; James & Teng, *Appl.Biol.* **4**:201, 1979; Bos, *Crop Protect.* **1**:263, 1982, viruses; Wood & Jellis 1984.

loti, *Cercospora*, *Stemphylium*.

lotus streak Yamashita et al. 1978; *Ann.phytopath.Soc.Japan* **51**:627, 1985; bacilliform particle assoc., *c.*300–340 × 90 nm, *Nelumbo nucifera* [65,4446].

lowbush blueberry, *Vaccinium angustifolium*.

LP 50, the median latent period q.v.

lucerne, *Medicago sativa*; Graham et al. 1979; Harvey, *N.Z.J.exp. Agric.* **10**:317, 1982, disease assessment; Hancock, *PD* **67**: 1203, 1983, seedlings; *Medicago* q.v. [62,1087; 63,3427].

—— **Australian latent v.** Taylor & Smith 1971; Jones & Forster, *Descr.V.* 225, 1980; Nepovirus, isometric *c.*25 nm diam., transm. sap, seedborne in lucerne, common in crops in Australia, New Zealand; Forster & Morris-Krsinich, *AAB* **107**:449, 1985, distinct str. in white clover in New Zealand, differs in host reactions [66,228].

—— —— **symptomless v.** Remah, Jones & Mitchell, *AAB* **109**:307, 1986; isometric *c.*26–28 nm diam., contains RNA, transm. sap to 11 of 22 plant spp. but only induced symptoms in 3 *Chenopodium*

spp. and *Gomphrena globosa*, seedborne in *C. quinoa*; its properties distinguish LASV from lucerne Australian latent v. and from all recognised Comoviruses and Nepoviruses, closest affinities are with rubus Chinese seedborne v., strawberry latent ringspot v., poss. arracacha B v. [66,2413].

—— **enation v.** Blattný 1959 as lucerne pappilosity; Bos 1964; Alliot & Signoret 1972; Alliot et al. *C.r.hebd.Séance Acad.Sci.* D. **274**:1974, 1972; Leclant et al., *Annls.Phytopath.* **5**:441, 1973; Phytorhabdovirus, particles 250 × 82–89 nm, transm. *Aphis craccivora*; Europe, Morocco [44,1452; 52,171–2; 54,2865].

—— **transient streak v.** Blackstock 1974; Forster & Jones, *Descr.V.* 224, 1980; Sobemovirus, isometric 27–28 nm diam., transm. sap; Australia, Canada, New Zealand; Jones et al., *J.gen.Virol.* **64**:1167, 1983, circular RNA, poss. satellite; Paliwal, *Can.J.Pl. Path.* **5**:75, 1983; **6**:1, 1984, in E. Canada, seed transm. in *Melilotus alba*, host reactions differ; Rao & Hiruki, *PD* **69**:610, 1985, in Canada (Alberta) [62,4349; 63,712, 3800; 65,2875].

—— **vein yellowing** van der Want & Bos, *Tijdschr.PlZiekt.* **65**:73, 1959; = bean leaf roll v. [39:113].

—— **virescence** Autonell & Faccioli, *Phytopath.Mediterranea* **19**:157, 1980; MLO assoc., Italy [63,197].

—— **white necrotic line pattern** (pea early browning v.), Gibbs & Harrison, *AAB* **54**:1, 1964 [44,382].

—— **witches' broom** Edwards, *J.Aust.Inst.agric.Sci.* **1**:31, 1935; see tomato big bud [14:516].

—— **yellowing** Cheo & Tsai, *Acta phytopath.sin.* **5**:7, 1959; transm. sap, China [39:138].

lucidum, *Ganoderma*.

luffa, see loofah.

luminescence, virus identification; Trubitsyn & Reifman, *Phytopath.Z.* **113**:51, 1985 [65,103].

lunata, *Ceuthospora*, *Curvularia*; **lunatus**, *Cochliobolus*.

luohanguo blistered leaves Lin & Chow 1984; MLO assoc., China (N. Guangxi), *Siratia grosvenori* [64,2784].

lupini, *Diaporthe*, see *D. woodii*.

lupinus leaf curl Verma, *Gartenbauwissenschaft* **39**:55, 1974; transm. *Bemisia tabaci*, India (N.), *L. hartwegii* [53,4492].

—— **witches' broom** Ulrychová, Jokeš & Kynčlová, *Biologia Pl.* **22**:363, 1980; MLO assoc., Czechoslovakia [60,3210].

luteobubalina, *Armillaria*.

*****Luteoviruses**, type barley yellow dwarf v., isometric 30 nm diam., 1 segment linear RNA, sedimenting at 115–118 S, thermal inactivation 65–70°, transm. aphid spp.; Rochow & Duffus in Kurstak 1981; Francki et al., vol. 1, 1985.

lychnis mosaic Gol'din, Sergeeva & Chernoivanova 1968; USSR, *L. chalcedonica* [**47**,3471].

—— **ringspot v.** Bennett, *Phytop.* **49**:706, 1959; Gibbs et al., *Virology* **20**:194, 1963; Lane, *ibid* **58**:323, 1974; Hordeivirus, rod *c.*125 nm long, transm. sap, seed, pollen; USA (California), *L. divaricata* [**39**:291; **43**,75; **55**,1693].

lycomarasmin, toxin from *Fusarium*, see fusaric acid.

lycopersici, *Didymella, Phoma, Pyrenochaeta, Septoria, Stemphylium.*

Lycopersicon esculentum, tomato.

Lygus pratensis, transm. rape savoying.

lyophilisation, as for freeze drying q.v.

lysis, (1) the enzymatic dissolution of all or part of a one or many celled structure; (2) dissolution of a bacterium infected with phage, after FBPP; for

mycelium in soil, Ko & Lockwood, *Phytop.* **60**:148, 1970.

lysogenic bacterium, a bacterium infected with phage and which is integrated with the bacterial genome. All progeny are likewise infected and can form phage, by occasional cells undergoing lysis spontaneously or after application of inducing agents, e.g. ultraviolet light. Lysogenic bacteria are not sensitive to the phage they carry, after FBPP.

lysogenisation, the infection of a bacterium by phage to form a lysogenic clone or subclone of descendants. Lysogenisation occurs some time after, or some bacterial generations after, the initial infection, FBPP.

lysosome, a body bounded by a single membrane and that contains more than one acid hydrolytic enzyme; Wilson, *A.R.Phytop.* **11**:247, 1973, a concept for plant pathology.

lythri, *Hainesia.*

M

macdonaldii, *Phoma*.
macerate, to soften by soaking, to cause disintegration of tissues; not the same as comminute which means to reduce to small fragments, see FBPP; Chesson, *J.appl.Bact.* **48**:1, 1980, maceration in relation to postharvest & plant material handling.
machamiento, see soybean proliferation.
maclura mosaic v. Pleše & Miličić 1973; Koenig, Lesemann & Pleše, *Descr.V.* 239, 1981; Potyvirus, filament 670 × 13–16 nm, transm. sap, *Myz.pers.*, non-persistent, Yugoslavia, *M. pomifera*.
macrocyclic, of Uredinomycetes, the 2 primary spore stages are present, i.e. telial with teliospores, aecial with aeciospores; if they are on one host, or host group, the life cycle is said to be autoecious; if each stage is on a host(s) that are taxonomically wide apart then the rust is said to be heteroecious.
Macrohyporia Johansen & Ryv. 1979; Polyporaceae; *M. extensa* (Peck) Ginns & Lowe, *CJB* **61**:1673, 1983; cubical butt rot of hardwoods in N. America (E.); Murrill, *Mycologia* **12**:110, 1920; Davidson & Campbell, *ibid* **46**:234, 1954, as *Poria cocos* Wolf [**34**:5; **63**,5135].
*****Macrophoma** (Sacc.) Berl. & Vogl. 1886; Coelom.; Sutton 1980 considered it synonymous with *Sphaeropsis*. *M. castaneicola* Kobayashi & Oishi, *Trans.mycol.Soc.Japan* **20**:429, 1979; con. aseptate, hyaline, mostly 19–22.5 × 6.5–7.5 μm; on *Castanea crenata* causing black root rot. *M. zeae*, anam. of *Botryosphaeria zeae*, is a *Dothiorella* [**60**,2207].
Macrophomina Petrak 1923; Coelom.; Sutton 1980; conidiomata pycnidial, pycnidia brown, thick walled; con. aseptate, hyaline, cylindrical to fusiform, comparatively large; in culture usually only sclerotia are formed.
M. phaseolina (Tassi) Goid. 1947; Holliday & Punithalingam, *Descr.F.* 275, 1970; con. 16–24 × 5–9 μm, sclerotia black, smooth, hard, up to 1 mm diam. Punithalingam, *Nova Hedwigia* **36**:249, 1982, reviewed the taxonomic status, gave a full description including conidiation, an account of a conidial appendage, and a synonymy of 13 names. He considered that transfers to *Tiarosporella* Höhnel 1919 by van der Aa in 1977, 1981, were unacceptable and that the 2 genera were distinct. The pathogen is plurivorous, causing ashy stem blight or charcoal rot; roots and stems show a destruction of the cortex; a root

inhabitant, widespread in warmer areas; the fungus usually invades immature, unthrifty, damaged or senescent tissues; plants are generally attacked as seedlings, when mature and at flowering; when conditions are hot and dry severe losses can occur. Infection arises from sclerotia which can survive for a few years in roots, survival is lower in soil, and the fungus can be seedborne. It has been implicated in more diseases than was justified, hence S.D. Garrett's 1956, *Biology of root infecting fungi*, reference to its imposter role.

Holliday 1980; Short et al., *Phytop.* **70**:13, 1980, survival in soil & soybean residue; Cottingham, *PD* **65**:355, 1981, numbers of sclerotia in soil; Yang & Owen, *Phytop.* **72**:819, 1982, assoc. with stem weevil larvae, *Cylindrocopturus adspersus*, in sunflower; Yang et al., *ibid* **73**:1467, 1983, transm. by stem weevil; Punithalingam, *Nova Hedwigia* **38**:339, 1983, nuclei; Kunwar et al., *Phytop.* **76**:532, 1986, in soybean seed [**60**,566; **61**,100, 7117; **63**,1151, 1907; **65**,6253].
macropodii, *Didymella*.
Macroposthonia curvata (Raski) de Grisse & Loof 1965; Loof, *Descr.N.* 58, 1974; damages roots of carnation and peach.
M. sphaerocephala (Taylor) de Grisse & Loof 1965; Orton-Williams, *Descr.N.* 28, 1973.
M. xenoplax (Raski) de Grisse & Loof 1965; Orton-Williams, *Descr.N.* 12, 1972; plurivorous, reported to cause ring nematode decline of carnation.
Macropsis triamaculata, transm. peach yellows.
Macrosiphum euphorbiae, transm. bean leaf roll, freesia mosaic, iris mild mosaic, narcissus latent, narcissus white streak, pea pimple pod, potato V, subterranean clover stunt, tomato yellow top, zucchini squash yellow mosaic.
macrospora, *Alternaria*, *Lirula*, *Mycosphaerella*, *Phytophthora*, *Sclerophthora*, *Stenocarpella*;
macrosporum, *Fusicladium*, *Synchytrium*;
macrosporus, *Protomyces*.
Macrosteles fascifrons, transm. alsike clover proliferation, aster yellows, clover phyllody, gladiolus grassy top, limonium yellows, oat blue dwarf, *Spiroplasma citri*, *S. phoeniceum*, strawberry witches' broom.
M. orientalis, transm. aster yellows, bupleurum yellows, chrysanthemum witches' broom, cryptotaenia witches' broom, oenanthe yellows, strawberry witches' broom.

M. sexnotatus, transm. delphinium yellows, gladiolus grassy top, gladiolus witches' broom, oat dark green dwarf.

macrostoma, *Phoma*.

Macrotyloma uniflorum, horsegram; see *Vigna*, Verdcourt.

maculans, *Leptosphaeria, Phloeospora, Taphrina*.

macularis, *Sphaerotheca, Venturia*.

maculata, see *Stagonospora arenaria*; **maculatum**, see *Diplocarpon mespili*.

Madagascar periwinkle, *Catharanthus roseus*.

Magnaporthe Krause & R. Webster, *Mycologia* **64**:110, 1972; Phyllachoraceae; Monod transferred the genus to *Phragmoporthe* q.v.; no stroma, ascomata dark, globose, long neck slightly or not protruding beyond leaf sheath surface; asci float freely, deliquescing; asco. long fusiform, curved, 2 septate.

M. grisea, anam. *Pyricularia grisea* q.v.

M. rhizophila Scott & Deacon, *TBMS* **81**:77, 1983; anam. *Phialophora* which may be confused with anam. of *Gaeumannomyces graminis*; asco. 24–28 × 6–8 μm; cereal roots in South Africa [63,93].

M. salvinii (Catt.) Krause & R. Webster; Ellis & Holliday, *Descr.F.* 344, 1972, *as Leptosphaeria salvinii* Catt.; anam. *Nakataea sigmoidea* Hara 1939; asco. av. 52 × 8.7 μm, con. almost always 3 septate, 40–83 × 11–14 μm; Tsuda et al., *TBMS* **78**:515, 1982, teleom. formation in culture; sclerotia are ± spherical. black, mostly 200–300 μm diam., often called *Sclerotium oryzae*. The genus was erected to accommodate this pathogen because it has unitunicate asci; *Leptosphaeria* has bitunicate asci. Rice stem rot, infection is prob. mostly from the sclerotia; and predisposing factors, e.g. a tendency to lodge may be important; sclerotia are viable for years and float free on the water; cultivation methods affect their numbers; variations in virulence are found; Bockus et al., *Phytop.* **68**:417, 1978; **69**:389, 862, 1979, competitive saprophytic ability low, effects of rice residue disposal methods on disease incidence, complete destruction desirable in USA (California), sclerotia have a half life of 1.9 years at 18 cm soil depth; Holliday 1980; Ou 1985 [**58**,689; **59**,2186, 5134].

magnoliae, *Cercosporidium, Phaeoisariopsis*.

mahonia mosaic Kochman & Stachyra, *Roczn.Nauk roln.* A **77**:297, 1957, Poland [**37**:340].

maize, *Zea mays*; Nishihara, *Bull.natn.Grassland Res.Inst.* 1:59, 1972, Japan, bibliography; Dodd, *PD* **64**:533, 1980, stress & stalk rots; Shurtleff 1980; Gordon et al. ed., *Southern Co-op.Ser.Bull.* 247, 1981, Ohio agric.Res.developm.Centre; *Proc.Int.Maize Virus Colloquium Workshop*,

Wooster, Ohio, 1983; Nault in Plumb & Thresh 1983, MLO & virus origins; Norton, *PD* **67**:253, 1983, nematodes; Brakke, *A.R.Phytop.* **22**:77, 1984, mutation & virus infection; Payak & Sharma, *Trop.Pest Management* **31**:302, 1985.

— **American leaf fleck** Stoner, *Phytop.* **42**:683, 1952; transm. *Myz.pers., Rhopalosiphum padi, R. maidis*, persistent, USA (California), infects *Phalaris stenoptera*, Harding grass [**32**:554].

— **Bulgarian leaf fleck** Atanasoff, *Phytopath.Z.* **52**:89, 1965; **56**:25, 1966; transm. aphid, seed [**44**,2474; **46**,110].

— **bushy stunt** Bradfute, Nault & Robertson 1977; MLO assoc., distribution similar to that of *Spiroplasma kunkelii* q.v. for references and with which MBS has been compared; transm. *Dalbulus eliminatus, D. maidis, Graminella nigrifrons*, infects teosinte.

— **chlorotic dwarf v.** Rosenkranz 1969; Gingery et al., *Descr.V.* 194, 1978; Gingery et al. in Kurstak 1981; Francki et al., vol. 2, 1985; isometric *c.*30 nm diam., contains ssRNA, transm. *Graminella nigrifrons, G. sonora*, semi-persistent, USA (S.), narrow host range in Gramineae, main overwintering host Johnson grass, a form of *Sorghum halepense*.

— — **mottle v.** Castillo & Hebert 1974; Gordon et al., *Descr.V.* 284, 1984; isometric *c.*30 nm diam., contains RNA, transm. sap, chrysomelid beetles, 6 spp.; Argentina, Mexico, Peru, USA; but economically important only in limited coastal areas of Peru, USA (Kansas, Nebraska); Gramineae, can cause severe stunt and premature death of maize.

— — **stripe** = maize stripe v.

— **Colombian stripe** Martínez-López, Cujía & Luque, *Fitopatología* **9**:93, 1974; prob. str. of maize rayado fino v. [**54**,4461].

— **dwarf mosaic v.**, filament, transm. sap, aphid spp., serologically related to sugarcane mosaic v. In 1964 a virus from maize was found to infect Johnson grass, a form of *Sorghum halepense*, and sorghum; this was later designated MDMV, and then later still that it be called SCMV; unfortunately both virus names have been used for the many strs. Only strs. A and B are known to be naturally transm. to sorghum, fide Toler, *PD* **69**:1011, 1985, who described MDMV as causing the most important virus disease of sorghum; Gordon et al. 1983, see maize; Madden in Plumb & Tresh 1983; McDaniel & Gordon, *PD* **69**:602, 1985, str. infecting oat [**65**,2797].

— **gooseneck stripe v.** Kulkarni 1972; isometric 35–40 nm diam., transm. *Peregrinus maidis*; Kenya, Tanzania; sorghum naturally infected [**52**,3908 f].

— **hoja blanca** = maize stripe v.

— **Iranian mosaic** v. Izadpanah et al., *Phytopath.Z.* **107**:283, 1983; Phytorhabdovirus, 150–220 × 70–90 nm, transm. *Ribautodelphax notabilis* [**63**,1233].

— **lethal necrosis** (maize chlorotic mottle v. + either maize dwarf mosaic v. or wheat streak mosaic v.), Niblett & Claflin, *PDR* **62**:15, 1978 [**57**,4910].

— **line** v. Kulkarni, *AAB* **75**:205, 1973; isometric 28 and 34 nm diam., transm. *Peregrinus maidis*, persistent, E. Africa [**53**,1788].

— **malformation** Klinkong & Sutabutra in Gordon et al. 1983, see maize; isometric 27 nm diam., transm. sap, Gramineae, causes large chlorotic streaks in sugarcane in Thailand.

— **mosaic** v. Kunkel 1921–2, 1927; Herold, *Descr.V.* 94,1972; Phytorhabdovirus, *c.*242 × 48 nm in sections, 225 × 90 in leaf dip, transm. *Peregrinus maidis*, persistent, infects sorghum, wild grasses; Falk & Tsai, *Phytop.* **73**:1536, 1983, characterisation; *ibid* **75**:852, 1985, detection & multiplication in *P. maidis* [**63**,1752; **65**,113].

— **mottle and chlorotic stunt** v. Rossel & Thottappilly in Gordon et al. 1983, see maize; isometric *c.*40 nm diam., transm. *Cicadulina triangula*, Nigeria.

— **rayado Colombiano** v. Martínez-Lopez & Rico de Cujia, *Fitopat. Colombiana* **6**:57, 1977; isometric *c.*30 nm diam., transm. leafhopper, serologically distinct from maize rayado fino v. [**57**,4440].

— — **fino** v. Ancalmo & Davis 1961; Gámez, *Descr.V.* 220, 1980; isometric *c.*31 nm diam., contains ssRNA, transm. *Baldulus tripsaci*, *Dalbulus* spp., *Graminella nigrifrons*, *Stirellus bicolor*, persistent, *D. maidis* is the main vector, Gramineae, rather narrow host range, tropical America; León & Gámez, *J.gen.Virol.* **56**:67, 1981, characteristics, resemblance to oat blue dwarf v.; Gingery et al., *Phytop.* **72**:1313, 1982, purification & properties; Gámez in Plumb & Thresh 1983; Rivera & Gámez, *Intervirology* **25**:76, 1986, multiplication in *D. maidis* [**61**,3439; **62**,1480; **65**,4886].

— **red leaf and red stripe** Šutić 1974; Kloeppler et al. 1982; poss. mollicute; Yugoslavia, USA (California) [**55**,1176; **61**,4893].

— **Rio IV** v. Nomé et al., *Phytopath.Z.* **101**:7, 1981; isometric 55–60 nm diam., transm. *Delphacodes*, Argentina; stunt, sterility, minute enations on leaf veins, ear proliferation [**61**,692].

— **rough dwarf** v. Biraghi 1949, 1952; Lovisolo, *Descr.V.* 72, 1971; Phytoreovirus, subgroup Fijivirus; isometric *c.*70 nm diam., transm.

planthoppers, including *Laodelphax striatellus* which transm. in nature, Gramineae; Asia, Mediterranean, S. America; has caused epidemics Milne & Lovisolo, *Adv.Virus Res.* **21**:267, 1977, review; Caciagli et al., *AAB* **107**:463, 1985, detection in plant & vector; Caciagli & Casetta, *ibid* **109**:337, 1986, in *L. striatellus* in relation to vector infectivity [**65**,5444; **66**,2824].

— **sterile stunt** v. Greber, *Aust.J.agric.Res.* **33**:13, 1982; Phytorhabdovirus (str. of barley yellow striate mosaic v.), Milne et al., *Intervirology* **25**:83, 1986; Australia, transm. *Sogatella longifurcifera*, the main natural vector [**61**,5683].

— **streak** v. Storey 1925; Bock, *Descr.V.* 133, 1974; Geminivirus, type, isometric, pairs of particles 20–30 nm, transm. *Cicadulina*, 6 spp., *C. mbila* is the most important, fide Shurtleff 1980; Africa (S. of Sahara), Asia (S.E.); Gramineae including maize, millet, sugarcane, wheat, and wild grasses; causes serious diseases, less severe as altitude increases; Damsteegt, *PD* **67**:734, 1983, for hosts [**62**,4884].

— **stripe** v. Shephard 1929; Storey 1936; Gingery, *Descr.V.* 300, 1985; filament, length undetermined, 3 nm diam., in a group with rice grassy stunt, rice hoja blanca, rice stripe; transm. *Peregrinus maidis*, persistent, transovarial; in a few spp. of Gramineae, tropics, can cause serious loss.

— **stunt** (*Spiroplasma kunkelii*).

— **stunting** v. Greber, *Aust.Pl.Path.Soc.Newsl.* **6**(1):18, 1977; Phytorhabdovirus, transm. *Peregrinus maidis*, ? a form of maize mosaic v. [**56**,5608].

— **subtle mosaic**, Louie et al. in Gordon et al. 1983, see maize.

— **tassel abortion** Kulkarni 1972; transm. *Malaxodes farinosus*, Kenya [**52**,3908 f].

— **tip necrosis** Stoner, Gustin & MaComb, *PDR* **51**:705, 1967; transm. sap, USA (S. Dakota [**47**,150].

— **vein enation** Ahlawat & Raychaudhuri, *Curr.Sci.* **45**:273, 1976; transm. *Cicadulina mbila*, persistent, India (N.), ? a form of maize streak v. [**55**,5162].

— **white leaf** v. Trujillo, Acosta & Piñero, *PDR* **58**:122, 1974; isometric, *c.*55–60 nm diam., transm. *Peregrinus maidis*, Venezuela; *Rottboellia exaltata*, a natural host; see maize gooseneck stripe and maize line also attributed to isometric particles transm. by *P. maidis* [**53**,2535].

— — **line mosaic** v. Boothroyd & Israel 1980; de Zoeten & Reddick, *Descr.V.* 283, 1984; isometric 35 nm diam., contains a single ssRNA species, transm. in soil poss. by a fungus, parts of Europe and USA, occurs naturally in a few grasses, maize cobs are deformed and contain few kernels, poss. assoc. with wet soils; Gingery & Louie, *Phytop.* **75**:870, 1985, assoc. satellite v., isometric 17 nm diam. [**65**,193].

majalis, *Ascochyta*.

Malaxodes farinosus, transm. maize tassel abortion, melinis dwarf.

malayensis, *Cercospora*.

Malaysian wilt, coconut; see palm lethal yellowing.

malformation (*Gibberella fujikuroi*) mango, etiology poss. not entirely certain.

malformin, from *Aspergillus niger*, distorts plant growth; Wood 1967.

mali, *Cylindrocarpon, Diplocarpon*.

malicorticis, *Cryptosporiopsis, Pezicula*.

mallet wound (*Ceratocystis fimbriata*) almond.

mallotivora, *Erwinia*.

mal seccin, toxin formed by *Deuterophoma tracheiphila*; **mal secco** (*D. tracheiphila*) citrus.

Malus, apple.

malvaceae infectious chlorosis Flores & Silberschmidt, *Phytopath.Z.* **60**:181, 1967; Jeske et al., *ibid* **89**:289, 1977; **97**:43, 1980; transm. *Bemisia tabaci*, sometimes called abutilon mosaic q.v.; Abouzid & Jeske, *J.Phytopath.* **115**:344, 1986, geminate particles [**47**,1056; **57**,1219; **59**,5662; **65**,5426].

malvacearum, *Puccinia*.

malva mottle M.I.C. & F.S. Henriques, *CJB* **64**:85, 1986; bacilliform particles assoc., *c*.300 × 75 nm, Portugal, refer to malva vein clearing v. [**65**,3664].

malvarum, *Colletotrichum*.

malva veinal necrosis v. Costa & Kitajima, *Bragantia* **29**, Nota 11:L1, 1970; rod 525 × 15 nm, transm. sap, Brazil (S.) [**51**,1235].

malvaviscus mosaic Herbas, *Turrialba* **24**:298, 1974; transm. sap, Brazil, *M. arboreus* [**54**,3972].

mamillatum, *Pythium*.

mammatoxin, and other toxic metabolites from *Hypoxylon mammatum*; *Populus tremuloides* is highly sensitive; Schipper, *Phytop.* **68**:866, 1978; Stermer et al., *ibid* **74**:654, 1984 [**58**,1970; **63**,5570].

mammatum, *Hypoxylon*.

mancozeb, maneb and Zn complex, protectant fungicide.

mandarin, see *Citrus*.

maneb, protectant fungicide, widely used, broad spectrum.

mangiferae, *Elsinoë, Oidium, Pestalotiopsis, Sphaceloma*.

mango, *Mangifera indica*; Thompson, *Trop.Agric.Trin.* **48**:63, 71, 1971, storage & transport; Javaid, *Trop.Pest Management* **31**:33, 1985, disease control in Zambia; Lim & Khoo 1985, diseases & disorders in Malaysia [**65**,4005].

mangold, a cv. of *Beta vulgaris*, beet.

Manihot esculenta, cassava, manioc, tapioca.

manihoticola, *Periconia, Sphaceloma, Uromyces* see *U. manihotis*.

manihotis, *Diaporthe, Mycosphaerella, Phomopsis, Uromyces*.

manihotis-catingae, *Uromyces* see *U. manihotis*.

Manila hemp, *Musa textilis*, see banana.

manioc, see cassava.

manshurica, *Peronosopora*.

maple, *Acer*.

—— **leaf perforation** Smolák, *Ochr.Rost.* **22**:173, 1949; Šubíková, *Biologia Pl.* **15**:166, 1973; transm. sap, Czechoslovakia [**30**:13; **54**,82].

—— **mosaics** Atanasoff, *Phytopath.Z.* **8**:197, 1935; Bulgaria; Szirmai, *Acta phytopath.Acad.Sci.hung.* **7**:197, 1972; transm. sap, *Trialeurodes vaporariorum*, Hungary [**14**:462; **52**,3080].

maracuja mosaic v. Fribourg, Koenig & Lesemann, *Phytop.* **77**:486, 1987; Tobamovirus, rod 304 × *c*.18 nm, transm. sap, *Passiflora edulis*, Peru (N.), systemic infections only in *Nicotiana benthamiana* and *P. edulis*, local lesions in 25 other spp. in 9 plant families [**66**,3914]

***Marasmiellus** Murrill 1915; Tricholomataceae; largely saprophytic, Pegler, *Kew Bull.* addit.ser. 6, 1977; Holliday 1980; Jackson & Firman, *Pl.Path.* **31**:187, 1982, seedborne marasmioid fungi in coconut; Latterell & Rossi, *PD* **68**:728, 1984, *Marasmiellus* sp. causing maize borde blanco, Central America [**61**,7125; **63**,5401].

M. cocophilus Pegler 1969; coconut lethal bole rot and basal stem break, seedlings; Bock et al., *AAB* **66**:453, 1970, E. Africa; Jackson & Firman, *Inform.Circ.S.Pacific Comm.* 83, 1979 [**50**,2416; **59**,3353].

M. inoderma (Berk.) Singer 1955; on several, mostly tropical, crops; Allen, *Agric.Gaz.N.S.W.* **81**:524, 1970, banana sheath rot; Sabet et al., *TBMS* **54**:123, 1970, maize root rot & wilt [**49**,2466; **50**,1312].

M. mesosporus Singer in Singer et al., *Mycologia* **65**:469, 1973; blight of *Ammophila breviligulata*, beach grass, USA (N. Carolina); Lucas et al., *PDR* **55**:582, 1971 [**51**,1589; **52**,4116].

M. scandens (Massee) Dennis & Reid, *Kew Bull.* **2**:289, 1957; thread blights of tropical crops, now mainly unimportant, see *Marasmius*.

M. semiustus (Berk. & M.A. Curtis) Singer 1968: pre-emergence shoot rot of coconuts, Singh et al., *Perak Plrs'.Assoc.J.* page 66, 1980 [**61**,5903].

M. troyanus (Murrill) Dennis, *Kew Bull.*addit.Ser. 3:31, 1970; plantain, banana, dry rot; Ordosgoitty et al., *Agron.trop.* **24**:33, 1974 [**54**,2367].

Marasmius Fr. 1835; Tricholomataceae; differs in

part from *Marasmiellus* in the stiff stipe, like horse hair, the tough pileus which dries out and can revive when wet, and the black rhizomorphs. The spp. are often mentioned in the older literature in the tropics as causing horse hair, brownish, or thread, whitish, blights. They formed as a superficial growth, sometimes causing leaf fall, on the foliage of perennial crops which were planted after felling primary forest. In well managed crops the fungi should cause no damage. Some spp. still require transfer to *Marasmiellus* and prob. other related genera; Petch, *Ann.R.bot.Gdns.Peradeniya* 6:43, 1915; 9:1, 1924; Briton-Jones & Baker, *Trop.Agric.Trin.* 11:55, 1934; Dennis & Reid, *Kew Bull.* 2:287, 1957; Holliday 1980; Turner 1981 [4:66; 13:539; 37:458].

M. graminum (Lib.) Berk. var. *brevispora* Dennis, *TBMS* 34:416, 1951; Pont, *Qd.J.agric.Anim.Sci.* 30:225, 1973, on maize; Muchovej & Couch, *Pl.Path.* 33:589, 1984, on senescent leaves of *Poa pratensis* [54,142].

M. oreades, see fairy rings, Drew Smith, *PD* 64:348, 1980.

M. palmivorus Sharples 1936; the sp. is a *Marasmiellus*; oil palm bunch rot, rubber white fan blight and pre-emergent shoot rot of coconut, Tey & Chan 1980; Turner 1981 [61,354].

marbled gall (*Physoderma alfalfae*) lucerne.

marcescens, *Serratia*.

marchitez, oil palm, see *Phytomonas*.

marginalis, *Pseudomonas*.

marginal leaf scorch (*Ascochyta heveae*) rubber; —— **necrosis** (*Pseudomonas marginalis* pv. *marginalis*) many crops; —— **scorch**, lettuce, Cl toxicity; —— **yellowing**, tomato, with blossom end rot q.v., Adams & Ho. 1985 [65,4094].

marginata, *Nummulariella*.

marigold mottle v. Naqvi, Hadi & Mahmood, *PD* 65:271, 1981; poss. Potyvirus, filament 650–700 nm long, transm. sap, *Myz.pers.*, India (Uttar Pradesh), *Tagetes erecta* [61,1249]. —— **phyllody** Sharma et al., *Int.J.Trop.Pl.Dis.* 3:45, 1985; MLO assoc., India, *Tagetes erecta* [65,3960].

market diseases, see postharvest loss.

Markham, Roy, 1916–79, born in England; Univ. Cambridge, Plant Virus Research Unit, Cambridge, director 1959; John Innes Institute, director 1967; original work on the physicochemical properties of plant viruses; FRS 1956. *Nature Lond.* 285:57, 1980; John Innes Annual Report for 1979:15, 1980; *Biogr.Mem.Fel.R.Soc.* 28:319, 1982.

marrow, see *Cucurbita*.

marsh spot, Mn deficiency; Glassock, *AAB* 28:316,

1941; Piper, *J.agric.Sci.* 31:448, 1941, pea; Hewitt, *Nature, Lond.* 155:22, 1945, bean [21:179; 24:173].

Marssonina Magnus 1906; Coelom.; emended by Sutton & Webster, *TBMS* 83:63, 1984; conidiomata acervular, conidiogenesis holoblastic; con. 1–2 septate, straight or curved, obpyriform, fusiform or Y shaped; Cellerino, *Cellulosa Carta* 30:3, 1979, review, spp. on poplar; Spiers & Hopcroft, *CJB* 61:3529, 1983, fine structure of conidial ontogeny; Spiers, *Eur.J.For.Path.* 14:202, 1984, host specificity & symptoms [60,4680; 63,3055; 64,1775].

M. brunnea, teleom. *Drepanopeziza punctiformis*; **Marssonina castagnei**, teleom. *Drepanopeziza populi-albae*; **M. coronaria**, teleom. *Diplocarpon mali*; **M. fragariae**, teleom. *D. earlianum*; **M. juglandis**, teleom. *Gnomonia leptostyla*; **M. kriegeriana**, teleom. *Drepanopeziza triandrae*; **M. populi**, teleom. *D. populorum*; **M. rosae**, teleom. *Diplocarpon rosae*; **M. salicicola**, teleom. *Drepanopeziza sphaeroides*; **M. tremulae**, poss. teleom. *D. tremulae*.

masked virus, one carried by a plant which does not show symptoms of its presence, FBPP.

Mason, Edmund William, 1890–1975; born in England, Univ. Cambridge, Birmingham; Imperial Bureau of Mycology, Kew, now the CAB International Mycological Institute, 1921–60. Mason was the first mycologist at the newly created bureau, see E.J. Butler, and the founder of the IMI herbarium; he set new standards for mycological herbaria; the work on Hyphomycetes had a great influence on their reclassification; his knowledge on this group and the Ascomycetes was profound; an obituarist stated: 'He will long be remembered as one of the original thinkers in mycological science'; Linnean Gold Medal 1961. *Bull.Br.mycol.Soc.* 9:114, 1975; *Kavaka* 3:145, 147, 1975; *TBMS* 66:371, 1976.

Massee, George Edward, 1850–1917; born in England; at Royal Botanic Gardens, Kew from 1893; he ranks with Berkeley and Cooke as one of the foremost mycologists in England; wrote: *Textbook of plant diseases*, 1899; *Textbook of fungi*, 1906; *Diseases of cultivated plants and trees*, 1910; Ramsbottom, in his obituary, *TBMS* 5:469, 1917, wrote: '...if he had had any capacity whatever for taking pains he would have been a genius.'

mass median diameter, the figure dividing a total volume of spray into two equal parts; one half of the mass of spray is contained in droplets of smaller diam. than MMD, and the other half is contained in droplets of larger diam., FBPP.

Mastigomycotina, the first division of the

Eumycota; thallus unicellular or mycelial, spores and/or gametes motile; Buczaki 1983.

Mastigosporium Riess 1852; Hyphom.; Carmichael et al. 1980; conidiomata synnematal, conidiogenesis holoblastic, conidiophore growth percurrent; con. multiseptate, hyaline; leaf spots on temperate grasses; Hughes, *Mycol.Pap.* 36, 1951; Austwick, *TBMS* **37**:161, 1954; Sampson & Western 1954; O'Rourke 1976; Smiley 1983.

mathematical modelling, of crop disease; Gilligan ed., *Adv.Pl.Path.* **3**, 1985.

Matsumoto, Takashi, 1891–1968; born in Japan, Univ. Tohoku Imperial, California Berkeley, Washington; chair 1928 at Univ. Taihoku Imperial, now the National Taiwan Univ.; diseases of tropical crops, especially those caused by viruses. *Phytop.* **58**:1325, 1968.

Matsumuratettix, transm. sugarcane grassy shoot; *M. hiroglyphicus*, transm. sugarcane white leaf.

matthiola green petal Gourret, *J. Microscopie* **9**:807, 1970; MLO assoc. [**51,184**].

maturity bronzing, banana; assoc. with hot, humid and overcast conditions; poss. due to excessive turgor pressure, Australia (Queensland); Campbell & Williams, *Aust.J.Exp.Agric.Anim.Husb.* **16**:428, 1976; **18**:603, 1978 [**56,1214**; **58,2329**].

Mauginiella Cav. 1925; Hyphom.; conidiomata hyphal, conidiogenesis holoarthric; con. aseptate, hyaline; *M. scaettae* Cav., date palm inflorescence rot, not reported from Americas; Nicot, *Rev.Mycol.* **37**(3), *Fiche Phytopath.Trop.* 24, 1972; Sigler & Carmichael, *Mycotaxon* **4**:349, 1976, Hyphom. with arthroconidia; von Arx et al., *Sydowia* **34**:42, 1981 [**61,7126**].

maximensis, *Claviceps.*

maydis, *Bipolaris, Hyalothyridium, Monographella, Peronosclerospora, Phyllachora, Phyllostica, Physoderma, Stenocarpella, Ustilago.*

Mayer, Adolf, 1843–1942, born in N. German Confederation; Univ. Heidelberg, Ghent, Halle; Agricultural Experiment Station, Wageningen, director 1876. In 1886 he wrote a paper on the mosaic disease of tobacco, now known to be caused by tobacco mosaic v., *Phytopath.Classics* 7, 1942, English translation by J. Johnson. Mayer showed that the filtered juice from diseased tobacco was infectious but thought the cause was bacterial. This was some 6 years before the work of Ivanovski and 12 years before that of Beijerinck on the same disease. Ainsworth 1981.

McAlpine, Daniel, 1849–1932; born in Scotland, Univ. London, Royal School of Mines; professor, Veterinary College, Edinburgh, Watt Herriot College; went to Australia in 1884, Univ. Melbourne and plant pathologist, Department of

Agriculture, Victoria from 1890; a pioneer of the subject in Australia; wrote books on the diseases of citrus, potato and stone fruit; studied cereal rusts; monographs on the Uredinales 1906, and the Ustilaginales 1910, of Australia. *Aust.Pl.Path.Newsl.* **5**:11, 1976.

McKay, Robert, 1889–1964; born in Ireland, Royal College of Science; chair, plant pathology, Univ. College, Dublin 1945; wrote handbooks on diseases of beet, cereals, crucifers, onion, potato, tomato; Medal of Honour, Horticultural Society of Ireland 1948; Boyle Medal, Royal Dublin Society 1957. *Nature Lond.* **203**:124, 1964; *Proc.R.Irish Acad.* B **76**:404, 1976.

McKinney, Harold Hall, 1889–1976; born in USA, Univ. Michigan State, Wisconsin; cereal virologist, gave the first demonstration of a soilborne virus, wheat soilborne mosaic v. *Phytop.* **67**:429, 1977.

meadii, *Phytophthora.*

meadow grass, *Poa pratensis.*

mealie variegation Fuller 1910 (maize streak v.), fide Damsteegt, see MSV.

mealy bugs, see vector.

—— **pod** (*Trachysphaera fructigena*) cacao.

measles, apple; type 1 assoc. with high Mn; type 2 due to B deficiency; Carroll, *J.agric.Qd.* **100**:29, 1974; internal bark necrosis, apple; assoc. with high Mn; Berg & Clulo, *Science N.Y.* **104**:265, 1946 [**26**:16; **53,4044**].

measurement of disease, see phytopathometry.

mechanical transmission, transmission of a virus from an infected host to a healthy one without the intervention of a vector; either by physical contact and friction of one plant against the other or experimentally, e.g. by rubbing sap from an infected plant on the leaves of a healthy one, FBPP; abbreviated here as sap transm.

median effective concentration, EC 50; the concentration of a material required to bring about a prescribed effect, e.g. retardation of growth, in 50% of the individuals in a population of a given sp., FBPP.

—— —— **dose,** ED 50; the dose of a material required to bring about a prescribed effect in 50% of the individuals of a population of a given sp., FBPP.

—— **latent period,** LP 50; the time taken for 50% of a population of viruliferous vectors to complete their latent period q.v., FBPP.

—— **lethal concentration,** LC 50; the concentration of toxicant required to kill 50% of a large group of individuals of one sp., FBPP.

—— —— **dose,** LD 50; the dose of a toxicant killing 50% of a large group of individuals of one sp., FBPP.

medicaginis, *Leptotrochila, Phoma.*

Medicago, *M. sativa*, alfalfa, lucerne q.v.; Chilton, *Misc.Publ.USDA* 499, 1943, fungi on *Medicago*, *Melilotus, Trifolium*; Bretag, *TBMS* **85**:329, 1985, fungi assoc. with *Medicago* spp. root rots in Australia [**64**,5437].

medicinal plants, Chattopadhyay 1967.

mediterraneum, *Hypoxylon*; **medusae**, *Melampsora*.

Megadelphax sordidula, transm. phleum green stripe.

megakarya, *Phytophthora*; **megalacanthum**, *Pythium*.

Megaloseptoria Naumov 1925; Coelom.; Sutton 1980; conidiomata pycnidial, conidiogenesis enteroblastic phialidic; con. filiform, multiseptate, hyaline; *M. mirabilis*, teleom. *Gemmamyces picea*.

megasperma, *Bremiella, Phytophthora*.

Mehta, Karamchand, 1892–1950; born in India, Government College, Lahore; Agra College and Univ.; an authority on temperate cereal rusts in India. *Indian Phytopath*. **3**:1, 1950.

Meinecke, Emilio Pepe Michael, 1869–1957; born in USA, Univ. Freiberg, Leipzig, Bonn, Heidelberg in central Europe; in 1898 he became an assistant to Hartig at Univ. Munich; Univ. La Plata, Argentina 1907–9; USDA from 1910; a pioneer of forest pathology; his classic *Forest pathology in forest regulation*, 1916, had an important influence on practice in N. America. *Phytop*. **47**:633, 1957.

melaleuca yellows Klein, Dabush & Bar-Joseph, *Phytoparasitica* **7**:169, 1979; agent like a bacterium assoc., 950–1400 × 400–600 nm, Israel, *M. armilaris* [**60**,346].

Melampsora Castagne 1843; Uredinales; telia remain covered by host tissue, forming a crust of a single layer of teliospores, aseptate, sessile; erect, capitate paraphyses interspersed in the uredia; some spp. are heteroecious with telia on *Populus* and *Salix*, aecia on angiosperms or conifers; many of the spp. are on poplar and willow; Peace 1962; Taris, *Ann.Épiphyt*. **19**:5, 1968, on poplar; Hiratsuka & Kaneko, *Rep.Tottori mycol.Inst*. 20:1, 1982, on willow Desprez-Loustau, *Eur.J.For.Path*. **16**:360, 1986, on pines in Europe [**48**,293; **62**,3275; **66**,1635].

M. euphorbiae Castagne, see spurge.

M. farlowii (Arthur) J.J. Davis 1915; Arthur 1934; autoecious, microcyclic, telia on *Tsuga canadensis*, hemlock rust, N. America, Hepting & Toole, *Phytop*. **29**:463, 1939 [**18**:719].

M. larici-populina Kleb. 1902; Walker, *Descr.F*. 479, 1975; heteroecious, telia on *Populus*, aecia on *Larix*; distinguished from other spp. on *Populus* by apically thickened uredial paraphyses, apically smooth, elongated uredospores, 30–44 × 14–19 μm; predominantly epiphyllous telia; absent from

N. America. The relatively recent introduction of this rust, and *Melampsora medusae*, to Australia and New Zealand has resulted in severe damage to poplars, especially to Lombardy poplar, *P. nigra* var. *italica*; Chandrashekar & Heather, *Euphytica* **29**:401, 1980; **30**:113, 1981; *Phytop*. **71**:421, 1981; *TBMS* **77**:375, 1981; **78**:381, 1982, races, effects of light & temp. on host reactions, stability of host resistance; Heather & Chandrashekar, *Aust.For.Res*. **12**:231, 1982, pathosystem [**60**,3362; **61**,409, 3699, 5983; **63**,268].

M. lini (Ehrenb.) Desm. 1850; Laundon & Waterston, *Descr.F*. 51, 1965; autoecious, macrocyclic, *Linum*, flax or linseed rust. The many races of this sp. were used by Flor to develop his gene for gene concept q.v. Gold & Littlefield and Gold & Statler, *CJB* **57**:629, 1979; **61**:308, 1983, fine structure of all states, telia formation & germination of teliospores; Coffey in Callow 1983; Saharan & Singh, *Indian Phytopath*. **38**:25, 1985, epidemiology [**58**,5388; **62**,2503; **66**,624].

M. medusae Thüm. 1878; Walker, *Descr. F*. 480, 1975; heteroecious, telia on *Populus*, aecia on conifers especially *Larix* and *Pseudotsuga*, less common on *Pinus*; distinguished from other spp. on *Populus* by uniformly thickened uredial paraphyses, obovate to oval uredospores 26–35 × 15–19 μm; spiny but with a smooth, equatorial patch, widespread; can cause severe damage to poplars and some conifers; Widin & Schipper, *Can.J.For.Res*. **10**:257, 1980, epidemiology in USA (N. central); Prakash & Heather, *AAB* **108**:403, 1986, effects of temp. on resistance to races in a poplar cv. [**60**,3364; **66**,336].

M. occidentalis H.S. Jackson, *Phytop*. **7**:354, 1917; Ziller 1974; heteroecious, telia on *Populus*, aecia on conifers, cottonwood rust, N. America (W.); causes discolouration and necrosis in conifers, Douglas fir and larch are more susceptible than pine, fir and spruce; premature defoliation in *Populus trichocarpa*; Hsiang & van der Kamp, *Can.J.Pl.Path*. **7**:247, 1985, variation in virulence & resistance in black cottonwood [**65**,1540].

M. populnea (Pers.) Karsten 1879; heteroecious, telia on *Populus*, aecia on *Larix* or *Pinus*, uredospores 15–25 × 11–18 μm, the forms or races on larch and pine are often referred to 2 spp.: *Melampsora larici-tremulae* Kleb. 1897, larch; *M. pinitorqua* Rostrup 1889, pine; fide Wilson & Henderson 1966. Although, morphologically, the 2 rusts are very similar, Longo et al., *Eur.J.For.Path*. **15**:432, 1985, upheld them as distinct. Poplars can be severely defoliated and the stage on pine can cause considerable damage, twisting rust of Scots pine; von Weissenberg &

Kurkela ed., *Folia Forestalia* 422, 1980, resistance in pines; Martinson, *Eur.J.For.path.* **15**:103, 1985, effect on growth & development of Scots pine [**60**,3984-91; **64**,5433; **65**,2443].

M. ricini Noronha 1952; Punithalingam, *Descr.F.* 171, 1968; castor rust, little investigated and therefore presumed to be not important, aecia are unknown.

Melampsorella Schröter 1874; Uredinales; telia not erumpent, intra-epidermal; teliospores laterally united, aseptate, 1 cell in depth, sessile; heteroecious, telia on dicotyledons, aecia on *Abies*.

M. caryophyllacearum Schröter; Wilson & Henderson 1966; Ziller 1974; telia on *Cerastium* and *Stellaria*, aecia on *Abies*; fir broom rust, can cause conspicuous, yellowish witches' brooms on fir: Singh, *Eur.J.For.Path.* **8**:25, 1978, on *A. balsamea* in Canada (Newfoundland) [**57**,4159].

Melampsoridium Kleb. 1899; Uredinales; Wilson & Henderson 1966; telia subepidermal, teliospores compacted laterally; single layer, aseptate; uredia characteristically ostiolate, with the ostiolar cells extended into elongated apices like spines; heteroecious, telia on dicotyledonous trees, aecia on *Larix*; Kaneko & Hiratsuka, *Trans.mycol.Soc.Japan* **22**:463, 1981, speciation based on position of uredospore germ tubes [**62**,76].

M. betulinum Kleb., fide Boerema, *Neth.J.Pl.Path.* **76**:165, 1970; telia on *Betula*, aecia on *Larix*, teliospores 30–52 × 8–16 μm, birch rust, causes yellow mottling on birch, can be damaging in nurseries and landscapes, survives on the telial host, alternation of hosts does not always occur, and the rust may not be found on larch; F. & H. Roll-Hansen, *Eur.J.For.Path.* **10**:382, 1980; **11**:77, 1981, uredospore chains, *Melampsoridium alni* is conspecific; Dooley, *PD* **68**:686, 1984, temp. & germination of uredospores which took place beneath dormant bud scales [**49**,3133; **61**,2476; **63**,5564].

Melanconiales, Coelom.; formerly used for Deuteromycotina with acervular conidiomata, not now accepted.

Melanconidaceae, Diaporthales or Sphaeriales; ascomata often large, oblique or horizontal, erumpent through a stromatic disc, asco. mainly broad, pigmented; *Cryptosporella, Melanconis*.

melandrium yellow fleck v. Hollings & Horváth 1978; *Descr.V.* 236, 1981; Bromovirus, isometric *c*.25 nm diam.; transm. sap, 173 spp. in 7 plant families, Hungary, *M. album*.

melanins, dark brown to black pigments in organisms; not essential for growth but they

enhance survival and competitiveness; Bell & Wheeler, *A.R.Phytop.* **24**:411, 1986, in fungi. **melanocephala**, *Puccinia*.

Melanopsichium Beck 1894; Ustilaginales; sori conspicuous, ± agglutinated in a slimy mass at maturity, forming into hard masses like galls, mostly in the inflorescence; for references see Spooner, *TBMS* **85**:540, 1985.

M. eleusinis (Kulk.) Mundkur & Thirum. 1946; ustilospores av. 9.5 μm diam., epispore densely pitted, infected grains of finger millet, galls 6–7 times the size of healthy grain; Kulkarni, *AAB* **9**:184, 1922; Thirumalachar & Mundkur, *Phytop.* **37**:481, 1947 [**2**:308; **26**:542].

melanose (*Diaporthe citri*) citrus.

melanosis (*Pseudomonas cichorii*) wheat.

Melasmia Lév. 1846; Coelom.; Sutton 1980; conidiomata stromatic, conidiogenesis enteroblastic phialidic; con. aseptate, hyaline; *M. acerina*, teleom. *Rhytisma acerinum*.

melia decline Vazquez et al. 1983; mollicute assoc., Argentina, *M. azedarach* [**64**,324].

meliae, *Pseudomonas*.

melia yellows Muñoz, Nome & Kitajima, *Fitopat.Brasileira* **12**:95, 1987; MLO assoc., Argentina (Córdoba), *M. azedarach*, umbrella tree.

Melilotus see *Medicago*

melilotus mosaic v. Naqvi & Mahmood, *Phytopath.Z.* **93**:249, 1978; filament 200–600 nm long, transm. sap, India (Uttar Pradesh) [**58**,3355].

melinis dwarf Kulkarni & Sheffield 1967; Kulkarni, *Phytop.* **59**:1783, 1969; transm. *Malaxodes farinosus*, E. Africa, *Melinis minutiflora*, causes a maize stunt [**47**,1379c; **49**,1686].

melissa mosaic (tobacco rattle v. + tomato black ring v.), Schmidt & Schmelzer, *Arch.Gartenb.* **24**:209, 1976; Germany, *M. officinalis* [**55**,5871].

mellea, *Armillaria*.

Meloderma Darker 1967; Hypodermataceae or Rhytismataceae; *M. desmazieresii* (Duby) Darker; Minter & Gibson, *Descr.F.* 569, 1978; anam. *Leptostroma strobicola* Hilitzer 1929; asco. 25–40 × 4 μm in a gelatinous sheath; pine needle blight.

Meloidodreita kirjanovae Poghossian 1966; Siddiqi, *Descr.N.* 113, 1985; root endoparasite, on *Mentha*; Israel, USSR (S.)

Meloidogyne, root knot nematodes; *c*.30 spp., collectively causing severe damage by causing root galls in many important crops, particularly in warmer areas. But 4 spp.: *M. arenaria, M. hapla, M. incognita, M. javanica*, account for nearly all infestations of cultivable land; Sasser et al., *A.R.Phytop.* **21**:271, 1983, international project.

M. acronea Coetzee 1956; Page, *Descr.N.* 114, 1985; damages cotton, sorghum; Africa (E. and S.).

M. arenaria (Neal) Chitwood 1949; Orton-Williams, *Descr.N.* 62, 1975; cosmopolitan, groundnut root knot nematode, damages this and other crops, causes important synergistic and harmful effects with fungus pathogens.

M. chitwoodii Golden et al. 1980; Jepson, *Descr.N.* 106, 1985; on potato, temperate cereals, N. America (N.W.).

M. exigua Goeldi 1887; Cain, *Descr.N.* 49, 1974; can cause yellowing and leaf fall in coffee, also on banana, red pepper, tea, watermelon.

M. graminicola Golden & Birchfield 1965; Mulk, *Descr.N.* 87, 1976; on Gramineae, may damage rice.

M. hapla Chitwood 1949; Orton-Williams, *Descr.N.* 31, 1974; northern root knot nematode.

M. incognita (Kofold & White) Chitwood 1949; Orton-Williams, *Descr.N.* 18, 1973; major damage to crops, warmer areas; often synergistic with fungus pathogens; see *Fusarium oxysporum* f. sp. *lycopersici*, Hirano.

M. javanica (Treub) Chitwood 1949; Orton-Williams, *Descr.N.* 3, 1972.

M. naasi Franklin 1965; *Descr.N.* 19, 1973; mainly on Gramineae, root galls, cooler areas.

melon, cantaloupe, muskmelon; *Cucumis melo*; Lecoq & Pitrat in Plumb & Thresh 1983, control of aphid borne viruses.

melonis, *Diaporthe*, *Phytophthora*.

melon leaf variegation Rubio-Huertos & Peña-Iglesias, *PDR* 57:649, 1973; bacilliform particle assoc. av. 320 × 60 nm, Spain [53,1606].

*—— **necrotic spot** v. Kishi 1960; Hibi & Furuki, *Descr.V.* 302, 1985; isometric c.30 nm diam., transm. sap, seed; in soil by *Olpidium radicale* zoospores and retained by the resting spores; Cucurbitaceae, narrow host range, causes diseases in greenhouse cucumbers, melons; Greece (Crete), Japan, Netherlands, USA; Tomlinson & Thomas, *AAB* 108:71, 1986, detected the virus causing a disease on cucumber grown hydroponically in England (N.E.) [66,376].

—— **rugose mosaic** v. Jones, Angood & Carpenter, *AAB* 108:303, 1986; prob. Tymovirus, isometric c.32 nm diam., contains ssRNA, transm. sap, People's Democratic Republic of Yemen [66,377].

—— **stunt and blister** v. Nameth, Dodds & Paulus, abs. *Phytop.* 73:793, 1983; poss. Potyvirus, filament 750 × 15 nm, transm. sap, *Myz.pers.*, USA (California).

Melophia Sacc. 1884; Coelom.; see phellophagy.

melting out (*Curvularia* spp.) bentgrass; (*Drechslera*

poae) Kentucky blue grass; for references: *Cochliobolus lunatus*, Muchovej.

membracids, see vector.

Mentha, mint; *M. piperata*, peppermint.

menthae, *Puccinia*.

mercurialis v. Phillips & Brunt 1981; bacilliform c.30–60 × 20 nm, transm. sap; and filament, England, *M. perennis* [61,4585].

Meria Vuill. 1896; Hyphom.; Carmichael et al. 1980; conidiomata hyphal. *M. laricis* Vuill., causing necrosis and death of needles of *Larix decidua*, simple conidiophores emerge through the stomata, each cell bearing a succession of con. on a sterigma near its apex, con. 8–10 × 2–3 μm; a disease of nurseries; Peace & Holmes, *Oxf.For.Memoirs* 15, 1933; Batko, *TBMS* 39:13, 1956, in Britain; Skarmoutsos & Millar, *Eur.J.For.Path.* 12:73, 1982, invasion through aphid feeding sites & other fungi [13:280; 36:146; 61,6623].

merismoides, *Fusarium*.

meristem tip culture, see tissue culture.

Merlinus brevidens (Allen) Siddiqi 1970; Siddiqi, *Descr.N.* 8, 1972; causing stunt in barley, wheat in USA (Oklahoma).

mesa central stunt (*Spiroplasma kunkelii*) maize.

mesosporus, *Marasmiellus*.

mespili, *Diplocarpon*, *Entomosporium*.

metabasidium, the developmental stage of a basidium in which meiosis occurs.

metalaxyl, systemic fungicide, particularly active against Oomycetes.

metal toxicity, in plants; Foy et al., *A.Rev.Pl.Physiol.* 29:511, 1978.

Metcalf, Haven, 1875–1940; born in USA, Univ. Harvard, Indiana, Nebraska; early studies on forest diseases in N. America, including chestnut blight (*Cryphonectria parasitica*) and white pine blister rust (*Cronartium ribicola*). *Phytop.* 31:289, 1941.

metham-sodium, soil fumigant.

methi mosaic wilt v. Bhaskar & Summanwar, *Indian Phytopath.* 35:688, 730, 1982; isometric 44 nm diam., transm. sap, *Trigonella foenum-graecum*, India (N.) [64,239–40].

methods, see: aerial photography, air spora, electron microscopy, epidemiology, genetic engineering, phytopathometry, preservation of fungi, serology, soilless culture, tissue culture, volumetric spore trap; *A guide to the use of terms in plant pathology*, *Phytop. Pap.* 17, 1973; Burchill ed., *ibid* 26, 1981; Daniels & Markham 1982; Johnson & Berger, *Phytop.* 72:1014, 1982, a comment on statistics; Schans et al., *ibid* 72:1582, 1982, fungi in seed with fluorescence microscopy; Zeyen. *A.R.Phytop.* 20:119, 1982, X-ray microanalysis; Johnston & Booth 1983; Lindow & Webb, *Phytop.* **73**:520, 1983, measuring leaf disease by microcomputer digitised video image analysis; Preece ed., *Methods in plant pathology*, vol. 1, by Hill, plant virology, 1984; vol. 2, by Lelliot & Stead, plant bacteriology, 1987; Teng & Rouse, *PD* **68**:539, 1984, computers; Dhingra & Sinclair, *Basic plant pathology methods*, 1985; Gilligan ed., *Adv.Pl.Path.* **3**:1985, mathematical modelling of crop disease; Raju & Olsen, *PD* **69**:189, 1985, indexing in ornamentals; Blazquez & Edwards, *AAB* **108**:243, 1986, spectral reflectance in diseased & healthy watermelon leaves; Christie & Edwardson, *PD* **70**:273, 1986, detection of virus inclusions by light microscopy.

methyl arsenic sulphide, fungicide seed dressing.

—— **isothiocyanate**, soil fumigant.

metiram, protectant fungicide, broad spectrum.

Metopolophium dirhodum, transm. barley yellow dwarf, rice yellows; *M. festucae*, transm. festuca mottle.

metrosideri, *Diaporthopsis*.

Metrosideros polymorpha, ohia dieback in USA (Hawaii), Mueller-Dombois 1985 [**65**,1541].

Metschnikowiaceae, Endomycetales; mycelium septate, often well developed, asci terminal or lateral, subglobose to cylindrical, often multispored, asco. acicular to fusiform; *Eremothecium, Nematospora*.

mexicana, *Ephelis, Phytophthora*.

Meyen, Franz Julius Ferdinand, 1804–40; born at Tilsit, Prussia, medical degree; chair of botany, Berlin; wrote *Pflanzenpathologie* 1841; Whetzel 1918.

michiganensis, *Clavibacter*; **michotii**, *Paraphaeosphaeria*.

microbodies, in plant pathogenic fungi, physiological function; Maxwell et al., *A.R.Phytop.* **15**:119, 1977.

microcephala, *Claviceps*.

microclimate, see ecoclimate.

microconidium, see pycniospore.

microcyclic, a life cycle in the Uredinales where one of the main spore stages, usually the aecial, is absent; it follows that the rust is always autoecious q.v.

Microcyclus Sacc. 1904; Dothidiaceae; Sivanesan 1984; ascomata erumpent, multilocular, pulvinate, on a hypostroma, usually on leaves, asco. 1 septate.

M. ulei (Henn.) v. Arx in E. Müller & v. Arx 1962; anam. *Fusicladium macrosporum* Kuijper 1912; Holliday, *Descr.F.* 225, 1970; asco. hyaline, cells unequal, the large one, with a more acute apex, lies towards the ascus base, 12–20 × 2.5 μm; the anam. causes distinctive and conspicuous olive green, powdery lesions on young rubber leaves; con. typically 1 septate, becoming greyish or olivaceous, proximal cell has a thickened, truncate end, and an extremely characteristic half turn or twist, 23–65 × 4–10 μm. This fungus is an obligate pathogen of rubber, and other *Hevea* spp.; it can completely defoliate trees of any age and cause destruction of large rubber plantings. It occurs on wild *Hevea* in tropical S. America, and is restricted to the mainland of tropical America, the islands of Haiti and Trinidad. *Microcyclus ulei* has been the main cause of the failure to develop a large rubber industry in this region for > 60 years.

There was a succession of epidemics in 3 areas: 1914–23, Guianas; 1930–43, Brazil; 1935–41, Central America. In *c*.1910 it became clear to the pathologists in British Guiana, now Guyana, that rubber cultivation was uneconomic. But the evidence of early epidemics was ignored by the advisors of Henry Ford who began huge plantings in Brazil *c*.17 years later. In 1919 Ford had said that history was bunk, *The Oxford dictionary of quotations*, 1953. Perhaps, if he had thought otherwise, he would not have presided over what was to become one of the most costly failures of an agricultural project, due to a fungus pathogen, in the world. By the mid 1940s Ford had abandoned his attempts to grow rubber in Brazil; he died in 1947. Both spore stages of the pathogen are air dispersed, the con. have a forenoon peak *c*.10.00 hours; the peak for the asco. is *c*.05.00 hours. Where susceptible rubber is grown, in areas where *M. ulei* is present, defoliation is inevitable; there are a few unimportant areas where the climate is unfavourable for disease development. Resistance is known but is not very effective because of the appearance of races. Fungicide control has recently become possible in Brazil; but whether the economy of the vast, susceptible rubber plantings of S.E. Asia would allow such control, should the fungus spread to this area, is problematical. Preventive measures, both research and quarantine, and led by the Malaysian Rubber Research and Development Board, have been in progress for > 25 years.

Stahel, *Meded.Dep.Landb.Suriname* 34, 111 pp., 1917, as *Melanopsammopsis ulei* gen.nov.; the teleom. was first described as *Dothidella ulei* Henn. 1904; Langford, *Tech.Bull.USDA* 882, 31 pp., 1945; Hilton, *J.Rubb.Res.Inst.Malaya* **14**:287, 1955; Holliday, *Phytopath.Pap.* 12, 1970, monograph; Holliday 1980; Chee & Holliday, Monograph 13, Malaysian Rubb.Res.Developm.Board, 1986; Dean, *Brazil and the struggle for rubber*, 1987.

Microdochium H. Sydow 1924; emended Sutton et al., *CJB* **50**:1899, 1972; Hyphom.; Carmichael et al. 1980; conidiomata sporodochial, conidiogenesis holoblastic, conidiophore growth sympodial; con. aseptate to multiseptate, hyaline, in some spp. the con. have appendages; Sutton & Hodges, *Nova Hedwigia* **27**:215, 1976.

M. bolleyi (Sprague) de Hoog & Hermanides-Nijhof 1977; con. *c*. 10 μm long, chlamydospores 9–11 μm diam.; at stem base of temperate cereals, especially wheat; flax root rot; Reinecke, *Z.PflKrank.PflSchutz.* **85**:679, 1978; Murray & Gadd, *TBMS* **76**:397, 1981, weakly pathogenic on barley; Reinecke & Fokkema, *ibid* **77**:343, 1981; antagonism to *Pseudocercosporella herpotrichoides*; Black & Brown, *Pl.Path.* **35**:592, 1986, assoc. with flax roots, N. Ireland [**58**,3245; **61**,199, 2814; **66**,1941].

M. fusarioides D. Harris, *TBMS* **84**:358, 1985; con. usually 3 septate, 30–60 μm long, poss. chlamydospores, from oospores of *Phytophthora syringae* in apple leaves, UK.

M. nivale, or *Fusarium nivale*, teleom. *Monographella nivalis*, Samuels & Hallet, *TBMS* **81**:473, 1983.

M. oryzae, teleom. *Monographella albescens*.

M. panattonianum (Berl.) B. Sutton et al., *TBMS* **86**:620, 1986; con. hyaline, smooth, fusiform, 1 septate, 12.5–15.5 × 2.5–3.5 μm; lettuce ringspot, poss. seedborne, N. temperate; Stevenson, *J. Pomol.* **17**:27, 1939, occurrence & spread in England; Couch & Grogan, *Phytop.* **45**:375, 1955, etiology & hosts in USA; Moline & Pollack, *ibid* **66**:669, 1976, morphology & as a postharvest pathogen; Marte & Cappelli, *Riv.Patol.veg.* **15**:117, 1979, severe loss in Italy; Jones, *Tests Agrochem. Cvs.* 7, *AAB* **108** suppl.:56, 1986, fungicides [**18**:569; **35**:67; **55**,6041; **59**,6047; **65**,6277].

M. tabacinum, see *Fusarium tabacinum*.

Micronectriella cucumeris, see *Fusarium tabacinum*, anam. fide Booth 1971; for teleom. of *F. tabacinum*, see Gams & Gerlach, *Persoonia* **5**:177, 1968, in *Plectosphaerella*.

micro-organism, see organism.

microsperma, *Diplodina*.

Microsphaera Lév. 1851; Erysiphaceae; ascomata with repeatedly, dichotomously branched appendages which form a compact head on a long stalk, several asci, on certain woody plants; Braun, *Nova Hedwigia* **39**:211, 1984.

M. alphitoides Griffon & Maubl. 1912; Robertson & Macfarlane, *TBMS* **29**:219, 1946; asco. 18–24 × 6–13 μm, con. 25–37 × 15–22 μm; oak mildew, Phillips & Burdekin 1982 for other *Microsphaera* spp. on *Quercus*; can cause

significant damage to young trees; Edwards & Ayres, *New Phytol.* **89**:411, 1981, cell death & cell wall papillae in 3 *Quercus* spp. differing in resistance [**26**:363; **61**,1939].

M. begoniae Sivan., *TBMS* **56**:304, 1971; anam. poss. *Oidium begoniae* var. *macrosporum* Mendonça & de Sequeira 1962, not *O. begoniae* Puttemans; asco. 18–23 × 11.5–14 μm, con. 40–65 × 11–15 μm, on *Begonia*. This sp. is like *Microsphaera hypophylla* Nevodovskii but the appendages are septate in *M. begoniae* and aseptate in *M. hypophylla* [**50**,3816].

M. diffusa Cooke & Peck 1871–2; Paxton & Rogers, *Mycologia* **66**:894, 1974; asco. 18 × 9 μm, con. 28–51 × 17–21 μm; soybean, other legumes; recently found to cause problems on soybean in USA (midwest, S.E.), mostly in cooler than normal seasons; McLaughlin et al., *Phytop.* **67**:726, 1977, morphology; Mignucci & Chamberlain, *ibid* **68**:169, 1978, anam. & hosts; Mignucci & Lim, *ibid* **70**:919, 1980, adult plant resistance; Dunleavy, *PD* **64**:291, 1980, loss [**54**,1957; **57**,868, 5213; **60**,568; 5192].

M. grossulariae (Wallr.) Lév. 1851; Sivanesan, *Descr.F.* 252, 1970; asco. 18–28 × 10–16 μm, European gooseberry mildew.

M. penicillata (Wallr.:Fr.) Lév.; Mukerji, *Descr.F.* 183; 1968; asco. 18–23 × 1–12 μm, alder, lilac, other plants; Morrall & McKenzie and Kharbanda & Bernier, *Can.J.Pl.Sci.* **57**:281, 745, 1977, on broad bean as var. *ludens*, identity, morphology, susceptibility in 5 cvs., use of benomyl [**56**,5887; **57**,1906].

Micrutalis, transm. tomato pseudo curly top.

mid-crown chlorosis, oil palm, assoc. with Cu deficiency; —— **yellowing**, oil palm, K deficiency, Turner 1981.

mid-season bunch rot (*Botryotinia fuckeliana*) grapevine, Australia; Nair & Parker, *Pl.Path.* **34**:302, 1985 [**64**,4435].

mid-vein necrosis (*Pseudomonas syringae* pv. *papulans*) apple.

mildew, popularly a visible fungus or mould growth on any substrate or surface; a growth on a plant surface, often on a leaf and usually a powdery or true mildew, i.e. a sp. in the Erysiphaceae; a downy or false mildew is a sp. in the Peronosporaceae; a dark or black mildew, one in the Capnodiales or Meliolales.

milk vetch dwarf v. Inouye 1964; Inouye et al., *Ann.Phytopath.Soc.Japan* **34**:28, 1968; Ohki et al., *ibid* **41**:508, 1975; prob. Luteovirus, spherical *c*.26 nm diam., transm. *Acyrthosiphon pisum*, *Aphis craccivora*, persistent, legumes, yellow dwarf of broad bean and pea [**45**,2694a; **47**,1712; **55**,3367].

Millardet, Pierre Marie Alexis, 1838–1902; born in

France, doctorate in medicine, Univ. Nancy and Bordeaux, chairs of botany; his perception and development of the use of Cu as an extremely effective fungicide was, as Large 1940 said, ' ... to be compared in significance only with Pasteur's triumph over the anthrax of sheep.' Others before Millardet had noticed the effects of the metal on fungi; but it is to him that we owe the development of the first universal fungicide, Bordeaux mixture q.v. He was walking in a Medoc vineyard in 1882 and noticed that the grapevines near the path were free from downy mildew (*Plasmopara viticola* q.v.). The growers used to sprinkle Cu and lime on the plants to prevent pilfering; where this had been done the vines were free from mildew. From this single observation sprang the voluminous literature describing the action and use of copper and lime against plant diseases caused by fungi: *Phytop.* **4**:1, 1914, Whetzel 1918; *Phytopath.Classics* 3, 1933, English translation by F.J. Schneiderhan.

milleranderage, or hen and chickens, grapevine, B deficiency; Scott, *Soil Sci.* **57**:55, 1944 [**23**:206].

milleri, *Mycosphaerella*.

millets, *Echinochloa frumentacea*, barnyard or Japanese; *Eleusine coracana*, finger; *Panicum miliaceum*, common or proso; *Paspalum scrobiculatum*, kodo; *Pennisetum americanum*, bulrush, cat tail, pearl or spiked; *Setaria italica*, foxtail; Ramakrishnan 1963.

Mills period, the period of apple leaf wetness satisfying requirements for leaf infection by *Venturia inaequalis* asco. at stated temps., Mills & Laplante, *Ext.Bull.Cornell agric.Exp.Stn.* 711, 1954, after FBPP; Jones & Croft, *PD* **65**:223, 1981, apple scab monitoring in USA (Michigan).

milo (*Periconia circinata*) sorghum.

mimics, see disease lesion mimics.

*mimosa.

mimosine, a non-protein amino acid that is highly concentrated in seeds and foliage of *Leucaena* and *Mimosa*; toxic to animals, fungi and plants; Serrano et al., *Aust.J.biol.Sci.* **36**:445, 1983 [**63**,2177].

minitans, *Coniothyrium*.

minor, *Rosellinia*, *Sclerotinia*, *Tilletiopsis*.

mint bolting and witches' broom Surgucheva & Protsenko 1971; Elmendorf, *Res.Bull.Univ.Wisconsin-Madison* R2901, 1977; poss. MLO, USSR.

minus, *Cyclaneusma*.

mirabilis, *Megaloseptoria*, **Phytophthora*.

—— **malformation** Lovisolo 1969; Italy [**48**,2707j].

—— **mosaic v.** Brunt et al. 1970; Brunt & Kitajima, *Phytopath.Z.* **76**:265, 1973; see dahlia mosaic v.,

Richins & Shepherd; Caulimovirus, isometric *c*.50 nm diam., transm. sap, aphid, USA (Illinois), *M. jalapa*, *M. nyctaginea* [**50**,1555u; **53**,194].

—— **witches' broom** Gosh et al., *Sci.Cult.* **41**:334, 1975; poss. MLO, India (N.) [**55**,2264].

mirabilissima, *Cumminsiella*.

miscanthi, *Peronosclerospora*.

miscanthus streak v. Yamashita 1983; Yamashita et al., *Ann.phytopath.Soc.Japan* **51**:582, 1985; Geminivirus, isometric, particle pairs 30 × 18 nm, *M. sacchariflorus* [**65**,4445].

mistletoes, plant of several genera which are parasitic on others; Hawksworth & Wiens, *A.R.Phytop.* **8**:187, 1970, *Arceuthobium*, dwarf mistletoes on Cupressaceae and Pinaceae.

mites, see vector; Moser, *TBMS* **84**:750, 1985, carrying asco. of coniferous blue stain fungi [**64**,5145].

Miuraea Hara 1948; Hyphom.; Deighton, *Mycol.Pap.* 133, 1973; con. borne as terminal blastospores, old conidial scars unthickened; *M. persica*, teleom. *Mycosphaerella pruni-persicae*.

miyabeanus, *Cochliobolus*.

Moko (*Pseudomonas solanacearum*) banana.

Molières decline, see cherry.

molinia streak v. Huth, *Acta biol.iugosl.* B **11**:195, 1974; Querfurth & Bercks, *Phytopath.Z.* **85**:193, 1976; isometric *c*.28 nm diam., transm. sap, Germany, *M. caerulea* [**55**,1308, 5598].

Moller, William John, 1936–81, born in Australia, Univ. Adelaide, California Davis, South Australian Department of Agriculture, specialising in diseases of grapevines and fruit trees. He confirmed that *Eutypa armeniacae* q.v. caused canker and dieback in grapevine; a disease attributed to *Phomopsis viticola* q.v. for > 60 years in N. America. He explained the separate roles for these fungi while studying grapevine dead arm. *Australasian Pl.Path.* **10**:63, 1981.

Mollicutes, the single class of the Tenericutes, see Procaryotae; there is one order, the Mycoplasmatales, 2 of the 3 families contain plant pathogens: Mycoplasmataceae and Spiroplasmataceae; their taxonomy is tentative. Very small procaryotes; no cell wall, enclosed in a plasma membrane; pleomorphic, gliding or rotary motion may occur; Gram negative, require fatty acids and sterols for growth, colonies with a characteristic fried egg appearance, tolerant of penicillin but not of tetracycline; see major texts and: Nienhaus & Sikora, *A.R.Phytop.* **17**:37, 1979; Behncken in Fahy & Persley 1983; Maniloff, *A.Rev.Microbiol.* **37**:477, 1983; Bové, *A.R.Phytop.* **22**:361, 1984.

molya (*Heterodera avenae*) barley, wheat.

Momordica charantia, bitter gourd.

momordicae, *Leandria*.
momordica mosaic Nagarajan & Ramakrishnan,
Proc.Indian Acad.Sci. B 73:30, 84, 1971; transm.
aphid, India (Tamil Nadu); Del Prado 1966,
Surinam [**45**,3457; **51**,3003; **52**,1316].
—— **witches' broom** Chou et al., as for lagenaria
witches' broom; Kitajima et al., *Fitopat.Brasileira*
6:115, 1981, MLO assoc., Brazil, Taiwan; Lin &
Choa 1982, in China on *M. grosvenori* [**61**,329;
63,296].
mompa, *Helicobasidium*.
monesis, *Chrysomyxa*.
Monilia Bonorden 1851; Hyphom.; Carmichael
et al. 1980; conidiomata hyphal, conidiogenesis
holoblastic, con. aseptate, hyaline, maturing
acropetally; teleom. in *Monilinia*; *Monilia cinerea*,
M. fructigena and *M. kusanoi* have teleom. in
Monilinia laxa, *M. fructigena* and *M. kusanoi*,
respectively. It should be noted that the serious
pathogen of cacao, *Moniliophthora roreri*, is not a
Monilia although it was originally so described.
Moniliaceae, Hyphom.; hyphae and/or con. are
hyaline or brightly coloured in mass, hence
moniliaceous, cf. Dematiaceae.
moniliforme, *Fusarium*.
moniliformin, toxin from *Gibberella fujikuroi*; Cole
et al., *Science N.Y.* **179**:1324, 1973 [**52**,2261].
moniliformis, *Ceratocystis*.
Monilinia Honey 1928; Sclerotiniaceae; asco.
aseptate, hyaline; sclerotium of the hollow
sphaeroid type; Byrde & Willetts, *The brown rot
fungi of fruit, their biology and control*, 1977; these
fungi, often placed in *Sclerotinia*, are now
generally accepted in *Monilinia*. The con. of *M.
fructicola*, *M. fructigena* and *M. laxa* vary in size
from 13–21 × 9–13 μm, those of the second sp.
being larger than those of the other two; see *M.
vaccinii-corymbosi*, Batra, for spp. on Ericaceae;
Harada, *Bull.Fac.Agric.Hirosaki Univ.* 27:30,
1977, spp. in Japan; Willetts et al.,
J.Gen.Microbiol. **103**:77, 1977; taxonomy &
extracellular cell degrading enzymes; von Arx,
Sydowia **34**:13, 1981, anam.; Willetts & Harada,
Mycologia **76**:314, 1984, apothecial formation
with particular reference to work in Japan
[**57**,2431; **59**,1623; **63**,3218].
M. aucupariae (Ludwig) Whetzel 1945; Harada &
Kudo, *Trans.mycol.Soc.Japan* **17**:126, 1976;
asco. 11–16 × 6–8 μm, con. 8–14 × 6–11 μm;
only on *Sorbus commixta*, blight of leaves and
young shoots, immature fruit rot; Europe,
Japan.
M. fructicola (Winter) Honey; Mordue, *Descr.F.*
616, 1979, *as Sclerotinia fructicola* (Winter) Rehm;
asco. 6–15 × 4–8 μm; most *Prunus* spp.; blossom
wilt, fruit rot, twig blight and canker of stone

fruit, also quince, infrequently in apple; essentially
in the New World. The fruit pustules are *c*.0.4 mm
diam., con. form long, unbranched germ tubes,
often not branching after 20 hours, on potato
dextrose agar there are abundant rings of
conidiophores, stroma formation sometimes
extensive; Byrde & Willetts, under genus; Biggs &
Northover, *Can.J.Pl.Path.* **7**:302, 1985, on peach,
inoculum sources [**65**,1395].
M. fructigena Honey ex Whetzel; Mordue, *Descr.F.*
617, 1979, as *Sclerotinia fructigena* Aderh. &
Ruhl.; asco. 9–12.5 × 5–6.8 μm; most *Prunus* spp.,
fruit rot can be severe, spur blight and canker,
pome and stone fruit, hazel nut drop; more
widespread than *Monilinia fructicola*, especially in
Asia and Europe; con. are mostly larger than
those of *M. fructicola* and *M. laxa*. The fruit
pustules are *c*.1 mm diam., long germ tubes
formed by con. before branching, on potato
dextrose agar rings of conidiophores are
sometimes formed, stroma infrequent; Byrde &
Willetts, under genus; Korf & Kohn, *Mycotaxon*
9:521, 1979, author citation of anam.
M. johnsonii (Ell. & Ev.) Honey 1936; Bond,
TBMS **44**:613, 1961; asco. 10.5–14 × 5–6 μm, con.
13–20 × 11–17 μm; *Crataegus* spp., hawthorn leaf
blotch and blossom blight; poss. Old and New
World forms; Dennis, *Pl.Path.* **14**:46, 1965, hosts
[**44**,2506].
M. kusanoi (Takahashi) Yamamoto 1977;
Kusunoki et al., *Bull.Forestry Forest
Prod.Res.Inst.* 328:1, 1984; asco. 12–15 × 5.5–
7 μm, con. 9–11 × 8–11 μm; flowering cherry
blight, Japan [**64**,1150].
M. laxa (Aderh. & Ruhl) Honey ex Whetzel;
Mordue, *Descr.F.* 619, 1979, as *Sclerotinia laxa*
Aderh. & Ruhl., asco. 7–19 × 4.5–8.5 μm; *Prunus*
spp., blossom and twig blights of pear and quince,
fruit rot, apricot canker. The fruit pustules are
c. 0.4 mm diam., con. form at 5° and above, a
lower temp. than for *Monilinia fructicola* and *M.
fructigena*; the germ tubes of the con. branch
close to the spore, and after *c*.20 hours it is often
difficult to identify the original germ tube because
of the scorpioid form; on potato dextrose agar
the colony margins are lobed, conidiophore
development sparse, stroma not often formed;
Byrde & Willetts, under genus.
M. linhartiana (Prill. & Delacr.) Buchw. 1949; fide
Cannon et al. 1985; Dennis, *Mycol.Pap.* 62:141,
1956; asco. 11–12 × 6–7.5 μm; quince fruit rot, the
disease can be serious in Turkey; Altinyay, *Bitki
Koruma Bült*.suppl. 1, 1972; **15**:139, 1975 [**52**,446;
55,3222].
M. oxycocci (Woronin) Honey 1928; fide Cannon
et al. 1985; asco. large 12–19 × 6–10 μm, small

9–14 × 4.5–7.5 µm; con. 16–28 × 11–22; *Vaccinium*, cranberry hard rot, cotton ball or tip blight.

M. vaccinii-corymbosi (Reade) Honey 1936; asco. 16–18 × 9–10 µm, con. 26–28 × 19–21 µm; *Vaccinium corymbosum*, blueberry mummy berry, an important disease in Canada, USA; Milholland, *Phytop.* 67:848, 1977, asco. infection, sclerotial germination & histology; Batra, *Mycologia* 75:131, 1983, *Monilinia* spp. on Ericaceae, morphology & biology [57,1794; 62,3133].

Moniliophthora H. Evans et al., *CJB* 56:2530, 1978; hyphae hyaline, septate, septa without clamp connections but with dolipores; con. catenate, globose to ellipsoid, thin and thick walls, formed basipetally; teleom. unknown. The genus was erected to accommodate the important pathogen of cacao, *Monilia roreri* Cif. 1933, which was found to have basidiomycete characteristics. This original name is, quite incorrectly, still being used by some in the Americas. The con. are not formed in chains with the youngest at the tip as in *Monilia*, i.e. acropetally, but in the reverse manner, i.e. the youngest con. are at the base of the chain, or basipetally. Also *Monilia* has teleom. in *Monilinia*.

M. roreri (Cif.) H. Evans et al.; con. easily separable, thick walls, pale yellow brown in mass, globose to subglobose, mostly 8–15 µm diam., sometimes ellipsoid 8–20 × 5–14 µm; cacao frosty pod rot; an obligate pathogen which, with *Crinipellis perniciosa*, causes the other of the 2 major diseases of the crop in parts of S. America. The fungus is still confined to the N.W. of S. America, Costa Rica, Panama; it is found on the eastern side of the Andean cordillera in Ecuador. Frosty pod rot caused epidemics on cacao in the 19th century in Colombia and Ecuador; and the fungus is presumed to be present on wild *Theobroma* in this area of the sub-continent. The disease is found over a wide climatic range, particularly with respect to rainfall, and is a potential threat to cacao in the Amazon basin and Bahia in Brazil. The con. are air dispersed and the only known infective propagules. They form white, cream to brown, powdery blooms on infected pods. Infection takes place on pods 1–3 months old; con. on pods still on the tree remain viable for 9 months. The disease is not restricted to wet environments since sporulation can occur in the dry season. The fungus survives on mummified pods between crops. Certain cultural measures can reduce incidence; Holliday 1980; Evans, *Phytopath.Pap.* 24, 1981; *Cocoa Grower's Bull.* 37:34, 1986 [61,150; 66,1818].

Monilochaetes Halsted ex Harter 1916; Hyphom.;

conidiomata synnematal, con. aseptate, pigmented, not in chains, 12–20 × 4–7 µm; the genus, not accepted by Carmichael et al. 1980, was erected to accommodate a pathogen of sweet potato, *M. infuscans*, see Harter, *J.agric.Res.* 5;791, 1916, causing scurf. The disease appears to be only important in USA where it disfigures the tubers; Holliday 1980; Lawrence et al., *Phytop.* 71:312, 1981, histology [61,1521].

monoceras, *Setosphaeria*, see genus.

monocerin, toxin formed by *Setosphaeria turcica*; Robeson & Strobel, *Agric.Biol.Chem.* 46:2681, 1982 [63,1764].

monoclinous, antheridium on oogonial stalk, cf. androgynous and diclinous.

monoclonal antibody, continuous cultures of fused cells secreting antibody of predefined specificity, Köhler & Milstein, *Nature Lond.* 256:495, 1975; Sander & Dietzgen, *Adv.Virus Res.* 29:131, 1984, against plant viruses; Halk & de Boer, *A.R.Phytop.* 23:321, 1985, use of in plant disease; Lin, *Pl.Prot.Bull.Taiwan* 29:91, 1987, in procaryotes, review.

monocyclic diseases, those that arise from inoculum that does not increase within the crop during a single growing season. Where inoculum does increase during a season, as progressively increasing numbers of plants become infectious, the disease is called polycyclic; Vanderplank, *Host–pathogen interactions in plant disease*, 1982; Thresh, *Adv.appl.Biol.* 8:1, 1983.

Monographella Petrak 1924; Amphisphaeriaceae; asci interspersed with apically free paraphyses, amyloid; von Arx, *TBMS* 83:373, 1984, on relationships with other genera & anam. [64,86].

M. albescens (Thüm.) Parkinson, Sivan. & C. Booth, *TBMS* 76:64, 1981; anam. *Microdochium oryzae* (Hashioka & Yokogi) Samuels & Hallett 1983, see *M. nivale*; Sivanesan, *Descr.F.* 729, 1982; asco. mostly 3 septate, 14–30 × 3.5–7.5 µm; con. mostly 1 septate, 11–16 × 3.5–4.5 µm; rice leaf scald or brown leaf spot, seedborne; Holliday 1980, as *Rhynchosporium oryzae*; Naito, *Bull.Tohoku Natn.agric.Exp.Stn.* 66:101, 1982, life history; Kim et al., *Korean J.Pl.Prot.* 23:126, 1984; Thomas, *Mycologia* 76:1111, 1984; Mia et al., *TBMS* 84:337, 1985, biology on seed, infection, damage, detection, survival & storage; Ou 1985 [60,5438; 62,200; 64,1592, 2538, 4312].

M. maydis E. Müller & Samuels, *Nova Hedwigia* 40:114, 1984; anam. in *Microdochium*; on maize in Mexico; the typical tar spot lesions contain only *Phyllachora maydis*, but larger lesions contain both fungi. *Monographella maydis* is found in symptomless green leaves, but with *P. maydis* it seems to become virulent.

M. nivalis (Schaffnit) E. Müller 1977; anam. *Microdochium nivale* (Fr.) Samuels & Hallett, 1983 see *M. nivale*; Booth, *Descr.F.* 309, 1971, as *Micronectriella nivalis* (Schaffnit) C. Booth; anam. *Fusarium nivale* (Fr.) Ces.; asco. mostly 1 septate, 10–17 × 3.5–4.5 μm; con. var. *nivale* mostly 1 septate, 13–20 × 2.5–3.3 μm; var. *majus* mostly 3 septate, 19–30 × 3.5–4.5 μm; temperate cereals, pre-emergence blight, foot or root rot, head blight; pink snow mould of turf; Subramanian & Bhat, *Revue Mycol.* **42**:293, 1978, morphology; Perry and Al-Hashimi & Perry, *TBMS* **86**:287, 373, 1986, pathogenicity to spring barley, survival & saprophytism in soil [65,4385, 4387].

monokaryon, monocaryotic, of fungi, having genetically ± identical, haploid nuclei, cf. dikaryon.

Monosporascus Pollack & Uecker, *Mycologia* **66**:348, 1974; Sordariales; ascomata globose, smooth, asci unitunicate, mostly with 1 asco., 56–90 × 30–55 μm; asco. aseptate, black, smooth, 30–50 μm diam. *M. cannonballus* Pollack & Uecker came from melon roots in USA (Arizona); most ascomata formed at 28–29° in vitro. *M. eutypoides* (Petrak) v. Arx, see Hawksworth & Ciccarone, *Mycopathologia* **66**:147, 1978, has asci with 2 asco. which are not > 45 μm diam. This latter sp. was first described as *Bitrimonospora indica* Sivan., Talde & Tilak, *TBMS* **63**:595, 1974; the asci contained 1–3 asco. Both spp. cause melon collapse or root rot; Reuveni et al., *Phytop.* **73**:1223, 1983; *TBMS* **80**:354, 1983, in Israel, most in vitro growth at 30°; Krikun, *Phytoparasitica* **13**:225, 1985, *M. eutypoides* distribution & poss. effects of soil temp. & fertilisers with irrigation in Israel; Uematsu et al., *Ann.phytopath.Soc.Japan* **51**:272, 1985, in Japan [58,4340; 62,4047; 63,1474; 65,961, 3048].

Monostichella salicis, teleom. *Drepanopeziza salicis*. **montanum**, *Peridermium*.

Moore, Frances Joan Harvey, 1920–86; born in England, Univ. Exeter, London Univ. College; in Ministry of Agriculture, UK, from 1948; an authority on diagnosis, disease assessment, research and development; OBE 1981. *Pl.Path.* **35**:259, 1986; *TBMS* **89**:141, 1987.

Moore, Walter Cecil, 1900–67; born in England, Univ. Cambridge; Plant Pathology Laboratory, Ministry of Agriculture, UK, 1925–62, head 1949. He had a widespread influence on advisory services and assisted in setting up the European Plant Protection Organisation; CBE 1955; wrote: *Diseases of bulbs*, 1939; edn. 2, 1979, ed. by J.S.W. Dickens; *Diseases of cereals*, 1945; *British parasitic fungi*, 1959. *AAB* **61**:167, 1968; *TBMS* **52**:353, 1969.

morbosa, *Apiosporina*.
mori, *Aecidium, Mycosphaerella, Stigmina*.
moricola, *Cristulariella, Phyllactinia*.
mors-uvae, *Sphaerotheca*.
Mortierella Coemans 1863; Mortierellaceae; sporangiophore arising directly from ordinary mycelium. *M. bainieri* Constantin 1889 causes mushroom shaggy stipe, the common cultivated sp.; Kuhlman, *Mycologia* **64**:325, 1972, zygospores; Fletcher 1984.
Mortierellaceae, Mucorales; sporangia and sporangiola typically with columella absent or rudimentary; zygospores warty, on opposed suspensors; *Mortierella*.
Morus, mulberry.
mosaic, a leaf symptom in which numerous, small areas of discolouration stand out against a background of a different tint, tending to have a clearly defined boundary delineated by veins. A pattern of green, angular areas against a predominantly yellow background may result. Alternatively the areas bounded by the veins may be chlorotic giving a yellow on green mosaic. If discrete areas of colour coalesce later a mottle symptom may result, FBPP.
mottle, a leaf symptom in which small but numerous areas of discolouration, commonly chlorotic, irregularly shaped and without sharply defined boundaries, stand out against a background of a different tint; the pattern is not related to the vein network, FBPP.
mottled needle cast (*Rhabdocline pseudotsugae*) Douglas fir; —— **rot** (*Fomes fomentarius*) beech, birch; —— **stripe** (*Pseudomonas rubrisubalbicans*) sugarcane.
mottle necrosis (*Pythium ultimum*) sweet potato; —— **scab** (*Elsinoë theae*) tea.
mould, a microfungus having an obvious mycelial or spore mass on a substrate; sometimes used as a part of a disease name, e.g. tobacco blue mould (*Peronospora tabacina*); Illman, *Mycologia* **62**:1214, 1970, mould for a fungus, mold for a shape.
mouldy core, apple, fungi infect open calyx cvs. and colonise the fruit core; Ellis & Barrat, *PD* **67**:150, 1983 [62,3104].
—— **rot** (*Ceratocystis fimbriata*) rubber.
mountain ash infectious variegation Baur 1907; Germany; Cooper 1979, in *Sorbus*.
—— —— **ringspot** Kegler, *Phytopath.Z.* **37**:214, 1959; Germany, *Sorbus aucuparia* [39:354].
Mucoraceae, Mucorales; sporangia columellate, specialised sporangiola absent, zygospores smooth to warty, and borne on opposed or side to side suspensors which may or may not have appendages; *Gilbertella, Rhizopus*.

Mucorales, Zygomycetes; sporangia multispored or few to one spore, sexual reproduction by zygospores; Hesseltine & Ellis in Ainsworth et al. IV B, 1973.

mucronatum, *Phragmidium*.

Mucuna aterrima, Bengal bean; *M. deeringiana*, velvet bean.

mucunae, *Uromyces*.

mukwa wilt, in the timber tree, *Pterocarpus angloensis*, prob. caused by *Fusarium oxysporum*, Zambia; Piearce, *PANS* 25:37, 1979 [58,5041].

mulberry, *Morus*.

— **curly little leaf** Zaprometov 1945; Chaduneli, *Zashch.Rast.Mosk.* 13:46, 1968; Giorgadze & Tulashvili 1973; transm. *Hishimonus sellatus*, USSR [25:17; 47,1977; 53,4527].

— **dwarf** 1868, Ishjima & Ishiie in Maramorosch & Raychaudhuri 1981; MLO, transm. *Hishimonides sellatiformis, Hishimonus sellatus*; Japan, Korea, prob. China, poss. USSR; one of the most destructive diseases of mulberry; Miyashita & Yamada, *Bull.seric.Exp.Stn.* 28:741, 1982, prevalence & control in Japan; Kim et al., *Korean J.Pl.Path.* 1:184, 1985, transm. to 5 herbaceous plants by *H. sellatus* [63,4004; 66,667].

— **latent v.** Tsuchizaki, *Ann.phytopath.Soc.Japan* 42:304, 1976; Carlavirus, filament *c.*700 nm long, transm. sap, Japan [56,1211].

— **leaf scorch** Kostka et al. 1982, *PD* 70:690, 1986; caused by a xylem limited bacterium q.v., Gram negative [66,669].

— **mosaic** Ho & Li, *Lingnan Sci.J.* 15:67, 1936; Raychaudhuri et al., *Indian Phytopath.* 15:187, 1962; transm. sap, *Rhopalosiphum maidis*; China, India, Thailand [15:386; 43,1362].

— — **dwarf** Zhang 1983; filament assoc., indicator *Vigna sinensis*, China (Zhejiang) [64,4421].

— **ringspot v.** Tsuchizaki, Hibino & Saito 1971; Tsuchizaki, *Descr.V.* 142, 1975; Nepovirus, isometric 22–25 nm diam., transm. sap, *Longidorus martini*, detected in soybean seed, Japan, mulberry is prob. the only natural host.

— **yellow net** Raychaudhuri, Chatterjee & Dhar, *Indian Phytopath.* 14:94, 1961; India (W. Bengal) [41:338].

mulches, in disease control, Palti 1981; in control of whitefly vectors of plant viruses, Cohen in Harris & Maramorosch 1982.

multicomponent virus, one whose genome is divided into 2 or more parts; a genome may be defined as the minimum complete set of genetic material; van Kammen, *A.R.Phytop.* 10:125, 1972; Fulton, *ibid* 18:131, 1980; Jaspars, *Adv.Virus Res.* 19:37, 1974; Reijnders, *ibid* 23:79, 1978.

multiline, a mixture of similar lines q.v. grown in fixed proportions with the intention of making it possible to substitute improved lines for less effective ones as the opportunity and need arises, FBPP. In plant pathology a multiline cv. is a mixture of isolines that differ by single, major genes for reaction to a pathogen, fide Browning & Frey in Jenkyn & Plumb 1981, and *A.R.Phytop.* 7:355, 1969; Groth, *Phytop.* 66:937, 1976, multilines & super races, a simple model; Marshall, *Theor.Appl.Genetics* 51:177, 1978; Wolfe, *A.R.Phytop.* 23:251, 1985, multiline cvs. & var. mixtures.

mummification, dry decay, skin becoming hard, dry and wrinkled.

mummy berry (*Monilinia vaccinii-corymbosi*) highbush blueberry.

Mundkur, Balchandra Bhavanishankar, 1896–1952; born in India, early education in Madras and then at Iowa State College of Agriculture; Indian Agricultural Research Institute at Pusa and then at Delhi; well known for his work on the rusts and smuts of India; wrote *Fungi and plant disease*, 1949; *The Ustilaginales of India*, 1952, with M.J. Thirumalachar. *Indian Phytopath.* 5:1, 1953.

mung bean, see *Vigna*.

— — **leaf curl** Nene 1972; Ghanekar & Beniwal, *Indian Phytopath.* 28:527, 1975; transm. sap [57,3713].

— — **mosaic** Singh & Varma, *Proc.natn.Acad.Sci.India* B 47:33, 1977; transm. sap, seedborne [59,6041].

— — **Philippine mottle v.** Talens, *Philipp.Phytopath.* 14:58, 1978; isometric 28–30 nm diam., transm. sap [59,1018].

— — **Taiwan mottle v.** Sun, Lai & Yan, abs. *Proc.Am.phytopath.Soc.* 4:91, 1977; poss. Potyvirus, filament 750 × 13 nm, transm. sap.

— — **witches' broom** Benigno, *Philipp.Phytopath.* 15:86, 1979; poss. MLO, Philippines [60,590].

— — **yellow mosaic v.** Nariani 1960; Honda & Ikegami, *Descr.V.* 323, 1986; Geminivirus, isometric, paired particles 30 × 18 nm, transm. sap, *Bemisia tabaci*, persistent; India, Pakistan, Sri Lanka, Thailand; causes severe losses in yields of mung bean and black gram; largely confined to Leguminosae.

— — — **mottle** Abu Kassim, *Malaysian Agric.J.* 53:29, 1981; transm. sap, Malaysia (W.) [62,1304].

Murphy, Hickman C., 1902–68; born in USA, Univ. West Virginia, Iowa; with F. Meehan described in 1947 one of the first examples of a host specific toxin, victorin q.v., formed by *Cochliobolus victoriae* in a blight of the oat cv. Victoria. *Phytop.* 59:525, 1969.

Murphy, Paul Aloysius, 1887–1938; Royal College of Science, Dublin; Imperial College, London; chair of plant pathology in 1927; notable early work on virus diseases, particularly of potato in Ireland; John Snell Memorial Medal, National Institute of Agricultural Botany, 1927; Boyle Medal, Royal Dublin Society, 1933. *Proc.R.Irish Acad.* B **76**:403, 1976.

Musa, banana *q.v.*, plantain; *M. textilis*, abaca, Manila hemp.

musae, *Cladosporium, Colletotrichum, Cordana, Cylindrocarpon, Guignardia, Haplobasidion, Mycosphaerella, Pseudocercospora, Uromyces, Veronaea*.

musarum, *Phyllosticta*.

mushroom, generally the fruit body or basidioma of an agaric or a bolete, especially an edible one. Several spp. are cultivated, especially in the N. temperate zone; the common one grown is mostly *Agaricus brunnescens* Peck; Sinden, *A.R.Phytop.* **9**:411, 1971, ecological control of pathogens & weed moulds; Frost & Passmore, *Phytopath.Z.* **98**:272, 1980, viruses; Fletcher 1984; Figueiredo & Mucci, *Biológico* **51**:93, 1985.

musicola, *Mycosphaerella, Phyllachora*; **musiva**, *Septoria*.

Muskett, Arthur E., 1900–84; born in England, Imperial College of Science, London; Queen's Univ. and Department of Agriculture, Belfast, chair of plant pathology 1945; extensive work in Northern Ireland on apple, flax, oat, potato, ryegrass; a very strong advisory sense; with J.P. Malone devised the Ulster method for testing flax seed for contaminating parasitic fungi; wrote: *Diseases of the flax plant*, 1949, with J. Colhoun; *Seedborne fungi*, 1964, with Malone; *Mycology and plant pathology in Ireland*, 1976. *AAB* **107**:353, 1985; *TBMS* **84**:575, 1985.

muskmelon, see melon.

—— **vein necrosis v.** Freitag 1952; Freitag & Milne, *Phytop.* **60**:166, 1970; Carlavirus, filament 674 × 15 nm, transm. sap, *Myz.pers.*, non-persistent, USA (California); only *Cucumis* in Cucurbitaceae; infected legumes systemically [**31**:368; **49**,2268].

—— **yellows v.** Lot, Delecolle & Lecoq, *Acta Hortic.* 127:175, 1982; filament *c*.1000 nm long, transm. *Trialeurodes vaporariorum*, France.

—— **yellow stunt** = zucchini squash yellow mosaic v., Lecoq et al., *PD* **67**:824, 1983 [**62**,5023].

mustard, black, brown, Ethiopian, see *Brassica*; white, *Sinapis alba*.

—— **black malformation v.** Avgelis & Quacquarelli, *Phytopath. Mediterranea* **12**:48, 1973; isometric *c*.30 nm diam., transm. sap, Italy; Hsu &

Civerolo, abs. *Phytop.* **71**:882, 1981, isometric mostly 33–36 nm diam., USA [**54**,262].

—— **mosaics** Azad & Sehgal, *Indian Phytopath.* **12**:45, 1959; Sharma, *ibid* **26**:346, 1973; Quacquarelli & Avgelis, *Phytopath. Mediterranea* **13**:160, 1974; transm. sap, aphid spp., non-persistent; India (N.), Italy [**39**:358; **54**,1483; **55**,4905].

mutila, *Diplodia*.

mycelial cord, or strand; a discrete, filamentous aggregation of hyphae which, unlike a rhizomorph, has no apical meristem; often formed by fungi infecting woody plants; Garrett 1970; Thompson & Rayner, *TBMS* **78**:193, 1982, prefer 'cord' to 'strand'.

—— **neck rot** (*Botrytis byssoidea*) onion, *Allium*.

mycelium, a mass of hyphae, thallus of a fungus; Gregory, *TBMS* **82**:1, 1984.

Mycena (Pers.) Roussel 1806; Tricholomataceae; pileus campanulate, generally small.

M. citricolor (Berk. & M.A. Curtis) Sacc. 1887; pileus 1.5–2.5 mm diam., forms a gemmifer which consists of a pedicel and a head; the latter is the gemma, an oblate spheroid *c*.0.36 mm diam., and is the large infective, airborne unit; the fungus is plurivorous; but, as causing cock's eye of coffee, a serious defoliation may result; tropical and subtropical America; Schieber in Kranz et al. 1977; Holliday 1980.

mycocecidium, a plant gall caused by a fungus.

Mycocentrospora Deighton, *Taxon* **21**:716, 1972; Hyphom.; Ellis 1971; Deighton *Mycol.Pap.* 124, 1971, as *Centrospora* Neergaard 1942, not *Centrospora* Trevisan 1845; Ellis 1976; conidiomata hyphal, conidiogenesis holoblastic, conidiophore growth sympodial; con. multiseptate, filiform, broader cells pale brown, the presence of a basal appendage is not a constant character.

M. acerina (Hartig) Deighton; Sutton & Gibson, *Descr.F.* 537, 1977; con. mostly 150–200 × 8–15 μm, base truncate, sometimes with a lateral, septate, downwardly directed appendage, 30–150 × 2–3 μm; Constantinescu, *Revue Mycol.* **42**:105, 1978; plurivorous, but particularly causing carrot liquorice rot, celery black crown rot, parsnip canker; O'Neill, *Pl.Path.* **34**:632, 1985, described a leaf spot and dieback of anemone seedlings. It has a low optimum temp. for growth, *c*.17°; spreads in soil and water; causes one of the most important disease rots of carrot and celery in long term, cold storage, and has recently been intensively studied in UK on the former crop. The fungus can also cause delayed emergence and damping off in carrot. It is a primary coloniser of root surfaces; in some soils there is a negligible decline in viability over 2 years. The pigmented mycelium and chlamydospores remain attached to the roots; these are the inoculum sources for infection during storage. Susceptibility in carrot increases with time in storage, wound depth and age at harvest. The disease is assoc. with the formation of the antifungal compound falcarindiol.

Davies, Day, Garrod and Lewis & Wall: *Physiol.Pl.Path.* **13**:241, 1978; *New Phytol.* **83**:463, 1979; *TBMS* **72**:515, 1979, falcarindiol, identification, action, site in carrot root; *TBMS* **74**:587, 1980; **75**,163, 207, 1980; **77**:139, 369, 1981, infection of carrot including leaves & stored roots, chlamydospore biology in soil, fungus biology on periderm & wounded tissue; *AAB* **95**:11, 1980; **99**:35, 1981, age of carrot roots at harvest & infection in store, wound healing in roots [**57**,4841; **58**,2007; **59**,1484, 2977; **60**,1124, 2848, 3398, 4176; **61**,1993, 2525, 3129; **65**,1320].

M. camelliae, see *Cercoseptoria ocellata*.

Mycogone Link 1809; Hyphom.; Carmichael et al. 1980; conidiomata hyphal, conidiogenesis holoblastic; con. solitary, 1 septate, hyaline.

M. perniciosa (Magnus) Delacr. 1900; Brady & Gibson, *Descr.F.* 499, 1976; con. upper cell warty, 16–31 × 13–30 μm, lower cell ± smooth, 8–17 × 8–13 μm; mushroom wet bubble, can cause severe loss in the common cultivated mushroom; Fletcher 1984; Holland et al., *TBMS* **85**:730, 1985, con. germination [**65**,2118].

M. rosea Link; Brady & Gibson, *Descr.F.* 500, 1976; con. upper cell strongly warty, 21–35 × 21–23 μm, lower cell ± smooth, 9–23 × 10–20 μm, colonies are a brighter colour than those of *Mycogone perniciosa*; causes a disease similar to mushroom wet bubble but not given by Fletcher 1984.

*mycoherbicide, a fungus deliberately used to cause disease in weeds with the object of controlling them. TeBeest & Templeton, *PD* **69**:6, 1985, delineate 2 strategies in their use. Classical: a fungus is released into a weed population, e.g. *Puccinia chondrillina* to control *Chondrilla juncea*, rush skeleton weed. Inundative: a massive, usually annual, release of a fungus into specific, weed infested crops, e.g. the f.sp. *aeschynomene* of *Glomerella cingulata* to control *Aeschynomene virginica*, northern jointvetch, in rice and soybean; and *Phytophthora palmivora* to control *Morrenia odorata*, strangler or milkweed vine, in citrus; Zettler & Freeman, *A.R.Phytop.* **10**: 455, 1972, aquatic weeds; Templeton et al., *ibid* **17**:301, 1979; Hasan, *Rev.Pl.Path.* **59**:349, 1980; Charudattan & Walker, *Biological control of*

weeds with plant pathogens, 1982; Templeton & Greaves, *Trop.Pest Management* **30**:333, 1984.

Mycoleptodiscus Ostazeski, *Mycologia* **59**:970, 1967, published 1968; Hyphom.; Ellis 1976; conidiomata sporodochial, conidiogenesis enteroblastic phialidic, sclerotia sometimes formed; con. hyaline, smooth 0–2 septate with a simple setula at the apex and often also at the base; Sutton & Hodges, *Nova Hedwigia* **27**:693, 1976.

M. indicus (Sahni) Sutton, *TBMS* **60**:528, 1973; con. 11–18.5 × 4.5–7.5 μm, with an unbranched setula or appendage 1–10 μm long and sometimes a basal appendage av. 3 μm long; sporodochia dark brown, of confluent cells 30–100 μm diam.; on monocotyledons in the tropics; causes black crust of vanilla in Brazil (Bahia), leaf, fruit and stem lesions, severe damage; Bezerra & Ram, *Fitopat.Brasileira* **11**:717, 1986.

mycology, the study of fungi q.v.

mycoparasite, a fungus parasitic on another fungus; see hyperparasite; Barnett & Binder, *A.R.Phytop.* **11**:273, 1973.

Mycoplasmataceae, Mollicutes; cells not helical, non-motile or a gliding motility; *Mycoplasma* spp. cause disease in animals; similar organisms, the MLOs or mycoplasmas, have been described causing disease in plants although fully adequate proof of pathogenicity is still lacking. They are found in the phloem, causing symptoms generally known as yellows, and in their insect vectors. The diseases were originally thought to be caused by viruses; the apparent causal agents passed through filters that retained bacteria; but Japanese workers, *Ann.phytopath.Soc.Japan* **33**:259, 1967, showed otherwise. Other symptoms include gross changes in flower structure, phyllody and colour, and leaf proliferation. These organisms lack a spiroplasma specific antigen; see Townsend 1983 under *Spiroplasma* which is in the Spiroplasmataceae, the other family in the Mollicutes q.v. that contains plant pathogens; see major texts and: Ploaie & Maramorosch, *Phytop.* **59**:536, 1969; Davis & Whitcomb, *A.R.Phytop.* **9**:119, 1971; Hull, *Rev.Pl.Path.* **50**:121, 1971; *PANS* **18**:154, 1972; Maramorosch, *A.Rev.Microbiol.* **28**:301, 1974; McCoy, *PD* **66**:539, 1982, use of tetracycline antibiotics; Shiomi & Sugiura, *Ann.phytopath.Soc.Japan* **50**:149, 1984, 12 MLO diseases in Japan considered to be caused by closely related agents, classified into 3 strs.; Lee & Davis, *A.R.Phytop.* **24**:339, 1986, prospects for in vitro culture [**47**,657; **48**,2840; **64**,553].

Mycoplasmatales, the one order of the Mollicutes q.v.

mycorrhiza, a symbiotic, non-pathogenic or feebly pathogenic, association of a fungus and the roots of a plant; some prefer an interpretation of mycorrhiza as one of limited parasitism rather than one of symbiosis; for interactions with plant pathogens and disease control chemicals: Zak, *A.R.Phytop.* **2**:377, 1964; Marx, *ibid* **10**:429, 1972; Wilcox, *ibid* **21**:221, 1983; Trappe et al., *ibid* **22**:331, 1984; Dehne, *Phytop.* **72**:1115, 1982, Perrin, *Eur.J.For.Path.* **15**:372, 1985.

mycosis, infection by a fungus or a disease so caused; it is generally used for infections of man and other animals, and is best restricted in this way. In plant pathology the term is most commonly found in tracheomycosis for a disease where the pathogen is confined, more or less, to the vessels and assoc. cells. The FBPP list prefers vascular wilt for this diseased condition; although tracheomycosis is more specific, in showing the disease to be caused by a fungus, whilst vascular wilt is less specific.

Mycosphaerella Johanson 1884; Dothideaceae; ascomata usually very small, mostly immersed in the host tissue, commonly in leaves, ±separate, asco. 1 septate, hyaline or nearly so. The spp. of this large genus include many important pathogens, see Sivanesan 1984 who keyed out some spp. by their anam. in > 10 genera and there are many more anam. names; Holliday 1980; Evans, *Mycol.Pap.* 153, 1984, on pines; Park & Keane, *TBMS* **79**:95, 101, 1982; **83**:93, 1984, on eucalyptus [**61**,7182–3; **63**,5141; **64**,328].

M. aleuritis Ou 1940; as *aleuritidis*; anam. *Pseudocercospora aleuritis* (Miyake) Deighton 1976; asco. 9–15 × 2.5–4 μm, con. 4–12 septate, 40–130 × 3–5 μm; the anam. of *Mycosphaerella websteri* Wiehe 1953 is a *Cercospora*, con. 1–18 septate, mostly 100–140 μm long; tung leaf spots.

M. allii-cepae Jordan, Maude & Burchill, *TBMS* **86**:392, 1986; anam. *Cladosporium allii-cepae* (Ranojević) M.B. Ellis 1976; Kirk, *Descr.F.* 679, 1980; asco. 25–40 × 10–20 μm; con. mostly 1 septate, 65–95 × 13–17 μm; onion leaf blotch. It had been considered that this fungus was the sole cause of the serious outbreak of *Allium* leaf blotch in England and Ireland. But cross inoculations with isolates from leek and onion showed that cross infection did not take place, i.e. the isolates were host specific. Morphological differences between the conidial states were found and 2 taxa are recognised, see *C. allii* for which a teleom. has not apparently been described; Hall & Kavanagh, *Pl.Path.* **33**:147, 1984, growth & sporulation in vitro [**63**,4668; **65**,4654].

M. anethi, prob. anam. *Cercosporidium punctum* q.v.

*M. arachidis** Deighton 1967; anam. *Cercospora arachidicola* Hori 1917; Mulder & Holliday, *Descr.F.* 411, 1974; asco. 7–16 × 3–4 μm; con. up to 12 septate, 35–110 × 3–6 μm, smooth; groundnut early leaf spot; often assoc. with *Mycosphaerella berkeleyi* on the same host, Holliday 1980 described both fungi and the diseases caused on this crop. In *M. arachidis* the conidiophores occur on both surfaces, the leaf lesions are red brown to black on the upper surface and lighter shades of brown on the lower one; the spots tend to be larger than those caused by *M. berkeleyi* and the dark stroma is absent; Cole, *Pl.Path.* **31**:355, 1982, interactions with *Didymosphaeria arachidicola*; Alderman & Beute, *Phytop.* **76**:715, 1986, effects of temp. & moisture on conidial germination & germ tube growth [**62**,2730; **66**,402].

M. areola Ehrlich & Wolf 1932; anam. *Ramularia gossypii* (Speg.) Cif. 1962; Mulder & Holliday, *Descr.F.* 520, 1976; asco. 12–16 × 3–4 μm; con. 1–3 septate, 14–30 × 4–5 μm, grey or areolate mildew of cotton; Sivanesan 1984 accepted the anam. and teleom. connection.

M. berkeleyi W.A. Jenkins 1938; anam. *Cercosporidium personatum* (Berk. & M.A. Curtis) Deighton 1967; Mulder & Holliday, *Descr.F.* 412, 1974; asco. 10–20 × 3–4 μm; con. mostly 3–4 septate, 30–50 × 6.5–7.5 μm; groundnut late leaf spot; often assoc. with *Mycosphaerella arachidis* on the same host. In *M. berkeleyi* the conidiophores occur very largely on the lower surface, they form characteristic dark brown, raised, stromatic areas which are usually arranged concentrically. Both pathogens are often studied together; both cause defoliation but differ in their effects on groundnut depending on geographical areas and time during the growing seasons; seed transm. is apparently unimportant; Cook, *Phytop.* **71**:787, 1981, susceptibility; Nevill, *AAB* **99**:77, 1981, components of resistance; Nevill 1982, inheritance of resistance, genetic model & selection [**61**,2028, 2562; **62**,1297].

M. brassicicola (Duby) Lindau 1897; anam. *Asteromella brassicae* (Chev.) Boerema & v. Kest. 1964; Punithalingam & Holliday, *Descr.F.* 468, 1975; asco. 18–23 × 3–5 μm, spermatia 3 × 1 μm; ringspot of brassicas, generally restricted to cool regions, seedborne; Hartill & Sutton, *AAB* **96**:153, 1980, inhibition of asco. germination on young leaves; Dixon 1981 [**60**,4017].

M. caricae H. & P. Sydow 1913; anam. *Phoma caricae-papayae* (Tarr) Punith., *TBMS* **75**:340, 1980; Punithalingam, *Descr.F.* 634, 1979, as *P. caricae* (Pat.) Punith.; asco. 7.5–15 × 3–5 μm, con. 7–12 × 3–4 μm; a member of a papaya stem end disease complex, and on fruit; Chowdhury, *TBMS* 33:317, 1950; Chau & Alvarez, *Phytop.* 69:500, 1979, asco. biology, diurnal periodicity, max. concentration 03.00–05.00 hours, USA (Hawaii) [30:479; 59,2843; 60,4297].

M. caryigena Demaree & Cole, *J.agric.Res.* 44:145, 1932; anam. *Pseudocercosporella caryigena* (Ell. & Ev.) Sivan. 1984; asco. 10–15 × 3.5 μm; con. 2–6 septate, 25–55 × 4–7 μm; pecan downy spot; Converse, *PDR* 42:393, 1958, evidence for races; Cole 1969, life history & control; Goff et al., *Phytop.* 77:491, 1987, seasonal spore dispersal, factors affecting formation & germination of con. [11:551; 37:610; 50,3207d; 66,3977].

M. cerasella, see *Cercospora circumscissa*.

M. chrysanthemi, not the cause of chrysanthemum ray blight, see *Didymella ligulicola*.

***M. citri** Whiteside 1972; anam. *Stenella citri-grisea* (Fisher) Sivan. 1984; Sivanesan & Holliday, *Descr.F.* 510, 1976; asco. 6–12 × 2–3 μm; con. mostly 3–6 septate, 6–50 × 2–4.5 μm; known in USA (Florida) as a cause of citrus greasy spot or black melanose; the disease has been reported from other parts of the Americas. In Japan a similar, or identical, disease is caused by *Mycosphaerella horii* whose asco. are larger and show a constriction at the septum. *M. citri* has been assoc. with a rust mite, *Phyllocoptora oleivora*; Whiteside, *Phytop.* 71:1108, 1981, discussed the anam. as *Cercospora citri-grisea* Fisher 1961, later placed in synonymy by Sivanesan; Whiteside, *PD* 66:687, 1982, chemical control; Mabbett & Phelps, *Trop.Pest Management* 29:137, 1983, Cu deposits & reduced volume sprays [61,2290; 62,206; 63,637].

M. contraria Hansf. 1941; anam. *Pseudocercospora contraria* (H. & P. Sydow) Deighton 1976; Little, *Descr.F.* 915, 1987, anam. only; asco. 8–12 × 2–3 μm, con. 32–55 × 4–8 μm; *Dioscorea* leaf spot; Emua & Fajola, *PD* 67:389, 1983, chemical control in Nigeria with *Didymosphaeria donacina* [62,3732].

M. convallariae McKeen & Zimmer, *CJB* 42:667, 1964; anam. *Ascochyta majalis* Massalonga 1899–1900; asco. 13–22 × 5–8 μm, con. 15–25 × 3.5–6 μm; lily of the valley leaf blotch; Jenkins, *Phytop.* 32:259, 1942, anam. [21:335; 43,2935].

M. cruenta Latham, *Mycologia* 26:525, 1934; anam. *Pseudocercospora cruenta* (Sacc.) Deighton 1976; Mulder & Holliday, *Descr.F.* 463, 1975, anam. only as *Cercospora cruenta* Sacc.; asco. 11–19.5 × 3–4 μm, con. 35–154 × 3–4.5 μm; cowpea leaf spot, on legumes; can cause severe defoliation, fide Allen 1983 [14:280].

***M. cryptica** (Cooke) Hansf. 1956; anam.

Colletogloeum nubilosum Ganapathi & Corbin, *TBMS* 72:237, 1979, who refer to the teleom. as *Mycosphaerella nubilosa* (Cooke) Hansf. 1956. Park & Keane, *ibid* 79:95, 101, 1982, placed *M. nubilosa* sensu Ganapathi & Corbin as a synonym of *M. cryptica*, and described the differences between Hansford's 2 spp.; asco. of *M. cryptica* have obtusely rounded ends and a clear constriction at the septum, slightly smaller than those of *M. nubilosa*, whose asco. lack a clear constriction and taper at one end. Both fungi cause leaf lesions on *Eucalyptus* in Australia. *M. cryptica* causes small, circular lesions on *E. globulus*; but on mature leaves of other eucalypts it causes large, blighting lesions. *M. nubilosa* was found only causing large, blighting lesions on juvenile leaves of *E. globulus* and 3 other eucalypt spp. *M. parva* Park & Keane 1982 was described as a saprophyte from lesions in assoc. with *M. nubilosa* [59,932; 61,7182–3].

M. dearnessii Barr 1972; a synonym of *Scirrhia acicola* q.v., fide Punithalingam & Gibson 1973 and Sivanesan 1984; Evans, see *Mycosphaerella*, described the fungus as *M. dearnessii*.

M. dendroides (Cooke) Demaree & Cole, *J.agric.Res.* 40:785, 1930; poss. anam. *Cercospora halstedii* Ell. & Ev. This may be a *Pseudocercospora*, fide Chupp 1954; asco. 18–26 × 4–6 μm; con. 1–8 septate, 35–83 × 4.5–7 μm; pecan leaf blotch [9:751; 33:635].

M. dianthi (Burt) Jørstad 1945; anam. *Cladosporium echinulatum* (Berk.) de Vries 1952; asco. 20–30 × 7–9 μm; con. commonly 2–4 septate, distinctly and densely echinulate, mostly 40–50 × 12–15 μm; carnation ringspot or fairy ringspot; Benlloch 1947, severe damage to outdoor carnations in Spain [29:155].

M. fijiensis Morelet 1969; anam. *Paracercospora fijiensis* (Morelet) Deighton 1979; Mulder & Holliday, *Descr.F.* 413, 1974; asco. 11–16 × 5–2.5 μm; con. 1–10 septate 20–132 × 2.5 μm; Stover, *PD* 64:750, 1980, reviewed the banana leaf spots caused by this sp., its var. *difformis* and *Mycosphaerella musicola*; the current anam. names were not given and the teleom. are, morphologically, very similar. The con. have thickened scars and the conidiophores are mostly on the lower leaf surface, emerging singly or in small groups. Banana black leaf streak was described from Fiji *c.*1964; it is widespread in S.E. Asia and the Pacific islands; disease outbreaks have occurred in Africa. The airborne asco. rather than the con. are prob. the more important inoculum. In Fiji, where epidemic levels have been reached, *M. fijiensis* has tended to replace *M. musicola*; fungicide spraying intervals are shorter compared with those for *M. musicola*; Long, *TBMS* **72**:299, 1979, disease in W. Samoa; Holliday 1980 [**59**,840; **60**,3855].

M.f. var. difformis Mulder & Stover, *TBMS* **67**:82, 1976; anam. *Paracercospora fijiensis* var. *difformis* (Mulder & Stover) Deighton 1979; asco. 14–20 × 4–6 μm; con. multiseptate, 51–108 × 3.5–7 μm; banana black Sigatoka, see Stover above; morphologically very similar to *Mycosphaerella fijiensis*, but a dark stroma may form with small groups of conidiophores which can also grow singly. The geographical distribution is similar to that of banana black leaf streak, but in 1973 an epidemic of banana black Sigatoka arose in Honduras; since then the var. has spread through most of Central America. It was reported from Australia (Queensland) in 1982, the first occurrence of *M. fijiensis* in the country.

M. fragariae (Tul.) Lindau 1897; anam. *Ramularia brunnea* Peck 1876; Sivanesan & Holliday, *Descr.F.* 708, 1981; asco. 11–14.5 × 2–3 μm; con. mostly 1 septate, 14–45 × 2–3 μm; strawberry white spot, now less important with the development of resistant cvs.

M. gibsonii H. Evans, *Mycol.Pap.* 153:61, 1984; anam. *Cercoseptoria pini-densiflorae* (Hori & Nambu) Deighton 1976, transferred to *Pseudocercospora* Deighton, *TBMS* **88**:390, 1987; Mulder & Gibson, *Descr.F.* 329, 1972, as *Cercospora pini-densiflorae* Hori & Nambu 1917; asco. 8.5–11 × 2–2.8 μm, con. 20–68 × 2.5–4.5 μm; pine brown needle blight, can be a serious disease of plants at the late nursery stage in Japan and Taiwan. In Tanzania the pathogen has caused severe defoliation in young plantings; Itô, *Bull.Govt.For.Exp.Stn. Tokyo* 246:21, 1972; Suto, *J.Jap.For.Soc.* **61**:180, 1979; *Bull.Shimane Prefect.For.Exp.Stn.* 32:1982 [**51**,4421; **59**,942; **63**,917].

M. gossypina (Atk.) Earle 1900; anam. *Cercospora gossypina* Cooke 1883; Little, *Descr.F.* 914, 1987, anam. only; asco. 15–18 × 3–4 μm; con. up to 12 or more septate, 60–300 × 2.5–4 μm; cotton leaf spot is of minor importance, the fungus infects plants under stress or in late season decline.

M. graminicola (Fuckel) Schröter 1894; anam. *Septoria tritici* Roberge 1842; Sutton & Waterston, *Descr.F.* 90, 1966; Bissett, *Fungi Canadenses* 244, 1983; both as anam. only; asco. 9–16 × 2.5–4 μm; con. 2–3 septate, 43–80 × 1.5–2 μm; wheat speckled leaf blotch; a major disease that is often studied with the one caused by *Leptosphaeria nodorum* q.v. for reviews; seedborne, on Gramineae; Forrer & Zadoks, *Neth.J.Pl.Path.* **89**:87, 1983, yield reduction; Coakley et al. and Eyal et al., *Phytop.* **75**:1245,

1456, 1985, prediction of disease severity & virulence patterns; Gaunt et al. and Thomson & Gaunt, *Ann.Bot.* **58**:33, 39, 1986, assessment in winter wheat & effects on yield in New Zealand [**63**,112; **65**,2274–5, 4856–7].

M. henningsii Sivan., *TBMS* **84**:522, 1984, for *Mycosphaerella manihotis* Ghesq. & Henrard 1924; anam. *Cercosporidium henningsii* (Allescher) Deighton 1976; Little, *Descr.F.* 912, 1987; asco. 18–20 × 4–4.5 μm; con. 3–8 septate, 38–85 × 5–7 μm; cassava brown leaf spot, see *Cercospora vicosae*; Madeuwesi, *Niger.J.Pl.Prot.* **1**:29, 1975; Teri et al., *Fitopat.Brasileira* **6**:341, 1981, epidemiology [**54**,5673; **61**,4560].

M. holci Tehon 1937; Anahosur & Sivanesan, *Descr.F.* 583, 1978; asco. 12–18 × 4.5–6.5 μm; *Phoma sorghina* q.v. was said to be the anam. but White & Morgan-Jones, *Mycotaxon* **18**:5, 1983, stated that there was no evidence for this. A minor leaf and inflorescence spot is caused, also a stem canker; mostly on Gramineae [**63**,2531].

M. horii Hara 1917; asco. 7.9–15.9 × 2.3–3.7 μm; Japanese citrus greasy spot, see *Mycosphaerella citri* and *Septoria citri*; Yamada and Tanaka & Yamada, *Bull.Tokai-Kinki Natn.agric.Exp.Stn.* Hort.Div. 1 & 3, 1952, 1956.

M. juglandis Kessler, *Mycologia* **76**:363, 1984; anam. *Cylindrosporium juglandis* Wolf 1914; asco. 8.2–16.4 × 2–2.5 μm; con. up to 7 septate, 20–55 × 1.7–2.2 μm; walnut leaf spot; Kessler, *PD* **69**:1092, 1985, premature leaf fall & chemical control [**63**,3592; **65**,4117].

M. laricina (Hartig) Migula, *Kryptogamen-Flora* **3**(3):301, 1913; asco. 7–15 μm long, fide Hiley, *The fungal diseases of the common larch*, 1919, as *Sphaerella laricina* Hartig; larch needle cast, Europe, absent from British Isles, *Larix decidua*; Patton & Spear, *PD* **67**:1149, 1983, USA (Iowa, Wisconsin), comparison with *Mycosphaerella laricis-leptolepidis* Ito, Sato & Ota 1957, in Japan, *Index Fungi* **2**:393, 1958; Peace 1962 [**38**:343, **63**,3602].

M. laricis-leptolepidis, see above.

*****M. linicola** Naumov 1926; anam. *Septoria linicola* (Speg.) Garassini 1938; Sivanesan & Holliday, *Descr.F.* 709, 1981; asco. 13–17 × 2.5–4 μm; con. aseptate, 17–40 × 1.5–3 μm; flax pasmo, causes leaf fall and stem infection, can cause lodging, seedborne; severe losses in yields of fibre and oil can occur.

M. macrospora (Kleb.) Jørstad 1945; anam. *Cladosporium iridis* (Fautrey & Roum.) de Vries 1952; Ellis & Waller, *Descr.F.* 435, 1974, anam. only; asco. 25–45 × 7–14 μm; con. mostly 2–3 septate, 35–70 × 13–25 μm, densely echinulate; iris blotch, also on freesia, gladiolus, *Hemerocallis*,

narcissus; Moore 1979; Hurley et al. 1982, as
Didymellina macrospora Kleb. [**61**,6426].

M. manihotis, see *Mycosphaerella henningsii*.

M. milleri Hodges & Haasis, *Mycologia* **56**:53,
1964; anam. *Cercosporidium magnoliae* (Ell. &
Harkn.) Sivan. 1984; asco. 21.1 × 3.5 μm; con.
mostly 2 septate, 23–45 × 5–6.5 μm; magnolia
angular leaf spot.

M. mori (Fuckel) Wolf 1935; anam. *Phloeospora
maculans* (Béreng.) Allescher in Rabenh. 1900;
asco. 12–14 × 3.5–4 μm; con. 2–4 septate,
29–55 × 4–5 μm; mulberry leaf spot; Cass-Smith &
Stewart, *J.Dep.Agric.W.Aust.* **24**:69, 1947, as
Septogloeum mori (Lév.) Briosi & Cavara;
Stewart, *J.R.Soc.W.Aust.* **34**:87, 1950 [**26**:551;
29:517].

M. musae (Speg.) H. Sydow & Sydow 1914; Pont,
Qd.J.agric.Sci. **17**:273, 1960; asco. 12–15 × 3–
4 μm, smaller than those of *Mycosphaerella
musicola*; banana leaf speckle; Stover,
Trop.Agric.Trin. **46**:325, 1969, *Mycosphaerella*
spp. & banana leaf spots; Allen & Peasley,
Australasian Pl.Path. **10**:24, 1981, chemical
control by aerial spraying [**41**:240; **49**,521;
61,2976].

M. musicola Leach ex Mulder in Mulder & Stover
TBMS **67**:77, 1976; anam. *Pseudocercospora
musae* Deighton 1976; Mulder & Holliday,
Descr.F. 414, 1974, anam. as *Cercospora musae*
Zimm. 1902; asco. 14–18 × 3–4 μm; con. 3–5
septate, 10–100 × 2–6 μm; the conidial scars are
unthickened, compared with those of
Mycosphaerella fijiensis and its var. *difformis*.
Also, in contrast to these last 2 fungi, the
condiophores are abundant on both leaf surfaces;
they grow in dense fascicles on a dark stroma.
Sigatoka is a major leaf disease of banana; the
pathogen, first described from Indonesia (Java) in
1902, became epidemic in Fiji in 1912.
Consequently *M. musicola* spread rapidly, causing
serious losses in Australia, Africa, Central
America, West Indies, particularly in Jamaica, in
the 1920s and 1930s. The asco. are airborne and
cause a characteristic leaf tip spotting. The con.
are waterborne and cause line spotting, mostly
towards the leaf base; they constitute the major
inoculum source. Control with fungicides is
largely obligatory; Sigatoka is severe under humid
conditions at 23–28°. Until the late 1950s the
large areas of cultivation in Central America were
sprayed with Bordeaux mixture. These elaborate
operations provide a classic example of the big
scale use of the fungicide. Contemporary control
techniques include the use of aircraft. Leach,
Banana leaf spot (Mycosphaerella musicola) *on the
Gros Michel variety in Jamaica*, Govt.Printer,

Kingston, 1946; Meredith, *Phytopath.Pap.* 11,
1970, monographs; Stover 1972 and under *M.
fijiensis*; Holliday 1980 [**26**:250].

M. nubilosa, and *Mycosphaerella parva*, see *M.
cryptica*; *M. pini*, see Evans, *Scirrhia pini*.

M. pinodes (Berk. & Blox.) Vestergr. 1896; anam.
Ascochyta pinodes L.K. Jones 1927;
Punithalingam & Holliday, *Descr.F.* 340, 1972;
Sivanesan 1984 described the fungus as *Didymella
pinodes* (Berk. & Blox.) Petrak 1924; asco.
8–18 × 4–8 μm, con. 8–19 × 3–5 μm; pea leaf, stem
and pod spot, and foot rot; because of the severe
foot rot syndrome this fungus causes a more
damaging disease than *A. pisi* with which it
occurs; seedborne, races; Dixon 1981; Hagedorn
1984; Sherf & MacNab 1986.

M. platanifolia (Cooke) Wolf, *Mycologia* **30**:62,
1938; anam. *Cercospora platanicola* Ell. & Ev.
1887; asco. 8–10 × 4–4.5 μm; con. 3–5 septate,
30–60 × 3–4 μm; also described is *M. stigmina-
platani* Wolf, *ibid*, page 60; anam. *Stigmina
platani* (Fuckel) Sacc. 1886; asco. 17–19 ×
6–7 μm; con. several septa, 15–40 × 7–10 μm; wall
dark, thick rough; both fungi from leaf spots of
Platanus occidentalis, USA; Itô, *J.Jap.For.Soc.*
41:229, 1959, incorrectly as *C. platanifolia*
[**17**:492; **40**:129].

*M. pomi, poss. anam. *Cylindrosporium pomi*
q.v.

M. populi (Auersw.) Schröter 1894; anam. *Septoria
populi* Desm. 1843; asco. 38–45 × 4–5 μm; con. 1
septate, 30–46 × 3–4 μm; Europe. *Mycosphaerella
populicola* G.E. Thompson, *Phytop.* **31**:251, 1941;
anam. *S. populicola* Peck 1886; asco. 22–32 ×
6–7 μm; con. 2–5 septate, 45–80 × 3–4.5 μm. *M.
populorum* Thompson, *ibid*, page 246; anam. *S.
musiva* Peck 1884; asco. 12–38 × 3–5.6 μm; con.
1–4 septate, 28–54 × 3.5–4 μm; N. America;
poplar leaf spots, and canker caused by *M.
populorum*; Bier, *Can.J.Res.C* **17**:195, 1939;
Waterman, *Phytop.* **36**:148, 1946, canker; Long
et al. 1986; Spielman et al., *PD* **70**:968, 1986
[**18**:770; **25**:426; **65**,6203; **66**,1620].

M. pruni-persicae Deighton, *TBMS* **50**:328, 1967;
anam. *Miuraea persica* (Sacc.) Hara 1948; asco.
12–20 × 2.5–3.5 μm, con. 17–86 × 2.5–7 μm; frosty
mildew of peach; Higgins & Wolf, *Phytop.*
27:690, 1937 [**16**:759].

M. puerariicola Weimer & Luttr., *Phytop.* **38**:350,
1948; anam. *Pseudocercospora puerariicola*
(Yamam.) Deighton 1976; asco. 14–23 × 2.5–
4.5 μm; con. 1–16 septate, 25–126 × 3–5.5 μm, on
Pueraria thunbergiana, kudzu.

M. pyri (Auersw.) Boerema, *Neth.J.Pl.Path.*
76:166, 1970; anam. *Septoria pyricola* (Desm.)
Desm. 1850; asco. 7–9 × 3–4 μm; con. 1–2 septate,

40–58 × 3–5 μm, pear leaf fleck; Anderson 1956 as *Mycosphaerella sentina* (Fr.) Schröter.

M. rabiei Kovachevski 1936; or *Didymella rabiei* (Kovach.) v. Arx 1962, poss. teleom. of *Ascochyta rabiei* q.v. [**15**:700].

M. rosicola B.H. Davis ex Deighton, *TBMS* **50**:328, 1967; anam. *Cercospora rosicola* Pass. 1877; asco. 13–17 × 4–5.3 μm; con. 3–8 septate, 40–64 × 3–4 μm; rose leaf spot; Davis, *Mycologia* **30**:282, 1938; Boelema, *Phytophylactica* **5**:7, 1973 [**17**:753; **53**,979].

M. sesame Sivan., *TBMS* **85**:397, 1985; anam. *Cercospora sesami* Zimm. 1904; Waller & Sutton, *Descr.F.* 627, 1979; asco. 11.5–14 × 4–5.5 μm; con. 7–15 septate, 90–150 × 3–4 μm; sesame leaf and pod spot, internally seedborne.

M. sesamicola Sivan., as above; anam. *Cercoseptoria sesami* (Hansf.) Deighton 1976; transferred to *Pseudocercospora* by Deighton, *TBMS* **88**:390, 1987; asco. 13.5–16.5 × 3.5–5 μm, con. up to 13 septate, 80–240 × 1.5–4 μm; sesame brown leaf spot; Sivanesan, loc.cit., keyed out *Cercospora* spp. and allied genera found on sesame; Orellana, *Phytop.* **51**:89, 1961; Schmutterer & Kranz, *Phytopath.Z.* **54**;193, 1965 [**40**:620; **45**,2205].

M. stigmina-platani, see *Mycosphaerella platanifolia*.

M. tassiana (de Not.) Johanson 1884; anam. *Cladosporium herbarum* (Pers.) Link ex Gray 1821; asco. 15–30 × 4.5–9.5 μm, con. 5–23 × 3–8 μm; plurivorous, frequently a secondary invader; Barr, *Mycologia* **50**:501, 1958, life history & morphology; Petrie & Vanterpool, *Can.Pl.Dis.Surv.* **58**:77, 1978, on Cruciferae in Canada (W.) [**58**,4074].

M. togashiana Ito & Kobayashi, *Bull.Govt.For.Exp.Stn.Meguro* 59:23, 1953; anam. *Pseudocercospora salicina* (Ell. & Ev.) Deighton 1976; asco. 12–17 × 2.8–4 μm; con. 1–8 septate, 24–56 × 2–5 μm; poplar leaf spot, *Populus simonii*; cf. *Mycosphaerella populi* [**34**:497].

M. tulipifera (Schwein.) Higgins, *Am.J.Bot.* **23**:598, 1936; anam. *Phaeoisariopsis liriodendri* (Ell. & Harkn.) Morgan-Jones & Brown, *Mycotaxon* **4**:493, 1976; asco. 9.5–16.8 × 2.3–3.5 μm; con. mostly 1 septate, 14–34 × 5–7 μm; angular leaf spot of *Liriodendron tulipifera*, tulip tree [**16**:286].

M. ulmi Kleb. 1902; anam. *Phloeospora ulmi* (Fr.) Wallr. 1833; asco. 20–28 × 3–4 μm; con. 3–5 septate, 30–60 × 4.5–6 μm; a minor leaf spot of elm.

M. zeae-maydis Mukunya & Boothroyd, *Phytop.* **63**:530, 1973, q.v. for other *Mycosphaerella* spp. on maize; anam. *Phyllosticta maydis* Arny & Nelson 1971, Sivanesan 1984 considered that the anam. is more like a *Phoma*; asco. 13–20 ×

5–6 μm, con. 8–20 × 3–7 μm; maize yellow leaf blight; the leaf lesions resemble those caused by *Cochliobolus heterostrophus*; Jiménez-Díaz 1976, review; Jiménez-Díaz & Boothroyd, *Phytopath.Mediterranea* **18**:3, 1979, asco. & epidemics [**51**,2430; **53**,510; **58**,673; **62**,4678].

mycostasis, synonymous with fungistasis q.v.

mycotoxicoses, literally poisoning by fungi which form toxins; preferred usage limited to the poisoning of animals including man, see toxin.

Mycovellosiella Rangel 1917; Hyphom.; Ellis 1971, 1976; Deighton, *Mycol.Pap.* 137, 1974; 144, 1979; Carmichael et al. 1980; conidiomata hyphal, conidiogenesis holoblastic, conidiophore growth sympodial; con. multiseptate, mostly 1–3 septate, pigmented; the distinctive characteristics are the thickened conidial scars; and the formation of an assurgent or repent, secondary, external mycelium, on the hyphae of which the conidiophores are borne terminally and as lateral branches; the spp. generally cause leaf spots; Holliday 1980.

M. cajani (Henn.) Rangel ex Trotter 1931; Waller & Sutton, *Descr.F.* 628, 1979; con. 19–45 × 4.5– 6.5 μm, on pigeon pea; chemical control has been described from India; Deighton 1974, see above, erected the vars. *cajani*, *indica* and *trichophila*.

M. concors (Caspary) Deighton 1974; Kirk, *Descr.F.* 724, 1982; con. 14–57 × 3.5–6 μm; on potato.

M. koepkei (Krüger) Deighton 1979; Mulder & Holliday, *Descr.F.* 417, 1974 as *Cercospora koepkei* Krüger; con. 21–44 × 4–6.5 μm; sugarcane yellow spot, has caused epidemics in Australia (Queensland), poss. races; Sivanesan & Waller 1986.

M. nattrassii Deighton 1974; Waller & Sutton, *Descr.F.* 629, 1979; con. 23–50 × 5.8–8 μm; circular, yellow patches on eggplant leaves.

M. oryzae (Deighton & Shaw) Deighton 1979; con. mostly 1 septate, 15–40 × 2.5–4.8 μm; rice white leaf streak; Deighton & Shaw, *TBMS* **43**:516, 1960, as *Ramularia oryzae*; see Sutton & Shahjahan, *Sphaerulina oryzina* [**40**:104].

M. phaseoli (Drummond) Deighton 1974; con. 0–1 septate, 4–21 × 3–5.5 μm; bean floury leaf spot; Deighton, *TBMS* **50**:123, 1967, as *Ramularia phaseoli* (Drummond) Deighton; Holliday 1980 [**46**, 2136].

M. puerariae Shaw & Deighton, *TBMS* **54**:327, 1970; con. 14–70 × 3.5–6 μm; yellow leaf mould of *Pueraria lobata* [**49**,2901].

M. vaginae (Krüger) Deighton 1979; Kirk, *Descr.F.* 725, 1982; con. 30–55 × 4–5 μm., sugarcane red leaf sheath spot, preferred to red leaf sheath rot; symptoms should not be confused with those caused by *Glomerella tucumanensis*; Sivanesan & Waller 1986.

mycoviruses, viruses that occur in fungi, first described by Hollings, *Nature Lond.* **196**:962, 1962. Hollings, *Adv.Virus Res.* **22**:1, 1978; *PD* **66**:1106, 1982; Ghabrial, *A.R.Phytop.* **18**:441, 1980; Ushiyama, *Trans.mycol.Soc.Japan* **21**:383, 1980; Teakle in Buczacki 1983; Buck et al., *Intervirology* **22**:17, 1984.

Myndus crudus, poss. transm. palm lethal yellowing; *M. taffini*, see foliar decay of coconut.

myosotis ringspot Cadman, *J.hort.Sci.* **31**:111, 1956; transm. sap, Scotland, ringspots caused in *Nicotiana rustica* and *Petunia* [**35**:618].

Myriangiaceae, Dothideales; ascomata often stromatic, superficial or erumpent, asci in monascus scattered locules, often at different levels; *Elsinoë*.

Myriogenospora Atk. 1894; Clavicipitaceae; very close to *Balansia*, with similar anam.; Diehl, *Balansia*; systemic in grasses; Luttrell & Bacon, *CJB* **55**:2090, 1977, morphology & taxonomy; Rykard et al., *Phytop.* **75**:950, 1985, host relations of *M. atramentosa* & *B. epichloë*. Tangle top of grasses and sugarcane leaf binding are caused by *M. atramentosa* (Berk. & M.A. Curtis) Diehl 1950, see Abbott & Tippett, *Phytop.* **31**:564, 1941. Sivanesan & Waller 1986 attribute leaf binding, a rare condition, to *M. aciculispora* Vizioli 1929 [**20**:596; **57**,1241; **65**,762].

Myriophyllum brasiliense, parrot feather, a weed problem in irrigation systems in USA (California); Bernhardt & Duniway, *PD* **68**:999, 1984, rot and stem rot (*Pythium carolinianum* Matthews) [**64**,1400].

Myriosclerotinia Buchw. 1947; Sclerotiniaceae; in culms of Cyperaceae, Gramineae, Juncaceae where the sclerotia form; anam. *Myrioconium* Sydow 1912; Hyphom.; *Myriosclerotinia borealis* (Bubák & Vleugel) Kohn, *Phytop.* **69**:885, 1979, black sclerotia spherical, oval or like flakes, 0.5–7 mm long; snow scald of grass; Årsvoll 1976,

sporulation, asco. germination & pathogenesis; Ozaki, *Bull.Hokkaido Prefect.agric.Exp.Stns.* 42:55, 1979, on cocksfoot as *Sclerotinia borealis* [**56**,1162; **60**,359].

myriotylum, *Pythium*.

myrobalan, cherry plum, *Prunus cerasifera*.

—— **latent ringspot v.** Dunez et al. 1971; Dunez & Dupont, *Descr.V.* 160, 1976; Nepovirus, isometric 28 nm diam., transm. sap, France, causes short internodes and rosetting in peach, enations on leaves of sweet cherry cv. Bing; Gallitelli et al., *J.gen.Virol.* **53**:57, 1981, comparison with Nepoviruses [**60**,6348].

Myrothecium Tode 1790; Hyphom.; conidiomata sporodochial or synnematal, conidiogenesis enteroblastic, phialidic; con. aseptate, pigmented, black in mass; Tulloch *Mycol.Pap.* 130, 1972, monograph; Nguyen et al., *TBMS* **61**:347, 1973, seedborne spp. & pathogenicity [**52**,1040; **53**,2927].

M. roridum Tode:Fr. 1828; Fitton & Holliday, *Descr.F.* 253, 1970; conidial mass wet when young, drying to become hard, shiny, black, convex, with a white halo of mycelium; con. av. $7.2 \times 1.8 \,\mu m$; plurivorous, causing necrotic lesions above ground, shotholes in leaves, crown dieback and stem canker, damping off; infects roots and spreads through the soil in which it is a saprophyte, sometimes seedborne, see roridin E; Holliday 1980; Chase, *PD* **67**:668, 1983, on leaves of ornamentals [**62**,4320].

Myxomycota, forming a plasmodium, i.e. a multinucleate mass, motile, amoeboid, no firm wall, characteristic of the growth phase, status as fungi uncertain; only the Plasmodiophoromycetes concern plant pathologists.

Myzus ascalonicus, transm. shallot latent; *M. macrosiphum*, transm. teasel mosaic.

M. persicae, green peach aphid, abbreviated here as *Myz.pers.*, widespread, polyphagous, transm. > 100 plant viruses, see vector; frequently used in virus transmission studies; Gibbs & Harrison 1976; see insects.

N

Nacobbus aberrans (Thorne) Thorne & Allen 1944; Stone & Burrows, Descr.N. 119, 1985; false root knot nematode, root galling of beet, potato, tomato.

nailhead canker (Nummulariella marginata) apple, — spot (Alternaria tomato) tomato.

nakanishikii, Puccinia.

Nakataea Hara 1939; Hyphom.; Ellis 1971; prob. = Pyricularia; N. sigmoidea, teleom. Magnaporthe salvinii.

names, see terminology, disease and plant names.

nandina mosaic v. Moreno, Attathom & Weathers, abs. Proc.Am.phytopath.Soc. 3:319, 1976; prob. Potexvirus, filament 460–560 nm long, transm. sap, USA (California), N. domestica, N. nana-purpurea; Zettler et al., Acta Hortic. 110:71, 1980.
— stem pitting v. Ahmed, Christie & Zettler 1982, Phytop. 73:470, 1983; poss. Closterovirus, filament 696–830 nm long, USA, N. domestica [62,3887].

narcissicola, Botryotinia, Botrytis.

Narcissus, N. pseudonarcissus, daffodil; N. jonquilla, jonquil; Moore 1979; Brunt, Acta Hortic. 110:23, 1980, viruses; Mowat, ibid 109:461, 1980, virus epidemiology; Monogr.Br.Crop.Prot.Council 33:193, 1986, production free from viruses in Scotland.

narcissus chlorotic spotting van Slogteren, Daffodil Tulip Yrbk. 12:19, 1946.
— degeneration v. Brunt 1969; Daffodil Tulip Yrbk. 36:29, 1970; Potyvirus, filament c.750 nm long, transm. Myz.pers., Isles of Scilly, N. tazetta cv. Grand Soleil d'Or.
— greys (narcissus yellow stripe v.).
— latent v. Brunt & Atkey 1967; Brunt, Descr.V. 170, 1976; AAB 87:355, 1977; Carlavirus, filament c.650 × 13 nm, transm. sap, Acyrthosiphon pisum, Aphis gossypii, Macrosiphum euphorbiae, Myz.pers; non-persistent, common in cvs. of narcissus and bulbous iris; see iris mild yellow mosaic v. which is a synonym [57,2957].
— late season yellows v. Brunt, Daffodil Tulip Yrbk. 36:18, 1970; poss. Potyvirus, Britain [50,2316].
— mosaic v. van Slogteren & de Bruyn Ouboter 1946; Mowat, Descr.V. 45, 1971; Potexvirus, filament c. 550 × 13 nm, transm. sap readily, prob. widespread; Maat 1976, str. from Nerine manselli; Koenig, Phytopath.Z. 84:193, 1975, relations with Potexviruses; Bancroft et al., J.gen.Virol. 50:451, 1980, structure [55,2170, 5786; 60,3615].

— mottle (tobacco rattle v.), van Slogteren 1958, see tulip mottle and stunt.
— tip necrosis v. Asjes 1972; Mowat et al., Descr.V. 166, 1976; poss. Tombusvirus, isometric c.30 nm diam., transm. sap, only known in narcissus, prob. widespread; Mowat et al., AAB 86:189, 1977, occurrence, purification & properties [57,636].
— white streak Chittenden 1933; McWhorter 1938; Haasis, Phytop. 29:890, 1939; Brunt, Daffodil Tulip Yrbk. 36:18, 1970; filament c.750 nm long, transm. sap, aphid spp., non-persistent. The evidence that this disease is caused by a virus is inadequate, W. P. Mowat, personal communication, 1986.
— yellow stripe v. Darlington 1908; Brunt, Descr.V. 76, 1971; Potyvirus, filament c.755 × 12 nm, transm. sap, 9 aphid spp., non-persistent, occurs naturally only in Amaryllidaceae, not serologically related to narcissus degeneration v. or narcissus white streak.

narmari, Leptosphaeria.

narrow brown leaf spot (Sphaerulina oryzina) rice.

nashicola, Venturia.

Nasonovia lactucae, transm. beet yellow stunt.

Nasturtium, N. officinale, watercress; the ornamental nasturtium is Tropaeolum majus.

nasturtium mosaic v. Jensen, abs. Phytop. 40:967, 1950; transm. sap, aphid spp., USA (California); Graca & Martin, Phytopath. Z. 88:276, 1977; rod, South Africa (Natal), Tropaeolum majus [56,4579].
— ringspot (str. of broad bean wilt v.), Doel, J.gen.Virol. 26:95, 1975 [54,2125].

natsudaidai dwarf, see citrus Satsuma dwarf.

nattrassii, Mycovellosiella.

near wilt (Fusarium oxysporum f.sp. pisi) pea.

necator, Sphaceloma, Uncinula.

Necator decretus Massee 1898; teleom. Corticium salmonicolor; Brooks & Sharples, AAB 2:58, 1915.

necatrix, Dematophora, Rosellinia.

neck rot, rotting of the 'neck' of a bulb or mangold, FBPP; (Botryotinia allii, B. porri, B. squamosa, Botrytis allii) onion.

necrophyte, an organism living on dead material, cf. necrotroph, saprophyte.

necrosis, death, usually of a clearly delimited part of a plant or part of a tissue, FBPP.

necrotic, dead; the terms holonecrotic, 'completely dead' and plesionecrotic 'nearly dead' have been used to describe, respectively, the central area of dead tissues and the surrounding zone of damaged, but not dead, tissues in leaf spots that are not delimited by cork barriers, FBPP.

—— **etch**, groundnut, genetic disorder; Hammons, *Peanut Sci.* **7**:13, 1980 [**60**,3425].

—— **fleck** (*Greeneria uvicola*) grapevine; —— **flecking**, potato, McKeen et al., *Can.Pl.Dis.Surv.* **53**:150, 1973 [**53**,4103].

—— **spot**, maize, genetic disorder; Mortimer & Gates, *Can.J.Pl.Sci.* **59**:147, 1979; symptoms like those caused by disease lesion mimics q.v. [**58**,5337].

—— —— **and blotch**, prune, prob. genetic disorder; Blodgett, *Phytop.* **30**:347, 1940 [**19**:480].

necrotroph, a fungus that kills tissues as it grows through them such that it is always colonising dead substrate, FBPP; also an organism that obtains nutrients from dead tissues of a living host, cf. saprophyte; perthophyte is synonymous.

Necroviruses, isometric particle, type tobacco necrosis v., fide *Descr.V.* set 19, 1985.

Nectria (Fr.) Fr. 1849; Hypocreaceae; the taxonomic monograph by Rossman, *Mycol.Pap.* 150, 1983, has been generally followed; 13 groups were described with the related genera: *Calonectria, Ophionectria, Paranectria, Scoleconectria, Trichonectria*. The original description did not specify asco. characteristics. Saccardo limited the spp. to those with 1 septate asco. Later segregations to other genera were often based on just asco. characteristics which are not now considered of primary importance; other characters include the anam. sp. and the colour reaction when ascomata are put in 3% KOH. Ascomata are superficial, sometimes with a stroma, bright in colour, usually red to orange and yellowish, rarely brown; asco. variable in shape and septation. Only a few spp. are important pathogens. They mostly cause cankers and diebacks of woody dicotyledons; invading through wounds; not particularly host specific.

N. cinnabarina (Tode:Fr.) Fr.; anam. *Tubercularia vulgaris* Tode: Fr. 1823; Booth, *Descr.F.* 531, 1977; asco. 1 septate, 12–20 × 4.5–6.5 *μ*m, sporodochia have a coral coloured layer hence the disease name, coral spot, mostly temperate, on woody dicotyledons; Bedker et al., *PD* **66**:1067, 1982, gave a recent account of a canker disease on *Gleditsia triacanthos*, honey locust.

*****N. coccinea** (Pers.: Fr.) Fr.; *Nectria coccinea* var. *faginata* Lohman, Watson & Ayers 1943; anam. in *Cylindrocarpon*, the sp. poss. *C. candidum* (Link) Wollenw. 1926 and the var., *C. faginatum* Booth 1966; Booth, *Descr.F.* 532, 533, 1977; ascomata 250–350 *μ*m diam. in groups of 5–30;

asco. 12–15 × 5–6 *μ*m; macroconidia 3–7 septate, 46–80 × 6–7 *μ*m. Var. *faginata* asco. 10.5–12.7 × 4.8–6.2 *μ*m, macroconidia 40–110 × 5–7 *μ*m; it is a geographical variant causing a similar disease on *Fagus grandiflora*, American beech. *N. coccinea* causes a serious disease on *F. sylvatica*, common or European beech, in Britain. It occurs in and outside Europe, prob. mostly as a saprophyte. Both fungi are assoc. with tarry spot, a bark stem necrosis, often called bark disease, and the scale insect *Cryptococcus fagisuga*. In the N. American situation the wounding caused by the insect allows fungal penetration; but in Europe the etiology appears to be more complex in that host stress factors play a role. The trees develop cankers and can be killed; Phillips & Burdekin 1982; Twery & Patterson, *Can.J.For.Res.* **14**:565, 1984, effects of disease on spp. composition & structure in hardwoods [**64**,1765].

N. ditissima Tul. 1865; coccinea group, fide Booth, *Mycol.Pap.* 73, 1959; ascomata with smooth 'varnished' walls, in groups of 5–30; asco. 14–21 × 5–8 *μ*m; on bare wood and cankered bark, causes deeply sunken cankers of beech; Philips & Burdekin 1982; Perrin & Gerwen, *Eur.J.For.Path.* **14**:170, 1984, variability in pathogenicity [**64**,793].

N. fuckeliana C. Booth 1959, *Descr.F.* 624, 1979; anam. *Cylindrocarpon cylindroides* Wollenw. var. *tenue* Wollenw. 1928; coccinea group, ascomata 300–400 *μ*m diam. on a well developed stroma, in groups of 10–100; asco. 13–16 × 5–6 *μ*m; macroconidia 3–7 septate, 33–85 × 4–7 *μ*m; wound pathogen, spruce bark necrosis, dieback in Pinaceae.

*****N. galligena** Bresad. in Strasser 1901; Booth, *Descr.F.* 147, 1967; anam. *Cylindrocarpon mali* (Allesch.) Wollenw. 1928; coccinea group, ascomata with rough walls, in groups of 2–5 on a common stroma; asco. 14–22 × 6–9 *μ*m; macroconidia 1–4 or more septate, from 10–28 × 4–5 *μ*m to 46–65 × 4–7 *μ*m; causes the important European canker of apple, particularly severe in wetter climates, also on pear and other trees. The form causing ash canker is prob. a f.sp. and distinct from the one on apple; Swinburne, *Rev.Pl.Path.* **54**:787, 1975; Flack & Swinburne, *TBMS* **68**:185, 1977, hosts, inoculations & ff.sp.; van der Scheer, *Meded.Proefst.Fruiteelt Wilhelminadorp* 18, 1980, review; McCracken & Cooke, *AAB* **107**:417, 1985, chemical control [**56**,4595; **66**,239].

N. haematococca, anam. *Fusarium solani* q.v.

N. inventa Pethybr., *TBMS* **6**:107, 1919; anam. a *Verticillium*; Hughes, *Mycol.Pap.* 45, 1951; asco.

9–10 × 4–5 μm; Murray, *Eur.J.For.Path.* **8**:65, 1978, anam. assoc. with stem lesions of *Acer pseudoplatanus*; Tsuneda & Skoropad, *TBMS* **74**:501, 1980, anam. as a mycoparasite [57,5696; 60,3390].

N. neomacrospora C. Booth & Samuels, *TBMS* **77**:645, 1981; Booth, *Descr.F.* 623, 1979, as *Nectria macrospora* (Wollenw.) Ouellette 1972; anam. *Cylindrocarpon cylindroides* Wollenw. 1913; asco. 16–22 × 5–7 μm; macroconidia 3–7 septate, 38–45 × 4.5–6 μm to 70–90 × 6–7 μm; canker, Pinaceae; Ouellette & Bard, *PDR* **50**:722, 1966; Ouellette, *Eur.J.For.Path.* **2**:172, 1972, canker of balsam fir, *Abies balsamea* [46,765; 52,1695].

*****N. radicicola** Gerlach & Nilsson 1963; anam. *Cylindrocarpon destructans* (Zinssm.) Scholten 1964; Booth, *Descr.F.* 148, 1967; asco. 10–13 × 3–3.5 μm; macroconidia 1–3 septate, 45–62 × 6.5–7.5 μm; plurivorous, temperate, root rots, dry brown rot; storage rot, e.g. grapevine, narcissus, strawberry; Channon & Thompson, *Pl.Path.* **30**:181, 1981, parsnip canker in Scotland; Matuo & Miyazawa, *Ann.phytopath.Soc.Japan* **50**:649, 1984; root rot of ginseng, *Panax ginseng*, f.sp. *panacis* [61,5307; 65,5678].

N. rigidiuscula Berk. & Broome 1873, fide Rossman 1983; anam. *Fusarium decemcellulare* Brick 1908; Booth & Waterston, *Descr.F.* 21, 1964, as *Calonectria rigidiuscula* (Berk. & Broome) Sacc.; ascomata cream to yellow; asco. 3 septate, 22–28 × 7–10 μm; macroconidia on sporodochia, 7–10 septate, 50–65 × 5–7 μm; mostly saprophytic, sometimes weakly pathogenic, assoc. with capsids and other fungi in causing a cacao dieback; also the relatively minor cacao disease, cushion and green point gall, is apparently assoc. with infection by *Nectria rigidiuscula*; Holliday 1980.

N. striispora Ell. & Ev. in Smith 1893, fide Samuels, *CJB* **51**:1281, 1973; anam. *Calostilbella calostilbe* Höhnel 1919; Booth & Holliday, *Descr.F.* 392, 1973 as *Calostilbe striispora* (Ell. & Ev.) Seaver 1928; ascomata yellowish, asco. 1 septate, longitudinally striate when mature, becoming yellowish brown; con. borne in heads on stipes up to 10 mm long, 3 septate, 2 large, yellow brown, central cells, and 2 small, crescent shaped, hyaline, apical cells, 37–50 × 12–14 μm. This fungus, on several hosts, causes a somewhat unique, if very minor, disease, in that it kills the shade trees of cacao in Trinidad and parts of S. America, and not the crop itself. These trees are *Erythrina glauca*, swamp or Bocare immortelle, and *E.poeppigiana*, mountain or Anauca immortelle; also assoc. with banana bonnygate, a minor disease.

Nectriella Nitschke ex Fuckel 1870; Hypocreaceae;

separated from *Nectria* because the red, fleshy, nonstromatic ascomata are partially or wholly immersed in the host tissue, asco. 1 septate.

N. pironii Alfieri & Samuels, *Mycologia* **71**:1181, 1979, published 1980; anam. *Kutilakesa pironii* Alfieri, *Mycotaxon* **10**:217, 1979; *Kutilakesa* Subram. 1956 = *Sarcopodium* Ehrenb. 1818, see Sutton, *TBMS* **76**:97, 1981; asco. 5–16 × 2.5–4.5 μm, con. pale orange, aseptate, 6–8 × 2–2.5 μm; causing cankers and galls on ornamental and other plants, a wound pathogen; Alfieri et al., *PDR* **63**:1016, 1979; *Proc.Fla.Sta.hort.Soc.* **93**:218, 1980 [59,2624, 4630; 60,936; 61,2307].

needle blight, a general term for some foliage diseases in conifers; (*Davisomella ampla*) jack pine, (*Elytroderma torres-juani, Lophodermella conjuncta, L. sulcigena, Meloderma desmazieresii, Ploioderma lethale*) pines, (*Rhizosphaera kalkhoffi*) pine, spruce, (*R. pini*) fir, (*Rosellinia minor*) conifers.

—— **cast**, as for needle blight; (*Bifusella linearis, Lophodermium* spp.) pines, (*Mycosphaerella laricina, M. laricis-leptolepidis*) larch, (*Phaeocryptopus gaeumannii*) Douglas fir, (*P. nudus*) Douglas fir, fir, hemlock, (*Rhizosphaera oudemansii*) fir.

—— **cushion rust** (*Chrysomyxa abietis*) spruce; ——.
nematodes, *Longidorus*.

—— **rusts** (*Chrysomyxa*) spruce, (*Pucciniastrum*) conifers.

negro coffee mosaic v. Verma & Naizi, *Z.PflKrankh.PflSchutz.* **81**:608, 1974; poss. Potyvirus, filament 550–580 × 21 nm, transm. sap, *Aphis fabae* ssp. *solanella*, India, *Cassia occidentalis*; Misra et al. 1984 [54,5038; 65,326].

nelsonii, *Gymnosporangium*.

Nelson's rust (*Gymnosporangium nelsonii, G. scopulorum*) juniper.

*****nematodes**, see vector; these vermiform animals, 0.3–10 mm long, av. 1 mm, are only briefly treated. A summary of their biology and a bibliography was given by Williams & Bridge in Johnston & Booth 1983. Some nematodes are important plant pathogens in themselves; some have damaging synergistic effects with bacteria and fungi; some transmit plant viruses, Nepoviruses and Tobraviruses. In some genera there is a marked sexual dimorphism. The female swells and fills with eggs. These are extruded in *Meloidogyne* and galls are formed on roots; or retained in cysts as in *Globodera* and *Heterodera*. The parasitic forms have a stylet or mouth spear which is extruded to pierce plant cells. A water film on soil particles is needed for mobility. Nematodes can survive for long periods under dehydration or low temps.; thousands of individuals may be found in a few grams of roots. In habit they can be divided into above ground and root forms; the latter may be ectoparasites, i.e. not normally penetrating the roots; semiendoparasites, i.e. usually feeding with the front end of the body in the root; or endoparasites. In each of these 3 groups the spp. may be either active and migratory or sedentary. Nematode infection alone tends to debilitate not kill plants.

In *A.R.Phytop.*: Triantaphyllou & Hirschmann, **18**:333, 1980, morphology, cytogenetics, speciation & evolution; Giebel, **20**:257, 1982, resistance mechanisms; Zuckerman & Jansson, **22**:95, 1984, chemotaxis & host recognition; Freckman & Caswell, **23**:275, 1985, ecology; Jatala, **24**:453, 1986, biological control. Also: Brown, *AAB* **84**:448, 1976, assessment of damage; Maggenti, *A.Rev.Microbiol.* **35**:135, 1981, as plant parasites; Gommers, *Helminth.Abs.* B **50**:9, 1981, biochemical interactions with plants; Sidhu & Webster, *Bot.Rev.* **47**:387, 1981, genetics of plant parasitism; Stone, *PD* **68**:551, 1984, changing approaches in taxonomy; Siddiqi, *Tylenchida. Parasites of plants and insects*, 1985; Benson & Barker, *PD* **69**:97, 1985, on ornamentals; Kerry & Brown ed., *Principles and practice of nematode control in crops*, 1987.

Nematospora Peglion 1897; Metschnikowiaceae; asco. needle shaped, mostly tropical, assoc. with stainer bugs, Hemiptera, transm. on mouth parts from plant to plant. *N. coryli* Peglion 1901; Mukerji, *Descr.F.* 184, 1968; asco. 30–40 × 2–3 μm, plurivorous, yeast spot of beans, soybeans and many other seeds. *N. gossypii* Ashby & Nowell 1926; Mukerji, *Descr.F.* 185, 1968; asco. 25–37 × 2–5 μm; Batra retains the sp. in *Ashbya* Guillierm. 1928; both spp. cause stigmatomycosis of cotton bolls; Batra, *Tech.Bull. USDA* 1469, 1973, monograph.

Neoaliturus fenestratus, synonym *Circulifer*, transm. safflower phyllody; *N. opacipennis*, transm. beet curly top; *N. tenellus*, transm. beet curly top, horseradish geminivirus, horseradish virescence, *Spiroplasma citri*.

Neocosmospora E.F. Sm. 1899; Hypocreaceae; ascomata usually formed singly, not in a stroma, frequent in culture; asco. ±brownish, usually aseptate, strongly ornamented. *N. vasinfecta* E.F. Sm., asco. mostly 10–11 μm diam. or 13–15.5 μm diam., soil inhabitant which may cause diseases in nurseries; Gray et al., *PD* **64**:321, 1980, soybean stem rot; Cannon & Hawksworth, *TBMS* **82**:673, 1984, revision of genus [**60**,569; **63**,3777].

Neofabraea malicorticis, see *Pezicula malicorticis*.

neomacrospora, *Nectria*.

Neotoxoptera formosana, transm. garlic mosaic.

Nephotettix, transm. rice viruses: bunchy stunt, dwarf, gall dwarf, transitory yellowing, tungro isometric, waika; and rice yellow dwarf MLO.

Nepoviruses, type tobacco ringspot v., *c.* 21 definitive members, isometric *c.* 28–30 nm diam., angular outlines, 3 particle types sedimenting into top, middle and bottom components, 2 segments linear ssRNA, both genomes needed for infection, top component has no RNA, middle and bottom ones have different amounts, thermal inactivation 55–70°, longevity in sap a few days or weeks, transm. sap, in soil by nematodes, most by pollen and seed, wide host ranges, many infect perennials; Harrison & Murant, *Descr.V.* 185, 1977; Murant in Kurstak 1981; Francki et al., vol. 2, 1985.

nerine latent v. Hakkaart 1972; Carlavirus, filament *c.*670 nm long, Netherlands, *N. bowdenii* [**52**,2803d, 2958f].

—— **X v.** Maat, *Neth.J.Pl.Path.* **82**:95, 1976; Potexvirus, filament av. 540 nm long, transm. sap difficult, Netherlands; Phillips & Brunt, *Acta Hortic.* 110:65, 1980, from *Agapanthus praecox* [**55**,5786].

—— **yellow leaf stripe v.** Brunt 1976; poss. Potyvirus, filament *c.*750 nm long, transm. sap, England [**56**,961d].

nervisida, *Gnomonia*; **nervisidum**, *Leptothyrium*; **nervisequia**, *Lirula*.

Nesoclutha pallida, transm. paspalum striate mosaic.

Nesophrosyne orientalis, transm. soybean rosette, sweet potato witches broom; *N. ryukyuensis*, transm. sweet potato dwarf.

net blotch (*Didymosphaeria arachidicola*) groundnut, (*Leandria momordicae*) cucumber, (*Pleospora herbarum*) broad bean, (*Pyrenophora dictyoides*) temperate grasses, (*P. teres*) barley, Gramineae.

netted scab (*Streptomyces*) potato; Scholte & Labruyère, *Potato Res.* **28**:443, 1985, suggested that this name be used for the disease commonly called russet scab q.v. in Europe [**65**,4519].

nettle, see *Urtica*.

New Hebrides disease, also foliar decay q.v.

—— **Logan 64** = black raspberry latent v.

—— **York canker** (*Botryosphaeria obtusa*) apple.

nicotianae, *Cercospora*, *Phytophthora*.

*****Nicotiana tabacum**, tobacco.

nicotiana clevelandii tumour Nienhaus & Gliem, *Phytopath.Z.* **78**:367, 1973; transm. sap to tobacco, seed- and soilborne, USA (California) [**53**,3610].

—— **glauca mosaic mottling v.** Eskarous, Habib &

Sayed, *Egyptian J.Microbiol.* **17**:179, 1982; rod, transm. sap [**63**,3549].

— **glutinosa necrosis** Miyamoto et al. 1965; *Sci.Repts.Fac.Agric.Kobe Univ.* **10**:79, 1971, Japan [**54**,91].

— **rustica v.** Pimpale & Summanwar, *Indian Phytopath.* **35**:217, 1982; isometric 26–30 nm diam., transm. sap, *Myz.pers.*, India (N.) [**63**,5544].

— **velutina mosaic v.** Randles, Harrison & Roberts 1976; Randles, *Descr.V.* 189, 1978; Tobamovirus, rod, mostly 125–150 × 18 nm, transm. sap, seedborne, S. Australia.

— **8 and 12**, = tobacco streak v. and tobacco ringspot v., respectively.

nidus-avis, *Gymnosporangium*.

niger, *Aspergillus*.

nigra, *Arkoola*; **nigrescens**, *Verticillium*; **nigricans**, *Pseudocercospora*; **nigrificans**, *Phoma*; **nigrifluens**, *Erwinia*.

Nigrospora Zimm. 1902; Hyphom.; Carmichael et al. 1980; conidiomata synnematal, conidiogenesis holoblastic; con. aseptate, solitary, spherical or broadly ellipsoidal, shining black, smooth; violent discharge, a rare phenomenon in the Hyphom.; Webster, *New Phytol.* **51**:229, 1952; Ellis 1971. The commonest spp. are *N. oryzae*, teleom. *Khuskia oryzae*, and *N. sphaerica* (Sacc.) Mason, con. mostly 16–18 μm diam. Both are common in the tropics on many plants, largely saprophytic. The latter sp. causes banana squirter in Australia, rare elsewhere; Simmonds, *Qd.agric.J.* **40**:98, 1933; Allen, *Aust.J.exp.Agric.Anim.Husb.* **10**:490, 1970; Holliday 1980 [**13**:42; **51**, 1691].

Nilaparvata bakeri, transm. rice grassy stunt; *N. lugens*, transm. rice grassy stunt, rice ragged stunt, rice wilted stunt; *N. muiri*, transm. rice grassy stunt.

nipponica, *Pyrenochaeta*, see *P. oryzae*.

nitens, *Gymnoconia*, *Lophodermium*.

nitrogen oxides, Hand, *ADAS Q.Rev.* **33**:134, 1979, injury to greenhouse crops.

nivale, *Fusarium*, *Microdochium*; **nivalis**, *Monographella*.

nivea, *Cytospora*; **niveum**, *Cyclaneusma*, *Leucostoma*.

nobleae, *Drechslera*.

Nocardia Trevisan 1889; Bradbury 1986; Gram positive bacteria, non-motile. *N. vaccinii* Demaree & Smith, *Phytop.* **42**:249, 1952; Bradbury, *Descr.B.* 891, 1987; bud proliferating gall of blueberry, shoots are killed, symptoms near soil level, USA (N.E.), the only known plant disease caused by a member of this genus [**32**:25].

nodorum, *Leptosphaeria*, *Septoria*, *Stagonospora*.

nodulosa, *Bipolaris*; **nodulosus**, *Cochliobolus*.

nomenclature, 'The scheme (believed to be a system) by which names are attached to objects including micro-organisms', Cowan 1978; a part of taxonomy q.v.; see the preface in Peace 1962 and FBPP. It is important that plant pathologists give the name of the pathogen that is considered correct by a recognised taxonomic authority. There are international rules that govern the nomenclature of most organisms.

non-host resistance, see resistance to pathogens.

non-infectious bud failure, almond, genetic disorder, Kester in Gilmer et al. 1976; — **crinkle leaf**, plum cv. Santa Rosa, prob. genetic disorder, Pine & Cochran, *Phytop.* **50**:701, 1960; — **shothole**, plum cv. Beaty, prob. genetic disorder, Smith & Cochran, *ibid* **33**:1101, 1943 [**23**:113; **40**:370].

non-pathogen, a micro-organism that is not known to infect a plant or animal and cause a disease; not a synonym of saprophyte.

non-persistent viruses, the vector becomes infective immediately after virus uptake and retention of the virus by it is only for hours; hence nonpersistent transm., see persistent viruses; Raccah, *Adv.Virus Res.* **31**:387, 1986, epidemiology & control.

northern anthracnose (*Kabatiella caulivora*) *clover*; — **cereal mosaic v.** as given in *Descr.V.* index 1986, see cereal northern mosaic v.; — **leaf blight** (*Setosphaeria turcica*) maize, sorghum; — **poor root** (see *Pythium arrhenomanes*) sugarcane; — **root knot nematode** is *Meloidogyne hapla*.

Nothophacidium J. Reid & Cain, *Mycologia* **54**, 194, 1962; Dermateaceae; ascomata cup shaped on a basal stroma, excipulum fleshy soft, on conifer foliage. *N. phyllophilum* (Peck) Smerlis, *CJB* **44**:563, 1966; asco. 5.5–8.5 × 4.5–6.5 μm, secondary invader of needles of *Abies balsamea* previously assoc. with snow blight [**42**:223].

nothoscordum mosaic McKinney, *Phytop.* **40**:703, 1950; transm. sap, USA (Louisiana), *N. fragrans*, false garlic [**30**:122].

novae-zelandiae, *Armillaria*.

noxia, *Koleroga*; **noxius**, *Phellinus*.

nubilosa, *Mycosphaerella*, see *M. cryptica*; **nubilosum**, *Colletogloeum*; **nubilum**, *Verticillium*.

nucleocapsid, the nucleic acid component of a virus particle and its protein coat or shell, the capsid.

nudus, *Phaeocryptopus*.

numerical threshold of infection, the minimum number of propagules of a parasite required for infection of an organism to take place under favourable conditions, Gaümann 1950; see FBPP.

Nummulariella Eckblad & Granmo 1978; Xylariaceae; *N. marginata* (Fr.) Eckblad & Granmo, Anderson 1956 as *Nummularia discreta*; Whalley & Edwards, *TBMS* **85**:385, 1985, described the anam.; the fungus causes apple blister or nailhead canker.

nummularium, *Hypoxylon.*

nursery blight (*Elsinoë randii*) pecan, —— **leaf spot** (*Phyllosticta elettariae*) cardamom.

nutrient film technique, see soilless culture.

nutrients and disease, Huber & Watson,

A.R.Phytop. **12**:139, 1974, N; Hancock & Huisman, *ibid* **19**:309, 1981, nutrient movement in host & pathogen system; Graham, *Adv.Bot.Res.* **10**:221, 1983, nutrient stress & plant susceptibility, particularly trace elements.

Nysius, transm. centrosema mosaic.

O

oahuensis, *Puccinia.*

oak, *Quercus.*

— **leaf deformation and yellowing**, rod assoc., central Europe; Horváth et al., *Z.PflKrankh.PflSchutz.* **82**:498, 1975 [**55**,2902].

— — **scorch**, see elm leaf scorch.

— **mosaic and ringspot** Barnett, *PDR* **55**:411, 1971; Kim & Fulton, *ibid* **57**:1029, 1973, filament assoc., USA (Arkansas), *Quercus marilandica, Q. velutina* [**51**,715; **53**,3191].

— **stunt** Gregory & Seliskar, *PDR* **54**:844, 1970; USA (Alabama), *Quercus lyrata* [**50**,1402].

— **yellowing** Nienhaus & Yarwood, *Phytop.* **62**:313, 1972; rod assoc. like tobacco mosaic v. particle, other symptoms, Cooper 1979 [**51**,4389].

oat, *Avena*, see cereals.

— **blue dwarf v.** Goto & Moore 1952; Banttari & Zeyen, *Descr.V.* 123, 1973; isometric 30–33 nm diam., contains ssRNA, transm. *Macrosteles fascifrons*, persistent, N. America (Kansas to Manitoba); stunt, vein enations & sterility in barley and oat; stunt, leaf crinkle and reduced flower set in flax; Banttari, *Phytop.* **71**:1242, 1981, serological assay [**61**,4077].

— **chlorotic stripe v.** Haber & Gill 1984; poss. Geminivirus, transm. *Rhopalosiphum maidis*, Canada, fide Harrison, *A.R.Phytop.* **23**:55, 1985.

— **dark green dwarf** Agarkov, *Zashch.Rast.Mosk.* **14**:22, 1969; transm. *Macrosteles sexnotatus*, USSR (Ukraine) [**48**,2340].

— **golden stripe** = oat tabular v.

— **mosaic v.** Atkinson 1945; Hebert & Panizo, *Descr.V.* 145, 1975; Potyvirus, filament *c.*600–750 × 12–14 nm, transm. sap difficult, *Polymyxa graminis*, soilborne; Britain, USA, poss. New Zealand; restricted to *Avena*, stunt late in season.

— **necrotic mottle v.** Gill & Westdal 1966; Gill, *Descr.V.* 169, 1976; Potyvirus, filament *c.*720 × 11 nm, transm. sap, Canada, oat and other grasses, does not infect barley or wheat; Gill, *Can.J.Pl.Path.* **2**:86, 1980, properties of protein & nucleic acid [**60**,1407].

— **pseudo-rosette v.** Soukhov & Vovk 1938; Smith 1972; Borodina et al., *Biol.Nauki.* 2:22, 1982; rod *c.*420 × 67 nm, transm. leafhoppers, *Laodelphax striatellus* most important; Japan, USSR (W. Siberia), abnormally prolific tillering; ? same as northern cereal mosaic v. [**61**,6298].

— **red leaf** = barley yellow dwarf v.

— **sterile dwarf v.** Průša 1958; Boccardo & Milne, *Descr.V.* 217, 1980; Phytoreovirus, Fijivirus subgroup, isometric 65–70 nm diam., transm. planthoppers, *Javesella pellucida* most important; Europe, Gramineae only, dark green and distorted leaves; *Lolium* spp. show conspicuous enations on nodes of flowering stems and spikes.

— **tubular v.** Plumb et al., *Annls.Phytopath.* **9**:365, 1977; Catherall & Chamberlain 1977; rod *c.*150 and 300 nm long, 20 nm wide, transm. sap erratically; England, Wales; oat golden stripe is a synonym [**57**,1566e; **58** 4a,1114].

obesa, *Septoria.*

obligate parasite, an organism that occurs in an intimate assoc. with, and which is wholly dependent for its nutrition on, another living organism; parasite, parasitism, applies to a mode of existence, pathogen q.v. to causing a disease; obligate is the antonym of facultative; see root and shoot metabolism, Walters.

obliquus, *Inonotus*; **oblonga**, *Phomopsis.*

obscura, *Armillaria*; **obscurans**, *Phomopsis.*

obtusa, *Botryosphaeria.*

occidentale, *Cronartium, Peridermium*; **occidentalis**, *Albugo, Melampsora.*

occulta, *Urocystis.*

ocellata, *Cercoseptoria.*

Ocfemia, Gerardo Offimaria, 1891–1959; born in the Philippines, College of Agriculture, Univ. Philippines; notable early work in tropical E. Asia. *Phytop.* **50**:403, 1960.

ochracea, *Ustilaginoidea.*

odontoglossum ringspot v. Jensen & Gould 1951; Paul, *Descr.V.* 155, 1975; Tobamovirus, rod 300 × 18 nm, transm. sap, in cultivated orchids, see cymbidium mosaic v.; Edwardson & Zettler in van Regenmortel & Fraenkel-Conrat 1986.

odontonema yellow net Ramakrishnan & Subramanian, *Curr.Sci.* **30**:431, 1961; India (S.) *O. nitidum* [**41**:461].

Oecanthus niveus, see *Cryptosporiopsis.*

oeconomicum, *Aristastoma.*

oedema, the accumulation of water in leaves, followed by epidermal rupture; Sagi & Rylski, *Phytoparasitica* **6**:151, 1978 [**58**,6067].

Oedocephalum Preuss 1851; Hyphom.; Carmichael et al. 1980; conidiomata hyphal, conidiogenesis holoblastic; con. aseptate, botryose, hyaline; *O. lineatum*, teleom. *Heterobasidion annosum.*

oenanthe yellows Shiomi & Sugiura, *Ann.phytopath.Soc.Japan* **49**:367, 1983; MLO assoc., transm. *Macrosteles orientalis*; *O. javanica* [**63**,3159].

oenotherae, *Pezizella*.

Ogilvie, Laurence, 1898–1980; Univ. Aberdeen, Cambridge; Department of Agriculture, Bermuda 1923–28; Long Ashton Research Station, Univ. Bristol; wrote *Diseases of vegetables* 1941, *edn*. 6, 1969. *Bull.Br.mycol.Soc.* **14**:153, 1980.

Ohio corn stunt = maize chlorotic dwarf.

Oidiopsis Scalia 1902; Hyphom.; not recognised by Carmichael et al. 1980; teleom. *Leveillula*.

Oidium Link 1824; Hyphom.; Carmichael et al. 1980; conidiomata hyphal, conidiogenesis holoarthric; con. aseptate, hyaline; teleom. in Erysiphaceae; where ascomata are present an identification should be straightforward. But frequently, especially in the tropics, they are not, or may be unknown. Therefore attempts have been made at classification based on the anam. Some of the criteria used are given under the teleom. genera: *Erysiphe, Leveillula, Microsphaera, Phyllactinia, Podosphaera, Sphaerotheca, Uncinula*. Many *Oidium* spp. have been omitted because of uncertain identity and/or inadequate data on pathology.

O. aceris, teleom. *Uncinula bicornis*.

O. anacardii Noack 1898; con. 34–37 × 18–22 μm; on cashew, reported to be severe in Tanzania, destroying the inflorescences and requiring chemical control; Casulli and Castellani & Casulli, *Riv.Agric.subtrop.trop.* **73**:241, 1979; **75**:211, 259, 1981 [**60**,1101; **61**,3711–12].

O. arachidis Chorin 1961; con. 31–44 × 13–15.5 μm; groundnut in Israel; Chorin & Frank, *Israel J.Bot.* **15**:133, 1966 [**46**,3607].

O. begoniae Puttemans 1911; con. 20–36 × 13–17 μm; the anam. of *Microsphaera begoniae* is poss. *Oidium begoniae* var. *macrosporum* which has larger con.; serious on elatior begonias in USA; Quinn & Powell, *PD* **65**:68, 1981, **66**:718, 1982; *Phytop.* **72**:480, 1982, 2 races, chemical control, effects of temp., light & relative humidity [**60**,5958; **61**,6419; **62**,223].

O. bixae Viégas 1944; con. 24–32 × 12–19 μm; on *Bixa orellana* in Venezuela; Capretti, *Riv.Agric.subtrop.trop.* **55**:13, 1961 [**40**:701].

O. caricae Noack 1898; con. 23–25 × 14.5–20 μm; on papaya in Brazil; several other spp. have been described from this crop: *Oidium caricae-papayae* Yen 1966; con. 36–52.8 × 15.6–21.6 μm; *O. caricicola* Yen & Wang 1973; con. 25–33 × 16–21 μm; both in Taiwan; *O. indicum* Kamat 1955; con. 31–46 × 13.7–23.4 μm; in India. Little pathology has been done on these spp.; Boeswinkel, *Fruits* **37**:473, 1982, identity of *O. caricae* in New Zealand [**62**,3146].

O. euonymi-japonicae (Arcang.) Sacc. 1903; con. *c*.30–38 × 13–14 μm; poss. teleom. *Uncinula euonymi-japonicae* Hara 1921, on *Euonymus japonicus*. The powdery mildew on *E. europeus* has been reported to be a *Microsphaera* and may therefore be a distinct sp.; Childs, *Phytop.* **30**:65, 1940; Tokushige, *Ann.phytopath.Soc.Japan* **17**:61, 1953 [**19**:297; **33**:30].

O. heveae Steinm. 1925; Sivanesan & Holliday, *Descr.F.* 508, 1976; con. in basipetalous chains, very variable in size, 24–42 × 12–17 μm; the commonest cause of rubber secondary leaf fall and prob. confined to this host, prob. absent from tropical America; the fungus infects the young leaves as the tree refoliates after the natural defoliation; as a result further, damaging leaf fall takes place; the trees become debilitated; a much less damaging pathogen compared with *Microcyclus ulei*; fungicide treatment may be needed on high yielding and very susceptible clones; Holliday 1980.

O. lini Škorić 1926; Allison, *Phytop.* **24**:305, 1934; con. 23–41 × 12–18 μm; an important disease on linseed in India; Singh & Saharan, *Euphytica* **28**:531, 1979, inheritance of resistance [**13**:515; **59**,312].

O. mangiferae Berthet 1914; con. 33–43 × 18–28 μm; on mango, important and often requires fungicide control, young leaves, blossoms and fruit are attacked; Palti et al., *PDR* **58**:45, 1974 [**53**,2638].

O. tingitanium Carter, *Phytop.* **5**:195, 1915; con. 20–28 × 10–15 μm; Petch, *ibid*, page 350 considered the powdery mildew on citrus in Sri Lanka to be different; Boeswinkel, *Nova Hedwigia* **34**:731, 1981, identity of sp. in USA (California) [**61**,2882].

O. tuckeri, teleom. *Uncinula necator*.

oil palm, *Elaeis guineensis*; Turner & Bull 1967; Turner, *Phytopath.Pap.* 14, 1971; Aderungboye, *PANS* **23**:305, 1977; Turner 1981; Renard & Quillec, *J.Pl.Prot.Tropics* **1**:69, 1984.

—— **assoc. v.** Plumb & Dabek 1976; Plumb et al., *Trop.Agric.Trin.* **55**:59, 1978; 2 rods *c*.607 and 788 nm long, 1 isometric *c*.32 nm diam., Papua New Guinea [**56**,1852g; **57**,2619].

oils, used in control of pathogens; Cuille, *PANS* **11**:281, 1965, aerial application; Calpouzos, *A.R.Phytop.* **4**:369, 1966; Simons & Zitter, *PD* **64**:542, 1980, aphid borne viruses.

oilseed rape, Davies on diseases in Scarisbrick & Daniels ed., *Oilseed rape* 1986.

okra, lady's fingers, *Abelmoschus esculentus*; often as *Hibiscus* q.v.

—— **mosaic v.** Givord, Pfeiffer & Hirth 1972; Givord & Koenig, *Descr.V.* 128, 1974; Tymovirus, isometric c.28 nm diam., transm. sap, W. Africa, wide host range; Lana et al., *PDR* **58**:616, 1974; Lana & Bozarth, *Nigerian J.Pl.Prot.* **1**:82, 1975, Nigerian form, transm. sap, chrysomelid beetles, *Podagrica uniformis, Syagrus calcaratus*; Givord, *Annls. Phytopath.* **9**:53, 1977; Igwegbe, *PD* **67**:320, 1983, strs.; Atiri, *AAB* **104**:261, 1984, in Nigerian weeds [**54**,639; **55**,500; **57**,1552; **62**,2917; **63**,3156].

oleae, *Elsinoë, Sphaceloma.*

Olea europaea, olive.

oleaginea, *Spilocaea.*

oleander necrosis Hollings, Stone & Brunt 1966; England [**46**,1175i].

—— **witches' broom** Chen, *Mem.Coll.Agric.Natn.Taiwan Univ.* **12**:67, 1971; *Nerium indicum* [**53**,185].

Oliarus atkinsoni, transm. phormium yellow leaf.

oligandrum, *Pythium*; **oligocladum**, *Teratosperma.*

oligogenic resistance, see resistance to pathogens.

olive, *Olea europaea*; Martelli, *Inftore.fitopatol.* **31**(1–2):97, 1981, poss. viruses [**61**,331].

—— **infective yellows** Ribaldi 1959; Italy [**38**:357].

—— **latent ringspot v.** Savino et al. 1983; Gallitelli et al., *Descr.V.* 301, 1985; Nepovirus, isometric c.28 nm diam., transm. sap to 7 spp. in 5 dicotyledonous families, Italy (central), no symptoms in naturally infected olive.

—— —— **1 v.** Gallitelli & Savino, *AAB* **106**:295, 1985; isometric c.30 nm diam., contains a major ssRNA, transm. sap, Italy (Apulia), the only disease symptoms in olive are occasional fasciations and bifurcations of leaves and twigs [**66**,257].

—— —— **2 v.** Savino et al., *Proc. 6th Congr.Medit.Phytopath.Union*; quasi-isometric to bacilliform particles.

—— **narrow leaf** Marte et al., *PD* **70**:171, 1986; strawberry latent ringspot v. assoc., olive cv. Ascolana Tenera, Italy (central) [**65**,2924].

—— **sickle leaf** Ciferri et al. 1953; Thomas, *PDR* **42**:1154, 1958; Waterworth & Monroe, *ibid* **59**:366, 1975; Israel, Italy, USA (California) [**34**:605; **38**:157; **54**,4592].

—— **spherosis** Lavee & Tanni 1984; poss. virus, Israel [**64**,3909].

Olivea Arthur 1917; Uredinales; uredia densely paraphysate at periphery, being replaced by telia; aecia lacking peridia; Thirumalachar & Mundkur, *Indian Phytopath.* **2**:236, 1949; Ono & Hennen, *Trans.mycol.Soc.Japan* **24**:369, 1983, taxonomy of Chaconiaceae genera which have aseptate, sessile, thin walled probasidia [**63**,4788].

O. tectonae (T.S. & K. Ramakrishnan) Mulder 1973 in Mulder & Gibson, *Descr.F.* 365, 1973; teliospores aseptate, sessile; previously placed in

Chaconia, transferred to *Olivea* because telial paraphyses are present; aecidia unknown, presumably autoecious; teak rust, potentially important, may require chemical control on young trees, Asia (E.).

Ollarianus balli, transm. rhynchosia little leaf.

Olpidiaceae, Chytridiales; thallus not differentiated into a well developed hyphal system; sporangium inoperculate, opening by the deliquescence or rupture of one or more papillae; zoospores often with a conspicuous globule, germination monopolar; *Olpidium.*

*****Olpidium** (A. Braun) Schröter 1886; Olpidiaceae; sporangia generally scattered, sporangium not filling host cell and mostly forming 1 discharge tube, resting spore usually filling its container.

O. brassicae (Woronin) Dang. 1886; Barr, *Fungi Canadenses* 176, 1980; sporangia spherical, 10–30 μm diam., to ellipsoid ± the shape of the host cell, 35–80 × 12–110 μm; mostly 1–3 discharge tubes; resting sporangia mostly 9–30 μm diam., stellate; zoospores 3.5–5.5 μm diam., spherical. Sahtiyanci 1962 transferred the fungus to *Pleotrachelus* but *P. brassicae* was given as a synonym by Barr. *Olpidium brassicae* is plurivorous, in roots, not a destructive pathogen but important in the transm. of viruses, these include: lettuce big vein, tobacco necrosis, tobacco stunt; it is also implicated in carrot rusty root in Canada; Sampson, *TBMS* **23**:199, 1939, morphology; Temmink & Campbell, *CJB* **46**:951, 1968; **47**:227, 1969, fine structure & virus transm. hypotheses; Aist & Israel, *Phytop.* **67**:187, 1977, papilla formation & host penetration; Subrahmanyam & McDonald, *TBMS* **75**:506, 1980, on groundnut, hosts & morphology; Singh & Pavgi, *Indian Phytopath.* **34**:173, 1981, chemotaxis & root infection [**47**,3740; **48**,2210; **56**,5249; **60**,4770; **61**,4417].

O. radicale Schwartz & W. Cook 1928, fide Lange & Insunza, *TBMS* **69**:377, 1977, who placed *Olpidium cucurbitacearum* Barr 1968 as a synonym, and compared *O. radicale* with *O. brassicae*; the former sp. differs in the smooth resting sporangia, 1 discharge tube and larger zoospores 7–10 μm diam., elongate; plurivorous, in roots; transm. cucumber necrosis, melon necrotic spot, poss. red clover necrotic mosaic; Lange & Olson, *ibid* **71**:43, 1978, zoospore [**57**,2836; **58**,1072].

O. viciae Kusano 1912; blister or warty scab of broad bean, infects other legumes; Xin et al., *Acta phytopath.sin.* **14**:165, 1984, stated that this is a major disease of broad bean in China (N.W. Sichuan); but apparently little has been published on the biology and pathology of the fungus. *Olpidium trifolii* Schröter 1889 also occurs on legumes; it was considered distinct by Kusano, *J.Coll.Agric.Imp.Univ.Tokyo* **10**:83, 1929; **11**:359, 1932; *Jap.J.Bot.* **8**:155, 1936 [**8**:580; **11**:720; **15**:659; **64**,2817].

omnivora, *Phymatotrichopsis*.

oncidium mosaic Jensen, *Calif.Agric.* **6**(2):7, 1952; transm. sap, USA (California) [**32**:130].

*Oncobasidium Talbot & Keane, *Aust.J.Bot.* **19**:203, 1971; Ceratobasidiaceae; basidiomata white, as effuse, adherent patches; basidia holobasidiate, obovate, later elongating, becoming capitate clavate. The genus was erected to accommodate an obligate pathogen of cacao, *O. theobromae* Talbot & Keane; basidiospores broad ellipsoid, flattened on one side, 15–25 × 6.5–8.5 μm. It causes vascular streak dieback, and has been reported from: China (Hainan Island), India (S.), Indonesia (Sumatra), Malaysia (Sabah, Sarawak, W.), New Britain, Papua New Guinea, Philippines. This serious dieback should not be confused with other forms of cacao diebacks; these usually have a complex etiology, and are assoc. not only with pathogens but also with insects and physiological factors. The disease often occurs in younger plants which can be killed. But it may cause significant yield loss in mature trees. Airborne basidiospores infect young leaves, and the fungus spreads in the xylem of leaves and young stems. Basidiocarps form around the leaf petiole scars; the growing tip dies and leaves fall; Holliday 1980; Keane, *AAB* **98**:227, 1981, epidemiology; Prior, *TBMS* **78**:571, 1982; *Pl.Path.* **34**:603, 1985, basidiospore formation in callus culture & measures against spread [**60**,6378; **61**,5629; **65**,1150].

Oncometopia nigricans, transm. xylem limited bacteria q.v.

onion, see *Allium*.

—— **mosaic v.** Ryžkov & Vovk 1937; Protsenko & Legunkova, *Mikrobiologiya* **30**:165, 1961; Cheremushkina 1975; prob. a rod, transm. sap, *Aceria tulipae*, USSR [**17**:91; **40**:644; **60**,2908].

—— **proliferation** Petre & Ploaie 1971; MLO assoc., this disease is prob. the same as onion phyllody Kleinhempel et al. 1975 and onion yellows Chen 1979; L. Bos, personal communication, 1984 [**53**,4682; **55**,2606; **60**,1898].

—— **yellow dwarf v.** Melhus et al. 1929; Bos, *Descr.V.* 158, 1976; Potyvirus, filament *c.*775 nm long, transm. sap, aphid, non-persistent, *Allium* spp.

ononis yellow mosaic v. Gibbs 1964; Gibbs et al., *J.gen.Microbiol.* **44**:177, 1966; Tymovirus, isometric 25–30 nm diam., transm. sap, England, *O. repens* [**43**,2790b; **46**,78].

oogonium, female sex organ of Oomycetes.

Oomycetes, Mastigomycotina, the same as the Phycomycetes sensu stricto; thallus mainly aseptate, diploid, cell walls of glucan-cellulose and have no chitin; zoospore with 2 equal flagella, tinsel flagellum directed forwards, whiplash flagellum directed backwards; Cohen & Coffey, *A.R.Phytop.* **24**:311, 1986, systemic fungicides.

oospore, a resting spore, from a fertilised oosphere which arises from an oogonium.

Ootheca mutabilis, transm. cowpea mottle.

ooze, (1) to be extruded as a viscous fluid through an aperture in a plant structure, e.g. hydathode, lenticel, stoma, wound; see exude. (2) A viscous fluid so extruded, cf. exudate, FBPP.

operculate, of an ascus or sporangium, opening by an apical lid to discharge spores as in the asci of the Pezizales, cf. inoperculate.

ophiobolins, toxins from *Cochliobolus*; Wood et al. 1972; Durbin 1981.

Ophiognomonia (Sacc.) Sacc. 1899; Gnomoniaceae. *O. pseudoplatani* (v. Tubeuf) D. Barrett & Pearce, *TBMS* **76**:317, 1981; ascomata 150–320 μm diam., beaks eccentric, 260–330 μm long, projecting 160–220 μm outside leaf; asco. 45–65 × 0.5–1.5 μm; sycamore giant leaf blotch, *Acer pseudoplatanus*, Europe; Monod, see Gnomoniaceae, transfered the fungus to *Pleuroceras pseudoplatani* [**61**,2454].

Ophiomyia simplex, see *Fusarium oxysporum* f.sp. *asparagi*.

ophiospora, *Melophia*, see phellophagy.

Ophiostoma H. Sydow & Sydow 1919; see *Ceratocystis*; *C. ulmi*, the cause of Dutch elm wilt, is preferred to *O. ulmi* (Buisman) Nannf.

Ophiostomatales, Ascomycotina; 1 family Ophiostomataceae; ascomata spherical, perithecioid, terrestrial, with a long beak; asci evanescent, spherical; asco. small, becoming free, often discharged in a slimy mass; *Ceratocystis*.

opium poppy mosaic Türkoglu, *J.Turkish Phytopath.* **8**:77, 1979; Anand & Summanwar, *Indian Phytopath.* **34**:262, 1981; transm. sap, 3 aphid spp. [**59**,5306; **61**4285].

—— —— **yellowing** Rozsypal, *Česká Biol.* **6**:428, 1957; Czechoslovakia [**39**:32].

opposite leaf test, see local lesion assay.

opuntia Sammon's v. Sammons & Chessin, *Nature Lond.* **191**:517, 1961; prob. Tobamovirus; Brandes, *Mitt.biol.BdAnst.Berl.* 110, 1964; Brandes & Chessin, *Virology* **25**:673, 1965 [**41**:155].

—— **witches' broom** Lesemann & Casper 1970; Casper et al., *PDR* **54**:851, 1970; MLO assoc., Europe, *O. tuna* [**49**,2498; **50**,1269].

orange, orangelo, see *Citrus*.

—— **blotch**, oil palm, K deficiency, Turner 1981.

—— **graft transmissible dwarfing** Schwinghammer & Broadbent, *Phytop.* **77**:205, 210, 1987; detection of viroids by transm. to chrysanthemum, Australia [66,3828–9].

—— **hobnail canker** (*Endothia gyrosa*) oak; —— **rust** (*Coleosporium ipomoeae*) pine, sweet potato, (*Gymnoconia nitens*) *Rubus*; —— **spotting**, oil palm, genetic disorder, Turner 1981.

orbiculare, *Colletotrichum*.

orchard grass, see cocksfoot.

orchid, Burnett 1974; Bund et al., *Orchideën* **40**(4a) 1978 [57,3490].

—— **fleck** v. Doi et al. 1969; *Descr.V.* 183, 1977; prob. Phytorhabdovirus, bacilliform *c*.150 × 40 nm, less in thin sections; transm. sap difficult, Japan and elsewhere; *Cymbidium, Dendrobium, Oncidium*.

—— **mosaic** = cymbidium mosaic v.

—— **ringspot** Kitajima, Blumenschein & Costa, *Phytopath.Z.* **81**:280, 1974; assoc. particle 50–200 × 40 nm, Brazil (S.) [54,4527].

oreades, *Marasmius*, see fairy rings.

oregonensis, *Didymosphaeria*.

organism, one meaning of an organ is: a part or member of an animal, fungus or plant that is adapted by its structure for a particular function. Organs function as a whole and in an orderly manner, i.e. an organism, a micro-organism, a living, replicating body. It may or may not be accepted that viruses can be described as organisms in the generally accepted sense; although they are, as replicating entities, organised in some sense; see taxonomy.

orientalis, *Caliciopsis*.

*ornamentals, and see trees; Sweet, *Acta Hortic.* 59:83, 1976, viruses in Britain; Stahl & Umgeller, *Pflanzenschutz im Zierpflanzenbau*, 1976; Pirone, *Diseases and pests of ornamental plants*, edn. 5, 1978; Tattar, *Diseases of shade trees*, 1978; Wheeler in Spencer 1978 and 1981, downy & powdery mildews; Baker & Lindeman, *A.R.Phytop.* **17**:253, 1979; Cooper 1979; Hollings & Stone in Ebbels & King 1979, use of stock free from viruses; Moore 1979; *Tech.Communic.Int.Soc.hortic.Sci.* 110, 1980; Buczacki & Harris, *Collins guide to pests, diseases and disorders of garden plants*, 1981; Griffin, *Booklet, Min.Agric.Fish Fd.* UK, 2364, 1981, cut flowers; Lawson, *PD* **65**:780, 1981, control of virus diseases.

ornithogalum mosaic Smith & Brierley, *Phytop.* **34**:497, 1944; transm. sap, aphid spp., USA (Oregon) [23:439].

Orosius albicinctus, transm. potato Indian purple top roll, potato witches' broom, sesame phyllody, tomato marginal flavescence, vernonia phyllody.

—— **argentatus**, transm. peanut Indonesian mosaic, potato purple top wilt, tobacco yellow dwarf, tomato big bud.

—— **cellulosus**, transm. cotton virescence, sesame phyllody.

—— **lotophagorum ryukyuensis** transm. sweet potato witches' broom.

Orthezia insignis, transm. epiphyllum mosaic.

Orton, William Allen, 1877–1930; born in USA, USDA; important early work especially in developing resistant crop cvs., potato viruses, and problems in international plant quarantine. *Phytop.* **21**:1, 1931.

oryzae, *Bipolaris, Cercospora, Entyloma, Ephelis, Khuskia, Microdochium, Mycovellosiella, Nigrospora, Pyrenochaeta, Pyricularia, Rhizoctonia, Rhizopus, Sarocladium, Sclerotium, Septoria*.

oryzae-sativae, *Balansia, Claviceps, Phomopsis, Rhizoctonia*.

Oryza sativa, rice; see Chang in Simmonds 1976 for a list of *Oryza* spp.

oryzicola, *Leptosphaeria*; **oryzina**, *Sphaerulina*.

ostiole, the schizogenous cavity lined with paraphyses and ending in a pore in the papilla or neck of a perithecium; any pore by which spores are freed from an ascigerous or pycnidial structure.

oudemansii, *Rhizosphaera*.

Oulema lichenis, *O. melanopus*, transm. cocksfoot mottle, phleum mottle.

Ovulariopsis Pat. & Hariot 1900; Hyphom.; sometimes used as an anam. for *Phyllactinia* but this is prob. best rejected, fide Yarwood in Spencer 1978.

Ovulinia Weiss, *Phytop.* **30**:242, 1940; Sclerotiniaceae; con. single, i.e. not catenate; monotypic, *O. azaleae* Weiss, anam. *Ovulitis azaleae*; asco. 10–18 × 8.5–10 μm, con. aseptate, 40–60 × 21–36 μm; azalea petal blight; Hemmi & Akai, *Mem.Coll.agric.Kyoto* 80, Phytopath.Ser. 13, 1959; Coyier & Roane 1986 [**19**:412; **39**:314].

Ovulitis Buchw. ex Buchw. 1970; Hyphom.; Carmichael et al. 1980; conidiomata hyphal, conidiogenesis holoblastic; con. hyaline, aseptate, botryose; teleom. *Ovulinia azaleae*.

Owens, Charles Elmer, 1877–1957; born in USA, Univ. Indiana, Wisconsin; at Oregon Agricultural College 1912–47; wrote *Principles of plant pathology* 1928. *Phytop.* **48**:291, 1958.

oxalicum, *Penicillium*.

oxalis chlorotic ringspot and decline v. Coyier, *PD* **65**:275, 1981; filament *c*.800–900 × 18 nm, transm. root contact, USA (Washington State), *O. regnellii*, ornamental shamrock [**61**,1760].

—— **leaf distortion** Giri & Sharma, *Curr.Sci.*

43:451, 1974; India (Himachal Pradesh), *O. corniculata* [**54**,2121].

oxycarboxin, systemic fungicide used against rusts.

oxycocci, *Monilinia*.

Oxycoccus, see *Vaccinium*.

oxygen, Bergman, *Bot.Rev.* **25**:417, 1959, O deficiency as a cause of disorder; Stolzy et al. in Bruehl 1975, O diffusion & in the root zone.

oxysporum, *Fusarium*.

ozone, damage to plants, and see air pollution; Rich, *A.R.Phytop.* **2**:253, 1964; Rist & Lorbeer in Blakeman 1981, interactions, fungi & the leaf; Reich & Amundson, *Science N.Y.* **230**:566, 1985, effects of ambient levels on photosynthesis [**65**,132].

P

pachyrhizi, *Phakopsora*.
pachysandra line pattern (alfalfa mosaic v.),
Hershman & Varney, *PD* **66**:1195, 1982,
P.terminalis [62,1536].
padwickii, *Alternaria, Phacidiopycnis*.
paeoniae, *Botrytis*.
pale brown rot (*Pseudophaeolus baudonii*) eucalypts
and other plants.
pallidoroseum, *Fusarium*.
palmarum, *Pestalotiopsis*.
palmivora, *Phytophthora*; palmivorus, *Marasmius*.
palm lethal yellowing Fawcett 1891; this devastating
disease was known in 1834 in the Cayman
Islands. It is prob. caused by an MLO but, like so
many diseases said to have such an etiology,
Koch's postulates have not yet been satisfied. The
causal agent may be transm. by *Myndus crudus*,
Howard et al., *Trop.Agric.Trin.* **60**:168, 1983; it
infects coconut and *c*.30 other palm spp.
Hundreds of thousands of trees have been killed
in USA (Florida) since 1955, and in Jamaica since
1961. The disease occurs in the Caribbean region.
Other diseases of obscure etiology may be caused
by the same, or similar, causal agent(s); these are:
Awka, dry bud rot, Julia & Mariau, *Oléagineaux*
37:517, 1982; Cape St Paul wilt, Johnson &
Harries, *Ghana J.agric.Sci.* **9**:125, 1976;
Kaincopé, Kribi, all in W. Africa. In India Kerala
wilt, first reported > 100 years ago, Solomon
et al., *Z.PflKrankh.PflSchutz.* **90**:295, 1983; also
in E. Asia, Malaysian wilt and stem necrosis. In
E. Africa lethal yellowing, Nienhaus et al., *ibid*
89:185, 1982.
Beakbane et al., *J.hort.Sci.* **47**:265, 1972;
Heinze et al., *Phytopath.Z.* **74**:230, 1972; Plavšić-
Banjac et al., *Phytop.* **62**,298, 1972, MLO assoc.;
reviews: Tsai in Harris & Maramorosch 1980;
Maramorosch & Hunt and Tsai & Thomas in
Maramorosch & Raychaudhuri 1981; Howard,
F.A.O.Pl.Prot.Bull. **31**:103, 1983; McCoy ed.
Bull.Florida agric.Exp.Stns. 834, 1983 [**51**,1776,
4263; **52**,820; **59**, 2300; **61**,5171; **62**,2076, 4408,
4746].
— mosaic Mayhew & Tidwell, *PDR* **62**:803, 1978;
filament assoc. av. 686 × 13 nm, USA (California),
Washingtonia robusta [**58**, 3939].
palms, see coconut, date, oil.
pan, see betel.
panacis, *Cylindrocarpon*.
Panama wilt (*Fusarium oxysporum* f.sp. *cubense*)
banana.

panattonianum, *Microdochium*.
panax, *Alternaria*.
— necrosis Li et al. 1983; MLO assoc., China
[64,705].
— ringspot Aragaki, Murakishi & Hendrix,
Phytop. **43**:643, 1953; USA (Hawaii), *Nothopanax
guilfoylei*, synonym? *Polyscias guilfoylei* [**33**:425].
— yellow and bushy stunts Li et al., *Acta
Microbiol.sin.* **22**:291, 1982; China, *P. ginseng*.
pandemic, see epidemic.
panel necrosis (*Botryodiplodia theobromae*) rubber.
pangola grass, *Digitaria decumbens*.
— stunt v. Dirven & van Hoof 1960; Milne,
Descr.V. 175, 1977; Phytoreovirus, Fijivirus
subgroup; isometric 65–70 nm diam., transm.
Sogatella furcifera, Brazil, Fiji, Guyana, Peru,
Taiwan; Gramineae.
panici-frumentacei, *Ustilago*, see *U. crus-galli*.
panicle blight, rice, as for ear blight.
Panicum maximum, Guinea grass; *P. miliaceum*, see
millets.
panicum golden yellow Gill, Chong & Caetano,
Can.J.Pl.Path. **3**:129, 1981; isometric particle
assoc. 26 nm diam., Brazil (Rio Grande do Sul),
P. sabulorum [**61**,2929].
— mosaic v. Still & Pickett 1957; Niblett et al.,
Descr.V. 177, 1977; isometric 25–30 nm, usually
assoc. with a serologically distinct virus, like a
satellite, particle 15–18 nm diam., transm. sap,
Mexico, USA; Gramineae, causes St Augustine
grass decline; Berger & Toler, *Phytop.* **73**:185,
1983, quantitative immunoelectrophoresis
[62,3004].
— Philippines mosaic Celino & Martinez,
Philipp.Agric. **40**:285, 1956; transm. aphid spp.,
P. distachyum [**36**:646].
— streak Patel 1969; Kenya, *P. maximum* [**50**,5b].
pannosa, *Sphaerotheca*.
papaveracea, *Pleospora*.
Papaver somniferum, opium poppy.
papaya, *Carica papaya*; Frossard, *Fruits* **24**:473,
483, 1969; Srivastava & Tandon, *PANS* **17**:51,
1971, postharvest; Hunter & Buddenhagen,
Trop.Agric.Trin. **49**:61, 1972, in USA (Hawaii);
Krochmal, *Ceiba* **18**:19, 1974; Alvarez &
Nishijima, *PD* **71**:681, 1987, postharvest.
— apical drooping necrosis Wan & Conover,
Proc.Fla.Sta.hort.Soc. **94**:318, 1981; bacilliform
particles assoc. 180–254 × 87–98 nm, resembles the
virus of entry below [62,2053].
— — necrosis v. Lastra & Quintero, *PD* **65**:439,

1981; Phytorhabdovirus, 210–230 × 80–84 nm, transm. *Empoasca papayae*, Venezuela [61,1300].
— **bunchy top** Cook 1931; Bird & Adsuar, *J.Agric.Univ.P.Rico* **36**:5, 1952; Story & Halliwell, *Phytop.* **59**:1336, 1969; Haque & Parasram, *PDR* **57**:412, 1973; MLO assoc., transm. *Empoasca papayae, Solanasca stevensi*; W. Indies [**32**:389; **49**, 529; **53**,629].
— **chlorosis** Kulkarni & Sheffield 1968; Kulkarni & Wamagata 1969; transm. sap, E. Africa [**48**,7f; **49**,2286i].
— **distortion ringspot** = papaya ringspot v.
— **Isabela yellow mosaic** Adsuar, *J.Agric.Univ.P.Rico* **56**:397, 1972; transm. sap. Puerto Rico [**52**,3781].
— **leaf curl** Sen, Ganguly & Mallik, *Indian J.Hort.* **3**:38, 1946; transm. sap; —— **distortion** Garga, *Indian Phytopath.* **16**:31, 1963; transm. *Myz.pers.*; —— **reduction** Singh, *PDR* **53**, 267, 1969; transm. sap, *Myz.pers.*, non-persistent, all India [**26**:20; **43**,1366; **48**,2499].
— **lethal yellowing v.** Loreto et al., *Biológico* **49**:275, 1983; isometric 29–32 nm diam., transm. sap, Brazil (Pernambuco), restricted host range, a threat to other, more important growing areas of the country [**64**,2652].
— **mild mosaic** = papaya mosaic v.
— **mosaic v.** Conover 1962, 1964; Purcifull & Hiebert, *Descr.V.* 56, 1971; Potexvirus, filament *c*.530 nm long, transm. sap; Cheema & Reddy 1985 described transm. by *Rhopalosiphum maidis*; Erickson et al., *Virology* **90**:36, 47, 60, 1978; Abouhaidar & Bancroft, *ibid* **90**:54, 1978; Tollin et al., *ibid* **98**:108, 1979, structure; Singh, *Indian J.Entomol.* **34**:240, 1972, in India [**54**,1359; **58**,2333–6; **59**,2647; **65**,825].
*—— **ringspot v.** Jensen 1949; Purcifull et al., *Descr.V.* 292, 1984; Potyvirus, filament *c*780 × 12, transm. sap, many aphid spp., non-persistent; isolates of 2 types P and W. The first infects papaya, causing stunt and severe leaf distortion; this is a major and widespread disease of the crop. The second infects cucurbits, watermelon mosaic 1 v. is a synonym, causing mottle, leaf and fruit distortion; generally the host range of the virus is narrow; Quiot-Douine et al., *Phytop.* **76**:346, 1986, a serologically distinct str. from squash, earlier called squash striped v. [**65**,4786].
— **severe chlorosis and necrosis** (tomato spotted wilt v.), Gonsalves & Trujillo, *PD* **70**:501, 1986 [**65**,5624].
— **yellow crinkle** Morwood 1931; McKnight, *Qd.agric.J.* **69**:153, 1949; Greber, *Qd.J.agric.Anim.Sci.* **23**:147, 1966; MLO assoc., Australia (Queensland) [**11**:157; **29**:36; **46**,2483].

papillae, ± synonymous with callosity q.v. and lignituber q.v.
papuana, *Deightoniella*.
Paracarsidara concolor, transm. wissadula proliferation.
Paracercospora Deighton, *Mycol.Pap.* 144:47, 1979; Hyphom.; conidiomata hyphal, conidiogenesis holoblastic, conidiophore growth sympodial; con. pigmented, multiseptate, filiform; distinguished by the narrow, thickened rim to the conidial scars; in *Cercospora* the thickening extends over the whole of the scar except for the small, central pore.
*****P. fijiensis,** and var. *difformis*, teleom. *Mycosphaerella fijiensis* and its var. *difformis*.
paradoxa, *Ceratocystis, Chalara, Ustilago* see *U. crus-galli*.
paragynous, having the antheridium at the side of the oogonium, Pythiaceae, cf. amphigynous; Ho, *Mycopathologia* **83**:119, 1983, in *Phytophthora* [**63**,1128].
Paralongidorus maximum (Bütschli) Siddiqi 1964; Heyns, *Descr.N.* 75, 1975; polyphagous, causes curling and galling of root tips, above ground stunt and death, mostly W. Europe.
Paraphaeosphaeria O. Eriksson 1967; Phaeosphaeriaceae; Sivanesan 1984; Shoemaker & Babcock, *CJB* **63**:1284, 1985; asco. straight, 2–10 septate, dark brown, punctate, ends broadly rounded; anam. in *Coniothyrium* [**65**,83].
P. michotii (Westend.) O. Eriksson; Morgan-Jones, *Descr.F.* 144, 1967, as *Leptosphaeria michotii* (Westend.) Sacc.; anam. *Coniothyrium scirpi* Trail 1889; asco. 2 septate, 14–23 × 3.5–6 μm, con. 5–13.5 × 3–5 μm; sugarcane leaf blast, a minor disease, on Gramineae; Shoemaker & Eriksson, *CJB* **45**, 1605, 1967, the teleom., Holliday 1980 as *L. michotii*; Sivanesan & Waller 1986.
Paraphelenchus myceliophthorus J.B. Goodey 1958; Hooper & Clark, *Descr.N.* 115, 1985; on the common, cultivated mushroom.
Para rubber, see rubber.
parasexual cycle, described by Pontecorvo & Roper in 1952; genetic recombination in filamentous fungi based on mitosis; it involves fusion of haploid nuclei in a heterokaryon, mitotic crossing over followed by haploidisation; Pontecorvo, *A.Rev.Microbiol.* **10**:393, 1956; Tinline & MacNeill, *A.R.Phytop.* **7**:147, 1969, in plant pathogens.
parasite, see obligate parasite.
parasitica, *Cryphonectria, Eutypella, Peronospora, Phialophora*.
parastolbur, see potato stolbur.
Paratanus exitiosus, transm. beet yellow wilt.
Paratrichodorus allius, transm. tobacco rattle, American isolates.

P. **anemones**, transm. pea early browning, English isolates; tobacco rattle, European isolates.

P. **christiei** (Allen) Siddiqi 1974; Heyns, *Descr.N.* 69, 1975; migratory ectoparasite, many hosts, attacked roots may lack fine feeder roots and become stubby; transm. tobacco rattle, American isolates.

P. **minor** (Colbran) Siddiqi 1974; Hooper, *Descr.N.* 103, 1977; migratory root ectoparasite, reported to transm. tobacco rattle.

P. **nanus**, transm. tobacco rattle, Dutch isolates.

P. **pachydermus** (Seinhorst) Siddiqi 1974; de Waele et al., *Descr.N.* 112, 1985; root ectoparasite, mostly Europe; transm. pea early browning, Dutch isolates; tobacco rattle, Dutch isolates.

P. **porosus**, transm. tobacco rattle, American isolates.

P. **teres**, transm. pea early browning, Dutch isolates.

Paratylenchus bukowinensis Micoletzky 1922; Andrássy, *Descr.N.* 79, 1976; sedentary, root ectoparasite, damages roots of umbelliferous crops, e.g. carrot and parsley, Europe.

P. **microdorus** Andrássy 1959; *Descr.N.* 108, 1985; no damage to crops reported.

P. **projectus** Jenkins 1956; Loof, *Descr.N.* 71, 1975; pin nematode, attacks roots of many crops, may cause stunting.

parksianum, *Peridermium*.

parrot feather, *Myriophyllum brasiliense* q.v.

parsley, *Petroselinum crispum*.

—— **carrot leaf** (chicory yellow mottle v.).

—— **chlorosis**, chlorotic mottle and necrotic spot (celery mosaic v.).

—— **green mottle v.** Avgelis & Quacquarelli, *Phytopath.Mediterranea* 13:1, 1974; filament 685–758 nm long, Italy (Apulia) [**55**,4233].

—— **latent v.** Bos, Huttinga & Maat, *Neth.J.Pl.Path.* 85:125, 1979; spherical *c*.27 nm diam., contains RNA, transm. sap, seed; Europe [**59**,1820].

—— **line pattern** (chicory yellow mottle v.), Vovlas, *Phytopath. Mediterranea* 17:193, 1978 [**60**,3285].

—— **rhabdovirus** Tomlinson & Webb 1974; particles *c*.214 × 87 nm, England [**54**,1537e].

—— **5 v.** Frowd & Tomlinson, *AAB* 72:177, 1972; poss. Potexvirus, filament *c*.500 nm long, transm. by *Cavariella aegopodii* in the presence of carrot mottle v., Britain [**52**,2340].

parsnip, *Pastinaca sativa*.

—— **mosaic v.** Murant, Munthe & Goold 1970; Murant, *Descr.V.* 91, 1972; Potyvirus, filament 730–760 × 14 nm, transm. sap, *Cavariella aegopodii*, C. *theobaldi*, *Myz.pers.*, non-persistent.

—— **mottle** Watson & Serjeant 1962–63; England [**41**:696; **42**:651].

*—— **yellow fleck v.** Murant & Goold 1968; Murant, *Descr.V.* 129, 1974; isometric *c*.30 nm, contains RNA, transm. sap, *Cavariella aegopodii*; depends on a helper, anthriscus yellow v., for aphid transm.; Elnagar & Murant, *AAB* 84:153, 169, 1976, relations of both viruses and vector; van Dijk & Bos, *Neth.J.Pl.Path.* 91:169, 1985, causing dieback of carrot, chervil, coriander, dill; in Netherlands [**56**,1049–50; **65**,951].

parthenium phyllody and witches' broom Varma et al., *Curr.Sci.* 43:349, 1974; Phatak et al., *Phytopath.Z.* 83:10, 1975; MLO assoc., India (N.), *P. hysterophorus*; Padmanabhan 1984, transm. *Hishimonus phycitis* [**54**,1203; **55**,1399; **64**,5282].

partial paralysis, olive, see bark measles.

parva, *Botryosphaeria*, see *B. dothidea*; *Mycosphaerella*, see *M. cryptica*; **parvum**, *Fusicoccum*.

pasmo (*Mycosphaerella linicola*) flax.

***paspali**, *Ascochyta, Claviceps, Marasmiellus*.

paspali-thunbergii, *Sorosporium*.

paspalum striate mosaic v., particles like a Geminivirus, transm. *Nesoclutha pallida*, fide Francki et al. 1985.

passage, an inoculation of a host with a parasite, or a substratum with a saprobe, and its re-isolation. It is often used in the sense of increasing or decreasing the virulence of a pathogen following successive passages; or, in the case of a bacterium, in selecting a new population, after FBPP.

passerinii, *Diplodina, Septoria*.

passiflorae, *Alternaria, Septoria*, see *S. passifloricola*.

Passiflora edulis, passion fruit; purple form *edulis*, yellow form *flavicarpa*; *P. quadrangularis*, giant granadilla; *P. ligularis*, sweet granadilla.

passiflora chlorotic spot van Velsen, *Papua New Guin.agric.J.* 13:160, 1961; transm. *Aphis gossypii*, non-persistent [**41**:163].

—— **latent v.** Schnepf & Brandes, *Phytopath.Z.* 43:102, 1961; Brandes & Wetter, *ibid* 49:61, 1963; rod *c*.650 × 13 nm, transm. sap, *P. coerulea*, *P. suberosa* [**43**,2366].

—— **mosaic v.** Gopo & Cavill, *Trans.Zimbabwe Scient.Assoc.* 61:27, 1982; isometric 18–20 nm diam., transm. sap to *P. allardii* only; Protsenko & Tamrazyan 1977 reported a rod 700 nm long in USSR [**57**,2152; **61**,6429].

—— **yellow vein mosaic** Wilson & Satyarajan, *Sci.Cult.* 36:224, 1970; India (S.), *P. foetida* [**50**,1840].

passifloricola, *Septoria*.

passion flower mosaic Protsenko & Tamrazyan 1977; assoc. particles 700 nm long, USSR, *Passiflora hybrida* [**57**,2152].

— **fruit**, *Passiflora edulis*: Kitajima et al., *Fitopat.Brasileira* **11**:409, 1986, viruses & MLOs.
—— **Brazilian mosaic v.** Chagas et al., *Fitopat.Brasileira* **9**:241, 1984; isometric 24 nm diam., transm. sap, Brazil (São Paulo) [**64**,1671].
—— **mosaic v.** Martini, *AAB* **50**:163, 1962; Del Rosario et al., *Philipp.Agric.* **48**:95, 1964; Ong & Ting, *MARDI Res.Bull.* **1**(1): 33, 1973; rod, transm. sap. *Aphis gossypii*, Malaysia, Nigeria, Philippines [**41**:535; **44**,2754; **53**,1031].
—— **Peruvian ringspot** (tomato ringspot v.), Koenig & Fribourg, *PD* **70**:244, 1986; caused apical necrosis in *Chenopodium*, but no systemic infection in tomato [**65**,4008].
—— **rhabdovirus** Pares, Martin & Morrison, *Australasian Pl.Path.* **12**:51, 1983; bacilliform *c.*250 × 68 nm; symptoms of, and with particles of, passion fruit woodiness v. [**63**,3477].
—— **ringspot v.** Wijs, *AAB* **77**:33, 1974; *Neth.J.Pl.Path.* **80**:133, 1974; prob. Potyvirus, filament 810–830 × 15 nm, transm. sap, *Aphis gossypii*, *A. spiraecola*, Ivory Coast, serologically related to passion fruit woodiness v. [**53**,3568; **54**,1825].
—— **tip necrosis** (cucumber mosaic v. + a mild str. of passion fruit woodiness v.), Pares et al., *Australasian Pl.Path.* **14**:76, 1985 [**65**,4481].
—— **vein clearing** Kitajima & Crestani, *Fitopat.Brasileira* **10**:681, 1985; bacilliform particles assoc., Brazil [**65**, 4482].
—— **woodiness v.** McKnight 1953; Taylor & Greber, *Descr.V.* 122, 1973; Potyvirus, filament 750 nm long, transm. sap, aphid, non-persistent; Australia, Brazil, Kenya, South Africa, Surinam; infects 44 spp. dicotyledons in 5 families; occurs naturally in some tropical legumes, transm. sap to bean and cowpea.
—— **yellow mosaic v.** Crestani et al., *Phytop.* **76**:951, 1986; Tymovirus, isometric *c.*30 nm diam., transm. *Diabotrica speciosa*, Brazil (Rio de Janeiro) [**66**, 1081].
Pasteur, Louis, 1822–95; born in France, one of the most famous of all biologists and who made a host of discoveries. He made no direct contribution to plant pathology; but by his work, and that of Koch, they finally laid to rest the already largely discredited doctrine of spontaneous generation; and showed that the germ theory of disease was true. The Institute Pasteur was opened in Paris in 1888. Copley and Rumford Medals of the Royal Society, 1848 and 1874; Foreign Member 1869. Biographies: Ducleaux, *Pasteur. Histoire d'un esprit*, 1896, English translation by E.F. Smith & F. Hedges 1920; Valléry-Radot, *La vie de Pasteur*, 1900, English translation by Devonshire 1911; Dubos,

Louis Pasteur, freelance of science, 1951; *Microbiological Sci.* **4**:285, 1987.
pasteurianus, *Acetobacter*; **pastinacae** *Itersonilia*.
Pastinaca sativa, parsnip.
patch, see bare patch; —— **canker** (*Phytophthora palmivora*) durian, rubber.
patchouli mosaic v. Subba Rao & Nagar, abs. *Phytop.* **76**:1109, 1986; poss. Potyvirus, particle 710 nm long, transm. sap, *Myz.pers.*, *Pogostemon patchouli*.
patch yellow (prob. *Fusarium oxysporum*) oil palm.
pathogen, an organism, often a micro-organism, a virus or viroid, which causes disease. It can be applied to a sp., race, str. or isolate. A pathogen may be avirulent and a host may be tolerant of a pathogen. **Pathogenesis** is the sequence of processes from host and pathogen contact to the complete syndrome; **pathogenic** is having the characteristics of a pathogen, and **pathogenicity** having the characteristics to cause disease. Cowan 1978 stated: 'Pathogenicity has always been difficult to determine and the so-called Koch's postulates were formulated in an attempt to define it. But, like insanity, it continues to defy description'. Virulence should not be considered as synonymous with pathogenicity, see FBPP; this guide does not recommend the use of horizontal and vertical pathogenicity, fide Robinson 1969.
Pathogenesis: in *A.R.Phytop.*: Rohringer & Samborski **5**:77, 1967, aromatic compounds; Watson, **8**:209, 1970, changes in virulence & population shifts; Matta, **9**:387, 1971, penetration & immunisation of uncongenial hosts; Schneider, **11**:119, 1973, cytological & histological aberrations in woody plants after infection by bacteria, flagellates, MLOs & viruses; Simons, **17**:75, 1979, modification of host & parasite interactions through artificial mutagenesis; Loomis & Adams, **21**:341, 1983, integrative analysis of host & pathogen relations. Also: Billing & Preece in Rhodes-Roberts & Skinner ed., *Bacteria and plants*, *Soc.appl.Bact.Sympos.Ser.* 10, 1982, entry, establishment & spread; Vanderplank, *Host–pathogen interactions in plant disease*, 1982; Wolfe & Knott, *Pl.Path.* **31**:79, 1982, pathogen populations, constraints on analysis of variation in pathogenicity; Callow 1983; Lebeda, *Phytopath.Z.* **110**:226, 1984, theory of host & parasite specificity; Groth & Bushnell, *Genetic basis of biochemical mechanisms of plant disease*, 1985; Day, *Pl.Path.* **35**: 263, 1986, a genetical perspective; Goodman et al., *The biochemistry and physiology of plant disease*, 1986.
pathometry, see phytopathometry.
pathotoxin, see toxin.
pathotype, a subdivision of a sp. distinguished by

common characters of pathogenicity, particularly in relation to host range, FBPP. Pathovar and f.sp. are synonymous; they are used for bacterial and fungal pathogens, respectively.

patterned midsummer wilt, Hymenomycetes assoc., *Poa* spp., USA; Pennypacker et al, *PD* **66**:419, 1982 [**61**,6442].

Patterson, Flora Wambaugh, 1847–1928; born in USA, Antioch and Wesleyan colleges, Ohio; USDA for > 25 years, built up the national fungus collections. *Phytop.* **19**:877, 1928; *Mycologia* **21**:1, 1929.

paulliniae, *Septoria.*

Paulownia, Hsieh, *Bull.Taiwan For.Res.Inst.* 388, 1983 [**64**,1291].

paulowniae, *Physalospora.*

paulownia witches' broom Tokushige 1950–1; Doi & Asuyama in Maramorosch & Raychaudhuri 1981; MLO, transm. *Halyomorpha* spp.; 6 cultivated *Paulownia* spp. all suceptible, an important disease.

pawpaw, see papaya.

PC toxin, host specific toxin from *Periconia circinata*; Scheffer & Pringle, *Nature Lond.* **191**:912, 1961 [**41**:225].

pea, *Pisum sativum*; Hagedorn 1984; Tu, *PD* **71**:9, 1987, root rot complex in Canada (Ontario).

— **and broad bean yellow dwarf** (milk vetch dwarf v.).

— **cyst nematode**, *Heterodera goettingiana.*

— **dwarf mosaic v.** Inouye 1965; *Rev.Pl.Prot.Res.* **2**:42, 1969; isometric 25 nm diam., transm. aphid spp., Japan [**45**, 2694a; **49**,1512].

— **early browning v.** Bos & van der Want 1962; Harrison, *Descr. V.* 120, 1973; Tobravirus, rod *c.*21 nm diam., 2 predominant lengths *c.*105, 215 nm; transm. sap, most isolates, to 30 spp. of dicotyledons in 10 families; transm. soil inhabiting nematodes, *Paratrichodorus*, *Trichodorus* spp.; seedborne, W. Europe, infects lucerne; Lockhart & Fischer, *Phytop.* **66**:1391, 1976, Moroccan isolate [**56**,3770].

— **enation mosaic v.** Osborn 1935; Peters, *Descr.V.* 257, 1982; Francki et al., vol. 2, 1985; isometric particles of 2 kinds, both 28–30 nm diam., but each containing one of the 2 species of linear ssRNA that comprise the genome, both are needed for infection; transm. sap. aphid spp. including *Acyrthosiphon pisum*, *Myz.pers.*, persistent, poss. seedborne, widespread in N. temperate zone, causes mosaics in broad bean, pea, sweet pea; Getz. et al., *Phytop.* **72**:1145, 1982, aphid transm. model [**62**, 942].

— **false leaf roll** Thottappilly & Schmutterer, *Z.PflKrankh. PflPath.PflSchutz.* **75**:1, 1968;

transm. sap, *Myz.pers.*, seedborne, prob. soilborne [**47**,1714].

— **fizzle top** = pea seedborne mosaic v.

— **green mottle v.** Valenta et al., *Acta Virol.* **13**:422, 1969; Comovirus, isometric *c.*25 nm diam.

— **leaf roll** = bean leaf roll v.

— — **rolling** (pea seedborne mosaic v.), also leaf rolling mosaic, leaf roll mosaic.

— — **veinal necrosis** (plantago mottle v.), Provvidenti & Granett, *AAB* **82**:85, 1976 [**55**,3799].

— **mild mosaic v.** Clark, *N.Z.J.agric.Res.* **15**:846, 1972; Comovirus, isometric 27 nm diam., transm. sap, seedborne in pea, New Zealand, infects a few legumes [**52**,2070].

— **mosaic** = bean yellow mosaic v.; — **mottle** = clover yellow mosaic v.; — **necrosis** = clover yellow vein v., fide Bos et al., *Neth.J.Pl.Path.* **83**:97, 1977 [**56**,5501].

— **necrosis mosaic v.** Sharma & Gupta, *Acta Bot.Indica* **6** Suppl.: 60, 1978; isometric 28–36 nm diam., India (Uttar Pradesh) [**59**,1502].

— **pimple pod** Wade, *Tasm. J.Agric.* **22**:40, 1951; **30**:311, 1959; transm. sap difficult, *Macrosiphum euphorbiae*, *Myz.pers.*, persistent, Australia (Tasmania) [**31**:3].

— **PO streak** = broad bean wilt v.

— **rhabdovirus** Caner, July & Vicente, *Summa Phytopathologica* **2**: 264, 1976; bacilliform 240 × 45 nm, transm. sap, Brazil (Minas Gerais), dwarfing, chlorotic mottle and leaf deformation [**56**,4764].

*— **seedborne mosaic v.** Musil 1966; Hampton & Mink, *Descr.V.* 146, 1975; Potyvirus, filament 770 × 12 nm, transm. sap, aphid spp., including *Acyrthosiphon pisum*, *Aphis craccivora*, *Rhopalosiphum padi*, non-persistent; Hampton et al., *Neth.J.Pl.Path.* **87**:1, 1981, host differentiation & serology; Hampton, *Phytop.* **72**:695, 1982, lentil str. [**60**,6146; **61**,7262].

— — **symptomless** Kowalska & Beczner 1980; a str. of broad bean stain v., L. Bos, personal communication, 1984 [**60**, 5799].

— **spotted wilt** (tomato spotted wilt v.), Prasada Rao et al., *Indian Phytopath.* **38**:90, 1985 [**66**,770].

— **streak v.** Zaumeyer 1938; Bos, *Descr.V.* 112, 1973; Carlavirus, filament, 630 nm long, transm. sap, *Acyrthosiphon pisum*, non-persistent; Europe, N. America; clover, lucerne, pea are infected naturally.

— **stunt** (red clover vein mosaic v.), *Phytopath.News* **19**:99, 1985.

— **symptomless** Mahmood et al., *Neth.J.Pl.Path.* **78**:204, 1972; str. of red clover mottle v. [**52**,1319].

— **tip yellowing** (bean leaf roll v.).

— **top necrosis v.** Musil & Lešková, *Ochr.Rost.*
9:259, 1973; isometric *c*.26–28 nm diam., transm.
sap, aphid spp., non-persistent, Czechoslovakia,
from clover. Prasada Rao et al. described a pea
top necrosis caused by tomato spotted wilt v., see
pea spotted wilt.
— — **yellows** (bean leaf roll v.).
— **wart** Gerasenkova & Makasheva, *Biol.Nauki*
12:99, 1969; transm. sap, USSR [**48**,1387].
— **Wisconsin streak** = pea streak v.
— — **stunt** = red clover vein mosaic v.
— **Y v.** Štefanac, Grbelja & Erić, *Acta bot.croat.*
40:35, 1981; Cucumovirus, Yugoslavia (Bosnia)
[**61**,3757].
peach, *Prunus persica*; Gilmer et al. 1976.
— **abnormal ripening** Mezzetti,
Phytopath.Mediterranea **1**:166 1962; Italy [**42**:32].
— **asteroid spot** Cochran & Smith, *Phytop.*
28:278, 1938; Williams et al. in Gilmer et al.
1976; Mexico, USA [**17**:609].
— **Australian decline** Doepel, McLean & Goss,
Aust.J.agric.Res. **30**:1089, 1979; Australia (W.)
[**60**,1529].
— **bark and wood grooving** Rosenberger & Jones,
Phytop. **66**:729, 1976; USA [**56**,769].
— **blotch** Willison, *Phytop.* **36**:273, 1946; Europe,
N. America [**25**:458].
— **calico** Blodgett, *Phytop.* **34**:650, 1944; Wagnon
et al. in Gilmer et al. 1976; widespread [**24**:65].
— **chlorotic spot** Wood, *N.Z.J.agric.Res.* **18**:255,
1975; New Zealand [**55**,1352].
— **dark green sunken mottle** Fry & Wood 1969;
Stubbs & Smith 1971 (apple chlorotic leaf spot
v.), P.R. Fridlund, personal communication, 1983
[**49**,2918; **51**,1644].
— **enation v.** Kishi, Abiko & Takanashi,
Ann.phytopath.Soc.Japan **39**:373, 1973; isometric
c.33 nm diam., transm. sap [**53**,4052].
— **grafting node necrosis** Sadovskiĭ 1986; bacteria
assoc., USSR [**66**,1056].
— **Grecian decline** Tsialis & Rumbos, *Acta
phytopath.Acad.Sci.hung.* **15**:279, 1980; poss.
MLO, Greece [**61**,1819].
— **latent mosaic** Desvignes, *Acta
phytopath.Acad.Sci.hung.* **15**: 183, 1980; France
[**61**,1803].
— **leaf vein browning** Sadovskiĭ & Meshcheryakov,
Mikrobiol. Zhurnal **46**:24, 1984; bacterium assoc.,
USSR [**63**,3992].
— **little peach** Bennett 1927; Jones et al., *Phytop.*
64:755, 1974; MLO assoc., USA, see peach X
[**54**,921].
— **mosaic** Hutchins, *Science N.Y.* **76**:123, 1932;
Gilmer et al. 1976; transm. *Eriophyes insidiosus*,
prob. restricted to Mexico, USA; has been called
peach American mosaic, presumably to

distinguish it from similar symptoms that occur in
Europe.
— **mottle**, Helton in Gilmer et al. 1976, USA.
— **Muir dwarf** Horne 1920 (prune dwarf v.), P.R.
Fridlund, personal communication, 1983.
— **necrotic leaf spot** (prunus necrotic ringspot v.).
— **oil blotch**, as for peach star mosaic.
— **phony** Hutchins 1921, 1929; Nyland et al. and
Hopkins et al., *Phytop.* **63**:1275, 1422, 1973;
caused by a xylem limited, Gram negative
bacterium q.v. Wells et al. described such a
bacterium, see *Xylella*, obtained from a peach
phony source. Peach phony is a serious disease
restricted to USA, not in California. In 1890 it
was causing severe losses in the S.E. peach
orchards. The trees become stunted and do not
show the leaf scorch symptom which is commonly
assoc. with some of the diseases caused by such a
bacterium. Trees do not die but form only small,
distorted fruit; Weaver et al., *PD* **64**:485, 1980,
Sorghum halepense as a poss. source of inoculum
in peach orchards; Wells et al., *Phytop.* **70**:817,
1980; **71**:1156, 1981, bacteria in perennial weeds,
transm. from diseased peach to plum and vice
versa [**8**:388; **53**,1881, 3557; **60**,1998, 2671;
61,2960].
— **red suture**, an old condition in USA and not
now of importance, see Klos in Gilmer et al.
1976. Maclean et al. 1984 decribed a suture red
spot in Elberta peach induced by fluoride
[**64**,5445].
— **ringspot** (prunus necrotic ringspot v.).
— **rosette** McClintock, *J.agric.Res.* **24**:307, 1923;
Kirkpatrick et al., *Phytop.* **65**:864, 1975;
KenKnight in Gilmer et al. 1976; MLO assoc., in
Prunus spp. and other plants, has caused serious
losses in peach in USA (S.E.), the disease does
not apparently occur outside USA [**3**:46; **55**,2798].
— — **and decline** (prune dwarf v. + prunus
necrotic ringspot v.), Fry & Wood, *Orchard. N.Z.*
42:397, 1969; Stubbs & Smith, *Aust.J.agric.Res.*
22:771, 1971; Smith et al., *ibid* **28**:103, 115, 441,
1977 [**49**,2918; **51**,1644; **56**,5111–13].
— **rosette mosaic v.** Cation 1933, 1942; Dias,
Descr.V. 150, 1975; Nepovirus, isometric *c*.28 nm
diam., transm. sap, *Xiphinema americanum*, USA
(mostly Michigan), causes grapevine degeneration;
Dias & Allen, *CJB* **58**:1747, 1980,
characterisation of the single protein & 2 nucleic
acids; Allen et al., *Can.J.Pl.Path.* **4**:16, 1982;
6:29, 1984, in Canada (Ontario), transm.
Longidorus diadecturus, *X. americanum*,
comparisons [**60**,3070; **61**,6487; **63**,3989].
— **seedling chlorosis** Fry & Wood,
N.Z.J.agric.Res. **16**:131, 1973; New Zealand
[**52**,3761].

—— **star mosaic** Kishi, Takanashi & Abiko, *Bull.hort.Res.Stn.Japan* A 12:197, 1973 [**54**,1351].
—— **stem pitting** (tomato ringspot v.), Smith et al., *Phytop.* **63**, 1404, 1973 [**53**,4514].
—— **stunt** (prune dwarf v.).
—— **tree short life**, see short life.
—— **wart** Blodgett, *Phytop.* **33**:21, 1943; in Gilmer et al. 1976; USA [**22**:213].
—— **willow leaf rosette** Scaramuzzi, *Notiz.Mal.Piante* 15:40, 1951; Italy [**31**:191].
—— **X** Anon., *Bull.Conn.agric.Exp.Stn.* 318, 1936; Nasu et al., 1970; Granett & Gilmer, *Phytop.* **61**:1036, 1971; Kloepper & Garrott, *ibid* **73**:357, 1983; MLO assoc.; the disease, widespread in N. America, is called eastern X or western X depending on the region, and called cherry X; *Prunus* spp. and herbaceous hosts are affected; transm. *Colladonus* spp., *Euscelidius variegatus*, *Scaphytopius acutus*; Gilmer & Blodgett in Gilmer et al. 1976; Gold in Maramorosch & Harris 1979, see vector; Sinha & Chiykowski, *Can.J.Pl.Path.* **2**:119, 1980; **6**:200, 1984, transm. by *S. acutus*, purification & serological detection; Gold & Sylvester, *Hilgardia* **50**, 1982, strs. & leafhopper transm. in differing light & temp. [**16**:21; **50**,1679; **51**,476; **60**,4567; **62**,1107, 3129; **64**,1512].
—— **yellow bud mosaic** (tomato ringspot v.).
—— —— **leaf roll** (str of peach X), for references see Purcell et al. *PD* **65**:365, 1981 [**61**,313].
—— —— **mosaic**, as for peach star mosaic.
—— —— **mottle** Wood, *N.Z.J.agric.Res.* **18**:255, 1975; New Zealand, from Moorpart apricot [**55**,1352].
—— **yellows** E.F. Smith 1888; Pine & Gilmer in Gilmer et al. 1976; Jones et al., *Phytop.* **64**:1154, 1974, MLO assoc. The disease has been known in USA since 1791. It was the subject of classical studies by Smith and, like other yellows diseases, was later thought to be due to a virus infection. The disease is also called little peach; it is not now of economic importance. Kunkel in 1933 showed that the causal agent was transm. by *Macropsis trimaculata*, the plum leafhopper [**54**,1200].
peacock's eye (*Spilocaea oleaginea*) olive.
*__peanut__, see groundnut.
—— **Chinese mild mottle v.**, see groundnut Chinese mild mottle v.
—— **chlorotic ring mottle v.** Fukumoto et al., *Ann.phytopath.Soc. Japan* **52**:496, 1986, Potyvirus, S.E. Asia, distant serological relationship with peanut mottle v. and 3 other viruses [**66**,1694].
—— **green mosaic v.** Sreenivasulu et al., *AAB* **98**:255, 1981; Potyvirus, filament *c*.750 nm long, transm. sap, *Aphis gossypii, Myz.pers.*, non-

persistent, India (Andhra Pradesh), serologically unrelated to peanut mottle v. [**60**,6738].
—— **Indian clump v.** Reddy et al., *AAB* **102**:305, 1983; rod, mostly 184 and 249 nm long, *c*.24 nm diam., transm. sap, soil; India (Punjab), severe stunting caused reported first in 1977, serologically distinct from peanut West African clump v.; Mayo & Reddy, *J.gen.Virol.* **66**:1347, 1985, translation products of RNA [**62**,3361].
—— **Indonesian mosaic** Thung 1946–7; Bergman, *Tijdschr.PlZiekt.* **62**:291, 1956; transm. *Orosius argentatus* [**36**:448].
—— **mottle v.** Kuhn 1965; Bock & Kuhn, *Descr.V.* 141, 1975; Potyvirus, filament 740–750 nm long, transm. sap, aphid spp., non-persistent, seedborne in groundnut, infects bean, lupin, pea, soybean; Demski et al., *PD* **65**:359, 1981; **67**:166, 1983, in forage legumes & epidemic in lupin; Rajeshawari et al., *Pl.Path.* **32**:197, 1983, characterisation of an Indian isolate [**61**,759; **62**,3093, 4547].
—— **proliferation** Muthusamy & Subramanian, *Curr.Sci.* **54**:1193, 1985; poss. MLO, India [**65**,3077].
—— **root knot nematode**, *Meloidogyne arenaria*.
—— **stripe v.** Demski et al. *AAB* **105**:495, 1984; Potyvirus, filament *c*.752 nm long, transm. sap, *Aphis craccivora*, non-persistent, serologically unrelated to peanut green mosaic v. and peanut mottle v., USA, from groundnut seed originating in China; Demski & Lovell, *PD* **69**:734, 1985, virus & seed distribution in USA [**64**,3619].
*—— **stunt v.** Troutman 1966; Mink, *Descr.V.* 92, 1972; Boswell & Gibbs 1983; Cucumovirus, isometric *c*.30 nm diam., transm. sap, aphid spp., non-persistent, wide host range; causes: epinasty, leaf distortion and stunt in bean; chlorotic mottle and stunt in Burley tobacco; systemic mottle in clovers and lucerne; Ahmed, *AAB* **109**:439, 1986, effect on growth & yield of broad bean in Sudan; Xu et al., *Phytop.* **76**:390, 1986, serotypes [**65**,4790; **66**,2596].
—— **West African clump v.** Thouvenel, Germani & Pfeiffer 1974; Thouvenel & Fauquet, *Descr.V.* 235, 1981; rod *c*.190 and 245 nm long, *c*.21 nm diam., contains ssRNA, transm. sap, seed, soil; prob. vector *Polymyxa graminis*; causes stunt but a yellowing str. does not; *Sorghum arundinaceum* is a symptomless, natural host.
pear, *Pyrus*; see apple, Porott et al.
—— **American stony pit** Kienholz, *Phytop.* **29**:260, 1939 [**18**:463].
—— **bark measles** Millecan et al., abs. *Phytop.* **52**:363, 1962; USA (California).
—— —— **split** Kegler 1966 [**45**,3362o].
—— **blister canker** Cropley, *Tech.Communic.Commonw.Bur.hort. Plantation*

Crops 30:103, 1963; also **bark necrosis** Posnette & Cropley, *J.hortic.Sci.* 33:289, 1958; Pleše et al., *Annls. Phytopath.* Hors-ser.:265, 1971.

— **brown leaf spot** Hidaka, *Tidsskr.PlAvl.* 65:60, 1961; Japan [41:465].

— **bud drop** Morvan 1961; *Annls Épiphytes* 16 Hors ser. 1:153, 1965; France, symptoms on cvs. Beurré Hardy & Doyenne du Comice; Trifonov, *Annls.Phytopath.* Hors ser.:285, 1971; see pear blister canker, Pleše et al.

— **corky pit** Keane & Welsh, *PDR* 44:636, 1960; cv. Flemish Beauty, Canada [40:115].

— **decline** McLarty 1948; Blodgett et al. and Shalla et al. 1963; Raju et al., *Phytop.* 73:350, 1983; MLO assoc., transm. *Psylla pyricola*, see Jensen et al., *ibid* 54:1346, 1964; a destructive disease, first found in Canada (British Columbia), which has killed many mature trees in Europe and N. America [42: 473; 43, 149; 44,1156; 62,3124].

— **deformation** Trifonov 1974; Bulgaria [55,4768].

— **European stony pit** Kristensen, *Tech.Communic.Commonw.Bur.hort.Plantation Crops* 30:97, 1963; Kegler et al., *Acta Hortic.* 67:209, 1976; carnation ringspot v. occurs in pear with stony pit; Kegler et al., *Arch.Phytopath.PflSchutz.* 13:297, 1977; Richter et al., *ibid* 14:411, 1978; it is not clear as to whether only a single pathogen is the cause [57,1773e, 3518; 58,5425].

— **fissure bark** Pares & Hutton, *Agric.Gaz. N.S.W.* 72:414, 1961 [41:158].

— **freckle pit** Wilkes & Welsh, *Can.Pl.Dis.Survey* 45:90, 1965; on cv. Anjou in Canada (British Columbia), symptoms distinct from pear European stony pit and they are not found in cvs. which are sensitive to stony pit [45,501].

— **latent** Waterworth 1965; *PDR* 55:983, 1971; transm. sap [51, 1633].

— **moria** = pear decline.

— **necrosis** Bălăşçuţă et al., *An.Inst.Cerc.Prot.Pl.* 15:51, 1979; Romania, linear necrosis of phloem and xylem in pear and quince [60,2077].

— **necrotic spot** Kishi, Takanashi & Abiko, *Acta Hortic.* 67:269, 1976; Japan [57,1774e].

— **pollen and corolla**, 2 isometric, serologically indistinguishable, viruses were isolated from these parts by Tomlinson et al., *Phytop.* 58:1026, 1968; transm. sap, caused symptoms in *Chenopodium*, cowpea, cucumber; the corolla isolate was sap transm. to pear seedlings but caused no symptoms [47,3493].

— **red leaf** Millican, Gotan & Nichols, *Bull.Calif.Dep.Agric.* 52:166, 1963 [43,1982].

— — **mottle**, see pear vein yellows.

— **ring pattern mosaic** (apple chlorotic leaf spot

v.); = pear green ring mosaic, — mosaic, — ring mosaic, — ringspot mosaic.

— **rough bark** Kristensen, *Tech.Communic.Commonw.Bur.hort. Plantation Crops* 30:107, 1963; Thomsen, *Acta Hortic.* 44: 119, 1975; similar to pear blister canker in symptoms and cvs. affected [55,3639a].

— **sooty ringspot** = quince sooty ringspot.

— **vein yellows and red mottle** Posnette, *Tech.Communic.Commonw. Bur.hort.Plantation Crops* 30:93, 1963; Fridlund, *PDR* 60: 891, 1976; Lemoine, *Annls.Phytopath.* 11:519, 1979, symptoms in quince. All isolates of red mottle also induce vein yellows but some isolates of the latter do not cause red mottle symptoms. It is not clear whether 2 distinct pathogens are involved.

pearl millet, see millets; Williams in Raychaudhuri & Verma ed. 1984, resistance [64,1530].

— — **mosaic v.** Seth, Raychaudhuri & Singh, *Indian J.agric.Sci.* 42:322, 1972; Singh & Seth, *Acta bot.ind.* 8:154, 1980; filament *c.*729 nm long, transm. sap, *Aphis gossypii, Rhopalosiphum maidis, Sitobion avenae*, non-persistent, Gramineae [52,716; 60,5905].

pecan nut, *Carya illoensis, C. pecan*; Payne et al., *Sci.Educ. Admin.Agric.Manuals Southern Ser.USDA* 5, 1979; Littrell & Bertrand, *PD* 65:769, 1981, fungicides & foliar pathogens.

— **bunch** Cole, *Phytop.* 27:604, 1937; Seliskar et al., *ibid* 64: 1269, 1974, MLO assoc., USA (S.) [16,717; 54,2986].

pectinases, pectinolytic enzymes; Bateman & Millar, *A.R.Phytop.* 4:119, 1966; Wood 1967; Byrde, *TBMS* 79:1, 1982; Collmer & Keen, *A.R.Phytop.* 26:383, 1986.

Pectobacterium Waldee 1945; Gram negative rod, motile, flagella peritrichous, or non-motile; type sp. *P. carotovorum* (Jones) Waldee. A valid name but not widely accepted for bacteria now placed in *Erwinia*; Bradbury 1986.

P. cypripedii (Hori) Brenner et al. 1973; nomenclaturally acceptable but see *Erwinia cypripedii. Pectobacterium rhapontici* (Millard) Patel & Kulkarni 1951; see *E. rhapontici*.

pedalium phyllody Padmanabhan et al. 1978; *J.Indian Bot.Soc.* 61: 95, 1982; MLO assoc., *P. murex* [62,2014].

pedicel lameness (*Coniella fragariae*) grapevine.

pedicillata, *Setosphaeria*, see genus.

Pedilophorus, transm. turnip yellow mosaic.

Peel's brimstone, name given by the Irish for the coarse, yellow maize meal imported from N. America by Robert Peel for food relief after the failure of the potato crops in Ireland in 1845 and later; see Irish famines, *Phytophthora infestans.*

peg, pod and root necrosis (*Calonectria pyrochroa*) groundnut.

pelargonii-zonalis, *Puccinia.*

Pelargonium, the garden geraniums; Stone, *Acta Hortic.* 110:177, 1980, viruses; Giboni 1985 [65,1347].

pelargonium curly top Kemp, *Can.Pl.Dis.Surv.* 41:265, 1961; Canada (Ontario) [41:230].

—— **flower break v.** Stone & Hollings 1973; Hollings & Stone, *Descr.V.* 130, 1974; Tombusvirus, isometric *c.*30 nm diam., transm. sap, England.

—— **latent** Kivilaan & Scheffler, *Phytop.* 49:282, 1959; USA (Michigan) [38:698].

—— **leaf curl v.** Pape 1927; Hollings, *AAB* 50:189, 1962; Hollings & Stone, *ibid* 56:87, 1965 (tomato bushy stunt v.) [41:717; 45,74].

—— **line pattern v.** Stone & Hollings 1977; Stone and Pleše & Štefanac, *Acta Hortic.* 110:177, 183, 1980; isometric *c.*30 nm diam., contains ssRNA, transm. sap; Štefanac et al., *Phytopath.Z.* 105:288, 1982, intracellular changes [62,1533].

—— **mosaic** Jones, *Bull.Wash.agric.Exp.Stn.* 390, 1940; USA (Washington State) [20:304].

—— **ring pattern v.** Stone, *Acta Hortic.* 36:113, 1974; isometric, England, prob. described earlier as pelargonium vein netting q.v. or ringspot q.v. by Stone & Hollings [51,2m; 56,961a].

—— **ringspot v.** Hollings, *Pl.Path.* 6:17, 1957; isometric, England, see pelargonium ring pattern v. [36:528].

—— **vein clearing v.** Russo, Castellano & Martelli, *Phytopath.Z.* 96:122, 1979; bacilliform 230 × 70 nm, Italy (S.) [59,3277].

—— —— **netting v.,** see pelargonium ring pattern v.; Pettersson, *Växtskyddsnotiser* 40:102, 1976 [56,255].

—— **yellow net vein** Reinert, Hildebrand & Beck 1961; Kemp, *Can.Pl.Dis.Surv.* 46:81, 1966; Canada, USA [45,3560].

—— —— **spot** Schmidt 1954; transm. aphid spp., Austria [33:706].

*—— **zonate spot v.** Martelli & Cirulli 1969; Gallitelli et al., *Descr.V.* 272, 1983; isometric 25–35 nm diam., contains ssRNA, transm. sap, in pollen and seed of *Nicotiana glutinosa,* Italy (S.); causes chlorotic rings, line patterns, malformation and necrosis in tomato leaves, found in *Chrysanthemum segetum* and globe artichoke.

pelleting, coating seed with inert material, often incorporating pesticides, to ensure uniform size and shape, FBPP.

pellucid ringspot, maize, Africa and poss. Papua New Guinea; Tanner, *Rhod.J.agric.Res.* 11:83, 1973, gave evidence that the condition, earlier thought to be caused by a virus, was due to Zn deficiency [52,3678].

penicillariae, *Tolyposporidium.*

penicillata, *Microsphaera*; **penicillatum,** *Dendryphion.*

Penicillium Link 1809; Hyphom.; Carmichael et al. 1980; conidiomata hyphal, conidiogenesis enteroblastic phialidic, conidiophores end in a branched structure like a brush, the penicillus, with 1 or more whorls of branches; con. aseptate, small, globose; cosmopolitan moulds, frequently greenish colonies; wound pathogens, causing decays of fruits and vegetables, important in contaminating stored cereal grain; Mislivic & Tuite, *Mycologia* 62:67, 75, 1970, in stored maize; Moore 1979; Pitt, *The genus* Penicillium *and its teleomorphic states* Eupenicillium *and* Talaromyces, 1979; *A laboratory guide to common* Penicillium *spp.* 1985; Caldwell et al., *Phytop.* 71:175, 1981, pathogenicity of 15 spp. to maize ears; Onions et al., *Smith's introduction to industrial mycology,* edn. 7, 1981; Ramirez, *Manual and atlas of the penicillia,* 1982; Hill & Lacy, *TBMS* 82:297, 1984, 33 spp. from barley grain [49,2842; 61,212; 63,2283].

P. aurantiogriseum Dierckx 1901; bulb rots of hyacinth, tulip and others, Moore 1979, as *P. cyclopium* Westl.

P. corymbiferum Westl. 1911; bulb rots mostly ornamentals; Smalley & Hansen, *Phytop.* 52:666, 1962, garlic decay; Saaltink, *Neth.J.Pl.Path.* 74:85, 1968, entry through wounds; Chauhan & Saaltink, *ibid* 75:197, 1969, infection of hyacinth, effects of temp. & relative humidity; Moore 1979; Sutton & Wale, *Pl.Path.* 34:566, 1985, control in crocus [42:170; 47,2500; 48,1760; 65,1334].

P. dangeardii, teleom. *Talaromyces flavus,* see genus.

P. digitatum Sacc. 1886; Onions, *Descr.F.* 96,1966; green mould of citrus fruits, commonly studied with *Penicillium italicum* on the same crops. *P. digitatum* forms a pasty, wrinkled mycelium which shows much in advance of the green, conidial mass; the decay margin has an indefinite border. Both fungi are mainly postharvest, wound pathogens. Green mould spreads less by fruit contact. There is a very large literature on the postharvest application of chemicals for control; tolerance of benzimidazole fungicides is widespread; Holliday 1980; Gutter et al., *Phytop.* 71:482, 1981; *Phytopath.Z.* 102:127, 1981, chemical control in strs. tolerant of benzimidazoles; Barmore & Brown, *Phytop.* 72:116, 1982, spread during fruit contact; Wild, *AAB* 103:237, 1983, double tolerance of benomyl & quazatine; Bancroft et al., *PD* 68:24, 1984, Californian lemons, control strategies for thiabendazole tolerant strs. [61,1193, 2883, 4976; 63,1281, 2337].

P. expansum Link; Onions, *Descr.F.* 97, 1966; blue mould of apple, generally isolated from mouldy fruits; Koffmann et al. 1978; Rosenberger et al., *PDR* 63:1033, 1979, control of strs. tolerant of benomyl [**58**,281; **59**,5257].

P. funiculosum Thom 1910; pineapple interfruitlet corking, leathery pocket and fruitlet core rot; and see *Gibberella fujikuroi* for the last disease; also on gladiolus, see Jackson, *Phytop.* 52:794, 1962, corky core rot; Rohrbach & Pfeiffer, *ibid* 66:392, 1976, on pineapple; Lim & Rohrbach, *ibid* 70:663, 1980, etiology & role of fungus strs. on pineapple [**42**:265; **56**,312; **60**,2109].

P. gladioli McCulloch & Thom 1928; Onions, *Descr.F.* 98, 1966; storage rot of gladiolus corms and other flower bulbs.

P. hirsutum Dierckx 1901; bulb and rhizome rots of hyacinth, iris, lily, tulip and others.

P. italicum Wehmer 1894; Onions, *Descr.F.* 99, 1966; blue mould of citrus fruits, often studied with the economically more important *Penicillium digitatum*. The powdery mycelium on the fruit surface does not extend much beyond the blue, conidial mass; the decay margin is well defined; has a greater tendency than green mould to spread by fruit contact, hence the name blue contact mould used by Fawcett 1936; Holliday 1980.

P. oxalicum Currie & Thom 1915; causes postharvest decays; Ricci et al., *Trop.Agric.Trin.* 56:41, 1979, on yam; Windels & Kommedahl, *Phytop.* 72:541, 1982, on pea; Kaiser & Hannan, *PD* 68:806, 1984, control of seed rot & damping off (*Pythium ultimum*) of chickpea [**58**,3605; **61**,6664; **64**,868].

P. sclerotigenum Yamamoto 1955; decay of yam tubers; Ogundana et al., *TBMS* 54:445, 1970, in Nigeria; Moura et al., *Fitopat. Brasileira* 1:67, 1976, in Brazil; Plumbley et al., *AAB* 106:277, 1985, imazalil & control [**37**:130; **50**,454; **56**,3378; **66**,432].

P. waksmanii Zaleski 1927; Dewey et al., *TBMS* 81:433, 1983, poss. cause of disease in the cultivated red alga, *Eucheuma striatum*, Caroline Islands, see *Scopulariopsis* [**63**,1912].

Peniophora Cooke 1879; Corticiaceae; like *Corticium* but having cystidia; *P. gigantea* (Fr.) Massee 1892 is used for the control of *Heterobasidion annosum*.

penniseti, *Cercospora, Plasmopara, Pyricularia.*

Pennisetum americanum, see millets.

pennisetum chlorotic stunt Slykhuis, *F.A.O.Pl.Prot.Bull.* 10:1, 1962; Australia (New South Wales), *P. clandestinum*, kikuyu grass [**42**:187].

Pentalonia nigronervosa, transm. banana bunchy top, cardamom foorkey, cardamom mosaic.

pentas mild mosaic (tomato ringspot v.), Chu et al., *Pl.Path.* 32:353, 1983; Australia (S. Australia), *P. lanceolata*; — **mosaic** (ribgrass mosaic v.) [**63**, 1302].

Pentatrichnopus, transm. strawberry mild yellow edge, strawberry mottle; *P. fragaefolii*, transm. strawberry Bulgarian yellows, strawberry chlorotic spot, strawberry epinasty, strawberry necrosis, strawberry vein necrosis; *P. jacobi*, transm. strawberry pseudo mild yellow edge; see the subgenus *Chaetosiphon.*

peony chlorosis Pisi & Marani, *Inftore.fitopatol.* 26(11–12):11, 1976; Italy [**56**,3074].

— **leaf curl** Brierley & Lorentz, *PDR* 41:691, 1957; USA (Maryland) [**37**:170].

— **ringspot v.** Dufrénoy 1934; Green 1935; Chang et al., *Ann. phytopath.Soc.Japan* 42:325, 1976; rod, 2 modal lengths 50–80 nm and 170–190 nm, 25 nm diam., prob. a str. of tobacco rattle v. [**14**:199; **15**:99; **56**,1632].

pepino, *Solanum muricatum.*

— **latent v.** Thomas, Mohamed & Fry, *AAB* 95:191, 1980; Carlavirus, filament 660–680 nm long, transm. sap, *Myz.pers.*, in Chile, from cuttings in New Zealand [**60**,1539].

— **mosaic v.** Jones, Koenig & Lesemann, *AAB* 94:61, 1980; Potexvirus, filament normally 508 nm long, transm. sap, contact, Peru, infects wild potato and potato cvs. [**59**,5539].

pepo, *Rosellinia.*

*****pepper**, see black and red pepper; the latter has some qualifying names, e.g. cayenne, chilli, bell, paprika, pimiento and sweet, all in *Capsicum* spp. They are shrubby perennials originating in the New World, and are mostly cultivated as herbaceous annuals. They are not the true pepper which is *Piper nigrum*; a woody, perennial vine indigenous in India (S.W.) and also of extremely ancient cultivation. The virus and MLO diseases listed below refer to *Capsicum* only. The viruses of *Piper* spp. are virtually unknown.

— **green vein banding** López Cardet & Blanco, *Revta.Agric.Cuba* 5:30, 1972; transm. sap, *Myz.pers.* [**54**,1083].

— **mild mosaic v.** Debrot, Lastra & Ladera, *Agronomía trop.* 30:85, 1980; Ladera et al., *Phytopath.Z.* 104:97, 1982; Potyvirus, filament 714 nm long, transm. *Myz.pers.*, non-persistent, Venezuela; considered to be serologically distinct from other Potyviruses in *Capsicum* [**62**,520; **63**,5704].

— — **mottle v.** Wetter et al., *Phytop.* 74:405, 1984; Tobamovirus, rod, normal length 312 nm, transm. sap, Italy (Sicily); distinguished from other members of the group by host range, symptoms and prob. amino acid composition [63,4201].

— **mottle v.** Nelson & Wheeler 1972; Nelson et al., *Descr. V.* 253, 1982; Potyvirus, filament *c.*737 nm long, transm. sap, aphid spp., USA, Central America, Solanaceae; often found in mixed infections with potato Y v. and tobacco etch v.

— **necrotic ringspot** Pop, *Product.Veget.Hort.* **28**(12):6, 1979; 1980; transm. sap, Romania [60,5771, 6766].

— **Puerto Rican mosaic** Roque & Adsuar, *J.Agric.Univ.P.Rico* **25**:40, 1941; Pérez & Adsuar, *ibid* **39**:165, 1955; transm. sap, *Myz.pers.*, Puerto Rico [21:401; 35:348].

— **ringspot and ring pattern** (tobacco rattle v.), Kitajima & Costa, *J.gen.Virol.* **4**:177, 1969; Silva & Nogueira 1979 [48,2263; 59,4759].

— **severe mosaic v.** Feldman & Gracia, *Phytopath.Z.* **89**:146, 1977; Potyvirus, filament 760 × 13 nm, transm. sap, Argentina, causes a severe loss in yield, serologically distinct from pepper mottle v.; Feldman & Gracia 1985 [56,5905; 65,5828].

— **spot** (*Leptosphaerulina trifolii*) lucerne and other plants, (*Sclerotinia trifoliorum*) clover, (*Stemphylium sarciniforme*) clover; **peppery spot** (*Pseudomonas syringae* pv. *maculicola*) brassicas.

— **spotting**, or pepper spot, white cabbage; a disorder whose cause is apparently not precisely known; called black speck in USA, grey speck in the Netherlands and spotted necrosis in USSR; Cox, *ADAS Q.Rev.* **25**:81, 1977. There is no evidence that a virus infection is assoc. with the condition, see turnip mosaic v., Walkey & Webb for larger necrotic lesions and vein streaking in cabbage.

— **veinal mottle v.** Brunt & Kenten 1971; *Descr.V.* 104, 1972; Potyvirus, filament *c.*770 × 12 nm, transm. sap, aphid, non-persistent, Ghana, causes a severe loss in yield; Lana et al., *Phytop.* **65**:1329, 1975; Lamptey & Bonsi, *Acta Hortic.* 53:227, 1977, the similar pepper veinal mosaic; Igwegbe & Waterworth, *Phytopath.Z.* **103**:9, 1982; Nigerian strs. & on eggplant; Givord, *PD* **66**:1081, 1982, in weed *Physalis angulata* in Ivory Coast; Sastry 1982, also in India [55,3386; 57,2395c; 61,5424, 6576; 62,838].

— — **necrosis** (potato Y v.), Rana et al., *Phytopath. Mediterranea* **10**:119, 1971 [51,1006].

— **vein banding** Dale, *AAB* **41**:240, 1954; transm. sap, *Aphis gossypii*, Trinidad, prob. like pepper Puerto Rican mosaic [34:278].

— **yellows** Iengo et al., *Inftore.Fitopatol.* **36**(3):35, 1986; MLO assoc., Italy [65,6222].

peppermint, *Mentha piperata*.

— **mosaic** Kacharmazov & Tanev, *Rasteniev.Nauki* **14**:114, 1977; Bulgaria [57,3554].

— **pale spot** Neubauer, *Preslia* **34**:193, 1963; Czechoslovakia [43,183].

percurrent, of a conidiogenous cell, growing straight on through the open end left by the secession of the preceding conidium.

Peregrinus maidis, transm. several maize viruses.

perenne, *Exobasidium*.

perennial canker, infection and host responses continue for > 1 season. Because no new wood is formed under such cankers the lesions appear to penetrate the wood deeply. This is due to the continued growth outside the limits of the lesion. Infection normally continues to be restricted mainly to the bark, after FBPP. — — (*Hypoxylon mammatum*) poplar, (*Pezicula malicorticis*) apple, pear (*Therrya piceae*) spruce.

— **ryegrass**, see ryegrass.

Perenniporia Murrill 1942; Polyporaceae; *P. fraxinophila* (Peck) Ryv., *Norwegian J.Bot.* **19**:143, 1972; Riffle et al., *PD* **68**:322, 1984, white mottle rot of heartwood of *Fraxinus pennsylvanica* in USA (Nebraska) [63,3570].

perfect state, the teleom. of a fungus; see fungi and holomorph.

pergularia mosaic v. Eranna & Nayudu, *Indian J.Microbiol.* **15**:191, 1975; Rajyalakshmi et al., *Curr.Sci.* **53**:332, 1984; filament *c.*730 nm long, transm. sap, *P. minor* [56,4148; 64,1148].

Periconia Tode 1791; Hyphom., Ellis 1971, 1976; Carmichael et al. 1980; conidiomata hyphal, conidiogenesis holoblastic; con. aseptate, pigmented, botryose, echinulate or verruculose; Bunning & Griffiths, *TBMS* **82**:397, 1984, development of conidium.

P. circinata (Mangin) Sacc. 1906; Ellis, *Descr.F.* 167, 1968, conidiophores circinate, con. 15–22 μm diam., milo of sorghum, a root and crown rot. *Periconia macrospinosa* Lefebvre & A.G. Johnson, *Mycologia* **41**:417, 1949, was also described from sorghum, con. 18–35 μm diam., Ellis loc.cit. 168; it is not pathogenic. Most disease is caused by the *P. circinata* isolates that form a host specific toxin. Early selection for resistance showed that host susceptibility was controlled by a dominant, major gene; Dunkle, *Phytop.* **69**:260, 1979, heterogeneous host reaction to toxin; Holliday 1980; Wolpert & Dunkle, *ibid* **70**: 872, 1980, partial characterisation of toxin; Dunkle & Wolpert, *Physiol.Pl.Path.* **18**:315, 1981, independence of symptoms & electrolyte leakage induced by toxin [59,1234; 60,4423; 61,704].

P. manihoticola (Vincens) Viégas 1955; con. 25–45 μm diam., leaf spot and blight of rubber, prob. only causes minor damage to seedlings; Stevenson

& Imle, *Mycologia* **37**:576, 1945, as the synonym *Periconia heveae* new sp.; Chen & Lee 1979 on cassava [**25**:9; **60**,2323].

periderm brown scorch (*Pythium megalacanthum*) Japanese radish.

Peridermium (Link) Schmidt & Kunze 1817; Uredinales; pycnia and aecia only, latter are white or yellow to orange, flattened or columnar blisters; a white membrane, the peridium is characteristic and encloses the aeciospores; telia in *Cronartium* and other rust genera, on gymnosperms.

P. appalachianum, teleom. *Cronartium appalachianum*.

P. bethelii Hedgcock & Long 1913; fide Hawksworth et al., *PD* **67**:729, 1983; on *Arceuthobium americanum*, dwarf mistletoe on *Pinus contorta*, lodgepole pine; apparently sometimes confused with the anam., *Peridermium pyriforme*, of *Cronartium comandrae*; *P. bethelii* has smaller, less pyriform aeciospores, 23.6–47.1 × 19.2 × 29.3 μm [**62**,5002].

P. cedri (Barclay) Sacc. 1886; telial host unknown, aecia on *Cedrus deodara*, causes witches' broom, India and Pakistan; Troup, *Indian Forester* **38**:222, 1912; **40**:469, 1914.

P. cerebrum, the following *Peridermium* spp. have teleom in *Cronartium*: **P. cerebrum** in *C. quercuum*, **P. comptoniae** in *C. comptoniae*, **P. cornui** in *C. flaccidum*, **P. himalayense** in *C. himalayense*, **P. occidentale** in *C. occidentale*, **P. pyriforme** in *C. comandrae*, **P. stalactiforme** in *C. coleosporioides*, **P. strobi** in *C. ribicola*.

P. coloradense, teleom. *Chrysomyxa arctostaphyli*; **Peridermium parksianum**, teleom. *C. piperiana*.

P. filamentosum Peck 1882; sometimes included in *Cronartium coleosporioides*.

P. harknessii, and **Peridermium pini**, see *Endocronartium*.

P. montanum, western form, teleom. *Coleosporium asterum*.

perilla mottle v. Lee et al., *Ann.phytopath.Soc.Japan* **46**:672, 1980; Potyvirus, filament 760 × 13 nm, transm. sap, *Myz.pers.*, non-persistent, *P. frutescens* var. *acuta* [**61**,1078].

periodicals, and other special publications, see bibliography.

periplasm, of Peronosporales, the outer layer of an antheridium or oogonium.

peristrophe mosaic Verma, Singh & Padma, *Acta phytopath.Acad. Sci.hung.* **10**:235, 1975; transm. *Aphis gossypii*, non-persistent, *P. bicalyculata*, India [**55**,4575].

perithecium, an ascoma that is shaped like a flask, sub-globose.

peritrichous, of rod shaped, motile bacteria where

the flagella tend to be inserted at any point around the cell.

periwinkle, *Vinca major*, *V. minor*; *Catharanthus roseus*, Madagascar periwinkle.

— **chlorotic stunt** Zaidi, Singh & Srivastava, *Curr.Sci.* **47**: 927, 1978; Zaidi et al., *Indian J.Pl.Path.* **2**:149, 1984; transm. aphid spp., more efficiently by *Aphis gossypii*, non-persistent, *Catharanthus roseus* [**58**,4971; **64**,4963].

— **wilt** McCoy et al., *PDR* **62**:1022, 1978; Davis et al., *Phytop.* **73**:1510, 1983; caused by a xylem limited, Gram negative bacterium; see *Xylella fastidiosa* and xylem limited bacteria; transm. *Homalodisca* sp., *Oncometropia nigricans*; did not cause symptoms of Pierce's leaf scald in grapevine; found in *Catharanthus roseus* and serologically related to the bacteria causing peach phony and plum leaf scald [**58**,5400; **63**,1827].

Perkinsiella, transm. sugarcane Fiji.

permeability changes, induced by a pathogen; Wheeler & Hanchey, *A.R.Phytop.* **6**:331, 1968; Bång 1983 [**62**,4110].

permitted tolerance, (1) the maximum content of toxicant allowed, e.g. of lead or arsenic, in foodstuffs for human consumption. (2) The maximum content of diseased plants allowed in a consignment moving in international or interstate trade, FBPP.

perniciosa, *Crinipellis*, *Mycogone*; **perniciosum**, *Sclerotium*.

Peronophythora Chen 1961 ex Ko et al., *Mycologia* **70**:381, 1978; the single, monotypic genus in the Peronophythoraceae; transitional between *Peronospora* and *Phytophthora*. *Peronophythora litchi*, authority as for genus; sporangia papillate, 27.7–33.3 × 18.5–22.2 μm, zoospore release as in *Phytophthora*; oogonia 27.8–33.3 × 24.1–29.6 μm, amphigynous, homothallic; blossom blight and fruit rot of *Litchi chinensis*; China, Papua New Guinea, Taiwan; Ann & Ko, *ibid* **72**:611, 1980, oospore germination; Kao & Leu, *ibid*, page 737, sporangium germination; Ho et al., *ibid* **76**:745, 1984, justification for retention in the genus; Chi et al., *Acta phytopath.sin.* **14**(2):133, 1984, infection [**42**:271; **58**,99; **60**,2697, 3854; **64**,689, 1198].

Peronophythoraceae, Peronosporales; distinguished from Pythiaceae by the generally determinate sporangiophores which are differentiated from the mycelium; *Peronophythora*.

Peronosclerospora (Ito) Shirai & K. Hara 1927, fide Shaw & Waterhouse, *Mycologia* **72**:425, 1980; Peronosporaceae. In 1913 Ito divided *Sclerospora* q.v. into 2 sections: those with a sporangium which germinated by zoospores, Eusclerospora; and those with a conidium germinating by a germ tube, *Peronosclerospora*. The latter became the new genus; it has few other morphological differences from *Sclerospora* which corresponds to Ito's Eusclerospora. The spp. are found on Gramineae in the sub-tropics and tropics, causing disease, downy mildews, particularly on maize, millets, sorghum, sugarcane. Infection takes place in the just germinated or young plant, sometimes directly from oospores in the soil. Later infection is from con. and through the leaves; these spores are air dispersed early in the day, mostly an hour or so before sunrise. The spread distance is generally short since the con. are extremely prone to desiccation. In very young plants, which can be killed, the infection becomes systemic; but in older ones it is localised. These fungi may be seedborne; internal infection of seed seems less common than external contamination. The oospores are the long survival units. Other hosts may be important in the epidemiology of a sp. in any crop.

Safeeulla 1976; Shaw, *Mycologia* **70**:594, 1978; Holliday 1980, as *Sclerospora*; Spencer 1981; Bonde, *Trop.Pest Management* **28**:49, 1982, epidemiology; Bonde et al., *Phytop.* **74**:1278, 1984, isozyme analysis; Sivanesan & Waller 1986; Duck et al., *ibid* **77**:438, 1987, sporulation in 3 spp. on maize [**58**,1075; **60**,738; **64**,1086].

P. heteropogoni Siradhana et al., *Curr.Sci.* **49**:316, 1980; con. 14.3–22.4 × 14.3–20.4 μm, oospores tuberculate, 24.5–36.7 μm diam., on *Heteropogon contortus*, infects maize but not sorghum in India (Rajasthan) [**60**,255].

P. maydis (Racib.) C.G. Shaw 1978; con. usually 23 × 18 μm, oospores unknown, the con. are smaller than those of *Peronosclerospora philippinensis*; Java downy mildew of maize, one of the important pathogens on the crop in Asia; Takahashi et al., *Phytop.* **71**:1133, 1981, distribution in shoot tips; Mikoshiba, *Tech.Bull.Trop.Agric.* 16, 1983 [**61**,2833, **64**,170].

P. miscanthi (T. Miyake) C.G. Shaw; con. 37–49 × 14–23 μm, oospores mostly 44–47 μm diam.; sugarcane leaf splitting, Fiji, Philippinnes, Taiwan, poss. Papua New Guinea; a minor disease where the leaves are converted into bundles of fibres, like whips.

P. philippinensis (Weston) C.G. Shaw; Holliday, *Descr.F.* 454, 1975, as *Sclerospora philippinensis* Weston; con. 27–39 × 17–21 μm, larger than those of *Peronosclerospora maydis*; oospores av. 19 μm diam., relatively small compared with those of other members of the genus and spp. in *Sclerospora*; restricted to parts of tropical E. Asia; a serious disease, Philippine downy mildew of maize; in this country this disease and Java

downy mildew were both called sleepy disease and were confused in the early literature; Josue & Exconde 1979, virulence; Kaneko & Aday 1980, inheritance of resistance; Bonde & Peterson, *Phytop.* **73**:875, 1983, grass hosts with *P. sacchari* [**60**,3699, 4412; **63**,128].

P. sacchari (T. Miyake) Shirai & K. Hara 1927, fide Sivanesan & Waller 1986; in *Mycologia* **70**:595, 1978 is *Peronosclerospora sacchari* (T. Miyake) C.G. Shaw comb.nov.; Mukerji & Holliday, *Descr.F.* 453, 1975, as *Sclerospora sacchari* Miyake; con. mostly 36 × 18 μm, oospores large, av. 50 μm diam., compared with other members of the genus; this downy mildew attacks both maize and sugarcane, passing from one crop to the other; on the latter the disease is called cane dew or leaf stripe; the distribution is limited to parts of S.E. Asia, including India; the pathogen has been virtually eradicated from Australia (Queensland); on maize the disease is serious in Taiwan; Bonde & Peterson, *PD* **65**:739, 1981, hosts in grasses; Poon et al., *Ann.phytopath.Soc. Japan* **48**:153, 162, 1982; *Pl.Prot.Bull.Taiwan* **24**:111, 121, 1982, history & disease patterns in Taiwan; Singh & Lal, *Trop.Pest Management* **31**:327, 1985, chemical control [**61**,3045, 7153–4; **62**, 1198–9; **65**,3273].

P. sorghi (Weston & Uppal) C.G. Shaw; Kenneth, *Descr.F.* 451, 1975, as *Sclerospora sorghi* Weston & Uppal; Francis & Williams, *ibid* 761, 1983; con. 15–23 × 15–19 μm, oospores 25–43 μm diam.; widespread on sorghum, and on maize; an important disease which has been intensively studied in India and USA, seedborne, races; Frederiksen, *PD* **64**:903, 1980, review; Rajasab et al., *Proc.Indian Natn.Sci.Acad.* B **46**:552, 1980, systemic infection; Prabhu et al., *ibid* **49**:459, 1983, seed penetration; Shetty & Safeeulla, *Proc.Indian Acad.Sci.* **90**:45, 465, 1981, conidial state, infection & environment; Tuleen et al., *Phytop.* **70**:905, 1980, cultural control; Craig and Janke et al., *ibid* **73**:1177, 1674, 1983, oospore infection, effects of deep tillage & roguing on oospore populations & disease incidence; Craig & Frederiksen, *PD* **67**:278, 1983, pathotypes; Schuh et al., *Phytop.* **77**:125, 1987, soil, inoculum & disease [**60**,4424,5903–4; **62**,3000; **63**, 1236, 1765, 4901; **66**,3790].

*Peronospora Corda 1837; Peronosporaceae; primary emergent conidiophore narrow, usually 8–10 μm across, dichotomously branched, branches ± reflexed, terminal ones sharp pointed, habit more graceful than in *Plasmopara*; con. have uniformly thick walls, germinating only by a germ tube; oogonial wall smooth. The distribution is temperate and warm temperate, hosts are mostly dicotyledons. Like other genera in the family, where the asexual spore does not form zoospores, the rate and distance of spread within a crop are much greater than in the genera where zoospores are formed. The dispersal peak for the con. is a few hours after sunrise, i.e. later than for *Peronosclerospora*; Holliday 1980; Spencer 1981.

P. anemones Tramier 1963; Francis & Griffin, *Descr.F.* 684, 1981; con. 20–28 × 16.6–23.6 μm, oospores 27–40 μm diam.; poss. widespread on anemone.

P. antirrhini Schröter 1874; Francis, *Descr.F.* 685, 1981; con. 23–30 × 14–18 μm, oospores 24–28 μm diam.; antirrhinum, poss. seedborne.

P. arborescens (Berk.) Casp. 1855; Francis, *Descr.F.* 686, 1981; con. 18–26 × 16–20 μm, oospores 42–48 μm diam.; opium poppy, poss. seedborne; Thakore et al., *Indian Phytopath.* 36:462, 1983, yield loss & chemical control [65,5670].

P. cytisi Rostrup 1892; Francis & Berrie, *Descr.F.* 762, 1983; con. 21–35 × 16–23 μm; oospores 28–35 μm diam.; laburnum, Europe.

P. destructor (Berk.) Casp. ex Berk. 1860; Palti, correct author, *Descr.F.* 456, 1975; con. 22–40 × 18–29 μm, relatively large for the genus; oospores 30–44 μm diam.; onion, *Allium* spp., prob. seedborne, severe downy mildew in cooler climates, infection can be systemic and arises from diseased bulbs; Virány in Spencer 1981; Hildebrand & Sutton, *Phytop.* 72:219, 1982, weather & epidemiology in Canada; Leach et al., *ibid* pages 881, 1052, conidial discharge; Bashi & Aylor, *ibid* 73:1135, 1983, survival of detached con.; Hildebrand & Sutton, *ibid* 74: 1444, 1984; *Can.J.Pl.Path.* 6:119, 127, 1984, weather & microclimate effects on sporulation, infection & conidial survival; Smith et al., *PD* 69:703, 1985; chemical control [61, 5413, 7269; 62,824; 63,1127, 5249–50; 64,1870; 65,1023].

P. dianthi de Bary 1863; Francis, *Descr.F.* 763, 1983; con. 25–30 × 15–18 μm, oospores 32–40 μm diam.; *Dianthus chinensis*.

P. dianthicola Barthelet 1946, no Latin diagnosis; Francis, *Descr.F.* 764, 1983; con. av. 23 × 17.8 μm, oospores 39 μm diam., carnation; America and Europe.

P. ducometi Siemaszko & Jankowska 1929; con. 23–25 × 15–16 μm; Tanaka 1934 has con. 20–30 × 14–18 μm, oospores 25–30 μm diam., as *Peronospora fagopyri* new sp. which is prob. a synonym; the fungus is distributed with crop, on buckwheat, disease incidence can be high in Canada, Zimmer did not find oospores, *PDR*

62:471, 1978; *Can.Pl.Dis.Surv.* 64:7, 25, 1984 [13:762; 58, 680; 64,1973, 2987].

P. farinosa (Fr.) Fr. 1849; Francis & Byford, *Descr.F.* 765, 1983, f.sp. *betae*; con. av. 25 × 19 μm, oospores 30–36 μm diam., also f.sp *spinaciae*. Both downy mildews of beet and spinach are seedborne and races occur on the latter crop; Byford in Spencer 1981; Inaba et al., *PD* 67:1139, 1983; *Phytop.* 74:214, 1984, seed transm. & heterothallism in f.sp. *spinaciae*; Frinking et al. and Frinking & Linders, *Neth.J.Pl.Path.* 91:215, 1985; 92:97, 107, 1986, oospore formation in f.sp. *spinaciae*, pathosystems on spinach & *Chenopodium album*, f.sp. *chenopodii* [63,3088, 3624; 65,2489; 66,365–6].

P. hyoscyami de Bary; considered by some to be the preferred name for the tobacco blue mould pathogen, *Peronospora tabacina*, which is usually used; as *P. hyoscyami* f.sp. *tabacina*; Skalický, *Acta Univ.Carol.*Biol.suppl. 2: 38, 1964; Shepherd, *TBMS* 55:253, 1970 [50,964].

P. lamii A. Braun in Rabenh. 1857; Francis, *Descr.F.* 688, 1981; con. 18–26 × 16–22 μm, oospores 30–38 μm diam.; on *Lamium*; has damaged *Ocimum basilicum*, sweet basil, in Uganda, and *Satureja hortensis*, savoury summer.

P. manshurica (Naoumov) H. Sydow in Gäumann 1923; Francis, *Descr.F.* 689, 1981; con. 20–26 × 18–21 μm; oospores 30–50 μm. Soybean downy mildew is a major and widespread disease, seed infested with oospores give rise to systemically infected seedlings, > 32 races have been delineated; Dunleavy in Spencer 1981; Sinclair 1982; Inaba et al., *Ann.phytopath.Soc.Japan* 45:468, 1979; 46,480, 533, 1980; 48:585, 1982; 49:252, 554, 1983, sexual stage biology; Lim et al., *PD* 68:71, 1984, race 33 in USA (Illinois) [59,4843; 60,5190–1; 62,2726; 63,318, 3138].

P. parasitica (Pers.:Fr.) Fr. 1849; con. 24–27 × 15–20 μm, oospores 30–40 μm diam.; plurivorous in Cruciferae and an important brassica downy mildew in cooler regions, seed transm. is in doubt; pathogenicity varies at the host spp. and generic level; but races, as determined by host cvs. within one sp. have prob. not been demonstrated; Holliday 1980; Channon in Spencer 1981; Kluzewski & Lucas, *Pl.Path.* 31:373, 1982; *TBMS* 81:591, 1983; rape isolates, infection in resistant & susceptible *Brassica* spp., oospore formation; Hartmann et al., *Can.J.Pl.Path.* 5:70, 1983, effects of water & temp. on con. formation & germination [62,2682; 63,956, 1458].

*P. sparsa Berk. 1862; Francis, *Descr.F.* 690, 1981; con. 18–24 × 16–20 μm, oospores 22–30 μm diam.;

on rose; Tate, *N.Z.J. Exp.agric.* **9**:371, 1981, on boysenberry in New Zealand [**61**,4245].

P. tabacina Adam 1933; the preferred name for some is *Peronospora hyoscyami* q.v.; con. 13–19 × 16–29 μm, oospores av. 46 μm diam.; tobacco blue mould is an old and still a major disease. Chemical control is obligatory when the crop environment is favourable for infection. Until 1958 the pathogen was confined to America and Australia, *Nicotiana* spp. are indigenous in both areas. In that year *P. tabacina* was reported from S. England and it consequently spread rapidly through the susceptible crops of Europe, N. Africa and W. Asia, causing enormous loss; it reached Iran in 1961–2. The Sahara has prevented spread to central and southern Africa; and the fungus is absent from most of the tropics where temp. conditions are unfavourable. It requires cool < 16° and moist conditions for epidemic spread which often begins in the seedbed if chemical control is ineffective. The young plants can be systemically infected. Davis et al., *PD* **65**:508, 1981, described an epidemic spread from Cuba and Jamaica in January 1980 northwards through N. America and reaching Canada (Ontario) in August. Another epidemic occurred in 1979 in Ontario. This was described graphically by McKeen, *PD* **65**: 8, 1981, who stated that it was '…an elegant example of the interactions of an aggressive and variable fungal pathogen, a susceptible host plant, the vagaries of the environment and the stupidity of man'. Enormous financial losses were incurred by growers, insurers and the state treasury. The episode is one more example in the history of plant pathology where even the most sophisticated measures against disease can break down.

Lucas 1975; Holliday 1980; Schiltz & Viennot-Bourgin in Spencer 1981; Gayed, *Lighter* **50** (1 & 3):5, 14, 1980; **52**(2):19, 1982, spread & loss in N. America; Aylor & Taylor, *Phytop.* **73**:525, 1983, escape of con. from a tobacco field; Rotem & Aylor, *ibid* **74**:309, 1984, autumn inoculum in USA (Connecticut); Nesmith, *PD* **68**:933, 1984, N. American warning system; Davis & Main, *ibid* **70**:490, 1986, applying atmospheric trajectory analysis to epidemiology [**59**,5357; **60**,1069, 6078; **61**,867, 7160; **62**,3649; **63**,4069; **64**,1250].

P. trifoliorum de Bary 1863; Francis, *Descr.F.* 768, 1983; con. 24–29 × 18–21 μm, oospores 20–30 μm diam.; on clover, lucerne and other forage legumes, temperate regions, host specific forms are known; Stuteville in Spencer 1981; Skinner & Stuteville, *Phytop.* **75**:119, 1985, inheritance of resistance, gene for gene relationship [**64**,2606].

P. valerianellae Fuckel in Sacc. **7**:253, 1888; con.

17–20 × 15–17 μm, on corn salad or lamb's lettuce, *Valerianella locusta*, seedborne; Champion & Mecheneau, *Seed Sci.Technol.* **7**:259, 1979 [**59**,1984].

P. viciae (Berk.) Casp. 1855; Mukerji, *Descr.F.* 455, 1975; con. 15–30 × 15–20 μm, oospores 25–37 μm diam.; causes a major pea disease, also on broad bean and *Lathyrus*, seedborne, systemic infection of seedlings, at least 8 races; some taxonomic confusion surrounds the nomenclature, see the account by Dixon in Spencer 1981. In USA the pathogen is referred to as *Peronospora pisi* Sydow. The fungi are morphologically the same and therefore *P. viciae*, based on *Botrytis viciae* Berk. 1846, is correct; Heydendorf & Hoffmann, *Z.PflKrank.PflSchutz.* **85**:561, 1978, 4 races; Holliday 1980; Dickinson & Singh, *Pl.Path.* **31**:333, 1982, colonisation & sporulation; Hagedorn 1984 [**58**,1446; **62**,2706].

Peronosporaceae, Peronosporales; the downy mildews, obligate parasites, sporangiophores or conidiophores, branched, growth determinate, differentiated from the mycelium, emerging singly or grouped, mostly through the stomata; sporangia or con. formed singly at branch tips, periplasm persistent and conspicuous, haustoria variable, usually branched; Spencer 1981; *Bremia, Bremiella, Peronosclerospora, Peronospora, Plasmopara, Pseudoperonospora, Sclerospora*.

Peronosporales, Oomycetes; zoospores not dimorphic, flagella 'laterally' attached; Waterhouse in Ainsworth et al. IVB, 1973; see Dick, under *Verrucalvus*, who reclassified the order; Shaw in Buczacki 1983.

perplexans, *Itersonilia*, see *I. pastinacae*.

Persea americana, avocado.

perseae, *Sphaceloma*.

persica, *Miuraea*; **persicae**, *Physalospora*; **persicaria**, *Gilbertella*.

persimmon, *Diospyros*.

—— mosaic v. Herbas, *Turrialba* **19**:480, 1969; isometric, transm. sap, Brazil (São Paulo); also a leaf fall, Mezzetti 1950, Italy [**29**:627; **49**,2942].

persistence of viruses, the length of time a virus is retained by a vector after being acquired from an infected plant. Persistence is a crucial feature which has been used to establish 3 categories with which other characteristics of transm. are assoc., i.e. viruses are called: persistent, semipersistent or nonpersistent q.v., according to whether they are retained by feeding vectors for weeks, days or hours, respectively, after FBPP. Persistent viruses are usually assoc. with highly specific virus and vector relationships; an uptake from the phloem requires *c*.15 minutes. There are 2 patterns: circulative and propagative. In the first the latent period is in hours and the virus is retained for some weeks. In the second this period is in days and the virus is retained continuously, and for life if it multiplies in the vector; hence persistent transm.; Sylvester, *A.Rev.Entomol.* **25**:257, 1980.

personata, *Cytospora.*

personatin, toxin from *Mycosphaerella berkeleyi*; Ramanujam & Swamy, *Phytopath.Mediterranea* **23**:63, 1984 [**64**,1851].

personatum, *Cercosporidium*; **persoonii,** *Leucostoma.*

perthophyte, a synonym of necrotroph.

Pesotum Crane & Schoknecht 1973; Hyphom.; Carmichael et al. 1980; conidiomata synnematal, conidiogenesis holoblastic, conidiophore growth sympodial; con. aseptate, hyaline; teleom. in *Ceratocystis*; *P. ulmi*, teleom. *C. ulmi.*

pest, an animal, usually an insect, mite or nematode, which damages animals and plants, or their products. Its use to cover other organisms, pathogens in a strict sense, is best avoided.

Pestalotia de Not. 1841, 1839 fide Sutton 1980; Coelom.; see Sutton, *CJB* **47**:2083, 1969, for a history of this genus and its relationships with *Pestalotiopsis.*

Pestalotiopsis Stey. 1949; Coelom.; conidiomata acervular; con. multiseptate, usually 5 cells, the 3 median ones are brown and the 2 end ones hyaline; simple or branched appendages; the basal cell has a shorter, usually simple, appendage; for *Pestalotia* versus *Pestalotiopsis* see Sutton 1980. The taxonomic uncertainty hardly affects plant pathologists since none of the spp. cause serious diseases. These are grey leaf spots, cankerous fruit lesions and diebacks. Mordue described 7 spp. in *Descr.F.* and referred to others: *Pestalotiopsis dichaeta* (Speg.) Stey. 1949; 675, 1980, on *Araucaria* and other hosts; *P. funerea* (Desm.) Stey.; 514, 1976, on conifers; *P. guepini* (Desm.) Stey.; 320, 1971, various hosts; *P. mangiferae* (Henn.) Stey.; 676, 1980, on mango and other hosts; *P. palmarum* (Cooke) Stey.; with Holliday, 319, 1971, on palms; *P. psidii* (Pat.) Mordue 1976; 515, 1976, on guava; *P. theae* (Saw.) Stey.; with Holliday, 318, 1971, on tea and other plants.

Pestalozzia, a later spelling of *Pestalotia.*

pesticide, a toxic chemical used against pests; sometimes, inadvisedly, used in an all embracing sense to include all chemicals used against bacteria, fungi and plants as well.

petal blight (*Itersonilia perplexans*) chrysanthemum.

Petch, Tom, 1870–1948; born in England, Univ. London; mycologist, Royal Botanic Gardens, Peradeniya, Ceylon, now Sri Lanka 1905–24; founder director, Tea Research Institute of Ceylon, 1925–8. Ainsworth, in his 1976 appreciation, called him one of the outstanding British workers of his generation in this field in the tropics. Petch's prolific output included work on the pathology of cacao, coconut, rubber and tea; the flora and fungi of Ceylon, and entomogenous fungi; wrote: *The physiology and diseases of* Hevea brasiliensis, 1911; *Diseases and pests of the rubber tree*, 1921; *Diseases of the tea bush*, 1923; with G.R. Bisby, *The fungi of Ceylon*, 1950, this last work includes 80 contributions by Petch. *Trop.Agriculturist* **104**:218, 1949; *Nature Lond.* **163**:202, 1949; *TBMS* **67**:179, 1976.

Pethybridge, George Herbert, 1871–1948; born in England, Univ. College of Wales, Royal College of Science, Dublin; Department of Agriculture, Ireland 1908–23; Ministry of Agriculture, UK 1923–36; original work on potato, flax and other crop diseases; Boyle Medal of the Royal Dublin Society 1921, OBE 1937. *Nature Lond.* **161**:1002, 1948; *TBMS* **33**:161, 1950; *Proc.R.Irish Acad.* B **76**:401, 1976.

petiole collapse, carrot, Ca deficiency; —— **crack,** marrow, parsnip, B deficiency; Scaif & Turner 1983.

Petri, Richard Julius, 1852–1921, curator of the Hygiene Museum, Berlin; one of Robert Koch's assistants who, in an advance on Koch's culture plates, devised the famous, and still universally used, Petri dish.

Petri dish, the simplest piece of microbiological equipment for culturing micro-organisms under sterile conditions, i.e. in pure culture. It consists of a bottom dish, flat, round, transparent, shallow with vertical sides; and an upper, identical dish of a slightly larger diam. which overlaps when placed over the lower one. Traditionally made of glass and then sterilised; now clear plastic, pre-sterilised dishes are very often used.

petroleum oils, see oils.

Petroselinum crispum, parsley.

petunia asteroid mosaic v. Lovisolo 1954; with Bode & Völk, *Phytopath.Z.* **53**:323, 1965; a str. of tomato bushy stunt v. [**45**,769].

—— **mosaics** Johnson, *Phytop.* **16**:141, 1926; USA (Wisconsin); Misra & Chenulu, *Indian Phytopath.* **19**:19 1966; transm. *Aphis gossypii*, Raizada et al., *Curr.Sci.* **48**:551, 1979; India, all transm. sap [**5**:509; **46**,3467; **59**,3286].

—— **mottle** Naqvi & Mahmood, *Geobios* **3**:149, 1976; *Curr.Sci.* **45**:707, 1976; transm. sap, aphid spp., non-persistent, India [**56**,2096, 2543].

—— **ringspot** (broad bean wilt v.), see nasturtium ringspot, Doel.

—— **vein clearing v.** Lesemann & Casper, *Phytop.* **63**:1118, 1973; prob. Caulimovirus, isometric 43–46 nm diam., ? Germany [**53**,1846].

— **yellow mottle** Rubio & Rosell, *Microbiología esp.* **12**:105, 1959; transm. sap, Spain [**39**:713].

Pezicula Tul. & C. Tul. 1865; Dermateaceae; ascomata ± stipitate, light coloured, usually pruinose, erumpent on wood or bark of shrubs or trees, asco. often 1–3 or more septate at maturity, hyaline; Groves, *Can.J.Res.* C **17**:125, 1939; Dennis, *Kew Bull.* **29**:158, 1974 [**18**:761].

P. alba Guthrie, *TBMS* **42**:504, 1959; anam. *Phlyctema vagabunda* Desm. 1847; asco. becoming 3–5 septate, 20–30 × 7–10 μm; con. 15–18 × 2.5–3 μm; apple bitter rot; Kennel & Weiler, *Z.PflKrankh.PflSchutz.* **91**:552, 1984; causing apple red lenticel [**39**:475; **64**,2629].

P. corticola (Jørg.) Nannf. 1932; anam. *Cryptosporiopsis corticola* (Edg.) Nannf. 1934; asco. becoming 1–3 septate, 17–29 × 7–9 μm; con. 30–32.5 × 9.5–10.5μm; apple surface canker.

P. malicorticis (H.S. Jackson) Nannf. 1932; poss. preferred name is *Neofabraea malicorticis* (Cordley) H.S. Jackson; anam. *Cryptosporiopsis malicorticis* (Cordley) Nannf.; asco. 11–23 × 4–9 μm, con. 11.5–16 × 3–4 μm; apple perennial canker, fruit bitter rot or ripe spot, pear bull's eye rot; intensively investigated in Europe, but is widespread; it is poss. that the teleom. would be better in *Neofabraea*, Sutton 1980; the anam. is often called *Gloeosporium perennans* Zeller & Childs, a synonym; for the teleom. see Wilkinson and Sharples, *TBMS* **28**:77, 1945; **42**:507, 1959; Corke, *J.hort.Sci.* **31**:272, 1956; **34**:85, 1959, seasonal variation & sporulation, infection of pruning wounds; Olsson, *Meddn.St.Växtsk.Anst.* **13**:104, 185, 1965, biology; Bompeix & Bondoux, *Annls.Phytopath.* **6**:1, 1974, pathogenicity; Bompeix, *Fruits* **28**:757, 863, 1973, factors affecting disease; Senula & Ficke, *Arch.Phytopath.PflSchutz.* **21**:183, 1985, bark necrosis, differences in virulence, most infection in autumn & winter through pruning cut & bark wounds [**25**:170; **36**:193, **38**:527; **39**:475; **45**,1842; **53**,3085; **54**,4003; **64**,5008].

*Pezizales, Ascomycotina; ascomata usually apothecioid, fleshy; asci unitunicate, apex not thickened, dehiscence typically by an operculum or slit; asco. aseptate, commonly sculptured.

Pezizella Fuckel 1870; Helotiaceae, tribe Helotioideae fide Dennis, *British Ascomycetes*, 1968.

P. oenotherae (Cooke & Ell.) Sacc. 1889; Sutton & Gibson, *Descr. F.* 535, 1977, as anam. *Hainesia lythri* (Desm.) Höhnel 1906; strawberry black lesion root rot, fruit rot and leafy spot; con. aseptate, 5–7.5 × 1.5–2 μm. Mass 1984 calls the disease tan brown rot, the teleom. is put in *Discohainesia* Nannf. 1932; on many plants;

Lundquist & Foreman 1986, leaf blight of eucalyptus seedlings in South Africa [**66**,326].

Pfeffinger, see cherry.

*Phacidiaceae, Helotiales; ascomata in a carbonaceous stroma which splits like teeth or a slit; like Rhytismataceae but asci amyloid; *Phacidium*.

Phacidiopycnis Potebnja 1912; Coelom.; Sutton 1980; conidiomata stromatic, conidiogenesis enteroblastic phialidic; con. aseptate, hyaline.

P. padwickii (Kheswalla) Sutton 1980; con. irregular, 10.5–14.5 × 4.5–12.5 μm, chickpea foot rot in India; Holliday 1980 as *Operculella* Kheswalla 1941; Singh & Bedi, *Indian Phytopath.* **32**:491, 1979 [**60**,4121].

P. pseudotsugae, teleom. *Potebniamyces coniferarum*.

P. tuberivora (Güssow & Foster) Sutton 1980; Punithalingam, *Descr.F.* 823, 1985; con. fusiform, 9.5–12 × 4.5–5.5 μm; potato tuber stem end hard rot, causes storage loss; Güssow & Foster, *Can.J.Res.* **6**:253, 1932, as *Phomopsis tuberivora*; Foster & Macleod, *ibid* **7**:520, 1932; Lester, *Pl.Path.* **5**:114, 1956 [**11**:671; **12**:241; **36**:121].

Phacidium Fr. 1823; Phacidiaceae; ascus pore amyloid; asco. aseptate, hyaline; on conifer needles.

P. coniferarum, see *Potebniamyces coniferarum*.

P. infestans P. Karsten 1888; Minter & Millar, *Descr.F.* 652, 1980; asco. 12–28 × 5–8 μm; on *Pinus sylvestris*, pine snow blight; differs from *Phacidium pini-cembrae* in the slightly larger asco. on av., ascomata on abaxial needle surface and always 8 asco. in an ascus; Turkey, USSR, other parts of Europe; reports of *P. infestans* on *Abies* and *Picea* in N. America are considered to be of another sp., *P. abietis*.

P. pini-cembrae (Rehm) Terrier 1942; Millar & Minter, *Descr.F.* 653, 1980; asco. 15–30 × 4–8 μm; on *Pinus cembra*, *P. siberica*, pine snow blight; ascomata on adaxial needle surface, sometimes asci have only 1–5 asco.; central Europe.

Phaeocryptopus Naumov 1915; Venturiaceae; ascomata borne singly on a stroma, usually dark, up to 150 μm diam.; asco. 1 septate; Butin, *Phytopath.Z.* **68**:269, 1970 [**50**,1435].

P. gaeumannii (Rhode) Petrak 1938; asco. 11–15 × 3.5–5 μm; Swiss needle cast of Douglas fir; Boyce, *Phytop.* **30**:649, 1940; Michaels & Chastagner, *PD* **68**:939, 942, 1984, distribution in USA (N.W.), disease severity & asco. release [**19**:736; **64**,1313–14].

P. nudus (Peck) Petrak 1938; asco. 11–15.5 × 4–5.5 μm; needle cast of *Abies*, *Pseudotsuga* and *Tsuga*; Uozumi, *J.Jap.For.Soc.* **41**:243, 1959 [**39**:508].

Phaeocytostroma Petrak 1921; Coelom.; Sutton, *Mycol.Pap.* 97, 1964; 1980; conidiomata stromatic, conidiogenesis enteroblastic phialidic; con. aseptate, pigmented; on grasses.

P. ambiguum (Mont.) Petrak 1927; con. 12–15 × 5.5–6.5 μm; causing a minor maize stalk rot; Smiljaković et al., *Zašt.Bilja* 30:47, 1979 [59,263].

P. iliau, teleom. *Clypeoporthe iliau.*

P. sacchari (Ell. & Ev.) B. Sutton 1964; Sutton & Waterston, *Descr.F.* 87, 1966; con. 11–14.5 × 3.5–5 μm; sugarcane rind or sour rot, assoc. with a stalk rot of mature canes growing under unfavourable conditions, canes that are over mature or weakened; Sivanesan & Waller 1986.

Phaeoisariopsis Ferraris 1909; Hyphom.; Ellis 1971, 1976; conidiomata synnematal, conidiogenesis holoblastic, conidiophore growth sympodial; con. filiform, pigmented, 3 or more septa; Emechebe, *AAB* 97:257, 1981, citrus brown spot in Nigeria [60,5448].

P. griseola (Sacc.) Ferraris; *Descr.F.* 847, 1986, no author; con. 3–6 septate, 30–70 μm long, 5–8 μm wide at the base; angular leaf spot of bean, Lima bean and other legumes, pathogenic variation, seedborne; Hocking, *PDR* 51:276, 1967, new virulent form in Tanzania; Holliday 1980; Schwartz et al., *PD* 65:494, 1981, loss in Colombia; Allen 1983 [46,2343; 61,927].

P. liriodendri, teleom. *Mycosphaerella tulipifera.*

P. magnoliae (Ell. & Harkn.) Jong & Morris 1968; Morgan-Jones & Brown, *Mycotaxon* 4:494, 1976; con. mostly 2 septate, 23–42.5 × 5–6.5 μm; Hodges & Haasis, *Mycologia* 54:448, 1962, magnolia leaf spot, as *Cercospora magnoliae* Ell. & Harkn. 1881, emended [42:497].

Phaeolus (Pat.) Pat. 1900; Hymenochaetaceae; basidiomata large, imbricate with a thick, short stipe; hymenophore tubular, context fleshy, brittle when dry.

P. schweinitzii Pat.; decay, butt or cubical rot of conifers; mostly when these follow hardwoods or in second rotation conifers; Phillips & Burdekin 1982; Barrett & Grieg and Barrett, *Eur.J.For.Path.* 15:412, 417, 1985, soil infestation by basidiospores, in soil around uninfected *Picea sitchensis* in UK (S.); Rossnev, *ibid* 15:66, 1985, on *Pinus peuce* [64,4532; 65,2465, 2479].

Phaeoramularia Muntañõla 1960; Hyphom.; Ellis 1971, 1976; Deighton, *Mycol.Pap.* 144, 1979; not recognised by Carmichael et al. 1980; conidiomata hyphal, conidiogenesis holoblastic, conidiophore growth sympodial; con. hyaline or olivaceous, 0–several septate.

P. capsicicola (Vassiljevsky) Deighton, *TBMS* 67:140, 1976, con. very commonly 1 septate but up to 2–5 septate, 12–92 × 3–7 μm; velvet spot of red pepper, can cause defoliation; Figueiredo et al., *Biológico* 49:45, 1983, severe in Brazil (São Paulo) when the crop is grown in polyethylene tunnels, chemical control [64,2872].

Phaeosphaeriaceae, Dothideales or Pleosporales; *Paraphaeosphaeria.*

phage, see bacteriophage.

Phakopsora Dietel 1895; Uredinales; teliospores aseptate, sessile, irregularly arranged in subepidermal crusts, sometimes combined with *Physopella*, aecidia unknown; Holliday 1980.

P. gossypii (Arthur) Hirats. 1955; Punithalingam, *Descr.F.* 172, 1968; teliospores 24–32 × 10–14 μm; one of the cotton rusts which can cause defoliation. It has been little investigated and therefore appears unimportant. The other rusts on the crop are *Puccinia cacabata* q.v. and *P. schedonnardi* Kellerm. & Sw. 1888, fide Arthur 1934. The latter has telia on Malvaceae and aecidia on Gramineae.

P. pachyrhizi H. Sydow & Sydow 1914; Anahosur & Waller, *Descr.F.* 589, 1978; microcylic, teliospores 18–30 × 6–12 μm, soybean rust, on many legume genera which include: *Cajanus, Canavalia, Crotalaria, Dolichos, Lupinus, Pachyrhizus, Phaseolus, Pueraria, Vigna.* The pathogen is widespread and has caused considerable damage in E. Asia. But USA, the largest producer of soybean, remains free from infection. Moderate temps. 18–26° are the most favourable for disease development. There are at least 4 races; and strs, from E. Asia can be more virulent than those from Africa, S. America and the West Indies; Bromfield, *Mongr.Am.phytopath.Soc.* 11, 1984; Burdon & Speer, *Euphytica* 33:891, 1984, differential hosts for race identification; Kuchler et al., *Phytop.* 74:916, 1984, economic consequences if spread to USA takes place [64,379; 65,2096].

phalaenopsis bacilliform = orchid fleck v.

phaseoli, *Elsinoë, Mycovellosiella, Phomopsis, Phytophthora, Synchytrium.*

phaseolina, *Macrophomina*; **phaseoli-radiati**, *Synchytrium.*

phaseolorum, *Ascochyta, Diaporthe.*

phaseolotoxin, or phaseotoxin, from *Pseudomonas syringae* pv. *phaseolicola*; see bacteria, Strobel 1977; Mitchell, *A.R.Phytop.* 22:225, 1984.

Phaseolus coccineus, runner bean; *P. lunatus*, Lima or butter bean; *P. vulgaris*, bean q.v.; for taxonomy see *Vigna*, Verdcourt.

Phellinus Quélet 1886; Hymenochaetaceae; Gilbertson, *Mycotaxon* 9:51, 1979, in N. America; basidiomata perennial, spp. formerly in *Fomes* or *Poria*; they cause a white rot decay of

living trees, mostly of the dead heartwood; a few
are active pathogens [59,665].

P. igniarius (L.:Fr.) Quélet; Pegler & Waterston,
Descr.F. 194, 1968; white heart rot of
dicotyledonous trees, a soft decay.

P. kawakamii Larsen, Lombard & Hodges,
Mycologia **77**:346, 1985; assoc. with a white
pocket rot of *Acacia* and *Casuarina* in USA
(Hawaii) [64,5507].

P. noxius (Corner) G. Cunn. 1965; Pegler &
Waterston, *Descr.F.* 195, 1968; plurivorous on
trees in the tropics; brown root rot, particularly
of rubber, and oil palm upper stem rot. The
rhizomorphs on the roots form a continuous skin,
brown internally, with the adhering soil particles
giving a characteristic rough appearance. This
distinguishes the fungus in the field from
Ganoderma philippii and *Rigidoporus lignosus*, 2
pathogens with comparable behaviour patterns
and with which *Phellinus noxius* has been studied
on rubber; Holliday 1980; Turner in Kranz et al.
1977, 1981; Bolland, *Aust.J.For.* **47**:2, 1984, root
rot of *Araucaria cunninghamii*, hoop pine, in
Australia (Queensland); Neil, *Eur.J.For.Path.*
16:274, 1986, on *Cordia alliodora* in Vanuatu
[64,5528; 66,1623].

P. pini (Thore:Fr.) Pilát 1942; red ring rot, also:
honeycomb, pecky rot, red heart, ring scale or
white speck; an important conifer heart rot in N.
America, also in N. Europe and Asia; Boyce
1961; Hepting 1971.

P. pomaceus (Pers.) Maire 1933; Pegler &
Waterston, *Descr.F.* 196, 1968; plum heart rot,
trees in Rosaceae, mostly *Prunus*.

P. robineae (Murr.) A. Ames 1913; on black locust;
Riffle et al., *PD* **69**:116, 1985, on *Robinia
pseudoacacia* in USA (Oklahoma) [64,2744].

P. robustus (P. Karsten) H. Bourdot & Galz. 1925;
Pegler & Waterston, *Descr.F.* 197, 1968; oak
yellow trunk rot, temperate conifers and
hardwoods, prob. does little damage.

phellophagy, attacking cork; Speer, *Mycotaxon*
21:235, 1984, *Melophia ophiospora* (Lév.) Sacc.,
from *Quercus suber* [64, 1286].

Phenacoccus aceris, transm. cherry little cherry.

phenolics, and plant disease, Friend 1981 [61,542].

2-phenylphenol, postharvest fungicide, e.g. on citrus.

phialidic, of conidiogenesis, where the conidium is
delimited by a new wall which is not derived from
any existing wall or layers of the conidiogenous
cell.

Phialophora Medlar 1915; Hyphom.; Ellis 1971,
1976; conidiomata hyphal, conidiogenesis
phialidic, flask shaped phialides have a collarette;
con. aseptate, hyaline, slimy; soil fungi; Schol-
Schwarz, *Persoonia* **6**:59, 1970; Cole & Hendrick,

Mycologia **65**:661, 1973, 6 wood inhabiting spp.
assoc. with blueing of wood in N. America; Gams
& Holubová-Jechová, *Stud.mycol.* 13, 1976, &
other genera; Walker, *Rev.Pl.Path.* **54**:113, 1975,
take all, Gramineae, see *P. radicicola*, Walker.

P. asteris (Dowson) Burge & Isaac 1974;
Hawksworth & Gibson *Descr.F.* 505, 1976; con.
in older cultures mostly $4-10 \times 2.3-5 \mu m$; aster
and sunflower wilt or yellows; Hoes, *Phytop.*
62:1088, 1972, on sunflower, leaf necrosis much
less than that caused by *Verticillium dahliae*, in
Canada; Tirilly & Moreau, *Bull.Soc.mycol.Fr.*
92:349, 1976, f.sp. *helianthi* [52,2343; 56,2620].

P. cinerescens (Wollenw.) v. Beyma 1940;
Hawksworth & Gibson, *Descr.F.* 503, 1976;
phialides in large clusters, con. in older cultures
$4-8 \times 2.5-3.5 \mu m$; carnation wilt or fan mould,
pathogenic variation.

P. graminicola (Deacon) J. Walker, *Mycotaxon*
11:90, 1980; see *Phialophora radicicola*, Walker;
sensu Scott, *TBMS* **55**:163, 1970, not *P.
radicicola* Cain, con. $5-11 \times 1.5-2.5 \mu m$, straight
or curved; Gramineae, studied in Britain since it
may protect roots against attack by
Gaeumannomyces graminis vars. *avenae* and *tritici*;
Smiley et al., *Phytop.* **75**:1160, 1985, described an
assoc. with patch of *Poa pratensis* [65,2343].

P. gregata (Allington & Chamberlain) W. Gams
1971; con. $3.5-4.8 \times 2.3-2.5 \mu m$; brown stem rot of
soybean and adzuki bean, systemic infection, the
soybean defoliating str. causes the typical leaf wilt
and necrosis; isolates differ in virulence and those
from these 2 hosts tend to be more virulent on the
host of their origin; the toxins formed by the
fungus are gregatins; Abel, *Rev.Pl.Path.* **56**:1065,
1977; Kobayashi et al., *Ann.phytopath.Soc.Japan*
45:409, 1979; **46**:253, 256, 1980; *PD* **67**:387,
1983 [59,4376; 60,4078, 4117; 62,3690].

*P. parasitica Ajello, Georg & Wang 1974;
Hawksworth & Gibson, *Descr.F.* 504, 1976; con.
$2.5-6 \times 1-2 \mu m$; Hawksworth et al., *TBMS*
66:427, 1976; assoc. with wilts and diebacks of
mature plants including: apricot, date palm,
grapevine, *Quercus virginiana* [56,626].

P. radicicola Cain 1952; see Walker in Asher &
Shipton ed., *Biology and control of take-all*, 1981,
for misapplications of name, history, morphology
& taxonomy.

P. tracheiphila, teleom. *Pseudopezicula tracheiphila*,
see *Pseudopeziza tracheiphila*.

P. zeicola Deacon & Scott, *TBMS* **81**:256, 1983,
q.v. for a comparison with similar fungi; con. of 2
types: rounded ends, not strongly curved,
$6-20 \times 1.5-6 \mu m$; falcate, $5-9 \times 1-1.5 \mu m$; assoc.
with a maize root and stalk rot, France and South
Africa, prob. widespread [63,1757].

philippii, *Ganoderma*.
Philippine downy mildew (*Peronosclerospora philippinensis*) maize.
philippinensis, *Peronosclerospora*.
phlei, *Cladosporium*.
phleichrome, non-host specific toxin formed by *Cladosporium phlei* q.v.
phleum green stripe Bremmer, *Ann.Agric.Fenn.* **13**:129, 1974; Heikinheimo & Raatikainen, *ibid* **15**:34, 1976; transm. *Megadelphax sordidula*, Finland, cereals [**55**,2181; **56**,608].
— **mottle v.** Catherall, *Pl.Path.* **19**:101, 1970; Benigno & A'Brook, *ibid* **21**:142, 1972; *AAB* **72**:43, 1972; Catherall & Chamberlain, *ibid* **87**:147, 1977; isometric *c*.30 nm diam., contains RNA, transm. sap, *Oulema lichensis, O. melanopus*, Gramineae; strs. are: brome stem leaf mottle, cocksfoot mild mosaic, festuca mottle, holcus transitory mottle [**50**,1277; **52**,1072, 1581; **57**,1756].
phloem canker (*Erwinia rubrifaciens*) European walnut.
— **limited bacteria**, see xylem limited bacteria; these organisms appear to resemble those limited to the xylem but they have been much less studied. They are assoc. with disease symptoms which include: flower virescence, leaf curl, premature death, stunt, yellows; some are transm. by leafhoppers; Hopkins, *A.R.Phytop.* **15**:277, 1977, referred to such bacteria assoc. with: citrus greening, clover club leaf, dodder stunt, potato leaflet stunt, rugose leaf curl, sida little leaf.
— **necrosis** (*Phytomonas*) coffee.
Phloeospora Wallr. 1833; Coelom.; Sutton 1980; conidiomata acervular, conidiogenesis holoblastic, conidiophore growth percurrent, con. multiseptate; **P. maculans**, teleom. *Mycosphaerella mori*; **P. ulmi**, teleom. *M. ulmi*.
phlox phyllody Misra, Sharma & Cousin, *Int.J.Trop.Pl.Dis.* **3**:7, 1985; MLO assoc., India, *P. drummondii*; Zajak 1978, Poland [**58**,5909; **65**,3957].
Phlyctema Desm. 1847; Coelom.; Sutton 1980; conidiomata stromatic, conidiogenesis enteroblastic phialidic; con. aseptate, hyaline; **P. vagabunda**, teleom. *Pezicula alba*.
Phoenix dactylifera, date palm.
Phoma Sacc. 1880; Coelom.; Sutton 1980; conidiomata pycnidial, wall thin; conidiogenesis enteroblastic phialidic; con. typically aseptate, hyaline. This taxonomically difficult genus has been studied in depth by Boerema and associates, references in *TBMS* **67**:289, 1976. There are few workable morphological criteria to delimit spp.; in vitro characteristics are not entirely satisfactory. *Phoma* should not be confused with *Ascochyta*;

the latter has typically 1 septate con. Many spp. are of little or no importance to plant pathologists; several are saprophytic and are found in soil; teleom. in *Didymella, Leptosphaeria, Mycosphaerella, Pleospora, Pyrenochaeta*; Holliday 1980; Johnston, *N.Z.J.Bot.* **19**:173, 1981, on grasses & pasture legumes.
P. aquilina, see *Pteridium aquilinum*.
P. betae, teleom. *Pleospora betae*.
P. caricae-papayae, teleom. *Mycosphaerella caricae*.
P. chrysanthemicola Hollós 1907; emended by Schneider & Plate, *Phytopath.Z.* **67**:97, 1970; con. 5–6 × 1.8–2 μm; chrysanthemum collar rot, a disease of the cortex; Kemp, *Can.J.Pl.Sci.* **38**:464, 1958; Hawkins et al., *Pl.Path.* **12**:21, 1963; Schneider & Boerema, *Phytopath.Z.* **83**:239, 1975 [**38**:211; **42**:465; **49**,2501; **55**,1283].
P. complanata (Tode:Fr.) Desm. 1851; con. 5.5–9 × 3.5 μm; parsnip canker, prob. seedborne; Cerkauska, *Can.J.Pl.Path.* **7**:135, 1985; **9**:63, 1987 [**64**,5538].
***P. destructiva** Plowr. 1881, emended Jamieson, *J.agric.Res.* **4**:19, 1915; con. 3.5–5 × 2 μm; close to *Phoma lycopersici* whose con. are larger and there are in vitro differences, see teleom. *Didymella lycopersici* of the latter. *P. destructiva* causes a potato leaf spot not unlike the spots caused by *Alternaria solani*, and a tomato fruit rot, see phomenone; Tisdale, *Bull. Fla.agric.Exp.Stn.* 308, 1937 [**17**:139].
P. epicoccina, see *Epicoccum purpurascens*.
P. eupyrena Sacc. 1879; con. 3–5 × 1.5–2 μm; assoc. with a needle cast of *Abies magnifica* and Douglas fir; Kliejunas et al., *PD* **69**:773, 1985, in a nursery, USA (California) [**65**,921].
P. exigua Desm. 1849; con. 5.5–10 × 2.5–3.5 μm, colonies with a scalloped or lobed margin, NaOH turns agar blue green; a general, often a wound, pathogen which also exists as ± specialised, pathogenic strs. The most important of these is var. *foveata*, causing potato gangrene which is a tuber rot in storage; var. *exigua* may also cause gangrene but is less pathogenic. Other vars. are: *diversispora*, causing severe damage to bean crops in Germany and the Netherlands in 1979, seedborne; *inoxydabilis* on *Vinca* and other plants; *lilacis* on lilac causing damping off; *linicola*, damping off and foot rot of flax and other plants; *sambuci-nigrae* on *Sambucus nigra*. *Phoma exigua* was described by von Aderkas & Brewer, *Can.J.Pl.Path.* **5**:164, 1983, as causing a midrib rot of *Matteuccia struthiopteris*, ostrich fern, which is a spring vegetable in Canada (New Brunswick) and USA (Maine). The croziers or fiddle heads become unsaleable; see potato, Boyd 1972; Rich 1983; Tichelaar, *Neth. J.Pl.Path.*

80:1691, 1974, differences between vars. *exigua* & *foveata*; Boerema et al., *Phytopath.Mediterranea* 18:105, 1979; *Z.PflKrank.PflSchutz.* 88:597, 1981, vars. [54,2426; 61,3145; 62,4699; 63,2638].
P. glumarum Ell. & Tracy 1888, fide Boerema et al., *Persoonia* 6:174, 1971; con. 3–4 μm long; incorrectly called *Phyllosticta glumarum* (Ell. & Tracy) Miyake 1910; common in tropics and subtropics, rice glume blight; Singh et al., *Indian Phytopath.* 31:419, 1978 [59,3756].
P. lingam, teleom. *Leptosphaeria maculans.*
P. lycopersici, teleom. *Didymella lycopersici.*
P. macdonaldii, teleom. *Leptosphaeria lindquistii.*
P. macrostoma Mont. 1849; con. 5.5–10.5 × 2.5–3.5 μm, very variable; Sidhu & Singh, *PDR* 63:878, 1979, leaf spot of jujube, *Zizyphus mauritiana*, in India (Punjab) [59,2274].
*P. medicaginis Malbr. & Roum. 1886; var. *medicaginis*; con. 6–11.5 × 2–3.5 μm, often becoming 1 septate, chlamydospores unicellular, colonies uniform, sparse formation of crystals on malt agar; lucerne spring black stem, N. temperate, seedborne. Var. *pinodella* (L.K. Jones) Boerema 1965; Punithalingam & Gibson, *Descr.F.* 518, 1976; con. 7–10 × 2.5–4 μm, often becoming 1 septate, chlamydospores unicellular, colonies variable, abundant formation of crystals on malt agar; clover and pea black stem or summer black stem; a less severe pathogen than 2 others on pea: *Ascochyta pisi* and *Mycosphaerella pinodes*, with which *Phoma medicaginis* var. *pinodella* has often been studied; Holliday 1980; Fukushiro & Furata, *Bull.Chogoku Natn.agric.Exp.Stn.* E 22:21, 1985, on lucerne in Japan.
P. nigrificans, teleom. *Didymella macropodii.*
P. sclerotioides Preuss ex Sacc. 1892; Boerema & Loerakker, *TBMS* 84:297, 1985; Boerema & van Kesteren, *Persoonia* 11:317, 1981, taxonomy & morphology, often called *Plenodomus meliloti* in plant pathology literature; scleroplectenchymatous pycnidia, con. 4.5–6.5 × 2–3 μm; on roots of forage legumes, assoc. with root rots in temperate regions with severe winters; Harris, *Australasian Pl.Path.* 15:14, 1986, assoc. with root diseases of cereals, legumes & weeds in Australia (S.) [66,1405].
P. sorghina (Sacc.) Boerema, Dorenbosch & v. Kest., *Persoonia* 7:134, 1973; Punithalingam, *Descr.F.* 825, 1985; con. 4–7 × 2 μm; *Phoma insidiosa* Tassi; Punithalingam & Holliday, *Descr.F.* 333, 1972, is a synonym; *Mycosphaerella holci* q.v. is not the teleom. and *P. glumarum* q.v. is a misapplied name; on grasses, seedborne, causing damping off and rice glume blight, common in tropics and sub-tropics.
P. telephii (Vestergr.) v. Kest., *Neth.J.Pl.Path.*

78:117, 1972; con. 4–8 × 2–3 μm; purple blotch of *Sedum* [52,162].
P. tracheiphila (Petri) Kantschaveli & Gikashvili 1948; *Index Fungi* 4:417, 1976; this transfer was made from *Deuterophoma tracheiphila* Petri and was accepted by Sutton 1980. As there may be doubt as to whether the fungus, that causes the serious citrus mal secco, is a *Phoma*, the disease is described under Petri's name which is often used.
P.t.f.sp. chrysanthemi Baker et al., *CJB* 63:1730, 1985; chrysanthemum, in the vascular system, does not infect rough lemon or sour orange; was common in USA (California) 1948–56, but is now controlled by planting healthy cuttings in fumigated soil; Taylor, *Aust.J.Exp.Agric.Husb,* 290, 1962, in Australia [41:715; 65,745].
P. valerinellae Gindrat, Semecnik & Bolay 1966; con. 4.5–5.5 × 1.5–2.5 μm; damping off of corn salad or lamb's lettuce, *Valerianella olitoria*, Europe; Boerema & de Jong, *Phytopath.Z.* 61:362, 1968; Vegh et al., *Revue hort.* 184:39, 1978; Nathaniels, *Pl.Path.* 34:449, 1986 [57,3729].
P. wasabiae, see internal black rot.
phomenone, toxin formed by *Phoma destructiva*; Bousquet, *Annls. Phytopath.* 5:289, 1973; Riche et al., *Tetrahedron Letters* 32:2765, 1974; Bottalico et al., *Phytopath.Mediterranea* 22:116, 1983; Iacobellis & Bottalico and Capasso et al., *ibid* 24:307, 311, 1985 [54,1315; 55,5043; 63,1956; 66,1137–8].
*Phomopsis (Sacc.) Bubák 1905; Coelom.; Sutton 1980 who stated that the genus requires revision; conidiomata stromatic, conidiogenesis enteroblastic phialidic; con. aseptate, hyaline, of 2 types: alpha or A, fusiform, straight; beta or B, filiform, straight or, more often, hamate i.e. curved at one end; anam. in Diaporthaceae; Punithalingam, *TBMS* 60:157, 1973; 63:229, 1974; 64:427, 1975, gave some spp. descriptions; some cause cankers.
P. alnea (Sacc.) Höhnel 1906; A con. 7–10 × 2–3 μm, B con. 20–26 × 1 μm; oak & Dorset, *PD* 67:691, 1983, causing a basal stem canker of *Alnus glutinosa* in USA (Kentucky) [62,4473].
P. anacardii Early & Punith. 1972; Punithalingam, *Descr.F.* 826, 1985; A con. 6–8 × 2–2.5 μm, B con. 22–26 × 0.5 μ; cashew dieback, prob. a minor disease.
P. capsici, teleom. *Diaporthe capsici.*
P. caricae-papayae Petrak & Cif. 1930; Punithalingam, *Descr.F.* 827, 1985; A con. 5–7 × 2–2.5 μm, B con. 20–27 × 0.5 μm; papaya fruit and stem rot, severe losses reported from India, *Phomopsis papayae* Gonz., Frag. & Cif. was not validily published.
P. castanea, teleom, *Amphiporthe castanea.*

P. cinerascens (Sacc.) Trav. 1906; poss. teleom. *Diaporthe cinerascens* Sacc. 1875; assoc. with a canker of *Ficus benjamina*, weeping fig, in Canada (Newfoundland); Hampson, *Can. Pl.Dis.Surv.* **61**:3, 1981 [**61**,3100].

P. citri, teleom. *Diaporthe citri*.

P. cucurbitae McKeen 1957; Punithalingam & Holliday, *Descr.F.* 469, 1975; A con. 8–12 × 2.5–3 μm, B con. 18–26 × 1 μm; resembles anam. of *Diaporthe melonis*; cucumber black rot, first described from Canada; Atkinson, *Can.J.Pl.Sci.* **60**:747, 1980, control with soil drenches [**60**,535].

P. elaeagni (Carter & Sacamano) R.H. Arnold & Carter, *Mycologia* **66**:193, 1974; A con. 5.5–11 × 1.5–2 μm, B con. 15–20 × 0.7–1 μm; canker of *Elaeagnus angustifolia*, Russian olive; Canada and USA; Carter & Sacamano, *ibid* **59**:535, 1967, as *Fusicoccum elaeagni*; Maffei & Morton, *PD* **67**:964, 1983, in S.E. Michigan [**46**,3113; **63**,909].

P. foeniculi Manoir & Vegh, *Phytopath.Z.* **100**:329, 1981; A con. 8–10 × 2–2.5 μm, B con. 18–25 × 1–1.5 μm; fennel decline in France [**61**,342].

P. gardeniae Buddin & Wakef. January 1938; Hansen & Barrett, *Mycologia* **30**:18, February 1938; A con. 6.8–12.3 × 2.7–4.3 μm, B con. 18.2–27.2 × 1.4–1.8 μm; gardenia canker; Rocca de Sarasola & Sarasola, *Fitopatología* **13**:44, 1978 [**17**:397; **58**,1805].

P. helianthi, teleom. *Diaporthe helianthi*.

P. ipomoeae-batatas Punith., *Descr.F.* 739, 1982; A con. 4–8 × 2.5–3.5 μm, B con. unknown; sweet potato leaf spot; this reference has comments on other similar fungi on this crop.

P. juniperivora Hahn 1920; Punithalingam & Gibson, *Descr.F.* 370, 1973; A con. 8–10 × 2–3 μm, B con 20–30 × 0.5–1 μm; juniper blight, conifers, mostly studied on *Juniperus virginiana*, eastern red cedar, largely a disease of young plants; Holliday 1980.

P. lebiseyi, teleom. *Cryptodiaporthe lebiseyi*.

P. leptostromiformis, teleom. *Diaporthe woodii*.

P. lokoyae, teleom. *Diaporthe lokoyae*.

P. longicolla Hobbs, *Mycologia* **77**:542, 1985; A con. 5–9.5 × 1.5–3.5 μm, soybean seed decay; Gleason et al., *Phytop.* **77**:371, 1987, serological detection in seed [**66**,4024].

P. manihotis, teleom. *Diaporthe manihotis*.

P. oblonga, see *Ceratocystis ulmi*.

P. obscurans (Ell. & Ev.) B. Sutton, *TBMS* **48**:615, 1965; Sutton & Waterston, *Descr.F.* 227, 1970; A con. 5.5–7.5 × 1.5–2 μm, B con. unknown; strawberry leaf blight and ripe fruit rot; Howard & Albregts, *PDR* **56**:23, 1972 [**51**,3462].

P. oryzae-sativae Punith., *Nova Hedwigia* **31**:882,

1979, published 1980; Punithalingam, *Descr.F.* 665, 1980; A con. 11–14.5 × 2.5–3.5 μm, B con. unknown; rice collar rot; *Phomopsis oryzae* has A con. 8–11 × 2–3 μm, B con 20–30 × 0.5 μm; on rice grain; *P. oryzae-sativae* has prob. been misdetermined as *Ascochyta oryzae* Catt., Punithalingam, *Mycol.Pap.* 142, 1979, who stated that *Ascochyta* on rice needs study.

P. phaseoli, teleom. *Diaporthe phaseolorum*.

P. psidii Nag Raj & Ponnappa 1974; *Index Fungi* **4**:316, 1975; A con. 7 × 2 μm, B con. 24 × 1.5 μm; Lim & Razak, *Fitopat. Brasileira* **11**:227, 1986, stylar end ring rot of guava [**66**,1075].

P. sclerotioides v. Kest., *Neth.J.Pl.Path.* **73**:115, 1967; Punithalingam & Holliday, *Descr.F.* 470, 1974; A con. 7–10 × 2.5–3.5 μm, B con. unknown, abundant sclerotia formed in culture; cucumber black rot, a virulent, primary pathogen of the crop under glass [**46**,3612].

P. sojae, see *Diaporthe phaseolorum*.

P. tanakae, teleom. *Diaporthe tanakae*.

P. theae Petch 1925; Punithalingam & Gibson, *Descr.F.* 330, 1972; A con. 5–9 × 2–2.5 μm, B con. 20–30 × 0.5–1 μm; tea collar and branch canker; E. Africa, India, Sri Lanka; generally worse where ecological conditions are suboptimal for the crop; Holliday 1980.

P. vaccinii, teleom. *Diaporthe vaccinii*.

P. vexans (Sacc. & Sydow) Harter 1914; Punithalingam & Holliday, *Descr.F.* 338, 1972; A con. 5–9 × 2.5 μm, B con. 20–30 × 0.5–1 μm; eggplant tip over, also stem blight or canker, leaf blight or spot, fruit rot; seedborne; Holliday 1980.

P. viticola Sacc. 1915; Punithalingam, *Descr.F.* 635, 1979; A con. 6–10 × 2.5–3 μm, B con. 18–30 × 0.5–1 μm; Shear in 1911 described a teleom. as *Cryptosporella viticola* but this has not been confirmed; grapevine cane and leaf spot. Confusion had arisen between this fungus and *Eutypa armeniacae* q.v. The latter causes a grapevine dieback, usually called dead arm, dying arm or excoriosis; this disease had been, until recently, attributed to *Phomopsis viticola*. The attribution is incorrect; Moller & Kasimatis, *PD* **65**:429, 1981; Cucuzza & Sall, *ibid* **63**;794, 1982, chemical, treatment [**61**,1311; **62**, 337].

phony peach, see peach phony.

phormium yellow leaf Boyce et al., *N.Z.J.Sci.Technol.* A **33**:76, 1951; **34**, suppl. 1, 1953; Ushiyama, *N.Z.J.Bot.* **7**:363, 1969; MLO assoc., transm. sap, *Oliarus atkinsoni*, New Zealand [**31**:606; **33**:229; **49**,1667].

Phorodon humuli, transm. hop American latent, hop latent.

photinicola, *Spilocaea*.

photosynthesis, effects of pathogens on: Habeshaw in Wood & Jellis 1984.

Phragmidium Link 1816; Uredinales; teliospores large, 1–10 cells with 2 or more germ pores in all of them, no conspicuous outer hygroscopic layer, pedicels hyaline, usually long and often swelling at the lower end; autoecious on Rosaceae and those given here are all macrocyclic; Hiratsuka et al. 1980, taxonomic revision of spp. in Japan; Bedlan 1984, *mucronatum* and *potentillae* groups [**60**,4291; **64**,2587].

P. bulbosum (Strauss) Schlecht. 1824; Laundon & Rainbow, *Descr.F.* 203, 1969; aeciospores verrucose with broad, shallow warts; blackberry leaf rust.

P. mucronatum (Pers.) Schlecht., Laundon & Rainbow, *Descr.F.* 204, 1969; telia black, aeciospores echinulate; rose rust, N. temperate, may cause defoliation in the older, susceptible cvs.: Shattock & Bhatti, *Pl.Path.* **32**:61, 67, 1983, detrimental effects & chemical control [**62**,2518–19].

P. rosae-pimpinellifoliae Dietel 1905; Laundon & Rainbow, *Descr.F.* 205, 1969; telia chestnut brown, rose rust.

P. rosae-setigerae Dietel; Laundon & Rainbow, *Descr.F.* 206, 1969; rust on setigera rose and hybrids.

P. rubi-idaei (DC.) Karsten 1879; Laundon & Rainbow, *Descr.F.* 207, 1969; raspberry western yellow or cane rust; N. temperate, Australia, New Zealand; can be damaging in USA (N.W.); in 1977 severe outbreaks arose in Scotland, particularly on cvs. Glen Clova and Malling Delight. Some cvs. are completely resistant; there is evidence for variations in pathogenicity in UK populations; Anthony et al., *Pl.Path.* **34**:510, 521, 1985; *AAB* **110**:263, 1987, life history, cv. & fungus interaction, effect of cane management techniques [**65**,1406–7; **66**,3436].

P. tuberculatum Müller 1885; Laundon & Rainbow, *Descr.F.* 208, 1969; telia black, aeciospores verrucose with short warts, rose rust.

P. violaceum (C.F. Schultz) Winter 1880; Laundon & Rainbow, *Descr.F.* 209, 1969; aeciospores echinulate, blackberry leaf rust; for references to, and the work on, the biological control of the European blackberry aggregate in Australia, see Bruzzese & Hasan, *Pl.Path.* **35**:413, 1986; *AAB* **108**:527, 585, 1986 [**65**,6290–1].

Phragmobasidiomycetidae, Hymenomycetes; metabasidium divided by primary septa, usually cruciate or horizontal.

Phragmoporthe Petrak 1934; Pseudovalsaceae; Monod, *Beihefte Sydowia* 9, 1983, transferred *Magnaporthe* Krause & Webster q.v. to this genus

and made these new combinations: *P. grisea* (Hebert) Monod and *P. salvinii* (Catt.) Monod.

phycomycetes, trivial term for 'lower fungi'.

Phyllachora Nitschke ex Fuckel 1870; Phyllachoraceae; ascomata embedded in a dark stroma or clypeus perforated by the ostioles; upper surface black, smooth; embedded in leaf tissues; the tar leaf spots, mostly on Gramineae and prob. none of much economic importance. Those spp. that may be damaging include: *P. cynodontis* Niessl, on Bermuda grass; *P. huberi* Henn., on rubber in tropical America; *P. lespedezae* (Schwein.) Sacc. on the cover crop *Lespedeza stipulacea*; *P. maydis* Maubl. on maize. Other spp. include *P. musicola* Booth & D. Shaw, causing banana black cross, and *P. sacchari* Henn. on sorghum and sugarcane, Anahosur & Sivanesan, *Descr.F.* 588, 1978. *Catacauma torrendiella* Batista 1948, prob. a *Phyllachora*, coconut verrucosis, has warranted attempts at chemical control, Oliveira et al., *Fitopat. Brasileira* **9**:521, 1984. Cannon described 9 other spp., *Descr.F.* 902–10, 1986; Holliday 1980; Smiley 1983 [**64**,2680].

Phyllachoraceae, Polystigmatales; *Gibellina*, *Glomerella*, *Magnaporthe*, *Phyllachora*.

Phyllactinia Lév. 1851; Erysiphaceae; ascomata have long, straight, unbranched, equatorial appendages with a basal swelling; several asci; con. borne singly, pointed; mycelium partly endophytic; speciation is apparently confused; most spp. prob. on temperate hardwoods. *P. corylea* (Pers.) P. Karsten, ash, birch, hazel and other trees; the mulberry pathogen has also been referred to this sp.; Clark & Ankora, *CJB* **47**:1289, 1969; *TBMS* **57**:162, 1971, development, con. release & germination. *P. dalbergiae* Pirozynski 1965, causes defoliation of *Dalbergia sissoo*; Mukerji, *Descr.F.* 186, 1968, conidiophores spirally coiled at the base. *P. guttata* (Wallr.:Fr.) Lév., common on *Corylus*, Kapoor, *Descr.F.* 157, 1967; Cullum & Webster, *TBMS* **68**:316, 1977; **72**:489, 1979, dehiscence & dispersal of ascomata. *P. moricola* (Henn.) Homma; a mulberry powdery mildew; Itoi et al., *Bull.sericult.Exp.Stn* **17**:321; 1962; Spencer 1978 [**43**,1158; **49**,7; **51**,1140; **56**,4362; **59**,1424].

Phyllocoptes fructiphilus, transm. rose rosette.

Phyllocoptora oleivora, see *Mycosphaerella citri*.

phyllody, the transformation of floral organs into structures like leaves.

phyllomania, abnormal formation of leaves.

phyllophilum, *Nothophacidium*.

Phyllosticta Pers. 1818; Coelom.; conidiomata pycnidial, conidiogenesis holoblastic; con. aseptate, hyaline, solitary, with a persistent mucilagenous sheath and an appendage, teleom. *Guignardia*: van der Aa, *Stud.mycol.* 5, 1973; Punithalingam & Woodhams, *Nova Hedwigia* **36**:151, 1982, appendage.

P. ampelicida, teleom. *Guignardia bidwellii*.

P. citricarpa, teleom. *Guignardia citricarpa*.

P. elettariae Chowdhury 1958; *Index Fungi* **2**:516, 1963; con. 2.3–7.3 × 2.8–4.2 μm; cardamom nursery leaf spot; Naidu, *J. Plantation Crops* **9**:23, 1981, India [**61**,2994].

P. elongata, teleom. *Guignardia vaccinii*.

P. maydis, teleom. *Mycosphaerella zeae-maydis*; **Phyllosticta musarum**, teleom. *Guignardia musae*.

P. sphaeropsoidea, teleom. *Guignardia aesculi*.

P. vaccinii Earle 1897; con. 10 × 7 μm; previously considered to be the anam. of *Guignardia vaccinii* q.v., Weidemann who described the correct anam. as *Phyllosticta elongata*. *P. vaccinii* differs in that the pycnidial diam. of 120 μm is greater and the ostiole is well differentiated, the con. length/ breadth ratio is 1.4, appendages are shorter and no teleom. is known; it causes a leaf spot and fruit rot of cranberry in USA (Massachusetts, New Jersey).

Phyllotreta, transm. turnip crinkle, turnip yellow mosaic.

Phymatotrichopsis Henneb., *Persoonia* **7**:199, 1973; Hyphom.; conidiomata hyphal, terminal and subterminal cells inflated, globose; conidiogenous cells holoblastic; con. aseptate, hyaline, botryose with a broad base and a frill of attachment, 5 μm diam. or 6–8 × 5–6 μm. *P. omnivora* (Duggar) Henneb. is based on *Phymatotrichum omnivorum* Duggar 1916. *Phymatotrichum* Bonorden 1851 = *Botrytis*, fide Hennebert, *ibid* page 183. The existence of a teleom. is doubtful; this has been described as *Trechispora brinkmanii* (Bresad.) Rogers.

The fungus causes Texas root rot of cotton and has been reported on > 2000 dicotyledons; monocotyledons are not significantly affected. The disease was first reported in 1880; and the pathogen is restricted to USA (S.W., mostly Texas and southerly areas of bordering states) and Mexico (central, N.). It is a soilborne, facultative saprophyte that apparently existed in isolated pockets under the natural grassland. *P. omnivora* is restricted by cool temps. to the N. and S., and to the E. by soil type. Percy, *PD* **67**:981, 1983, mapped the reported and theoretical distribution and obtained a good fit. After cultivation, notably with cotton, the disease began to erupt; significant annual losses to this crop in Texas still occur. The fungus is found to a soil depth of > 240 cm. Infection of living roots arises from sclerotia at a depth of 45–75 cm; these structures survive for at

least 10 years. Infection is also through overwintering as simple rhizomorphs. Host penetration is followed by plant death. In wet weather fungus mats bearing con. are formed on the soil surface. The epidemiological role of these spores, if any, is unknown. Disease patches spread for 2–10 m a year; the pathogen being active in heavy soils with a pH of > 7 and at high, summer temps. Sclerotia are not formed in sandy, acidic soils; and therefore survival from one year to the next does not take place.

Deep ploughing may delay disease and long rotations with sorghum can be effective in control. Cotton cvs. with a seedling tolerance of cool soils can be used. Soils with long histories of Texas root rot, where control has not been poss., are used for livestock. Streets & Bloss, *Monogr.Am.Phytopath.Soc.* 8, 1973; Lyda, *A.R.Phytop.* **16**:193, 1978; Jeger & Lyda, *AAB* **109**:523, 1986 [**63**,439; **66**,2390].

physalis mild chlorosis MacKinnon, *CJB* **43**:509, 1965; transm. *Myz.pers.*, Canada (New Brunswick), *P. floridana*, a component of physalis yellow net [**44**,2755].

—— **mosaic v.** Peters & Derks, *Neth.J.Pl.Path.* **80**:124, 1974; Tymovirus, isometric 27 nm diam., transm. sap. USA (Illinois), *P. subglabrata* [**54**,1194].

—— **mottle** Chávez & Rodríguez Montessoro 1984; transm. sap to 7 spp. Solanaceae, seedborne, Mexico, *P. ixocarpa* [**66**,1725].

—— **shoestring v.** Verma & Chowdhury, *Indian J.Pl.Path.* **2**:95, 1984; rod *c.*415 nm long, India, *P. minima* [**64**,5284].

—— **vein blotch** MacKinnon & Lawson, *CJB* **44**:1219, 1966; transm. *Myz.pers.*, persistent, Canada (New Brunswick), *P. floridana* [**46**,284].

—— **yellow mottle v.** Moline & Fries, abs. *Phytop.* **62**:1109, 1972; isometric 29 nm diam., transm. sap, USA, *P. angulata* [**52**,2820].

—— —— **net** (physalis mild chlorosis + physalis yellow speck).

—— —— **speck** MacKinnon 1965; MacKinnon & Lawson, *CJB* **44**:795, 1966; transm. *Myz.pers.*; Canada (New Brunswick), *P. floridana*, a component of physalis yellow net [**45**,1287q, 3104].

Physalospora Niessl 1876; Amphisphaeriaceae; ascomata fleshy, glabrous or setose, asco. > 20 μm long, close to *Glomerella* which has asco. < 20 μm long; Holliday 1980.

P. paulowniae Itô & Kobayashi, *Rep.For.Exp.Stn.Tokyo* **49**:85, 1951; asco. 21–27 × 6–11 μm; assoc. with dieback of *Paulownia tormentosa*, Japan [**31**:153].

P. persicae Abiko & Kitajima,
Ann.phytopath.Soc.Japan 36:264, 1970; asco.
15–32.5 × 12.5 μm; peach blister canker
[50,1884].
P. psidii Stevens & Peirce, *Indian J.agric.Sci.* 3:913,
1933; asco. 30–37 × 13–16 μm, guava dieback
[13:269].
P. rhodina, anam. *Botryodiplodia theobromae* q.v.
P. zeicola Ell. & Ev. 1890; Eddins & Voorhees,
Phytop. 23:63, 1933; asco. 18–20 × 8–10 μm, poss.
anam. *Diplodia frumenti* Ell. & Ev. 1886; Eddins,
ibid 20:733, 1930; maize ear and stalk rot [10:96;
12:366].
physiologic disease, disorder q.v. is preferred.
—— race, abbreviated here as race; a taxon of
parasites, particularly fungi, characterised by
specialistion to different cvs. of one host sp., FBPP
q.v. This definition may be considered inadequate
by some without the insertion of pathogenicity
which is the main, or usually only, differing
characteristic between such races. Races and ff. sp.
do not differ in morphology. The origin of the term
goes back at least to Eriksson in 1894. He
demonstrated the existence of different parasitic
forms in cereal rusts specialised for host spp. now
known as formae speciales, see forma specialis,
singular. Stakman in 1913 showed that within these
ff. sp. there existed races. Thus ff.sp. and races
represent 2 groups that differ in levels of parasitic
specialisation. Parlevliet, *EPPO Bull.* 15:145,
1985, defined race as a population in which all
individuals carry the same combination of
virulence genes. Terminology is still confused and
perhaps this is to be expected in the still evolving
study of resistance, Robinson 1969, and where
more unified concepts are still being sought.
Bacteriologists use pathovar or pathotype which
are both comparable in definition to f. sp. and
race; Caten in Wolfe & Caten 1987.
physiotype, 'a population of a pathogen in which
all individuals have a particular character of
physiology, but not of pathogenicity, in common,'
Robinson 1969, see resistance to pathogens; same
as physiodeme which would appear to be the
preferred usage, FBPP.
Physoderma Wallr. 1833, fide Karling, *Lloydia*
13:29, 1950; Physodermataceae q.v. the genus
being coextensive with the family; 2 independent
forms: a monocentric, epibiotic zoosporangium;
and a polycentric, endobiotic system bearing
septate, turbinate cells on which many, dark, thick
walled, resting spores are formed; these spores
germinate to form zoospores; the vegetative
system is predominantly rhizoidal and aseptate;
parasites of vascular plants. Karling united
Urophlyctis q.v. with *Physoderma*; Sparrow

distinguished between them; Walker, see *P.
alfalfae*, followed Karling; Holliday 1980 [30:125].
P. alfalfae (Pat. & Lagerh.) Karling; Walker,
Descr.F. 751, 1983; lucerne crown wart or
marbled gall; very restricted in distribution and
only likely to be a problem in low wet land.
P. leproides (Trabut) Karling; Walker, *Descr.F.*
752, 1983; on sugarbeet, large galls on roots and
smaller ones on leaves and stems; Whitney &
Duffus 1986 as *Urophlyctis leproides* (Trabut)
Magnus.
P. maydis (Miyabe) Miyabe 1909; Walker, *Descr.F.*
753, 1983; maize brown spot, a widespread
disease which has been studied in USA where it is
of only minor importance.
Physodermataceae, Blastocladiales; as for
Physoderma, Lange & Olson, *TBMS* 74:449, 1980
[60,3056].
Physopella Arthur 1906; Uredinales; Cummins &
Ramachar, *Mycologia* 50:741, 1958; teliospores
aseptate, sessile in subepidermal crusts, sometimes
in chains; the genus has been combined with
Phakopsora; Cummins 1971; Holliday 1980
[38:354].
P. ampelopsidis (Dietel & Sydow) Cummins &
Ramachar; Punithalingam, *Descr.F.* 173, 1968 q.v.
for other rusts on grapevine, *Vitis*; microcyclic,
teliospores 20–30 × 12–15 μm, uredospores
18–28 × 12–17 μm; grapevine leaf rust; Leu &
Wu, *Pl. Prot.Bull.Taiwan* 25:167, 1983, infection
[63,204].
P. zeae (Mains) Cummins & Ramachar; Laundon
& Waterston, *Descr.F.* 5, 1964; microcyclic;
teliospores 22–34 × 12–18 μm, uredospores
22–28 × 16–20 μm; on maize, economically less
important than the other 2 maize rusts: *Puccinia
polysora* and *P. sorghi*; but has caused damage on
the Pacific side of Central America and in
Venezuela; Bonde et al. *Phytop.* 72:1489, 1982,
leaf penetration; Heath & Bonde, *CJB* 61:2231,
1983, fine structure of uredia [62,1489; 63,582].
physostegia virescence and proliferation Giunchedi
& Poggi Pollini, *Phytopath.Mediterranea* 25:151,
1986; MLO assoc., Italy (Emilia-Romagna), *P.
virginiana*, false dragon head flower.
phytarbovirus, plant viruses transm. persistently by
leafhoppers, aphids and other arthropods;
Whitcomb & Davis, *A.Rev.Entomol.* 15:405, 1970.
phytiatry, the treatment of plant diseases; the term
usually implies the use of chemical methods for
preventing or eliminating infection, FBPP; its use
in English is uncommon, see chemotherapy.
phytoalexin, a substance which inhibits the growth
of a micro-organism, i.e. a pathogen in a living
plant, and which is formed only locally and only
when these 2 organisms interact. It is absent from

non-infected host tissue at concentrations which could exert an inhibitory effect on the pathogen; it is not specific; Müller & Börger 1940, see Müller, *Aust.J.biol.Soc.* 11:275, 1958. Several hundred phytoalexins have been reported; they are not given in the order; Bailey & Mansfield ed., *Phytoalexins*, 1982; Darvill & Albersheim, *A.Rev.Pl.Physiol.* 35:243, 1984; Ebel, *A.R.Phytop.* 24:235, 1986; Dixon 1986; Keen 1986; Bailey in Day & Jellis 1987 [65,4680, 4800].

phytomed, Blumenbach & Laux 1986, a thesaurus of alphabetical lists of terms in German and English in plant pathology and used in the phytomed database [65,5356].

Phytomonas, flagellate Protozoa; first found in plants in 1909; they occur in several plant families and are spread by insects. Their implication as etiological agents in disease was discovered by Stahel in 1931 but his work was neglected for c.35 years, Vermeulen, *Neth.J.Pl.Path.* 74:202, 1968. Stahel transm. the protozoa through grafts in work on coffee phloem necrosis where the leaves fall gradually after yellowing and the plant dies in 3–12 months. But there is a more acute form of phloem necrosis. Old diseases of coconut in Surinam and Trinidad, i.e. bronze leaf, Cedros and coronie wilts, have a similar etiology; now called heart rot. So does a damaging disease of oil palm called marchitez or sudden wither which is found in northern areas of S. America. Insects have been implicated in the spread of the marchitez pathogen. *P. staheli* R. McGhee & A. McGhee, *J.Protozool.* 26:348, 1979, was described from sieve tubes of coconut and oil palm; it may be the cause of these palm diseases. These protozoa are apparently restricted to tropical America. Kitajima et al., *Phytop.* 76:638, 1986, reported a *Phytomonas* from the lactiferous ducts of cassava and which was assoc. with chochamento das raizes, empty roots, an important disease in Brazil (Espirito Santo). Davis et al., *ibid* 77:177, 1987, grew one of these organisms, *P. davidi*, for the first time on a defined medium; McCoy & Martinez-Lopez, *PD* 66:675, 1982; Dollet, *A.R.Phytop.* 22:115, 1984 [48,809; 60,2140; 62,345; 66,428,3711].

phytoncide, a non-specific term applied to any substance which confers resistance by preventing growth of the attacking organism; too all embracing to be useful, after FBPP.

phytopathogen, an organism which causes disease in plants.

phytopathometry, the measurement of plant disease, Large 1966; required mainly for assessing the effects, e.g. crop loss, of disease on a plant or crop; in screening for chemical control and host resistance. Measurements can be made using grading methods, standard diagrams and field keys. For more precision these methods are used on plants growing in controlled environments; Large, *A.R.Phytop.* 4:9, 1966, measurement; Dimock, *ibid* 5:265, 1967, controlled environment; James, *Can.Pl.Dis. Surv.* 51:39, 1971, disease assessment keys & diagrams; *Plant growth stages and disease assessment keys*, Min.Agric.Fish.Fd., UK; Horsfall & Cowling, *Plant disease, an advanced treatise*, vol. 2, 1978; Hebert, *Phytop.* 72:1269, 1982, the rationale for the Horsfall & Barrat plant disease assessment scale; Seem, *A.R.Phytop.* 22:133, 1984.

*Phytophthora de Bary 1876; Pythiaceae; sporangia ovoid or obpyriform with a distinct apical emission zone, zoospores formed before emission, quick dispersal, any vesicle formed is quickly evanescent, cf. *Pythium*; oogonia smooth, occasionally ornamented, never spiny; oospores mostly aplerotic. Six morphological groups or characteristics are recognised: I and II sporangial apex markedly papillate with a thickening at least 4 μm deep; III and IV apex less papillate; I and IV exit pores of sporangia 7 μm diam. or less; V and VI apex not papillate, exit pores 12 μm diam. or more; sporangia proliferate internally in V and VI; antheridia are mostly paragynous in I, III and V; homothallic spp.; and mostly amphigynous in II, IV and VI, physiologically heterothallic. The genus is diploid with meiosis occurring in the gametangia. The spp. said to be caducous, i.e. sporangia are detached with a pedicel of ±characteristic length, have papillate sporangia. Those that are not caducous include all the non-papillate spp. Waterhouse, in Gregory ed., Phytophthora *disease of cocoa*, 1974, tabulated the spp. into caducous, most shed in water only; and persistent, i.e. not shed. The characteristics of caducity apparently need more study. *P. infestans* and *P. phaseoli* are the only truly airborne spp., with sporangia that drop readily in air. Other spp. said to have caducous sporangia, e.g. *P. megakarya* and *P. palmivora*, are not truly airborne and are not caught in volumetric traps to any great extent. In these 2 groups the characteristics of caducity have differences which have not been adequately explained.

The optimum temps. for growth in vitro are reflected in the geographical distribution. The spp. fall into 3 groups: 15–22°, 20–28°, 25–32°, abbreviated for each sp. as LT, MT, HT; the group number I–VI is also given. The spp. are either plurivorous or specialised for hosts. Infection is followed by a rapid necrosis of the plant organ attacked. The complete death of a plant, including woody perennials, may result from canopy destruction, extensive root necrosis or stem cankers and cortical rots. Some spp. have a significant saprophytic phase and some are water moulds, and not plant pathogens to any significant degree. Some crops, e.g. apple, cacao, citrus and rubber, are attacked by several spp. which cause similar disease syndromes. There has been no monograph on one of the most important phytopathogenic genera of fungi for > 55 years. This largely reflects the difficulties inherent in work on this very adaptable and variable genus. One of the problems has been the failure to use uniform techniques. It has, for example, been known for nearly 25 years that *Phytophthora* spp. can complete their life cycles on a defined medium. Yet taxonomists, although recognising the importance of standard, defined media for comparative studies, have failed to adopt this and other strictly defined techniques. Brasier in Erwin et al. gave an account of problems and prospects.

Blackwell, *Mycol.Pap.* 30, 1949, terminology; Waterhouse, *ibid* 122, 1970, the original descriptions of spp.; Newhook et al., *ibid* 143, 1978, key; Ribeiro, *A. source book of the genus* Phytophthora, 1978; Holliday 1980; Ho, *Mycologia* 73:705, 1981; *Mycopathologia* 79:141, 1982, synoptic keys, cluster analysis & the taxonomic groups; Allen 1983; Erwin et al. ed., Phytophthora: *its biology, taxonomy, ecology and pathology*, 1983.

P. arecae (Coleman) Pethybr. 1913; Stamps, *Descr.F.* 833, 1985; short sporangial pedicel, oogonia av. 30 μm diam., group II, HT; not extensively studied, mostly reported from India causing koleroga or mahali of *Areca catechu*; it infects the nuts.

P. boehmeriae Saw. 1927; Stamps, *Descr.F.* 591, 1978; short sporangial pedicel, oogonia av. 27 μm diam., group II, MT; on *Boehmeria nivea* and other plants, a little known sp.

P. botryosa Chee, *TBMS* 52, 428, 1969; Stamps, *Descr.F.* 835 1985; short sporangial pedicel, characterised by the clumped sporangia, group II, MT; rubber pod rot and abnormal leaf fall, can cause black stripe; Andaman Islands, Malaysia (W.), Thailand; under monsoon rainfall conditions can cause leaf canopy epidemics [48,3135].

P. cactorum (Lebert & Cohn) Schröter 1886; Waterhouse & Waterston, *Descr.F.* 111, 1966; short sporangial pedicel, oogonia mostly 25–32 μm diam., group I, LT; plurivorus, mostly Rosaceae, temperate areas, soilborne, fruits in contact with the soil become infected; mostly and fully studied causing apple collar or crown rot,

with *Phytophthora citricola* and *P. syringae*; and strawberry crown and fruit leather rot; APPLE: Sewell & Wilson, *AAB* 74:149, 159, 1973, seasonal effects on infection & disease in England, effect of stock on scion var. resistance; Sewell et al., *ibid* 76:179, 1974, seasonal varations in soil activity; Ellis et al., *PD* 70:24, 1986, control; Utkhede, *Phytoprotection* 67:1, 1986, review; STRAWBERRY: Harris & Stickels, *Pl.Path.* 30:205, 1981, in the greenhouse in England; Grove et al., *Phytop.* 75:165, 611, 700, 1985, on fruit in USA, effects of temp. & wetness on sporulation, splash dispersal [53, 217–18, 3547; 61,5123; 64,3137, 5030; 65,296, 5027, 6108].

P. cajani Amin, Baldev & Williams, *Mycologia* 70:174, 1978 = *Phytophthora drechsleri* f.sp. *cajani* [57,4722].

P. cambivora (Petri) Buisman 1927; Waterhouse & Waterston, *Descr.F.* 112, 1966; oogonia av. 43 μm diam., group VI, MT; plurivorous, often trees, root rots, one cause of chestnut ink; Reichard & Bolay 1986 described a severe attack on this tree in Switzerland; Wicks & Lee, *Aust.J.Exp.Agric.* 25:705, 1985; *Aust. J.agric.Res.* 37:277, 1986, almond crown rot in S. Australia, other spp. of genus [65,3997; 66,1151, 1535].

P. capsici Leonian 1922; Stamps. *Descr.F.* 836, 1985; oogonia av. 30 μm diam., group II, HT, see *Phytophthora MF*4; on many plants, but particularly causing stem and fruit rots of red pepper, soft rots of cucurbit and tomato fruit, and on eggplant. No other *Phytophthora* sp. causes so much damage to red pepper; not all cvs. have adequate resistance and infection can lead to the complete collapse of mature plants; there are differences in pathogenicity to various hosts; the inherent characteristics of the fungus have been intensively studied; Holliday 1980; Steekelenburg, *Neth.J.Pl.Path.* 86:259, 1980; Schlub, *J.agric.Sci.* 100:7, 1983, on red pepper [60,1748; 62,2244].

*P. cinnamomi Rands 1922; Waterhouse & Waterston, *Descr.F.* 113, 1966; oogonia av. 40 μm diam. group VI, MT; in warm temperate and subtropical regions; the fungus does not cause disease problems in the lowland tropics; plurivorous but few monocotyledons are attacked; main hosts are: avocado, conifers, eucalypts and other Australian native, woody spp., woody ornamentals. Before the fungus had been described by Rands dying patches of the dry forests of *Eucalyptus marginata* in S.W. Australia had been reported. But only c.40 years later was there good evidence for a pathogen, later shown to be *Phytophthora cinnamomi*, as the cause. There followed an explosive pandemic in the natural vegetation of this region of Australia and in

Victoria. This destruction by a pathogen over
large areas of natural plant communities is
unique. The disease is known as jarrah dieback.
The spread of this soilborne fungus was c.400 m/
year downslope and c.175 m/year on flat terrain.
Severe root rots of avocado, one of this crops
most important diseases, and woody ornamental
are caused; and chestnut ink with P. cambivora.
Zentmyer et al., Mycologia 71:55, 1979, temp.,
nutrition & oospore formation; Zentmyer,
Monogr.Am.phytopath.Soc. 10:1980; Phytop.
71:925, 1981; Zentmyer & Guillemet, PD 65:475,
1981, poss. pathotypes; Weste on jarrah dieback
in Erwin et al., under genus; Whiley et al.,
Aust.J.Exp.Agric. 26:249, 1986, chemical control
in avocado [58,5711; 61,586; 66,2441].
P. citricola Saw. 1927; Waterhouse & Waterston,
Descr.F. 114, 1966; oogonia mostly 27–32 μm
diam., group III, MT; plurivorous, diseases
reported include; citrus brown rot, hop black rot,
raspberry root rot, rhododendron dieback, tomato
basal rot; isolates from hop vary in virulence;
Holliday 1980; Matheron & Mircetich, Phytop.
75:970, 973, 977, 1985, pathogenicity to walnut
stocks with other Phytophthora spp. [65,2456–8].
P. citrophthora (R.E. Smith & E.H. Smith) Leonian
1925; Waterhouse & Waterston, Descr.F. 33,
1964; oogonia unknown, group II, MT; this
fungus has been largely studied as the most
frequent cause of important citrus diseases: collar
rot, foot rot or gummosis of the main stem and
crown roots, rot of smaller roots, leaf and twig
blight, and fruit brown rot. The other spp. of the
genus which are assoc. with all or some of these
diseases are: citricola, hibernalis, nicotianae var.
parasitica, palmivora. All these fungi cause similar
symptoms which may vary according to host and
growing conditions. They vary in their optimum
temp. requirements; and this affects their
geographical distribution and economic
importance. During long periods of wet weather
extensive damage to citrus can be caused and seed
can be infected. Resistant rootstocks should be
used. Phytophthora citrophthora occurs on many
other crops, and may be one pathogen in the
cacao black pod complex in Brazil, see P.
palmivora, bark canker, black pod; woody
Rosaceae are often attacked.
    Chitzanidis in Kranz et al. 1977; Ridings et al.
1979 on a pathotype in the control of Morrenia
odorata, milkweed vine, a serious weed in citrus in
USA (Florida); Holliday 1980; Solel, PD 67:878,
1983, aerial versus ground chemical control of
citrus brown rot; Tuzco et al., ibid 68:502, 1984,
resistance in citrus rootstocks; Kellam &
Zentmyer, Phytop. 76:159, 1986, on cacao with

other Phytophthora spp.; Mycologia 78:351, 1986,
single oospore isolates from cacao in Brazil with
P. capsici [59,5419; 63,631, 4435; 65,3205;
66,117].
P. clandestina Taylor, Pascoe & Greenhalgh,
Mycotaxon 22:80, 1985; sporangia papillate,
deciduous, pedicel 1–6 μm long; oogonia av. 30
μm diam., abundant in roots; antheridia
paragynous or amphigynous, with digitate
processes; in Australia, causing damping off and
tap root rot of subterranean clover; Greenhalgh &
Taylor, PD 69:1002, 1985, widespread in Victoria
State; Wong et al. TBMS 86:479, 1986;
J.Phytopath. 116:67, 1986; Can.J.Microbiol.
32:553, 1986; effects of soil temp., moisture &
other fungus pathogens; soil behaviour; growth
in vitro & survival [65,3968, 4457, 6091; 66,3662].
P. colocasiae Racib. 1900; oogonia av. 29 μm
diam., group IV, HT; this very characteristic
tropical sp., with its narrowly elongate sporangia,
has an extremely limited host range and a
restricted distribution in E. and S. Asia, and
Oceania; spp. of Alocasia, Colocasia, Xanthosoma
are attacked and a damaging leaf collapse is the
result; where these edible aroids form a part of
the staple diet the disease is economically
important; other Araceae may be very resistant;
Holliday 1980; Gollifer et al., AAB 94:379, 1980,
inoculum survival; Jackson et al., ibid 96:1, 1980,
control by spacing & chemical [59,5538;
60,2930].
P. cryptogea Pethybr. & Lafferty 1919; Stamps,
Descr.F. 592, 1978; oogonia 20–32 μm diam.,
group VI, MT; close to Phytophthora drechsleri;
plurivorous on temperate crops, mostly foot and
root rots; Kröber, Phytopath.Z. 102:219, 1981,
found differences of pathogenicity; causes tulip
shanking and may be assoc. with P. erythroseptica
in potato tuber pink rot; Michell et al., Phytop.
68:1446, 1978, watercress stem & root rot,
zoospore numbers & infection; Bumbieris,
Aust.J.Bot. 27:11, 1979, soil biology; MacDonald,
Phytop. 74:621, 1984, salinity & effects on
chrysanthemum roots in a hydroponically grown
crop; Ho & Jong, Mycotaxon 27:289, 1986,
considered the same as P. drechsleri; Stirling &
Irwin, Pl.Path. 35:527, 1986, root rot of guar in
Australia (Queensland) [58,4145; 59,157; 61,4179;
63,4458; 66,1373, 2148].
P. drechsleri Tucker 1931; Stamps, Descr.F. 840,
1985; oogonia 36 μm diam., group VI, HT; close
to Phytophthora cryptogea but from which it
differs in the larger sporangia and oospores, and
higher opt. temp. requirements; some authorities
would merge the 2 spp.; plurivorous, diseases
caused include: sugarbeet tap root rot, cucurbit

necroses of stem, leaves, fruit and roots, green death in Iran; stem canker of *Albizia chinensis*, a coffee shade tree, in *Stizolobium* and cassava root rot; the chlamydospores have been found in weeds. Most work on the pathogen has been done in a root rot of safflower in USA (California), especially in irrigated cultivations; root infection causes above ground collapse; Holliday 1980; Kannaiyan et al., *Mycologia* 72:169, 1980, f.sp. *cajani* causing pigeon pea stem blight; Sharma et al., *PD* 66:22, 1982, inheritance of resistance in pigeon pea; Alavi et al., *Pl.Path.* 31:221, 1982, green death of cucurbits in Iran [59,6044; 61,5385; 62,448].

P. erythroseptica Pethybr. 1913; Stamps, *Descr.F.* 593, 1978; oogonia 30–35 μm diam., group VI, MT; potato tuber pink rot, tomato buckeye fruit rot, tulip shanking; Seemüller et al., *NachrBl.dt.PflSchutzdienst* 38:17, 1986, causing raspberry root rot [65,5612].

P. fragariae Hickman 1940; oogonia 39 μm diam., group V, LT; causes the extremely important strawberry red core, sometimes called red stele; loganberry can become diseased, and *Geum*, *Potentilla*, and *Rubus* spp. may spread inoculum; more disease is found in acid soils and there is a long soil survival as oospores; at least 15 races occur but a better differentiation is needed; infection is through the actively growing roots and the red discolouration may extend into the crown; plants affected by red core are often found in patches of stunted and dying plants in an infested field; Montgomerie, *Hort.Rev.Commonw.Bur.hortic.Plantation Crops* 5, 1977, review; in Ebbels & King 1979, legislative control; Wicks, *PD* 67:1225, 1983, chemical control; Maas 1984; Duncan, *TBMS* 85:455, 585, 1985, oospore germination in vitro; effects of fungicides on oospore survival, infectivity & germination; Kennedy et al., *Pl.Path.* 35:344, 1986, virulence of single zoospore isolates [63,3467; 65,297, 1959; 66,665].

P. heveae Thompson 1929; Stamps, *Descr.F.* 594, 1978; oogonia 25–28 μm diam., group II, MT; the pathology of this sp. is little known; although it has been intensively studied in vitro since oogonia are formed abundantly in culture. It has been isolated from: avocado, Brazil nut, cacao, cashew, coconut, eucalyptus, guava, mango, red pepper, rhododendron, rubber and from soil; the geographical distribution is widespread but records are few.

P. hibernalis Carne 1925; Waterhouse & Waterston, *Descr.F.* 31:1964; oogonia av. 35 μm diam., group IV, LT: citrus brown rot; one of the spp. of the genus attacking citrus but its temp.

requirements limit the sp. to the cooler growing areas, fruits, stems and leaves can be infected; Holliday 1980.

P. ilicis Buddenhagen & Young, *Phytop.* 47:100, 1957; oogonia av. 21 μm diam., group IV, LT; holly twig blight, USA (Oregon, Washington State), Ticknor et al. 1980 reported on resistance in spp. other than *Ilex aquifolium* and in hybrids [36:471; 60,5568].

****P. infestans** (Mont.) de Bary 1876; Stamps, *Descr.F.* 838, 1985; group IV, LT; sporangia typically airborne and dropping off, with a short pedicel, more readily than any other sp. of the genus, except for *Phytophthora phaseoli* which is very similar morphologically; oogonia 38 μm diam. Potato blight, late blight in N. America, is found in all cool, wet regions; a similar disease is caused on tomato. *P. infestans*, arguably the most studied of all plant pathogens, has been used intensively in work on the biochemistry and genetics of host and parasite reactions, Clarke in Callow 1983. The potato, domesticated in the altoplano of Bolivia and Peru thousands of years ago, was brought to Spain c.1570. But the fungus, first called *Botrytis infestans* in 1845 by Montagne, prob. did not reach Europe and N. America until 1842. Its origin was poss. in Mexico where there is a race population and where the 2 compatibility types exist. Elsewhere only the A1 type was known until the A2 type was recorded in Scotland and Switzerland; Hohl & Iselin, *TBMS* 83:529, 1984; Malcolmson, *ibid* 85:531, 1985. The first major epidemics on potato began in Canada (E.) and USA (N.E.) in 1843, and in Europe 2 years later; history by Bourke, *Nature Lond.* 203:805, 1964. The disastrous social and economic consequences of such epidemics in Ireland have been fully documented, see Irish famines. Blight is most severe at 13–20°, and where mists and rain are frequent. Spread by the sporangia and zoospores in potato fields is extremly rapid. An annual epidemic, causing total loss, can arise from one diseased tuber per square km. The pathogen overwinters in tubers, i.e. seed, ground keepers and cul piles, and on tomato in a summer gap. Persistence in soil is c.11 weeks. The extreme, genetic variation is asexual in origin; certain races pass readily from potato to tomato whose seed may spread inoculum, Vartanian & Endo, *Phytop.* 75:375, 1985. Control, apart from sanitary measures, is with fungicides and forecasting is widely used, see Beaumont period, blitecast. The long investigations to find lasting resistance of the race specific type has been unsuccessful since *P. infestans* soon adapts to and overcomes each newly found source of such

resistance. But non-specific resistance is known and will prob. become commercially satisfactory.

Butler & Jones 1949; Walker 1969; Lapwood & Hide in Western 1971; Holliday 1980; Kolbe, *Pflschutz.Nachr.Bayer* **35**:247, 1982–3, forecasting; Yamamoto, *Shokubutsu Byogai Kenkyu* **9**:1, 1982, resistance; Hooker 1981; Rich 1983; Skidmore et al., *Pl.Path.* **33**:173, 1984, a poss. mechanism for oospore formation in the field; Campbell et al., *TBMS* **84**:533, 1985, formation of oospores in vitro in A1 isolates [**44**,800; **63**,4531; **64**,288, 3520, 5120; **65**,345].

P. inflata Caroselli & Tucker, *Phytop.* **39**:485, 1949; oogonia av. 34 μm diam., group III; elm pit canker, Canada, USA [**28**:600].

P. iranica Ershad, *Mitt.biol.BundAnst.Ld-u.Forstw.* **140**:63, 1971; oogonia av. 34 μm diam., group I; on eggplant, Iran [**51**,198].

P. katsurae Ko & Chang, *Mycologia* **71**:841, 1979; Ko & Arakawa, *Trans.mycol.Soc.Japan* **21**:215, 1980; Stamps, *Descr.F.* 837, 1985; oogonia with warty protruberances, 27 × 25 μm; group II, MT; chestnut trunk rot; *Phytophthora cambivora* and *P. cinnamomi* also attack chestnut; pathogenic to oak; Australia (Queensland), Ivory Coast, Japan, Papua New Guinea, Taiwan, USA (Hawaii); *Castanea crenata* and *C. sativa* are susceptible, *C. dentata* and *C. mollissima* are resistant; Uchida 1976 as the synonym *P. castaneae* Katsura & Uchida [**58**,2951; **59**,2349; **60**,4257].

P. lateralis Tucker & Milbrath, *Mycologia* **34**:97, 1942; oospores 35–40 μm diam., group V, LT; root rot of Lawson cypress, *Chamaecyparis lawsoniana*; Canada (British Columbia) USA (California, Oregon, Washington State); Trione, *Phytop.* **49**:306, 1959; **64**:1531, 1974 [**21**:276; **39**:58; **54**,3011].

P. macrospora, see *Sclerophthora macrospora*; the type sp. of *Sclerophthora* is best placed in *Phytophthora* fide Ito & Tanaka 1940.

P. meadii McRae 1918; Stamps, *Descr.F.* 834, 1985; oogonia av. 33 μm diam., group II, HT; abnormal leaf fall and black stripe of rubber in India and Sri Lanka. This crop is the only one on which serious diseases are caused; although the fungus has been recorded from pineapple in USA (Hawaii). It has been considered by some to be synonymous with certain other tropical members of group II; but Dantanarayana et al., *TBMS* **82**:113, 1984, found it to be distinct. *Phytophthora meadii* is the principal pathogen of the genus on rubber in Sir Lanka. Abnormal leaf fall occurs in regions where there is heavy, monsoon rain and the disease is also caused by *P. botryosa* and *P. palmivora*. Damaging attacks can occur in S.W. India in particular; also in Indonesia (Sumatra)

and parts of Malaysia. Abnormal leaf fall of rubber is caused by infection of mature leaves. It contrasts with secondary leaf fall of the same crop and which is caused by infection of young leaves by *Glomerella cingulata* and *Oidium heveae*; Peries in Kranz et al. 1977; Holliday 1980; Rajalakshmy, *TBMS* **85**:723, 1985, mating types [**63**,1936; **65**,2005].

P. megakarya Brasier & Griffin, *TBMS* **72**:137, 1979; Stamps, *Descr.F.* 832, 1985; group II, MT; a segregate from *Phytophthora palmivora* and see *P.* MF4. It has 5–6 large chromosomes, a sporangial pedicel 10–30 μm long, a slightly lower growth optimum temp., and a limited distribution, W. Africa. It causes one form of cacao black pod and is epidemic on Nigerian cacao where it has been very fully investigated. The fungus survives the dry season on the feeding roots and poss. as encysted zoospores in the soil. During the wet season zoospores spread in water droplets to the lower pods on the tree. Infections then increase via secondary sources of inoculum, e.g. infected pods and ant tents. Cacao black pod is caused by other *Phytophthora* spp. in W. Africa, S. America and S.E. Asia. Epidemiological work, on the lines done in W. Africa, is needed elsewhere. Behaviour patterns will differ, e.g. in Ghana *P. palmivora* becomes systemic in the flower cushion and invades the pod via the peduncle. *P. megakarya* seldom shows this infection pattern. In Brazil other spp. of the genus have been implicated in this disease. Gregory & Madison ed., *Phytopath. Pap.* 25, 1981; Gregory et al., *Cocoa Growers Bull.* 35:5, 1984.

P. megasperma Drechsler 1931; Waterhouse & Waterston, *Descr.F.* 115, 1966; group V, MT; the sp. is taxonomically complex. Hansen et al., *TBMS* **87**:557, 1986, examined 93 isolates and divided them into 9 sub-groups on the basis of: mean in vitro growth rate/day, mean oogonial diam., chromosome number, mean relative DNA (% of *Phytophthora infestans*); classical morphology and electrophoretic patterns were also used. The sub-groups are, based on pathogenicity: clover, Douglas fir, lucerne, rosaceous fruit trees, soybean, and a major group on a broad range of hosts sub-divided into 4. Two ecological groups were recognised; one on herbaceous legumes, and the other on many diverse hosts including woody plants. The first has an obligate, gene for gene pathosystem; the second has a generalised pathogenicity. Whether some forms of the fungus merit a new specific rank remains to be determined. Faris et al., *CJB* **64**:262, 1986, on the basis of protein patterns, distinguished 2 groups in 26 lucerne isolates: (1) minimum growth at 5°, optimum 25–30° and maximum 35°; highly pathogenic, small oogonia; (2) growth temps. < 5°, 20°, 30°, respectively, less pathogenic, large oogonia.

The fungus has been very largely studied as a pathogen of lucerne and soybean on which it causes stem and root rots. In soybean 24 races have been delineated in USA and pathogenicity characteristics examined in considerable depth; for reviews: Holliday 1980; Sinclair & Shurtleff 1982; Paxton in Callow 1983; Schmitthenner, *PD* **69**:363, 1985; also: Buzzell et al., *Phytop.* **72**:801, 1982, host & pathogen interactions on soybean hypocotyls; Wilkinson & Millar, *ibid* **72**:790, 1982, soil temp. & moisture effects on lucerne root rot; Jimenez & Lockwood and Förster et al., *ibid* **72**:662, 1982; **73**:442, 1983, oospore germination; Faris et al., *Can.J.Pl.Path.* **5**:29, 1983, isolates of differential pathogenicity to lucerne; Stack & Millar, *Phytop.* **75**:1393, 1398, 1985, colonisation of organic matter & soil survival [**61**,7067, 7237, 7240; **62**,3778, 4930; **65**,2353–4, 4460; **66**,2253].

P. melonis Katsura, *Trans.mycol.Soc.Japan* **17**:238, 1976; oogonia 27.5 35 µm diam., group VI, HT; foot rot of cucumber, on other Cucurbitaceae; China, Japan, Taiwan; Lu & Gong 1982; Lin & Wu, *Pl.Prot.Bull.Taiwan* **27**:257, 1985, disease patterns & control [**57**,2024; **63**,968; **65**,2072].

P. mexicana Hotson & Hartge, *Phytop.* **13**:520, 1923; oogonia 28–40 µm diam., group II, MT; on tomato causing a wilt and fruit decay Kröber et al., *Phytopath.Z.* **107**:244, 1983, stem base rot of *Dieffenbachia maculata* [**3**:373; **63**,1297].

P. MF4, see *Phytophthora palmivora* and Brasier & Medeiros, *TBMS* **70**:295, 1978. This morphological form of the genus, still not named, is one segregate from *P. palmivora*, the other is *P. megakarya*. They constitute what used to be called the atypical and other strs. of *P. palmivora* sensu stricto as originally described; both cause important diseases. The earlier work on *P. palmivora* sensu lato, i.e. before the recognition of the 2 segregates in 1977, did not always give adequate morphological details. It may therefore be difficult, if not impossible, to determine which of these 2 spp. or MF4 was the pathogen prob. concerned. Like *P. palmivora*, MF4 has small, 9–12 chromosomes, but differs in its narrow, long pedicel, mostly 20–150 µm, and in the tapered base to the sporangium. *P. megakarya* has larger, fewer chromosomes and a shorter pedicel.

An atypical form of *P. palmivora* was intensively investigated by Holliday & Mowat, *Phytopath. Pap.* 5, 1963. It causes foot rot of black pepper in Malaysia (Sarawak). Historical records indicate that the disease is an old one; its etiology was first determined by Muller in 1936 in Sumatra. The Sarawak and Sumatran isolates cannot be compared but there seems little doubt that they are the same. The fungus causes the most severe disease of the crop and severely limits production in Brazil, India, Indonesia and Malaysia. The black pepper isolates have now been shown to be the same as, or very similar to, MF4 which is therefore widespread. The original segregation of the form was based on isolates from cacao only.

MF4 has been considered to be the same as *P. capsici*, Alizadeh & Tsao, *TBMS* **85**:47, 71, 1985; Tsao et al., *F.A.O.Pl. Prot. Bull.* **33**:61, 1985. But this view has not yet been generally accepted; G.M. Waterhouse, personal communication, 1987. One of the characteristics of MF4 from black pepper is that the sporangia are difficult to dislodge; unlike the spp., including *P. capsici*, that have caducous sporangia that fairly readily fall away from the sporangiophore when placed in water. Furthermore, the pathogenicity characteristics of *P. capsici* and MF4 are very different. The former causes diseases on several crops in the Cucurbitaceae and Solanaceae; it is particularly important on *Capsicum*. The latter is restricted to the Piperaceae and causes diseases only on *Piper betle* and *P. nigrum*, Turner, *TBMS* **57**:61, 1971; Alconero et al., *Phytop.* **62**:144, 1972, described black pepper foot rot in Brazil and Puerto Rico [**51**,558, 3483; **57**,4380; **64**,4741–2, 5053].

P. nicotianae Breda de Haan 1896; Waterhouse & Waterston, *Descr.F.* 34, 35, 1964; sometimes incorrectly referred to *Phytophthora parasitica* Dastur which is illegitimate, fide Waterhouse 1963; oogonia 24–26 µm diam., group II, HT. This sp. is very like *P. palmivora* with which it has not been very closely compared in recent years. The morphological differences are few; the sporangial pedicel of *P. nicotianae* is 2 µm long whilst that of *P. palmivora* is more variable, mostly 5 µm. Waterhouse in Gregory 1974, see *P. palmivora*, called *P. nicotianae* sporangia caducous, shed in water only; Trichilo & Aragaki, *Mycologia* **74**:927, 1982, called them non-caducous. *P. nicotianae* var. *parasitica* (Dastur) Waterhouse *Mycol.Pap.* 92:14, 1963, like *P. palmivora*, is plurivorous; but its pathogenicity is more catholic in that it causes diseases in both temperate and tropical areas; it also has a higher optimum temp. for growth, grows well at 35–37.5°, and has a wider temp. range. Var. *parasitica* does not cause any significant diseases on black pepper, cacao, coconut and rubber, cf. *P. palmivora*. It causes tomato fruit buckeye and is one of the members of the genus attacking citrus. Var. *nicotianae* is largely distinguished by its specificity to tobacco, causing black shank. Black shank has been mostly studied in USA; its damaging effects are increased by root knot nematodes; there are 3 races of var. *nicotianae* on tobacco.

Weststeijn, *Neth.J.Pl.Path.* **79**, suppl. 1, 1973, on tomato; Lucas 1975; Holliday 1980; Tsao et al., *TBMS* **75**:153, 1980, oospores in vitro; Davis, Csinos & Minton and Ioannou & Grogan, *PD* **66**:218, 1982; **67**:204, 1983; **68**:429, 1984, chemical control in citrus, tobacco, tomato; Jacobi et al. and Campbell et al., *Phytop.* **73**:139, 1983; **74**:230, 1984, effects of rain & temp. on black shank, quantitative analysis of its spread; Hoy et al., *ibid* **74**:474, 1984, irrigation & buckeye; Sherf & MacNab, 1986, on eggplant & tomato [**53**,279; **60**,2419; **61**,5762; **62**,2312, 3230–1; **63**,3018, 4076, 4564].

P. palmivora (E. Butler) E. Butler 1919; sensu stricto fide Brasier & Griffin, *TBMS* **72**:111, 1979; Stamps, *Descr.F.* 831, 1985; oospores av. 22–24 μm diam.; oogonia mostly 23–35 μm diam. fide Waterhouse in Gregory ed., Phytophthora *disease of cocoa*, 1974, q.v. for an account of the sp. sensu lato; group II, HT. *Phytophthora palmivora* has 9–12 small chromosomes like those of *P.* MF4, the sporangial pedicel is characteristically mostly 5 μm long. The sp. is based on Butler's fungus from Palmyra palm in India, see Ashby, *TBMS* **14**:18, 1929; it is plurivorous, widespread in the wet tropics and economically very important.

Over many years this sp. acquired accretions of other forms with morphological, cultural and prob. pathological differences. The result was an extremely confused situation, particularly with respect to the isolates from cacao which is the most important crop attacked. A disentanglement was begun by Turner although the significance of his work was overlooked, *TBMS* **43**:665, 1960; **44**: 409, 1961; *Phytop.* **51**:161, 1961. In 1975 Sansome et al., *Nature Lond.* **255**:704, described chromosome size differences amongst isolates from cacao. One group was later described as *P. megakarya*; another group is *P. palmivora* and the third is *P.* MF4 q.v. The differences between these 3 groups were described by Brasier in Gregory & Madison, *Phytopath.Pap.* 25, 1981. Like most pathogenic members of the genus *P. palmivora* is splash dispersed and survives in soil, Onesirosan, *Phytop.* **61**:975, 1981. It may be assoc. with other spp. in the genus on the same crop and causing similar diseases. Those which have been adequately studied are: cacao black pod, bark or stem and cushion canker, citrus fruit rot and

gummosis, papaya root rot, rubber black stripe, patch canker and abnormal leaf fall. Lesser diseases include betel wilt and decline, durian patch canker, palm bud rots. C.P.A. Bennett and his colleagues described a premature nutfall of coconut in Indonesia (Sulawesi), personal communication 1986; the condition is also caused by other pathogens and poor growing conditions. An epidemiological study of *P. palmivora*, similar to the one recently done for *P. megakarya* in Nigeria, on cacao is prob. needed; particularly with respect to S. America.

Chee, *Rev.appl.Mycol.* **48**:337, 1969, hosts; Gregory loc.cit.; Thorold 1975; Asare-Nyako in Kranz et al. 1977; Holliday 1980, review treatments; Harris et al., *TBMS* **82**:249, 1984, from coconut roots; Pereira and Mabbett, *Cocoa Growers Bull.* 36:23, 1985; 37:24, 1986, cacao black pod including chemical control; see black pod and bark canker [**8**:526; **40**:350, 521; **41**:86; **51**,1284; **54**,3815; **58**,4309; **63**,2452].

P. parasitica Dastur 1913, see *Phytophthora nicotianae*.

P. phaseoli Thaxter 1889; oogonia mostly 30 μm diam., group IV, LT, oospores in the host especially the pods; morphologically very near *Phytophthora infestans*, with truly airborne sporangia, but restricted to *Phaseolus*; studied only in USA (E. and central) and on Lima bean; the fungus has also been reported from: Italy, Mexico, Philippines, Puerto Rico, Sri Lanka, USSR (Caucasus), Zaire; races occur and disease forecasting is established; Sherf & MacNab 1986.

P. porri Foister 1931; Stamps, *Descr.F.* 595, 1978; oogonia mostly 38–39 μm diam., group III, MT; originally described as causing leek white tip, a leaf dieback in England and Scotland. Later it was shown to be pathogenic to other *Allium* spp. The fungus can cause a cabbage rot in storage and an isolate from *Campanula* did not infect leek. In Australia (S.) a low temp. form caused a lettuce stem rot, Sitepu & Bumbieris, *Australasian Pl.Path.* **10**:59, 1981. This form did not cause disease in either cabbage or leek. In Canada (Alberta) *Phytophthora porri* was described from stored carrots, Ho, *Mycologia* **75**:747, 1983. Carnation, chrysanthemum, gladiolus, tulip are attacked. The apparent pathogenic specialisation needs further study [**61**,4530; **63**,957].

P. primulae Tomlinson, *TBMS* **35**:233, 1952; Stamps, *Descr.F.* 839, 1985; oogonia av. 37 μm diam., group III, LT; primula brown core and root rot; in England on asparagus and parsley; the fungus occurs in Denmark and New Zealand; Clarkson & Phillips, *Tests Agrochem.Cvs* 8, *AAB* **110**, suppl.:64, 1987, chemical control in parsley.

P. pseudotsugae Hamm & Hansen, *CJB* **61**:2630, 1983; sporangiophore simple, unbranched; oogonia av. 35 μm diam., group I, MT; from roots of Douglas fir, USA (Oregon, Washington State) [**63**,1996].

P. quininea Crandall, *Mycologia* **39**:220, 1947; oogonia 68–83 μm diam., group V, MT; root and collar rot of *Cinchona*, Peru [**26**:467].

P. richardiae Buisman 1927; oogonia 34–38 μm diam., group VI, MT; root rot of *Calla* and arum lily, *Zantedeschia aethiopica*; Europe (N.W., central), Philippines, USA [**6**:616].

P. sinensis Yu & Zhuang, *Mycotaxon* **14**:183, 1982; oogonia av. 24 μm diam., group VI, HT; assoc. with a cucumber blight in China [**61**,5315].

P. syringae (Kleb.) Kleb. 1909; Waterhouse & Waterston, *Descr.F.* 32, 1964; oogonia av. 34 μm diam., group III, LT; Harris & Cole, *TBMS* **79**:527, 1982, found an optimum temp. of 12.5–15° for oospore germination. *Phytophthora syringae* is a widespread pathogen causing lilac twig blight and wilt, diseases of woody Rosaceae, particularly apple collar and fruit rots, on citrus. Sewell & Wilson, *AAB* **53**:275, 1964, described lethal collar rot of apple in England and compared *P. syringae* with *P. cactorum*. Apple fruit rot had hitherto been considered unimportant in England but in the 1970s heavy losses were reported in stored apples in Kent. The fruit was being infected via soil splash in the field around harvest time; fallen apple leaves were colonised. The horticultural practices of lower trees on dwarfing rootstocks and bare soil cultivation had led to this situation. This is an excellent example of how cultural change in crops may lead to heavy, unexpected loss if the pathogen situation is not fully considered. Infection of crops by *Phytophthora* spp. through soil splash is a common phenomenon. It has often been reported, in the tropics and sub-tropics, on black pepper, cacao and citrus.

On apple in England: Upstone and Upstone & Gunn, *Pl.Path.* **27**:24, 30, 1978; Harris, *AAB* **91**:309, 331, 1979; **107**:179, 1985; *TBMS* **85**:153, 1985. Also: Bostock & Doster, *PD* **69**:568, 1985, assoc. with pruning wound cankers on almond [**43**,2319; **58**,1310–11, 5944–5; **62**,2035; **64**,5004; **65**,2901; **66**,1043].

P. verrucosa Alcock & Foister, *Trans.bot.Soc.Edinb.* **33**:65, 1940; oogonia 37 μm diam., group V; tomato toe rot; Howells, *Scot.J.Agric.* **19**:47, 1936, and on primula; Byford 1958 reported the fungus from dahlia; England, Scotland [**15**:614; **20**:91; **37**:538].

P. vesicula Anastasiou & Churchland, *CJB* **47**:252, 1969 q.v. for differences from other aquatic

Phytophthora spp.; oogonia av. 46 μm diam., group III. Fell & Master, *ibid* **53**:2908, 1975, assoc. with degrading mangrove litter with 4 new spp. of the genus; Pegg et al., *Australasian Pl.Path.* **9**(3):6, 1980, on a sp. like *P. vesicula* & assoc. with mangrove death in Australia (Queensland) [**55**,3475; **60**, 4688].

P. vignae Purss, *Qd.J.agric.Sci.* **14**:141, 1957; also in *Bull.Div.Pl.Ind.Qd.* 107.; oogonia av. 32 μm diam., group VI, HT; cowpea stem rot; Australia (Queensland), India, Japan; 4 races in Australia; Kitazawa et al., *Ann.phytopath.Soc.Japan* **45**:406, 1979, ff.sp. for adzuki bean & cowpea; Ishiguro & Ui, *ibid* **47**:213, 1981, factors affecting oospore germination; Holliday 1980 [**37**:433; **59**,4375; **61**,2711].

Phytoreoviruses, type sub-group 1 as for group, wound tumour v.; type sub-group 2 Fijiviruses, sugarcane Fiji v.; type sub-group 3, rice ragged stunt v. The group has *c*.11 definitive members; isometric, double shell particles *c*.70 nm diam., 10–12 segments dsRNA, transm. hemipterous insects, leafhoppers and planthoppers; but not by sap, except rarely by needle puncture; often confined to phloem where hyperplasia occurs; Boccardo & Milne, *Descr.V.* 294, 1984; Francki et al., vol. 1, 1985.

Phytorhabdoviruses, type sub-group 1, lettuce necrotic yellows v.; type sub-group 2, potato yellow dwarf v.; the types differ in their protein molecular weights; cacao swollen shoot v., and other viruses similar to it, may fall into a third group. The group has *c*.37 definitive members; bacilliform, 135–380 × 70–95 nm, enveloped, 1 segment linear ssRNA, sedimenting at 1000–1200 S, thermal inactivation 45–60°, longevity in sap 1–3 days, transm. sap and insects, persistent; Francki et al. in Kurstak 1981; Peters, *Descr.V.* 244, 1981.

phytosanitary certificate, a certificate of health of plants or plant commodities, FBPP; a model certificate is outlined by Johnston & Booth 1983.

phytosanitation, measures requiring removal or destruction of infected or infested plant material likely to be a source of pathogens or pests, FBPP.

phytostilbenes, diphenylethylenes; Hart & Shrimpton, *Phytop.* **69**: 1138, 1979, role in resistance of wood to decay; Hart, *A.R.Phytop.* **19**:437, 1981, role as post-infection inhibitory compounds.

phytotoxin, see toxin.

piaropi, *Cercospora*.

Picea, spruce.

piceae, *Gemmamyces*, *Therrya*.

Pierce's grapevine leaf scald Pierce 1892; Goheen

et al., *Phytop.* **63**:341, 1973; Hopkins & Mollenhauer, *Science N.Y.* **179**:298, 1973; Davis et al., *ibid* **199**:75, 1978; caused by a xylem limited bacterium q.v., see *Xylella*. Periodic epidemics are caused and the pathogen can limit cultivation in USA (California, S.E.). The causal agent was transm. by Hewitt in 1939; but it was to take 34 years before a bacterium of this type was shown to be assoc. with the disease. A few years later it became the first disease to be known to be caused by such a bacterium. Elsewhere Pierce's leaf scald is found only in Costa Rica and Mexico. The bacterium infects several plants, including weeds near crops. It is apparently serologically different from the similar bacterium which causes peach phony; Hopkins, *Phytop.* **74**:1395, 1984; **75**:713, 1985, bacterial virulence & characteristics; Tyson et al., *ibid* **75**:264, 1985, scanning electron microscopy; Purcell & Frazier, *Hilgardia* 53(4), 1985, habits & dispersal of leafhopper vectors [**52**,3011; **53**,635; **64**,1679, 5044; **65**,309, 1977].

Piesma cinerea, transm. beet savoy; *P. quadratum*, transm. beet latent rosette, beet leaf curl, spinach witches' broom.

piesmids, see vector.

pigeon pea, *Cajanus cajan*; Kannaiyan et al., *Trop.Pest Management* **30**:62, 1984 [**64**,866].

—— —— **bushy canopy**, and mid foliar vein yellowing; Licha-Baquero and McCoy et al., respectively; see pigeon pea witches' broom.

—— —— **mosaics**, Newton & Peiris, *F.A.O.Pl.Prot.Bull.* **2**:17, 1953, Sri Lanka; Bisht & Banerjee, *Labdev.J.Sci.Technol.* **3**:271, 1965; Singh & Mall, *Curr.Sci.* **45**:635, 1976; last reference sap transm., India [**33**:581; **47**,81; **56**,2779].

—— —— **proliferation**, see pigeon pea witches' broom.

—— —— **rosette** Maramorosch, Kimura & Nene, *F.A.O.Pl.Prot.Bull.* **24**:33, 1976; MLO assoc., India (Andhra Pradesh) [**56**,2778].

—— —— **sterility mosaic** Capoor 1950; *Indian J.Agric.Sci.* **22**: 271, 1952; transm. *Aceria cajani*, S.E. Asia [**29**:405; **32**:629].

—— —— **witches' broom** Maramorosch et al., *F.A.O.Pl.Prot.Bull.* **22**:32, 1974; Vakili & Maramorosch, *PDR* **58**:96, 1974; Licha-Baquero & Maramorosch, *J.Agric.Univ.P.Rico* **64**:424, 1980; McCoy et al., *PD* **67**: 443, 1983; MLO assoc., poss. transm. *Empoasca* spp., W. Indies, USA (Florida). A plant rhabdovirus reported in work on this disease is considered to be assoc. with other symptoms. It is prob. pigeon pea proliferation listed by Boswell et al., *Rev. Pl.Path.* **65**:221, 1986 [**53**,2762; **54**,3073; **60**,4113; **62**,3712].

—— —— **yellow mosaic** Newton & Peiris, transm. sap, Sri Lanka see pigeon pea mosaics.

pigweed mosaic v. Singh et al., *Phytopath.Z.* **75**:82, 1972; filament 700–725 × 14 nm, transm. aphid, non-persistent, India (N.), *Amaranthus* [**52**,2206].

pinastri, *Leptostroma*, *Lophodermium*.

pincushion scab (*Elsinoë* sp.) Proteaceae, Benić & Knox-Davies 1983 [**64**,218].

pine, see *Pinus*; assoc. virus particles, Biddle & Tinsley, *Nature Lond.* **219**: 1387, 1968, Scots pine, Britain; see Cooper 1979 for conifers and assoc. viruses; **pinea**, *Brunchorstia*, *Caliciopsis*, *Diplodia*.

pineapple, *Ananas comosus*: Rohrbach & Apt, *PD* **70**:81, 1986, nematodes & other pathogens.

—— **chlorotic leaf spot** Kitajima et al., *Phytopath.Z.* **82**:83, 1975; bacilliform particles assoc., Brazil [**54**,4587].

—— **disease** (*Ceratocystis paradoxa*) sugarcane.

—— **yellow spot** (tomato spotted wilt v.).

pine cone rusts, *Cronartium conigenum*, *C. strobilinum*.

pine wood nematode, *Bursaphelenchus xylophilus*.

pini, *Cytospora*, *Dasyscyphus*, *Discosia*, *Endocronartium*, *Lachnellula*, *Mycosphaerella*, *Peridermium*, *Phellinus*, *Ramichloridium*, *Rhizosphaera*, *Scirrhia*.

pini-cembrae, *Phacidium*.

pinicola, *Ascocalyx*, *Atropellis*, *Crumenulopsis*, *Leptomelanconium*.

pini-densiflorae, *Cercoseptoria*.

piniphila, *Atropellis*; **piniphilum**, *Digitosporium*.

pinitorqua, *Melampsora*, see *M. populnea*.

pink crust (*Corticium salmonicolor*) woody, very largely tropical crops, see disease. Although usually called pink disease I have suggested that the use of the word disease in a description of a disease is undesirable. Petch, *The physiology and diseases of* Hevea brasiliensis, 1911, used incrustation in describing the disease, and it is in rubber that infection by *C. salmonicolor* has been most studied; hence pink crust. Pink disease is also used for the effects caused by fungus infection of turf grasses e.g. *Laetisaria fuciformis*, *Limonomyces culmigenus*, *L. roseipellis*.

—— **eye** (*Pseudomonas fluorescens*) potato; —— **grain** (*Erwinia rhapontici*) wheat; —— **mould** (*Trichothecium roseum*) apple; —— —— **rot** (*Trichothecium*) several crops, e.g. on fruit of apple, pear, tomato, on fruit after harvest; —— **patch** (*Laetisaria fuciformis*) turf grasses; —— **root** (*Pyrenochaeta terrestris*) onion, *Allium*; —— **rot** (*Phytophthora cryptogea*, *P. erythroseptica*) potato; —— **seed** (*Erwinia rhapontici*) wheat; —— **snow mould** (*Monographella nivalis*) temperate cereals and grasses.

pin nematode, *Paratylenchus projectus.*
pinnule blight (*Corticium anceps*) *Pteridium aquilinum.*
pinodella, var. of *Phoma medicaginis.*
pinodes, *Ascochyta, Didymella, Mycosphaerella.*
piñon blister rust (*Cronartium occidentale*) piñon pines.
Pinus, pine; Peterson & Jewell, *A.R.Phytop.* 6:23, 1968, American stem rusts; Gibson, *Commonw.For.Rev.* 49:267, 1970, *P. patula*; Gilbertson, *Fungi that decay ponderosa pine*, Univ. Arizona, 1974; Hiratsuka & Powell 1976, Canadian stem rusts; Gibson, *Diseases of forest trees widely planted as exotics in the tropics and southern hemisphere*, part II, *the genus* Pinus, 1979; Rowan, *PD* 65:53, 1981, chemical control of *P. taeda* seedling root rot; Krause, *Phytop.* 72:382, 1982, salt spray injury; Barnard et al., *PD* 69:196, 1985, fungi assoc. with root disease of *P. clausa* [58,402].
Piper betle, betel, pan; *P. nigrum*, black pepper.
piperiana, *Chrysomyxa.*
Piptoporus (Bull.) P. Karsten 1881; Polyporaceae; *P. betulinus* (Bull.) P. Karsten, MacDonald, *AAB* 24:289, 1937, red brown cubical stem rot of birch, the commonest rot of the European birch, *Betula pendula*, B. pubescens [16:716].
piri, Elsinoë.
piricularin, toxin from *Pyricularia oryzae*; Tarmari & Kaji in 1954, fide Ou 1985.
pirina, *Venturia*; pirinum, *Sphaceloma.*
pirolata, *Chrysomyxa*; pironii, *Nectriella.*
pisi, *Ascochyta, Erysiphe*; pisicola, *Fusicladium*; pisi-sativi, *Uromyces.*
pistachio, *Pistacia vera*; Ashworth et al., *Phytop.* 75:1084, 1985, nutrient disorders [65,2464].
— rosette Kreutzberg 1940; transm. seed, *Liothrips pistaciae*, Afghanistan, Iran, USSR (Tadjikistan, Uzbekistan) [21:210].
pisum bacilliform v. Caner, July & Vicente, *Summa Phytopath.* 2:264, 1976; particle *c.*280 × 65 nm, transm. sap to *Datura stramonium* and *Nicotiana glutinosa* [56,4764].
Pisum sativum, pea.
pit (*Streptomyces ipomoeae*) sweet potato; — canker (*Phytophthora inflata*) elm.
pitch canker (*Gibberella fujikuroi*) pines.
Pithomyces Berk. & Broome 1873; Hyphom.; Ellis 1971, 1976; Carmichael et al. 1980; conidiomata hyphal, conidiogenesis holoblastic; con. solitary, multiseptate or muriform, pigmented, usually echinulate or verruculose. *P. chartarum* (Berk & M.A. Curtis) M.B. Ellis 1960; Sutton & Gibson, *Descr.F.* 540, 1977; con. 18–29 × 10–17 μm; teleom. *Leptosphaerulina chartarum* Roux, *TBMS* 86:320, 1986; asco. usually 3 transverse septa, 1

longitudinal septum, smooth, 23–27 × 12 μm; largely saprophytic, causes a mycotoxicosis, facial eczema of sheep, can cause necroses of rice and sorghum leaves.
pith necrosis (*Pseudomonas corrugata*) tomato; — — and crown death, strawberry, Scotland, Mason & Rath, *Hort.Res.* 18:13, 1978, related to increasing plant age and soil parental material [57,5602].
pittieriana, *Puccinia.*
pitting (*Pyricularia grisea*) banana.
pittosporum rough bark Thomas & Baker, *Phytop.* 37:192, 1947; Alfieri & Seymour, *Proc.Fla.St.hort.Soc.* 83:438, 1970; USA (California) [26:341].
— vein yellowing v., or vein clearing, Plavšic-Banjac, Miličić & Erić, *Phytopath.Z.* 86:225, 1976; Rana & Franco, *Phytopath. Mediterranea* 18:48, 1979; Marani & Bertaccini, *ibid* 19:57, 1980; bacilliform 340–400 × 8 nm, transm. sap [56,1631; 62,4707].
plane, *Platanus*; in N. America the planes are usually called sycamore.
Planococcoides njalensis, transm. cacao swollen shoot; see *Adansonia.*
Planococcus citri, transm. cacao swollen shoot, cacao mottle leaf, grapevine A; *P. ficus*, transm. grapevine A; *P. kenyae*, transm. the 2 cacao viruses.
Plantago, Hammond, *Adv.Virus Res.* 27:103, 1982, a host for economically important viruses.
plantago American mottle v. Granett 1972, *Phytop.* 63:1313, 1973; Tymovirus, isometric av. 26 nm diam., transm. sap to antirrhinum, pea, *Plantago*; causes pea leaf veinal necrosis q.v., USA [53,2483].
— Argentinian v. Gracia, Koenig & Lesemann, *Phytop.* 73:1488, 1983; Potexvirus, filament 538 μm long, transm. sap, wide host range [63,2114].
— English mottle Hitchborn, Hills & Hull, *Virology* 28:768, 1966; bacilliform particles assoc., mostly 330 × 63 nm [45,2771].
— mosaic (ribgrass mosaic v.).
— severe mottle v. Rowhani & Peterson, *Can.J.Pl.Path.* 2:12, 1980; Potexvirus, filament, modal length 536 nm, transm. sap, Canada (Quebec) [59,5663].
plantain X v. Hammond 1977; Hammond & Hull, *Descr.V.* 266, 1983; Potexvirus, filament *c.*570–580 × 12 nm, transm. sap, Britain.
— yellows Staniulis & Genyte, *Phytopath.Z.* 86:240, 1976; USSR (Lithuania) [56,1058].
— 4–8 Hammond, *Pl.Path.* 30:237, 1981, England [61,5439].
planta macho, see tomato.
plantarii, *Pseudomonas.*

plant breeding, see resistance to pathogens; ——
exudate, see exudate.

—— **hormones**, in plant disease, Mahadevan, *Growth regulators, micro-organisms and diseased plants*, 1984.

—— **names**, the most widely used books were: Airy Shaw 1966, Purseglove 1968, 1972, Howes 1974, Simmonds 1976. Howes has a bibliography of other works. Crop names have been corrected but for plant names in general the one given by the author(s) is used. Common plant names are only inserted in the order where there is some literature on pathology, where it is a crop name or where the name is used to designate a pathogen, e.g. MLO, virus or viroid; see disease names.

—— **physiology**, effect of disease on, Ayres ed., *Soc.exp.Biol.* Seminar Ser. 11, 1981.

—— **quarantine**, regulation of the movement of living plants or their parts and plant products between political boundaries and within political territories. It involves legislation, inspection, treatment, certification, post and intermediate quarantine, and international co-operation; Prentice in Western 1971; Hewitt & Chiarappa ed., *Plant health and quarantine in international transfer of genetic resources*, 1977; Ebbels & King 1979; Mathys & Baker, *A.R.Phytop.* **18**:85, 1980; Waterworth & White and Dowling et al., *PD* **66**:87, 345, 1982; Johnston & Booth 1983.

—— **reoviruses**, and plant rhabdoviruses, see Phytoreoviruses and Phytorhabdoviruses, respectively.

—— **transport systems**, effects of pathogens on, Farrar in Wood & Jellis 1984.

—— **tumour**, see tumour.

plaque, a restricted, clear area in a lawn of confluent bacterial growth in culture, resulting from lysis of the bacteria in that area by a phage, FBPP. The term has been used in mycology, e.g. Riegman & Wessels, *TBMS* **75**:325, 1980.

plasmatocecidia, galls caused by MLO or viruses; Paclt, *Z.PflKrankh.PflSchutz.* **88**:152, 1981 [**60**,5820].

plasmid, 'A small, autonomously replicating molecule of covalently closed circular DNA that is devoid of protein, and which is not essential for the survival of its host. Plasmids have been found only in bacteria'; fide Diener & Prusiner in Maramorosch & McKelvey 1985; Nester & Kosuge, *A.Rev.Microbiol.* **35**:531, 1981, plasmids specifying plant hyperplasias; Panopoulos & Peet, *A.R.Phytop.* **23**:381, 1985, molecular genetics of plant pathogenic bacteria & their plasmids.

Plasmodiophora Woronin 1877; Plasmodiophoromycetes.

P. brassicae Woronin; Buczacki, *Descr.F.* 621,

1979; cysts or resting sporangia not united into cytosori but lying free in the host cell, minutely spiny; on Cruciferae, clubroot or finger and toe of brassicas is an important disease with a long history of investigation and control is not entirely satisfactory; the symptoms are gross root malformations. The life history is not completely known. A haploid, uninucleate resting spore in soil germinates into a biflagellate, uninucleate, primary zoospore; this injects a root hair with a plasmodium which becomes multinucleate by mitosis, forms zoosporangia and uninucleate, biflagellate, secondary zoospores. These are released into the soil but later phases are not clear. They may pair, fuse and form a secondary, binucleate plasmodium which becomes multinucleate by mitosis. Karyogamy is followed by meiosis, formation of resting spores which are released after root disintegration, soil survival is 18 years or more.

Infection does not usually kill the plant but it induces wilt, stunting and physical instability. The fungus is very variable in pathogenicity. The pathotypes or races have been called populations since they prob. lack homogeneity. A complex host and pathogen situation has led to the adoption of a European clubroot differential set, Toxopeus et al., *TBMS* **87**:279, 1986. Control by applying lime has long been practised since the disease is more severe on acid soils. Chemical treatment of transplants and soil can be satisfactory. High resistance is found in *Raphanus*. Colhoun, *Phytopath.Pap.* 3, 1958, monograph; Jönsson, *Acta Agric.Scand.* **28**:261, 1975, bibliography; Mazin, *Mikol.i Fitopatol.* **10**:446, 1976; Crute et al., *Pl.Br.Abs.* **50**:91, 1980, variation in pathogen & resistance; Buczacki 1983; Campbell et al., *Phytop.* **75**:665, 1985, control with lime in USA (California); Crute, *Adv.Pl.Path.* **5**:1, 1986, relationships of pathogen & hosts [**65**,422; **66**,1659].

Plasmodiophoromycetes, Plasmodiophorales, Myxomycota; obligate plant parasites, growth phase a minute intra-cellular plasmodium; *Ligniera, Plasmodiophora, Polymyxa, Sorosphaera, Spongospora*; Karling, *The Plasmodiophorales*, edn. 2, 1968; Waterhouse in Ainsworth et al. IV B, 1983; Fraser & Buczacki, *TBMS* **80**:107, 1983, ribosomal RNA molecular weights & affinities with other taxonomic groups [**62**,2892].

Plasmopara Schröter 1886; Peronosporaceae; sporophore variable but typically narrow, usually 8–10 μm wide; its secondary and later branches at right angles to the axis, straight tips blunt; spore wall poroid, germination usually by zoospores; Spencer 1981.

P. halstedii (Farlow) Berl. & de Toni 1888; sporangia 18–30 × 14–25 μm, oospores 30–32 μm diam.; sunflower false or downy mildew, on Compositae, the fungus is found on wild Helianthus in N. America; a major, widespread disease but not reported from Australasia. The young plant is infected and the pathogen becomes systemic, seedborne, latent infection of roots, oospore survival in soil is prob. in years; Holliday 1980; Sackston in Spencer 1981; Carson, PD 65:842, 1981, races; Melero-Vara et al., ibid 66:132, 1982, control with metalaxyl; Gray & Sackston, CJB 63:1725, 1985, infection by 3 races [61,2382, 5164; 65,842].

P. penniseti, Kenneth & Kranz, TBMS 60:591, 1973; sporangia 19–23.7 × 14.2–17 μm, oospores unknown; pearl millet leaf brown stripe, Ethiopia [52,4057].

P. viticola (Berk. & Curtis ex de Bary) Berl. & de Toni 1886; sporangia 17–25 × 10–16 μm, oospores 28–40 μm diam. Grapevine downy mildew has an important niche in the history of plant pathology. It is indigenous on Vitis in N. America. Stock of these plants was imported by Europe in c.1870 in the hope of finding resistance to Viteus vitifoliae, an aphid which infested the leaves and roots. This aphid, also a native of N. America, had been brought to Europe a few years before. The introduction of rootstocks to be used against the aphid led also to the introduction of Plasmopara viticola. This is one of the early examples of man's spread of a fungus pathogen into a susceptible and extremely important economic crop. Berkeley & Curtis's fungus was described in 1848. But Heald in 1926 stated that the fungal cause of the disease had already been recognised by Schweinitz in 1834. This date is the one usually given for Schweinitz's Synopsis fungorum in America Boreali in Trans.Am.Phil.Soc. 4:141, it should be 1832, fide Rogers, Mycologia 36: 526, 1944.

The spread of P. viticola in France led to the momentous discovery of the fungicide bouille bordelaise or Bordeaux mixture by Millardet in 1882. Epidemics of grapevine downy mildew had raged in France in the years following 1878; Bordeaux gave eventual control. The fungus overwinters mainly as oospores in fallen leaves, and primary infections in spring arise from this inoculum. Later infections come through sporangia which have a dispersal peak at c.1800 hours. Zoospore liberation has an optimum temp. of 22–25°; host penetration is mainly through the stomata. For control chemicals are applied c.5–6 times, based on forecasts, in late spring and in summer; Lafon & Bulit and Viennot-Bourgin in Spencer 1981; Gehmann et al.,

Z.PflKrankh.PflSchutz. 94:230, 1987, temp. & oospore formation.

platani, Gnomonia; platanicola, Cercospora; platanifolia, Mycosphaerella.

Platanus, planes.

platanus leaf scorch = sycamore leaf scorch.

Plectomycetes, a class proposed for Ascomycotina with ± globose, non-ostiolate ascomata but now recognised to be heterogeneous.

*Plectophomella Moesz 1922; Coelom.; Sutton 1980; conidiomata stromatic, conidiogenesis enteroblastic phialidic; con. aseptate, hyaline, P. ulmi (Verrall & May) Redfern & B. Sutton, TBMS 77:383, 1981; conidiomata 115 μm diam., in longitudinal fissures, con. 2.5–4 × 1–1.5 μm; Verrall & May, Mycologia 29:321, 1937, as Dothiorella ulmi; elm dieback or wilt, USA. P. concentrica Redfern & Sutton loc.cit.; conidiomata up to 800 μm diam. in concentric zones, con. 2.5–4 × 1–1.5 μm; canker or dieback of Ulmus glabra in Britain [16:782; 61,3664].

Plectosphaerella Kleb. 1930; Trichosphaeriaceae, see Fusarium tabacinum.

Pleiochaeta (Sacc.) S. Hughes 1951; Hyphom.; Ellis 1971; Carmichael et al. 1980; conidiomata hyphal, conidiogenesis holoblastic, conidiophore growth sympodial; con. multiseptate, hyaline to pigmented; apical cells bear several long, hyaline appendages which may branch, Holliday 1980.

P. setosa (Kirchner) S. Hughes; Pirozynski, Fungi Canadenses 12, 1974; Ellis & Holliday, Descr.F. 495, 1976; con. mostly 5 septate, 60–90 × 14–22 μm; lupin brown spot, roots are attacked, on bean and Crotalaria; leaf, stem and pod lesions, seedborne.

Plenodomus Preuss 1841 = Phoma; Boerema et al. TBMS 77:61, 1981, formally proposed Plenodomus to be a section of Phoma. Plenodomus destruens Harter, J.agric.Res. 1:251, 1913, was described as causing a sweet potato foot rot; for P. meliloti see Phoma sclerotioides [61,1596].

pleomorphic, pleomorphism, having more than one form; in biology showing different forms in the life cycle or life history of an organism, e.g. fungi, see holomorph. Polymorphism, virtually synonymous, is generally used in entomology; thus the 2 words have acquired a biological distinctness; Savile, Mycologia 61:1161. 1969.

Pleospora Rabenh. ex Ces. de Not. 1863; Pleosporaceae, Sivanesan 1984; ascomata scattered, unilocular, immersed, short papillate ostiole; asco. muriform, yellow brown to dark brown; anam. include: Alternaria, Diplodia, Phoma, Stemphylium; Simmons, Mycotaxon 25:288, 1986.

P. alfalfae Simmons, Sydowia 38:292, 1985,

published 1986; anam. *Stemphylium alfalfae* Simmons loc.cit.; asco. *c*.38–42 × 12–14 μm; con. *c*.32–35 × 16–19 μm. The sp. was erected in connection with recent work by Cowling et al., see *Pleospora herbarum*, and Irwin, *PD* **68**:531, 1984, on a lucerne disease [63, 4479].

P. allii, poss. anam. *Stemphylium vesicarium* q.v.

P. bjöerlingii Byford 1963; Booth, *Descr.F.* 149, 1967; Sivanesan 1984 as *Pleospora betae* (Berl.) Nowodowski 1915; anam. *Phoma betae* Frank 1892; asco. with 3 transverse septa, 18–25 × 7–10 μm; con. 5–9 × 3–5 μm; beet black leg, seedborne; Maude & Bambridge, *Pl.Path* **34**:435, 1985, found that red beet seed gave a high germination after storage for 13 years at 10° and 50% relative humidity; and infection declined from 27.5 to 4.5%. A seed soak in thiram or ethyl mercury phosphate improved germination and reduced mean infection from 17% to < 1%. Byford, *ibid* page 463, compared chemical treatments for improvement of sugarbeet seedling establishment; Dixon 1981; Sherf & MacNab 1986 [65,412, 1570].

P. drummondii Nirenberg & Plate, *Phytopath.Z.* **107**:365, 1983; anam. *Stemphylium drummondii*, same authors; asco. 43–59 × 20–26 μm, con. 33.8 × 22.6, verrucose; leaf spot of *Phlox drummondii*, previously ascribed to *Pleospora herbarum*.

P. herbarum (Pers.) Rabenh. ex Ces. & de Not. 1863, fide Cannon et al. 1985; Booth & Pirozynski, *Descr.F.* 150, 1967; Corlett et al., *Fungi Canadenses* 232, 1982; anam. *Stemphylium herbarum* Simmonds, *Sydowia* **38**:291, 1985, published 1986; asco. finally muriform, 26–45 × 14–20 μm; con. commonly with 3 transverse and 1–3 longitudinal septa, 26–42 × 24–30 μm; saprophytic and weakly pathogenic, particularly temperate and subtropical, seedborne, may vary in pathogenicity between isolates; Cowling et al., *Phytop.* **71**:679, 1981; **72**:26, 1982, biotypes on lucerne in N. America; see *Pleospora alfalfae*; Holliday 1980 [61,1264, 5060].

P. papaveracea (de Not.) Sacc. 1883; Sivanesan & Holliday, *Descr.F.* 730, 1982; anam. *Dendryphion penicillatum* (Corda) Fr. 1849; asco. 3 septate, with a longitudinal septum usually in the 2 central cells, 20–25 × 6–9 μm; con. usually 3 septate, 17–28 × 5–9 μm; opium poppy leaf blight, seedborne, studied mostly in E. and central Europe, and in Japan.

Pleosporaceae, Dothideales; *Cochliobolus, Didymella, Gemmamyces, Herpotrichia, Leptosphaeria, Pleospora, Pyrenophora, Setosphaeria*.

plerotic, of oospores, Pythiaceae, filling the oogonium.

plesionecrotic, see necrotic.

Pleuroceras pseudoplatani, see *Ophiognomonia pseudoplatani*.

Plioderma Darker, *CJB* **45**:1424, 1967; Hypodermataceae; ascomata mono-ascocarpus, basal tissue weakly developed, mostly elliptical; asci rather broad, asco. aseptate, like rods.

P. hedgcockii (Dearness) Darker; Cannon & Minter, *Descr.F.* 799, 1984; anam. *Leptostroma hedgcockii* Dearness 1928; asco. 27–35 × 7.5–10 μm; con. mostly 1 septate, septum 4–6 μm from base, 15–19 × 4.5–5.5 μm; pine needle blight, USA, may cause premature needle shed.

P. lethale (Dearness) Darker; Minter & Gibson, *Descr.F.* 570, 1978; asco. 20–30 × *c*. 4 μm, in a gelatinous sheath 4–6 μm wide; pine needle blight, USA (E.)

Plowright, Charles Bagge, 1849–1910; born in England; a doctor and surgeon; wrote: *A monograph of the British Uredineae and Ustilagineae*, 1889. *TBMS* **3**:231, 1910.

plum, *Prunus* q.v.

—— **American line pattern v.** Valleau 1932; Fulton, *Descr.V.* 280, 1984; Ilarvirus, quasi-isometric, sedimenting as 4 components *c*.26, 28, 31, 33 nm diam., transm. sap, N. America but may be more widely distributed, not serologically related to prunus necrotic ringspot v. or apple mosaic v.; these 3 viruses or their strs. may cause line pattern in plum.

—— **bark split** Posnette 1953; Posnette & Ellenberger, *AAB* **45**: 573, 1957; may be caused by apple chlorotic leaf spot v.; Dunez et al., *Annls.Phytopath.* **3**:523, 1971; *PDR* **56**:293, 1972 [33:95; 37:242; 51,4163–4].

—— **chlorosis and wilt** Ionică et al., *An.Inst.Cerc.Prot.Pl.* **14**:11, 1979; MLO assoc., Romania [59,1313].

—— **crinkle leaf**, prob. genetic disorder; Pine in Gilmer et al. 1976.

—— **decline** Goidànich 1933; Giunchedi et al., *Phytopath. Mediterranea* **17**:205, 1978, MLO assoc., Italy, *Prunus salicina* [60,3237].

—— **European line pattern** (apple mosaic v.).

—— **false crinkle**, not transm., prob. a disorder, New Zealand, fide Atkinson 1971.

—— **fruit crinkle** Chamberlain, Atkinson & Hunter, *N.Z.J.Agric. Res.* **2**:174, 1959; New Zealand (Auckland, Hawke's Bay), symptoms resemble those of plum pox, prob. eradicated, fide Atkinson 1971; *Prunus salicina* [38:413].

—— —— **spotting** Canova & Casalicchio, *Phytopath.Mediterranea* **2**:147, 1963; Italy [43, 779 k].

— **infectious chlorosis** Louw 1949, 1951, South Africa (Cape Province) [**28**:338; **31**:537].
— **latent**, or prunus latent; Proeseler, *Phytopath.Z.* **63**:1, 1968; transm. sap, *Vasates fockeui*, Germany [**48**,853].
— **leaf scald** Fernandez Valiela & Bakarcic 1954, 1963; Kitajima et al. and Raju et al., *Phytop.* **65**:476, 1975; **72**:1460, 1982; a xylem limited, Gram negative bacterium, see *Xylella*; pathogen very similar to others, see peach phony. The disease, first described from Argentina (Paraná river delta) > 45 years ago, has caused severe losses in Brazil (Paraná, Rio Grande do Sul, Santa Catarina); prob. also in São Paulo; French & Feliciano, *PD* **56**:515, 1982 [**35**:617; **44**,2548; **55**,320; **61**,6485; **62**,1570].
— **ochre mosaic** Blattný, *Phytopath.Z.* **49**:102, 1963 [**43**,2310].
— **pox** v. Atanasoff 1932; Kegler & Schade, *Descr.V.* 70, 1971; Potyvirus, filament *c.* 764 × 20 nm, transm. sap, aphid spp., non-persistent; Europe, Turkey; restricted host range, causes diseases in apricot, peach, plum; *EPPO Bull.* **4**(1), 1974; Adams in Scott & Bainbridge 1978, incidence & control in England.
— **pseudo pox** Caspar 1977, Germany [**57**,672].
— **punctiform mosaic** Ragozzino, Graziano & Bellone 1972, Italy (Campania) [**52**,2310].
— **ringspot mosaic** Blattný, *Ochr.Rost.* **14**:86, 1938; Czechoslovakia [**17**:543].
— **rusty blotch** Pine & Cochrane, *PDR* **44**:87, 1960; originally described as transmissable but this is prob. incorrect; USA (California), *Prunus salicina*; Pine in Gilmer et al. 1976 [**39**:478].
— **shot hole** Smith & Cochrane, *Phytop.* **33**:1101, 1943; Pine & Welsh in Gilmer et al. 1976; prob. genetic abnormality [**23**:113].
— **stripe mosaic** Blattný 1969; Czechoslovakia [**49**,1428].
— **white spot** Thomas & Rawlins, *Hilgardia* **12**:623, 1939; USA (California) [**19**:416].
Plumeria rubra, frangipani, temple tree.
poae, *Drechslera, Fusarium*; **poae-nemoralis**, *Puccinia*.
Poa pratensis, Kentucky blue grass, meadow grass.
poarum, *Puccinia*.
poa semi-latent v. Slykhuis, *Phytop.* **62**:508, 1972; Hordeivirus, rod 125–225 × 25 nm, transm. sap, *P. palustris*, other Gramineae including oat and wheat; Hunter et al., *Phytop.* **76**: 322, 1986, in a structural comparison with barley stripe mosaic v. confirmed that the 2 viruses are distinct Hordeiviruses [**65**,4779].
pocket plum (*Taphrina pruni*).
Podagrica uniformis, transm. okra mosaic.

pod and stem blight (*Diaporthe phaseolorum*) Lima bean, soybean.
Podosphaera Kunze 1823; Erysiphaceae; mycelium superficial or ectoparasitic, ascomata with dichotomously branched appendages and 1 ascus; con. in chains, ellipsoidal with fibrosin bodies; Spencer 1978.
P. clandestina (Wallr.) Lév. 1851; Khairi & Preece, *Descr.F.* 478, 1975; asco. 10–12 × 7–14 μm; con. 18–23 × 9–10 μm; hawthorn mildew, *Crataegus* spp., on pear, quince and other plants; poss. different pathogenic forms.
P. leucotricha (Ell. & Ev.) E. Salmon 1900; Kapoor, *Descr.F.* 158, 1967; asco. 22–26 × 12–15 μm; con. in long chains, 22–30 × 12–15 μm; distinguished from other *Podosphaera* spp. by the tuft of unbranched, apical appendages on the ascoma; basal appendages are present; apple powdery mildew, often of minor importance, but it requires many fungicide applications in Britain; leaves, young shoots, blossoms and fruit are attacked, defoliation may be caused; the fungus overwinters as dormant mycelium in fruit and vegetative buds; ascomata are formed but asco. are prob. not epidemiologically important. Burchill and Bull in Spencer 1978; Jeger & Butt, *F.A.O.Pl. Prot.Bull.* **32**:61, 1984, management of the disease with *Venturia inequalis*; Locke & Andrews, *Pl.Path* **35**:241, 1986; effects of fungicides on mildew, tree growth & cropping in cv. Cox's Orange Pippin; Jeger et al., Butt & Jeger and Jeger & Butt, *ibid* **35**:477, 491, 498, 1986, resistance, formation of con., epidemics in mixed cv. orchard [**64**,1644; **65**,5595; **66**,1980–2].
P. tridactyla (Wallr.) de Bary 1870; Mukerji, *Descr.F.* 187, 1968; asco. 18–30 × 12–15 μm, con. 22–45 × 14–20 μm; differs from *Podosphaera leucotricha* in having long, branched appendages; *Prunus* mildew.
pod rot and black stripe (*Phytophthora botryosa, P. heveae, P. palmivora*) rubber; — **twist** (*Pseudomonas flectens*) bean.
pogostemon mottle Kitajima & Costa, *Fitopat.Brasileira* **4**:55, 1979; bacilliform particles assoc., Brazil [**58**,5279].
— **yellow mosaic** Sastry & Vasanthakumar, *Curr.Sci.* **50**:767, 1981; transm. *Bemisia tabaci*, India (Mysore), *P. patchouly* [**61**,3583].
— — **spot** Roland, *Parasitica* **6**:8, 1950; transm. sap., Belgium, *P. patchouly* [**29**:619].
poinsettia, *Euphorbia pulcherrima*.
— **cryptic** v. Koenig & Lesemann, *PD* **64**:782, 1980; isometric 28 nm diam., USA [**60**,3205].
— **mosaic** v. R.W. & J.L. Fulton 1980; Koenig et al., *Descr.V.* 311, 1986; poss. Tymovirus, contains

ssRNA, isometric *c*.26 and 29 nm diam., sedimenting as 2 components, transm. sap, occurs naturally in *Euphorbia fulgens, E. pulcherrima*; experimental hosts *Euphorbia, Nicotiana*; poinsettias with the virus usually contain poinsettia cryptic v.

— **stunt** Davino & Cartia, *Tec.agric.Catania* **33**:81, 1981; Italy (Sicily) [**62**,2015].

poison hemlock, *Conium maculatum* q.v.

— — **ringspot** Freitag & Severin, *Hilgardia* **16**:389, 1945; transm. *Hyadaphis foeniculi*, USA (California) [**25**:289].

pokeweed, *Phytolacca americana.*

— **crinkle leaf** Lackey, *PDR* **49**:1002, 1965; USA (California), from *Adenostema fasciculatum* [**45**,1319].

— **mosaic v.** Woods 1902; Shepherd, *Descr.V.* 97, 1972; Potyvirus, filament av. 776 × 12–13 nm, transm. sap, *Myz.pers.*, non-persistent, N. America.

pokkah boeng (*Gibberella fujikuroi*) sugarcane.

polar, of bacterial flagella at one end or pole of the cell, a rod; sometimes called cephalotrichous.

polianthes leaf mottle Pearson & Horner, *Australasian Pl.Path.* **15**:39, 1986; poss. Potyvirus, filament *c*.733 nm long, New Zealand, *P. tuberosa* [**66**,1950].

Pollaccia Baldacci & Cif. 1937; Hyphom.; Ellis 1971, 1976; Sivanesan 1984; conidiomata sporodochial, conidiogenesis holoblastic, conidiophore growth percurrent; con. solitary, pigmented, 1–2 septate; teleom. *Venturia*; **P. balsamiferae, P. radiosa, P. saliciperda,** teleom. *V. populina, V. macularis, V. saliciperda,* respectively.

pollen, in the spread of viruses; Härdtl, *Gartenbauwissenschaft* **43**:34, 1978; Phatak, *Trop.Pest Management* **26**:278, 1980; Dunez, *Revue Cytol.Biol.Vég.Botaniste* **5**:21, 1982; Haight & Gibbs, *Pl.Path.* **32**:369, 1983, described the effect of viruses on pollen morphology; Mandahar & Gill, *Z.PflKrankh.PflSchutz.* **91**:246, 1984; Huang & Kokko, *Phytop.* **75**:859, 1985, described infection of pollen by *Verticillium albo-atrum* [**58**,3720; **63**,1165; **64**,96; **65**,254].

pollu (*Glomerella cingulata*) black pepper.

pollution, see air pollution.

polyblastis, *Botryotinia, Botrytis*; **polychrous,** *Lentinus.*

polycyclic diseases, see monocyclic diseases.

polygenic resistance, see resistance to pathogens.

polygoni, *Erysiphe*; **polymastum,** *Pythium*; **polymorpha,** *Xylaria.*

polymorphism, see pleomorphic which is the preferred term for fungi.

Polymorphum Chev. 1822; Coelom.; Sutton 1980; conidiomata stromatic, conidiogenesis holoblastic; con. aseptate, solitary; Hawksworth &

Punithalingam, *TBMS* **60**:501, 1973, typification; *P. quercinum*, teleom. *Ascodichaena rugosa.*

Polymyxa Ledingham 1939; Plasmodiophoromycetes; cysts 4–7 μm diam., mostly united in cytosori, variable in shape, little or no hypertrophy in hosts; sporangia relatively large, elongated, lobed, exit tubes long; Barr, *Can.J.Pl.Path.* **1**:85, 1979; *Fungi Canadenses* 199, 200, 1981 [**59**,3158].

P. betae Keskin 1964, on Chenopodiaceae, beet, spinach, *Chenopodium*; transm. beet necrotic yellow vein, Abe, *Rep. Hokkaido prefect.agric.Exp.Stns* 60, 1987, ecology & control as a vector of BNYV causing sugarbeet rhizomania, Japan; f.sp. *amaranthi* on *Amaranthus retroflexus. Polymyxa graminis* Ledingham mostly on Gramineae, including cereals; transm. barley yellow mosaic, oat mosaic, prob. peanut West African clump, poss. rice necrosis mosaic, wheat soilborne mosaic, wheat spindle streak mosaic; Abu & Ui, *Ann.phytopth.Soc.Japan* **52**:394, 1986, hosts of *P. betae* [**66**,1653].

polypodium leaf deformation Canova & Casalicchio, *Phytopath. Mediterranea* **2**:88, 1963; transm. sap, Italy [**43**,106].

Polyporaceae, Aphyllophorales; poroid form, basidiomata very variable but never clavarioid; basidiospores hyaline, rarely ornamented or amyloid; lignicolous or humicolous; *Fomes, Fomitopsis, Heterobasidion, Laetiporus, Lentinus, Macrohyporia, Perenniporia, Piptoporus, Polyporus, Poria, Rigidoporus*; Pegler, *Bull.Br.mycol.Soc.* suppl. 7(1) 1973; in Ainsworth et al., IVB, 1973; Gilbertson & Ryvarden, *North American polypores*, vol. 1, 1986.

Polyporus (Mich.) Fr. 1821, fide Pegler 1973; Polyporaceae; lignicolous, context whitish, pores not hexagonal, basidiospores not globose. *P. squamosus* Fr. 1821; saddle back fungus, Dryad's saddle; heart rot of temperate hardwoods, particularly in elm; Campbell & Munson, *AAB* **23**:453, 1936 [**16**:137].

Polyscytalum Reiss 1853; Hyphom.; Ellis 1971; Carmichael et al. 1980; conidiomata hyphal, conidiogenesis holoblastic, con. 0–1 septate, hyaline, catenate, acropetal.

P. pustulans (Owen & Wakef.) M.B. Ellis 1976; con. mostly aseptate, 6–18 × 2–3 μm; potato tuber skin spot; Hide et al., *AAB* **64**:265, 1969, time & infection, control in England; Lapwood & Hide in Western 1971 [**49**,1122].

polysora, *Puccinia*; **polyspora,** *Sydowia.*

Polystigmatales, Ascomycotina; ascomata perithecioid, ascohymenial, immersed, thin walls; asci unitunicate, with a narrow, apical annulus, not amyloid.

Polythrincium Kunze 1817; Hyphom.; Ellis 1971;
Carmichael et al. 1980; conidiomata hyphal,
conidiogenesis holoblastic; conidiophore growth
sympodial, large scars; con. 1 septate, solitary,
pigmented, cuneiform or pyriform; *P. trifolii*,
teleom. *Cymadothea trifolii*.

pomaceus, *Phellinus*.

pome fruit, see fruits.

pomi, *Cylindrosporium*, *Mycosphaerella*,
Schizothyrium, *Spilocaea*.

pomigena, *Gloeodes*.

Poncirus trifoliata, trifoliate orange.

popcorn (*Ciboria carunculoides*) mulberry.

*****poplar**, *Populus*.

— **decline** v. Martin, Berbee & Omuemu, *Phytop.*
72:1158, 1982; Potyvirus, filament, modal length
800–810 nm, transm. sap, USA (Wisconsin)
[**62**,1226].

— **mosaic** v. Atanasoff 1935; Biddle & Tinsley,
Descr.V. 75, 1971; Carlavirus, filament 626–
735 nm long, transm. sap; Canada, Europe;
growth reduced, poplar is the only natural host;
Boccardo & Milne, *Phytopath.Z.* **87**:120, 1976;
Cooper 1979; van der Meer et al., *Neth.J.Pl.Path.*
86:99, 1980; Cooper & Edwards, *AAB* **99**:53,
1981, detection, analysis, purification &
distribution in hybrid poplars; Cooper et al.,
Eur.J.For.Path. **16**:116, 1986, detection &
occurrence in England & Poland [**56**,2686;
60,1095; **61**,2467; **65**,6201].

— **necrotic leaf spot** Boyer, *CJB* **40**:1237, 1962;
Boyer & Navratil, *ibid* **48**:1141, 1970; transm.
seed, Canada (Quebec) [**42**:282; **51**,4396].

— **vein yellowing** Navratil, abs. *Phytop.* **71**:245,
1981; bacilliform particles assoc. *c.*460–540 ×
13 nm, Canada (British Columbia), USA
(Oregon).

poppy, *Papaver*; *P. somniferum*, opium poppy.

populea, *Cryptodiaporthe*; **populeum**, *Discosporium*;
populi, *Marssonina*, *Mycosphaerella*, *Septoria*;
populi-albae, *Drepanopeziza*; **populicola**,
Mycosphaerella, see *M. populi*, *Septoria*; **populina**,
Cryptosphaeria, *Taphrina*, *Venturia*; **populiperda**,
Septotina, *Septotis*; **populnea**, *Melampsora*;
populorum, *Drepanopeziza*, *Mycosphaerella*, see
M. populi.

Populus, *P. alba*, white poplar; *P. deltoides*, eastern
cottonwood; *P. grandidentata*, large toothed
aspen; *P. nigra*, black poplar; *P. tremula*,
European aspen; *P. tremuloides*, Canadian or
quaking aspen; *P. trichocarpa*, black cottonwood;
Thielges & Land 1976; Lindsey & Gilbertson,
Basidiomycetes that decay aspen in North America,
Bibliotheca Mycologica 63, 1978; Pinon,
Eur.J.For.Path. **14**:415, 1984, disease management
[**58**,5034].

populus bushy top v. Palukaitis et al., *Intervirology*
16:136, 1981; Tobamovirus, China, *P. tomentosa*
[**61**,5580].

Poria Pers. 1794; Polyporaceae; lignicolous,
basidiomata resupinate, context white becoming
yellowish; mostly saprophytic; Wright, *Mycologia*
56:692, 1964.

P. hypobrunnea Petch 1916; Pegler & Gibson,
Descr.F. 322, 1972; root and branch canker of
rubber and tea, on woody, tropical crops;
Holliday 1980 for *Poria hypolateritia* (Berk.)
Cooke and *P. vincta* (Berk.) Cooke.

poroid, like tubes, see Aphyllophorales.

porri, *Alternaria*, *Botryotinia*, *Botrytis*,
Leptotrochila, *Phytophthora*.

portulacae, *Dichotomophthora*.

postbloom fruit drop (*Glomerella cingulata*)
citrus.

postharvest loss, Coursey & Booth, *Rev.Pl.Path.*
51:751, 1972, tropics; Tuite & Foster,
A.R.Phytop. **17**:343, 1979, grain storage; Sommer,
PD **66**:357, 1982; fruit; Dennis 1983; Williams &
McDonald, *A.R.Phytop.* **21**:153, 1983, grain
moulds; Eckert & Ogawa, *ibid* **23**:421, 1985,
chemical control, sub-tropical & tropical fruits;
Wilson & Pusey, *PD* **69**:375, 1985, biological
control.

potato, *Solanum tuberosum*; see desert habitats;
Salaman, *The history and social influence of the
potato* 1949, revised impression, J.G. Hawkes ed.,
1985; Lapwood & Hide in Western 1971; Bokx,
Viruses of potatoes and potato seed production,
1972; Boyd, *Rev.Pl.Path.* **51**:297, 1972, storage;
Shepard & Claflin, *A.R.Phytop.* **13**:271, 1975,
seed potato certification; Hill in Scott &
Bainbridge 1978, viruses in seed tubers in Britain;
Brenchley & Wilcox, *Potato diseases*, 1979; Todd
& Howell in Ebbels & King, 1979, quarantine in
UK & W. Europe; Fribourg, *Fitopatología* **15**(2):
13, 1980, viruses in central & S. America; *AAB*
96:341, 1980, tuber damage; Thurston, *PD*
64:252, 1980, in developing countries; Hooker
1981; Geeson in Dennis 1983; Rich 1983. The
potato viruses that are distinguished by a capital
letter are placed alphabetically with the
abbreviation 'v.' for a virus placed last; although
the convention is to use e.g. potato virus A.

— **A** v. Murphy & McKay 1932; Bartels, *Descr.V.*
54, 1971; Potyvirus, filament *c.*730 × 15 nm,
transm. at least 7 aphid spp., non-persistent,
potato is the only natural host but the virus
infects many Solanaceae, can decrease yields by
up to 40 %.

— **ABC** (tobacco necrosis v.), Nordam,
Tijdschr.Pl Ziekt. **63**:237, 1957 [**37**:245].

— **acropetal necrosis** = potato Y v.

—— Andean calico (tobacco ringspot v.), Fribourg, *Phytop.* **67**:174, 1977 [**56**,5153].

—— —— latent (eggplant mosaic v.), Jones & Fribourg and Fribourg et al., *AAB* **86**:123, 373, 1977; Koenig et al., *Phytop.* **69**:748, 1979; transm. *Epitrix* sp., true seed; Andean region of S. America, wide host range [**56**,5154; **57**,752; **59**,3870].

—— —— mottle v. Fribourg, Jones & Koenig 1977; *Descr.V.* 203, 1979; Comovirus, isometric 26 nm diam., transm. sap in potato cvs., Andean region of S. America, narrow host range; Avila et al., *PD* **68**:997, 1984, str. B from Brazil causing severe crinkly mosaic in potato cv. Delta [**64**,1226].

—— aucuba mosaic v. Quanjer 1921; Kassanis & Govier, *Descr.V.* 98, 1972; Potexvirus, rod 580 × 11 nm, transm. sap, aphid spp., non-persistent, with helper including potato A v. or Y v., systemic infection in young *Capsicum* is lethal, of little economic importance; Mossop, *N.Z.J.agric.Res.* **25**:449, 1982, latent in *Cyphomandra betaceae* in New Zealand with tamarillo mosaic v. [**62**,1587].

—— black ringspot v. Salazar & Harrison 1977; *Descr.V.* 206, 1979; Nepovirus, isometric *c.*25 nm diam., RNA genome in 2 pieces, transm. sap, Peru, wide host range.

—— bouquet (tomato black ring v.).

—— C (potato Y v.).

—— calico str. of tobacco ringspot v. = potato black ringspot v.

—— Chilean v. Accatino, *Agricultura téc.* **26**:85, 1966; rod 230–260 × 15–20 nm, transm. *Myz.pers.* [**46**,1073].

—— Christmas tree Delgado Sánchez, *Agricultura téc. Mexico* **3**:382, 1975; poss. MLO [**55**,5324].

—— concentric necrosis (potato mop top v.).

—— corky ringspot = tobacco rattle v.

—— cyst nematodes, *Heterodera* or *Globodera pallida*, *H.* or *G. rostochiensis*; Trudgill, *AAB* **108**:181, 1986, yield loss, review & prospects in Britain.

—— deforming mosaic Calderoni et al. 1962; Delhey et al., *Potato Res.* **24**:123, 1981; Argentina [**62**,2100].

—— E = potato M v.

—— F and G = potato aucuba mosaic v.

—— French purple top roll Cousin, *Annls.Phytopath.* **7**:167, 1975; MLO assoc. [**55**,4825].

—— green dwarf (beet curly top v.).

—— hairy sprout Khanna, Central Potato Res. Inst., Simla, India 1973; plants showing marginal flavescence, purple top roll, witches' broom [**53**,1059].

—— haywire (aster yellows MLO), Semancik & Peterson, *Phytop.* **61**:1316, 1971 [**51**,2774].

—— Hopei vein banding Chi et al., *Acta phytopath.sin.* **7**:109, 1964; transm. aphid spp., China [**44**,2228].

—— Indian purple top roll Nagaich & Giri, *Am.Potato J.* **50**:79, 1973; Nagaich 1978; MLO assoc.; Singh et al., *Indian Phytopath.* **36**:646, 1983; **39**:14, 1986, transm. *Alebroides nigroscutellatus, Orosius albicinctus*; a serious disease in India [**54**,1402; **60**,2728; **65**,6158].

—— internal rust (tobacco rattle v.), Barchend & Heidel, *NachrBl.PflSchutzdienst.DDR* **39**:109, 1985; see internal rust spot.

—— late breaking Milbrath & English 1949; Raymer & Milbrath, *Phytop.* **50**:312, 1960; prob. aster yellows MLO [**28**:535; **39**:611].

—— leaflet stunt Klein, Zimmerman-Gries & Sneh, *Phytop.* **66**:564, 1976; assoc. organism in the phloem like a bacterium, Israel [**55**,5900].

—— leaf malformation Mallozzi & Barros, *Summa Phytopathologica* **10**:280, 1984; MLO assoc., Brazil (São Paulo) [**64**,4462].

—— —— roll v. Quanjer, van der Lek & Oortwijin Botjes 1916; Harrison, *Descr.V.* 291, 1984; Luteovirus, isometric *c.*24 nm diam., transm. aphid spp., persistent, *Myz.pers.* most efficient and important, a disease of great economic significance.

—— —— rolling mosaic = potato M v.

—— M v. Schultz & Folsom 1923; Wetter, *Descr.V.* 87, 1972; Carlavirus, filament *c.*650 × 12 nm, transm. sap, *Myz.pers.*, other aphid spp., non-persistent, narrow host range, mostly Solanaceae.

—— marginal flavescence Nagaich & Giri 1971; Nagaich et al., *Phytopath.Z.* **81**:273, 1974; MLO assoc., transm. *Orosius albicinctus*, India (N.) [**52**,1647; **54**,4607].

—— mop top v. Calvert & Harrison 1966; Harrison, *Descr.V.* 138, 1974; Furovirus, rod 100–150 or 250–300 × 18–20 nm, transm. sap, *Spongospora subterranea* f. sp. *subterranea*, survives in resting spores, soilborne, Europe (W.), S. America, prob. elsewhere if the fungus is present.

—— paracrinkle = potato M v.; parastolbur, see potato stolbur.

—— phloem necrosis = potato leaf roll v.

—— phyllody Nagaich 1978; Khurana et al., *Indian Phytopath.* **34**:532, 1981; MLO assoc., India (N.) [**60**,2728; **61**,6543].

—— pseudo-aucuba (tomato black ring v.).

—— purple top wilt Brentzel 1938; Norris, *Aust.J.agric.Res.* **5**:1, 1954; MLO assoc.; Shiomi & Sugiura, *Ann.phytopath.Soc.Japan* **50**:455, 1984; transm. *Scleroracus flavopictus* but not by *Macrosteles fascifrons* or *M. orientalis*. The last 2 insects transm. aster yellows MLO and 3 MLO

strs. in Japan. This disease is found in 4
Australian states. Harding & Teakle,
Aust.J.agri.Res. **36**:443, 1985, transm. the eggplant
little leaf MLO by grafting to tomato and big bud
symptoms resulted. Transm. to potato by grafting
and *Orosius argentatus* from tomato big bud
plants caused potato purple top symptoms
[**17**:480; **33**:621; **64**,2414; **65**,1466].
— **rot nematode**, *Ditylenchus destructor*.
— **rugose mosaic** (potato Y v. or potato Y v.
+ potato X v.).
— **S v.** de Bruyn Ouboter 1952; Wetter, *Descr.V.*
60, 1971; Rich 1983; Carlavirus, filament 650 ×
12 nm, transm. sap, some isolates by aphids, can
reduce yield by up to 20%. Slack, *PD* **67**:786,
1983, described an isolate which was unusual
because of its systemic invasion of *Chenopodium
quinoa*, transm. *Myz.pers.* [**62**,4971].
— **severe chlorosis** (cucumber mosaic v.),
Somerville et al. *PD* **71**:18, 1987; symptoms of
mosaic, shortened internodes, deformed leaves
and knobby tubers, a legume str. of CMV, USA
(California) [**66**,4427].
— — **mosaic** = potato Y v.
— **spindle tuber viroid** Martin 1922; Diener &
Raymer, *Descr.V.* 66, 1971; etiology determined
by Diener 1971, see viroid; Rich 1983. Infected
plants are smaller and more erect than healthy
ones, leaves darker green, branching angle more
acute, tubers are elongated and may be tapered at
both ends, eyes are more numerous and have
prominent bud scales; yields can be reduced by
> 60% by virulent strs. Transm. by sap, contact,
knife cut and by cultivation equipment; through
true seed and by insects. The viroid infects
tomato, causing epinasty, rugosity, lateral twisting
of leaflets and veinal necrosis; and other
Solanaceae. It has not been found in wild potato
spp. or local cvs. in the Peruvian Andes; occurs
in: Argentina, Canada, Europe (E.), USA (N.,
N.E.); detected in Australia in 1982; and may
occur in South Africa, causing tomato bunchy top
q.v. No high host resistance appears to have been
found; Harris et al. in Ebbels & King 1979,
problems with breeding material; Salazar et al. in
Maramorosch & McKelvey 1985, elimination by
cold treatment & meristem culture; Grasmick &
Slack, *CJB* **64**:336, 1986, effects on sexual
reproduction & transm. in true potato seed
[**64**,4505].
— **spraing**, a tuber syndrome, at first of raised
necrotic rings on the surface, with underlying arcs
of corky tissue; caused by potato mop top v. or
tobacco rattle v. on some potato cvs.; Calvert &
Harrison, *Pl.Path.* **15**:134, 1966 [**46**,409].
— **stem mottle** = tobacco rattle v.

— **stolbur**, MLO assoc., Ploaie & Maramorosch,
Phytop. **59**:536, 1969 [**48**,2840].
— **stunt** Cockerham & McGhee 1953, fide Smith
1972.
— **T v.** Salazar & Harrison 1977; *AAB* **89**:223,
1978; *Descr.V.* 187, 1978; apple stem grooving v.
group, contains RNA, filament *c.*640 × 12 nm,
transm. sap, seed, pollen in *Solanum demissum*;
Bolivia, Peru; limited host range, usually latent in
potato but may cause a mild leaf mottle [**58**,2371].
— **tuber blotch** = potato aucuba mosaic v.
— — **necrotic ringspot** Beczner et al., *Potato Res.*
27:339, 1984; assoc. with infection by potato Y v.
[**65**,4032].
— **U v.** Jones, Fribourg & Koenig, *Phytop.*
73:195, 1983; prob. Nepovirus, isometric *c.*28 nm
diam., transm. sap, seed of *Chenopodium,
Nicotiana debneyi*, tobacco; hosts, 44 spp. in 7
plant families [**62**,3208].
— **V v.** Fribourg & Nakashima, *Phytop.* **74**:1363,
1984; Jones & Fribourg, *Descr.V.* 316, 1986;
Potyvirus, filament *c.*760 nm long, transm. sap,
*Brachycaudus helichrysi, Macrosiphum euphorbiae,
Myz.pers., Rhopalosiphoninus latysiphon*, non-
persistent; France, Netherlands, Peru, UK;
Solanaceae; natural infection only in potato, on
inoculation necrotic spots, severe systemic necrosis
[**64**,1709].
— **vein yellowing** Alba 1950; Silberschmidt,
Phytop. **44**:415, 1954; Colombia, Ecuador
[**34**:241].
— **wild mosaic v.** Jones & Fribourg, *Phytop.*
69:446, 1979; Potyvirus, filament *c.*735 nm long,
transm. sap, *Myz.pers.*, Peru, from *Solanum
chancayense*, infected some Solanaceae but not 13
potato cvs. [**59**,2884].
— **witches' broom** Hungerford & Dana 1924;
Harrison & Roberts, *AAB* **63**:347, 1969, in
Scotland, MLO assoc., see potato stolbur;
Nagaich et al., *Phytopath.Z.* **81**:273, 1974,
transm. *Orosius albicinctus*, India, see potato
Indian purple top roll [**4**:55; **48**,3623; **54**,4607].
— **X v.** Smith 1931; Bercks, *Descr.V.* 4, 1970;
Rich 1983; Potexvirus, type, filament *c.*515 × 12
nm; transm. sap and by contact, very contagious,
may be no obvious symptoms or a leaf mottle;
natural hosts, potato, tomato; PVX + tomato str.
of tobacco mosaic v. causes tomato double streak;
isolates differing in virulence are common; Tollin
et al., *J.Gen.Virol.* **49**:407, 1980, particle
structure; Adams et al., *Pl.Path.* **35**:517, 1986,
effect of temp. on infection of potato cvs. with
different combinations of hypersensitivity genes
[**60**,1594; **66**,2032].
— **Y v.** Smith 1931; Bokx & Huttinga, *Descr.V.*
242, 1981; Potyvirus, type, filament 730 × 11 nm,

transm. sap, aphid spp., particularly *Myz.pers.*, non-persistent, can cause economic loss in *Capsicum*, tobacco, tomato; most potato cvs. free from PVY; symptoms depend on cv.; leaf tip mottle, necrotic streaks on leaf veins, necrosis on petiole and stem, leaves wither and droop, stunt; strs. differ in virulence; Katis & Gibson, *Potato Res.* **28**:65, 1985, transm. cereal aphids [65,4038].

—— **yellow dwarf v.** Barrus & Chupp 1922; Black, *Descr.V.* 35, 1970; Rich 1983; Phytorhabdovirus, bacilliform 380 × 75 nm, transm. sap, difficult, and leafhoppers: New York str. by *Aceratagallia sanguinolenta*; New Jersey str. by *Agallia constricta, A. quadripunctata*; other strs.; Canada (E.), USA (N.E.); Falk & Weathers, *Phytop.* **73**:81, 1983, comparison of serotypes [62,1854].

—— —— **mosaic v.** Roberts, Buck & Coutts, abs. *PD* **70**:603, 1986; poss. Geminivirus, transm. sap to *Lycopersicon, Nicotiana, Petunia*, and by *Bemisia tabaci*, Venezuela.

—— —— **necrosis** (potato aucuba mosaic v.).

—— —— **spot** Bonde & Merriam, *Phytop.* **44**:608, 1954; Janke & Ramson 1960; Germany, USA (Maine) [**34**:313; **40**:556].

Potebniamyces Smerlis 1962; Dermataceae or Hypodermataceae; small, fleshy stroma beneath bark, ascomata cup shaped, asco. aseptate.

P. **balsamicola** Smerlis, *CJB* **40**:352, 1962; asco. 8–15 × 3–5 μm; con. 6–11 × 2.5–5 μm; assoc. with a twig and branch blight of *Abies balsamea*, balsam fir [**41**:622].

P. **coniferarum** (Hahn) Smerlis, fide Punithalingam & Gibson, *Descr.F.* 517, 1976, as anam. *Phacidiopycnis pseudotsugae* (M. Wilson) Hahn 1957; Cannon et al. 1985 accepted *Phacidium coniferarum* (Hahn) DiCosmo 1983 for the teleom.; asco. 10–18 × 3–6 μm; con. 4–7 × 2–3 μm; on branches and leaves of conifers causing twig dieback and basal canker, particularly in *Larix, Pinus, Pseudotsuga*; Hahn, *Mycologia* **49**:227, 1957, as *Phacidiella coniferarum* Hahn [**36**:625].

P. **pyri** (Berk. & Broome) Dennis 1978, anam. *Phacidiopycnis malorum* Potebnia, fide Cannon et al. 1985; asco. 15–23 × 8–11 μm, con. 9–13 × 6–9 μm; apple and pear bark canker; Wollenweber, *Angew.Bot.* **19**:131, 1937, black cortical blight of quince [**16**:690].

Potexviruses, type potato X v., *c.* 22 definitive members, filament 470–580 × 11–13 nm, 1 segment linear ssRNA, sedimenting at 114–130 S, thermal inactivation 60–80°, longevity in sap usually 14–21 days, transm. sap readily, by contact; fairly narrow host ranges; Koenig, *Descr.V.* 200, 1978; Purcifull & Edwardson in Kurstak 1981; Francki et al., vol. 2, 1985; Short & Davies, *AAB* **110**:213,

1987, host ranges, symptoms caused & amino acid composition of 8 members [66,2750].

potexvirus sieg Koenig & Lesemann, *Phytopath.Z.* **112**:105, 1985; filament, transm. sap. Germany (river Sieg) [64,3758].

Potyviruses, type potato Y v., prob. > 50 members, the largest group of plant viruses; divided into subgroups 1, 2, 3, 4, transm. by aphids, fungi, mites, whiteflies, respectively; the 1st subgroup is much the largest; filament 680–900 × 12 nm, 1 segment linear ssRNA, sedimenting at *c.*150 S, thermal inactivation usually 55–60°, longevity in sap usually 2–4 days, transm. sap, mostly aphids, non-persistent, induce characteristic inclusion bodies, called pinwheels, in the cytoplasm; Hollings & Brunt, *Descr.V.* 245, 1981 and in Kurstak 1981; Sako & Ogata, *Virology* **112**:762, 1981, helper factors & aphid transm.; Francki et al., vol. 2, 1985; Katis et al., *Pl.Path.* **35**:152, 1986, interference between Potyviruses during aphid transm. [**61**,596; **65**,5419].

powdery dry rot (*Fusarium solani* var. *coeruleum*; *F. trichothecioides*) potato; —— **mildew**, see mildew; —— **scab** (*Spongospora subterranea* f.sp. *subterranea*) potato.

pox (plum pox v.) apricot, peach, plum, (*Streptomyces ipomoeae*) sweet potato.

pratensis, *Leptosphaeria*.

Pratylenchoides crenicauda Winslow 1958; Siddiqi, *Descr.N.* 38, 1974; on grasses.

Pratylenchus, root lesion nematodes, often cause severe damage, particularly in warmer regions.

P. **brachyurus** (Godfrey) Filipjev & Schuurmans Stekhoven 1941; Corbett, *Descr.N.* 89, 1976; root endoparasite, tropics, can cause extensive damage to a root system.

P. **coffeae** (Zimmermann) Filipjev & Schuurmans Stekhoven; Siddiqi, *Descr.N.* 6, 1972.

P. **fallax** Seinhorst 1968; *Descr.N.* 100, 1977; migratory endoparasite of roots, causes patchy growth in barley.

P. **goodeyi** Sher & Allen 1953; Machon & Hunt, *Descr.N.* 120, 1985; migratory endoparasite, important on banana, > half the root system may be destroyed, Canary Islands, E. Africa.

P. **loosi** Loof 1960; Seinhorst, *Descr.N.* 98, 1977; migratory endoparasite, roots, can be serious on tea in Sri Lanka.

P. **neglectus** (Rensch) Filipjev & Schuurmans Stekhoven; Townsend & Anderson, *Descr.N.* 82, 1976; feeds externally and then becomes a migratory endoparasite; damages roots of bean, maize, tobacco, causing above ground symptoms of stunt, wilt and yellowing.

P. **penetrans** (Cobb) Chitwood & Oteifa 1952; Corbett, *Descr.N.* 25, 1973; mainly cooler regions,

> 350 hosts, in conifer and fruit nurseries, assoc. with soil sickness and replant diseases.

P. pratensis (de Man) Filipjev 1936; Loof, *Descr.N.* 52, 1974.

P. scribneri Steiner in Sherbakoff & Stanley 1943; Loof, *Descr.N.* 110, 1985; migratory root endoparasite; important on soybean, bean top growth may be suppressed; Sinclair 1982.

P. thornei Sher & Allen 1953; Fortuner, *Descr.N.* 93, 1977; on roots, can cause stunting of wheat.

P. vulnus Allen & Jensen 1951; Corbett, *Descr.N.* 37, 1974; mostly on roots of woody plants, including fruit crops, causes stunting.

P. zeae Graham 1951; Fortuner, *Descr.N.* 77, 1976; migratory root endoparasite; causes disease in cotton, maize, rice, sugarcane, tobacco.

predisposition, disease proneness of a plant, see stress; prob. best used for a condition of the plant which operates before infection by a pathogen; and it affects the susceptibility of the host rather than directly affecting the pathogen, fide Tarr 1972; Yarwood in Heitefuss & Williams 1976.

premature death (*Leptosphaeria lindquistii*) sunflower, Donald et al., *PD* 71:466, 1987; — **desiccation**, oil palm, prob. assoc. with K deficiency, Turner 1981; — **fruit abortion** (*Choanephora cucurbitarum*) okra, Adebanjo, *J.Pl.Prot.Tropics* 2:131, 1985; — **leaf fall** (*Pseudocercospora anacardii*) cashew; — **needle cast** (*Naemacyclus minor*) pine; — **nut fall** (*Phytophthora palmivora*) coconut [66,4608].

prepenetration period, the period between the time that a vector is placed on a plant and the time that the host tissues are penetrated, FBPP.

preservation of fungi, Smith & Onions, *The preservation and maintenance of living fungi*, 1983; *TBMS* 81:535, 1983, comparison of techniques; Smith, *ibid* 79:415, 1982; 80:333, 360, 1983, freeze drying & liquid N; Dahmen et al., *Phytop.* 73:241, 1983, liquid N.

Prévost, Isaac Bénédict, 1755–1819; born in Switzerland, lived in France; his careful work, including inoculations, on *Tilletia tritici* and published in 1807, 'constitute the first experimental evidence for the pathogenicity of a micro-organism', Ainsworth 1981. Prévost, and Tillet in 1755, carried out the first clear demonstrations that the theory of spontaneous generation was untenable. This was nearly 40 years before Berkeley considered that a fungus caused potato blight and c.70 years before Koch's work on anthrax; *Phytopath.Classics* 6, 1939, English translation by G.W. Keitt; Keitt, *Phytop.* 46:2, 1956.

Prillieux, Edouard Ernest, 1829–1915; born in France, chair at Institute National Agronomique,

Paris, 1888; initiated plant pathology teaching in France; wrote: *Maladies des plantes agricoles et des arbres fruitiers et forestiers causées par des parasites végétaux*, 2 vol., 1895–7; Whetzel 1918; Ainsworth 1981.

primula American mosaic Tompkins & Middleton, *J.agric.Res.* 63: 671, 1941; transm. sap, USA (California) [21:201].

primulae, *Phytophthora, Ramularia*.

primula green petal Mokrá 1964; Stevens & Spurdon, *Pl.Path.* 21:195, 1972; MLO assoc. [44,3073; 52,1933].

—— **hardy** Uschdraweit & Valentin, *Phytopath.Z.* 36:122, 1959; Germany [39:314].

—— **Italian mosaic v.** Lisa & Lovisolo, *Acta Hortic.* 59:167, 1976; poss. Potyvirus, filament modal length 760 nm, transm. *Myz.pers.*, wide host range, causes flower colour break [56,2097].

—— **mottle v.** Singh et al., *Indian Phytopath.* 23:148, 1970; filament c.720–745 nm long, transm. sap, *Aphis gossypii, Myz.pers.*, non-persistent, India (N.) [49,3346].

—— **phyllody** Kleinhempel, Müller & Spaar 1972; MLO assoc. [52,658].

—— **yellow vein necrosis** Kochman & Stachyra, *Roczn.Nauk roln.A* 77:297, 1957, Poland [37:340].

prion, a minute infectious entity containing protein and which is resistant to procedures that modify or hydrolyse nucleic acids; whether or not prions contain nucleic acid remains to be established. Prions, presumably, replicate within cells; this mechanism is unknown. They have only been found in animals where they can cause disease; Diener & Prusiner in Maramorosch & McKelvey 1985.

privet, *Ligustrum*; Cooper 1979, viruses.

probineb, broad spectrum, protectant fungicide.

probing, used of aphids, and some other vectors with stylets, for the brief, initial penetration of superficial cells. This should be distinguished from feeding which usually involves penetration to the vascular tissues, after FBPP.

procaryote, prokaryote, a single cell, or simple association of cells; smallest 0.2–10 μm diam., defined by cellular not organismal characteristics, no nuclear membrane, cell division not accompanied by changes in texture or staining properties, no microtubule or spindle system. The Procaryotae, cf. eucaryote, have 4 divisions. The Firmicutes and Gracilicutes constitute the bacteria; they have rigid cell walls; the latter are Gram negative and have no endospores. The former are Gram positive and endospores may be present. The Tenericutes have non-rigid cell walls; one class, the Mollicutes. Mendosicutes are primitive forms not assoc. with disease; the cell wall has no conventional peptidoglycan; Starr et al. 1981; Mount & Lacy ed., *Phytopathogenic prokaryotes*, 1982; Krieg 1984.

procera, *Verticicladiella*.

prochloraz, broad spectrum, protectant fungicide with eradicant properties.

procymidone, protectant fungicide.

prolata, *Setosphaeria*, see genus.

proliferation, elongation of shoots and roots after they have normally ceased to elongate.

propagative viruses, see persistent viruses.

propagule, that form or part of an organism by which it may be dispersed or reproduced, FBPP.

prophage, the form in which phages are perpetuated in lysogenic bacteria, FBPP.

proso millet, see millets.

Prospodium Arthur 1907; Uredinales; autoecious, Bignoniaceae and Verbenaceae, warmer areas of America. *P. bicolor* F.A. Ferreira & Hennen, *Mycologia* **78**:801, 1986; macrocyclic, on *Tabebuia serratifolia*, Brazil, Trinidad; damages nursery plantings and younger trees in Brazil [66,1622].

proteamaculans, *Serratia*.

protectant, a substance that protects an organism against infection, FBPP.

protection, the action of a str. or isolate in protecting against infection and invasion by a second str. or isolate. Cross protection indicates that the protective action is reciprocal, FBPP; Fulton, *A.R.Phytop.* **24**:67, 1986; use for control of viruses.

proteins, in pathogenesis, changes in diseased plants; their appearance may be assoc. with the development of acquired resistance; Uritani, *A.R.Phytop.* **9**:211, 1971; *Neth.J.Pl.Path.* **89**(6), 1983.

prothiocarb, systemic fungicide, applied to soil, active against Oomycetes.

Protomyces Unger 1833; Protomycetaceae; resting spores, or chlamydospores, intercalary with smooth walls, parasitic, causing galls on stems, leaves and fruits of Compositae and Umbelliferae. *P. macrosporus* Unger; resting spores 50–70 μm diam., coriander stem gall; Holliday 1980, and the related genus *Protomycopsis* Magnus 1905; Valverde & Templeton, *PD* **68**:716, 1984, leaf gall on the weed *Torilis japonica* in USA (Arkansas) [63,5718].

Protomycetaceae, Taphrinales; mycelium forming resting spores with thick walls, germinating to form a spore sac, a synascus which is multinucleate and forms endospores; the spore sac is often taken to be a group of fused asci.

protoplasts, see tissue culture.

prototroph, prototrophic, micro-organims with no

nutritional needs beyond that of the wild type, cf. auxotroph.

protozoa, see *Phytomonas*.

provirus, the DNA form of an RNA virus. The provirus may be integrated with the host chromosomes and transm. from one generation to another; it does not contain protein, fide Diener & Prusiner in Maramorosch & McKelvey 1985.

pruinosa, *Encoelia*, see *Cenangium singulare*.

prune, see *Prunus*.

—— **bark split** (apple chlorotic leaf spot v.), Dunez et al., *Acta Hortic.* 44:81, 1975 [55,3638b].

—— **brown line** (tomato ringspot v.).

—— **diamond canker** Smith 1932, *Phytop.* **31**:886, 1941; USA (California) [**12**:180; **21**:83].

—— **dwarf** v. Thomas & Hildebrand 1936; Fulton, *Descr.V.* 19, 1970; Ilarvirus, isometric 19–20 nm diam., to bacilliform up to 73 nm long, transm. sap, seed, pollen in *Prunus*, experimental host range fairly wide; Kunze et al., *Phytopath.Z.* **110**:251, 1984, characterising isolates [**64**,250].

prunella mosaic Liro 1930; transm. sap, aphid spp., Finland, *P. vulgaris* [**10**:687].

pruni, *Taphrina*; **pruni-persicae**, *Mycosphaerella*; **pruni-spinosa**, *Tranzchelia*.

Prunus, *P. americana*, American plum; *P. amygdalus*, almond; *P. armeniaca*, apricot; *P. avium*, sweet cherry; *P. cerasifera*, cherry plum, myrobalan; *P. cerasus*, sour cherry; *P. domestica*, European plum, prune; *P. insititia*, damson; *P. persica*, peach; *P. salicina*, Japanese plum; *P. virginiana*, chokecherry; see fruits.

prunus chlorotic spot Callahan 1962; transm. sap, Canada (New Brunswick) [**42**:345].

—— **isometric** v. Kleinhempel et al., *Zentbl.Bakt.ParasitKde.* 2 **126**:659; 1971; particle 41 nm diam., transm. sap, Germany [**51**,4115].

—— **Krassa Severa ringspot** Basak & Millikan 1968, Poland [**49**,2522w].

—— **Krikon stem necrosis** Desvignes & Savio, *Annls.Phytopath.* hors sér.:75, 1971; seedborne, France [**51**,2633f].

—— **necrotic ringspot** v. Cochrane & Hutchins 1941; Fulton, *Descr.V.* 5, 1970; Ilarvirus, isometric *c*.23 nm diam., transm. sap, widespread in temperate regions, fairly wide host range; Barbara et al., *AAB* **90**:395, 1978, detection & serotyping of strs. in hop & plum; Cole et al., *Phytop.* **72**:1542, 1982, on pollen; Mink in Plumb & Thresh 1983, poss. role of honey bees in long distance spread; Schimanski & Fuchs, *Zentbl.Mikrobiologie* **139**:649, 1984, transm. apricot seed; Smith & Skotland, *PD* **70**:1019, 1986, in hop [**58**,3173; **62**,2039; **64**,5447; **66**,2453].

—— **rasp leaf** (tomato ringspot v.).

—— **ringspot** (prunus necrotic ringspot v.).

— **stem pitting** Lott et al., *Can.Pl.Dis.Surv.*
42:229, 1962; (tomato ringspot v.), Smith et al.,
Phytop. **63**:1404, 1973; Mircetich & Moller in
Klement, see apoplexy [**53**,4514].
—'**yellow bud mosaic** (tomato ringspot v.).
— — **mosaic** Králiková 1956; Bulgaria [**37**:411].
psalliotae, *Verticillium*.
Psammotettix alienus, transm. gramineae chlorosis,
wheat dwarf, wheat yellow stunting; *P. striatus*
transm. winter wheat Russian mosaic.
pseudobifrons, *Ciborinia*; see *C. whetzelii.*
***Pseudocercospora** Speg. 1910; Hyphom.; Ellis 1971,
1976; Sivanesan 1984, *Mycosphaerella* teleom.;
conidiomata hyphal, conidiogenesis holoblastic,
conidiophore growth sympodial; con. filiform,
pigmented, solitary, obclavate near base, hilum
unthickened, multiseptate; Holliday 1980.
P. abelmoschi (Ell. & Ev.) Deighton 1976; Waller
& Sutton, *Descr.F.* 625, 1979; con. 33–58 × 4.5–8
µm; okra leaf mould, *Hibiscus* spp., frequent but
a minor disease.
P. aleuritis, teleom. *Mycosphaerella aleuritis.*
P. anacardii E. Castell. & Casulli,
Riv.Agric.subtrop.trop. **75**:103, 1981, *Index Fungi*
5:136, 1982, premature leaf fall of cashew in
Tanzania [**61**,2473].
P. contraria, teleom. *Mycosphaerella contraria;*
Pseudocercospora cruenta, teleom. *M. cruenta.*
P. fuligena (Roldan) Deighton, *Mycol.Pap.*
140:144, 1976; Mulder & Holliday, *Descr.F.* 465,
1975, as *Cercospora fuligena* Roldan 1938; con.
15–120 × 3.5–5 µm; tomato leaf spot, mostly
tropics; a potentially serious disease which needs
further study.
P. musae, teleom. *Mycosphaerella musicola.*
P. nigricans (Cooke) Deighton, *Mycol.Pap.*
140:149, 1976; con. 3–5 septate, 30–65 × 4–6 µm;
fide Ellis 1976 as *Cercospora nigricans* Cooke
1883; on *Cassia*; Hofmeister & Charudattan, PD
71:44, 1987, causing leaf chlorosis, necrosis and
fall in *C. obtusifolia*, sicklepod, weed in USA
(S.E.); pathogenic only to this plant and not to 23
crop and weed spp. in 10 families [**66**,4613].
P. pini-densiflorae, teleom. *Mycosphaerella gibsonii.*
P. psophocarpi (Yen) Deighton, *Mycol.Pap.*
140:151, 1976; con. av. 28–33 × 5–6 µm; winged
bean leaf spot, can cause complete defoliation;
Price & Munro, *TBMS* **70**:47, 1978, in Papua
New Guinea; Price et al., *AAB* **101**:473, 1982,
incidence & distribution with *Synchytrium
psophocarpi* & *Oidium* sp. [**57**,4250; **62**,2227].
P. puerariicola, teleom. *Mycosphaerella
puerariicola*; **Pseudocercospora salicina**, teleom. *M.
togashiana.*
P. taiwanensis (Matsumoto & Yamamoto) Yen
1981; con. 3–15 septate, 65–275 × 2–4 µm; a

secondary invader on sugarcane leaf blight lesions
caused by *Leptosphaeria taiwanensis*. It was at first
thought to be both the cause of this leaf blight
and the anam. of *L. taiwanensis* q.v.; this was
disproved by Hsieh 1979.
P. timorensis (Cooke) Deighton 1976; Little,
Descr.F. 918, 1987; con. 3–8 septate, 50–100 × 3–
4 µm; leaf spot of sweet potato and other
Ipomoea spp.
Pseudocercosporella Deighton, *Mycol.Pap.* 133:38,
1973; Hyphom.; Carmichael et al. 1980;
conidiomata hyphal, conidiogenesis holoblastic,
conidiophore growth sympodial; con. filiform,
hyaline, slightly tapered towards the hilum, not
acicular; very similar to *Cercoseptoria* and
Pseudocercospora.
P. capsellae (Ell. & Ev.) Deighton, as above, page
42; con. mostly 3 septate, 30–90 × 2–3 µm;
brassica white spot, poss. seedborne, chemical
control may be needed; Holliday 1980.
P. caryigena, teleom. *Mycosphaerella caryigena.*
P. herpotrichoides (Fron) Deighton, *Mycol.Pap.*
133:46, 1973; Booth & Waller, *Descr.F.* 386,
1973; con. 3–7 septate, 26.5–47 × 1–2 µm; eyespot
of temperate cereals; causes foot rot, lodging and
post-emergent death; the fungus is found on
many wild and cultivated grasses in cooler
regions; con. are splash dispersed. The disease
tends to be most severe on winter wheat in
Europe, in wet conditions, on heavy soils and
where conditions are conducive for lodging. There
is a long survival in soil and crop debris.
Experimentally eyespot is most severe at a temp.
regime of 10° night and 15° day compared with
one of 5° and 10°.

Bateman & Taylor, *TBMS* **67**:95, 1976, infection
of wheat seedlings; Scott & Hollins, *Pl.Path.*
27:125, 1978, prediction of yield loss; Hollins &
Scott, *AAB* **95**:19, 1980, correlation of infection
with number of wet days/week; Fitt &
Bainbridge, *Phytopath.Z.* **106**:214, 1983; Fitt &
Nijman, *Neth.J.Pl.Path.* **89**:198, 1983; both on
dispersal of con.; Higgins, *TBMS* **82**:443, 1984,
germination of con. & initial growth on wheat;
Higgins & Fitt, *Phytopath.Z.* **111**:222, 1984,
formation of con., pathogenicity, most con. in
England in February; Fitt et al., *TBMS* **88**:149,
1987, pathogenicity of rye & wheat types;
Siebrasse & Fehrmann, *Z.PflKrankh.PflSchutz.*
94:137, 1987, model for chemical control
[**56**,2031; **58**,4785; **60**,807; **62**,3493; **63**,2824, 2826;
64,2465; **66**,3754].
Pseudococcus longispinus, transm. colocasia bobone.
Pseudocochliobolus Tsuda, Ueyama & Nishihara,
Mycologia **69**:1117, 1977, published 1978 =
Cochliobolus fide Sivanesan 1984.

pseudohispidus, *Inonotus*.
Pseudomonas Migula 1844; bacteria; Gram negative rods, single or in chains of a few cells, no sheaths or resting stages, polar flagella single or many, rarely non-motile; 4 homology groups have been separated using ribosomal RNA hybridisation; the spp. have been divided into fluorescent and non-fluorescent groups; Schroth et al. in Starr et al. 1981; Fahy & Lloyd and Hayward in Fahy & Persley 1983; Palleroni in Krieg 1984, Bradbury 1986; Vivian in Day & Jellis 1987, genetic control of pathogenicity.

P. agarici Young, *N.Z.J.agric.Res.* **13**:977, 1970; Bradbury, *Descr.B.* 892, 1987; drippy gill of the common cultivated mushroom, *Agaricus brunnescens*; Australia (New South Wales), Irish Republic, New Zealand, UK, prob. more widespread [50,2629].

P. amygdali Psallidas & Panagopoulos 1975; Bradbury, *Descr.B.* 556, 1977; almond, perennial cankers on branches, 3–5 to 15–20 cm long, may be girdling.

P. andropogonis (Smith) Stapp 1928; Bradbury, *Descr.B.* 372, 1973; 1, rarely 2, polar flagella; leaf stripes in Gramineae and leaf spots in legumes, other hosts; > 1 serological group; Caruso, *PD* **68**:910, 1984, on chickpea; Moffet et al., *Pl.Path.* **35**:34, 1986, new hosts in Australia (E.) & isolate characterisation; Hayward in Fahy & Persley 1983 [64,874; 65,3687].

P. avenae Manns 1909; Bradbury, *Descr.B.* 371, 1973, as *Pseudomonas alboprecipitans* Rosen 1922; brown stripe, Gramineae, causes yellowish or brownish leaf streaks; Hayward in Fahy & Persley 1983; Shakya et al., *Phytopath.Z.* **114**:256, 1985, on rice, widespread [65,2321].

P. betle (Ragunathan) Săvulescu 1947; betel leaf spot; India, Malaysia (W.), Mauritius, Sri Lanka; poss. a *Xanthomonas*; Bradbury 1986.

P. caricapapayae Robbs, *Revta.Soc.Brasiliera Agron.* **12**:73, 1956; Robbs 1960; papaya leaf spot, Brazil (S.E.) [40:75].

P. caryophylli (Burkholder) Starr & Burkholder 1942; Bradbury *Descr.B.* 373, 1973; carnation stem crack, a vascular wilt, *Dianthus* spp.; Hayward in Fahy & Persley 1983.

P. cattleyae (Pavarino) Săvulescu 1947; orchid brown spot; Hayward in Fahy & Persley 1983.

P. cepacia ex Burkholder, *Phytop.* **40**:115, 1950; Palleroni & Holmes, *Int.J.syst.Bact.* **31**:479, 1981; slippery or sour skin of onion, prob. widespread; a bulb scale rot; Hayward in Fahy & Persley 1983 [29:489].

*****P. cichorii** (Swingle) Stapp 1928; Bradbury, *Descr.B.* 695, 1981; dark brown to blackish spots in many plants, dark brown streaks in tomato,

lettuce varnish spot, wheat melanosis; Bazzi et al., *Phytopath.Z.* **111**:251, 1984; Piening & MacPherson, *Can.J.Pl.Path.* **7**:168, 1985; Chase, *Pl.Path.* **36**:219, 1987, leaf & petiole rot of *Ficus lyrata* in USA (Florida) [64,2864, 5362; 66,4324].

P. corrugata Roberts & Scarlett 1981; Scarlett et al., *AAB* **88**:105, 1978; Bradbury, *Descr.B.* 893, 1987; tomato pith necrosis, causes a pith collapse, bacteria exude from stem wounds; Denmark, England, Italy, Germany, New Zealand, South Africa, Sweden, USA (California, Florida) [57,2655].

P. ficuserectae Goto, *Int.J.syst.Bact.* **33**:546, 1983; twig dieback of *Ficus erecta*, Japan [63,270].

P. flectens Johnson, *Qd.J.agric.Sci.* **13**:127, 1956; pod twist; bean, *Macroptilium atropurpureum*, yound pods may wither and drop, Australia (New South Wales) [36:805].

P. fluorescens (Trevisan) Migula 1895; saprophytic or weakly parasitic; *Pseudomonas tolaasii* may be a form of this sp.; Wong et al., *J.appl.Bact.* **52**:43, 1982, who described mushroom ginger blotch; Bradbury 1986 tabulated the characteristics of 5 biovars [61,3804].

P. fuscovaginae Miyajima, Tanii & Akita, *Int.J.syst.Bact.* **33**:656, 1983; rice leaf sheath brown rot; Burundi, Japan; Tani et al., *Ann.phytopath.Soc.Japan* **42**:540, 1976; Zeigler & Alvarez, *PD* **71**:592, 1987 [56,4050; 63,151].

P. gladioli Sereini 1913. Pv. *gladioli* causes lacquer scab or scab of gladiolus bulbs, a serious disease, Moore 1979; other flower bulb crops may be attacked; causes leaf spots and blights of ferns, Chase et al., *PD* **68**:344, 1984. Pv. *alliicola* (Burkholder) Young et al. 1978, causes a rot of inner bulb scales of onion and dry, necrotic leaf spots; Burkholder, *Phytop.* **32**:141, 1942; Starr & Burkholder, *ibid*, page 598, as *Pseudomonas alliicola*; Hayward in Fahy & Persley 1983 [21,325; 63,3386].

P. glumae Kurita & Tabei 1967; rice grain rot, over half the grains in the panicle can be attacked; Bradbury 1986.

P. marginalis (Brown) Stevens 1925; Bradbury 1986 described 3 pvs. Pv. *alfalfae* (Shinde & Lukezic) Young et al. 1978; root browning and stunt of lucerne in USA (N.E.), *Phytop.* **64**:865, 1974. Pv. *marginalis*; a leaf marginal necrosis in many crops, a soft rot in storage. Pv. *pastinacae* (Burkholder) Young et al. 1978; a firm, dark brown rot of upper roots of parsnip and a soft rot of petiole bases, USA (N.E.), *Phytop.* **50**:280, 1960 [39:648; 54,2314].

P. meliae Ogimi 1981; Bradbury 1986; galls on trunks and branches of *Melia azedarach*, China

berry, nim tree, pride of Persia, Persian lilac and other names; Japan, Taiwan.
P. plantarii Azegami et al., *Int.J.syst.Bact.* **37**:144, 1987; rice seedling blight in Japan.
P. rubrilineans (Lee et al.) Stapp 1928; Bradbury, *Descr.B.* 127, 1967; single polar flagellum; sugarcane red stripe and top rot; the leaf stripes are long, narrow, sharply defined and dark red; top rot, which may or may not occur with the leaf stripes, causes a greater loss; Hayward in Fahy & Persley 1983; Sivanesan & Waller 1986.
P. rubrisubalbicans (Christopher & Edgerton) Krasil'nokov 1949; Bradbury, *Descr.B.* 128, 1967; several polar flagella; sugarcane mottle stripe, stripes paler than those caused by *Pseudomonas rubrilineans*, little loss caused; references as above.
*****P. solanacearum** (Smith) Smith 1914; Hayward & Waterston, *Descr.B.* 15, 1964; the literature on this pathogen is very large and the host range is extremely wide, see Bradbury 1986 who described the bacterium as very heterogeneous. Three races have been delineated: (1) on solanaceous crops, intense tyrosinase reaction; (2) banana and *Heliconia*, no tyrosinase; (3) potato, weak tyrosinase reaction; 13 pathotypes have been distinguished. The bacterium is widespread in warmer regions and causes vascular wilts; it is soilborne but can contaminate seed and be found on foliage. The main diseases are: Moko of bananas, Stover 1972; tomato wilt, Anon 1983; Granville wilt of tobacco, Lucas 1975; potato brown rot or slime, Hooker 1981, Rich 1983; also: Buddenhagen & Kelman, *A.R.Phytop.* **2**:203, 1964; Harris in Kranz et al. 1977; Moffett et al., *AAB* **98**:403, 1981, seed & soil, sources of inoculum & foliage colonisation; Hayward in Fahy & Persley 1983; Woods, *Phytop.* **74**:972, 1984, banana, atypical symptoms induced by afluidal variants [**61**,2600; **64**,259].
P. syringae van Hall 1902; a very wide host range, causes necroses of plant organs above ground; Bradbury 1986 described 46 pvs.; Fahy & Lloyd in Fahy & Persley 1983.
P. s. pv. aceris (Ark) Young, Dye & Wilkie in Young et al., *N.Z.J.agric.Res.* **21**:153, 1978. *Acer* leaf spot; *Phytop.* **29**:968, 1939, USA (California) [**19**:173].
P. s. pv. antirrhini (Takimoto) Young et al., antirrhinum leaf spot; Moffett 1966; Simpson et al., *Pl.Path.* **20**:127, 1971 [**46**,2440; **51**,1555].
P. s. pv. apii (Jagger) Young et al.; celery leaf spot; *J.agric.Res.* **21**:185, 1921; may cause defoliation; Thayer & Wehlburg, *Phytop.* **55**:554, 1965, in USA (Florida) [**44**,2940].
P. s. pv. aptata (Brown & Jamieson) Young et al.; beet leaf spot, other crops; *J.agric.Res.* **1**:189,

1913; Ark & Leach, *Phytop.* **36**:549, 1946, seedborne [**26**:40].
P. s. pv. atrofaciens (McCulloch) Young et al.; basal glume rot of barley, oat, wheat; *J.agric.Res.* **18**:543, 1920; Wilkie, *N.Z.J.agric.Res.* **16**:155, 1973 [**53**,121].
P. s. pv. atropurpurea (Reddy & Godkin) Young et al.; grasses; poss. differs from pv. *coronafaciens* only in host range, forms coronatine; *Phytop.* **13**:75, 1923 [**2**:357].
P. s. pv. avellanae Psallidas 1984, pathotype str. not yet designated, canker of hazelnut; Psallidas, *EPPO Bull.* **17**:257, 1987.
P. s. pv. berberidis (Thornberry & Anderson) Young et al.; *Berberis* leaf spot, can cause leaf fall; *J.agric.Res.* **43**:29, 1931; Roberts & Preece, *J.appl.Bact.* **56**:507, 1984 [**11**:109; **63**,4950].
P. s. pv. cannabina (Šutić & Dowson) Young et al.; hemp, water soaked lesions on leaves, spreading and becoming dark; *Phytopath.Z.* **34**:307, 1959 [**38**:408].
P. s. pv. coronafaciens (Elliott) Young et al.; Bradbury, *Descr.B.* 235, 1970, as *Pseudomonas coronafaciens* (Elliot) Stevens; halo blight of oat, forms the non-specific toxin coronatine; *J.agric.Res.* **19**:139, 1920.
P. s. pv. eriobotryae (Takimoto) Young et al,; loquat bud blight and twig canker; *J.Pl.Prot. Tokyo* **18**:349, 1931, fide Bradbury 1986.
P. s. pv. garcae (Amaral, Teixeria & Pinheiro) Young et al.; coffee, twig blight and dieback, Elgon or Solai dieback of this crop in Kenya is a major disease there; called halo blight in Brazil where the bacterium was originally described; *Arq.Inst.biol.São Paulo* **23**:151, 1956; Costa et al., *Phytopath.Z.* **28**:427, 1957, in Brazil; Ramos, *PDR* **63**:6, 1979, epidemiology; Ramos & Kamidi, *PD* **65**:581, 1981, seasonal periodicity & distribution in Kenya; Kairu et al., *Pl.Path.* **34**:207, 1985, control with Cu [**36**:643; **37**:42; **58**,5870; **61**,1739; **64**,4332].
P. s. pv. glycinea (Coerper) Young et al.; leaf spot, a common disease of soybean; *J.agric.Res.* **18**:179, 1919; Kennedy & Ercolani, *Phytop.* **68**:1196, 1978, epiphytic multiplication; Peña et al., *Acta agron.Palmira* **32**:53, 1982, races; Gnanamanickam et al., *CJB* **61**:3271, 1983, toxins; Park & Lim, *Phytop.* **75**:520, 1985, overwintering; *PD* **70**:214, 1986, effect on yield [**58**, 3557; **63**,2073, 3139; **64**,5528; **65**,3594].
P. s. pv. helianthi (Kawamura) Young et al.; *Helianthus* leaf spot; *Ann.phytopath.Soc.Japan* **4**:25, 1934; Thornberry & Anderson, *Phytop.* **27**:946, 1937 [**14**:314; **17**:98].
P. s. pv. lachrymans (Smith & Bryan) Young et al.; Bradbury, *Descr.B.* 124, 1967, as *Pseudomonas*

lachrymans (Smith & Bryan) Carsner; cucumber angular leaf spot, on other cucurbits; can cause significant loss; Komoto & Kimura, *Bull.Chugoku Natn.agric.Exp.Stn.*E. 21:1, 1983, seed transm.; Kritzman & Zutra, *Phytoparasitica* **11**:99, 1983, survival; Mabbett & Phelps, *Trop.Pest Management* **30**:444, 1984, chemical control; Sherf & MacNab 1986 [**63**,1476–7; **64**,2792].

P. s. pv. lapsa (Ark) Young et al.; stalk rot of maize and sorghum; Bradbury 1986.

P. s. pv. maculicola (McCulloch) Young et al.; peppery spot of cauliflower and broccoli, on other brassicas; Bradbury 1986; Sherf & MacNab 1986, as *Pseudomonas maculicola* (McCulloch) F.L. Stevens.

P s. pv. mellea (Johnson) Young et al; Wisconsin leaf spot of tobacco; *J.agric.Res.* **23**:481, 1923 [**3**:68].

P. s. pv. mori (Boyer & Lambert) Young et al.; on mulberry, cankers are caused on shoots which are killed; Takahashi, *JARQ* **14**:41, 1980; Bradbury 1986 [**60**,2688].

P. s. pv. morsprunorum (Wormald) Young et al., Bradbury, *Descr.B.* 125, 1967, as *Pseudomonas morsprunorum* Wormald; *Prunus* stem canker, especially cherry and plum; Crosse, *A.R.Phytop.* **4**:291, 1966; see Garrett, pv. *syringae*.

P. s. pv. myricae Ogimi & Higuchi, *Ann.phytopath.Soc.Japan* **47**:443, 1981; *Myrica rubra*, galls up to several cm diam. on trunks and branches [**61**,5265].

P. s. pv. panici (Elliott) Young et al.; *Panicum miliaceum*, on barley, foxtail millet, rice; brown streaks on leaves, sheaths and culms, upper part of plant may die; *J.agric.Res.* **26**:151, 1923.

P. s. pv. papulans (Rose) Dhanvantari, *N.Z.J.agric.Res.* **20**:557, 1977; apple, blister spot of fruit, small cankers or rough bark on branches, midvein necrosis on leaves of cv. Mutsu; Rose, *Phytop.* **7**:198, 1917; Burr & Hurwitz, *PDR* **63**:157, 1979; Bedford et al., *Can.J.Pl.Path.* **6**:17, 1984; Bonn & Bedford, *ibid* **8**:167, 1986 [**57**,4542; **59**,830; **63**,3980; **66**,1993].

P. s. pv. passiflorae (Reid) Young et al., passion fruit grease spot, necrotic spots on leaves, stems and fruit; *N.Z.J.Sci.Technol.* A **20**:260, 1938, Bradbury 1986 has 1939 [**18**:331].

P. s. pv. persicae (Prunier et al.) Young et al.; peach, leaf spot, canker, fruit gummosis, dieback and death, France; *Annls.Phytopath.* **2**:181, 1970 [**50**,787b].

P. s. pv. phaseolicola (Burkholder) Young et al., Hayward & Waterston, *Descr.B.* 45, 1965, as *Pseudomonas phaseolicola* (Burkholder) Dowson; bean halo blight and on other legumes; a serious, seedborne pathogen, particularly in cooler

climates; Allen 1983; Fahy & Lloyd in Fahy & Persley 1983; Sherf & MacNab 1986.

P. s. pv. philadelphi Roberts, *J.appl.Bact.* **59**:283, 1985; leaf spot of *Philadelphus*; *P. coronarius*, mock orange or syringa [**65**,1348].

P. s. pv. photiniae Goto, *Ann.phytopath.Soc.Japan* **49**:457, 1983; leaf spot of *Photinia glabra* [**63**,3061].

P. s. pv. pisi (Sackett) Young et al.; Bradbury, *Descr.B.* 126, 1967; as *Pseudomonas pisi* Sackett; pea blight, a serious seedborne pathogen; Hagedorn 1984; Sherf & MacNab 1986.

P. s. pv. primulae (Ark & Gardner) Young et al., leaf spot of *Primula* spp., large parts of the leaves can become necrotic; *Phytop.* **26**:1050, 1936 [**16**:256].

P. s. pv. ribicola (Bohn & Maloit) Young et al.; leaf spot and defoliation of *Ribes aureum*, *J.agric.Res.* **73**:281, 1946, USA (Wyoming) [**26**:66].

P. s. pv. savastanoi (Smith) Young et al.; causing the widespread olive knot; galls or knots form on trunks, branches, roots and even leaves; they reach a large size in a few months. This pv. is heterogeneous. Janse, fide Bradbury 1986, erected *Pseudomonas syringae* ssp. *savastanoi* and distinguished 3 pvs. within it: *fraxinii, nerii, oleae*; these names were not validly published. Bradbury loc.cit. considered that the erection of this ssp. is at least premature. This and the 'rather clumsy nomenclature' caused him to maintain *P. syringae* pv. *savastanoi*. Natural hosts include: *Forsythia intermedia, Fraxinus* spp., *Ligustrum* spp., *Nerium oleander, Olea europaea* and vars. *oleaster* and *sativa*.

P. s. pv. sesami (Malkoff) Young et al.; Bradbury, *Descr.B.* 696, 1981; dark brown to black leaf spots, tending to be limited by veins, on sesame, seedborne, 2 races.

P. s. pv. striafaciens (Elliott) Young et al.; Bradbury, *Descr.B.* 238, 1970, as *Pseudomonas striafaciens* (Elliott) Starr & Burkholder, oat stripe blight; Fahy & Lloyd in Fahy & Persley 1983.

P. s. pv. syringae van Hall 1902; Hayward & Waterson, *Descr.B.* 46, 1965; Fahy & Lloyd in Fahy & Persley 1983, Bradbury 1986. The pathogen has a very wide host range. Some of the diseases caused are: canker, blossom and bud blight, shoot wilt of stone fruit; blast and blossom blight of pear; blister bark and blossom blight of apple, blast or black pit of citrus, lilac blight, bean brown spot. The pv. occurs as an epiphyte on many unrelated plants. The symptoms on apple and pear should not be confused with those caused by *Erwinia amylovora*. One of the 58 synonyms listed by Bradbury loc.cit. is

Pseudomonas oryzicola Klement 1955, Bradbury, *Descr.B.* 236, 1970, causing rice sheath blight. The diseases on stone fruit, prob. the most serious, were reviewed by Cross, *P. syringae* pv. *morsprunorum* q.v; another review is by Cameron, *Tech.Bull.Ore.agric.Exp.Stn.* 66, 1962, fide Hayward & Waterston loc.cit.; for toxins see syringomycin; Klement, *EPPO Bull.* 7:57, 1977, on apricot; Latorre et al., *PD* 69:409, 1985, cherry, isolation & epiphytic populations; Gaudet & Kokko, *Can.J.Pl.Path.* 8:208, 1986, sorghum stunt, lesions on root & coleoptiles; Garrett, *Pl.Path.* 35:114, 1986, cherry, effect of stock on scion susceptibility to pvs. *morsprunorum* and *syringae*; Mansvelt & Hattingh, *PD* 70:403, 1986, apple [**64**, 4410; **65**,3999, 5032; **66**,1900].

P. s. pv. tabaci (Wolf & Foster) Young et al.; Bradbury, *Descr.B.* 129, 1967, as *Pseudomonas tabaci* (Wolf & Foster) Stevens; tobacco wildfire. The bacterium is found naturally on many plants but causes no other diseases, except a minor one on soybean, on bean and cowpea. Blackfire of tobacco was originally attributed to *Pseudomonas angulata* (Fromme & Murray) Stevens. This bacterium is now known to be a form of *P. s.* pv. *tabaci* which does not form a toxin. The symptoms of wildfire on leaves are characteristic, with broad chlorotic haloes to the leaf spots. The haloes are caused by toxin forming strs.; see tabtoxin. The strs. which do not form the toxin cause angular leaf spots, blackfire; these spots are darker than those caused by the strs. which form the toxin, are bounded by the veins and have no chlorotic haloes. Lucas 1975; Allen 1983; Fahy & Lloyd in Fahy & Persley 1983; Nishimura et al. 1985; Bradbury 1986; Deall & Cole, *Pl.Path.* 35:74, 1986 [**65**,886, 4085].

P. s. pv. tagetis (Hellmers) Young et al.; leaf spot of *Tagetes* spp., *Helianthus* spp., sunflower and apical chlorosis of Jerusalem artichoke, forms tagetitoxin; Shane & Baumer, *PD* 68:257, 1984; Rhodehamel & Durbin, *ibid* 69,589, 1985, hosts; Laberge & Sackston, *Phytoprotection* 67:117, 1986 [**63**,3691; **65**,2629; **66**,2159].

P. s. pv. theae (Hori) Young et al.; tea shoot blight; China, Japan; Horikawa, *Proc.Kansai Pl.Prot.Soc.* 27:7, 1985; Ando et al., *Ann.phytopath.Soc.Japan* 52:478, 1986 [**65**,361; **66**,1589].

P. s. pv. tomato (Okabe) Young et al.; tomato speck or small leaf spot, on *Capsicum*, seedborne; Getz et al., *Phytop.* 73:36, 39, 1983, tomato fruit development stage & susceptibility; fine structure of infection; McCarter et al., *ibid* 73:1393, 1983, survival; Mitchell et al., *Physiol.Pl.Path.* 23:315, 1983, toxins; Pyke et al., *N.Z.J.exp.Agric.* 12:161,

1984, control in seed; Bashan et al., *Ann.Bot.* 55:803, 1985, leaf morphology & infection; Leite & Mohan, *Fitopat.Brasileira* 10:541, 1985; Jardine & Stephens, *PD* 71:405, 1987, chemical control [**62**,2157–8; **63**,1959, 1961; **64**,2162, 4498; **65**,4547].

P. tolaasii Paine 1919; Bradbury, *Descr.B.* 894, 1987; mushroom brown blotch, *Agaricus* spp., for mushroom ginger blotch see Wong et al., *Pseudomonas fluorescens*; Fletcher 1984; Wong & Preece, *J.appl.Bact.* 58, 259, 269, 275, 1985, chemical control; Goor et al. 1986, diversity of *Pseudomonas* spp. on common mushroom [**64**,4077–9; **66**,1709].

P. viridiflava (Burkholder) Dowson 1939; Bradbury, *Descr.B.* 895, 1987; on leaves, blossoms and fruits of many plants; prob. a common epiphyte, entering through wounds as a secondary organism, or when a plant is under stress.

Pseudoperonospora Rostov. 1903; Peronosporaceae; Waterhouse & Brothers, *Mycol.Pap.* 148, 1981; differs from *Peronospora* in that the vegetative reproductive unit is a sporangium forming zoospores; *Peronospora* forms con.; differs from *Plasmopara* in that the angle of branching is more often acute and the branch tips are pointed; Spencer 1981.

P. cubensis (Berk. & Curtis) Rostov.; Palti, *Descr.F.* 457, 1975; sporangia 20–40 × 14–25 μm; oospores 22–42 μm diam., described from many regions; cucurbit downy mildew is important on *Citrullus*, *Cucumis*, *Cucurbita*; often destructive in warmer conditions; sporangia air dispersed with a forenoon peak; the shortest dew period for zoospore infection is 2 hours at 20°; carry over is mostly through mycelium in the host. There is strong evidence for the existence of races but a more precise differentiation is needed. Further work on the occurrence of oospores is also desirable; Holliday 1980; Palti & Cohen, *Phytoparasitica* 8:109, 1980; Cohen in Spencer 1981; Inaba et al. 1986; Sherf & MacNab 1986 [**60**,536; **66**,2123].

P. humuli (Miyabe & Takah.) G. Wilson 1914; Francis, *Descr.F.* 769, 1983; sporangia 22–30 × 16 μm, oospores 25–40 μm diam.; hop downy mildew is one of the most serious diseases of the crop; localised lesions are caused on leaves, flowers, cones and shoots are infected systemically. Each season infection arises from overwintering mycelium in rootstocks which become debilitated; the role of the oospores in the epidemiology is not precisely understood. Zoospores infect through the stomata; infection needs a minimum of 1.5 hours for leaves and 3 hours for shoots. Sporangia show the expected diurnal release pattern with a forenoon peak. Differences in pathogenicity prob. occur between isolates; the position on races appears unclear; Ratushina et al. 1985 refer to 16 races in USSR. Control is largely through fungicides and based on disease prediction; Royle & Kremheller in Spencer 1981; Skotland & Johnson, *PD* 67:1183, 1983; control in USA [**63**,3491; **65**,4490].

Pseudopezicula, see *Pseudopeziza tracheiphila*.

Pseudopeziza Fuckel 1870; Dermataceae; on leaves and herbaceous stems, ascomata ± smooth, arising from a stroma, not forming rhizomorphs, excipulum nearly or wholly lacking, asco. aseptate; Schüepp, *Phytopath.Z.* **36**:224, 1959.

P. tracheiphila Müller-Thurgau, *Zentbl.Bakt.ParasitKde.* 2 **10**:57, 1903, also pages: 8, 48, 81, 113; anam. *Phialophora tracheiphila* (Sacc. & Sacc.) Korf; asco. 18–22 × 4.5–11 μm. Korf et al., *Mycotaxon* **26**:457, 1986, transferred the fungus to a new genus *Pseudopezicula*, page 462, in the Peziculoideae; the new combination *P. tracheiphila* has 8 spored asci; the American form was designated a new sp. *P. tetraspora* which has 4 spored asci; on grapevine leaves, rotbrenner, red fire or red rot; central Europe, Brazil (Rio Grande do Sul), Jordan, Tunisia, Turkey; Ochs 1960; Kundert Bolay 1968; Kundert 1970; Neider, *Pflanzenschutzberichte* **47**:19, 1986, chemical control [**40**:452; **47**,2539; **49**,2569; **65**,6133; **66**,264].

P. trifolii (Biv.–Bern.) Fuckel 1870, fide Cannon et al. 1985; asco. 8–12 × 4–5 μm. Booth & Waller, *Descr.F.* 636–7, 1979, described the 2 spp. *Pseudopeziza medicaginis* (Lib.) Sacc. 1889 and *P. trifolii*. Boerema & Verhoeven, *Neth.J.Pl.Path.* **85**:170, 1979, accepted the treatment of this complex by Schüepp, see genus, into 5 ff. sp.; they referred to Schmiedeknecht 1964 who maintained a differentiation into 3 spp. The fungi cause leaf spots on herbage legumes in temperate areas. The 5 ff. sp. fide Boerema & Verhoeven are: *medicaginis-lupulinae* (Schmiedeknecht) Schüepp, on *Medicago*, lucerne is resistant; *medicaginis-sativae*, same authors, ± specific on lucerne but may infect some other *Medicago* spp.; *meliloti* (Sydow) Schüepp, on *Melilotus, Trigonella*; *trifolii-pratensis* Schüepp, on *Trifolium* but white clover is not susceptible; *trifolii-repentis* Schüepp, on white clover but other clovers are susceptible; Semeniuk, *Phytopath.Z.* **110**:281, 290, 1984, asco. germination, infection & dispersal with peak after sunrise [**64**,2972–3].

Pseudophaeolus Ryv. 1975; Aphyllophorales; *P. baudonii* (Pat.) Ryv.; van der Westhuizen, *Descr.F.* 442, 1975, as *Polyporus baudonii* Pat.

1914; Ofosu-Asiedu, *TBMS* **65**:285, 1975, root rot of *Eucalyptus citriodora* in Ghana; Rattan & Pawsey, *Trop.Pest Management* **27**:225, 1981, root rot and death of tea in Malawi [**55**,2909; **61**,1374].

pseudoplatani, *Ophiognomonia, Pleuroceras*.

Pseudoseptoria Speg. 1910; Coelom.; Sutton 1980; conidiomata pycnidial, conidiogenesis holoblastic, conidiophore growth percurrent; con. falcate, fusiform, aseptate; on Gramineae. *P. donacis* (Pass.) B. Sutton 1977; Punithalingam & Waller, *Descr.F.* 400, 1973, as *Selenophoma donacis* (Pass.) Sprague & A.G. Johnson 1940; con. 16–30 × 2–3 μm; halo spot of temperate cereals and grasses; has caused severe defoliation of *Dactylis glomerata* and *Phleum pratense*.

Pseudospiropes M.B. Ellis, *Dematiaceous Hyphomycetes*: 258, 1971; Hyphom.; conidiomata hyphal, conidiogenesis holoblastic, conidiophore growth sympodial; con. pigmented, multiseptate. *P. elaeidis* (Steyaert) Deighton, *TBMS* **85**:739, 1985; Mulder & Holliday, *Descr.F.* 464, 1975, as *Cercospora elaeidis* Steyaert 1948; con. 43–130 × 5–8 μm; oil palm freckle. The fungus was transferred on the grounds that it differed from other *Cercospora* spp. with pigmented con. in that the conidial scars were situated on short protruberances rather than lying flat.

pseudothecium, a stromatic ascoma which has asci in numerous, unwalled locules as in the loculascomycetes.

pseudotsugae, *Caliciopsis, Corniculariella, Dasyscyphus, Dermea, Durandiella, Phacidiopycnis, Phytophthora, Rhabdocline*.

Pseudotsuga menziesii, Douglas fir.

Pseudovalsaceae, Diaporthales; ascomata erect, beaks usually central, erumpent; *Phragmoporthe*.

psidii, *Pestalotiopsis, Phomopsis, Puccinia*.

Psidium guajava, guava.

psophocarpi, *Pseudocercospora, Synchytrium*.

Psophocarpus tetragonolobus, Goa or winged bean.

psychromorbidus, *Coprinus*.

psychrophil, an organism which can grow at low temps.; see snow moulds; Deverall in Ainsworth & Sussman, vol. 3, 1968, on psychrophilic fungi.

Psylla pyricola, transm. pear decline.

psyllids, see vector.

Psylliodes, transm. turnip crinkle, turnip yellow mosaic; *P. affinis* transm. solanum dulcamara mottle.

ptelea yellow spotting, arabis mosaic v. and cherry leaf roll v. detected; Schmelzer & Stahl, *Acta phytopath.Acad.Scient.hung.* **7**:221, 1972; Schmelzer, *Z.Bact.ParasitKde.* 2, **127**:140, 1972 [**51**,4751; **52**,2880].

Pteridium aquilinum, bracken, a serious weed. Gregor, *Phytopath.Z.* **8**:401, 1935, described a disease, later called pinnule blight, in Scotland and caused by *Corticium anceps* (Bres. & H. Sydow) Gregor, *Annals.Mycol.* **30**:463, 1932; the fungus is found elsewhere in Europe. Irvine et al., *AAB* **110**:25, 1987, reported on curl tip assoc. with *Ascochyta pteridis* Bres. and *Phoma aquilina* Sacc. & Penzig. The former appeared to infect undamaged plants to cause fleck symptoms assoc. with curl tip. The latter caused curl tip when inoculated through wounds, McElwee 1983, fide Irvine et al. But it caused no damage when con. were applied to intact tissue; Burge et al. 1986, as *A. aquilina* [**14**:797; **66**,2655, 3185].

Puccinia Pers. 1801; Uredinales; telia and globose pycnia subepidermal, teliospores typically 2 celled, truly pedicillate, on Angiosperms, autoecious or heteroecious, macrocyclic or microcyclic, > 3000 spp.; with *Fusarium* and *Phytophthora* one of the most important pathogenic genera; Butler & Jones 1949; Dickson 1956; Wilson & Henderson 1966; Hooker, *A.R.Phytop.* **5**:163, 1967, genetics & expression of resistance; Walker 1969; Cummins 1971, 1978; Holliday 1980; Dixon 1981; Allen 1983.

P. allii Rud. 1829; Laundon & Waterston, *Descr.F.* 52, 1965; autoecious, macrocyclic; teliospores 36–65 × 18–28 μm; *Allium*, see *Descr.F.* for other rusts on the genus; prob. less important in the tropics than elsewhere, transm. seed, isolates show variation in virulence but no races differentiated; Jones, *Tests Agrochem.Cvs.* 6 *AAB* **106** suppl.:52, 54, 1985, fungicides; Uma & Taylor, *TBMS* **87**:320, 1986, teliospores in England.

P. antirrhini Dietel & Holway 1897; Sivanesan, *Descr.F.* 262, 1970; microcyclic, teliospores 30–55 × 18–19μm; on antirrhinum; infection by uredospores; the fungus is an early example of the rapid spread of an air dispersed pathogen. The disease was reported from USA (California) in 1896; it was first found in Europe in 1931 and spread throughout the continent reaching N. Africa and W. Asia in 1936–7; prob. seedborne, 2 races; Gawthrop & Jones, *Ann.Bot.* **47**:197, 1981, review of resistance, chronology of genetic change in the pathogen's virulence [**60**,4498].

P. apii Desm. 1823; Laundon & Rainbow, *Descr.F.* 284, 1971; autoecious, macrocyclic, teliospores 30–50 × 20–26 μm; celery; prob. of little importance.

P. arachidis Speg. 1884; Laundon & Waterston, *Descr.F.* 53, 1965; microcyclic, teliospores 38–42 × 14–16 μm; *Arachis* spp., the only rust on groundnut. It has been known for a long time in tropical America and was not considered

economically important, although some damage occurred in the West Indies and USA (S.). More recently *Puccinia arachidis* has spread in Africa, Asia and Australasia; there has been considerable damage to groundnut in India. Spread is mostly through the uredospores which have a diurnal periodicity with a peak at noon; the fungus may be found on seed; Cook, *Phytop.* **70**:822, 826, 1980, susceptibility & relationship with leaf wettability; Mayee, *Indian Bot. Reptr.* 1:75, 1982, review; Mallaiah & Rao, *TBMS* **78**:21, 1982, air dispersal; Subrahmanyam et al., *Phytop.* **73**:253, 726, 1983; *PD* **67**:209, 1983; **69**:813, 1985, resistance, effect of host genotype on uredospore formation & germination, review; Savary, *Neth.J.Pl.Path.* **92**:115, 1986; **93**:15, 25, 1987, uredospore dispersal, plant age & development of primary disease gradients, decrease in susceptibility related to host development & leaf age [**60**,4088–9; **61**,4492; **62**,3372, 3374, 4551; **66**,401].

P. asparagi DC. 1805; Waterston, *Descr.F.* 54, 1965; autoecious, macrocyclic, teliospores 30–50 × 19–26 μm; asparagus, can be serious; Blanchette et al., *PD* **66**:904, 1982, resistance; Johnson, *Phytop.* **76**:208, 1986, slow rusting [**62**,518; **65**,3647].

P. cacabata Arthur & Holway 1925; Mulder & Holliday, *Descr.F.* 294, 1971; heteroecious, macrocyclic, teliospores 27 × 39 × 19–26 μm; telia on *Bouteloua* and *Chloris*, aecia on *Gossypium*; southwestern cotton rust; S. America, USA, West Indies; the telial stage is more widespread than the aecial one; significant damage has been caused in Arizona and Mexico.

P. calcitrapae DC. 1805; Wilson & Henderson 1966; autoecious, macrocyclic, teliospores 24–50 × 16–27 μm; on *Arctum, Carduus, Carlina, Centaurea, Cirsium*; Politis et al., *Phytop.* **74**:687, 1984, as *Puccinia carduorum* Jacky which is a synonym, evaluation for control of musk thistle, *Carduus nutans*, infected globe artichoke; Politis & Bruckart, *PD* **70**:288, 1986, conditions required for infection of musk thistle [**63**,5714; **65**,5349].

P. canaliculata (Schwein.) Lagerh. 1894; Arthur 1934; heteroecious, macrocyclic, teliospores 39–64 × 15–21 μm; telia on *Cyperus*, aecia on *Ambrosia trifida, Xanthium*; Phatak et al., *Science N.Y.* **219**:1446, 1983; Callaway et al., *PD* **69**:924, 1985; poss. control of *C. esculentus* and *C. rotundus*, yellow & purple nut sedges, respectively [**62**,2755; **65**,4236].

P. carduorum, see *Puccinia calcitrapae*.

P. caricina DC. 1815; heteroecious, macrocyclic, teliospores 35–66 × 14–23 μm; telia on *Carex*, aecia on *Ribes, Urtica* and other plants; Wilson &

Henderson 1966 described 15 vars. for Britain; var. *pringsheimiana*, teliospores 50–75 × 15–20 μm, causes gooseberry rust which is of little importance and controlled by removal of the telial hosts.

P. carthami Corda 1840; Punithalingam, *Descr.F.* 174, 1968; autoecious, macrocyclic, teliospores 36–44 × 24–30 μm; safflower rust, see *Descr.F.* for rusts on *Carthamus*, seedborne, at least 4 races, a severe disease in USA (W.); Holliday 1980.

P. chondrillina Bubák & Sydow 1901; Arthur 1934; autoecious, macrocyclic, teliospores 30–39 × 19–23 μm; on *Chondrilla* spp. in Mediterranean region and Asia (W., central), and on *C. juncea*, skeleton weed, in USA (E.). The weed was accidentally brought to Australia at the beginning of this century; it became damaging in the wheat cultivations of E. Australia and chemical control failed. The rust was therefore introduced for control; some success has been achieved but several forms of the weed exist and further strs. of the fungus have been introduced. The weed also became serious in USA (W.) *c.*1960; Hasan and Hasan & Wapshere, *AAB* **72**:257, 1972; **74**:323, 1973; Cullen et al., *Nature Lond.* **244**:462, 1973; Groves & Williams, *Aust.J.agric.Res.* **26**:975, 1975; Hasan, *AAB* **99**:119, 1981; Emge et al. and Adams & Line, *Phytop.* **71**:839, 1981; **74**:742, 745, 1984 [**52**,1379; **53**,35, 390; **55**,2614; **61**,2049, 3211; **63**,5715–16].

P. chrysanthemi Roze 1900; Punithalingam, *Descr.F.* 175, 1968; microcyclic, teliospores 35–57 × 20–25 μm; chrysanthemum black or brown rust; see *Descr.F.* for rusts on *Chrysanthemum*.

P. cordiae (Henn.) Arthur, fide Briton-Jones, *Mem.Imp.Coll.Trop.Agric.Trin.* 3, 1930; teliospores 27–48 × 18–30 μm; causing canker of cypre, *Cordia alliodora* in Trinidad, West Indies [**10**:215].

P. coronata Corda 1837; heteroecious, macrocyclic, teliospores 36–65 × 14–19 μm, apex characteristically flattened with digitate projections; telia on Gramineae, aecia on *Rhamnus*, including *R. catharticus*, buckthorn, and *Frangula alnus*, alder buckthorn; var. *avenae* and ff. sp. cause the important crown rust of oat, particularly in Canada and USA (N.), many races, spread over short distances is prob. more important than that over long ones; Simmons, *Monogr.Am.phytopath.Soc.* 5, 1970; Manners in Western 1971; Kopec et al., *PD* **67**:98, 1983, on perennial ryegrass [**62**,2022].

P. cynodontis Lacroix 1859; Mulder & Holliday, *Descr.F.* 292, 1971; heteroecious, macrocyclic,

teliospores 28–60 × 15–25 μm; telia on *Cynodon dactylon*, aecia on *Plantago*, races.

P. graminis Pers. 1801; Wilson & Henderson 1966; Cummins 1971; heteroecious, macrocyclic, telia black, teliospores 35–60 × 12–22 μm; on wheat, temperate cereals, other Gramineae; aecia on *Berberis*, especially *B. vulgaris*; wheat black stem rust. The rust that so troubled the earliest civilisations of Eurasia. It was accurately described by Fontana and Targioni-Tozzetti in Italy in 1767. De Bary in Germany, 1865–6, demonstrated the complete life cycle; he called the phenomenon of alternate hosts heteroecism. French legislation in 1660 showed that the European common barberry was by then suspect as the source of infection for wheat. More than 300 years later Roelfs, *PD* **66**:177, 1982, was describing the effects on the rust of barberry eradication in USA. In 1894 Eriksson determined the existence of ff.sp.; and some 19 years later Stakman and others demonstrated that the f.sp. *tritici* existed as many races.

Six ff.sp. were given in 1966: *agrostidis*, *Agrostis*; *avenae*, oat and grasses; *phlei-pratensis*, *Phleum pratense*; *poae*, *Poa*; *secalis*, barley, rye and grasses; *tritici*, barley, rye, wheat and grasses. The last and most important f.sp. exists in > 200 pathogenic forms or races. It has a temp. optimum of *c.*20° and is generally only severe in the warmer growing areas, Australia, India, N.Africa, N.America. The rust still limits production. It may overwinter on barberry or as uredospores which overwinter in, and spread from, other areas. Thus in USA epidemics in northerly areas may result from spread originating further south. In India *Puccinia graminis* f.sp. *tritici* oversummers in the foothills of the Himalayas and the southerly hills. In 1916 an epidemic in USA reduced wheat yields by *c.*65% and a barberry eradication campaign began. Roelfs loc.cit. described its beneficial effects: delayed disease onset, virulence reduction in spring inoculum, fewer and stabilisation of races. There is now virtually no spread from barberry to commercial fields. Clark et al., *Can.J.Pl.Path.* **8**:193, 1986, gave an account of *P.g.*f.sp. *avenae* and its races in Canada 1968–83, and the results from barberry eradication. Continual selection is required to keep host resistance levels adequate.

Butler, *Fungi and disease in plants*, 1918; Eriksson, *Fungus diseases of plants*, edn. 2, 1930; Heald 1933; Butler & Jones 1949; Stakman, *AAB* **42**:22, 1955; Dickson 1956; Stakman & Harrar, *Principles of plant pathology*, 1957; Walker 1969; Manners in Western 1971; Wiese 1987; Bushnell & Roelfs ed. *The cereal rusts*, 1984; Knott,

Phytop. **76**:1149, 1986, genetic structure of populations.

P. granularis Kalchbr. & Cooke & 1882; Sivanesan, *Descr.F.* 263, 1970; autoecious, macrocyclic, teliospores 43–80 × 22–23.5 μm, pelargonium rust, Africa.

P. helianthi Schwein. 1822; Laundon & Waterston, *Descr.F.* 55, 1965; autoecious, macrocyclic, teliospores 40–60 × 18–30; see *Descr.F.* for rusts on *Helianthus*; sunflower rust causes an important disease, it occurs on Jerusalem artichoke, races, seed dispersal prob. not significant; Holliday 1980; Kochman & Goulter, *Australasian Pl.Path.* **13**:3, 1984; Yang, *Ann.phytopath.Soc.Japan* **52**:248, 1986, new race overcoming resistance to other races in N. America & Argentina [**64**,703; **66**,1095].

P. hordei Otth 1871; Wilson & Henderson 1966; heteroecious, macrocyclic, teliospores 40–54 × 15–24 μm, high proportion are aseptate, mesospores 25–45 × 16–24 μm; on barley, brown rust or leaf rust; aecia on *Ornithogalum*; widespread with crop, locally important, > 50 races; Dickson 1956; Manners in Western 1971; Mathre 1982; Gareth Jones & Clifford 1983.

P. horiana Henn. 1901; Punithalingam, *Descr.F.* 176, 1968; microcyclic, teliospores 32–45 × 12–18 μm; chrysanthemum white rust; see *Descr.F.* for rusts on *Chrysanthemum*; spread by basidiospores, a widespread but restricted distribution, absent from Australia and eradicated from England and Wales; Dickens in Ebbels & King 1979, biology & eradication from UK; Leu et al., *Pl.Prot.Bull.Taiwan* **24**:9, 1982; Walker, *TBMS* **81**:664, 1983, distribution & spread [**61**,6423; **63**,1293].

P. iridis Rabenh. 1844; Laundon & Rainbow, *Descr.F.* 285, 1971; heteroecious, macrocyclic, teliospores 30–50 × 13–21 μm; aecia on *Urtica*; see *Descr.F.* for rusts on *Iris*; some host specificity.

P. kuehnii E. Butler 1914; Laundon & Waterston, *Descr.F.* 10, 1964; microcyclic, teliospores 25–40 × 10–18 μm; a less important sugarcane rust than *Puccinia melanocephala* q.v. for a morphological distinction in the uredospores; prob. restricted to Asia and Australasia.

P. leveillei Mont. 1852; Sivanesan, *Descr.F.* 264, 1970; microcyclic, teliospores 26–34 × 16–22 μm; geranium rust; see *Descr.F.* for rusts on *Geranium* spp.

P. malvacearum Mont.; Sivanesan, *Descr.F.* 265, 1970; Parmelee & deCarteret, *Fungi Canadenses* 171, 1980; microcyclic, only teliospores known, 35–75 × 13–26 μm; hollyhock rust, on *Althaea*, *Malva*; see *Descr.F.* for rusts on Malvaceae. The rust was described from Chile in 1852 and

Australia in 1857; it spread rapidly in Europe after being reported from Spain in 1869.

P. melanocephala H. Sydow & Sydow 1907; Laundon & Waterston, *Descr.F.* 9, 1964, as *Puccinia erianthi* Padw. & Khan; microcyclic, teliospores 36–55 × 18–24 μm. This sugarcane rust causes a much more severe disease than does *P. kuehnii* on the same crop. A wrong identification can be made since both may usually be found in the field as uredia only. Mordue, *TBMS* **84**:758, 1985, described uredospore ornamentation as a character which had no intermediates. Uredospores of *P. melanocephala* are densely echinulate, the spines are 1–5.5 μm apart; those of *P. kuehnii* are sparsely so, spines mostly 3–4 μm apart. The former rust, originally only in China and India, has now spread to most sugarcane cultivations and caused damaging epidemics. It appears to be still absent from S. America south of the river Amazon; Holliday 1980; Liu, *Sugcane Path.Newsl.* 26:26, 1981; Purdy et al., *PD* **67**:1292, 1983; **69**, 689, 1985; Sivanesan & Waller 1986 [**65**,876].

P. menthae Pers. 1801; Laundon & Waterston, *Descr.F.* 7, 1964; Wilson & Henderson 1966 called it a collective sp., several vars. have been proposed; autoecious, macrocyclic, teliospores 22–30 × 17–24 μm; on many genera of Labiatae; mint rust can cause severe crop loss; races, the form on *Mentha piperata* is not pathogenic to *M. spicata* and vice versa. The only other rust on *Mentha* is *Puccinia angustata* Peck 1873, fide Arthur 1934; Laundon & Waterston loc.cit. It is heteroecious, macrocyclic, with telia on Cyperaceae and aecia on Labiatae; Harvey, *Australasian Pl.Path.* **8**:44, 1979, detrimental effect on oil yield in New Zealand; Beresford, *N.Z.J.agric.Res.* **25**:431, 1982, 3 races in New Zealand [**59**,2868; **62**,1600].

P. nakanishikii Dietel 1905; Cummins 1971; microcyclic, teliospores 33–44 × 20–25 μm; rust of citronella and lemon grasses, *Cymbopogon nardus* and *C. citratus*, causing damage in Sri Lanka, Bandara, *PD* **65**:164, 1981 [**61**,1330].

P. oahuensis Ell. & Ev. 1895; Mordue & Crichett, *Descr.F.* 516, 1976, microcyclic, teliospores 27–46 × 16–22 μm; pangola grass rust, may be economically important, see *Descr.F.* for other rusts on *Digitaria*.

P. pelargonii-zonalis Doidge 1926; Sivanesan, *Descr.F.* 266, 1970; microcyclic, teliospores 36–50 × 16–24 μm; see *Descr.F.* for other rusts on *Pelargonium*.

P. pittieriana Henn. 1904; Laundon & Rainbow, *Descr.F.* 286, 1971; microcyclic, teliospores 23–38 × 17–25 μm; potato and tomato rust,

Solanum spp.; Central and S. America; the only rust on tomato; damage has been caused on potato in Colombia and Ecuador, and heavy infection of *S. demissum* in Mexico. The reason that the rust has not followed these crops elsewhere may be because of the absence of overwintering or carryover hosts.

P. poae-nemoralis Otth 1871; Wilson & Henderson 1966; Cummins 1971, as *Puccinia brachypodii* var. *poae-nemoralis* (Otth) Cummins & H.C. Green 1966; heteroecious, macrocyclic, teliospores 30–45 × 16–22 μm; on temperate grasses; aecial stage rare, on *Berberis*.

P. poarum Niels 1877; Wilson & Henderson 1966; Cummins 1971; heteroecious, macrocyclic, teliospores 30–45 × 16–22 μm; on temperate grasses including *Agrostis, Festuca, Phleum, Poa*; aecia on *Tussilago farfara* and other genera; Whipps & Lewis, *TBMS* **82**:455, 1984, infection & development of aecial state, effect of oxycarboxin on fungus growth [**63**,3160].

P. polysora Underw. 1897; Laundon & Waterston, *Descr.F.* 4, 1964; microcyclic, teliospores 35–50 × 16–26 μm, uredospores 28–38 × 22–30 μm; southern rust of maize, on *Erianthus, Euchlaena, Tripsacum*; see *Descr.F.* and Cummins 1971 for other maize rusts. This rust, presumably indigenous in parts of America, was confined to this continent until 1948–9 when it appeared in W. Africa. Its consequent rapid spread across to the Indian ocean seaboard in a susceptible crop caused very heavy losses; *Puccinia polysora* is favoured by high temps. *P. sorghi*, on maize, is macrocyclic, has smaller and darker uredospores and the uredia are common on both leaf surfaces; Holliday 1980; Shurtleff 1980.

P. psidii Winter 1884; Laundon & Waterston, *Descr.F.* 56, 1965; microcyclic, teliospores 30–48 × 19–22 μm; Myrtaceae; Central, S. America, USA (Florida), West Indies. In the mid 1930s the rust caused severe losses on pimento or allspice, *Pimenta dioica*, in Jamaica. In Brazil (Minas Gerais) in the 1980s losses were reported on a provenance of *Eucalyptus grandis* where *Puccinia psidii* is common on wild *Eugenia jambos*, rose apple. The rust infects guava. Evidence from Jamaica indicates that the forms on guava, pimento and *Eugenia* behave as different races; Holliday 1980; Dianese et al., *PD* **68**:314, 1984, on eucalyptus [**63**,3576].

P. purpurea Cooke 1876; Laundon & Waterston, *Descr.F.* 8, 1964; microcyclic, teliospores 40–60 × 25–32 μm; sorghum, for other rusts on sorghum see Tarr 1962 and *Descr.F.*; the disease is little noticed until the crop is maturing; provided the cvs. have some resistance there is

unlikely to be any significant damage; Holliday 1980.

P. recondita Roberge ex Desm. 1857, fide Cummins 1971 who gave > 50 synonyms, and treated the fungus as a sp. complex whose morphological variability is extreme. Wilson & Henderson 1966 described 11 ff.sp.; the 2 of economic importance, but particularly the second, are: f.sp. *recondita*, rye leaf rust; and f.sp. *tritici*, wheat brown or leaf rust. Less commonly the rye and wheat forms are treated as distinct spp., Savile, *Fungi Canadenses* 309, 1986; heteroecious, macrocyclic, teliospores 36–65 × 13–24 μm; Cummins gave 5 families for the aecial hosts; f.sp. *tritici* has aecia on *Anchusa* and *Thalictrum* spp., and *Isopyrum fumarioides*. But, unlike *Puccinia graminis* f.sp. *tritici*, the aecial hosts are epidemiologically unimportant.

Brown rust is prob. the most important of the wheat rusts and can be more damaging than black stem rust, since it is significant as a pathogen over a wider climatic range. The uredial pustules, usually on leaves, are cinnamon brown compared with the yellowish or orange of the other temperate cereal rusts. Brown rust is most severe in the warmer, more humid cultivation areas, but it can also be severe in N.W. Europe. The uredia carryover infection; the uredospores can survive relatively low temps. There are > 200 races. Dubin & Torres, *A.R.Phytop.* **19**:41, 1981, gave a graphic account of an epidemic in 1976–7 on wheat in N.W. Mexico. This was due to a breakdown of the resistance in the most widely planted cv. and weather. The serious losses which would have been caused were effectively reduced by the large scale use of 2 systemic fungicides.

Chester, *The nature and prevention of the cereal rusts as exemplified in the leaf rust of wheat*, 1946; Butler & Jones 1949; Dickson 1956; Manners in Western 1971; Gareth Jones & Clifford 1983; de Milliano et al., *Neth.J.Pl.Path.* **92**:49, 1986, testing wheat for race non-specific resistance; Hassan et al., *TBMS* **86**:365, 1986, summer & winter survival & infection by soilborne uredospores; Wiese 1987 [**65**,4360–1].

*P. sorghi Schwein 1832; Laundon & Waterston, *Descr.F.* 3, 1964, q.v. for other maize rusts; Parmelee & Savile, *Fungi Canadenses* 302, 1986; heteroecious, macrocyclic; telia on maize, *Euchlaena mexicana, E. perennis*, teliospores 35–50 × 16–23 μm, uredospores 24–29 × 22–29 μm; aecia on *Oxalis*, restricted to temperate areas. This rust has teliospores with a thicker apical wall and long pedicels compared with *Puccinia polysora*, and has lower temp. optima; uredospores do not overwinter; at least 14 races; Holliday 1980; Pataky, *Phytop.* **76**:702, 1986;

Headrick & Pataky, *ibid* **77**:454, 1987, partial
resistance; *PD* **70**:950, 1986, night temp., mist &
infection [**66**,183, 1441,3777].
*P. striiformis Westend. 1854; Mulder & Booth,
Descr.F. 291, 1971; Savile, *Fungi Canadenses* 250,
1981; microcyclic; teliospores 30–70 μm long,
upper cells 16–24 μm wide, basal cells 9–12 μm
wide. Wheat stripe or yellow rust, f.sp. *tritici*; on
barley f.sp. *hordei* and on *c.*230 grass spp. This is
the most important wheat rust of the cooler parts
of Europe and other areas in the world. It was the
first disease in which it was shown, by Biffen in
1904, that plant resistance to a pathogen was
genetically controlled and could be inherited as a
single Mendelian factor. The rust,
characteristically, forms small uredia in long lines,
mostly on the leaves; it overwinters in the
dicaryotic stage and has lower temp. optima than
the rusts on other temperate cereals; there are
> 60races, barley and rye can be damaged.
    Zadoks, *Tijdschr.PlZiekt.* **67**:69, 1961, review
for N.W. Europe; Manners in Western 1971;
Rapilly, *A.R.Phytop.* **17**:59, 1979, epidemiology;
Johnson et al., *TBMS* **58**:475, 1972, nomenclature
for wheat races; Priestley & Doling, *ibid* **63**:549,
1974, aggressiveness of isolates; Priestley in Scott
& Bainbridge 1978, detection of increased
virulence to wheat; Gopalan & Manners, *ibid*
82:239, 1984, effects of temp., rain, relative
humidity, leaf senescence on uredospore
germination; Osman-Ghani & Manners, *Pl.Path.*
34:75, 1985, partial resistance in barley;
McGregor & Manners, *ibid*, page 263, effects of
light & temp. on growth & sporulation on wheat;
Dubin & Stubbs, *PD* **70**:141, 1986, epidemic on
barley in S. America; Gaunt & Cole, *ibid* **71**:102,
1987, in New Zealand [**51**,4782; **54**,1679; **63**,2262;
64,3819, 4242; **65**,2782].
P. substriata Ell. & Barth. 1897; heteroecious,
macrocyclic; the sp. was split into 4 vars. by
Ramachar & Cummins 1965, Cummins 1971;
indica and *penicillariae* are the common rusts on
millet, Laundon & Waterston, *Descr.F.* 6, 1964.
The first, Indian form, has aecia on eggplant,
teliospores 51–71 × 17–20; it appears to be
confined to India although aecia on eggplant are
known from elsewhere. The second form is found
in Africa, India, Sri Lanka; aecia unknown,
teliospores 44–58 × 24–27 μm. Var. *imposita*, in the
Western hemisphere has aecia on *Solanum* spp.,
teliospores 38–50 × 23–28 μm, on *Digitaria* spp.
Var. *insolita* occurs in equatorial Africa on
Panicum and *Setaria* spp., teliospores
30–37 × 17–20 μm. Eboh, *TBMS* **87**:476, 1986,
described var. *decrospora* from Nigeria on
Pennisetum americanum, long teliospores

67–115 × 19–29 μm. Millet rust can cause extensive
leaf necrosis; most work has been done in India;
the evidence for the existence of races appears
inadequate; Holliday 1980.
P. thaliae Dietel 1899; Sivanesan, *Descr.F.* 267,
1970; microcyclic, rust of *Canna* spp., teliospores
40–80 × 14–20 μm.
Pucciniastrum Otth 1861; Uredinales; Wilson &
Henderson 1966; Ziller 1974; heteroecious,
macrocyclic; telia and uredia on angiosperms,
aecia on Pinaceae; telia not erumpent, but intra-
or subepidermal, teliospores mostly aseptate, in a
1 cell thick layer, divided by vertical septa into
several cells, Arthur 1934 has a key.
P. americanum (Farlow) Arthur 1920; Laundon &
Rainbow, *Descr.F.* 210, 1969; telia on *Rubus*,
teliospores 15–30 × 15–22 μm; aecia on *Picea*; late
leaf or late yellow rust of raspberry, white spruce
needle rust, N. America; may cause premature
defoliation of *P. engelmanii* and *P. glauca*, and
raspberry cvs.
P. areolatum (Fr.) Otth 1863; telia on *Prunus*,
teliospores 22–30 × 8–14 μm; aecia on spruce,
cone rust, N. temperate zone (E.), can cause seed
loss in *Picea*.
P. boehmeriae (Dietel) Sydow 1903; begonia rust,
teliospores may be absent, aecial stage reported
on *Abies firma*; Kakishima et al.,
Ann.phytopath.Soc.Japan **51**:623, 1985, on
Begonia spp. [**65**,4437].
P. epilobii Otth 1861; telia on *Epilobium* spp.,
teliospores 2 celled 18–28 × 7–15 μm, 4 celled
20–30 μm diam.; uredospores occur on *Fuchsia*
spp. but prob. not teliospores; aecia on *Abies*,
fireweed rust, can damage cones and needles.
Another rust on cultivated fuchsias is *Uredo*
fuchsiae (Cooke) Henn. 1903, found on plants
native to Australia. It is distinguished from
Pucciniastrum epilobii by the larger, bright orange
sori, absence of a peridium and larger,
conspicuously echinulate uredospores; McNabb &
Laurenson, *N.Z.J.agric.Res.* **8**:336, 1965, uredial
stage on fuchsia; Hiratsuka et al., *CJB* **45**:1913,
1967, inoculated *A. lasiocarpa* with teliospores
from *E. angustifolium* & showed a distinction
from *P. goeppertianum* in the seasonal
development of the aecial state; Gardner, *PDR*
63:136, 1979, in USA (Hawaii) [**44**,2508; **47**,448;
59,321].
P. goeppertianum (Kühn) Kleb. 1904; telia on
Vaccinium, teliospores 18–30 × 10–14; aecia on
Abies; damages fir needles and causes witches'
brooms on blueberry, N. temperate zone; van
Sickle, *PDR* **57**:608, 1973, loss in blueberry in
Canada (Maritime Provinces); *Can.J.For.Res.*
4:138, 1974, growth loss in balsam fir; *CJB* **53**:8,

1975; **55**:745, 1977, basidiospore formation, discharge & seasonal periodicity & infection of *A. balsamea* [**54**,2492, 3519; **56**,5726].

P. vaccinii (Winter) Jørstad 1952; telia on *Vaccinium* and other Ericaceae, teliospores 14–17 × 7–10 μm; aecia on *Tsuga*; hemlock rust, may defoliate *Vaccinium*, N. temperate zone but aecia occur only in N. America, fide Wilson & Henderson 1966.

puerariae, *Mycovellosiella*; **puerariicola**, *Mycosphaerella*, *Pseudocercospora*.

pulicaris, *Gibberella*.

pummelo, see *Citrus*.

pumpkin, see *Cucurbita*.

—— **Indian mosaics** Hariharasubramanian & Badami, *Phytopath.Z.* **51**:274, 1964, transm. *Aphis cytisorum*; Ghosh & Mukhopadhyay 1979, enation and mild mosaic; Singh, *Z.PflKrankh.PflSchutz.* **89**:79, 1982, transm. *A. craccivora*, non-persistent [**44**,1738; **58**,5078; **61**,4422].

—— **yellows** Ragozzino, *Inftore.fitopatol.* **28**(9):13, 1978; MLO assoc., Italy [**58**,3528].

—— **yellow vein mosaic** Varma, *Curr.Sci.* **24**:317, 1955; Capoor & Ahmad, *Indian Phytopath.* **28**:241, 1975; transm. *Bemisia tabaci*, semi-persistent, India [**35**:69; **56**,1798].

punctiformis, *Drepanopeziza*.

Punctodera punctata (Thorne) Mulvey & Stone 1976; Webley & Lewis, *Descr.N.* 102, 1977; temperate cereals and grasses, prob. does no damage to crops.

punctum, *Cercosporidium*.

pure culture, see axenic culture.

purple blight, pea, Mn toxicity; —— **blotch** (*Alternaria porri*) onion, *Allium*, (*Phoma telephii*) *Sedum*, (*Septocyta rubrum*) blackberry; —— **leaf spot** (*Stagonospora arenaria*) cocksfoot; —— **seed stain** (*Cercospora kikuchii*) soybean; —— **spot** (*Stemphylium vesicarium*) asparagus.

purpurascens, *Epicoccum*.

purpurea, *Claviceps*, *Puccinia*; **purpureum**, *Chondrostereum*, *Helicobasidium*, see *H. brebissonii*.

pustulans, *Polyscytalum*.

pustular canker, prune cv. Tragedy, genetic disorder; Pine & Cochrane, *PDR* **47**:521, 1963.

pustule, a spot like a blister, on a leaf, stem or fruit and from which erupts a fruiting structure of a fungus, after FBPP; (*Xanthomonas campestris* pv. *glycines*) soybean.

putrefaciens, *Fusicoccum*.

pycnidium, a conidioma which is globose to lageniform, ostiolate, brown, a wall of 2–3 cells thick, but can be thicker, inner surface is lined with conidiogenous cells.

pycniospore, of rusts, a spore from a pycnium or spermatium; not a pycnidiospore which is a conidium from a pycnidium and is obsolete. Pycniospores, or sometimes spermatia, carry mating factors as in heterothallic fungi, e.g. the + and − thalli in rusts, stage 0 in the life cycle. Each pycnium is either + or − and the 2 mycelia from the pycniospores form a dikaryon which forms the binucleate aeciospores.

pycnothyrium, a conidioma which is superficial, flattened, like a shield, brown or darker, with ± radiate upper and sometimes lower walls.

pyracanthae, *Spilocaea*.

pyracarbolid, systemic fungicide active against *Rhizoctonia*, rusts and smuts.

pyramidalis, *Grovesinia*.

pyrazophos, systemic fungicide for powdery mildews.

Pyrenochaeta de Not. 1849; Coelom.; Sutton 1980; pycnidia setose, setae abundant around the ostiole but less so over the rest of the pycnidium, thin walled; conidiogenesis enteroblastic phialidic, con. aseptate, ellipsoid; Schneider & Schwarz, *Mitt.Biol.Bundes Land.u.Forstw.* 189, 1979, monograph; Holliday 1980.

P. glycines Stewart, *Mycologia* **49**:115, 1957; con. 4.5–7.5 × 2–3 μm, sclerotial state *Dactuliophora glycines* Datnoff et al., *TBMS* **87**:297, 1986 q.v. for morphology of *Pyrenochaeta glycines*; soybean red leaf blotch, causes leaf fall and a severe disease of the crop in Africa (E., S. and central); soilborne sclerotia prob. initiate infection; Hartman et al. and Datnoff et al. *PD* **71**:113, 132, 1987 [**66**,4025,4027].

P. lycopersici Schneider & Gerlach 1966; Punithalingam & Holliday, *Descr.F.* 398, 1973; con. 4.5–8 × 1.5–2 μm; tomato corky root or brown root rot, soilborne, on tobacco and other plants, prob. absent from the tropics. The pathogen, long known as the grey sterile fungus, was first isolated *c*.1929 and shown to cause a cortical rot of tomato roots in 1944; in Europe it is found usually in greenhouses; Ebben & Last in Bruehl 1975; Ball, *TBMS* **73**:363, 366, 1979, microsclerotia; Campbell et al., *PD* **66**:657, 1982, on field tomato in USA (California) & soil fumigation; McGrath & Campbell, *ibid* **67**:1245, 1983, inducing sporulation; Hockey & Jeves, *TBMS* **82**:151, 1984, isolation & identification; Polley, *Pl.Path.* **34**:502, 1985, tomato yield loss [**59**,3896–7; **62**,389; **63**,1957, 3559; **65**,1518].

P. oryzae Shirai ex Miyake 1910; Punithalingam, *Descr.F.* 666, 1980; con. 4–6 × 1.5–2 μm, setae 50–140 × 4–5 μm; rice sheath blotch. *Pyrenochaeta nipponica* Hara 1918, also on rice, con. 3–4 × 1–1.5 μm, setae 45–75 × 2.5–3 μm.

P. terrestris (Hansen) Gorena, Walker & Larson 1948; Punithalingam & Holliday, *Descr.F.* 397, 1973; con. 4–7 × 1.5–2 μm, setae 60–180 μm long; onion pink root, *Allium* spp., soilborne, seedlings may collapse, pink pigment forms in culture; Nishio & Kusano, *JARQ* **13**:37, 1979, assoc. with upland rice soil sickness in Japan; Katan et al., *Phytoparasitica* **8**:39, 1980, control by solar heating, polyethylene mulch in Israel; Levy & Gornik, *ibid* **9**:51, 1981, infection of onion cvs. [59,2751; **60**,1186, 6198].

pyrenocines, toxins from *Pyrenochaeta terrestris*; Sato et al., *Agric.Biol.Chem.* **43**:2409, 1979; Sparace et al., *Physiol.Molec.Pl.Path.* **28**:381,1987 [66,2635].

pyrenomycetes, a group of Ascomycotina where the ascoma is perithecioid and asci are unitunicate; the class is not now accepted.

Pyrenopeziza Fuckel 1870; Dermataceae; ascomata on herbaceous tissue but not grasses or sedges, hairs hyaline if present, erumpent, no stroma or hyphal strands assoc.; Gremmen, *Fungus* **28**:37, 1958; Hütter, *Phytopath.Z.* **33**:1, 1958, keys [38:68, 581].

*P. brassicae Sutton & Rawlinson in Rawlinson et al., *TBMS* **71**:426, 1978; anam. *Cylindrosporium concentricum* Grev. 1823; Sutton & Gibson, *Descr.F.* 536, 1977; asco. 13.5–15.5 × 2.5–3 μm, aseptate or 1 septate; con. 10–16 × 3–4 μm; light leaf spot of brassicas has recently become serious in New Zealand and UK, warranting fungicide control. Con. are splash dispersed and remain infective for at least 10 months. Asco. are released after wetting but there is no clear diurnal periodicity in the field; effective seed transm. appears in doubt although the fungus was obtained from seed after up to 2 months from harvest. There is variation in virulence but no races defined.

    Cheah et al., *N.Z.J.Bot.* **18**:197, 1980, first occurrence in New Zealand; *N.Z.J.agric.Res.* **24**:391, 1981, chemical control; *TBMS* **79**:536, 1982, asco. release; Maddock et al., *ibid* **76**:371, 1981, resistance; **77**:153, 207, 1981, con. survival & in seed, teleom.; Hartill & Cheah, *N.Z.J.agric.Res.* **27**:441, 1984, disease development; Rawlinson et al., *J.agric.Res.* **103**:613, 1984, chemical control; Cheah & Hartill, *N.Z.J.agric.Res.* **28**:567, 1985, disease cycle in New Zealand; McCartney et al., *J.agric.Res.* **107**:299, 1986, dispersal of asco. from oilseed rape [58,4075; **60**,2841; **61**,454, 1986–7, 6030; **62**,2191; **64**,834, 2206; **65**,3040; **66**,369].

Pyrenophora Fr. 1849; Pleosporaceae or Pyrenophoraceae; Sivanesan 1984; ascomata on herbaceous stems, no stroma, often with dark brown bristles on upper surface, asco. generally < 50 μm long, dictyoseptate or muriform, smaller on av. than those of *Pleospora*.

P. avenae Ito & Kuribay. 1930; Ellis & Waller, *Descr.F.* 389, 1973, anam. only, *Drechslera avenae* (Eidam) Scharif 1963; asco. 35–75 × 17–30 μm, con. 30–170 × 11–22 μm; oat leaf stripe, blotch or spot, seedborne, forms pyrenophorin q.v.

P. bromi (Died.) Drechsler 1923; anam. *Drechslera bromi* (Died.) Shoem. 1959; asco. 48–69 × 14–23 μm; con. 100–250 × 14–26 μm; brown leaf spot of *Bromus inermis*; Frauenstein, *Phytopath.Z.* **44**:1, 1962; *NachrBl.dt.PflSchutzdienst.Berl.* **16**:90, 1962; *Züchter* **32**:265, 1962 [41:719; **42**:129, 390].

P. dactylidis Ammon, *Phytopath.Z.* **47**:256, 1963; anam. *Drechslera dactylidis* Shoem. 1962; asco. mostly 50–55 × 20–30 μm, con. 57–143 × 12–17 μm; leaf spot and blotch of *Dactylis glomerata*; Zeiders, *PD* **64**:211, 1980 [43,398; **59**,5029].

P. dictyoides Paul & Parbery 1968; Ellis & Waller, *Descr.F.* 493, 1976, anam. only, *Drechslera dictyoides* (Drechsler) Shoem. 1959; asco. 33–55 × 15–24 μm, con. mostly 50–90 × 15–17 μm; net blotch of *Festuca*, on *Lolium*, *Phleum*, *Poa*; Cromey & Cole, *Pl.Path.* **34**:83, 1985, infection [64,3872].

P. graminea Ito & Kuribay. 1930; Ellis & Waller, *Descr.F.* 388, 1973, anam. only, *Drechslera graminea* (Rabenh. ex Schlecht.) Shoem. 1959; asco. 45–75 × 20–33 μm; con. mostly 50–80 × 14–23 μm, barley leaf stripe, on other temperate cereals, variation in virulence but no races defined, seedborne, of limited importance, has a pycnidial state; Smedegård-Peterson, *Friesia* **10**:61, 1972, teleom.; Arnst et al., *N.Z.J.agric.Res.* **21**:697, 1978, in New Zealand; Magnus 1979, yield loss in Norway; Tekauz & Chiko, *Can.J.Pl.Path.* **2**:152, 1980, in Canada; Gordon et al., *PD* **69**:474, 1985, chemical seed treatment [52,3651; **58**,3800; **59**,4549; **60**,4396; **64**,5371].

P. lolii Dovaston, *TBMS* **31**:251, 1948; Ellis & Holliday, *Descr.F.* 492, 1976; anam. only, *Drechslera siccans* (Drechsler) Shoem. 1959; asco. 46–67 × 16–22 μm, con. mostly 60–100 × 16–18 μm; brown blight of temperate grasses; Wilkins, *Euphytica* **22**:106, 1973, commonest *Drechslera* sp. on *Festuca* and *Lolium* in Wales; Cook, *AAB* **81**:251, 1975; loss in *L. multiflorum* experiments; Lam, *TBMS* **83**:305, 1984, more frequent on *L. multiflorum* than on *L. perenne* in England & Wales; Scholes & Shattock, *New Phytol.* **98**:377, 1984, Cd increased leaf spot [53,587; **55**,1600; **64**,227, 1158].

P. teres Drechsler, the name usually used, see below.

P. trichostoma (Fr.) Fuckel 1870, fide Cannon et al. 1985; Ellis & Waller, *Descr.F.* 390, 1973, anam. only, *Drechslera teres* (Sacc.) Shoem. 1959, as *Pyrenophora teres* Drechsler 1923; asco. 36–65 × 14–28 μm, con. mostly 90–120 × 19–21 μm; barley net blotch, on several grass genera, can be destructive, seedborne, prob. races, con. have a diurnal periodicity with a daytime peak, asco. are prob. important inoculum; Shipton et al., *Rev.Pl.Path.* **52**:269, 1973; Jordan et al., *Pl.Path.* **30**:77, 1981; **33**:547, 1984; **34**:200, 1985, life history, infection, effects of straw disposal & cultivation, dry matter & yield loss; Martin & Clough and Martin, *Can.J.Pl.Path.* **6**:105, 1984; **7**:83, 1985, con. dispersal & weather; disease progress & yield loss as affected by fungicides on seed; Shaw, *Pl.Path.* **35**:294, 1986, effects of temp. & leaf wetness on con. & infection; Deadman & Cooke, *AAB* **110**:33, 1987, effects on growth & yield [**61**,4876; **64**,580, 1561, 3827, 4261; **66**,558, 2819].

P. tritici-repentis Drechsler 1923; Ellis & Waller, *Descr.F.* 494, 1976, anam. only, *Drechslera tritici-repentis* (Died.) Shoem. 1959; asco. mostly 42–55 × 14–22mm, con. mostly 70–120 × 14–17 μm; yellow leaf spot of temperate cereals and grasses, seedborne, con. have a diurnal periodicity with a peak at *c*.1200 in Canada; Gruen, Howard, Morrall and Platt, *CJB* **53**:1040, 2345, 1975; **55**:254, 1977; *Can.J.Pl.Path.* **2**:53, 58, 1980, epidemiology, formation & liberation of con.; Rees & Platz, *Aust.J.Exp.Agric.Anim.Husb.* **19**:369, 1979, occurrence & control in N.E. Australia; Loughman & Deverall, *Pl.Path.* **35**:443, 1986, infection of resistant & susceptible wheat cvs. [**55**,779, 2767; **56**,3872; **59**,1215; **60**,2411–12; **66**,1865].

Pyrenophoraceae, Dothideales; *Cochliobolus*, *Pyrenophora*, *Setosphaeria*.

pyrenophorin, toxin or antibiotic from *Pyrenophora avenae*; Ishibashi, *J.Agric.Chem.Soc.Japan* **35**:257, 1961, fide Graniti & Puglia, *Phytopath.Mediterranea* **23**:39, 1984 [**64**,1568].

pyrethrum, *Chrysanthemum cinerariifolium*.

pyri, *Mycosphaerella*, *Potebniamyces*; **pyricola**, *Septoria*.

Pyricularia Sacc. 1880; Hyphom.; Ellis 1971, 1976; Carmichael et al., 1980; conidiomata hyphal, conidiogenesis holoblastic, conidiophore growth sympodial; con. usually 2 septate, pigmented, hilum often protruberant; see *P.grisea* for a teleom. and Yaegashi & Udagawa 1978 who illustrated con. of *P. grisea* from *Digitaria sanguinalis*, *P. oryzae* from rice and *Pyricularia* sp. from *Eleusine coracana*, *E. indica*, *Eragrostis curvula*, *Phalaris arundinacea*; Yaegashi,

Ann.phytopath.Soc.Japan **44**:626, 1978, isolates from *Eleusine*, *Eragrostis*, asco. segregation for pathogenicity; Holliday 1980 [**59**,271].

P. angulata Hashioka, *Trans.mycol.Soc.Japan* **12**:127, 1971; con. 18–28 × 5–9 μm; on Cannaceae, Musaceae, Zingiberaceae; banana blast, Wan Gyu Kim et al., *Korean J.Pl.Path.* **3**:2, 1987.

P. curcumae Asuyama 1963, a Latin description was given by Rathaiah, *PD* **64**:104, 1980; con. 16–23 × 7–9 μm; this paper described turmeric blast in India where severe infection was found in 1977; the leaf spots are ringed with black sclerotia, 100–250 μm diam. [**59**,1831].

P. grisea (Cooke) Sacc. 1886; differs little from *Pyricularia oryzae* in morphology. The teleom. was described by Hebert 1971 and transferred independently to *Magnaporthe grisea* by Barr, *Mycologia* **69**:954, 1977, and Yaegashi & Udagawa, *CJB* **56**:181, 1978; asco. 3 septate, mostly 18–23 × 5–7 μm; it was fully described by Yaegashi, *Bull.Tohoku Natn.agric.Exp.Stn.* 63:49, 1981, using isolates from 35 spp. in 20 genera of Gramineae. This name is usually retained for the forms on grasses, other than rice, causing grey leaf spot; pitting or Johnston fruit spot of banana is also attributed to it; Atilano & Busey, *PD* **67**:782, 1983, on St Augustine grass [**61**,1595; **62**,4925].

P. oryzae Cavara 1891; Subramanian, *Descr.F.* 169, 1968; con. 17–23 × 8–11 μm; Ou 1985 gave a full account of rice blast, this crop's most serious fungus disease. *Pyricularia oryzae* shows extreme variation, both in its behaviour on rice cvs. and in infection of other grasses. It is morphologically indistinguishable from *P. grisea*; but this earlier name, perhaps unsatisfactorily, is used for the forms that do not occur on rice. A teleom., see *P. grisea*, has been described from cultures by mating isolates of either fungus from several plants including rice; but not if they came from rice only; Kato & Yamaguchi, *Ann.phytopath.Soc.Japan* **48**:607, 1982.

Blast, found wherever rice is cultivated, was referred to in a Chinese book of 1637. Seedlings, plants at tillering and the panicles are all severely attacked and heavy losses can follow. The dry con. are wind and splash dispersed, with a peak at 00.01–04.00 hours; their survival is very short in the field. The most rapid infection, with a favourable dew period, is at 24–26°. In cooler regions temp. is critical but in the tropics the limiting factors are dew, rain and relative humidity. The fungus is seedborne; races have been known since 1922. Extreme variation in pathogenicity is found, and race patterns differ with geographical area. Distinct, differential host sets have been devised for Colombia, India, Japan, Korea, Philippines, Taiwan, USA. An international set of 8 cvs. characterises 32 race groups. Ahn & Ou, *Phytop.* **72**:279, 282, 1982, found that cvs. which have resistance to many races behave as if the resistance was non-specific. The cultural, chemical and resistance control measures that are applied depend very much, as in many other diseases of plants, on socio-economic factors. In Japan chemicals are widely used in sophisticated systems, see blastcast. They are used in some S. American countries where there are large scale cultivations. But in much of tropical Asia such methods are often inapplicable on economic grounds.

See literature under rice and: *The rice blast disease*; *Proc.Sympos.Int.Rice Res.Inst.* 1963, published 1965; *Horizontal resistance to the blast disease of rice*; *Proc. Seminar Centro Agricultura Tropicale* 1971, published 1975; Holliday 1980; Ou, *PD* **64**:439, 1980, control; *A.R.Phytop* **18**:167, 1980, pathogen variability & host resistance; Kingsolver et al., *Bull.Penn.agric.Exp.Stn.* 853, 1984, epidemiology [**61**,4941–2; **62**,2473; **64**,4917].

P. penniseti Prasada & Goyal, *Curr.Sci.* **39**:287, 1970; con. av. 27.5 × 9.2 μm, longer than those of *Pyricularia oryzae*, *Pennisetum americanum* [**50**,661].

P. setariae Nisikado 1917; con. 19–30 × 9–15 μm; *Setaria italica*; a form of it prob. occurs on *Eleusine coracana*, blast of finger millet, and *Pennisetum americanum*; prob. cannot be morphologically distinguished from *Pyricularia grisea*; Ramakrishnan 1963.

P. zingiberi Nisikado 1917; Nisikado, *Jap.J.Bot.* **3**:239, 1927; con. 12–27 × 6–9 μm; ginger blast, appears to be specific on *Zingiber*; Rathaiah, *Indian Phytopath.* **32**:321, 1979; Sasaki & Honda, *Ann.Rep.Soc.Pl.Prot.N.Japan* 36:134, 1985, control with ultraviolet absorbing film [**6**:637; **60**,3296; **65**,1991].

pyridinitril, protectant fungicide, fruit trees and vegetables.

pyriforme, *Geniculodendron*, *Peridermium*.

pyrochroa, *Calonectria*; **pyrorum**, *Fusicladium*.

Pyrus, pear; *P. communis*, common or European; *P. serotina*, Japanese; *P. ussuriensis*, Chinese.

Pythiaceae, Peronosporales; facultative or obligate pathogens or saprophytic; sporangiophores or conidiophores usually undifferentiated from the mycelium, branched; indeterminate, i.e. resuming growth after formation of a sporangium or conidium, either from below or within the previous empty sporangium; periplasm thin or absent; *Phytophthora*, *Pythium*, *Sclerophthora*, *Trachysphaera*.

pythiophyla, *Sclerophoma*.

Pythium Pringsh. 1858; Pythiaceae; sporangia filamentous, inflated, hyphal, spherical, ovoid or occasionally obpyriform; contents emerge, virtually undifferentiated, into a spherical, evanescent vesicle, usually through a discharge tube, division into zoospores follows; antheridia always paragynous, oogonia smooth or spiny, oospore plerotic or aplerotic, mostly homothallic. *Phytophthora* differs morphologically in that the sporangia have a distinct, apical emission zone, thickened or papillate; the contents differentiate into zoospores within the sporangium and disperse directly; if a vesicle is formed it is extremely evanescent. Some have thought that these 2 genera should be combined but the current view is otherwise. Whilst they have similarities as pathogens, their differences in this respect are highly significant. *Pythium* spp. are soil and water inhabitants, saprophytes and general pathogens with limited host specificities. Several spp. often occur together in one disease syndrome, e.g. in damping off, root rots and plant declines; see replant effect. They can attack those parts of plants close to soil level; van der Plaats-Niterink, *Stud.Mycol.* 21, 1981, monograph; Holliday 1980; Robertson, *N.Z.J.Bot.* **18**:73, 1980; Hendrix & Campbell in Buczacki 1983; Ali-Shtayeh, *TBMS* **85**:761, 1985, diam. of oogonia & oospores as taxonomic criteria; Sivanesan & Waller 1986 [**65**,1773].

P. aphanidermatum (Edson) Fitzp. 1923; Waterhouse & Waterston, *Descr.F.* 36, 1964; oogonia av. 23 μm diam., smooth, stalks straight, oospores aplerotic; warmer regions, plurivorous, cottony blight of turf grasses, cottony leak of cucurbits; Ruben et al., *Phytop.* **70**:54, 1980, oospore germination; Stanghellini, *ibid* **72**:1481, 1982, inoculum density in irrigated beet fields [**59**,5602; **62**,1679].

P. arrhenomanes Drechsler 1928; Waterhouse & Waterston, *Descr.F.* 39, 1964; oogonia av. 32.5 μm diam., oogonia smooth, many antheridia/ oogonium; on > 30 genera in Gramineae, seedling blight and root rot; Egan and others 1984, a poss. factor in the northern poor root syndrome in sugarcane in Australia [**64**,1735–41].

P. butleri Subramaniam 1919; Waterhouse & Waterston, *Descr.F.* 37, 1964; = *Pythium aphanidermatum* fide van der Plaats-Niterink 1981.

P. debaryanum Hesse 1874; this name is frequent in the literature but the identity is in doubt, see van der Plaats-Niterink 1981; the fungus has been confused with *Pythium intermedium* de Bary, *P. irregulare* Buisman, *P. ultimum* Trow.

P. deliense Meurs 1934; Waterhouse & Waterston, *Descr.F.* 116, 1966; oogonia av. 21.9 μm diam., stalks curved towards antheridia, warmer regions, plurivorous, causes tobacco stem burn.

P. graminicola Subramaniam 1928; Waterhouse & Waterston, *Descr.F.* 38, 1964; oogonia av. 32.5 μm diam.; root rots of Gramineae, often assoc. with *Pythium arrhenomanes*.

P. irregulare Buisman 1927; sporangia seldom formed, oogonia av. 18.5 μm diam., spiny, variable in size; commonly pathogenic to seedlings.

P. mamillatum Meurs 1928; Waterhouse & Waterston, *Descr.F.* 117, 1966; sporangia globose, ovoid or broadly ellipsoid; oogonia av. 16 μm diam., spiny; commonly pathogenic to seedlings.

P. megalacanthum de Bary 1881; oogonia large, av. > 50 μm diam., spiny; Wiersema 1955 attributed flax scorch to this fungus in the Netherlands; a flax disease was also attributed to *Pythium buismaniae* van der Plaats-Niterink 1981. Wakaida et al. 1973 described periderm brown scorch of Japanese radish [35:296; 53,1986].

P. myriotylum Drechsler 1930; Waterhouse & Waterston, *Descr.F.* 118, 1966; oogonia smooth, often entangled with many antheridial stalks, oospores plerotic, clusters of 'appressoria' in vitro, 28 μm diam.; warmer areas, seedling rots; Frank in Kranz et al. 1977; Csinos, *CJB* 57:2059, 1979, toxin [59,2335].

P. okanoganense Lipps, *Mycologia* 72:1172, 1980; oospores av. 20 μm diam.; barley and wheat snow rot, a snow mould; Japan, USA; Lipps & Bruehl and Lipps, *Phytop.* 70:723, 794, 1980; Takamatsu & Ichitani and Ichitani et al., *Ann.phytopath.Soc.Japan* 52:82, 209, 1986 [60,3670–1; 66,520, 922].

P. oligandrum Drechsler 1930; Waterhouse & Waterston, *Descr.F.* 119, 1966; sporangia in complexes, terminal clusters, linear or branched arrangements; oogonia av. 25 μm diam., with long, slender spines, plurivorous, causes damping off. An aggressive hyperparasite of other fungi and therefore of poss. use in biological control; Veselý, *Phytopath.Z.* 90:113, 1977, on *Pythium*

spp.; Al-Hamdani & Cooke, *TBMS* 81:619, 1983, on *Rhizoctonia solani*; Al-Hamdani et al., *Pl.Path.* 32:449, 1983, control of *P. ultimum* on cress; Lutchmeah & Cooke, *TBMS* 83:696, 1984; *Pl.Path.* 34:528, 1985, antagonism, pelleting seed with *P. oligandrum* to control damping off; Walther & Gindrat, *J.Phytopath.* 119:167, 1987, control of damping off in sugarbeet [57,2682; 63,1132, 1463; 64,936; 65,1564].

P. polymastum Drechsler 1939; oogonia av. 53 μm diam., spiny; Vanterpool, *CJB* 52:1205, 1974, on crucifers; Coplin et al., *PD* 64:63, 1980, on lettuce [54,594].

P. splendens Braun 1925; Waterhouse & Waterston, *Descr.F.* 120, 1966; hyphal swellings large, often 30–40 μm diam., oogonia av. 29 μm diam., smooth; oospores aplerotic, one of the few heterothallic sp.; seedling diseases, assoc. with oil palm blast; Bolton, *Can.J.Pl.Path.* 3:177, 1981, evidence for pathogenic strs. [61,2904].

P. sulcatum Pratt & Mitchell, *CJB* 51:334, 1973; oogonia av. 16.5 μm diam., smooth; assoc. with carrot rusty root complex in N. America, Wisbey et al., *Can.J.Pl.Sci.* 57:235, 1977; Kemp & Barr, *Phytopath.Z.* 91:203, 1978; causing brown blot root rot of carrot in Japan, Nagai et al. and Watanabe et al., *Ann.phytopath.Soc.Japan* 52:278, 287, 1986 [56,5260; 57,5740; 66,1194–5].

P. ultimum Trow 1901; sporangia globose, monoclinous antheridia, like sacs; oogonia av. 21.5 μm diam., smooth; one of the most studied of all the members of the genus. Recent work has been mostly on soil survival and pathogenicity to chickpea, cotton, pea, soybean; Holliday 1980; Rush et al., *Phytop.* 76:1330, 1986, on wheat, effects of tillage & chaff on inoculum density in the field [66,4138].

P. violae, see cavity spot.

P. zingiberum Takahashi, *Ann.phytopath.Soc.Japan* 18:115, 1954; van der Plaats-Niterink 1981 finds this fungus, as *Pythium zingiberis*, doubtful and poss. a synonym of *P. volutum* Vanterpool & Truscott 1932; oogonia av. 30 μm diam., smooth; causes ginger and mioga rhizome rot in Japan; Ichitani et al. *ibid* 46:435, 539, 1980; 47:151, 1981; 48:674, 1982 [60,5070–1; 61,3013; 62,2586].

Q

quail pea mosaic v. Moore 1973; Moore & Scott, *Descr.V.* 238, 1981; Comovirus, isometric *c.*30 nm diam., transm. sap, beetles; USA (Arkansas), Central America; *Strophostyles helvola*, bean, soybean.

Quanjer, Hendrick Marius, 1879–1961; born in Netherlands, Agricultural Univ. Wageningen, pioneering studies on potato disease caused by viruses, particularly potato leaf roll. *Tijdschr.PlZiekt.* **67**:36, 1961.

quarantine, see plant quarantine.

quercina, *Chalara, Erwinia, Gnomonia*; quercinum, *Amphicytostroma, Colpoma, Polymorphum*.

Quercus, oak.

quercus, *Fusicoccum*; quercus-falcatae, *Elsinoë*; quercuum, *Cronartium*.

quick decline, see citrus tristeza v.

quinacetol sulphate, protective fungicide for seedborne pathogens.

quince, *Cydonia oblonga*.

— bark necrosis Posnette & Cropley, *J.hort.Sci.* **33**:289, 1958; prob. due to the agent which causes apple rubbery wood [**38**:152].

— deformation Christoff, *Phytopath.Z.* **8**:285, 1935; Scaramuzzi, *ibid* **30**:259, 1957; some isolates of pear vein yellows and red mottle can cause deformation; Fleisher et al., *Acta Hortic.* 44:123, 1975, described a non-transm. form of deformation which is prob. a genetic disorder [**14**:640; **37**:290; **55**,3639b].

— Japanese mosaic Docea, Pop & Frătilă 1962; Romania [**42**:267].

— ringspot Issa, *Biológico* **25**:64, 1959; Brazil (São Paulo) [**38**:609].

— sooty ringspot Posnette, *J.hort.Sci.* **32**:53, 1957; Posnette & Cropley, *Tech.Communic.Commonw.Bur.Hortic.Plantation Crops* **30**:109, 1963; Refatti & Osler, *Riv.Patol.Veg.* **9**:45, 1973; latent in apple and pear cvs. [**36**:330, **53**,1878].

— stunt Posnette & Cropley, as above, page 113; a disease complex which is latent in pear and some quince cvs.; it is thought to be caused by several agents.

— yellow blotch, see quince bark necrosis.

— — leaf Docea & Frătilă 1964, Romania [**44**, 3105].

— — mosaic Nagaich & Vashisth, *Indian Phytopath.* **15**:222, 1962; India (Himachal Pradesh, Jammu & Kashmir) [**43**:1349].

quininea, *Phytophthora*.

Quinisulcius capitatus (Allen) Siddiqi 1971; Jairajpuri, *Descr.N.* 111, 1985.

quinmethionate, protectant fungicide for powdery mildews.

quinqueseptata, *Calonectria*, see *C. pyrochroa*.

quintozene, soil and seed application, particularly against Basidiomycetes.

quisqualis, *Ampelomyces*.

R

rabiei, *Ascochyta*, *Mycosphaerella*.

race, an abbreviation for physiologic race q.v.

— specific and non-specific resistance, see resistance to pathogens.

radicale, *Olpidium*; radicicola, *Lagena*, *Nectria*, *Phialophora*.

radicin, toxin formed by *Alternaria radicina*; Durbin 1981; Robeson & Strobel, *Phytochemistry* 21:1821, 1982; *AAB* 107:409, 1985; Tal et al., *Phytochemistry* 24:729, 1985, from *A. helianthi* & other toxins [61,6525; 64,3164; 66,270].

radicina, *Alternaria*.

radiobacter, *Agrobacterium*, see genus; radiosa, *Pollacia*.

radish, *Raphanus sativus*; Rod, *Tests Agrochem.Cvs.* 6, *AAB* 106, suppl.:64, 1985, effects of fungicides on yield & germination characteristics of radish seed.

— enation mosaic = radish mosaic v.

— mosaic v. Tompkins 1939; Campbell, *Descr.V.* 121, 1973; Comovirus, isometric *c*.30 nm diam., transm. sap, beetles; occurs naturally only in crucifers; Plakolli & Stefanac, *Phytopath.Z.* 87:114, 1976, isolate relationships [56,2716].

— P and R = turnip mosaic v.

— phyllody Misra & Gupta, *Proc.Indian Acad.Sci.* B 85:319, 1977; MLO assoc., India (Jaipur) [57,1887].

— stunt Isiyama & Misawa, *Ann.phytopath.Soc.Japan* 12:116, 1943; transm. *Lipaphis erysimi*, *Myz.pers.*; Japan (Sizuoka); infects Cruciferae, *Capsicum*, pea, spinach [30:354].

— wild rhabdovirus Kitajima & Costa, *Fitopat.Brasileira* 4:55, 1979; Brazil [58,5279].

— yellow edge v. Natsuaki et al. 1979; Natsuaki, *Descr.V.* 298, 1985; isometric *c*.30 nm diam., contains dsRNA, a cryptic virus q.v.; transm. seed, in naturally infected radish, most Japanese cvs. are symptomless, pollen and ovule transm. to seed in crosses of infected radish and healthy Chinese cabbage, common in Japan, found in seed from widely separated countries.

— yellows = beet western yellows v.

Radopholus similis (Cobb) Thorne 1949; Orton Williams & Siddiqi, *Descr.N.* 27, 1973; root endoparasite, warmer regions, causes extensive cavities in the root cortex, many hosts; diseases caused or assoc. with: avocado and tea declines, banana toppling, black pepper yellows, citrus spreading decline; banana and citrus forms.

ragweed stunt Timmer et al., *Phytop.* 73:975, 1983; xylem limited bacterium, transm. *Homalodisca coagulata*, *Oncometopia nigricans*; USA, *Ambrosia artemesiifolia* [63,1535].

Ramakrishnan, Krishnaier, 1920–78; born in India, Univ. Madras, chair plant pathology, Agricultural College and Research Institute, Coimbatore; a notable student of fungi and teacher. *Kavaka* 6:71, 1978.

Ramichloridium Stahel ex de Hoog 1977; Hyphom.; = *Periconiella* fide Ellis, *Mycol.Pap.* 111, 1967; 1976; but see de Hoog et al., *TBMS* 81:485, 1983, who defined the genus, gave a key which included related taxa, and described *R. pini* de Hoog & Rahman which causes a dieback of *Pinus contorta* in Scotland [63,1434].

ramie, *Boehmeria nivea*; Sarma, *Trop.Pest Management* 27:370, 1981.

Ramularia Unger 1833; Hyphom.; Carmichael et al. 1980; conidiomata hyphal, conidiogenesis holoblastic, conidiophore growth sympodial or con. catenate, acropetal; hyaline, cylindrical, aseptate to multiseptate; cause leaf spots, the conidiophores emerge through the stomata, most spp. of little or no economic importance; Hughes, *TBMS* 32:34, 1949, history; Wilson, *Mycopathologia* 95:41, 1986, 10 spp. [28:457].

R. beticola Fautrey & Lamb 1897; con. 0–2 septate, 8.2 × 1.5 μm; beet leaf spot; Byford, *J.agric.Sci.* 85:369, 1975; *AAB* 82:291, 1976, in England, premature defoliation & chemical control [55,1520, 3334].

R. brunnea, teleom. *Mycosphaerella fragariae*.

R. coriandri Moesz & Smarods 1930; coriander leaf spot is of some significance in USSR; Andreeva 1980, Savenko & Pin'kovskiĭ 1981 [59,5858; 60,6508].

R. foeniculi Sibilia, *Boll.Staz.Patol.veg.Roma* 12:233, 1932; con. av. 45.5 × 6.5 μm; can be severe on fennel; Prasad et al., *Curr.Sci.* 30:65, 1961 [12:244; 40:620].

R. gossypii, teleom. *Mycosphaerella areola*.

R. primulae Thüm. 1878; Adebayo, *TBMS* 61:159, 1973; con. av. 27 × 3.5 μm, primula leaf spot; Raabe & Hurlimann, *PDR* 55:586, 1971, chemical control [51,1568; 53,978].

R. rhei Allescher 1896; rhubarb leaf and stalk spot; Ormrod et al., *Can.Pl.Dis.Surv.* 65:29, 1985, described damage to the crop in Canada (British Columbia) and chemical control [66,2157].

R. vallisumbrosae Cavara 1899; con. mostly 3 septate, av. 56 × 2 μm; narcissus white mould; Gregory, *TBMS* **23**:24, 1939; Moore 1979 [**18**:598].

Ramulispora Miura 1920; Hyphom.; Carmichael et al. 1980; conidiomata sporodochial; conidiophores emerging through stomata or from epidermal cells, unbranched or very short branches, ± fasciculate; con. multiseptate, hyaline filiform; sclerotia usually present, superficial, cf. *Gloeocercospora sorghi* which has immersed sclerotia; Rawla, *TBMS* **60**:283, 1973; Holliday 1980 [**52**,3572].

R. sorghi (Ell. & Ev.) Olive & Lefèbvre 1946; Anahosur, *Descr.F.* 585, 1978; con. formed in pink, gelatinous masses, 3–8 septate, 36–90 × 2–3 μm, with lateral branches, sclerotia black, 100–230 × 53–190 μm; differs from *Ramulispora sorghicola* in the larger leaf spots and the abundant non-setose sclerotia; sorghum sooty stripe, seedborne.

R. sorghicola E. Harris 1960; Anahosur, *Descr.F.* 586, 1978; see above; sorghum common leaf spot, a lesser pathogen compared with *Ramulispora sorghi*.

randii, *Elsinoë*; **ranunculi**, *Erysiphe*.

ranunculus breaking Devergne, Cardin & Marais, *Annls.Phytopath.* 1 hors sér.:321, 1969; France (S.E.) [**49**,910f].

— **mottle** Laird & Dickson, *PDR* **49**:449, 1965; transm. sap, USA (California), limited host range [**44**,2816].

rape, see *Brassica* and oilseed rape; Walsh & Tomlinson, *AAB* **107**:485, 1985, viruses in UK [**65**,5219].

— **Chinese mosaic** v. Chang et al., *Scientia Sin.* **13**:1421, 1964; Tobamovirus, fide Francki et al., vol 2, 1985.

— **green petal** Horváth, *Acta phytopath.Acad.Sci.hung* **4**:363, 1969; *Növenytermelés* **19**:49, 1970; poss. MLO [**49**,2215, 3041].

— **savoying** Kauffmann 1936; transm. sap, *Lygus pratensis*, Germany [**16**:10].

raphani, *Alternaria. Aphanomyces.*

Raphanus sativus, radish.

raphani, see *Rubus*, black raspberry.

— **bushy dwarf** v. Cadman 1961; Murant, *Descr.V.* 165, 1976; isometric c.33 nm diam., contains ssRNA, transm. sap; pollen and seed in raspberry; causes no symptoms in many raspberry cvs. and many herbaceous plants; one of several viruses assoc. with bushy dwarf or symptomless decline of cv. Lloyd George; Jones et al., *AAB* **100**:135, 1982, poss. cause of raspberry yellows; Barbara et al., *ibid* **105**:49, 1984; **106**:75, 1985,

isolates differing in infection of *Rubus*, occurrence in England [**61**,5117; **64**,252; **65**,2903].

— **chlorotic net** = raspberry vein chlorosis v.

— **curly dwarf** Prentice & Harris, *J.hort.Sci.* **25**:122, 1950; England [**29**:315].

— **Himalayan mosaic** Azad & Sehgal, *Indian Phytopath.* **11**:159, 1958; India (Simla Hills), *Rubus ellipticus* [**38**:605].

— **leaf curl** Bennett, *Phytop.* **20**:787, 1930; Stace-Smith, *CJB* **40**:651, 1962; transm. *Aphis idaei, A. rubicola, A. rubifolii*, persistent, N. America, A. and B strs. [**10**:195; **42**:34].

— — **mottle** Cadman, *AAB* **38**:801, 1951; transm. *Amphorophora* spp., prob. semi-persistent; Jones 1982 compared RLM with black raspberry necrosis v. q.v. and raspberry leaf spot [**31**:439].

— — **spot** Cadman, *AAB* **39**:501, 1952; transm. *Amphorophora* spp., prob. semi-persistent; Jones 1982 compared RLS with black raspberry necrosis v. q.v. [**32**:682].

— **line pattern** Basak, *Bull.Acad.pol.Sci.* **19**:681, 1971; transm. sap, Poland [**51**,4196].

— **red calico** Johnson, *PDR* **56**:779, 1972; USA (Washington State) [**52**,2690].

— **ringspot** v. Cadman 1956; Murant, *Descr.V.* 198, 1978; Nepovirus, isometric c.28 nm diam., transm. sap, seed, *Longidorus* spp.; many hosts, e.g. narcissus, grapevine, red currant, causing spoon leaf; strawberry; disease outbreaks often as obvious patches; Rana et al., *Phytopath.Z.* **112**:222, 1985, 2 serologically distinct strs. from artichoke [**64**,4084].

— — **and decline** (tomato ringspot v.).

— **Scottish leaf curl** (raspberry ringspot v.).

— **Tweddell** (cherry rasp leaf v.).

— **veinbanding mosaic** (black raspberry necrosis v. + rubus yellow net v.).

— **vein chlorosis** v. Cadman 1952; Jones et al., *Descr.V.* 174, 1977; Phytorhabdovirus, bacilliform c.430–500 × 65–80 nm, transm. *Aphis idaei*, persistent; Murant & Roberts, *Acta Hortic.* **95**:31, 1980, particles in *A. idaei* [**60**,3848].

— **yellow blotch** (raspberry ringspot v.).

— — **dwarf** = arabis mosaic v.

— — **mosaic** = rubus yellow net v.

— **yellows** Cadman, *AAB* **39**:495, 1952; Scotland, see raspberry bushy dwarf v. [**32**:682].

— **yellow spot** Basak 1974, fide Jones, *Acta Hortic.* 129:41, 1982.

rathayi, *Clavibacter*.

ratoon stunting (*Clavibacter xyli* ssp. *xyli*) sugarcane.

rauvolfia bunchy top Varadarajan, *Indian Phytopath.* **20**:255, 1967; India (Gujarat), *R. serpentina* [**47**,2799].

Ravn, Frederick Kolpin, 1873–1920; born in

Denmark, Univ. Copenhagen, chair of plant pathology, Royal Veterinary and Agricultural High School, in Copenhagen; he followed E. Rostrup who had been appointed to the first professorial chair in this discipline in the world; one of the founders of plant disease work in the country and an authority on cereal pathology. *Phytop.* **11**:1, 1921.

Rawlins, Thomas Ellsworth, 1894–1972; born in USA, Univ. California Berkeley, Wisconsin; with C.M. Tompkins, he introduced carborundum as an abrasive in inoculations of plants with viruses. *Phytop.* **64**:573, 1974.

ray blight (*Didymella ligulicola*) chrysanthemum; — **speck** (*Stemphylium floridanum*) chrysanthemum.

rayssiae, *Sclerophthora*.

razor strop fungus, *Piptoporus betulinus*.

Ré, Filipo. 1763–1817; born in Italy, agriculturist and botanist, chairs at Univ. Bologna, Modena; wrote *Saggio teorica practico sulle mallatie delle piante* 1807, edn. 2, 1817; English translation by M.J. Berkeley, *Gdnrs.Chron.* 1849:228 to 1850:469; Whetzel 1918; Ainsworth 1981.

Recilia dorsalis, transm. rice dwarf, rice orange leaf; *R. trifasciata*, transm. cotton virescence.

recognition, an early, specific event that triggers a rapid, overt response by a host plant; it either facilitates or impedes further growth of a pathogen; Sequeira, *A.R.Phytop.* **16**:453, 1978; Etzler, *Phytop.* **71**:744, 1981; Mahadevan et al., *Curr.Sci.* **51**:263, 1982; Keen & M.J. Holliday in Mount & Lacy 1982, procaryotae q.v.; Daly, *A.R.Phytop.* **22**:273, 1984; Etzler, *A.Rev.Pl.Physiol.* **36**:209, 1985.

recondita, *Puccinia*.

red band needle blight (*Scirrhia pini*) pine, conifers. — **beet**, a cv. of *Beta vulgaris*, see beet. — **brown cubical heart rot** (*Laetiporus sulphureus*) tree wood; — — **leaf sheath** (*Mycovellosiella vaginae*) sugarcane. — **clover**, see *Trifolium*; Akita, *Bull.Natn.Grassl.Res.Inst.* 20:93, 1981, viruses in Japan [**61**,5053]. — — **cryptic v.**, see cryptic viruses. — — **enation v.** Musil & Leskova 1980; isometric 28 nm diam., Europe, fide Boswell & Gibbs 1983. — — **mild mosaic v.** Gerhardson, *Phytopath.Z.* **89**:116, 1977; isometric 27–28 nm diam., transm. sap, *Acyrthosiphon pisum*, *Myz.pers.*, semi-persistent; Sweden [**57**,196]. — — **mosaic** Vela & Rubio-Huertos, *Phytopath.Z.* **79**:343, 1974; bacilliform and rod particles assoc., Spain [**54**,1320]. — — **mottle v.** Sinha 1960; Valenta & Marcinka, *Descr.V.* 74, 1971; Comovirus, isometric *c*.30 nm

diam., transm. sap, *Apion africanum*, *A. varipes*; Europe, restricted host range; Tomenius & Oxelfelt, *J.gen.Virol.* **61**:143, 1982, fine structure of pea leaf cells infected with 3 strs. [**61**,6662]. — — **necrosis** (bean yellow mosaic v.). * — — **necrotic mosaic v.** Musil & Matisova 1967; Hollings & Stone, *Descr.V.* 181, 1977; Dianthovirus, isometric *c*.30 nm diam., transm. sap, soil; Europe, many herbaceous plants can be infected, *Melilotus officinalis* is naturally so; Bowen & Plumb, *AAB* **91**:227, 1979, occurrence & effects; Gould et al., *Virology* **108**:499, 1981, bi-partite genome [**58**,5916; **60**,4885]. — — **rough vein** Musil & Kvičala, *Phytopath.Z.* **77**:189, 1973; MLO assoc., Czechoslovakia [**53**,1858]. — — **vein mosaic v.** Osborn 1937; Varma, *Descr.V.* 22, 1970; Carlavirus, filament *c*.645 nm long, transm. sap, seed, aphid spp. non-persistent, most hosts are legumes; Bos et al., *Neth.J.Pl.Path.* **78**:125, 1972, str. from pea; Weber et al., *Phytopath.Z.* **103**:1, 1982, particle diversity & RNA duality [**52**,1318; **61**,4756]. — **core** or red stele (*Phytophthora fragariae*) strawberry. — **currant**, *Ribes rubrum* and other spp. — — **double flower** Rakús, *Ochr.Rost.* **9**:143, 1973; Czechoslovakia [**53**,1883]. — — **leaf pattern** Frazier et al. 1970 [**51**,3455]. — — **ringspot and spoon leaf** (raspberry ringspot v.). — — **yellow leaf spot** van der Meer in Frazier et al. 1970 [**51**,3455]. — — **yellows** Rakús, Králik & Brčák, *Ochr.Rost.* **10**:307, 1974; MLO assoc. [**54**,5000]. — **fire** (*Pseudopeziza tracheiphila*) grapevine; — **heart** (*Torula ligniperda*) *Betula papyrifera*; — **leaf** (*Exobasidium vaccinii*) *Vaccinium*; — — **blotch** (*Pyrenochaeta glycines*) soybean; — — **sheath spot** (*Mycovellosiella vaginae*) sugarcane; — **leg** (*Botryotinia fuckeliana*) lettuce; — **lenticel** (*Pezicula alba*) apple. — **pepper**, *Capsicum*, see pepper; the prefix pepper to all virus, or suspected MLO and virus, names means *Capsicum*, not the true pepper which is *Piper nigrum*; Boswell et al., *Fmr's Bull. USDA* 205, 1952; Villalón, *PD* **65**:557, 1981, breeding for resistance to viruses; Martelli & Quacquarelli, *Acta Hortic.* 127:39, 1983, viruses with tomato; Horváth, *Acta phytopath.entomol.hung.* **21**:29, 35, 59, 1986. — **raspberry calico** Johnson, *PDR* **56**:779, 1972; USA (Washington State) [**52**,2690]. — **ring** (*Rhadinaphelenchus cocophilus*) coconut; — — **rot** (*Phellinus pini*) conifers; — **root rot** (*Ganoderma philippii*) rubber, woody crops; —

rot (*Glomerella tucumanensis*) sugarcane; — rust (*Cephaleuros virescens*) plurivorous; — sheath rot (*Deightoniella torulosa*) abaca; — shoot (*Exobasidium perenne*) cranberry; — stalk rot and leaf spot (*Colletotrichum graminicola*) sorghum; — stripe (*Pseudomonas rubrilineans*) sugarcane; — thread (*Laetisaria fuciformis*) turf grasses; — top (*Diplodia pinea*) pine, conifers.

reflective mulch, a polyethylene, straw or other material that is placed on the soil surface around the crop to control airborne insect vectors.

regrowth dieback, eucalyptus, related to summer drought, Australia (Tasmania); see *Armillaria hinnulea*; West, *AAB* 93:337, 1979 [59,3424].

rehmiana, *Sphaerulina*.

Reichert, Israel, 1891–1975, born in Poland, went to Palestine 1908, Univ. Berlin 1916–20; a versatile biologist and an originator of mycology and plant pathology in the region that is now largely Israel; Gold Medal, Mediterranean Phytopathological Union 1969. *Phytoparasitica* 4:71, 1976.

reilianum, *Sporisorium*.

remote sensing, see stress, Jackson.

reniform nematode, *Rotylenchus reniformis*.

reoviruses, see Phytoreoviruses.

repens, *Sphaerostilbe*.

replant effect, severe reduction in rates of root and shoot growth in a second planting of a perennial crop following the same crop, or one that is closely related, on the same site. It is suggested that one uses 'specific replant disease' and 'specific replant disorder' when the effect is attributable to a parasitic organism(s) or a non-parasitic factor, respectively. 'Specific replant effect' is used when the cause is undetermined, after FBPP. These effects on pome and stone fruits were reviewed by Traquair, *Can.J.Pl.Path.* 6:54, 1984.

residue, the amount of a crop protectant remaining in or on plant tissue after a given time, especially at harvest, FBPP.

resin bleeding (*Crumenulopsis sororia*) pines; — canker (*Cronartium flaccidum*) Scots pine; — top (*C. flaccidum*, *Endocronartium pini*) Scots pine.

resinosis, an abnormal exudation of resin from conifers, FBPP.

resistance to pathogens, a plant suppresses or retards invasion by a potential pathogen; resistance and pathogen virulence are complementary. The single most important control measure and under genetic control. In practice it very largely implies the study of infection by specialised, or obligate type, pathogens. Therefore 2 phenomena have perhaps been inadequately studied, although understandably so. First, that most plants are

resistant to most pathogens, i.e. are not hosts; Heath, *A.R.Phytop.* 18:211, 1980. Second, that many pathogens are unspecialised and cause disease in many botanically unrelated plants. Some resistance shows in situations that are relatively simple, e.g. the prevention of infection by a pathogen by a leaf cuticle, stem bark or root suberisation. These can operate against all types of pathogens. Work is primarily directed towards understanding the reaction phenomena when an obligate pathogen penetrates not a normally susceptible plant but one that has some resistance, especially that of the type called hypersensitivity q.v. Such high host resistance involves inducible defence mechanisms, e.g. phytoalexin synthesis by the host, deposition of material like lignin, accumulation of proteinase inhibitors, and increases in the activity of certain hydrolytic enzymes.

The experimental work on the genetics of resistance goes back 84 years to Biffen who showed that its inheritance was Mendelian. Thereafter, for many years, investigations were almost wholly done on a high resistance, or virtual immunity, and controlled in the plant crop by one or a few genes. Such resistance is called oligogenic. Its disadvantage was that selection in the pathogen led to new virulence genes arising and the consequent total breakdown of such resistance. Thus began the continuous cycle of breeding and selection for new resistance. Most of this work, in a sophisticated sense, has been in cereals, potatoes and some legumes. Some specialised pathogens do not mutate to form new virulent forms at all readily; these are said to be immobile. Another pattern of resistance came to be recognised. It showed in such phenomena as reduced growth and sporulation of a pathogen. Such incomplete, but nevertheless significant resistance, was under polygenic control. It was not readily broken down by a pathogen but caused difficulties for breeders. It came to be called durable resistance or, at an earlier time, field resistance.

These 2 patterns, not always clearly differentiated, were analysed by Vanderplank in the 1960s, *Disease resistance in plants*, edn. 2, 1984. There developed a complex, confused, and prob. excessive terminology, most of which is not noted in the book; Robinson, *Rev.appl.Mycol.* 48:593, 1969; *Rev.Pl.Path.* 50:233, 1971; 52:483, 1973. The patterns of resistance were briefly considered by Wood, *Proc. Indian Acad.Sci.*, *Pl.Sci.* 93:195, 1984 and Simmonds 1985. The extreme types are: race specific or oligogenic and race non-specific or polygenic; these are the vertical and horizontal

resistances of Vanderplank. Two further patterns
are pathotype or race non-specific, oligogenic; and
interaction resistance, where several major genes
operate, heterogeneous, oligogenic. Durable
resistance in the host occurs when the pathogen is
immobile.
MECHANISMS: Sequeira, *A.Rev.Microbiol.*
37:51, 1983; Bailey & Deverall, *The dynamics of
host defence*, 1984; Harris & Frederiksen,
A.R.Phytop. **22**:247, 1984; Fraser ed., *Mechanisms
of resistance to plant diseases*, 1985.
DURABILITY: Lamberti et al. ed., *Durable
resistance in crops*, 1983; Johnson, *A.R.Phytop.*
22:309, 1984. BREEDING: Russell, *Plant
breeding for pest and disease resistance*, 1978;
Simmonds, *Principles of crop improvement*, 1979;
F.A.O. Pl.Prot.Bull. **31**:2, 1983; **33**:13, 1985;
Robinson, *A.R.Phytop.* **18**:189, 1980;
Buddenhagen, *ibid* **21**:385, 1983; Wenzel, *ibid*
23:149, 1985; Singh, *Breeding for resistance to
diseases and insect pests*, 1986. OTHER
ASPECTS: Shepard, *A.R.Phytop.* **19**:145, 1981,
protoplasts as sources; Kiyosawa, *ibid* **20**:93,
1982; Ouchi, *ibid* **21**:289, 1983, induction; Bruehl,
Phytop. **73**:948, 1983, non-specific resistance to
soilborne fungi; Kuć, *Fitopat. Brasiliera* **10**:17,
1985, induced systemic resistance & interferons;
Smedegaard-Peterson & Tolstrup, *A.R.Phytop.*
23:475, 1985; Wolfe, *ibid* **23**:251, 1985, control
with multi-cvs. & cv. mixtures; Johnson in Day &
Jellis 1987.

respiration, role in diseased plants, Smedegaard-
Petersen in Wood & Jellis 1984.

resting spore, of a fungus, a spore or a sporangium,
usually with a thickened wall, and which can lie
dormant in host debris or soil for years, e.g.
chlamydospore, oospore.

resupinate, of basidiomata in the macro-fungi, lying
flat on the substrate with the hymenium on the
upper surface.

reticulatum, *Exobasidium*.

Rhabdocline H. Sydow 1922, emended Parker &
Reid, *CJB* **47**:1533, 1969; Hypodermataceae; asci
8 spored, asco. ovoid to clavate, becoming 1
septate, strongly constricted in the middle, hyaline
[**49**,877].

R. pseudotsugae H. Sydow, ssp. *pseudotsugae*;
Millar & Minter, *Descr.F.* 651, 1980; asco.
13–19 × 5–8 μm, apical cell brown, basal
cell ± hyaline; mottled needle cast of Douglas fir,
poss. races, forms of host differ in susceptibility;
Kurkela 1981, growth reduction in Douglas fir;
McDowell & Merrill, *PD* **69**:715, 1985,
occurrence in USA (Pennsylvania) with
Rhabdocline weirii ssp. oblonga [**61**,7200; **65**,931].

rhabdoviruses, see Phytorhabdoviruses.

Rhadinaphelenchus cocophilus (Cobb) J.B. Goodey
1960; Brathwaite & Siddiqi, *Descr.N.* 72, 1975;
attacks coconut roots in the West Indies, S. and
Central America, causing the serious red ring;
palms are killed in a few months; the nematodes
can be found in stems up to *c*.8 feet above soil
level. Oil palm is also infected but the status of
any disease on this crop is unclear. *R. cocophilus*
has also been implicated in little leaf of both
palms; Dean,
Tech.Communic.Commonw.Inst.Helminthology 47,
1979; Griffith, *PD* **71**,193, 1987.

rhapontici, *Erwinia*.

rhei, *Ramularia*.

Rheum rhaponticum, rhubarb.

rhipsalis mosaic Pape, *Gartenwelt* **36**:707, 731,
1932; Germany [**12**:570].

Rhizina Fr. 1815; Helvellaceae; ascomata attached
to soil or wood by many cylindrical and branched
structures like roots. *R. undulata* Fr.; Booth &
Gibson, *Descr.F.* 324, 1972; asco. aseptate,
24–40 × 9–11 μm, group dying of conifers; disease
outbreaks are found around old fire sites or where
soil has been heated in other ways, e.g. asphalt
road construction, temps. of ± 40° stimulate the
asco. to germinate; Holliday 1980; Phillips &
Burdekin 1982.

rhizobitoxine, toxin from *Rhizobium japonicum*;
Owens & Wright, *Pl.Physiol.* **40**:927, 931, 1965;
Strobel 1977, see bacteria; Chakraborty &
Purkayastha, *Can.J.Microbiol.* **30**:285, 1984
[**45**,676; **63**,5220].

Rhizobium, a genus of root nodule bacteria which
fix atmospheric N; found in the root nodules of
leguminous plants; Vance, *A.Rev.Microbiol.*
37:399, 1983; Djordjevic et al., *A.R.Phytop.*
25:145, 1987.

Rhizoctonia DC. 1816; Agonomycetales, forms
sclerotia of uniform texture; the cells are
essentially similar and, while the outer ones may
be darker and thicker walled, there is no obvious
differentiation into a rind and a medulla. Sclerotia
that are differentiated, like *Sclerotium*, can
therefore be readily excluded from *Rhizoctonia*.
Hyphae emanate from the sclerotia and the
mycelium is assoc. with the roots of living plants;
the spp. have basidial states in *Athelia*,
*Botryobasidium, Ceratobasidium, Helicobasidium,
Thanatephorus, Uthatobasidium, Waitea*; see *T.
cucumeris* whose sclerotial state *R. solani* is the
most important member of the genus; Tu &
Kimbrough, *Bot.Gaz.* **139**:454, 1978, taxonomy;
Ogoshi et al., *Trans.mycol.Soc.Japan* **20**:33, 1979;
24:79, 1983; *J.Fac.Agric.Hokkaido Univ.* **61**:244,
1983; anastomosis groups, binucleate isolates,
Ceratobasidium; Martin & Lucas, *Phytop.* **74**:170,

1984, on turf grasses in USA; Wong &
Sivasithamparam and Wong et al., *TBMS* **85**:21,
156, 1985, on subterranean clover in Australia
[**58**,3714; **59**,2076; **62**,3787; **63**,452, 2924;
64,4985–6].

R. carotae Rader 1948; Mordue, *Descr.F.* 408,
1974; distinguished from *Rhizoctonia solani* by the
presence of clamp connections; hyphal diam.,
particularly in sclerotia, less than in the *R. solani*
group, and the pigmentation is paler; carrot crater
rot in cold storage, soilborne, spreads on and
between storage crates on which it can persist
from one season to the next.

R. cerealis, teleom. *Ceratobasidium cereale*.

R. crocorum, teleom. *Helicobasidium brebissonii*.

R. fragariae Husain & McKeen, *Phytop.* **53**:533,
541, 1963; assoc. with strawberry degeneration or
root rot, at low soil temp. in Canada; Wilhelm et
al., *ibid* **62**:700, 1972 in USA; Kohmoto et al.
1981 [**42**:694; **52**,1211; **61**,3551].

R. oryzae Ryker & Gooch, *Phytop.* **28**:238, 1938;
the sclerotial masses in vitro are of indefinite size
and shape compared with the round, regular
sclerotia of *Rhizoctonia solani*; prob. of little
importance as the cause of rice sheath spot; see
Oniki et al., *R. zeae*; Ou 1985 [**17**:622].

R. oryzae-sativae (Saw.) Mordue, *Descr.F.* 409,
1974; on rice; morphologically close to
Rhizoctonia solani.

R. solani, teleom. *Thanatephorus cucumeris*.

R. tuliparum Whetzel & Arthur 1924; Mordue,
Descr.F. 407, 1974; grey bulb rot, temperate,
ornamental, bulb crops; Europe, N. America,
USSR; germination of sclerotia is vigorous in
England in winter and spring but not in summer;
mycelium grows through the soil for at least
10 m; fungus primarily a shoot pathogen; long
soil survival, outbreaks of disease occurred in a
field which had not been planted with a
susceptible crop for 6 years; Gladders & Coley-
Smith, *TBMS* **68**:115, 1977; **71**:129, 1978;
72:251, 1979; **74**:579, 1980; Javed, *Trans.
mycol.Soc.Japan* **19**:47, 1978; Coley-Smith et al.,
Pl.Path. **28**:128, 1979 [**56**,3432; **58**,1256, 1813;
59,806, 5234; **60**,2064].

R. zeae Voorhees, *Phytop.* **24**:1299, 1934; root and
ear rots of maize; Sumner & Bell, *ibid* **72**:86,
1982, on maize in USA with *Rhizoctonia solani*;
Martin & Lucas, *PD* **67**:676, 1983, on turf grasses
in USA, prob. more severe in warm seasons,
differences in virulence; Oniki et al.,
Trans.mycol.Soc.Japan **26**:189, 1985, some isolates
& some of *R. oryzae* formed a teleom. *Waitea
circinata* [**14**:232; **61**,4895; **62**,4337; **65**,587].

rhizogenes, *Agrobacterium*.

rhizomania (beet necrotic yellow vein v.) beet.

rhizome rot (*Botryotinia convoluta*) iris, (*Erwinia
chrysanthemi* pv. *paradisiaca*) banana.

rhizomorph, an aggregation of hyphae, like a root,
having a well defined apical meristem and often
differentiated into a rind of small dark cells which
surround a core of elongated, colourless cells;
Garrett 1970.

rhizophila, *Magnaporthe*.

Rhizopus Ehrenb. 1821; Mucoraceae; Kirk, *Taxon*
35:375, 1986; sporangiophores on arching stolons,
usually borne opposite tufts of rhizoids and
typically unbranched, sporangia not in umbels;
common moulds and important organisms in food
spoilage; can be distinguished from the
ecologically similar *Mucor* by the stolons and
rhizoids; Holliday 1980; Onions et al., *Smith's
introduction to industrial mycology*, edn. 7, 1981;
Schipper & Stalpers, *Stud.Mycol.* 25, 1984;
Seviour et al., *TBMS* **84**:701, 1985, taxonomy.

R. oryzae Went & Prinsen Geerligs 1895; Lunn,
Descr.F. 525, 1977; a common saprophyte causing
soft rots of fruits and vegetables, barn or pole rot
of tobacco, beet root rot, yam tuber rot and
assoc. with sunflower head rot.

R. sexualis (Smith) Callen 1940; Lunn, *Descr.F.*
526, 1977; soft rot of fruit, particularly of
strawberry.

R. stolonifer (Ehrenb.:Fr.) Lind 1913; Lunn,
Descr.F. 524, 1977; whiskers or leak of fruits and
vegetables; sweet potato soft rot, assoc. with
cotton boll rot; Roy, *TBMS* **77**:434, 1981, jack
fruit, *Artocarpus heterophyllus*, rot &
disseminating insects, as *Rhizopus artocarpi*
Racib., a synonym; Sonoda, *PD* **64**:296, 1980,
soft rot of greenhouse cucumbers [**60**,530;
61,3207].

Rhizosphaera Mangin & Hariot 1907; Coelom.;
Sutton 1980; conidiomata pycnidial,
conidiogenesis enteroblastic, phialidic; con.
aseptate, hyaline, comparatively large, on conifer
needles.

R. kalkhoffii Bubák 1914; Diamandis & Minter,
Descr.F. 656, 1980; con. $4.5–9.5 \times 2.5–4.5 \ \mu m$; pine
and spruce needle blight, prob. geographical
variation in pathogenicity.

R. oudemansii Maubl. 1907; con. $9–16 \times 5.5–9 \ \mu m$;
Martínez & Ramírez, *Mycopathologia* **83**:175,
1983, assoc. with the needle cast of *Abies pinsapo*
in Spain [**63**,3601].

R. pini (Corda) Maubl.; Diamandis & Minter,
Descr.F. 657, 1980; con. $15–30 \times 7–12 \ \mu m$; fir
needle blight.

rhizosphere, the region around plant roots; Ferriss,
Phytop. **71**:1229, 1981; **73**:1355, 1983; Curl, *PD*
66:624, 1982; Drury et al., *Phytop.* **73**:1351,
1983; Funck-Jensen & Hockenhull,

Växtskyddsnotiser **48**:49, 1984; Gangawane 1985; Foster, *A.R.Phytop.* **24**:211, 1986; Schippers et al., *ibid* **25**:339, 1987 [**65**,4753].

rhodina, *Physalospora*.

Rhodococcus Zopf 1891; bacteria; Bradbury 1986; Gram positive, aerobic, pleomorphic cells, often initially forming a 'mycelium' that soon breaks into irregular elements, non-motile, no spores.

R. fascians (Tilford) Goodfellow, *Syst.appl.Microbiol.* **5**:225, 1984; Bradbury, *Descr.B.* 121, 1967, as *Corynebacterium fascians* (Tilford) Dowson; fasciation of sweet pea, leafy gall of many plants; *J.agric.Res.* **53**:383, 1936; Lacey, *AAB* **26**:262, 1939; Elia et al., *Phytopath.Z.* **110**:89, 1984 [**16**:102; **18**:596; **63**:5277].

Rhododendron, Coyier & Roane 1986.

rhododendron necrotic ringspot Coyier et al., *Phytop.* **67**:1090, 1977; filament assoc. 460–540 × 13 nm, Canada (British Columbia), USA (Oregon) [**57**,2553].

rhoeo mosaic (tobacco mosaic v.) Thompson & Corbett, *PD* **69**:356, 1985 [**64**,4369].

Rhopalosiphoninus latysiphon, transm. potato V v.

Rhopalosiphum maidis, transm. barley mosaic, barley yellow dwarf, cardamom Indian streak, cereal leaf spot, guinea grass mosaic, maize American leaf fleck, mulberry mosaic, oat chlorotic stripe, wheat yellow leaf. *R. padi*, transm. barley yellow dwarf, festuca necrosis, lolium mottle, maize American leaf fleck, pea seedborne mosaic, ryegrass chlorotic streak, tradescantia leaf distortion.

rhubarb, *Rheum rhaponticum*; Tomlinson & Walkey, *AAB* **59**:415, 1967, viruses in Britain; Walkey et al., *Pl.Path.* **31**:253, 1982, reinfection & yield of virus tested stock [**46**,3302; **62**,517].

— **light green spot** (turnip mosaic v.), Walkey et al., see above, cv. Timperley Early.

— **mosaic** (arabis mosaic v.), fide Francki et al., vol. 2, 1985.

— **ringspot** Yale & Vaughan, *Phytop.* **44**:118, 1954; transm. sap, USA (Oregon) [**34**:73].

— **1 and 2 v.** MacLachlan 1958; *Can.J.Pl.Sci.* **40**:104, 1960, both rods 478 × 15 nm, transm. sap, Canada (Ontario) [**38**:377; **39**:762].

rhynchosia little leaf Dabek, *AAB* **103**:431, 1983; MLO assoc., transm. *Ollarianus balli*, Jamaica, weed *R. minima* [**63**,2113].

— **mosaic** Bird 1962; Bird & Sánchez, *J. Agric.Univ.P.Rico* **55**:461, 1971; Kvíčala, *Zentbl.Bakt.ParasitKde.* 2 **133**:451, 1978; transm. *Bemisia tabaci*, *R. minima*, Cuba, Puerto Rico [**51**,4745; **58**,5757].

Rhynchosporium Heinsen ex Frank 1901; Hyphom.;

Carmichael et al. 1980; conidiomata hyphal; con. 1 septate, hyaline, curved, beaked.

R. secalis (Oudem.) J.J. Davis 1919; Owen, *Descr.F.* 387, 1973; con. 11–35 × 3–3.5 μm; barley scald, rye and other Gramineae, the leaves and sheaths are mostly attacked, a major disease in cooler and wetter areas, seedborne but its importance not always clear, carryover on crop debris, a weakly competitive saprophyte and prob. not soilborne to any great degree, variable pathogenicity, splash dispersed con., races on barley in USA (California), forms rhynchosporoside; Shipton et al., *Rev.Pl.Path.* **53**:839, 1974; Mathre 1982; Mayfield & Clare, *Aust.J.agric.Res.* **35**:789, 799, 1984, survival over summer, effects of stubble treatments & sowing sequences; Khan, *Aust.J.exp.Agric.* **26**:231, 1986, chemical control [**64**,2006–7; **66**,2328].

rhynchosporoside, toxin from *Rhynchosporium secalis*; Auriol et al., *Proc.natn.Acad.Sci.U.S.A.* **75**:4339, 1978; Mazars et al., *Phytopath.Z.* **107**:1, 1983 [**58**,4358; **62**,4270].

Rhytisma Fr. 1819, fide Cannon et al. 1985; Rhytismataceae; asco. filiform to slightly clavate. *R. acerinum* (Pers.) Fr., Cannon & Minter, *Descr.F.* 791, 1984; anam. *Melasmia acerina* Lév. 1846; asco. 60–80 × 1.5–2.5 μm, con. 6–10 × *c.*1 μm; maple tar spot, rarely serious although unsightly, sensitive to SO₂ levels in the atmosphere; Hudler et al., *PD* **71**:65, 1987, reported severe defoliation of *Acer platanoides* in USA (New York State), their fungus differed morphologically from *R. acerinum*; Duravetz & Morgan-Jones, *CJB* **49**:1267, 1971, ascoma development in *R. acerinum* & *R. punctatum* (Pers.) Fr. [**66**,4481].

Rhytismataceae, Rhytismatales; ascomata carbonaceous, spreading, multiloculate, asci lacking a pore, not amyloid; asco. aseptate, sheathed, often included in Hypodermataceae; *Cyclaneusma, Didymascella, Lirula, Lophodermella, Lophodermium, Meloderma, Naemacyclus, Rhytisma, Therrya*; Dicosmo et al., *Mycotaxon* **21**:1, 1984, on Phacidiaceae q.v.; Minter & Cannon, *TBMS* **83**:65, 1984, asco. discharge [**63**,4787].

Rhytismatales, Ascomycotina; ascomata apothecioid or long, narrow; immersed in a stroma or host tissue, especially leaves; exposed by rupture, asci unitunicate, annulus small if present.

Ribautodelphax notabilis, transm. maize Iranian mosaic.

ribavirin, see antiviral compounds.

Ribes, black and red currants; *R. grossularia*, gooseberry.

ribes ringspot (tobacco rattle v.), Schmelzer 1970 [50,115].
— yellows Králík & Brčák, *Biologia Pl.* 17:214, 1975; Czechoslovakia, *R. houghtonianum* [55,329].
ribesia, *Dothiora*.
ribgrass mosaic v. Holmes 1941; Oshima & Harrison, *Descr.V.* 152, 1975; Wetter in van Regenmortel & Fraenkel-Conrat 1986; Tobamovirus, rod 300 × 18 nm, transm. sap, *Plantago lanceolata*; causes necrotic mosaic in tobacco and internal necrosis in tomato fruit; many minor variants.
ribicola, *Cronartium*.
ribis, *Botryosphaeria, Drepanopeziza, Gloeosporidiella*.
ribonucleic acid, commonly RNA; Chakravorty & Shaw, *A.R.Phytop.* 15:135, 1977, role in host & parasite specificity.
rice, *Oryza sativa*; Padwick, *Manual of rice diseases*, 1950; Ou & Jennings, *A.R.Phytop.* 7:383, 1969, resistance; Hashioka, *Riso* 18:279; 19:11, 111, 309; 20:235; 21:11, 1969–72; Ramakrishnan, *Diseases of rice*, 1971; Atkins, *Agric.Handb.* USDA 448, 1974; Magi, *Jap.Pestic.Inf.* 27:5, 1976, seed disinfection: *Pest control in rice, PANS Manual* 3, 1976; Khush, *Adv.Agronomy* 29:265, 1977, resistance; Yamaguchi, *Rev.Pl.Prot.Res.* 10:49, 1977, seed disinfection; Gangopadhyay, *Current concepts on fungal diseases of rice*, 1983; Ou 1985; Gangopadhyay & Padmanabhan, *Breeding for disease resistance in rice*, 1987.
— black streaked dwarf v. Kuribayashi & Shinkai 1952; Shikata, *Descr.V.* 135, 1974; Phytoreovirus, subgroup Fijiviruses; isometric *c.*75–80 nm diam., transm. *Laodelphax striatellus, Unkanodes albifascia, U. sapporana*, persistent; China (coastal regions) Japan, Korea; can cause serious damage to barley, maize, rice, wheat; Shikata & Kitagawa, *Ann.phytopath.Soc.Japan* 40:329, 1974; *Virology* 77:826, 1977, multiplication, properties & morphology; Lee & Kim, *Korean J.Pl.Path.* 1:190, 1985, effects on *L. striatellus* [54,4921; 56,5004; 66,988].
— bunchy stunt v. Xie et al., *Acta phytopath.sin.* 9:93, 1979; 12(4):16, 1982; 13(3):15, 1983; isometric, transm. *Nephotettix cincticeps, N. impicticeps*, China [59,4152; 62,2459; 63,2310].
— chlorotic streak Anjaneyulu et al., *Int.Rice Res.Newsl.* 5(3):12, 1980; transm. *Heterococcus rehi*, India (Cuttack) [60,2007].
— deformation v. Luk'yanchikov & Sorokina, *Zashch.Rast.Mosk.* 12:35, 1979; isometric *c.*30 nm diam., transm. *Schizaphis graminum*, USSR [60,2577].
— dwarf v. Takata 1895–6; Iida et al., *Descr.V.* 102, 1972; Phytoreovirus, subgroup 1, isometric

*c.*70 nm diam., transm. *Nephotettix cincticeps, N. nigropictus, Recilia dorsalis*, persistent; Japan, Korea; Gramineae; Uyeda & Shikata, *Ann. phytopath.Soc.Japan* 48:295, 1982, particle structure; Xie et al., *Acta phytopath.sin.* 9:93, 1979, in China, like RDV but not serologically related [59,4152; 62,654].
— gall dwarf v. Omura et al. and Putta et al. 1980; Omura & Inoue, *Descr.V.* 296, 1985; Phytoreovirus, subgroup 1; isometric *c.*65 nm diam., transm. *Nephotettix* and *Recilia* spp., persistent, transovarial; Gramineae, only rice naturally infected. The 4, serologically distinct, rice viruses in the Phytoreovirus group can be identified by symptoms in the greenhouse at 24–30°. RGDV causes large numbers of small galls or enations after 20 days, they do not change in colour. Rice black streaked dwarf v. and rice ragged stunt v. cause a few elongated galls after 50–60 days. In the former their colour changes to dark brown; in the latter there is no colour change. Rice dwarf v. does not cause galls to form, only chlorotic flecks.
— giallume (barley yellow dwarf v.), Corbetta 1967; Amici et al., *Riv.Patol.veg.* 14:127, 1978 [58,3828].
— grassy stunt v. John 1965; Bergonia et al. and Rivera et al. 1966; Hibino, *Descr.V.* 320, 1986; rice stripe virus group; 4 species RNA, filament, many particles circular, modal contour length 950–1350 × 6–8 nm; transm. *Nilaparvata lugens* especially, also *N. bakeri, N. muiri*; persistent; confined to *Oryza*, economically important in the Asian tropics and subtropics, 3 variants in Taiwan: grassy stunt B and Y, wilted stunt; a str. 2 in the Philippines.
— hoja blanca v. Gracés 1940; Morales & Niessen, *Descr.V.* 299, 1985; rice stripe virus group, filament, length undetermined, 3–4 nm diam., transm. *Sogatodes cubanus, S. oryzicola*, persistent, transovarial, Gramineae; America, mainly tropics; particle morphology and vector relations similar to those of rice stripe v. q.v. Toriyama 1986, and maize stripe v.; rice grassy stunt v. also belongs to the rice stripe virus group.
— mentek = rice tungro v.
— necrosis mosaic v. Fujii et al. 1966–7; Inouye & Fujii, *Descr.V.* 172, 1977; Potyvirus, subgroup 2, filament 275 or 550 nm long, 13–14 nm diam., transm. sap difficult, soilborne, poss. transm. *Polymyxa graminis*, Japan, rice only; similar disease in India (Cuttack), Ghosh, *Curr.Sci.* 48:1045, 1979 [59,4145].
— orange leaf Rivera, Ou & Pathak, *PDR* 47:1045, 1963; Saito et al., *ibid* 60:649, 1976; Lin

et al. 1983; transm. *Recilia dorsalis*, parts of S.E. Asia [**43**,1631; **56**,2054; **65**,695].

— **penyakit merah** = rice tungro v.

— **ragged stunt v.** Hibino et al. 1977; Milne et al., *Descr.V.* 248, 1982; Phytoreovirus, subgroup 3, isometric 50 nm diam. with spikes and poss. a complete outer shell out to 65 nm diam., transm. *Nilaparvata lugens*, persistent, S.E. Asia, Japan; Gramineae; Hibino & Kimura, *Phytop.* **72**:656, 1982, detection in vectors: Kawano et al., *J.Fac.Agric.Hokkaido Univ.* **61**:408, 1984, particle structure & ds RNA [**61**,7011; **63**,3891].

— **root nematode**, *Hirschmanniella oryzae*.

— **rosette** = rice grassy stunt v.

— **stripe v.** Uyeda 1917; Kuribayashi 1931; Toriyama, *Descr.V.* 269, 1983; filament length undetermined, *c*.8 nm wide, contains ssRNA. This virus gives its name to a group which also contains maize stripe v., rice grassy stunt v., rice hoja blanca v.; transm. sap, difficult, *Laodelphax striatellus, Unkanodes albifascia, U. sapporana*, persistent; Asia with crop; Gramineae; Toriyama, *Microbiological Sci.* **3**:347, 1986, review of the group [**66**,2358].

— **stunt** (rice dwarf v.).

— **transitory yellowing v.** Chiu et al. 1965; Shikata, *Descr.V.* 100, 1972; Phytorhabdovirus, subgroup 2; bacilliform *c*.129 × 96 nm, dip preparations; 193 × 94 nm, leaf sections; transm. *Nephotettix cincticeps, N. nigropictus, N. virescens*, persistent; China, Taiwan; Chen & Chiu, *Pl.Prot.Bull. Taiwan* **22**:297, 1980, transm. [**61**,1165].

— **tungro v.** Rivera & Ou 1965; Gálvez, *Descr.V.* 67,1971, isometric; Ou 1985 fully discussed this important virus disease of rice in S.E. Asia, known under several names. But he did not refer to the work that characterises the causal agent(s) as 2 viruses, one bacilliform and the other isometric. The *Descr.V.* 1986 index places RTV, tentatively, under a maize chlorotic dwarf virus group under viruses with isometric particles; see rice waika v.; Ling in Plumb & Thresh 1983; Mukhopadhyay 1984 [**64**,1106].

— — **bacilliform v.** Saito et al., *Phytop.* **65**:793, 1975; particle 150–350 × 25 nm, see below [**55**,728].

— — **isometric v.** Hibino et al., *Phytop.* **68**:1412, 1978; **69**:1266, 1979; *PD* **67**:774, 1983; particle 30–33 nm diam., transm. *Nephotettix virescens*, the most important vector, other *Nephotettix* spp., non-persistent; RTIV can be transm. alone but rice tungro bacilliform v. is dependent on RTIV for its transm. [**58**,3829; **59**,5127; **62**,4896].

— **waika v.** Nishi, Kimura & Maejima, *Ann.phytopath.Soc.Japan* **41**:233, 1975; Doi et al., *ibid* **41**:228, 1975; Inoue, *PDR* **62**:867, 1978;

Inoue & Hirao, *Appl.Entomol.Zool.* **15**:433, 1980; isometric, transm. *Nephotettix* spp., semi- or non-persistent; Japan; this is apparently the S str. of rice tungro isometric v. referred to by Ou 1985 as rice tungro v. [**55**,2221–2; **58**,3279; **61**,3457].

— **wilted stunt** Chen 1978; *Int.Rice Res.Newsl.* **6**(1):13, 1981; transm. *Nilaparvata lugens*, persistent, Taiwan, lethal [**61**,711].

— **yellow dwarf** Iida & Shinkai 1950; Ou 1985; MLO assoc., transm. *Nephotettix* spp., Asian tropics.

— — **mottle v.** Bakker 1970; *Descr.V.* 149, 1975; Sobemovirus, isometric *c*.25 nm diam., transm. sap, chrysomelid beetles; Kenya, W. Africa.

richardiae, *Phytophthora*.

ricini, *Alternaria, Botryotinia, Botrytis, Melampsora*.

ricinodendron mosaic Deighton & Tinsley 1958; Brunt, *Trop.Agric.Trin.* **40**:325, 1963; Ghana, *R. heudelotii* [**43**,590].

Ricinus communis, castor.

Rickettsiales, Procaryotae; Gram negative, typical bacterial cell walls, no flagella, rods, cocci or pleomorphic, multiply only in a eucaryotic host cell, cause diseases in animals. Some bacteria which occur in diseased plants were originally described as like rickettsias but they are not related; see phloem limited bacteria and xylem limited bacteria.

rigidiuscula, *Nectria*.

Rigidoporus Murrill 1905; Polyporaceae; basidiomata woody, context thin, white, not stratified; pores white, discolouring at maturity; basidiospores smooth, hyaline.

R. lignosus (Klotzsch) Imazeki 1952; Pegler & Waterston, *Descr.F.* 198, 1968; white root rot of tropical crops, fans of mycelium, rhizomorphs white to yellowish. The fungus had been almost entirely studied on rubber in Indonesia, Malaysia (W.) and Sri Lanka. It is widespread in S.E. Asia, although not everywhere as a rubber pathogen; occurs in Africa but its identification in tropical America is uncertain. *Rigidoporus lignosus* caused enormous losses during the development of the crop in Indonesia and Malaysia; the major disease problem for many years. It has been frequently studied with *Ganoderma philippii* and *Phellinus noxius* which are both lesser pathogens but have similar behavioural patterns to those of *R. lignosus*. The 3 fungi can be distinguished in the field by their characteristic mycelial growths on the roots within 30 cm of soil level.

Host penetration is mostly through rhizomorphs during root contact but spore invasion of cut surfaces, e.g. trunk stumps, can take place. Infection patterns show variations depending on climate. In the Ivory Coast a deep,

100 cm infection of the tap root is usual. In Bangladesh and Burma infection is apparently not serious, prob. because of severe dry seasons. The roots develop an undifferentiated rot; 3-year-old trees can be killed in 6 months. The size of the infection source, i.e. the inoculum potential, is an important factor in disease spread. Infective sources may therefore be considered as potential rather than actual. Bancroft's work on this disease in 1912 is one of the first demonstrations of inoculum potential phenomena. In the early history of white root rot the fungus spread from the indigenous forest to the planted crop. But control problems now very largely relate to patterns where an existing woody crop is being replaced by the same one or another. Control depends on: reducing the chance of spore infection; promoting the rapid decay of woody debris on clearing existing crops; hastening decay of potential inoculum sources during early growth of the new crop by dissipating rhizomorphs; early detection of actual infective sources by tree inspection; and finally the prevention of epiphytic growth on roots and collars with fungicides; Fox in Baker & Snyder, 1965; Peries in Kranz et al. 1977; Holliday 1980.

R. ulmarius (Sow.:Fr.) Imazeki; Pegler & Waterston, *Descr.F.* 199, 1968; elm butt rot and some other temperate, hardwood trees.

R. zonalis (Berk.) Imazeki; Pegler & Waterston, *Descr.F.* 200, 1968; distinguished from *Rigidoporus lignosus* by the encrusted cystidia; white pocket rot of dicotyledonous trees, a ubiquitous saprophyte.

rim blight (*Ascochyta heveae*) rubber.

rind (*Phaeocytostroma sacchari*) sugarcane.

— **blemish**, citrus; wind scarring in Australia (New South Wales); Freeman 1976 [55,5023–4].

— **necrosis**, watermelon; poss. assoc. with bacteria; Kontaxis, *PDR* 60:122, 1976 [55,4924].

ring dying (*Rosellinia desmazieresii*) creeping willow.

— **neck**, avocado; water stress during fruit development; Whiley et al., *Aust.J.Exp.Agric.* 26:249, 1986 [66,2441].

— **nematode decline** (*Macroposthonia xenoplax*) carnation.

— **rot** (*Clavibacter michiganensis* ssp. *sepedonicus*) potato.

ringspot, usually on a leaf, a spot surrounded by a ring or rings which are a darker green, chlorotic or necrotic; the symptoms may be caused by a virus; (*Leptosphaeria sacchari*) sugarcane, (*Microdochium panattonianum*) lettuce, (*Mycosphaerella brassicicola*) brassicas, (*M. dianthi*) carnation, (*Pleospora herbarum*) lucerne, (*Stemphylium sarciniforme*) clover.

Rio Grande gummosis, citrus; an old disorder reported from USA (California, Florida) of unknown cause. Childs, *PDR* 62:390, 395, 1978, stated that it was still of some consequence in Florida on grapefruit and lemon; it was poss. related to salt concentration particularly KCl [58,234–5].

— — **stunt** (*Spiroplasma kunkelii*) maize.

ripe fruit rot (*Botryosphaeria dothidea, B. parva* assoc.) kiwi fruit.

— **rot** (*Glomerella cingulata*) grapevine.

Ritzema Bos, Jan, 1850–1928; born in the Netherlands, Agricultural School, Groningen; first director Willie Commelin Scholten Phytopathological Laboratory, Amsterdam, 1895; chair Univ. Agriculture, Wageningen 1918; laid the foundations of applied biology in the Netherlands. *AAB* 16:483, 1929.

RNA plant viruses, Goldbach, *A.R.Phytop.* 24:289, 1986, molecular evolution & genomes.

robineae, *Phellinus*.

robinia mosaic v. Atanasoff 1935; Schmelzer, *Descr.V.* 65, 1971; Boswell & Gibbs 1983 as black locust true mosaic v., Cucumovirus, isometric *c.*40 nm or 27–28 nm diam., transm. sap, aphid spp., non-persistent, central and S.E. Europe; *R. pseudoacacia*, false acacia, wide host range; Richter et al., *Arch.Phytopath.PflSchutz.* 15:1, 1979, serological relationships with cucumber mosaic v. [58,5281].

— **witches' broom** Waters 1898; Jackson & Hartley, *Phytop.* 23:83, 1933; Seliskar et al., *ibid* 63:30, 1973; MLO assoc.; Europe, USA, *R. pseudoacacia* [12:405; 52,3442].

Robusta coffee, see coffee; **robustus**, *Phellinus*.

rodmanii, *Cercospora*.

Roesleria Thüm. & Pass. 1877; Caliciaceae; fide Redhead; = *Coniocybe* Ach. 1816 in Cannon et al. 1985. *R. subterranea* (Weinm.) Redhead, *CJB* 62:2516, 1984; asco. 1 septate, previously considered to be aseptate, Europe and N. America on fruit tree roots and grapevine; Beckwith, *J.agric.Res.* 27:609, 1924, on grapevine as *R. hypogaea* Thüm. & Pass.; Deal et al., *Phytop.* 62;503, 1972, grapevine replant situation [3:567; 51,4239; 64,4167].

Rolfs, Peter Henry, 1865–1944; born in USA, Iowa State College, Univ. Florida, USDA; went to Brazil in 1921 to help in founding an agricultural college at Viçosa where he stayed for > 12 years; early work on tropical and subtropical crops. *Phytop.* 35:491, 1945.

rolfsii, *Corticium*.

root and collar rot (*Phytophthora quininea*) Cinchona.

— — **shoot metabolism**, effects of obligate fungus pathogens; Walters, *Biol.Rev.* 60:47, 1985.

—— —— soil inhabitants, Garrett, *Biology of root infecting fungi*, 1956, placed the parasitic growth habit of fungi on roots in 3 groups: on roots, in soil and mycorrhiza. The first consisted of forms which had some host specificity, and which could have a transitory or long, i.e. dormant, existence in soil in the absence of the host. The second consisted of plurivorous fungi on roots but which have an effective saprophytic existence. Garrett 1970, in a development of his earlier work, considered unspecialised root fungus parasites and specialised forms. The latter were divided into vascular wilt fungi and ectotrophic fungi; Baker & Snyder 1965; Toussoun et al. 1970; Bruehl 1975; Lynch in Rhodes-Roberts & Skinner ed., *Soc.appl.Bact.Sympos.Ser.* 10, 1982; Lockwood, *Phytop.* 76:20, 1986; see soilborne pathogens.

—— black stain (*Ceratocystis wageneri*).

—— exudates, see root and soil inhabitants, Lockwood.

—— graft transmission, tree pathogens; Epstein, *A.R.Phytop.* 16:181, 1978.

—— growth, effects of infection; Ayres in Wood & Jellies 1984.

—— knot nematodes, *Meloidogyne*, Sasser, *PD* 64:36, 1980; —— lesion nematodes, *Pratylenchus*.

—— parasitic weeds, *Orobanche*, *Striga* and others; Musselman, *A.R.Phytop.* 18:463, 1980.

—— plate rot (*Nectria radicicola*) narcissus.

—— rot (*Aphanomyces euteiches*) pea, (*Phytophthora cinnamomi*) forest trees, ornamental trees and shrubs, avocado, (*P. citricola*) raspberry, (*P. drechsleri*) safflower, (*P. lateralis*) Lawson cypress, (*P. palmivora*) papaya; for subterranean clover, several fungi, Barbetti et al., *Rev.Pl.Path.* 65:287, 1986.

rootstock necrosis, an incompatibility which causes declines of lemon on sour orange stock, and of Eureka lemon on sweet orange stock; Schneider, *Calif.Citrogr.* 41:117, 387, 1956 [35:365, 887].

roreri, *Moniliophthora*.

roridin E, a factor in the pathogenicity of *Myrothecium roridum* to *Cucumis melo*, fide Fernando et al., *TBMS* 86:273, 1986 [65,4605].

roridum, *Myrothecium*.

rosae, *Diplocarpon*, *Marssonina*, *Septoria*; **rosae-pimpinellifoliae**, *Phragmidium*; **rosae-setigerae**, *Phragmidium*.

rosarum, *Elsinoë*, *Sphaceloma*.

Rose, Dean Humbolt, 1878–1963; born in USA, Univ. Kansas, Washington; USDA, postharvest diseases. *Phytop.* 54:745, 1964.

rose, *Rosa*; Ikin & Frost, *Phytopath.Z.* 79:160, 1974, viruses in UK; Thomas, *A.Rep.Glasshse.Crops Res.Inst.*, for 1979:178, degeneration & dieback; *AAB* 94:91, 1980;

105:213, 1984, serological detection of viruses; epidemiology of arabis mosaic v., prunus necrotic ringspot v., strawberry latent ringspot v. in UK; Horst 1983 [54,893; 59,4214; 64,1151].

rosea, *Fomitopsis*, *Mycogone*; **roseipellis**, *Limonomyces*.

rose bud proliferation Bos & Perquin, *Neth.J.Pl.Path.* 81:187, 1975; Netherlands [55,1825].

—— colour and flower break Hunter, *N.Z.J.agric.Res.* 9:1070, 1966, in New Zealand; Farrar & Frost, *Pl.Path.* 21:97, 1972, in England [46,661: 52,430].

—— cowlforming Klášterský 1949, 1951; Czechoslovakia [29:28; 31:38].

—— curling Devergne & Goujon, *Annls.Phytopath.* 7:71, 1975; France [55,4156].

—— dieback, see rose, Thomas, report for 1979.

—— flower proliferation Gualaccini, *Boll.Staz.Patol.veg.Roma* 21:45, 1963; Italy [43,2940].

—— leaf curl Slack et al. 1975; *PDR* 60:178, 1976; USA (California) [55,4747].

—— little leaf Meyer, *S.Afr.J.agric.Sci.* 3:47, 1960; South Africa, prob. close to rose wilt.

—— mosaic, symptoms very variable, caused by several viruses, separately or in a combination, i.e. apple mosaic v., arabis mosaic v., prunus necrotic ringspot v., strawberry latent ringspot v. PNRSV is prob. the most important; in UK it delays the onset of flowering, reduces the size and number of blossoms, and increased the proportion of deformed ones; see rose, Thomas 1980, 1984; *AAB* 98:419, 1981; 100:129, 1982; *Pl.Path.* 33:155, 1984 [61,1762, 5034; 63,4465].

—— mottled mosaic Kirkpatrick, Lindner & Cheney, *PDR* 52:499, 1968; USA [47,2765].

—— pinch off Roberts 1962, fide Thomas, rose dieback.

—— rosette Thomas & Scott, *Phytop.* 43:218, 1953; Allington et al., *J.econ.Ent.* 61:1137, 1968; transm. *Phyllocoptes fructiphilus*; Crowe, *PD* 67:544, 1983, outbreak in USA (central states) [33:85; 48,1766; 62,3544].

—— spring dwarf Traylor, Wagnon & Williams, *PDR* 55:294, 1971; Slack et al., *ibid* 60:183, 1976; USA (California) [50,3856; 55,4748].

—— streak Brierley 1935; Brierley & Smith, *J.agric.Res.* 61:625, 1940; Klesser, *S.Afr.J.agric.Res.* 10:849, 1967; Schmelzer, *Phytopath.Z.* 58:92, 1967; Europe, South Africa, USA [14:363; 20:365; 46,1623; 47,1909].

—— stunt Hutton, *Gdnr's.Chron.* 168:31, 1970; see rose, Ikin & Frost; prob. close to rose wilt.

—— tobamovirus Hicks 1979; Hicks & Frost, *Pl.Path.* 33:581, 1984; transm. sap, serologically related to, but distinct from, the type str. of tobacco mosaic v.; UK [64,1621].

—vein yellowing (tobacco streak v.).

— wilt, see rose, Thomas, report for 1979; Hammett, *PDR* 55:916, 1971, distinguishing from *Verticillium* wilt [51,1571].

—witches' broom = rose rosette.

— yellow vein mosaic Sastry, *Indian Phytopath.* 19:316, 1966; India (N.) [47,230].

roselle, see *Hibiscus*.

Rosellinia de Not. 1844; Xylariaceae; ascomata superficial, ostiolate, often as dense swarms on a common mycelial mat or subiculum, subglobose, smooth, dark; asco. pigmented, aseptate, often with a minute, hyaline appendage; anam. *Dematophora*. The fungi are soil inhabitants; they often form mycelial sheets or strands which can spread from litter and attack the roots and lower stems of woody plants. In some spp. rhizomorphs form on roots. The tropical spp. are much less important than formerly when crops were planted on land which had been cleared from primary forest; Gibson in Kranz et al. 1977; Dargan & Thind, *Mycologia* 71:1010, 1979, in India (N.W. Himalayas); Holliday 1980 [59,3635].

R. arcuata Petch 1916; Sivanesan & Holliday, *Descr.F.* 353, 1972; hyphae of subiculum without swellings as in *Rosellinia necatrix*, asco. 30–48 × 3–6 μm; black root rot, tropical and subtropical, only severe on tea in parts of Asia; = *R. bothrina* (Berk. & Br.) Sacc. 1882, fide Francis, *Sydowia* 38:83, 1985.

R. bunodes (Berk. & Br.) Sacc. 1882; Sivanesan & Holliday, *Descr.F.* 351, 1972; asco. 80–120 × 5–9 μm; black root rot, tropical and subtropical; more widespread, and has been more damaging, than *Rosellinia pepo*.

R. desmazieresii (Berk. & Br.) Sacc. 1878; asco. 24–30 × 6.8–7.5 μm; Barrett & Payne, *TBMS* 78:566, 1982, gave an account of the morphology and ring dying of *Salix repens*, creeping willow, in England (N.W.) [61,5988].

R. herpotrichioides Hepting & Davidson, *Phytop.* 27:307, 1937; asco. 23 × 26 × 9–10 μm; on leaves and twigs of *Tsuga canadensis*, USA, reported on Douglas fir in Canada; see Francis, below [16:574].

R. minor (Höhnel) Francis, *TBMS* 87:397, 1986, asco. 20–25 × 5.5–7 μm, needle blights of conifers in Europe and N. America; Francis reviewed *Rosellinia* on conifers.

R. necatrix Berl. ex Prill. 1904; anam. *Dematophora necatrix* Hartig 1883; Sivanesan & Holliday, *Descr.F.* 352, 1972; asco. 30–50 × 5–8 μm; Cannon et al. 1985 did not accept this name; see Francis, *Sydowia* 38:75, 1985, for a discussion; white root

rot of temperate fruit crops, particularly apple, grapevine, mulberry; no well defined rhizomorphs; hyphae swollen near the septa, black sclerotia 130 × 98 μm formed in vitro; Sztejnberg et al., *Phytop.* 70:525, 1980; *PD* 64:662, 1980; 71:365, 1987, microsclerotia, hosts, control by solarisation; Sousa, *Eur.J.For.Path.* 15:323, 1985, host susceptibility & control [60,1926–7; 65,2357; 66,886].

R. pepo Pat. 1908; Booth & Holliday, *Descr.F.* 354, 1972; asco. 50–69 × 7–9 μm; like *Rosellinia bunodes* causes black root rot of tropical and subtropical crops, particularly in the W. Hemisphere, but appears to have a more limited distribution and host range.

rosellus, *Hypomyces*.

rosette, a symptom often denoting Zn deficiency; Bould et al. 1983.

rosetting, severe reduction of internode growth in a vertical axis without comparable reduction in the size of leaves, FBPP.

roseum, *Gliocladium*, *Trichothecium*.

rosicola, *Cercospora*, *Mycosphaerella*.

rostellata, *Gnomonia*, see *G. rubi*; rostrata, *Setosphaeria*; rostratum, *Exserohilum*.

Rostrup, Frederick George Emil, 1831–1907; born in Denmark, school teacher at Skaarup, Royal Veterinary and Agricultural High School, Copenhagen; held the first ever chair in plant pathology 1902; he built up an important herbarium of fungi which was catalogued by J. Lind in 1913; wrote *Plantepatologi*, 1902. Whetzel 1918, Ainsworth 1981.

rot, disintegration of tissue as a result of the action of invading organisms, usually bacteria or fungi; a disease so characterised, after FBPP.

rotbrenner (*Pseudopeziza tracheiphila*) grapevine.

rottboellia mosaic Celino & Martinez, *Philipp.Agric.* 40:285, 1956; Espeleta & Nuque, *Araneta J.Agric.* 6:45, 1959; transm. *Aphis* spp., Philippines, *R. exaltata* [36:646; 39:583].

—stunt Fajemisin et al. 1982; isometric particle assoc., transm. sap, Nigeria, *R. exaltata* [62,2992].

Rotylenchus buxophilus Golden 1956; Siddiqi, *Descr.N.* 55, 1974; migratory ectoparasite.

R. parvus (Williams) Sher 1961; Heyns, *Descr.N.* 83, 1976; little or no damage to plant roots.

R. reniformis Linford & Oliveira 1950; Siddiqi, *Descr.N.* 5, 1972; tropical and subtropical.

R. robustus (de Man) Filipjev 1936; Siddiqi, *Descr.N.* 11, 1972; mostly temperate; can cause stunting.

rough leaf spot (*Ascochyta sorghi*) sorghum.

rubber, Para rubber, *Hevea brasiliensis*; Sharples, *Diseases and pests of the rubber tree*, 1936; Wastie, *PANS* 21:268, 1975; *F.A.O.Pl.Prot.Bull.* 34:193,

1986; Rao, *Maladies of* Hevea *in Malaysia*, 1975;
Chee, *Micro-organisms associated with rubber*
(Hevea brasiliensis *Mull.Arg.*), *Rubb.Res.Inst.*
Malaysia, 1976; Holliday 1980, rubber diseases
are all caused by fungi; Nandris et al., *PD*
71:298, 1987, root diseases.
——**bark cracking** Bobilioff 1930; Peries, *PDR*
61:946, 1977; S.E. Asia [**57**,3076].
rubbery wood, a structural abnormality caused by a
virus; the branches become flexible and mature
trees show a weeping appearance.
rubescens, *Cytospora*.
rubi, *Agrobacterium, Cercosporella, Cylindrosporium,*
Gnomonia, Sphaerulina; **rubi-idaei**, *Phragmidium*;
ruborum, *Septocyta*.
rubrifaciens, *Erwinia*; **rubrilineans**, *Pseudomonas*;
rubrisubalbicans, *Pseudomonas*.
*****Rubus**, blackberry; *R. idaeus, R. strigosus*, red
raspberry; *R. occidentalis*, black raspberry;
raspberry viruses: Cadman *Hortic.Res.* **1**:47,
1961; Converse, *Agric.Handb.USDA* 310, 1966;
HortSci **12**:471, 1977; Putz, *Annls.Phytopath.*
1:275, 1969; Jones & Wood, *N.Z.J.agric.Res.*
22:173, 1979; Jones, *Bull.Scottish*
hortic.Res.Inst.Assoc. 19:15, 1981; Guy et al.,
Acta Hortic. 129:31, 1982; Stace-Smith, *PD*
68:274, 1984 [**59**,837; *Hortic. Abs.* **51**,8418].
rubus Chinese seedborne v. Barbara, Ashby &
McNamara, *AAB* **107**:45, 1985; poss. Nepovirus,
isometric *c*.30 nm diam., transm. sap and seed of
Chenopodium quinoa, Nicotiana bigelowii; from a
symptomless plant of a *Rubus* sp. grown from
seed from China [**66**,433].
—— **mosaic** Azad & Sehgal, *Indian Phytopath.*
11:159, 1958; India (N.), *R. ellipticus* [**38**:605].
——**sterility** Morris 1938; Hemphill 1962; in Frazier
1970 [**51**,3455].
—— **stunt** Prentice, *J.hort.Sci.* 26:35, 1950; Murant
& Roberts, *AAB* 67:389, 1971; MLO assoc.
[**30**:376; **50**,4001].
——**yellow net v.** Stace-Smith 1955; Stace-Smith &
Jones, *Descr.V.* 188, 1978; bacilliform
c.80–150 × 25–31 nm, transm. *Amphorophora* spp.,
prob. widespread with crop, occurs naturally in
Rubus; forms complexes with other viruses; with
black raspberry necrosis v. causes raspberry
veinbanding mosaic.
ruffles, tobacco, genetic disorder; Wolf & Sharp,
J.Elisha Mitchell scient.Soc. 68:85, 1952 [**32**:456].
rugby stocking, see colour banding.
rugosa, *Ascodichaena*.
rugose leaf curl Grylls, *Aust.J.biol.Soc.* 7:47, 1954;
Behncken & Gowanlock, *ibid* 29:137, 1976;
bacterium assoc., in clover phloem. The causal
agent was originally found in *Austroagallia*
torrida, the vector, in Australia and at first

thought to be a virus; transm. to 16 spp. in 8
plant families [**34**:515; **56**,272].
rugulosin, see diaporthin, McCarroll & Thor.
rumple, citrus; Knorr, *PDR* **47**:335, 1963, reported
the cause unknown; Terranova & Schuderi 1972
suggested a relationship with citrus impietratura;
Majorana & Continella 1983 reported graft
transm. [**42**:681; **53**,544; **63**,4430].
runner bean, *Phaseolus coccineus*.
—— —— **mosaic** Vashisth & Nagaich, *Indian*
Phytopath. **18**:311, 1965; transm. sap, seedborne,
India (Himachal Pradesh) [**46**,799].
russet crack (sweet potato feathery mottle v.) sweet
potato.
russeting, suberisation of epidermal cell walls
without hyperplasia of cells; Vogl et al. 1985,
russeting in the apple cv. Golden Delicious
[**65**,4468].
russett scab (*Streptomyces*) potato, poss. distinct
from potato common scab, see netted scab;
Harrison, *Am.Pot.J.* **39**:368, 1962; Bång, *Pot.Res.*
22:203, 1979, effect on yield [**42**:339; **59**,3876].
rust, a disease caused by, and a sp. in, the class
Urediniomycetes, order Uredinales; from the
predominant yellowish to orange and orange
brown of the spores.
rusty root (*Olpidium brassicae, Pythium* spp.)
carrot, see *P. sulcatum*.
—— **spot**, peach fruit; suspected to be caused by
Podosphaera leucotricha from nearby apple
orchards in USA; Ries & Royse, *Phytop.* **68**:896,
1978 [**58**,1856].
ruta ringspot Silberschmidt, *Biológico* **12**:219, 1946;
transm. sap, Brazil (São Paulo), *R. graveolens*
[**25**:558].
ryegrass, Italian, *Lolium multiflorum*; perennial, *L.*
perenne.
—— **bacilliform v.** Plumb, *Acta Biol.Iugoslavica* B
11:95, 1974; Plumb & James, *AAB* **80**:181, 1975;
particles *c*.210 × 67 nm, Britain, in symptomless
Lolium spp. [**54**,4539; **55**,1304].
—— **chlorotic streak** Clark & Christensen,
N.Z.J.agric.Res. **15**:179, 1972; transm.
Rhopalosiphum padi, persistent, New Zealand
[**51**,2608].
——**cryptic v.**, see ryegrass spherical v.
—— **enation** Huth, *NachrBl.PflSchutzdienstes* **27**:49,
1975; dwarfed; deep green, suppressed
inflorescence, Germany [**55**,777].
—— **mosaic v.** Bruehl, Toko & McKinney 1957;
Slykhuis & Paliwal, *Descr.V.* 86, 1972; Potyvirus,
subgroup 3, filament *c*.700 × 15 nm, transm. sap,
Abacarus hystrix, semi-persistent, Gramineae;
Gibson, *Pl.Path.* **30**:25, 1981, spread by vector &
prevention; Heard & Chapman, *AAB* **108**:341,
1986, pattern of local spread in mown grassland;

Catherall, *Pl.Path.* **36**:73, 1987, effect, with barley yellow dwarf v., on productivity of perennial & Italian ryegrasses [**61**,5047; **65**,5575; **66**,3409].
——**mottle v.** Toriyama, Mikoshiba & Doi, *Ann. phytopath.Soc.Japan* **49**:610, 1983; isometric, *c*.28 nm diam., contains RNA, transm. sap, Japan, in *Dactylis glomerata* and *Lolium multiflorum*, infects temperate cereals and grasses [**63**,4472].

—— **spherical v.** Plumb & James 1973; Cooper, *Annls.Phytopath.* **9**:261, 1977; Luisoni & Milne 1981; particle av. 29 nm diam., transm. seed, Europe; a cryptic v. q.v.; Milne, *Microbiologica* **3**:333, 1980, electron microscopy [**54**,4: **58**,1094; **62**,1081, 1542].
—— **streak** = brome mosaic v.; —— —— **mosaic** = ryegrass mosaic v.

S

sacchari, *Bipolaris, Cytospora, Leptosphaeria, Peronosclerospora, Phaeocytostroma, Phyllachora, Stagonospora.*

Saccharum, sugarcane.

saddle back fungus, *Polyporus squamosus.*

safflower, *Carthamus tinctorius*; Kolte, *Diseases of annual edible oilseed crops*, vol. 3, 1985.

— mosaics Thangamani et al., *Madras agric.J.* **57**:326, 1970; Chenulu et al., *Phytopath.Z.* **71**:129, 1971; Chauhan & Singh, *Indian Phytopath.* **32**:301, 1979; transm. sap, aphid spp., seed [**50**,1327; **51**,562; **60**,3286].

— phyllody Klein, *PDR* **54**:735, 1970; Zelcer et al., *Phytop.* **62**:1453, 1972; Raccah & Klein, *ibid* **72**:230, 1982; MLO assoc., transm. *Neoaliturus fenestratus,* Israel [**50**,1328; **52**,2966; **61**,5159].

saguaro cactus v. Milbrath & Nelson 1972; Nelson & Tremaine, *Descr.V.* 148, 1975; Tombusvirus, isometric *c.*32 nm diam., transm. sap readily, USA (Arizona), naturally only in *Carnegia gigantea*, no symptoms.

Saint Anthony's fire, see *Claviceps purpurea.*

— Augustine grass, *Stenotaphrum secundatum*; decline (panicum mosaic v.).

— Croix decline (prob. *Corynespora cassiicola*) papaya.

— John's wilt (*Fusarium oxysporum* f.sp. *pisi* and var. *redolens, F. solani* f.sp. *pisi*) pea.

saintpaulia leaf necrosis Čiamper & Dokoupil, *Acta Virol.* **18**: 355, 1974; bacilliform particles assoc., 2 types, 200–220 × 60–65 nm and shorter ones, *S. ionantha.*

— line pattern Rønde Kristensen et al. 1963; transm. sap, Denmark [**43**,1514].

Salaman, Redcliffe Nathan, 1874–1955; born in England, Univ. Cambridge, qualified in medicine but after an illness turned to studying potato genetics and became a world authority on this crop. He was the first to demonstrate the genetic control of resistance to *Phytophthora infestans* in 1908 and 1914. He originated the work to build up potato stocks free from viruses. He discovered potato paracrinkle, now known as potato M v., which was present in all plants of the potato cv. King Edward, Smith 1972; FRS 1935; wrote; *Potato varieties* 1926; *The history and social influence of the potato,* 1949, revised impression edited by J. G. Hawkes 1985.

Biogr.Mem.Fel.R.Soc. **1**:239, 1955.

salicella, *Cryptodiaporthe*; salicicola, *Marssonina*;

salicina, *Cryptodiaporthe* see *C. salicella, Pseudocercospora.*

saliciperda, *Pollaccia, Venturia*; salicis, *Discella, Drepanopeziza, Erwinia, Monostichella.*

salinity, and sodicity; effects on plant growth, Bernstein, *A.R.Phytop.* **13**:295, 1975; salt tolerance and crop production, Pasternak, *ibid* **25**:271, 1987.

Salix, willow.

salix chlorosis Svobodová, Blattný & Procházková 1962, Czechoslovakia, Hungary [**44**,865].

— witches' broom Westphal & Michler, *C.r.hebd.Séanc.Acad.Sci.* D **281**:403, 1975; bacilliform particles assoc. [**55**,3308].

— yellows Holmes, Hirumi & Maramorosch, *Phytop.* **62**:826, 1972; MLO assoc., USA (N.E.); causes witches' brooms in *S. rigida* [**52**,1283].

Salmon, Ernest Stanley, 1871–1959, Univ. London, Wye College 1906–39, hop breeding, wrote *A monograph of the Erysiphaceae*, 1900; Horace Brown Medal, Institute of Brewing, 1955. *Nature Lond.* **184**:1188, 1959.

salmonicolor, *Corticium.*

salt, see salinity.

salvia mosaic Gigante, *Boll.Staz.Patol.veg.Roma* 3 **13**:67, 1955; transm. sap, Italy [**36**:31].

— yellow vein mosaic Verma, *Gartenbauwissenschaft* **39**:565, 1974; transm. *Bemisia tabaci,* India (Andhra Pradesh) [**54**,3366].

— 1 Roland, *Parasitica* **6**:8, 1950; transm. aphid, Belgium [**29**:619].

salvinii,*Magnaporthe.*

sambucinum, *Fusarium.*

Sambucus, elderberry; sambucus ringspot and yellow net (cherry leaf roll v.).

Sampson, Kathleen, 1892–1980; born in England, Univ. London, Royal Holloway College; Univ. Leeds; Univ. College of Wales 1919–45; diseases of grasses; wrote, with J. H. Western, *Diseases of British grasses and herbage legumes*, 1941; edn. 2, 1954; with G. C. Ainsworth, *The British smut fungi (Ustilaginales)*, 1950. *TBMS* **75**:353, 1980.

sandal, Indian or true sandalwood, *Santalum album.*

— leaf curl mosaic Venkata Rao, *Indian Forester* **59**:772, 1933; India (Mysore) [**13**:338].

— spike Rao & Iyongar 1934; MLO assoc., detected in 1969. This disease of Indian sandalwood was first reported at the turn of the century in India; it occurs in the natural forests. The leaves are reduced in size and the internodes became shortened; death results in 2–3 years after the symptoms appear; transm. *Coelidia indica*, *Nephotettix virescens*; sap transm. to seedlings resulted in death; Raychaudhuri & Varma, *Rev.Pl.Path.* **59**:99, 1980; Nayar, *Eur.J.For.Path.* **10**:236, 1980; **11**:29, 1981; **14**:59, 1984, control, hosts, in vitro growth, sap transm.; Mohamed Ali et al., *Pl.Path.* **36**:119, 1987, symptom remission with tetracyclines [**13**:735; **60**,1650; **61**,2472; **63**,3594].

sand drawn, tobacco, Mg deficiency.

— **hill decline**, citrus, the same as citrus blight q.v.

Santalum album, sandal.

sapinea, *Sphaeropsis*.

saponaria leaf curl Azad, *Indian Phytopath.* **6**:141, 1953; India (N.), *S. vaccaria* [**34**:38].

Saprolegniales, Oomycetes; zoospores dimorphic or showing evidence of this; *Aphanomyces*: Dick in Ainsworth et al. IVB, 1973.

saprophyte, an organism obtaining nutrients from dead, organic matter; hence being saprophytic; cf. necrotroph which is considered synonymous with perthophyte; Garrett 1970 for behaviour of saprophytic fungi, i.e. colonisation, competitiveness and survival; see competitive saprophytic ability; preferred to saprotrophy; Hudson, *Fungal saprophytism*, *Inst.Biol.Stud.Biol.* 32, 1972.

sap streak (*Ceratocystis coerulescens*) sugar maple.

sarciniforme, *Stemphylium*.

sarcody, swellings above constricted, girdled areas of stems.

Sarcopodium Ehrenb. 1918; Hyphom.; Ellis 1976; conidiomata sporodochial, conidiogenesis enteroblastic phialidic, con. aseptate; see *Nectriella pironii*.

Sarcosomataceae, Pezizales; lignicolous, ascomata apothecioid, large, often stipitate, tough; asci suboperculate; *Urnula*.

Sarocladium W. Gams & D. Hawksw. 1976; Hyphom.; Carmichael et al. 1980; conidiomata hyphal, conidiogenesis enteroblastic phialidic, con. aseptate, hyaline; resembles *Gliocladium* and *Verticillium*.

S. attenuatum, authors as above; Brady. *Descr.F.* 674, 1980; con. 4.5–8 × 0.6–1 μm; differs from *Sarocladium oryzae* by the more regularly verticillate conidiophores and the longer con. with truncate ends; rice sheath rot, seedborne; Ngala, *Pl.Path.* **32**:289, 1983, causing rice grain spotting or dirty panicles [**63**,1272].

S. oryzae (Saw.) W. Gams & D. Hawksw.; Brady, *Descr.F.* 673, 1980; con. 3.5–7 × 0.8–1.5 μm; rice sheath rot, seedborne; Holliday 1980; Manibhushanrao et al., *Z.PflKrankh.PflSchutz.* **93**:319, 1986, review; Boa & Brady, *TBMS*

89:161, 1987, assoc. with bamboo blight in Bangladesh.

sasakii, *Thanatephorus*.

satellite virus, a virus or nucleic acid that does not multiply in cells without a specific helper virus; that is not necessary for the multiplication of the helper; and that has no appreciable sequence homology with the helper virus genome; Murant & Mayo, *A.R.Phytop.* **20**:49, 1982; Francki, *A.Rev.Microbiol.* **39**:151, 1985; Kassanis 1960–1, *Descr.V.* 15, 1970.

sativus, *Cochliobolus*.

Satsuma dwarf, see citrus Satsuma dwarf.

saubinettii, *Dichomera*.

savoying, leaf puckering due to different growth rates in adjacent tissue.

Săvulescu, Alice, 1905–70; born in Romania, Univ. Bucharest; Colombia, USA; Institute of Biology, Romania; mycologist and plant pathologist. *Rev.roum.Biol. bot.ser.* **15**:139, 1970.

—, Trajan, 1889–1963; Romanian mycologist noted for work on rusts and smuts; wrote, with O. Săvulescu, *Tratat de patologie*, 1959. *Sydowia* **18**:1, 1964.

scab, a discrete, superficial lesion involving localised, severe roughening or pitting; more commonly, abnormal thickening of the surface layers, with or without the development of cork; localised hyperplasia of the surface tissues; a disease characterised by lesions of one of these types, FBPP. There are many diseases called scab; they are mostly caused by fungi in these genera: *Cladosporium, Elsinoë, Fusicladium, Sphaceloma, Venturia*; and see *Streptomyces*.

scabby stripe, on radish roots; Hirokawa & Kawashima, *Proc.Assoc.Pl.Prot. Hokuriku* **28**:66, 1980, caused by soil fungi, chemical treatment, Japan [**61**,450].

scaettae, *Mauginiella*.

scald, lesions that suggest to the eye the apparent consequences of scalding with hot water. Such lesions are mainly bleached and may be partly translucent; they are not chlorotic in the usual sense of that term, after FBPP.

—, bean; Mo deficiency; Wilson, *Aust.J.Sci.* **11**:209, 1949 [**29**:396].

— (*Diplocarpon mespili*) Rosaceae, (*Rhynchosporium secalis*) barley.

scale rot (*Fusarium oxysporum* f.sp. *lilii*) lily.

scaly butt (citrus exocortis viroid).

scandens, *Marasmiellus*.

Scaphoideus littoralis, transm. grapevine corky bark, grapevine flavescence dorée; *S. luteolus*, transm. elm phloem necrosis; *S. titanus*, transm. grapevine flavescence dorée.

Scaphytopius acutus, transm. peach X, soybean bud

proliferation, *Spiroplasma citri*; *Scaphytopius fuliginosus* transm. soybean proliferation; *S. magdalensis*, transm. blueberry stunt; *S. nitridus*, transm. *Spiroplasma citri*; *Scaphytopius verecundus*; transm. blueberry stunt.

scarlet runner bean, see runner bean.

schefflera ringspot Gaard, de Zoeten & Heimann, abs. *Phytop.* **72**:1135, 1982; long rod assoc., USA.

Schilberszky, Károly, 1863–1935; born in Hungary, Univ. Budapest, held the first chair of plant pathology in the country. He first described the fungus that causes potato wart, as *Chrysophlyctis endobiotica* in 1896, now *Synchytrium endobioticum*: Gold Medal, Paris Exhibition 1900. *Acta phytopath.Acad.Sci.hung.* **1**:5, 1966.

schizanthus witches' broom Ahlawat & Chenulu, *Curr.Sci.* **52**:367, 1983; poss. MLO, transm. leafhoppers, India (Darjeeling, Sikkim Hills) *S. wisetonensis* [**62**,4706].

Schizaphis graminum, transm. barley yellow dwarf, cereal leaf spot, rice deformation.

Schizothyriaceae, Dothideales; ascomata pseudothecioid, opening by crumbling or splitting of shield of inordinately arranged cells, forming openings that resemble pores, slits or irregular clefts; *Schizothyrium*.

Schizothyrium Desm. 1849 or 1852; Schizothyriaceae; ascomata dark, shield a brown to black crust, crumbling or splitting at maturity, asci distributed individually.

S. pomi (Fr.:Mont) v. Arx 1959; anam. *Zygophiala jamaicensis* Mason in Martin 1945; asco. 10–14 × 3.5 μm; con. 13–20 × 4–6 μm; fly speck of apple and pear, on other fruit, carnation greasy blotch; fully described by Baker et al., *Phytop.* **67**:580, 1977; Nasu et al., *Ann.phytopath.Soc.Japan* **51**:536, 1985; **52**:445, 466, 1986, on apple, grapevine, Japanese persimmon, histology & fine structure [**57**,102; **65**,4339; **66**,1557–8].

Schlumbergera, see *Zygocactus*.

Schramm, Gerhard Felix, 1910–69; German biochemist, Univ. Göttingen, Munich; Max Planck Instititutes, Berlin, Tübingen. In 1956 he isolated, with Gierer, and simultaneously with the independent work by Fraenkel-Conrat in Berkeley, California, the infectious nucleic acid of tobacco mosaic v., fide Waterson & Wilkinson 1978.

Schultz, Eugene S., 1884–1969; born in USA, Univ. Wisconsin, Columbia; virus diseases of potato, recognition of latent infection. *Phytop.* **60**:745, 1970.

schweinitzii, *Phaeolus*.

scientific words, see terminology.

scirpi *Coniothyrium*.

Scirrhia Nitschke ex Fuckel 1870; Dothidiaceae; Sivanesan 1984; stroma dark with partly erumpent, unilocular ascomata along the apex, asco. 1 septate, hyaline; Holliday 1980.

S. acicola (Dearn.) Siggers 1939; anam. *Lecanostica acicola* (Thüm.) H. Sydow 1924; Punithalingam & Gibson, *Descr.F.* 367, 1973; see *Mycosphaerella*, Evans; asco. 9–16 × 2.5–4 μm, con. 1–3 septate, 15–35 × 3–4 μm; brown spot needle blight, *Pinus*; important on *P. palustris* in the nursery and in the early years after planting out.

S. pini Funk & Parker 1966; anam. *Dothistroma septospora* (Dorog.) Morelet var. *septospora* 1968; Punithalingam & Gibson, *Descr.F.* 368, 1973; Sivanesan 1984; Evans described the fungus under *Mycosphaerella* q.v. as *M. pini* Rostrup in Munk 1957; asco. 11–16 × 3–4 μm, con. 1–5 septate, mostly 3, 25–60 × 2–3 μm; red band needle blight of *Pinus*, a widespread and major disease of exotic pine crops; damaging to *P. radiata* in Africa, Chile, New Zealand; and to *P. contorta*, *P. nigra*, *P. ponderosa* in N. America. Cool, wet conditions are most favourable for disease outbreaks; some spp. become resistant with age, others remain highly susceptible or are resistant. *Scirrhia pini* forms the toxin dothistromin.

Gibson, *A.R.Phytop.* **10**:51, 1972; Edwards & Walker, *Aust.For.Res.* **8**:125, 1978, in Australia; Gilmour, *Eur.J.For.Path.* **11**:265, 1981, effect of season on infection in New Zealand; van der Pas, *N.Z.J.For.Sci.* **11**:210, 1981, reduced early growth rates in host; Wilcox, *ibid* **12**:14, 1982, inheritance of resistance; Woollons & Hayward and van der Pas et al., *N.Z.J.For.Sci.* **14**:14, 23,1984; chemical control; Butin, *Sydowia* **38**:20, 1985, formation of both states on *P. nigra* needles; Franich et al., *Physiol.Molec.Pl.Path.* **28**:267, 1986, tissue injury by dothistromin & correlation with resistance in *P. radiata* families [**59**,958; **61**,2489; **62**, 4483–4; **65**,2047–8, 4573; **66**,1171].

Scirtothrips dorsalis, transm. groundnut bud necrosis.

scitaminea, *Ustilago*.

Sciurus carolinensis, grey squirrel, see *Cryptostroma corticale*.

Sclerophoma Höhnel 1909; Coelom.; Sutton 1980; conidiomata stromatic, conidiogenesis enteroblastic phialidic; con. aseptate, hyaline, sometimes tapered to the base.

S. pythiophila, teleom. *Sydowia polyspora*.

S. semenospora Funk 1980, see *Xenomeris abietis*; con. 5–7 × 2–2.5 μm; poss. pathogenic on stressed Douglas fir.

S. xenomeria, teleom. *Xenomeris abietis*.

Sclerophthora Thirum., C. Shaw & Narasimhan 1953; Pythiaceae; sporangia usually ovoid or obpyriform with a distinct apical emission zone, a thickening or papilla; zoospores formed before emission, or nearly so, and dispersing quickly; antheridia always paragynous, oogonial wall fused to the plerotic oospore and a thick brown covering forms. The type sp. *S. macrospora* was first placed in *Sclerospora* and later in *Phytophthora*. *Sclerophthora* has sporangiophores and sporangia that are like *Phytophthora*, but the oospores differ in being plerotic with thick, brown walls; unlike *Phytophthora*, both *Sclerophthora* and *Sclerospora* are virtually restricted to Gramineae; Payak, *Indian Phytopath.* **23**:183, 1970, summarised the differences between these 3 genera. Newhook et al., *Mycol. Pap.* 143, 1978, prefer the type sp. as *P. macrospora* (Sacc.) Ito & Tanaka 1940, and placed the fungus in their group III; Kenneth in Spencer 1981.

S. macrospora (Sacc.) Thirum., C. Shaw & Narasimhan; oogonia mostly 57–73 × 63–75 μm; wide host range on temperate and tropical cereals and other grasses, yellow tuft; causes maize crazy top and rice yellow wilt; also on sugarcane; infection is often systemic and in this case the seedlings are first infected; Holliday 1980; Dernoeden & Jackson, *Phytop.* **70**:1009, 1980; *TBMS* **76**:337, 1981, infection of *Lolium perenne* & enhanced oospore gemination; Grisham et al., *PD* **69**:289, 1985, effect on growth & development of St Augustine grass [**60**,3821; **61**,2134; **64**,4383].

S. rayssiae Kenneth, Koltin & Wahl 1964; oogonia unevenly thickened 44–59 μm diam., oospores smooth, mostly 33 μm diam., smaller sporangia and oospores than those of *Sclerophthora macrospora*; originally described from barley in Israel, the fungus causes brown stripe of maize. The maize form, var. *zeae*, was separated on the grounds of larger sporangia and smaller oospores; seedborne, infects seedlings. The maize disease can be severe in India; Chamswarng et al., *Kasetsart J.* **10**:14, 1976, described infection of *Digitaria bicornis* in Thailand; Holliday 1980 [**58**,1273].

Scleroracus flavopictus, transm. potato purple top wilt; *S. striatulus*, transm. cranberry false blossom.

Sclerospora Schröter 1879; Peronosporaceae; primary aerial sporophore from host surface 10 μm or more broad, usually 15–25 μm, × 2–3 dichotomously branched in the upper part; sporangia usually non-papillate, forming zoospores; oogonial wall thick, rough or ornamented, oospore plerotic. In 1913 Ito noted that the spp. could be placed in one of 2 groups: those in which the asexual spore was a sporangium germinating by zoospores,

eusclerospora; and those in which this spore was a conidium germinating by a germ tube, peronosclerospora; see *Peronosclerospora*. All the spp. except one belong in the latter group which is now recognised as a distinct genus. *Sclerospora* is now therefore monotypic, the sole representative of Ito's eusclerospora group; Holliday 1980 described the *Peronosclerospora* spp. under *Sclerospora*; Spencer 1981.

S. graminicola (Sacc.) Schröter 1886; Kenneth, *Descr.F.* 452, 1975; Francis & Williams, *Descr.F.* 770, 1983; sporangia 14–31 × 12–21 μm; sometimes larger, elongated 28–57 × 17–26 μm; oospores smooth, 19–40 μm diam., thick walls; oogonial wall brownish, smooth or slightly uneven; causes green ear, major diseases in bulrush and foxtail millets, very damaging in India; on other Gramineae. The inflorescence is transformed into a loose, green head which is composed of structures like leaves; there are chlorotic leaf streaks, stunt and excessive tillering; seedborne, Shetty et al., *TBMS* **74**:127, 1980. The seedlings are invaded systemically from oospores in the soil; different pathogenic forms occur on the millets.

Singh & Williams, *Phytop.* **70**:1187, 1980, sporangia & epidemiology; Shetty & Ahmad, *Indian Phytopath.* **34**:307, 1981, 2 races; Ramesh & Safeeulla, *Ind.J.agric.Sci.* **53**:1081, 1983, zoospore infection of flowers; Ball and Ball & Pike, *AAB* **102**:257, 265, 1983; **104**:41, 1984, pathogenic variability in Indian & W. African isolates; Michelmore et al., *Phytop.* **72**:1368, 1982; Idris & Ball, *Pl.Path.* **33**:219, 1984, 2 mating types; Ball et al., *AAB* **108**:519, 1986, pathogenicity of African & Indian forms [**59**,4139; **60**,4958; **61**,5702; **62**,2450, 3002–3; **63**,3328, 4409; **65**,6002].

Sclerosporales, see *Verrucalvus*, Dick et al., for a reclassification of the Peronosporales.

sclerotigenum, *Penicillium*.

Sclerotinia Fuckel !870; Sclerotiniaceae; Kohn, *Mycotaxon* **9**:365, 1979, redefined the genus to include only those spp. which form tuberoid sclerotia not incorporating host tissue within the sclerotial medulla; develop an apothecial ectal excipulum of globose cells; and have no disseminative conidial state. Three well known spp. were retained in *Sclerotinia* and 25, out of 259 epithets, were included as imperfectly known. Kohn's treatment of the genus sensu stricto is followed here; Holliday 1980 described some spp. sensu lato, see Dennis, *Mycol.Pap.* 62, 1956, and other genera sometimes included are *Botryotinia*, *Ciborinia*, *Monilinia*, *Myriosclerotinia*, *Ovulinia*, all q.v.

Phytop. **69**:875, 1979, symposium on all aspects including a key to plant pathogens; Wong & Willets, *J.Gen.Microbiol.* **112**:29, 1979, cytology; Willets & Wong, *Bot.Rev.* **46**,101, 1980, biology & nomenclature; Scott, *TBMS* **77**:674, 1981, serological relationships; Cruickshank, *ibid* **80**:117, 1983, distinctions by pectic zymograms; Tariq et al., *ibid* **84**:381, 1985, cultural & biochemical characteristics [**59**,190, 1620; **61**,3356; **62**,2893; **64**,3342].

S. bulborum (Wakker) Sacc. 1889; ascomata disks 3–5 mm diam., asco. 16 × 8 μm, sclerotia up to 12 mm long; imperfectly known fide Kohn 1979, hyacinth black slime, uncommon; Gould & Russell, *PDR* **49**:443, 1965, Moore 1979 [**44**,2813].

S. homoeocarpa F.T. Bennet 1937; Mordue, *Descr.F.* 618, 1979; the name is uncertain fide Kohn 1979; ascomata disks 0.5–1.5 mm diam.; asco. *c.*16 × 6 μm; the flat sclerotia, like parchment, distinguish this sp. from *Sclerotinia sclerotiorum*; dollar spot of turf grasses, particularly golf greens; Smiley 1983; Hall, *Can.J.Pl.Sci.* **64**:167, 1984, weather & infection periods [**63**,2926].

S. minor Jagger, *J.agric.Res.* **20**:333, 1920; ascomata disks 1–9 mm diam., asco. mostly 12–16 × 6–8 μm, sclerotia black, irregular, up to 2 mm diam., often anastamosing to form flattened bodies several mm long. This sp., like *Sclerotinia sclerotiorum*, is plurivorous, it has smaller sclerotia and larger asco. But it has been mostly studied as the cause of lettuce drop especially in USA (California). Asco. have been implicated in spread but mycelial germination of sclerotia is prob. the most frequent source of infection. Disease incidence at harvest is positively correlated with the numbers of sclerotia that can cause infection; those > 1 cm from the main stem usually do not do so. Sclerotia can be spread with seed and crop debris; Imolehin et al., *Phytop.* **70**:1153, 1158, 1162, 1980, biology of sclerotia & of infection; Dillard & Grogan, *ibid* **75**:90, 1985; Patterson & Grogan, *PD* **69**:766, 1985, chemical control; Sherf & MacNab 1986 [**60**,6191–3; **64**,2862; **65**,1021].

*S. sclerotiorum (Lib.) de Bary 1884; Mordue & Holliday, *Descr.F.* 513, 1976; ascomata disks 1–10 mm diam., asco. 9–13 × 4–6.5 μm, sclerotia usually 5–15 × 3–5 mm but variable; larger sclerotia and smaller asco. than *Sclerotinia minor*; cottony or watery soft rot, in the field and postharvest; sometimes called white mould; the fungus attacks most vegetables, citrus, sugarbeet and sunflower under cool, moist conditions. Infection occurs either from directly germinating sclerotia in the soil or airborne asco. In the first

case the attack on the lower stem causes damping off and a wilt, spread by root contact, takes place. Asco., which can spread on pollen, infect unwounded, soft tissue; flowers being a frequent site of entry. Seed and weeds may be sources of inoculum, and the sclerotia ensure survival.

Purdy, *Phytop.* **69**:875, 1979; Holliday 1980; Akai, *Rep. Hokkaido Prefect.agric.Exp.Stns* 36, 1981; Steadman, *PD* **67**:346, 1983, bean; all reviews; Hims, *Pl.Path.* **28**:197, 1979, infection from weeds; Huang and Huang & Dueck, *Can.J.Pl.Path.* **2**:26, 47, 1980, mycoparasites & infection from sclerotia; Huang & Hoes and Weiss et al., *PD* **64**:81, 757, 1980, plant spacing, sclerotial position, temp., moisture & infection; Morrall & Dueck, *Can.J.Pl.Path.* **4**:161, 1982, rape, epidemiology; Caesar & Pearson, *Phytop.* **73**:1024, 1983, asco. survival; Lamarque and Payen, *EPPO Bull.* **13**(2):75, 277, 1983, sunflower, epidemiology, climate & forecasting; Sutton & Deverall, *Pl.Path.* **32**:251, 1983; **33**:377, 1984, asco. infection & phytoalexin; Holley & Nelson, *Phytop.* **76**:71, 1986, sunflower, plant population & inoculum; Yarden et al., *PD* **70**:738, 1986, chemical control in soil; Ben-Yephet et al., *Pl.Path.* **35**:146, 1986, lettuce, chemical control [**59**,1822, 5440, 5869; **60**,2713, 4057; **61**,2011; **62**,2583–4; **63**,986, 1478, 5622; **65**,5115, 5819; **66**,484].

S. trifoliorum Jakob Eriksson 1880; ascomata disks up to 8 mm diam., mostly *c.* 4 mm diam.; asco. 14–17 × 5–8 μm, dimorphic; sclerotia black, irregular, 2–20 mm long; causes one of the main diseases of red clover, and the fungus has caused severe damage and death in forage legumes in parts of Europe. Both asco. and mycelium from sclerotia cause infection. In the first case this causes pepper spotting on leaves followed by a quiescent stage for up to a few weeks. A later aggressive stage destroys the foliage; this change may be caused by the external environment. Internally seedborne and spread via sclerotia in seed. There is some evidence for the existence of races and a var. *fabae* with larger asco. was described by Keay in 1939; Sampson & Western 1954; Scott, *Bot.Rev.* **50**:491, 1984; Uhm & Fujii, *Phytop.* **73**:565, 569, 1983, asco. dimorphism & heterothallism; Scott & Fielding, *TBMS* **84**:317, 1985, pectolytic enzyme patterns induced by different hosts [**62**,3894–5; **64**,2591].

Sclerotiniaceae, Helotiales; ascomata apothecioid from sclerotia or stromata, ascus pore almost always amyloid; *Botryotinia, Ciboria, Ciborinia, Gloeotinia, Grovesinia, Monilinia, Myrosclerotinia, Ovulinia, Sclerotinia, Septotinia, Stromatinia*.

sclerotioides, *Phoma, Phomopsis*.

sclerotiorum, *Sclerotinia*.

sclerotium, a firm, frequently rounded mass of hyphal tissue, with or without host tissue, and bearing no spores; Coley-Smith & Cooke, *A.R.Phytop.* **9**:65, 1971, survival & germination; Willets, *Biol.Rev.* **46**:387, 1971; **47**:515, 1972, survival & morphogenesis; Chet & Henis, *A.R.Phytop.* **13**:169, 1975, morphogenesis; Insell et al., *CJB* **63**:2305, 1985, comparisons of sclerotia of 4 fungi [**65**,2666].

Sclerotium Tode 1790; Agonomycetales, the sclerotia are distinguished from those of *Rhizoctonia* in that they have a sharply defined rind of evenly thickened cells with pigmented walls, sometimes a narrow cortex of cells with thin walls and finally an inner medulla of interwined hyphae. Several spp. have been described on rice, Ou 1985. Mordue, *TBMS* **81**:654, 1983, described dolipore septa in *S. hydrophilum* Sacc. 1892 which therefore has basidiomycete affinities. A form from *Zizania aquatica* also showed such affinities, Punter et al., *Mycologia* **76**:722, 1984 [**64**,601].

S. cepivorum Berk. 1841; Mordue, *Descr.F.* 512, 1976; sclerotia black, nearly spherical, 200–500 μm diam., with the characteristic structure given for the genus and including the narrow cortex, a phialidic state present. Onion and *Allium* white rot is a major disease of cooler regions. Sometimes large sclerotial masses, like those of *Botrytis* spp. are formed, e.g. in Egypt. A conspicuous, white mycelium covers the roots and bulb base in which many sclerotia form; a soft decay continues in storage. The sclerotia germinate in natural soil only in response to exudate formed from onion roots not to that from roots of other plants. A 10 year survival of sclerotia generally precludes crop rotation as a control measure. But under long periods of winter soil saturation and/or flooding the survival rate was much less in Canada (British Columbia). There is less disease in leek compared with garlic and onion. Some *Allium* spp. and related genera show no disease since they do not stimulate sclerotial germination.

Entwistle & Munasinghe in Scott & Bainbridge 1978; Holliday 1980; Adams, *Phytop.* **71**:1178, 1981, inoculum density & disease; Utkhede, *Z.PflKrank.PflSchutz.* **89**:291, 1982, review; Georgy & Coley-Smith, *TBMS* **79**:534, 1982, sclerotial morphology; Utkhede & Rahe, *PD* **67**:153, 1983, chemical & biological control; Coley-Smith & Esler, *Pl.Path.* **32**:373, 1983; **33**:199, 1984, susceptibility & resistance; Leggett & Rahe, *AAB* **106**:255, 1985, sclerotial survival in Canada; Coley-Smith, *Pl.Path.* **35**:362, 370, 1986,

interactions with, & flavour compounds of, *Allium* cvs; Somerville & Hall, *PD* **71**:229, 1987, sclerotial germination & secondary formation [**61**,3184; **62**,2241, 3385; **63**,1524, 4669; **66**,806–7, 1251].

S. oryzae, a name often, but inappropriately, used for the sclerotia of *Magnaporthe salvinii*.

S. perniciosum van Slogteren & Thomas 1930; black sclerotia 1–2 mmm diam.; tulip smoulder [**10**:32].

S. rolfsii, sclerotial state of *Corticium rolfsii*.

S. wakkerii Boerema & Posthumus, *Neth.J.Pl.Path.* **69**:205, 1963; sclerotia 2–5 mm diam.; black leg of iris and tulip [**42**:766].

sclerotivorum, *Sporidesmium*.

Scolytus, see *Ceratocystis ulmi*.

scoparia witches' broom Hirumi & Maramorosch, abs. *Phytop.* **62**:670, 1972; MLO assoc., Togo, *S. dulcis*.

scoparium, *Cylindrocladium*.

scopolia vein clearing v. Protsenko & Khotin, *Dokl.Akad.Nauk SSSR* **215**:725, 1974; isometric *c.* 25 nm diam., transm. sap [**54**,4602].

Scopulariopsis Bainier 1907; emended Morton & Smith, *Mycol.Pap.* 86:17, 1963; Hyphom.; conidiomata hyphal, conidiogenesis holoblastic, conidiophore growth percurrent; con. aseptate, hyaline. *S. brevicaulis* (Sacc.) Bainier assoc. with a disease called ice-ice on the cultivated red alga *Eucheuma striatum*, East Caroline Islands; Dewey et al., *TBMS* **83**:621, 1984; see *Penicillium waksmanii* [**64**,1016].

scorch, any symptom such as a lesion or lesions that suggest the action of fire on the affected part, especially on green leaves or blossoms. The symptoms may have several causes and their observation provides clues as to the nature of the cause, after FBPP.

—— (*Didymella exitialis*) barley, wheat, (*Kabatiella caulivora*) clover, (see *Pythium megalacanthum*) flax; —— **bud and twig blight and canker** (*Apiognomonia veneta*) plane.

scorzonerae, *Alternaria*.

screening host, a plant in which one or more of the viruses of a mixture can be separated from others, FBPP.

Scrophularia californica, figwort.

scrophularia mottle v. Hein 1959; Bercks, *Descr.V.* 113, 1973; Tymovirus, isometric *c.* 25–27 nm diam., transm. sap, *Cionus*; rather wide host range, Germany, *S. nodosa*.

scurf (*Monilochaetes infuscans*) sweet potato.

scurfy root (*Thanatephorus cucumeris*) radish.

Scutelloma bradys (Steiner & Lettew) Andrássy 1958; Siddiqi, *Descr.N.* 10:1972; yam nematode, causes severe dry rot.

Scutellonema brachyurum (Steiner) Andrássy; Siddiqi, *Descr.N.* 54, 1974; root ectoparasite.

Scutylenchus quadrifer (Andrássy) Siddiqi 1979; Andrássy, *Descr. N.* 108, 1985; little data on damage to crops.

sea tangle, see laminaria coiling stunt.

seaweeds, pathology of, N. America (Pacific N.W.); Andrews, *CJB* 55:1019, 1977 [56,5541].

secalis, *Rhynchosporium*.

sechium witches' broom, see lagenaria and momordica witches' brooms.

secondary leaf fall (*Glomerella cingulata, Oidium heveae*) rubber.

seditiosum, *Lophodermium*.

seed infection index, a formula devised and used to compare susceptibility of bean cvs. to seed infection by fungi; Asmus & Dhingra, *Seed Sci.Technol.* 13:53, 1985 [64,3612].

—— pathology, healthy planting material is the first step towards disease control; the very large literature was last fully assessed by Neergaard 1977, also: Hewett in Scott & Bainbridge 1978 and Ebbels & King 1979, epidemiology & regulation by certification; Richardson 1979, 1981, 1983; Baker, *Seed Sci. Technol.* 8:575, 1980, flower seed; McGee and Schaad, *PD* 65:638, 1981; 66:885, 1982, including bacteria; Geng et al., *ibid* 67:236, 1983, quality control; *Phytop.* 73:314, 1983, seed deterioration; *Seed Sci.Technol.* 11(3,3a) 1983, symposia; Mathur, *ibid* 11:113, 1983, testing seed of tropical plants; *ibid* 12(1), 1984, congress; Kulik, *ibid* 12:831, 1984, detection; *ibid* 13:299, 1985, rules, seed testing; Jeffs ed. *Br.Crop Protect.Council Monogr.* 2, edn.2, 1986; Neergaard, *A.R.Phytop.* 24:1, 1986.

—— piece decay, potato; Miska & Nelson, *Can.Pl.Dis.Surv.* 55:126, 1975, bibliography.

—— treatment, coating or impregnating seeds with a chemical for protection against pathogens.

segetum, *Ustilago*.

Seimatosporium Corda 1833; Coelom.; Sutton 1980; conidiomata acervular to cupulate, conidiogenesis holoblastic, conidiophore growth percurrent; con. mostly 2–5 septate, pigmented, sometimes with a single or branched appendage from each end cell which may not be pigmented.

S. etheridgei Funk, *Eur.J.For.Path.* 8:56, 1978; con. 3–6 septate, 30–44 × 13–15 μm; assoc. with a hypertrophy of bark of *Populus tremuloides* main stems in Canada (British Columbia).

S. lichenicola, teleom. *Clethridium corticola*.

Seiridium Nees:Fr. 1821; Coelom.; Sutton 1980; conidiomata acervular, conidiogenesis holoblastic, conidiophore growth percurrent; con. 5 septate, 4 median cells thick walled, brown; apical and basal cells hyaline, either or both with a simple or

branched appendage. Boesewinkel, *TBMS* 80:544, 1983, reported on the confusion between the 3 spp. assoc. with a severe *Cupressus* canker. S. cardinale (Wagener) B. Sutton & Gibson, *Descr.F.* 326, 1972; con. 24–30 × 8–10.5 μm, appendages usually absent. S. cupressi, teleom. *Lepteutypa cupressi*. S. unicorne (Cooke & Ell.) B. Sutton 1975; con. 25–27 × 7.5–10 μm, appendages present, apical one frequently perpendicular; Motta, *Annali Ist sper. Patol.veg.Roma* 9:205, 1984; Xenopoulos & Diamandis, *Eur.J.For.Path.*15:223, 1985, both S. cardinale, seed infection, chemical control & spread in Greece [62,4474; 64,4002; 65,932].

Sekiguchi lesion, rice, genetic disorder induced by infection with *Cochliobolus miyabeanus* or *Pyricularia oryzae*, or by some chemicals; Marchetti et al., *Phytop.* 73:603, 1983 [62,3854].

Selby, Augustine Dawson, 1859–1924; born in USA, Univ. Ohio State, Washington, Columbia; at Ohio Agricultural Experiment Station for nearly 30 years; wrote *A condensed handbook of the diseases of cultivated plants*, 1900, revised 1910; the book was widely used outside Ohio State since it was the only general text then available. *Phytop.* 15:1, 1925.

selective medium, used to isolate a micro-organism, or a group of such organisms, from a sample which contains a heterogeneous mixture of them; the medium contains substances that inhibit the growth of unwanted micro-organisms. Sometimes the technique is used unnecessarily and perhaps sometimes undesirably. This is because the characteristics, both of the micro-organism and its environment, have not been adequately studied. Such a study will often show that isolation procedures can be much simpler and equally effective without the use of antibiotic materials; Tsao, *A.R.Phytop.* 8:157, 1970, for plant pathogenic fungi.

self-inhibitor, of fungus spores, see endogenous inhibitor.

semenospora, *Sclerophoma*.

Semiaphis heraclei, transm. carrot latent.

semi-persistent viruses, a rather long uptake of a virus by a vector from the phloem, but it becomes infective at once; retention of the virus by vector is in days.

semiustus, *Marasmiellus*.

senecio mosaics Jones, *Phytop.* 34:941, 1944; transm. sap, *Brachycaudus helichrysi*, seed; USA (Washington State); Singh, Verma & Padma, *Gartenbauwissenschaft* 40:67, 1975, transm. sap, India (N.) [24:150; 54,4530].

senescence, Farkas in Horsfall & Cowling, vol. 3, 1978.

senna coffee, see *Cassia*.

sensitive, reacting with severe symptoms to infection by a pathogen or injury by another agent, e.g. a toxin; the antonym is tolerant q.v., see FBPP.

Septocyta Petrak 1927; Coelom.; Sutton 1980; conidiomata stromatic, conidiogenesis holoblastic, conidiophore growth sympodial; con. 1–3 septate, filiform, hyaline.

S. ruborum (Lib.) Petrak 1968; Punithalingam, *Descr.F.* 667, 1980; con. mostly 20–30 × 1–1.5 μm, blackberry purple blotch.

Septoria Sacc. 1884; Coelom., Sutton 1980; conidiomata pycnidial, pycnidia simple with thin walls, con. filiform, hyaline, multiseptate. The genus has > 2000 described taxa and there is no practical system of identification and taxonomy. Relationships based on conidiogenesis have not been considered, hence heterogeneity. Some pathogenic spp. in one of 3 groups are: holoblastic, sympodial conidiogenesis, like the type sp. *S. cytisi* Desm., 1847, *S. chrysanthemella*, *S. helianthi*, *S. obesa*, *S. passerinii*; with phialides, *S. apiicola*, *S. tritici*: with simple holoblastic development and no sympodial or percurrent proliferation, *S. adanensis*, *S. glycines*, *S. lactucae*, *S. leucanthemi*, *S. lycopersici*, *S. socia*; Richardson & Noble, *Pl.Path.* 19:159, 1970, on temperate cereals; Holliday 1980; Constantinescu, *TBMS* 83:383, 1984, 6 spp. on Betulaceae [50,1699; 64,316].

S. adanensis Petrak 1953; Punithalingam, *Descr.F.* 136, 1967; con. 2–4 septate, 26–36 × 1.5–2 μm, blastospores, leaf spot of cultivated chrysanthemum; see *Descr.F.* for other *Septoria* spp. on *Chrysanthemum*.

S. apiicola Speg. 1887; Sutton & Waterston, *Descr.F.* 88, 1966; con. 1–5 septate, 22–56 × 2–2.5 μm, phialospores, celery late blight, an important disease, seedborne; Sheridan, *AAB* 57:75, 1966, sources of inoculum; Maude, *ibid* 65:249, 1970, control on seed with a thiram soak; Maude & Shuring, *Pl.Path.* 19:177, 1970, persistence on celery debris in soil; Vulsteke & Meeus, *Z.PflKrank.PflSchutz.* 93:237, 1986, chemical control [45, 1973; 49,2713; 50,2058; 65,5774].

S. avenae, teleom. *Leptosphaeria avenaria*.

S. cannabis (Lasch) Sacc. 1884; Punithalingam, *Descr.F.* 668, 1980; con. mostly 2–3 septate, 30–38 × 1.5–2 μm, blastospores; white leaf spot or leaf blight of common hemp, can be severe.

S. carthami Murashk. 1926; Punithalingam, *Descr.F.* 669, 1980; con. mostly 5–6 septate, 70–90 × 2–2.5 μm; safflower white leaf spot, see *Descr.F.* for other *Septoria* spp. on *Carthamus*, poss. seedborne.

S. chrysanthemella Sacc. 1895; Punithalingam, *Descr.F.* 137, 1967; con. mostly 4–9 septate, 36–65 × 1.5–2.5 μm; black leaf spot or blotch of chrysanthemum; see *Descr.F.* for other *Septoria* spp. on *Chrysanthemum*; Leu & Yang, *Pl.Prot.Bull.Taiwan* 25:23, 1983, seasonal incidence & chemical control [62,4324].

S. citri Pass. 1877; con. 14–15 × 2–3 μm; Wellings, *TBMS* 76:495, 1981, assoc. with citrus greasy spot & *Mycosphaerella* spp. from citrus [61,239].

S. cucurbitacearum Sacc. 1876; Punithalingam, *Descr.F.* 740, 1982; con. 3–6 septate, 36–62 × 1.5 μm; leaf spot, see *Descr.F.* for other *Septoria* spp. on cucurbits, can be severe; Bradshaw, *Pl.Path.* 33:135, 1984 [63,3107].

S. digitalis Pass., in *Syll.Fung.* 3:534, 1884; con. 25–30 × 1.5 μm; *Digitalis* spp.; Grzybowska 1976, 1977, in Poland, pathogenicity & seed treatment [58,1881–3].

S. elaeagni (Chev.) Desm., in *Syll.Fung.* 10:357, 1892; con. 0–6 septate, 16–28 × 2–3 μm; leaf spot of *Elaeagnus angustifolia*, Russian olive; Lorenzini et al., *Phytopath.Z.* 110:134, 1984, in Italy [64,1145].

S. gladioli Pass. 1874; con. usually 3 septate, 20–55 × 2–4 μm; gladiolus hard rot, black bodies, common on corm lesions and covering scales may be sclerotia; Stone, *TBMS* 41:505, 1958; Schenk, *Tijdschr.PlZiekt.* 66:205, 1960; Moore 1979 [38:261; 40:228].

S. glycines Hemmi 1915; Punithalingam & Holliday, *Descr.F.* 339, 1972; con. 2–4 septate, 30–50 × 1.5–2 μm; soybean brown spot, see *Descr.F.* for other *Septoria* spp. on *Glycine*; an important seedborne pathogen of the crop that has been fully studied recently, particularly in USA; Williams & Nyvall and Lim, *Phytop.* 70:900, 974, 1980; Pataky & Lim and Gray, *PD* 65:588, 1981; 67:525, 1983, effects on yield; Pataky & Lim, *Phytop.* 71:438, 1051, 1981, chemical control, effects of row width & plant growth habit; Ross, *ibid* 72:236, 1982; Peterson & Edwards, *PD* 66:995, 1982, effects of temp. & leaf wetness [60,4083, 5181; 61,487, 2020, 2555, 5364; 62,808, 3705].

S. helianthi Ellis & Kellerman 1883; Holliday & Punithalingam, *Descr.F.* 276, 1970; con. 3–5 septate, 50–85 × 2–3 μm, sunflower leaf spot, can cause damage; *Septoria helianthicola* Cooke & Harkn. has con. 30–35 × 1 μm; Carson, *PD* 71:548, 1987, effects on seed yield with *Alternaria zinniae*.

S. humuli Westend. 1845; Punithalingam, *Descr.F.* 829, 1985; con. 3–4 septate, 40–50 × 1.5–2 μm, hop leaf spot, may cause defoliation.

S. lactucae Pass. 1879; Punithalingam & Holliday,

Descr.F. 335, 1972; con. 1–3 septate, 25–40 × 1.5–2 μm; lettuce leaf spot, see *Descr.F.* for other *Septoria* spp. on *Lactuca*; seedborne.

S. leucanthemi Sacc. & Speg. 1878; Punithalingam, *Descr.F.* 138, 1967; con. 6–12 septate, 72–96 × 2.5–3 μm; leaf spot of chrysanthemum and Compositae, see *Descr.F.* for other *Septoria* spp. on *Chrysanthemum*.

S. linicola, teleom. *Mycosphaerella linicola.*

S. lycopersici Speg. 1882; Sutton & Waterston, *Descr.F.* 89, 1966; con. 2–6 septate, 52–95 × 2 μm; tomato leaf spot; and a form on potato, Piglionica et al., *Phytopath.Mediterranea* **17**:81, 1978, S. & Central America; seedborne, can become epidemic, crop debris prob. not important in spread; Marcinkowska, *Acta Agrobot.* **30**:341, 359, 373, 384, 1977, in Poland; Balakrishna & Rai, *Indian Phytop.* **31**:458, 1978, toxin; Barksdale, *PD* **66**:239, 1982, induced an epidemic & showed effectiveness of host resistance [**58**, 1957–60; **59**,3895; **60**,3313, **61**,5962].

S. musiva, teleom. *Mycosphaerella populorum.*

S. nodorum, teleom. *Leptosphaeria nodorum.*

S. obesa Syd. 1914; Punithalingam, *Descr.F.* 139, 1967; con. 5–9 septate, 51–91 × 2.5–3.5 μm; brown spot of cultivated chrysanthemum, see *Descr.F.* for other *Septoria* spp. on *Chrysanthemum.*

S. oryzae, teleom. *Leptosphaeria oryzicola.*

S. passerinii Sacc. 1884; Punithalingam & Waterston, *Descr.F.* 277, 1970; Bissett, *Fungi Canadenses* 243, 1983; con. 0–3 septate, 26–42 × 1.5–2 μm; barley speckled leaf blotch, important in Canada.

S. passifloricola Punith. *Descr.F.* 670, 1980; con. 0–3 septate, 14–22 × 1.5–2 μm; leaf, blossom, fruit and stem spot of passion fruit; *Septoria passiflorae* Louw 1941 is a synonym; *S. passiflorae* Syd. 1939 has con. 35–52 × 1.5–2 μm.

S. paulliniae Freire & Albuquerque, *Fitopat.Brasileira* **3**:302, 1978; con. 4–8 septate, 33–70 × 3.5–6.5 μm; black crust of guarana, *Paullinia cupana* var. *sorbilis*, Brazil [**58**,4940].

S. populi, teleom. *Mycosphaerella populi*; **Septoria populicola**, teleom. *M. populicola.*

S. pyricola, teleom. *Mycosphaerella pyri.*

S. rosae, teleom. *Sphaerulina rehmiana.*

S. socia Pass. 1879; Punithalingam, *Descr.F.* 140, 1967; con. 1–3 septate, 20–34 × 0.7–1 μm; has characteristically thin, short con.; leaf spot of ox-eye daisy, on other *Chrysanthemum* spp.

S. steviae Ishiba, Yokoyama & Tani, *Ann.phytopath.Soc.Japan* **48**:39, 1982; con. 0–5 septate, 25–63 × 1.5–2.8 μm; leaf spot of *Stevia rebaudiana* [**61**,5800].

S. tritici, teleom. *Mycosphaerella graminicola.*

S. vignae Henn. 1907; Punithalingam, *Descr.F.* 830,

1985; con. mostly 3 septate, 25–40 × 1–1.5 μm; cowpea leaf spot; may be the same as *Septoria vignicola* Vasant Rao, *Curr.Sci.* **32**:367, 1963, fide Punithalingam, con. 2–4 septate, 21–67 × 1.3–2.1 μm; Rawal & Sohi, *Indian J.Pl.Path.* **2**:59, 1984, loss [**43**,304; **64**,1856].

septospora, *Dothistroma.*

Septotinia Whetzel 1937 ex Groves & Elliot, *CJB* **39**:227, 1961; Sclerotiniaceae; ascomata arise from stromatised host tissue, con. mostly 2–4 septate.

S. populiperda Waterman & Cash, *Mycologia* **42**:377, 1950; anam. *Septotis populiperda* (Moesz & Smarods) Waterman & Cash; asco. 10–13 × 4–5 μm, hyaline; con.becoming 1–3 septate 12–15 × 6–9 μm; poplar leaf blotch, may cause premature defoliation [**30**:130].

Septotis Buchwald ex v. Arx 1970; Coelom.; conidiomata acervular, conidiogenesis holoarthric, con. 1–multiseptate; *S. populiperda*, teleom. *Septotinia populiperda.*

septum, of a fungus, a cell wall or partition; *Ainsworth & Bisby's dictionary of the fungi*, edn. 7, 1983.

sequoiae, *Cercospora.*

serial transmission, the transm. of a virus by a test vector, after a single acquisition access period, to 2 or more test plants in succession, FBPP.

serica, *Stromatinia.*

serology, see Elisa, methods; the science of serum reactions; the relations between proteins or polysaccharides, the antigens q.v., and antibody in the serum of infected or immunised, warm blooded animals. Serological reactions are those between antigen(s) and antibody(ies). The different kinds of reaction, e.g. agglutination, complement q.v. fixation and precipitation depend on the physical conditions of the test. Such serological tests may show differences between apparently similar organisms; the subdivisions so formed are called serotypes; Figueredo et al., *Summa Phytopathologica* **3**:233, 1977, plant pathogenic fungi; van Regenmortel, *A.R.Phytop.* **16**:57, 1978; *Serology and immunochemistry of plant viruses*, 1982; Clark, *A.R.Phytop.* **19**:83, 1981, immunosorbent assays in plant pathology; Torrance & Jones, *Pl.Path.* **30**:1, 1981, viruses.

serpens, *Geniculosporium, Hypoxylon, Verticicladiella.*

Serpula (Pers.) Gray 1821; Coniophoraceae; hymenium reticulately folded; Seehann, *Eur.J.For.Path.* **16**:207, 1986, *S. himantioides* (Fr.) Karsten causing butt rot of *Larix kaempferi* and *Pseudotsuga menziesii* [**66**,340].

Serratia Bizio 1823; bacteria; Gram positive rods, peritrichous flagella; few pathogens; Bradbury

1986 described *S. marcescens* Bizio, assoc. with a crown and root rot of lucerne, and *S. proteamaculans* (Paine & Stansfield) Grimont et al. 1978.

sesame, *Sesamum indicum*; Malaguti, *Revta Fac.Agron.Maracay* **7**:109, 1973, leaf pathogens; Mathur & Kabeere, *Seed Sci. Technol.* **3**:655, 1975, seedborne fungi in Uganda; Kolte, *Diseases of annual edible oilseed crops*, vol. 2, 1985; Maiti et al., *Trop. Pest Management* **31**:317, 1985 [**54**,524; **55**,2830; **65**,3438].

— **leaf curl** Wallace 1933; Vasudeva 1954; transm. *Bemisia tabaci*; India, Tanzania [**12**:552; **35**:279].

— **mosaic** Gangopadhyay, *Sci.Cult.* **33**:537, 1967; transm. sap, India (Bengal) [**47**,3179].

— **phyllody** Pal & Nath, *Indian J.agric.Sci.* **5**:517, 1935; Cousin et al. 1970; Murugesan et al., *Madras agric.J.* **60**:492, 1973; Desmidts & Laboucheix, *F.A.O.Pl.Prot.Bull.* **22**:19, 1974; MLO assoc., transm. *Orosius albicinctus, O. cellulosus* [**15**:396; **50**,2411; **54**,1832, 2394].

sesami, *Alternaria, Cercoseptoria, Cercospora, Mycosphaerella*; **sesamicola,** *Alternaria, Mycosphaerella, Synchytrium.*

Sesamum indicum, sesame.

sesbania mosaic v. Sreenivasulu & Nayudu, *Curr.Sci.* **51**:86, 1982; prob. Comovirus, isometric 28 nm diam., transm. sap, India (Andhra Pradesh), *S. grandiflora*; Singh & Srivastava 1985, seedborne in *S. sesban* [**61**,5592; **66**,1519].

setariae, *Bipolaris, Cochliobolus, Pyricularia*; **setariae-italicae,** *Uromyces.*

Setaria italica, see millets.

setosa, *Pleiochaeta.*

Setosphaeria Leonard & Suggs, *Mycologia* **66**:294, 1974; Pleosporaceae or Pyrenophoraceae; Sivanesan 1984; ascomata superficial, with simple, short, rigid setae surrounding the ostiolar region; asco. hyaline, usually 2–6 septate, smooth, with a mucilagenous sheath; anam. *Exserohilum*; Sivanesan, *Descr.F.* 885–8, 1986 described the spp.: *S. holmii, S. monoceras, S. pedicillata, S. prolata*; Holliday 1980.

S. rostrata Leonard 1976; anam. *Exserohilum rostratum* (Drechsler) Leonard & Suggs; Anahosur & Sivanesan, *Descr.F.* 587, 1978; asco. usually 3–5 septate, 29–85 × 9–21 μm; con. 6–16 septate, 30–190 × 10–29 μm; seedling blight, often on Gramineae.

S. turcica (Luttr.) Leonard & Suggs; anam. *Exserohilum turcicum* (Pass.) Leonard & Suggs; Ellis & Holliday, *Descr.F.* 304, 1971, anam. only as *Drechslera turcica* (Pass.) Subram. & Jain 1966; asco. mostly 3 septate, 40–78 × 12–18 μm; con. 4–9 septate, 50–144 × 18–33 μm; northern leaf

blight of maize; so called because, in USA, it has a more northerly distribution compared with *Cochliobolus heterostrophus* causing maize southern leaf blight; on other Gramineae, seed transm. prob. rare, con. dispersal shows a forenoon peak; C. M. Leach has described con. release in great detail; isolates from maize and sorghum are generally specific on their crop of origin; 3 races have been differentiated; forms the toxin monocerin; Smith & Kinsey, Perkins & Hooker and Jordan et al., *PD* **64**:779, 1980; **65**:502, 1981; **67**:1163, 1983, races; Levy & Cohen, *Phytop.* **73**:722, 1983, external environment & infection; Shenoi & Ramalingam, *Indian Phytopath.* **36**:700, 1983, on sorghum; Chang & Fan, *Bot.Bull.Acad.sin.* **14**:209, 1986, distinct race on maize & sorghum [**60**,3145; **61**,695; **62**,4676; **63**,3318; **65**,5998].

sett rot (*Gibberella fujikuroi*) sugarcane.

seven curls (*Glomerella cingulata*) onion, Remiro & Kimati, *Summa Phytopathologica* **1**:51, 1975 [**54**,4285].

sexualis, *Rhizopus.*

Seymour, Arthur Bliss, 1859–1933; born in USA, Univ. Illinois, Harvard; noted for his *Host index of the fungi of North America*, 1929. *Mycologia* **26**:279, 1934.

shaggy stipe (*Mortierella bainieri*) common mushroom.

Shalla, Thomas Allen, 1933–83; born in USA, Univ. Colorado State, California Davis; fundamental work in plant virology. *Phytop.* **73**:1602, 1983.

shallot, see *Allium.*

— **latent v.** Bos, Huttinga & Maat 1978; Bos, *Descr.V.* 250, 1982; Carlavirus, filament *c*.650 nm long, transm. sap, *Myzus ascalonicus*, non-persistent, widespread in shallot, causes a mild chlorotic streaking in leek, occurs in onion, host range limited.

— **yellows v.** Green 1945; Henderson, *Pl.Path.* **2**:130, 1953; Hollings 1968; rod 620–640 nm long, transm. sap, aphid spp., poss. non-persistent, Britain [**24**:133; **33**:401; **48**,1024s].

shallow bark canker (*Erwinia nigrifluens*) European walnut.

shanking (*Phytophthora cryptogea, P. erythroseptica*) tulip.

sharka (plum pox v.) apricot, peach, plum.

sharp eyespot (*Ceratobasidium cereale*) barley, oat, wheat.

shattering (*Bipolaris cactivora*) Easter cactus.

Shear, Cornelius Lott, 1865–1956; born in USA, Univ. Washington, USDA, a notable plant pathologist but best known for his taxonomic classic, with F.E. Clements, *The genera of fungi,*

1931. *Mycologia* **49**:283, 1957; *Phytop.* **47**:321, 1957; *Sydowia* **11**:1, 1957; *Taxon* **6**:7, 1957.
sheath blight (*Thanatephorus cucumeris, T. sasakii*) rice, Gangopadhyay & Chakrabarti, *Rev.Pl.Path.* **61**:451, 1982; Lee & Rush, *PD* **67**:829, 1983; (*T. cucumeris*) sorghum, O'Neil & Rush, *PD* **66**:15, 1982 [**62**,1495].
— **blotch** (*Pyrenochaeta oryzae*) rice.
— **nematode**, *Hemicycliophora arenaria.*
— **rot** (*Cytospora sacchari*) sugarcane, (*Marasmiellus inoderma*) banana, (*Sarocladium attenuatum, S. oryzae*) rice; — **spot** (*Rhizoctonia oryzae*) rice.
shellflower mosaic Verma & Niazi, *Gartenbauwissenschaft* **39**:51, 1974; transm. sap, *Lipaphis erysimi*, India (N.), *Moluccella laevis* [**53**,4828].
shelling, fruit falling prematurely.
shiraiana, *Ciboria.*
shoeflower, see *Hibiscus.*
shoestring, a symptom of extreme narrowing of the leaf, often caused by a virus; a name given to the rhizomorphs of *Armillaria mellea* and hence to the diseases caused by this fungus.
shoot blight (*Pseudomonas syringae* pv. *theae*) tea, (*Sirococcus strobilinus*) conifers.
short life, peach; used for a complex disease syndrome which ends in the death of trees in USA (S.E.). It appears in 3–4 years, particularly on sandy soils or where peach has been grown before, and *c.* spring time. Factors contributing to this collapse are: acid soil, nematode damage by *Criconemella* spp., freeze injury, canker (*Pseudomonas syringae* pv. *syringae*) and a too early pruning, i.e. in autumn; Ritchie & Clayton, Nyczepir et al. and Reilly et al., *PD* **65**:462, 1981; **67**:507, 1983; **69**:874, 1985; **70**:538, 1986, review; effects of *C. xenoplax*; incidence of *Criconemella* spp.; physiological environment, pathogens & culture.
— **spored rust** (*Chrysomyxa ledi* var. *rhododendri*) rhododendron.
shothole, a leaf disease symptom where a limited necrotic lesion falls away to leave a ± circular hole in the lamina; sometimes a sporulating fructification develops around the circumference, e.g. the ascomata of *Microcyclus ulei.*
siccans, *Drechslera.*
sickle leaf, cacao; Zn deficiency; Greenwood & Hayfron, *Emp.J.exp.Agric.* **19**:73, 1951 [**30**:459].
sickness (*Heterodera carotae*) carrot. Also used for a condition in lucerne prob. caused by pythiaceous fungi and the absence of effective *Rhizobium* strs.; in Canada; Damirgi et al. 1976, 1978; Faechner & Bolton, *Can.J.Pl.Sci.* **58**:891, 945, 1978; G.Y. & W.K. Tan 1986, *Rhizobium*

[**58**,1282, 3353; **66**,1515; abs. *Soils Fertilisers* **39**, 7151; **41**,6575].
sida little leaf Hirumi et al., abs. *Phytop.* **64**:581, 1974; assoc. bacteria in the phloem, Puerto Rico, *S. cordifolia.*
— **mosaic** Bird, Sánchez & Lopez-Rosa 1970; Bird et al., *J.Agric.Univ. P.Rico* **55**:461, 1971; transm. *Bemisia tabaci*, Puerto Rico [**51**,4745].
— **yellow vein mosaic** Nair & Wilson, *Agric.Res.J.Kerala* **7**:123, 1969; transm. *Bemisia tabaci*, India (Kerala) [**50**,1136].
siderophores, iron chelates, as factors in germination, infection and aggression of fungus pathogens; Swinburne in Blakeman 1981; Brown & Swinburne, *TBMS* **77**:119, 1981, effects on progressive lesions caused by *Colletotrichum musae* on banana fruit; *Physiol.Pl.Path.* **21**:13, 1982, effects on lesion development caused by *Botryotinia fuckeliana* on broad bean leaves; Lockwood & Schippers, *TBMS* **82**:589, 1984, as a factor in soil mycostasis; Slade et al., *J.Gen.Microbiol.* **132**:21, 1986, *C. acutatum*; Leong, *A.R.Phytop.* **24**:187, 1986, biochemistry & poss. role in control of plant pathogens [**61**,1866; **62**,472; **63**,3767; **65**,4002].
sieve tube necrosis, citrus; a condition that is found in some virus infections, e.g. citrus tristeza v.; in other cases the symptoms are not assoc. with, or caused by, a transm. agent; they are prob. inherited disorders or stock and scion incompatibilities; Allen et al., *Citrograph* **62**:79, 1977, a decline of Frost Nucellar lemon on *Citrus macrophylla* stock in USA (Arizona) [**56**,3568].
Sigatoka (*Mycosphaerella musicola*) banana.
sigla, letters, especially initials, or other characters used to denote words, see acronym. Used universally as abbreviations for plant viruses in scientific papers, e.g. CGMMV, cucumber green mottle mosaic virus; to avoid ambiguity extra letters are sometimes added to the initials, e.g. Ho, hop; Tu, turnip; Tob, tobacco; Tom, tomato. The excessive use of sigla is not always easy to avoid but it can be done and leads to a much more readable script.
sigmoidea, *Nakataea.*
Silberschmidt, Karl Martin, 1903–73; born in Germany, Univ. Ludovica Maximiliana, Munich, Berlin; head plant virology, Instituto Biológico, São Paulo, Brazil; an authority on this subject. *Biológico* **39**:109, 1973; *Arquivos Inst.Biológico* **40**:164, 1973; *Phytop.* **64**:159, 1974.
silky green McWhorter 1954; McWhorter & Cook, *PDR* **42**:51, 1958; this supposed virus was reported from pea in USA (N.W.) that was infected with pea enation mosaic v. The symptoms in broad bean resemble those caused by bean

yellow vein banding v. q.v. Silky green may be similar to BYVBV and depends on the assistor virus PEMV for aphid transm. [**34**:201; **37**:615].

silvering, or chimera, tomato; Grimbly & Thomas and Grimbly, *J.hort.Sci.* **52**:49, 469, 1977; **54**:247, 1979; **56**:65, 1981, chimerical structure, leaf distortion variant, elimination with a cytoplasmic variant [**56**,3232; **59**,1888; **60**,4664].

—— , *Cucurbita* leaves, disorder in summer squash in Israel; distinguished from silver mottling, a genetic characteristic; Paris et al., *Can.J.Pl.Sci.* **67**:593, 1987.

—— (*Curtobacterium flaccumfaciens* pv. *betae*) mangold, red beet.

silver leaf (*Chondrostereum purpureum*) apple, pear, plum; —— **scurf** (*Helminthosporium solani*) potato; —— **top** (*Fusarium poae*) temperate cereals and grasses.

Sinapsis alba, white mustard.

sinensis, *Phytophthora*.

single antibody dot immunoassay, for detection of plant viruses; Graddon & Randles 1986 [**66**,645].

singulare, *Cenangium*.

Sirex, wood wasps, see *Amylostereum*.

Sirococcus Preuss 1855; Coelom.; Sutton 1980; conidiomata stromatic, conidiogenesis enteroblastic phialidic, con. 1 septate, hyaline.

S. clavigignenti-juglandacearum Nair, Kostichka & Kuntz, *Mycologia* **71**:643, 1979; con. 9–17 × 1–1.5 μm; *Juglans cinerea*, butternut canker, first seen in 1967 in USA (Wisconsin) and now a lethal threat to this hardwood throughout its natural range in the eastern and midwestern states; Tisserat & Kuntz, *Can.J.For.Res.* **13**:1139, 1983; *Phytop.* **73**:1628, 1983; *PD* **68**:613, 1984, dispersal gradients of con., longevity of con., disease development [**63**,1986, 5152, 5574].

S. strobilinus Preuss; con. 10–15 × 2–3 μm; shoot blight of *Pinus contorta*, conifers, Europe, N. America; Schneider & Paetzholdt, *NachrbBl.dt.PflSchutzdienst.Stuttg.* **16**:73, 1964; Smith, *PDR* **57**:69, 1973; Wall & Magasi, *Can.J.For.Res.* **6**:448, 1976; correct name is *Sirococcus conigenus* (DC.) P. Cannon & Minter 1983, fide Punithalingam, *Pl.Path.* **37**:162,1988 [**43**,3344; **52**,3092; **56**,5226].

Sirosporium Bubák & Serebr. 1912; Hyphom.; Ellis 1971, 1976; conidiomata hyphal, conidiogenesis holoblastic, conidiophore growth sympodial; con. multiseptate, pigmented.

S. diffusum (Heald & Wolf) Deighton in Ellis 1976; con. 3–18 septate, 20–140 × 3–6 μm; pecan brown leaf spot, may need chemical control; commonly known as the synonym *Cercospora fusca* Rands, *J.agric.Res.* **1**:312, 1914.

sisal, *Agave*; Bock, *Wld.Crops* **17**:64, 1965.

Sitobion avenae, transm. barley yellow dwarf, pearl millet mosaic.

Sitona lineatus, transm. broad bean stain, echtes ackerbohnenmosaik.

skeleton weed, *Chondrilla juncea*.

skin spot (*Polyscytalum pustulans*) potato.

skyrin, toxin; see diaporthin, McCaroll & Thor; Gäumann & Obrist, *Phytopath.Z.* **37**:145, 1960.

sleepy (*Peronosclerospora maydis*, *P. philippinensis*) maize, (*Verticillium albo-atrum*) tomato.

slime (*Pseudomonas solanacearum*) potato.

—— **flux**, wetwood q.v.; a liquid or semi-liquid exudation from the inner bark or wood of large stems of hardwoods, preceded by an injury. It is found mostly in ornamental and shade trees and less often in forest trees. The phenomenon is one of 'bleeding' and results from the internal tree pressures which cause wetwood. If the flux is excessive the exuding sap becomes colonised by micro-organisms and ferments. The flow of this fermented sap then kills tissues around the injury which becomes enlarged; Ogilvie, *TBMS* **9**:167, 1924; Stautz, *Phytopath.Z.* **3**:163, 1931.

—— **spores**, fungus spores that become separated from the spore bearing structures with a slimy coat. They are water dispersed and therefore have epidemiological characteristics that differ fundamentally from dry or air dispersed spores. Spores removed from their substrate by rain splash may be either dry or slimy; Louis & Cooke, *TBMS* **84**:661, 1985, conidial matrix & spore germination in fungus pathogens [**64**,4735].

slippery skin (*Pseudomonas cepacia*) onion.

slip skin maceration (*Rhizopus* spp. assoc.) prune; Sholberg & Ogawa, *Phytop.* **73**:708, 1983 [**62**,4718].

slow decline (*Tylenchulus semipenetrans*) citrus.

—— **rusting**, used for a host's expression of a mostly race non-specific resistance to spp. in the Uredinales; Kulkarni & Chopra, *Z.PflKrankh.PflSchutz.* **87**:562, 1980, *Puccinia* on cereals; Wilcoxson, *Phytop.* **71**:989, 1981, genetics of in cereals; Rashid & Bernier, *Crop Protect.* **5**:218, 1986, *Uromyces viciae-fabae* on broad bean [**60**,1938; **65**,5267].

—— **wilt** (*Erwinia chrysanthemi* pv. *dianthicola*) carnation.

small sclerotial neck rot (*Botryotinia squamosa*) onion; —— **leaf spot** (*Pseudomonas syringae* pv. *tomato*) tomato.

Smith, Erwin Frink, 1854–1927; born in USA, Univ. Michigan, USDA; the founder of plant bacteriology and a famous protagonist in the arguments with the German Alfred Fischer as to whether bacteria could cause disease in plants;

some workers in Europe thought that they did not, *Phytopath.Classics* 13, 1981; wrote: *Bacteria in relation to plant disease*, 3 vol., 1905, 1911, 1914; *Introduction to bacterial diseases of plants*, 1920. *Phytop.* **17**:657, 1927; *J.Bact.* **15**:1, 1928; *A.R.Phytop.* **21**:21, 1983.

—, Kenneth Manley, 1892–1981; born in Scotland, Royal College of Science, Univ. Manchester; Potato Virus Research Station 1927, director 1939, later Plant Virus Research Unit. He had an immense influence on the establishment of plant virology; pioneer work on insect vectors of plant viruses, and insect viruses as Smith trained as an entomologist; FRS 1938, CBE 1956; his books include: *Plant viruses*, 1948, edn. 4, 1974; *A textbook of plant virus diseases*, 1937, edn. 3, 1972; with R. Markham, *Mumps, measles and mosaics*, 1954; *Viruses*, 1962; *Virus–insect relationships*, 1976. *Biogr.Mem.Fel.R.Soc.* **28**:451, 1982; *Adv. Virus Res.* **27**:ix, 1982.

—, Worthington George, 1835–1917; born in England, architect and artist on the staff of the *Gardeners' Chronicle*. He became interested in fungi and the plant diseases they caused and wrote *Diseases of field and garden crops* in 1884. His appearance, or notoriety, in plant pathological history is mostly due to his drawing of 1875 in which he illustrated his claim to have found the oogonia and oospores of the fungus causing potato blight; a claim which de Bary correctly rejected. Awarded 2 gold medals by the Royal Horticultural Society. *TBMS* **6**:65, 1917; *Gardnrs' chron.* **62**:180, 1917; Large 1940.

smoke, toxicity to fungi; Zagory & Parmeter, *Phytop.* **74**:1027, 1984 [**64**,910].

smoulder (*Botryotinia narcissicola*) narcissus.

smudge (*Colletotrichum circinans*) leek, onion.

smut, a disease caused by, and a sp. in, the class Ustilaginomycetes, order Ustilaginales; from the predominant very dark, almost black in mass, colour of the spores.

snow blights (*Phacidium* spp.) conifers.

— **moulds**, low temp. tolerant, fungus pathogens which grow on an unfrozen soil surface below a snow cover. They cause significant diseases of winter cereals, other grasses and some forage legumes; see *Acremonium boreale*, *Coprinus psychromorbidus*, *Myriosclerotinia borealis*, *Typhula* spp.; Jamalainen, *A.R.Phytop.* **12**:281, 1974, crop resistance; Wiese 1987; Smith, *Can.J.Pl.Path.* **3**:15, 1981, diagnosis, culture & pathogenicity; Smiley 1983 [**61**,157].

— **rot** (*Pythium okanoganense*) barley, wheat; — **scald** (*Discosia pini*) conifers.

Snyder, William Cowperthwaite, 1904–80; born in USA, Univ. California Berkeley and Davis,

Wisconsin; an authority on *Fusarium*, extensive work with N. Hansen; also soil environment and disease. *Phytop.* **71**:473, 1981; *A.R.Phytop.* **24**:27, 1986.

soapwort leaf curl Verma, *Gartenbauwissenschaft* **39**:567, 1974; transm. *Bemisia tabaci*, India, *Saponaria* [**54**,3367].

Sobemoviruses, type southern bean mosaic v., other members: cocksfoot mottle v., lucerne transient streak v., rice yellow mottle v., sowbane mosaic v., turnip rosette v., velvet tobacco mottle v.; isometric *c.* 28–30 nm diam., 1 segment linear ssRNA, sedimenting at 112–115 S, thermal inactivation 85–95°, longevity in vitro 28–140 days, transm. sap and beetles; Sehgal in Kurstak 1981; Francki et al., vol. 1, 1985.

socia, *Septoria*.

societies, for plant pathology, Johnston & Booth 1983; for mycology, *Ainsworth & Bisby's dictionary of the fungi*, edn. 7, 1983.

sodic soils, sodicity, containing excess exchangeable Na; see salinity.

soft rot, rotting of tissue, usually parenchyma, by action of a pathogen on the middle lamella of cell walls; cells are separated but retain their identity for a time, after FBPP; (*Ceratocystis paradoxa*) pineapple, (*Choanephora cucurbitarum*) general, (*Erwinia carotovora*) general, (*Phytophthora capsici*) cucurbits, (*P. megasperma*) carrot, (*Rhizopus* spp.) general.

Sogatella, transm. eleusine mosaic, streak and stunt; *S. furcifera* transm. pangola stunt; *S. kolophon* transm. digitaria striate, maize sterile stunt; *S longifurcifera* transm. echinochloa ragged stunt, maize sterile stunt.

Sogatodes cubanus, and *S. oryzicola* transm. rice hoja blanca.

soilborne pathogens, Baker & Snyder 1965; Garrett 1970; Toussoun et al. 1970; Bruehl 1975; Bowen & Rovira, *A.R.Phytop.* **14**:121, 1976, root colonisation; Shipton, *ibid* **15**:387, 1977, monoculture; Schmidt, *A.Rev.Microbiol.* **33**:355, 1979, initiation of root & microbial interactions; Grogan et al., *Phytop.* **70**:361, 1980; Gilligan, *A.R.Phytop.* **21**:45, 1983, modelling; Nicot et al., *Phytop.* **74**:1399, 1984; Campbell & Noe, *A.R.Phytop.* **23**:129, 1985, spatial analysis; Parker et al. ed., *Ecology and management of soilborne plant pathogens*, 1985; Sun & Huang, *PD* **69**:917, 1985, soil amendment & control; Gilligan, *Phytop.* **75**:61, 1985, probability models for host infection by fungi; Lockwood, *ibid* **76**:20, 1986, concepts & connections; Bruehl, *Soilborne plant pathogens*, 1987.

soil disinfestation, see control of plant diseases, Mulder.

—— **fungi**, see *Ainsworth & Bisby's dictionary of the fungi*, edn. 7, 1983.

soilless culture, hydroponics, growth of crops in a water nutrient film; Graves, *Hortic.Rev.* **5**:1, 1983; disease problems: Jenkins & Averre, *PD* **67**:968, 1983; Bates & Stanghellini and Stanghellini et al., *ibid* **68**:989, 1075, 1984; Gold & Stanghellini, *Phytop.* **75**:333, 1985; Lane & Eddy, *Z. PflKrankh. PflSchutz.* **92**:417, 1985; Paludan, *Tidsskr.PlAvl.* **89**:467, 1985; Price & Fox, *Aust.J.agric.Res.* **37**:65, 1986 [**63**,498; **64**,1327, 2203, 5168; **65**,186, 2223, 5777].

soil rot (*Streptomyces ipomoeae*) sweet potato.

sojae, *Phomopsis*; **sojina**, see *Cercosporidium sojinum*; **sojinum**, *Cercosporidium*.

Solai dieback, as for Elgon dieback.

solanacearum, *Pseudomonas*.

Solanasca stevensi, transm. papaya bunchy top.

solani, *Alternaria, Angiosorus, Fusarium, Helminthosporium, Rhizoctonia, Stemphylium*.

Solanum, *S. aviculare*, kangaroo apple; *S. melongena*, aubergine, eggplant; *S. muricatum*, melon pear, pepino; *S. tuberosum*, Irish potato, potato.

solanum apical leaf curling v. Hooker & Salazar, *AAB* **103**:449, 1983; particles *c*.17 × 52 nm of 3 quasi-isometric units in a straight chain, transm. graft only, occurs naturally in potato in Peru (Andes), infected *Datura* and tomato inter alia [**65**,864].

—— **big bud** Marwitz, Petzold & Roth, *Phytopath.Z.* **95**:305, 1979; MLO assoc., Ecuador, *S. marginatum* [**59**,1829].

—— **carolinense v.** Anderson 1960; Gooding & Sun, abs. *Phytop.* **62**:803, 1972; filament *c*.730 × 13 nm, transm. *Myz.pers.*, non-persistent, to burley tobacco, in tomato, USA [**40**:148; **52**,2461].

—— **dulcamara mottle v.** Gibbs et al. 1965–6; see ononis yellow mosaic v., Tymovirus, isometric 25–30 nm diam., transm. sap, seed, *Psylliodes affinis*, related to potato Andean latent v. and OYMV; Beczner et al., *Acta phytopath.Acad.Sci.hung.* **11**:245, 1976, Hungarian str. [**57**,1343].

—— **Indian mosaics**, several authors, *Indian J.expl.Biol.* **5**:120, 1967; **12**:356, 1974; *Indian Phytopath.* **24**:127, 1971; **27**:316, 1974; **28**:209, 1975; *Indian J.Mycol.Pl.Path.* **5**:86, 1975; filament 755 × 15 nm, transm. sap, aphid spp., *S. khasianum, S. torvum* [**51**,2756; **55**,3484; **56**,617, 1686, 2176].

—— **laciniatum bacilliform** Lim, Protsenko & Surgucheva 1980; particles assoc. 250 × 75 nm; Protsenko et al. 1980, MLO assoc.; Thomson, *N.Z.J.agric.Res.* **19**:521, 1976, viruses in *S. laciniatum* [**56**,3158; **60**,2716; **61**,350].

—— **leaf deformation and mottle** (cucumber mosaic v.), Camele et al., *Riv.Patol.Veg.* **22**:17, 1986; Italy; *S. pseudocapsicum*, Jerusalem cherry.

—— **nodiflorum mottle v.** Greber 1965, 1973; Greber & Randles, *Descr.V.* 318, 1986; Sobemovirus, isometric 28–30 nm diam., contains a linear, genomic RNA and a satellite, circular ssRNA, natural infections common in *S. nodiflorum* in Australia (E.), less common in 2 other *Solanum* spp., transm. sap, *Epilachna* spp.

—— **viroid**, as for columnea latent viroid q.v., *S. phureja, S. stenotomum*.

solar heating, or solarisation; a method of heating soil to control plant pathogens by applying polyethylene sheet on the soil surface as for a mulch; it improves crop growth, reduces weeds and is cheap; intensively used in Israel; Katan, *PD* **64**:450, 1980; *A.R.Phytop.* **19**:211, 1981; Porter & Merriman, *Pl.Path.* **34**:108, 1985, in Australia (Victoria); Stapleton & DeVay, *Crop Protect.* **5**:190, 1986 [**64**,3778; **65**,4747].

solidago distortion MacClement & Richards, *CJB* **34**:793, 1956; transm. sap, Canada (Ontario) [**36**:304].

solopathogen, when meiosis fails in the germinating ustilospore of *Ustilago maydis* the sporidia so formed are diploid and contain both mating type genes. Such sporidia can be pathogenic as a single spore and cause infection when inoculated singly. They are called solopathogens, or solopathogenic lines, R. Holliday, *Genetic Res.* **2**:204, 231, 1961. The term may be reasonably used for any fungus that shows a comparable phenomenon.

Sonchus, *S. oleraceus*, sowthistle.

sonchus Argentinian v. Vega et al., *Phytopath.Z.* **85**:7, 1976; bacilliform 270–350 × 45–57 nm, transm. sap, limited host range [**55**,3966].

—— **Indian yellow vein** Reddy & Janardhan, *Curr.Sci.* **44**:596, 1975; transm. *Uroleucon* (*Uromelan*) *sonchi*, considered different from sowthistle yellow vein v. [**55**,3965].

—— **mosaic v.** Qureshi, Naqvi & Mahmood, *Indian Phytopath.* **32**:264, 1979; rod 470–475 × 13–14 nm, transm. sap, *Myz.pers.*, India, *S. asper* [**60**,3465].

—— **yellow net v.** S.R. & R. G. Christie & Edwardson 1974; Jackson & S.R. Christie, *Descr.V.* 205, 1979, Phytorhabdovirus, bacilliform 248–294 nm, transm. sap, *Aphis coreopsidis*, USA, in weeds of Compositae, causes severe stunt in experimentally infected lettuce.

sooty bark (*Cryptostroma corticale*) sycamore; —— **blotch** (*Cymadothea trifolii*) clover.

—— **moulds**, Hughes, *Mycologia* **68**:693, 1976, used the term for any saprophytic fungus, usually with dark hyphae, which forms superficial, brown to black colonies on living plants. They are most noticeable on leaves, are often assoc. with insect secretions and can be detrimental to the plant's physiology; a mildew q.v. that is dark.

— stripe (*Ramulispora sorghi*) sorghum.

Sorauer, Paul Carl Moritz, 1838–1916; born in Germany, one of the main founders of plant pathology, his classic *Handbuch der Pflanzenkrankheiten* 1874, last edn. 1911, is still a standard text since it developed into a modern text of several vol. with contributions from many authors. *Ber.dtsch.bot.Ges.* **34**:(50)–(57), 1916; *Nature Lond.* **96**:600, 1916.

sorbi, *Eutypella*; **sordida**, *Valsa*.

soreshin (*Thanatephorus cucumeris*) cotton, tobacco.

sorghi, *Ascochyta, Cercospora, Claviceps, Gloeocercospora, Peronosclerospora, Puccinia, Ramulispora, Sphacelia, Sporisorium*; **sorghicola**, *Bipolaris, Ramulispora*; **sorghina**, *Phoma*.

sorghum, *Sorghum*; Johnson grass is a form of *S. halepense*; Tarr, *Diseases of sorghum, sudan grass and broomcorn*, 1962; Nishihara, *Bull.natn.Grassland Res.Inst.* **3**:134, 1973, bibliography; *Proc.Int.Workshop, Int.Crops Res.Inst.Semi-arid tropics* 1978, published 1980; Hepperly et al., *PD* **66**:902, 1982, fungicides & seedborne fungi; Frederiksen 1986.

— chlorosis Capoor, Rao & Varma, *Indian J.agric.Sci.* **38**:198, 1968; transm. *Peregrinus maidis*, India (Bombay) [**47**,3438].

— concentric ring blotch Gorter & Klesser, *S.Afr.J.agric.Sci.* **7**:329, 1964; transm. sap to maize and other Gramineae, South Africa (Transvaal) [**44**,690].

— mosaic v. Signoret, *Annls.Phytopath.* **2**:681, 1970; *PDR* **55**:1090, 1971; filament 750 nm long, transm. sap, France (S.) [**51**,319,2432].

— red stripe (sugarcane mosaic v.).

— stunt mosaic v. Mayhew & Flock, *PD* **65**:84, 1981; bacilliform 220 × 68 nm, transm. *Graminella sonora* to maize, sorghum, wheat, USA (California) [**60**,5901].

— xylem limited bacterium Weaver et al., *PD* **64**:485, 1980; related to bacteria of peach phony and Pierce's grapevine leaf scald [**60**,1998].

— yellow freckle Cherian & Kylasam 1937; transm. *Peregrinus maidis*, India (Tamil Nadu) [**17**:169].

— — stunt Zummo et al., *PDR* **59**:714, 1975; MLO assoc., USA [**55**,1775].

sorokiniana, *Bipolaris*; **sororia**, *Crumenulopsis*.

Sorosphaera Schröter 1886; Plasmodiophorales; cysts or resting sporangia united in loose or compact clusters or cytosori, predominately spherical to subspherical or ellipsoidal.

S. vascularum (Matz) M.T. Cook, *J.Dept.Agric.P.Rico* **21**:85, 1937; Karling 1968,

see Plasmodiophoromycetes, placed the organism under the original name *Plasmodiophora vascularum* Matz, *ibid* **4**:45, 1920, which is given as a synonym by Sivanesan & Waller 1986; causes dry top rot of sugarcane in parts of the W. hemisphere; references to the disease apparently cease after 1949 [**16**:409].

Sorosporium Rudolphi 1829; Ustilaginaceae; Mundkur & Thirumalachar 1952; Fischer 1953; sori with well defined columellae and peridia unlike in *Ustilago*, sori mostly in reproductive parts; ustilospores loosely united in powdery balls which are fragile and tend to fragment into single ustilospores at maturity, no sterile cells, germination as in *Ustilago*; Langdon & Fullerton, *Aust.J.Bot.* **23**:915, 1975 [**55**,3947].

S. ehrenbergii Kühn 1877, fide Thirumalachar & Neergaard 1978; Ainsworth, *Descr.F.* 76, 1965, Holliday 1980, as *Tolyposporium ehrenbergii* (Kühn) Pat. 1903; sori in ovaries, mostly 1–3 cm long, 0.5–1 cm wide, spore balls 45–200 μm diam., ustilospores 10–15 μm diam.; sorghum long smut; Africa, Asia; usually not > 10% of the flowers are affected, flowers infected by airborne basidiospores; var. *grandiglobum* Uppal & Patel 1943, has larger spore balls, more spores/ball and ustilospores 9.5–14.5 μm diam., on *Sorghum purpureo-sericeum*.

S. paspali-thunbergii (Henn.) S. Ito 1935; frequently called *Sorosporium paspali* McAlp. 1910; spore balls 30–50 μm diam., ustilospores 9–15 × 8–11 μm; head smut of kodo millet, the inflorescence is transformed into a sorus up to 8 cm long and 0.5 cm wide, cream coloured when young; seedborne; Sattar, *Bull.agric.Res.Inst.Pusa* 201, 1930 [**9**:774].

sour orange, see *Citrus*; — rot (*Geotrichum candidum*) fruits, vegetables, (*Phaeocytostroma sacchari*) sugarcane; — skin (*Pseudomonas cepacia*) onion.

South American leaf blight (*Microcyclus ulei*) rubber.

southern anthracnose (*Colletotrichum trifolii, Glomerella cingulata*) Leguminosae; — bean mosaic v., as given in *Descr.V.* set 1986, see bean southern mosaic v.; — cone rust (*Cronartium strobilinum*) evergreen oak, pine; — leaf blight (*Cochliobolus heterostrophus*) maize; — rust (*Puccinia polysora*) maize; — stem blight (*Corticium rolfsii*) many crops; — sunn hemp mosaic = sunn hemp mosaic v.

sowbane mosaic v. Bennett & Costa 1961; Kado, *Descr.V.* 64, 1971; Sobemovirus, isometric 26 nm diam., transm. sap, seed, prob. insects; causes chlorotic mottling in *Chenopodium* spp., but plants may be symptomless; Francki & Miles, *Pl.Path.* **34**:11, 1985, transm. on pollen surface;

Zebzami et al., *Phytop.* **77**:571, 1987, 3 serotypes [**64**,3777; **66**,4171].

sowthistle, *Sonchus oleraceus.*

—— **yellow vein v.** Duffus 1963; Peters, *Descr.V.* 62, 1971; Phytorhabdovirus, bacilliform *c.* 230 × 100 nm, transm. *Hyperomyzus lactucae*, persistent; Sylvester, *Virology* **56**:632, 1973; *J.econ.Ent.* **71**:17, 1978, effects on aphid & rate of transovarial passage [**53**,2917].

soybean, *Glycine max*; Costa, *Summa Phytopathologica* **3**:3, 1977; Takahashi et al., *Bull.Tohoku Nat.agric.Exp.Stn.* 62, 1980, viruses; Goodman et al., *PD* **65**:214, 1981, international soybean program; Kittle & Gray, *ibid* **66**:213, 1982, chemical control; Sinclair 1982; Koldenhoven et al., *ibid* **67**:1394, 1983, losses in USA in 1982; Calzolari, *Inftore.fitopatol.* **37**(3): 7, 1987 [**61**,6086].

—— **bud blight** (tobacco ringspot v.), Koehler 1944; Allington, *Phytop.* **36**:314, 1946; (tobacco streak v.), Costa et al. 1955; Costa & Carvalho, *Phytopath.Z.* **42**:113, 1961; (cowpea severe mosaic v.), Thongmeearkom et al. 1978; Anjos & Lin, *PD* **68**:405, 1984; these 3 viruses can each cause severe diseases in soybean [**24**:133; **25**:484; **41**:333; **63**,4631].

—— —— **proliferation** Derrick & Newsom, *PD* **68**:343, 1984; MLO assoc., transm. *Scaphytopius acutus*, USA (Louisiana), similar to soybean proliferation [**63**,3662].

—— **calico mosaic** (alfalfa mosaic v.), Almeida et al., *Fitopat. Brasileira* **7**:133, 1982 [**61**,6689].

—— **chlorotic mottle v.** Iwaki et al., *PD* **68**:1009, 1984; prob. Caulimovirus, isometric 50 nm diam., transm. sap, Japan, not serologically related to cauliflower mosaic v. [**64**,1368].

—— —— **spot** Inouye 1969 = soybean mild mosaic v. [**49**,1512].

—— **crinkle v.** Bock, Hollings & Ngugi 1971; filament 750 nm long, E. Africa [**51**,3056h].

—— —— **leaf v.** Iwaki et al., *PD* **67**:546, 1983; poss. Geminivirus, transm. *Bemisia tabaci*, Thailand [**62**,3703].

—— **dwarf v.** Tamada et al. 1969; Tamada & Kojima, *Descr.V.* 179, 1977; Luteovirus, isometric *c.* 25 nm diam., transm. *Aulacorthum solani*, persistent, Japan, dwarfing and yellowing strs., causes: yellows in bean, mild yellowing in pea; infects clover symptomlessly; Thottappilly et al. 1984, transm. *Bemisia tabaci*, persistent, Nigeria; Damsteegt & Hewings, *Phytop.* **77**:515, 1987, factors affecting aphid transm. in 2 strs. [**64**,2232].

—— **fleck** (tobacco rattle v.), Natsuaki et al., *Ann.phytopath.Soc.Japan* **46**:357, 1980 [**60**,4075].

—— **green vein** Owusu, *Ghana J.agric.Sci.* **4**:201, 1971; transm. sap to cacao, poss. seedborne, Ghana [**51**,4514].

——**Indonesian dwarf v.** Iwaki et al., *PD* **64**:1027, 1980; poss. Luteovirus, isometric *c.*26 nm diam., transm. *Aphis glycines*, persistent, no relationship with soybean dwarf v. [**60**,5182].

—— **mild mosaic v.** Takahashi, Tanaka & Tsuda, *Ann.phytopath.Soc.Japan* **40**:103, 1974; isometric 26–27 nm diam., transm. sap, *Myz.pers.*, seedborne, Japan [**54**,1066].

—— **mosaic v.** Gardner & Kendrick 1921; Bos, *Descr.V.* 93, 1972; Irwin & Schultz, *F.A.O.Pl.Prot.Bull.* **29**:41, 1981; Potyvirus, filament 750 × 15–18 nm, transm. sap, seed, *c.* 30 aphid spp., non-persistent; only a few aphid spp. are important, rather narrow host range, strs. differ in virulence; a major, worldwide disease of soybean; destructive in this crop especially in mixed infections with bean pod mottle v.; Schultz et al., *AAB* **103**:87, 1983, aphid acquisition & transm. [**62**,5032].

—— **proliferation** Granada 1976; *PDR* **63**:47, 1979; Fletcher et al., *PD* **68**:994, 1984; MLO assoc., transm. *Scaphytopius fuliginosus*, disease called machamiento or machismo; Colombia, Mexico [**59**,544; **64**,1369].

—— **rosette** Lo 1966; transm. *Nesophrosyne orientalis*, Taiwan [**46**,235c].

—— **rugose mosaic** Cheo & Tsai, *Acta phytopath.sin.* **5**:7, 1959; seedborne, China (N.) [**39**:138].

—— **stunt** (cucumber mosaic v.), Takahashi et al., *Bull.Tohoku Nat.agric.Exp.Stn.* 62:1, 1980; Hanada & Tochihara, *Phytop.* **72**:761, 1982, properties [**60**,2879; **61**,7234].

—— **witches' broom** Kulkarni & Sheffield 1966; prob. MLO assoc., E. Africa; ;Dhingra & Chenulu, *Curr.Sci.* **52**:603, 1983, transm. *Orosius* sp., India (N.), comments on this diseased condition in Indonesia and Japan [**46**,2941b; **63**,2604].

—— **yellow mottle v.** Bock, Hollings & Ngugi 1971, isometric 25–30 nm diam., E. Africa [**51**,3056h].

—— —— **vein v.** Senboku et al., *Trop.Agric.Res.Ser.* **19**:101, 1986; rod 500–550 × 15–20 nm, transm. sap, Thailand [**66**,3559].

—— **Z v.** Dale & Gibbs, *Intervirology* **6**:325, 1975–6; filament av. 610 nm long, contains RNA, Australia, causes rugose leaves and a mosaic [**56**, 1817].

sparaxis mosaic Smith & Brierley, *Phytop.* **34**:593, 1944; transm. *Myz.pers.*, USA [**23**:488].

sparmannia enation Uschdraweit, *Zentbl.Bakt.ParasitKde.* 2 **123**:348, 1969; *S. africana*, African hemp [**50**,130].

sparsa, *Peronospora.*

spartina mottle v. Jones, *AAB* **94**:77, 1980; poss. Potyvirus, filament 725 × 12 nm, transm. sap, England, *S. angelica* [**59**, 4086].

Spaulding, Perley, 1878–1960; born in USA; Univ. Vermont, Washington; USDA; an authority on forest pathology; his work culminated in 3 major, annotated lists of diseases of forest trees of N. America and the rest of the world: *USDA Handbooks* 100, 1956; 139, 1958; 197, 1961. *Phytop.* **51**:209, 1961.

spear rot, oil palm; Turner 1981, assoc. micro-organisms.

specificity, between parasite or pathogen and host, see recognition; Browder & Eversmeyer, *Phytop.* **76**:379, 1986, specificity, resistance & susceptibility.

specific replant effects, see replant effect.

speck (*Pseudomonas syringae* pv. *tomato*) tomato.

speckle (*Cladosporium musae*, *Veronaea musae*) banana; — **blotch** (*Leptosphaeria avenaria* f.sp. *avenaria*) oat; — **bottom**, potato, Cipar et al., *Pot.Res.* **17**:307, 1974, Zn deficiency, correction with soil fumigation [**54**,1415].

speckled leaf blotch (*Mycosphaerella graminicola*) wheat, (*Septoria passerinii*) barley; — **snow mould** (*Typhula* spp.) temperate cereals and other grasses; — **yellows**, beet, Mn deficiency, Bould et al. 1983.

Spegazzini, Carlos, 1858–1926; born in Italy, went to Argentina in 1880, professor of natural history in Buenos Aires; made a large contribution to the mycology of S. America. *Nature Lond.* **118**:704, 1926; Marchionatto, *Revta Fac.Agron.Univ.nac.La Plata* **25**:11, 1940.

spermatium, see pycniospore.

Spermospora R. Sprague 1948; Hyphom.; conidiomata hyphal, conidiogenesis holoblastic; con. 2–4 septate, solitary, hyaline; apical cell extended into a long, narrow, straight or curved rostrum, an unthickened hilum at the base; Deighton, *TBMS* **51**:41, 1968.

S. bromivora (Latch) Deighton, in Laundon, *N.Z.J.Bot.* **8**:64, 1970; con. usually 1–3 septate, 12–55 × 1.8–2.7 μm; Latch, *N.Z.J.agric.Res.* **8**:960, 1965, as *Cercosporella bromivora* new sp.; damaging *Bromus wildenowii* [**45**,816].

S. lolii MacGarvie & O'Rourke, *Irish J.agric.Res.* **8**:152, 1969; con. 2–6 septate, 40–70 × 3.5–4.8 μm; leaf spot of *Lolium perenne* [**48**,2439].

Sphacelia Lév. 1827; Hyphom.; Carmichael et al. 1980; conidiomata sporodochial; con. aseptate, single, hyaline; teleom. *Claviceps*; *S. sorghi*, teleom. *C. sorghi*.

Sphaceloma de Bary 1874; Coelom.; Sutton 1980; conidiomata acervular, conidiogenesis phialidic enteroblastic; con. usually aseptate, hyaline;

teleom. *Elsinoë*; diseases caused called scabs and most spp. have been little investigated as pathogens; Holliday 1980.

S. ampelinum, teleom. *Elsinoë ampelinum*; **S. australis**, teleom. *E. australis*; **S. batatas**, teleom. *E. batatas*.

S. cardamomi Muthappa, *Sydowia* **19**:145, 1965; con. 4.3–6.5 × 2 μm; cardamom, said to cause severe damage in India (Mysore); Naidu, *J. Plantation Crops* **6**:48, 1978, screening for resistance in India [**58**,820].

S. fawcettii, teleom. *Elsinoë fawcettii*, var. **scabiosa** (McAlp. & Tryon) Jenkins 1936; Sivanesan & Critchett, *Descr.F.* 437, 1974; con. 8–16 × 2–6 μm; Tryon's, or Australian, citrus scab; the con. are longer and the conidiophores more robust than those of *E. fawcettii*.

S. glycines Kurata & Kuribayashi, *Ann.phytopath.Soc.Japan* **18**: 120, 1954; con. 4.7–13 × 2.1–5.6 μm; on soybean; Tasugi & Mogi, *ibid* **23**:159, 1958, hosts [**37**:568; **39**:364].

S. iwatae, **Sphaceloma mangiferae**, **S. manihoticola**, **S. necator**, **S. oleae**, teleom. *Elsinoë iwatae*, *E. mangiferae*, *E. brasiliensis*, *E. veneta*, *E. oleae*, respectively.

S. perseae Jenkins, *Phytop.* **24**:84, 1934; con. hyaline aseptate, often 5–8 × 3–4 μm; other con. septate, often 12–20 μm long reaching 30–35 μm long, pigmented, can be several septate; germination of the larger con. by sprout cells or by germ tubes; avocado scab, may require chemical control; Jenkins, *J.agric. Res.* **49**:859, 1934; Frossard in Kranz et al. 1977 [**13**:386; **14**:459].

S. pirinum, **Sphaceloma rosarum**, **S. theae**, teleom. *Elsinoë piri*, *E. rosarum*, *E. leucospila*, respectively.

S. zorniae Bitanc. & Jenkins 1940; con. 3.5–8.5 × 2–4.5 μm; tropical forage legumes; S. America; Lenné. *PD* **65**:162, 1981, causing defoliation in Colombia [**61**,1266].

Sphacelotheca de Bary 1884; Ustilaginaceae; sori in ovaries bounded by fungus peridium of hyaline cells, over which are ovary wall tissues; dark ustilospores in chains, with disjunctors between them; columella of cells similar to those of the peridium; the components of the sorus develop from mycelium in host tissue at the sorus base; fide Langdon & Fullerton, *Mycotaxon* **6**:421, 1978, who defined the criteria for the genus. The structure and function of the columella are different from those of the columella in *Sporisorium*; sporogenesis in *Ustilago* differs from that in *Sphacelotheca*; Holliday 1980 [**57**,3339].

S. cruenta (Kühn) Potter 1912; Ainsworth, *Descr.F.* 71, 1965; ustilospores av. 7–8 μm diam.; sorghum loose smut; not to be confused with the 2 other sorghum smuts: *Sporisorium reilianum* and *S. sorghi*; in *Sphacelotheca cruenta* the plant is stunted and heading is premature, the membrane ruptures easily, the ustilospores are minutely echinulate and intermediate in size. Primarily seedborne, infection occurring in the germinating seed; but heads can be infected by airborne inoculum; in both cases a systemic infection results; ustilospores have a low survival rate in soil; at least 3 races; a standard seed treatment gives control.

S. destruens (Schlecht.) Stevenson & A.G. Johnson 1944; Ainsworth, *Descr.F.* 72, 1965; ustilospores smooth or slightly punctate, av. 9.6 μm diam.; head smut of common millet, seedborne, control by a seed treatment.

Sphaeriales, Ascomycotina; in part the pyrenomycetes, mostly with ostiolate, perithecioid ascomata, asci unitunicate.

sphaerica, *Nigrospora*; **sphaerioides**, *Drepanopeziza*; **sphaerococca**, *Tilletia*; **sphaeroidea**, *Beniowskia*.

Sphaeronema californicum Raski & Sher 1952; Machon, *Descr.N.* 70, 1975; in roots of *Umbellularia californica*, Californian laurel, USA.

Sphaeropsidales, Coelom.; traditionally used for the group with pycnidia, not now generally accepted.

Sphaeropsis Sacc. 1880; Coelom.; Sutton 1980; conidiomata pycnidial, conidiogenesis holoblastic; con. mostly 1 septate, pigmented, solitary; Sutton considered that *Macrophoma* is a synonym.

S. sapinea, see *Diplodia pinea*.

S. tumefaciens Hedges 1911; Holliday & Punithalingam, *Descr.F.* 278, 1970; con. mostly aseptate, 20–34 × 6–10 μm; knot of citrus and other plants; an old and not well known disease, the knots are growths like galls a few cm diam.; most recently investigated on woody ornamentals, the distribution is uncertain, most reports are from the Americas; Holliday 1980.

sphaeropsoidea, *Phyllosticta*.

Sphaerostilbe Tul. & C. Tul. 1861; = *Nectria*, fide Dingley 1951 and Rogerson 1970; not accepted by Cannon et al. 1985. *S. repens* Berk. & Br. 1873; Booth & Holliday, *Descr.F.* 391, 1973; asco. hyaline to light brown, rough walls, 1 septate, 18–20 × 8–9 μm; conidiophore synnematal; con. hyaline, aseptate, 14–26 × 6–9 μm; stinking root rot, from the characteristic sickly, sour smell of the attack on roots of woody plants in the wet tropics; sometimes called red or violet root rot; conspicuous white rhizomorphs, becoming dark brown to black, under the bark. The fungus can kill woody crops, e.g. tea, rubber and shade trees, but only under poorly aerated, almost waterlogged soil conditions.

Sphaerotheca Lév. 1851; Erysiphaceae; mycelium epiphytic, ascomata with simple appendages, 1 ascus; conidiophores with a straight base; con. catenate, ellipsoidal, with fibrosin bodies; Spencer 1978.

S. fuliginea (Schlecht.) Pollacci 1911; Kapoor, *Descr.F.* 159, 1967; ascomata 66–68 μm diam., asco. 17–22 × 12–20 μm, con. 25–37 × 14–25 μm; the teleom. may be rare and is often absent, it was most recently described by Grand, *Mycologia* **79**:484, 1987; causes the important cucurbit powdery mildew which develops under widely differing conditions; 3 races have been described but races 1 and 2 may be each a complex of races. The large literature was reviewed by Ballantyne, *Proc.Linn.Soc.N.S.W.* **99**:100, 1975; Sitterly in Spencer 1978; Holliday 1980; see *Ampelomyces* and *Tilletiopsis* for biological control; Abiko, *Ann.phytopath.Soc.Japan* **44**:612, 1978, 5 groups based on hosts; Abul-Hayja, *Z.PflKrankh.PflSchutz.* **89**:671, 1982, growth & sporulation on resistant & susceptible cucumber; Sowell, *PD* **66**: 130, 1982, shift from race 2 to race 1 on melon in USA [**59**,186; **61**,5317; **62**,2698].

S. humuli (DC.) Burr. 1887; Royle in Spencer 1978 considered that this is the correct binomial for the fungus causing hop powdery mildew; Cannon et al. 1985 placed it as a synonym of *Sphaerotheca macularis*; *S. humuli* is highly specialised on hop, causing an important disease; the mycelium perennates in dormant, aerial buds of rootstocks, 4 races; Liyanage & Royle, *AAB* **83**:381, 1976, overwintering [**56**,321].

S. macularis (Wallr.) Lind 1913; Mukerji, *Descr.F.* 188, 1968; ascomata 60–125 μm diam., larger than those of *Sphaerotheca fuliginea*, asco. 18–25 × 15–23 μm, con. 25–38 × 15–23 μm; on many plants but physiologically specialised; causes mildews in raspberry and strawberry; Peries, *AAB* **50**:211, 225, 1962, specialised form on the latter crop; the fungus does not cause hop powdery mildew, see Royle above [**42**:35].

S. mors-uvae (Schwein.) Berk & Curt. 1876; Purnell & Sivanesan, *Descr.F.* 254, 1970; asco. 20–30 × 12–15 μm; con. 26–30 × 17–20 μm; American gooseberry mildew; on *Ribes* spp., mostly gooseberry; Corke & Jordan in Spencer 1978.

S. pannosa (Wallr.) Lév. 1851; Mukerji, *Descr.F.* 189, 1968; asco. 20–30 × 12–17 μm, con. 20–35 × 14–20 μm; rose powdery mildew, perennates prob. mostly in dormant buds but not as ascomata; grows well only on young tissue; races, a form on peach; Price, *AAB* **65**:231, 1970, epidemiology & control; Wheeler in Scott &

Bainbridge 1978 and Spencer 1978; Bender & Coyier, *Phytop.* **74**:100, 1984, 5 races; *TBMS* **84**:647, 1985, heterothallism [**49**,2891; **63**,2209; **64**,4975].

Sphaerulina Sacc. 1878; Dothideaceae; Sivanesan 1984; asco. cylindrical to filiform, > 1 septum, septa equidistant.

S. oryzina Hara 1918; anam. *Cercospora oryzae* Miyake 1910; Mulder & Holliday, *Descr.F.* 420, 1974, anam. only; asco. 3 septate, 20–33 × 4–5 μm; con. 0–1 septate, 15–60 × 3–6 μm; rice narrow brown leaf spot; a relatively minor disease but severe losses can be caused; several races; Sutton & Shahjahan, *Nova Hedwigia* **35**:197, 1981, compared the anam. with *Mycovellosiella oryzae*; Estrada et al., *PD* **65**:793, 1981, races in the Philippines; Ou 1985 [**61**,2281].

S. rehmiana Jaap 1907; anam. *Septoria rosae* Desm. 1831; asco. 2–5 septate, 30–80 × 2.5–3.5 μm; con. 3–5 septate, 40–75 × 2–3 μm; rose leaf scorch and surface canker, Europe; Boerema, *Neth.J.Pl.Path.* **69**:76, 1963 [**42**:553].

S. rubi Demaree & Wilcox, *Phytop.* **33**:997, 1943; anam. *Cylindrosporium rubi* Ell. & Morgan 1885, fide Sivanesan 1984; asco. mostly 3 septate, 22–30 × 2.5–3.5 μm; con. 3–9 septate, 32–86 × 2.5 μm; raspberry leaf spot; Europe, N. America; Tsonkovski & Paneva 1980, biology & chemical control in Bulgaria [**23**:135; **60**,6559].

spices, Chattopadhyay 1967; Purseglove et al., *Spices*, 2 vol., 1981; see: betel, black pepper, cardamom, coriander, ginger, red pepper.

spicifer, *Cochliobolus*; **spicifera**, *Bipolaris*.

spike, see sandal; — **blight** (*Clavibacter tritici*) wheat; spiked millet, see millets.

Spilocaea Fr. 1819; Hyphom.; Ellis, 1971, 1976; conidiomata hyphal, conidiogenesis holoblastic, conidiophore growth percurrent; con. 1 septate, hyaline or pigmented.

S. eriobotryae (Cav.) S. Hughes, *CJB* **31**:563, 1953; con. 6–20 × 6–7 μm, fide Ferraris, *Annals. Mycol.* **7**:283, 1909; loquat black spot or scab; Raabe & Gardner, see *Spilocaea photinicola*, considered that the fungus is same as *S. pyracanthe*; occasional reports of the need for chemical control.

S. oleaginea (Castagne) S. Hughes, as above, page 564; con. mostly 21–26 × 10–12 μm; olive peacock's eye, causes leaf fall and may warrant chemical control but little experimental work on etiology seems to have been done; Tenerini & Loprieno, *Phytopath.Z.* **39**:101, 1960; Assawah, *Phytopath.Mediterranea* **6**: 144, 1967; Chen & Zhang, *Acta phytopath.sin.* **13**(1):31, 1983 [**40**:236; **47**,3515; **62**,3921].

S. photinicola (McClain) M.B. Ellis 1976; con. 0–1

septate, 22–29 × 8–11 μm; the name was cited as = *Spilocaea pyracanthae* by Raabe & Gardner, *Phytop.* **62**:914, 1972; but the con. are larger and often 1 septate; the av. length of con. given by McClain, *Phytop.* **15**:178, 1925, i.e. 22 μm, exceeds that of the largest con. in *S. pyracanthae*, see Ellis 1976; scab, on leaves and berries of *Heteromeles arbutifolia* [**4**:556; **52**,1563].

S. pomi, teleom. *Venturia inaequalis*.

S. pyracanthae (Otth) v. Arx 1957; con. aseptate, 14–21 × 7–10 μm; scab of *Pyracantha coccinea*, fire thorn; smaller con. than in *Spilocaea photinicola* q.v.

spinach, *Spinacea oleracea*; Sumner et al., *Phytop.* **66**:1267, 1976, root diseases; Bailiss & Okonkwo, *J.hort.Sci.* **54**:289, 1979, viruses [**56**,2715; **59**,3450].

— **blight** (cucumber mosaic v.), Smith 1972.

— **chlorotic spot** Gathuru & Bock 1974; transm. *Myz. pers.*, E. Africa [**54**,3094h].

— **latent v.** Stefanac 1978; Bos, *Descr.V.* 281, 1984; Ilarvirus, isometric *c.* 27 nm diam., transm. sap, frequently seedborne in many plants with no symptoms, prob. prevalent and of practical importance.

— **mosaic v.** Naqvi & Mahmood, *Indian Phytopath.* **28**:268, 1975; *Indian J.Pl.Path.* **3**:152, 1985; particle 660–670 nm long, transm. sap, *Myz.pers.* [**56**,1789; **65**,3033].

— **stunt v.** Halliwell et al., *Misc.Publs.Texas agric.Exp.Stn.* 1525, 1982; isometric particle, transm. sap to 13 spp. in Chenopodiaceae and Cucurbitaceae [**63**,2543].

— **temperate v.**, a cryptic v. q.v. and see beet temperate v., Natsuaki et al.

— **tobravirus** Bailiss & Okonkwo see spinach; rods 115 and 209 nm long.

— **witches' broom** Schmutterer & Lorra and Nienhaus & Schmutterer, *Z.PflKrankh.PflSchutz.* **83**:295, 641, 1976; bacterium assoc., transm. *Piesma quadratum*, Germany [**56**,442, 3750].

— **yellow dwarf v.** Severin & Little, *Hilgardia* **17**:555, 1947; Thomas et al., *Phytop.* **63**:538, 1973; rod 250 × 15 nm, transm. sap, *Myz.pers.*, USA (California, Texas) [**27**:345; **53**,305].

— — **mottle and spot** (tobacco rattle v.), Robinson et al. 1979; Kurpa et al., *AAB* **98**:243, 1981 [**59**,4002; **60**,6698].

Spinacia oleracea, spinach.

spindling sprout, potato; sprouts *c.* a quarter the diam. of normal ones, due to a lack of tuber vigour; in Israel, etiology unknown; Marco, *Phytopath.Mediterranea* **20**:181, 1981 [**63**,797].

spiral nematode, *Helicotylenchus dihystera*.

*****Spiroplasma** Saglio et al. 1973; Spiroplasmataceae; Bradbury 1986; cells pleomorphic, filaments

helical to branched or not helical, sometimes
spherical or ovoid; helical forms in vitro usually
3–5 μm long, 100–200 nm diam.; motile,
sometimes with rotary movements; no organelles
of locomotion, but intracellular fibrils present,
binary fission, facultatively anaerobic, cholesterol
required for growth; serotypes from ticks, insects
and plants, phloem or surface. In some forms
Spiroplasma appears indistinguishable from
organisms like MLO; but see Townsend,
J.gen.Microbiol. **129**:1959, 1983, for an antigen
specific for *Spiroplasma*. It is unlikely that the
MLO, that have not been grown in vitro, are
related to this genus; see major texts and
references under Mollicutes and
Mycoplasmataceae; Razin, *Microbiol.Rev.* **42**:414,
1978; Whitcomb, *A.Rev.Microbiol.* **34**:677, 1980;
A.Rev.Entomol. **26**:397, 1981; Markham &
Townsend, *Sci.Prog.Oxf.* **67**:43, 1981; Daniels,
A.R.Phytop. **21**:291, 1983.
S. **citri** Saglio et al.; causes citrus stubborn and
horseradish brittle root; infects many herbaceous
plants and exists as different serotypes. It was the
first plant mollicute to be grown in vitro and the
first one to be shown to cause disease in plants.
Citrus stubborn was reported from USA
(California) in 1915; in 1969 c. a million trees
were affected. The disease is widespread in the
Mediterranean countries. Trees become stunted,
leaves are small and show chlorotic mottling, fruit
is abnormally shaped and seeds abort. The
optimum temp. for in vitro growth is 32°;
symptoms are more severe at day/night temps of
35°/27° than at 27°/23°; transm. by *Macrosteles
fascifrons, Neoaliturus tenellus, Scaphytopius
acutus, S. nitridus*; Liu et al. *Phytop.* **73**:582, 585,
1983; O'Hayer et al., *AAB* **102**:311, 1983. Brittle
root was reported from USA c.1936 but its
etiology was not known for > 40 years. In Illinois
there have been epidemics; Raju et al. and
Fletcher et al., *Phytop.* **71**:1067, 1073, 1981;
reviews: Wallace in Reuther et al. 1978; Gumpf
& Calavan in Maramorosch & Raychaudhuri
1981; Bové, *F.A.O.Pl.Prot. Bull.* **34**:15, 1986; also
Archer et al., *Pl.Path.* **31**:299, 1982, detection by
Elisa; Fletcher, *Phytop.* **73**:354, 1983, distribution
in USA [**61**,2595–6; **62**,2520, 3189–90, 3808–9].
S. **cocos** Eden-Green & Waters, *J.gen.Microbiol.*
124:263, 1981; Eden-Green et al., *AAB* **102**:127,
1983; isolated from coconut but no evidence that
this organism causes palm lethal yellowing; prob.
another serotype of the *Spiroplasma citri* group
[**61**,352; **62**,2077].
S. **kunkelii** Whitcomb et al., *Int.J.syst.Bact.* **36**:170,
1986; causes maize stunt, Mesa Central or Rio
Grande maize stunt. An important disease in the

lowlands of Central and S. America; present in
USA (S.). A chlorosis is followed by a change
towards red at the tops of older leaves; successive
leaves show chlorotic spots which become stripes;
plants are stunted and bear many small ear
shoots. *Dalbulus maidis* is the most important
vector. Others, some experimental, are *D.
elimatus, Euscelidius variegatus, Graminella
nigrifrons, Stirellus bicolor*. Teosinte and *Zea* spp.
are infected; Davis et al., *Science N.Y.* **176**:521,
1972; Chen & Liao, *ibid* **188**:1015, 1975, Madden,
Nault et al., *Phytop.* **70**:659, 1980; **73**:1608, 1983;
74:977, 1984; *AAB* **105**:431, 1984; Alivizatos,
Phytopath.Z. **110**:148, 1984; Alivizatos &
Markham, *AAB* **108**:535, 545, 1986 [**51**,3973;
55,4058; **60**,2546; **63**,1748, 5332; **64**,160, 3834;
65,3866, 5987–8].
S. **phoeniceum** Saillard et al., *Int.J.syst.Bact.*
37:106, 1987; plant pathogenic, in Syria from
Catharanthus roseus, periwinkle; experimental
transm. to periwinkle using *Macrosteles fascifrons*.
Spiroplasmataceae, Mollicutes; cells helical during
logarithmic growth; with rotary, flexional and
translational motility; see major texts and
references under Mollicutes, Mycoplasmataceae,
Spiroplasma.
splashmeter, an apparatus for measuring upward
splash; use in cereals infected by *Leptosphaeria
nodorum, Mycosphaerella graminicola,
Rhynchosporium secalis*; Shaw, *Pl.Path.* **36**:201,
1987 [**66**,4185].
splendens, *Pythium*.
split seed, lupin, related to the effect of the gene
iucundus; Perry & Gartrell, *J.Agric.West.Aust.*
17:20, 1976 [**56**,276].
Spongospora Brunchorst 1887;
Plasmodiophoromycetes; Waterhouse in
Ainsworth et al. IVB, 1973; cysts mostly united in
a loose or compact cytosorus which is like a
sponge, obovoid, subspherical or irregular; traversed
by canals and fissures; obligate endoparasites,
often inducing hypertrophy in infected cells.
S. **subterranea** (Wallr.) Lagerh. 1891 f.sp. *nasturti*,
specific to watercress causing crook rot; transm.
watercress chlorotic leaf spot; Tomlinson *TBMS*
41:491, 1958; *NAAS q.Rev.* 49:13, 1960 [**38**:286].
S.s.f.sp. **subterranea** Tomlinson 1958; Hims &
Preece, *Descr.F.* 477, 1975; spore balls 19–85 μm
diam., resting spores or cysts 3.5–4.5 μm diam.;
potato powdery scab. The resting spores, which
remain viable in soil for c. 6 years, form primary
zoospores which infect root hairs. The
multinucleate plasmodium then formed produces
zoosporangia and secondary zoospores; secondary
zoospores are released in the cortex and a
secondary plasmodium forms a spore ball from

which resting spores reach the soil. Secondary zoospores may reinfect the host to form further motile spores. Powdery scab is a relatively minor disease and only severe under cool, wet soil conditions. Control measures include planting healthy tubers, long rotations and resistance. An important side effect of the fungus is that it transm. potato mop top v.; Hooker 1981; Rich 1983.

spongy rot (*Inonotus hispidus*) temperate hardwoods.

spora, see air spora.

sporangiospore, a fungus spore formed in a sporangium.

sporangium, a fungus structure which forms endogenous, asexual spores.

spore, of a fungus, see air spora, longevity of fungi, slime spores, volumetric spore trap; Cochrane in Horsfall & Dimond 1960 vol. 2; Allen and Staples & Yaniv in Heitfuss & Williams 1976, spore germination; *Ainsworth & Bisby's dictionary of the fungi*, edn. 7, 1983; —— **release**, see air spora; —— **trap**, see volumetric spore trap; —— **yield**, Johnson & Taylor, *A.R.Phytop.* **14**:97, 1976.

Sporidesmium Link 1809; Hyphom.; Ellis 1971, 1976; conidiomata hyphal, conidiogenesis holoblastic, conidiophore growth percurrent, con. multiseptate, pigmented, solitary.

S. sclerotivorum Uecker, Ayers & Adams, *Mycotaxon* **7**:276, 1978; con. mostly 6 septate, 60–92 × 6–8 μm, on sclerotia of *Sclerotinia sclerotiorum*; Ayers et al. and Barnett & Ayers, *Can.J.Microbiol.* **27**:664, 685, 1981, in vitro growth; Adams & Ayers, *Phytop.* **70**:366, 1980; **71**:90, 1981; **72**:485, 1982, destruction of sclerotia of *S. minor*, *S. sclerotiorum* & *Sclerotium cepivorum*; Bullock et al. and Adams, *ibid* **76**:101, 1986; **77**:575, 1987, haustorial formation in sclerotia & comparison of isolates [**60**,3041, 5736; **61**,2714–15, 6718; **65**,4752].

Sporisorium Ehrenb. in Link 1825; Ustilaginaceae; the characteristics of the columella differ from that in *Sphacelotheca*, q.v. Langdon & Fullerton.

S. reilianum (Kühn) Langdon & Fullerton, *Mycotaxon* **6**:452, 1978; Ainsworth, *Descr.F.* 73, 1965; as *Sphacelotheca reiliana* (Kühn) Clinton; sorghum and maize head smut; differs from *Sporisorium sorghi* and *Sphacelotheca cruenta*, both sorghum smuts, in the conspicuously echinulate ustilospores which are also larger, 9–12 μm diam. In head smut the plant is not stunted and heading is normal; the sorus, rupturing easily, is in the inflorescence which may be completely replaced by the smut; seedling infection from soilborne inoculum becomes systemic. There are 2 groups of races one infecting maize and one

infecting sorghum. Seed chemical treatment may give only partial control since infection can occur after germination in infested soil; Holliday 1980 as *Sphacelotheca reiliana*; Stromberg et al. *PD* **68**:880, 1984, expression of disease & resistance in maize; Matyac, *Phytop.* **75**:924, 1985, colonisation of maize; Matyac & Kommedahl, *ibid* **76**:487, 1986, ustilospore survival in soil; Stienstra et al., *PD* **69**:301, 1985, chemical control; Frederiksen 1986 [**64**,587, 4278; **65**,682, 5992].

S. sorghi Ehrenb. in Link 1825, fide Langdon & Fullerton 1978; Ainsworth, *Descr.F.* 74, 1965, as *Sphacelotheca sorghi* (Link) Clinton; sorghum covered smut, the ustilospores are minutely echinulate, 6–7 μm diam., smaller than those of *Sporisorium reilianum* and *Sphacelotheca cruenta*. In covered smut the plant is not stunted and heading is normal, the sorus membrane is rather permanent. One of the most important diseases of sorghum and serious where seed treatment is not used; infection takes place between sowing and emergence; there are at least 8 races; for the smuts on sorghum see Tarr, *Diseases of sorghum, Sudan grass and broom corn*, 1962; Holliday 1980 as *Sphacelotheca sorghi*; Frederiksen 1986; Ingold, *TBMS* **87**:474, 1986, the basidium.

sporodochium, a conidioma where the spore mass is supported by a superficial, pulvinate mass of short conidiophores.

sporostasis, inhibition of fungus spore germination; Robinson et al., *TBMS* **51**:113, 1968 [47,2052].

sporotrichioides, *Fusarium*.

spot anthracnose, adopted by Jenkins, *PDR* **31**:71, 1947, as a name for diseases caused by spp. of *Elsinoë* and *Sphaceloma*; these diseases are more usually called scabs; —— —— (*E. rosarum*) rose; —— **blotch** (*Cochliobolus sativus*) temperate cereals.

Sprague, Roderick, 1901–62; born in USA, Univ. Washington State, Cincinnati, Wisconsin; USDA 1936–47; Tree Fruit Experiment Station, Wenatchee; a noted authority on graminaceous fungi, wrote *Diseases of cereals and grasses in North America (fungi except smuts and rusts)*, 1950. *Mycologia* **54**:587, 1962.

spray, to apply water, or chemicals for disease control, a suspension or solution dispersed as droplets. The physical characteristics of a spray depend in part on the type of propellant, air or water, and droplet size. Sprays are applied to crops at different volumes/area, e.g. for high volume 60–100 gallons/acre to very low volumes 5–20 gallons. Spray droplets may be coarse, median diam. > 200 μm to a mist with droplets 50–100 μm diam.; see FBPP for assoc. terms.

spreader, or wetter, a substance added to a spray to

reduce surface tension of the applied droplets, and to help even spread and distribution over the target.

spreading decline (*Radopholus similis*) citrus.

spring beauty latent v. Valverde 1984; *Phytop.* **75**:395, 1985; Bromovirus, isometric 28 nm diam., transm. sap, USA (Arkansas), *Claytonia virginica* [**64**,5283].

— **blackstem** (*Didymella lethalis*, *Phoma medicaginis* var. *medicaginis*) clovers, lucerne; —
dead spot (*Cochliobolus spicifer*, *Leptosphaeria korrae*, *L. narmari*) Bermuda grass, turf grasses; — **dwarf** (*Aphelenchoides fragariae*) strawberry.

spruce, *Picea*.

— **deformation** Blattný 1948; Smolak 1948; Čech et al., *Phytop.* **51**:183, 1961; rod av. 625 nm long assoc., Europe [**40**:637].

— **forest decline**, Ebrahim-Nesbat & Heitefuss, *Eur.J.For.Path.* **15**:182, 1985; bacteria in tracheids, *Picea abies*, Germany [**64**,5156].

— **needle rusts** (*Chrysomyxa* spp.), — **bud or shoot rust** (*C. woroninii*).

— **variegated stunt** Černý et al., *Česká Mykol.* **31**:126, 1977; isometric particle assoc., Czechoslovakia [**57**,5712].

spur blight (*Didymella applanata*) red raspberry, *Rubus*.

spurge, *Euphorbia*; Bruckart et al., *PD* **70**:847, 1986, virulence of *Melampsora euphorbiae* on the weed in N. America [**66**,1270].

squamosa, *Botryotinia*, *Botrytis*; **squamosus**, *Polyporus*; **squarrosulus**, *Lentinus*.

squash, see *Cucurbita*, Zucchini.

— **American leaf curl v.** Flock & Mayhew, *PD* **65**:75, 1981; Cohen et al., *Phytop.* **73**:1669, 1983; Geminivirus, pairs isometric particles 38 × 22 nm, transm. *Bemisia tabaci*, persistent, USA (S.W.), *Cucurbita* spp., causes green mosaic and leaf distortion in bean [**60**,6131; **63**,2025].

— **mosaic v.** Freitag 1941; Campbell, *Descr.V.* 43, 1971; Comovirus, isometric *c.* 30 nm diam., transm. sap, seed, beetles; mostly on cucurbits; Lockhart et al., *PD* **66**:1191, 1982 [**62**,1690].

— **Philippines leaf curl** Benigno, *Philipp.Agric.* **61**:304, 1977–8; transm. *Bemisia tabaci*, semi-persistent, seed [**59**,2403].

— **striped mosaic** (papaya ringspot v.).

squirter (*Nigrospora sphaerica*) banana.

stachys stunt v. Pisi, Bellardi & Milne, *Phytopath.Mediterranea* **26**:7, 1987; prob. Potyvirus, particle 720–750 nm long, Italy (N.), *S. tubifera*.

stachytarpheta leaf distortion Wilson & Sathiarajan, *Sci.Cult.* **31**:251, 1965; India (Kerala), *S. indica* [**45**,533].

— **mosaic and rosette** Loos, *Trop.Agriculturist* **98**:8, 1942; Sri Lanka, *S. jamaicensis* [**22**:156].

stackburn (*Alternaria padwickii*) rice.

staghead (*Albugo candida*) turnip rape.

Stagonospora (Sacc.) Sacc. 1884; Coelom.; Sutton 1980; conidiomata pycnidial, conidiogenesis holoblastic; con. multiseptate, hyaline, solitary; Castellani & Germano, *Annali Fac.Sci.agr.Univ.Torino* **10**:1, 1975–7, on Gramineae [**59**,672].

S. arenaria Sacc.; con. mostly 3 septate, 25–60 × 2.5–5 μm; *Stagonospora maculata* (Grove) Sprague is prob. a synonym; purple leaf spot or purple brown blotch, reduces yield and quality of cocksfoot grass; Graham, *Phytop.* **42**:653, 1952, Sherwood and Sherwood et al., *ibid* **72**:146, 1982; **73**:173, 1983, histology, assessment of leaf area attacked; Berg et al. 1986, effects of temp. & photoperiod on resistance [**32**:566; **61**,5052; **62**,3087; **66**,1512].

S. avenae, see *Leptosphaeria avenaria*.

S. curtisii (Berk.) Sacc.; con. 0–5 septate, 4.5–30 × 3–8 μm; narcissus fire or leaf scorch, other temperate, ornamental bulb crops; Moore 1979.

S. foliicola (Bresad.) Bubák 1915, fide Sprague 1950; con. 5–9 septate, 35–80 × 4–6.5 μm; tawny blotch of reed canary grass, *Phalaris arundinacea*; Zeiders, *PDR* **59**:779, 1975, biology; Zeiders & Sherwood, *Crop.Sci.* **17**:594, 651, 1977, cultural effects on, & reactions of, host genotypes [**55**,2772; **57**,1761–2].

S. nodorum, see *Leptosphaeria nodorum*.

S. sacchari Lo & Ling 1950; Sivanesan, *Descr.F.* 776, 1983; con. mostly 3 septate, 36–49 × 8–11 μm; sugarcane leaf scorch, an apparently minor disease, mostly studied in the Philippines and Taiwan.

S. tainanensis, teleom. *Leptosphaeria taiwanensis*.

Stahel, Gerold, 1887–1955; born in Switzerland, Univ. Basel; in Surinam 1914–50, except for a brief period in 1921 as professor of phytopathology at Univ. Agriculture, Wageningen; botanist and plant pathologist. Stahel was prob. the most outstanding worker in his field in tropical America at the time. His monographs on *Crinipellis perniciosa* and *Microcyclus ulei* are meticulous models of research; both fungi still limit the production of cacao and rubber, respectively, in S. America. His work on coffee phloem necrosis, assoc. with protozoa, was neglected by others for *c.* 35 years, see *Phytomonas*; publications and obituary in *De West-Ind.Gids* 36, 1955; H. Vermeulen, personal communication, 1985.

Stakman, Elvin Charles, 1885–1979; born in USA,

Univ. Minnesota, director USDA Federal Cereal Rust Laboratory; classical studies on *Puccinia graminis* var. *tritici*; Emil Hansen Gold Medal and Prize; wrote, with J. G. Harrar, *Principles of plant pathology*, 1957. *Phytop.* **69**:195, 1979; *Kavaka* **7**:93, 1979; *Bull.Br.mycol.Soc.* **14**:152, 1980; *A.R.Phytop.* **22**:11, 1984; biography by C.M. Christensen 1984.

stalactiform blister rust (*Cronartium coleosporioides*) pine.

stalactiforme, *Peridermium.*

stalk break (*Sclerotinia sclerotiorum*) potato; — **rot** (*Diplodia maydis, Gibberella fujikuroi, G. zeae*) maize, (*G. fujikuroi*) sorghum, (*Phytophthora megasperma*) soybean; — **smut** (*Urocystis occulta*) rye.

stamen blight (*Hapalosphaeria deformans*) *Rubus.*

stand opening (*Inonotus tomentosus*) white spruce.

Stanley, Wendell Meredith, 1904–71; born in USA, Univ. Illinois, California Berkeley, Rockefeller Institute, Princeton; shared the Nobel Prize for chemistry in 1946 for the 'preparation of virus protein in a pure form' with J.B. Sumner and J.H. Northrop, fide Waterson & Wilkinson 1978. *Nature Lond.* **233**:149, 1971.

star grass, see Bermuda grass.

statice Y. v. Lesemann, Koenig & Hein, *Phytopath.Z.* **95**:128, 1979; poss. Potyvirus, filament 748 nm long, transm. sap, Germany, *Limonium sinuatum* [**59**,1283].

Stecklenberg (prunus necrotic ringspot v.) *Prunus* spp.

Stegophora H. Sydow & Sydow 1916; Gnomoniaceae; fide Barr 1978; ascomata usually single in substrate, stromatic tissues forming a clypeus, asco. ellipsoid, septum near base.

S. ulmea (Schwein.:Fr.) H. Sydow & Sydow; sometimes as *Gnomonia ulmea* (Schwein.:Fr.) Thüm.; the anam. was originally *Gloeosporium ulmeum* Miles, *Bot.Gaz.* **71**:161, 1921, and later transferred to *Cylindrosporella = Asteroma* fide Sutton 1977; the name is now *A. ulmeum* (Miles) B. Sutton 1980; asco. 10–13.5 × 3–6 μm; both macro- and micro-con. were described by McGranahan & Smalley, *Phytop.* **74**:1296, 1300, 1984; elm black spot, widespread in N. America where it causes an important leaf disease; Pomerleau 1938; Morgan-Jones, *Svensk.Bot.Tidskr.* **52**:363, 1958 [**18**:146; **64**, 1280–1].

stellaria ribbon leaf Schwarz, *Z.PflKrankh.PflPath.PflSchutz.* **66**:86, 1959; transm. sap, Germany, *S. media* [**38**:503].
— **yellows** Begtrup, see cirsium yellows.

stem and bulb nematode, *Ditylenchus dipasci*; —
— **fruit rot** (*Didymella lycopersici*) tomato; —

atrophy, grapevine, assoc. with high temp. at flowering, Theiler 1986; — **bleeding** (*Ceratocystis paradoxa*) coconut; — **blight** (*Alternaria godetiae*) Godetia, (*Diaporthe woodii*) lupin, (*Phytophthora drechsleri*) pigeon pea; — **and branch canker** (*Lepteutypa cupressi*) *Cupressus,* conifers; — **break** (*Alternaria helianthi*) sunflower, (*Colletotrichum curvatum*) sunnhemp; — **burn** (*Pythium deliense*) tobacco; — **canker** (*Diaporthe phaseolorum*) soybean, (*Leptosphaeria coniothyrium*) rose, (*Thanatephorus cucumeris*) bean, Hide et al., *AAB* **106**:413, 423, 1985, potato; — **collapse** (*Erwinia chrysanthemi* pv. *zeae*) cardamom; — **crack,** celery, B deficiency, (*Pseudomonas caryophylli*) carnation; — **end hard rot** (*Phacidiopycnis tuberivora*) potato; — **end rot** (*Alternaria solani*) tomato, (*Botryodiplodia theobromae*) avocado, citrus, (*Diaporthe actinidae*) kiwi fruit, (*D. citri*) citrus, (*Mycosphaerella caricae*) papaya; — **gall** (*Protomyces macrosporus*) coriander; — **necrosis** (see palm lethal yellowing) coconut; — **nematode,** *Ditylenchus angustus* on rice; — **rot** (*Fusarium merismoides*) tomato, (*Magnaporthe salvinii*) rice, (*Phytophthora vignae*) adzuki bean, cowpea; — **rust** (*Puccinia graminis*) wheat; — **streak necrosis,** potato, Mn toxicity, Berger & Gerloff, *Am.Pot.J.* **24**:156, 1947; Robinson & Callbeck, *ibid* **32**:418, 1955; — **tip necrosis** (soybean mosaic v.) soybean, Tu & Buzzell, *Can.J.Pl.Sci.* **67**:661, 1987 [**27**:40; **35**:389; **65**,5669; **66**,288–9].

Stemphylium Wallr. 1833; Hyphom.; Ellis 1971, 1976; conidiomata hyphal, conidiogenesis enteroblastic, monotretic, terminal percurrent; con. muriform, pigmented, solitary; distinguished from *Alternaria,* with which it has been confused, by the percurrent q.v. growth; some teleom. in *Pleospora,* see *Ulocladium,* Simmons 1967; Wiltshire, *TBMS* **21**:211, 1938, conceptions of the genus; Simmons, *Mycologia* **61**:1, 1969, anam. & teleom.; Holliday 1980; Irwin et al., *Aust.J.Bot.* **34**:281, 1986, Australian isolates [**18**:141; **48**,2220; **66**,2251].

S. cucurbitacearum Osner 1918 = *Leandria momordicae* q.v.

S. drummondii, teleom. *Pleospora drummondii.*

S. herbarum, teleom. *Pleospora herbarum.*

S. loti Graham, *Phytop.* **43**:578, 1953; con. 21–37 × 14–33 μm, on *Lotus corniculatus,* bird's foot trefoil, forms black, erumpent stromata on dead stems [**33**:356].

S. lycopersici (Enjoji) Yamamoto 1960; Ellis & Gibson, *Descr.F.* 471, 1975; con. usually constricted at the 3 major transverse septa, mostly 50–74 × 16–23 μm, length: breadth ratio 3:1 or more; tomato leaf spot, can cause defoliation; there are other hosts, e.g. carnation, chrysanthemum, *Kalenchoë*; Chau & Alvarez, *PD* **67**: 1279, 1983, described a postharvest fruit rot of papaya [**63**,3474].

S. sarciniforme (Cavara) Wiltshire 1938; Booth, *Descr.F.* 671, 1980; Corlet et al., *Fungi Canadenses* 233, 1982; con. sometimes constricted at the median, transverse septum, 30–50 × 22–33 μm; target or pepper spot of clover; one of the sp. of the genus which infects legumes, mostly leaf spots; the relative importance of the diseases caused seems unclear, the other spp. include: *Stemphylium globuliferum* (Vest.) Simmons, *Mycologia* **61**:14, 1969, con. 28–30 × 25–28 μm; *S. herbarum*, *S. loti*, *S. trifolii*; Zeiders, *Phytop.* **50**:757, 1960, *Stemphylium* on legumes; see genus, Irwin et al.

S. solani Weber, *Phytop.* **20**:516, 1930; Ellis & Gibson, *Descr.F.* 472, 1975; con. mostly constricted at the median septum only, 35–55 × 18–28 μm, length: breadth ratio 2:1; tomato grey leaf spot, mostly tropical and subtropical; this sp. differs from *Stemphylium lycopersici* in the con. length: breadth ratio and constriction at only 1 major transverse septum; poss. races; Holliday 1980 [**9**:812].

S. trifolii Graham *Phytop.* **47**:215, 1957; con. 27–50 × 11–21 μm, a greater length: breadth ratio than *Stemphylium herbarum* and *S. sarciniforme* and more distinct constrictions at the transverse septa; on clovers [**36**:593].

S. vesicarium (Wallr.) Simmons, *Mycologia* **61**:9, 1969; *Pleospora allii* (Klotzsch) Ces. & de Not., fide Cannon et al. 1985, may be the teleom.; con. 25–45 × 12–22 μm, length: breadth ratio 2:1; originally described from onion but causes disease in several crops including purple spot of young asparagus; Miller et al., *PDR* **62**:851, 1978, onion leaf blight; Lacy, *PD* **66**:1198, 1982, sand blast of asparagus, infection through wounds; Lamprecht & Knox-Davies, *Phytophylactica* **16**:189, 1984, on lucerne seed; Johnson & Lunden, *PD* **70**:419, 1986, wounding, wetness & asparagus infection; Brunelli et al. 1986, chemical control of pear brown rot; Falloon et al., *Phytop.* **77**:407, 1987, asparagus epidemiology, stomatal penetration; Lowe et al., *Aust.J.Exp.Agric.* **27**:59, 1987, with *Leptosphaerulina trifolii*, damage to lucerne & chemical control [**58**,3578; **62**,1747; **64**,2608; **65**,5034, 5330; **66**,4054].

stemphyloxins, toxins from *Pleospora herbarum*; Barash et al., *Science N.Y.* **220**:1065, 1983; Manulis et al., *J.Phytopath.* **115**:283, 1986 [**62**,3444; **65**,4090].

Stenella H. Sydow 1930; Hyphom.; Ellis 1971, 1976; conidiomata hyphal, conidiogenesis holoblastic, conidiophore growth sympodial, scars usually conspicuous; con. multiseptate, pigmented; similarities with *Cercospora*, *Cladosporium*, *Mycovellosiella*; Deighton, *Mycol.Pap.* 144:42, 1979; Mulder, *TBMS* **79**:469, 1982; see *Ramichloridium*, de Hoog et al.

S. citri-grisea, teleom. *Mycosphaerella citri*.

Stenocarpella Syd. 1917, fide Sutton 1980; Coelom.; conidiomata pycnidial, conidiogenesis enteroblastic phialidic; con. 0–3 septate, commonly 2 celled, pigmented.

S. macrospora (Earle) B. Sutton, *Mycol.Pap.* 141:202, 1977; con. 44–82 × 7.5–11.5 μm; *Stenocarpella maydis* (Berk.) B. Sutton 1980; con. 15–34 × 5–8 μm; Sutton & Waterston, *Descr.F.* 83, 84, 1966, as *Diplodia macrospora* Earle and *D. maydis* (Berk.) Sacc. These 2 fungi are 2 of those causing maize ear and stalk rots, leaf spots and seedling blights. In USA (S.) *S. maydis* is generally considered to be of more economic importance than *S. macrospora*, the lesser pathogen. But work by Latterall & Rossi, *PD* **67**:725, 1983, in Africa and S. America, has shown that *S. macrospora* can be an aggressive pathogen and, unlike *S. maydis*, can attack at all growth stages, and cause losses both in yield and in storage; both spp. can be seedborne; Holliday 1980, *S. maydis* as *D. maydis* [**62**,4885].

stenospila, *Bipolaris*.

Stenotaphrum secundatum, Saint Augustine grass.

stephanotis vein clearing Paludan, *St. plantepatol.Forsøg* 449: 11, 1970; Denmark, *S. floribunda* [**50**,3861].

Stereaceae, Aphyllophorales; thelephoroid, flattened basidiomata; appressed, effused-reflexed, often zoned, lignicolous; Talbot in Ainsworth et al., vol. IVB, 1973; *Amylostereum*, *Chondrostereum*, *Stereum*, *Veluticeps*.

Stereum Hill ex Pers. 1794; decays of temperate trees; Boyce 1961; Hepting 1971; Phillips & Burdekin 1982.

sterilisation, making free from living micro-organisms; see methods and *Ainsworth & Bisby's dictionary of the fungi*, edn. 7, 1983.

sterols, in growth and reproduction of fungi; Hendrix, *A.R.Phytop.* **8**:111, 1970; Elliot, *Adv.Microbial Physiol.* **15**:121, 1977.

Stevens, Frank Lincoln, 1871–1934; born in USA, Rutgers College, Univ. Chicago, Illinois, N. Carolina Agricultural Experiment Station, worked in Puerto Rico; wrote with J.G. Hall, *Diseases of economic plants*, 1910; *The fungi with cause plant disease*, 1913; *Plant disease fungi*, 1925. *Mycologia* **27**:1, 1935.

stevensii, *Botryosphaeria*; **steviae**, *Septoria*.

Stewart, Fred Carleton, 1868–1946; born in USA, Iowa State College, New York Agricultural Experiment Station, a notable authority on potato diseases. *Phytop.* **37**:687, 1947.

stewartii, *Erwinia*.

Stewart's wilt (*Erwinia stewartii*) maize.

sticker, a substance added to a plant protectant to increase its tenacity; one of high viscosity, and used to stick powered seed dressings on to seed or to increase retention on plant foliage.

stigmatomycosis, cotton; see *Eremothecium*, *Nematospora*.

Stigmina Sacc. 1880; Hyphom.; Ellis 1971, 1976; conidiomata sporodochial, conidiogenesis holoblastic, conidiophore growth percurrent; con. several celled to muriform, pigmented.

S. carpophila (Lév.) M.B. Ellis, *Mycol.Pap.* **72**:56, 1959; con. mostly with 3–7 transverse septa and occasionally with 1 or 2 oblique or longitudinal septa, mostly 30–60 × 9–18 μm, narrowing towards the base, pigmented, septa dark; on *Prunus*, causing dieback or shot hole; attacks buds, shoots, leaves and fruits. There are many references to chemical control in E. Europe; Wilson 1937; Atkinson 1971; Highberg & Ogawa, *PD* **70**:825, 828, 1986, chemical control on almond in USA (California) & survival in dormant buds [**17**:256; **66**,1062–3].

S. mori (Nomura) Shirata & Takahashi, *J.Sericultural Sci.Japan* **44**:411, 1975; *Ann.phytopath.Soc.Japan* **44**:190, 1978; mulberry branch blight [**55**,4208; **58**,2326].

stigmina-platani, *Mycosphaerella*.

Stigmina theae, see *Cercoseptoria ocellata*.

stilbenes, diphenylethylenes, mostly in bark, wood and leaves of forest trees; Hart, *A.R.Phytop.* **19**:437, 1981, role in decay & resistance to pathogens.

stilboides, *Fusarium*, *Gibberella*.

stinking root rot (*Sphaerostilbe repens*, a *Nectria*) woody crops; — **smut** (*Tilletia laevis*, *T. tritici*) wheat.

Stirellus bicolor, transm. maize rayado fino v., *Spiroplasma kunkelii*.

stock green petal Gourret, *J. Microscopie* **9**:807, 1970; MLO assoc., *Matthiola* [**51**,184].

stolbur, see potato stolbur.

stolonifer, *Rhizopus*.

stone fruit, see fruits.

Storey, Harold Haydon, 1894–1969; born in England, Univ. Cambridge; Amani Institute, Tanganyika, now Tanzania, 1928–41; this became the East African Agriculture and Forest Research organisation where Storey worked from 1948 to just before his death; directed industrial research in E. Africa in 1939–45. He did fundamental work,

especially on transm., on viruses of cassava, groundnut, maize, sugarcane, tobacco. He saved the tea industry of Nyasaland, now Malawi, from extinction by discovering the cause of tea yellows; one of the most eminent of plant pathologists in the tropics; FRS 1946, CMG 1949. *AAB* **64**:188, 1969; *Biogr.Mem.Fel.R.Soc.* **15**:239, 1969; *Nature Lond.* **222**:905, 1969.

Storey's bark (*Gibberella stilboides*) coffee.

straggling, the untidy criss cross of lodged barley or wheat stems caused by infection by *Pseudocercosporella herpotrichoides*.

strangles, the hypocotyls of singled beet plants become constricted at *c.* soil level and they fall; no particular pathogen is involved. Control is by decreasing exposure at singling or delaying singling until the 4th true leaf stage; Boyd, *Tech.Bull.Edinb.Sch.Agric.* 26, 1966 [**47**,3625].

strawberry, *Fragaria ananassa*; Aerts, *Neth.J.Pl.Path.* **80**:215, 1974, viruses; Wilhelm & Paulus, *PD* **64**:264, 1980, soil fumigation in USA (California); Maas 1984; see *Pentatrichnopus*.

— **band mosaic** Maassen & Németh. *Phytopath.Z.* **42**:57, 1961; Hungary [**41**:239].

— **blossom sterility** Khristov, *C.r.Acad.bulg.Sci.* **2**:241, 1969; Bulgaria.

— **Bulgarian yellows** Kacharmazov 1974 [**55**,4791].

— **chlorotic fleck** Horn & Carver, *PDR* **46**:762, 1962; USA (Louisiana) [**42**:333].

— — **spot** Maassen, *Phytopath.Z.* **57**:138, 1966; Germany [**46**,1671].

— **crinkle v.** Zeller & Vaughan 1932; Sylvester et al., *Descr.V.* 163, 1976; Phytorhabdovirus, bacilliform *c.* 90–380 × 69 nm, transm. *Chaetosiphon* spp., persistent, can be damaging with other viruses in disease complexes; Getz et al., *Phytop.* **72**:1441, 1982, latent period in vector [**62**,1580].

— **eastern vein banding** = strawberry vein banding v.

— **epinasty** Maassen, *Phytopath.Z.* **62**:343, 1968; transm. aphid, persistent, Germany [**48**,535].

— **feather leaf** McGrew 1970, see Frazier 1970 [**51**,3455].

— **green petal** Posnette 1952; *Pl.Path.* **2**:17, 1953; Frazier & Posnette, *AAB* **45**:580, 1957; Chiykowski, *CJB* **40**:1615, 1962; Beakbane et al., *J.gen.Microbiol.* **66**:55, 1971; MLO assoc., transm. *Aphrodes bicinctus*, *Euscelis plebeja*; Chiykowski & Craig, *Can.J.Pl.Sci.* **58**:467, 1978, plant & insect age & transm. [**32**:632; **37**:243; **42**:376; **50**,3912; **57**,5598].

— **latent A and B** = strawberry crinkle v.; latent C, see Maas 1984.

— — **ringspot v.** Lister 1964; Murant, *Descr.V.*

126, 1974; poss. Nepovirus, isometric *c*. 30 nm diam., contains RNA, transm. sap readily, *Xiphinema* spp., Europe (W.), wide host range, seedborne in several plants; infects blackberry; celery, causing strap leaf; cherry, currant, grapevine, narcissus, olive, peach, plum, raspberry, rhubarb, rose; Hicks et al., *Seed Sci. Technol.* **14**:409, 1986, effects on seed & seedlings of *Chenopodium quinoa* & parsnip [**66**,1662].

— leaf roll Berkeley & Plakidas, *Phytop.* **32**:631, 1942; Canada (New Brunswick, Ontario), USA (Maryland, New York); little economic importance [**21**:533].

— lethal decline Schwartze & Frazier, *PDR* **48**:833, 1964; Frazier & Jensen, *Phytop.* **60**:1527, 1970; resemblances to strawberry pallidosis; a disease resembling strawberry lethal decline was induced in strawberry by a Californian str. of peach western X; Canada (British Columbia), USA (Pacific N.W.) [**44**,1176; **50**,1310].

— — yellows Stubbs 1968; Greber & Gowanlock, *Aust.J.agric. Res.* **30**:1101, 1979; 2 types, MLO and bacterium assoc., Australia (Queensland) [**59**,5841].

— little leaf Shanmuganathan & Garrett, *Aust.Pl.Path.Soc. Newsl.* **5**(1) suppl. abs. 83, 1976, MLO assoc., Australia.

— mild yellow edge v. Prentice 1948; Martin & Converse, *Acta Hortic.* 129:75, 1982; *Phytopath.Z.* **114**:21, 1985; Yoshikawa et al., *Ann.phytopath.Soc.Japan* **50**:659, 1984; prob. Luteovirus, isometric 22–25 nm diam., transm. *Chaetosiphon fragaefolii*, persistent; causes severe disease in strawberry with other strawberry viruses [**27**:571; **65**,5614, 6121].

— mosaic and yellow crinkle (arabis mosaic v.).

— mottle Prentice, *AAB* **39**:487, 1952; Smith 1972; transm. sap, aphid spp. [**32**:573].

— necrosis Schöniger, *Phytopath.Z.* **32**:325, 1958; Maassen, *ibid* **41**:271, 1961; transm. sap, *Chaetosiphon fragaefolii*, non-persistent, Germany [**38**:20; **41**:161].

— necrotic shock (tobacco streak v.), Frazier et al. 1962; Stace-Smith & Frazier, *Phytop.* **61**:757, 1971; Johnson et al., *PD* **68**:390, 1984, transm. in strawberry seed [**42**:135; **50**,4010; **63**,4506].

— pallidosis Frazier & Stubbs, *PDR* **53**:524, 1969; Frazier, *ibid* **59**:40, 1975; transm. experimentally by *Coelidia olitoria*, N. America; may contribute to severe degeneration with strawberry crinkle v. and strawberry mild yellow edge v., or in other complexes [**48**,3581; **54**,3410].

— pseudo-mild yellow edge v. Frazier, *Phytop.* **56**:571, 1966; Yoshikawa & Inouye, *Ann.phytopath.Soc.Japan* **52**:643, 1986; poss. Carlavirus, filament 625 × 12 nm, transm.

Chaetosiphon fragaefolii, *Aphis gossypii*, in USA transm. was said to be semi-persistent [**45**,2914; **66**,1541].

— stunt Zeller & Weaver, *Phytop.* **31**:849, 1941; transm. *Chaetosiphon fragaefolii*, USA (Pacific N.W.); status dubious, not seen since 1941, fide Maas 1984.

— vein banding v. Frazier 1955; Frazier & Converse, *Descr.V.* 219, 1980; Caulimovirus, isometric 40–50 nm diam., transm. *Chaetosiphon* spp., semi-persistent, widespread, damaging only when certain other viruses are present.

— vein chlorosis = strawberry crinkle v.

— — clearing Canova & Tacconi, *Phytopath.Mediterranea* **4**:31, 1965; transm. sap, soil, Italy [**45**,2179].

— — necrosis Stingle & King, *Phytop.* **55**:1269, 1965; Maassen, *Phytopath.Z.* **62**:343, 1968; transm. sap, *Chaetosiphon fragaefolii*, semi-persistent; Germany, USA (Minnesota) [**45**,1108; **48**,535].

— witches' broom Zeller, *Phytop.* **17**:329, 1927; Huhtanen & Converse, *ibid* **61**:1137, 1971; Shiomi & Sugiura, *Ann.phytopath.Soc.Japan* **49**:727, 1983; MLO assoc., USA; and in Japan, transm. *Macrosteles fascifrons*, *M. orientalis* [**6**:673; **51**,1668; **63**,4507].

— yellow vein banding (strawberry vein banding v.).

— yellows diseases, see Maas 1984.

strawbreaker (*Pseudocercosporella herpotrichoides*) wheat, other temperate cereals, grasses; American usage; called eyespot in England; preferred usage here is true eyespot.

streak, a disease characterised by elongate lesions or areas of discolouration, usually of limited length, on leaves with parallel venation or on stems, cf. stripe, FBPP; — (*Xanthomonas campestris* pv. *translucens*) barley.

streptanthera mosaic, as for sparaxis mosaic.

Streptomyces Waksman & Henrici 1943; Actinomycetales; Bradbury 1986; well developed, branched, coenocytic mycelium, not readily fragmenting, usually < 1 μm wide; aerial mycelium forming chains of round, oval or cylindrical spores; Gram positive; distinguished from most fungi by the smaller mycelial diam. and sensitivity to antibacterial antibiotics but not to antifungal ones. For plant pathologists the most serious and widespread disease caused is potato common scab. The pathogen is often given as *S. scabies* (Thaxter) Waksman & Henrici 1948; but for taxonomic reasons this name is excluded from the Approved Lists of Bacteria and is therefore unacceptable. Bradbury accepted 18 spp. several of which are said to cause common scab;

there has been long disagreement as to whether the disease is caused by one or several *Streptomyces* spp. The severe symptoms on potato tubers are deep pits and raised excrescences which can cause severe losses. Other tuber crops are attacked, e.g. beet, carrot, mangold, parsnip, radish. *S. ipomoeae* (Person & Martin) Waksman & Henrici causes sweet potato scab. The disease has been called: soil rot, ground rot, pit or pox; feeding roots can be destroyed and the plants become stunted, dark lesions on the tubers; see potato; Trehan & Grewal 1980, cultural control of potato scab; Bradbury, *Descr.F.* 697, 1981, *S. ipomoeae*; Levick et al., *Phytop.* **75**:568, 1985, radish scab [**62**,369; **64**,5176].

stress, predisposition q.v.; effects of external environment on plants and disease; Yarwood in Horsfall & Dimond, vol. 1, 1959; *Phytop.* **63**:451, 1973, water; Schoeneweiss, *A.R.Phytop.* **13**:193, 1975; *PD* **65**:308, 1981, woody plants; Ayres, *A.R.Phytop.* **22**:53, 1984; Houston 1984; Blum 1985; Jackson, *ibid* **24**:265, 1986, remote sensing [**63**,3563; **64**,4670].

striatus, *Uromyces*; **strictum**, *Acremonium*.

Striga, see root parasitic weeds.

striiformis, *Puccinia, Ustilago*; **striispora**, *Nectria*.

stringy root (*Aphanomyces cochlioides*) beet.

stripe, a disease characterised by elongate lesions or areas of discolouration, of indefinite length, on leaves with parallel venation or on stems, cf. streak, after FBPP; —— (*Ustilago striiformis*) temperate grasses; —— **blight** (*Pseudomonas syringae* pv. *striafaciens*) oat; —— **canker** (*Phytophthora cinnamomi*) cinnamon; —— **rust** (*Puccinia striiformis*) barley, wheat, temperate grasses.

strobi, *Peridermium*; **strobicola**, *Leptostroma*; **strobilinum**, *Cronartium*; **strobilinus**, *Sirococcus*.

Stromatinia (Boud.) Boud. 1907; Sclerotiniaceae; stroma forms a thin crust on bulbs, corms or rhizomes, black microsclerotia present in host, but the ascomata do not arise from them but from the stroma on the host tissue.

S. gladioli (Drayton) Whetzel 1945; ascomata disks 3–7 mm diam., asco. 10–17 × 5.5–9.5 μm; gladiolus corm dry rot, prob. confined to the Iridaceae; minute sclerotia are just visible on leaf bases after the old corm has rotted; spread through the sclerotia which can survive in undisturbed soil for > 5 years and up to 10 years; Gould, *PDR* **42**:1011, 1958; Moore 1979; Jeves & Coley-Smith, *TBMS* **67**:419, 1976; **74**:13, 1980; persistence in soil, spread & sclerotial germination [**38**:148; **56**,4078; **59**,4208].

S. serica (Keay) Kohn, *Phytop.* **69**:886, 1979; asco. av. 20.5 × 10 μm, black sclerotia 2–10 × 2–5 mm;

on *Gypsophila*; Keay, *J.Bot.Lond.* **75**:130, 1937; *AAB* **26**:227, 1939 [**16**:679; **18**:628].

structures of viruses, Harrison in Mahy & Pattison ed., *Sympos.Soc.Gen.Microbiol.* 36, part 1, 1984.

stubborn (*Spiroplasma citri*) citrus.

stubby root nematodes, *Paratrichodorus*.

stub dieback (*Gibberella zeae*) carnation, Nelson et al. *Phytop.* **65**:575, 1975 [**54**,4954].

stump rot (*Phytophthora nicotianae* var. *parasitica*) lily, Moore 1979.

stunt (*Clavibacter xyli* ssp. *cynodontis*) Bermuda grass, (*Glomus*) tobacco, (*Thanatephorus cucumeris*) barley, Murray, *TBMS* **76**:383, 1981 [**61**,198].

stunted greening, beet, P deficiency; Whitney & Duffus 1986.

stunting of dodder Giannotti et al. 1970, 1974; bacterium assoc., fide Hopkins, *A.R.Phytop.* **15**:277, 1977.

stylar end ring rot (*Phomopsis psidii*) guava.

stylosanthes mild mottle Kulkarni et al. 1970; Kenya [**50**,2637f].

sub-clover mottle = pea dwarf mosaic v.

sublineolum, *Colletotrichum*; **substriata**, *Puccinia*; **subterranea**, *Roesleria, Spongospora*.

subterranean clover, see *Trifolium*; Barbetti & Sivasithamparam, *Rev.Pl.Path.* **65**:513, 1986, leaf disease fungi; Barbetti et al., *AAB* **109**:259, 1986, root rot fungi; Johnstone & McLean, *ibid* **110**:421, 1987, virus diseases [**66**,2904].

—— **mottle v.** Francki et al., *Pl.Path.* **32**:47, 1983; Sobemovirus, isometric *c*. 30 nm diam.; as well as ssRNA, particles also contain circular and linear RNAs like viroids of 2 size classes; transm. sap to subterranean clover, prob. seedborne, Australia (Western Australia, S.W.), the effects of infection are severe [**62**,2531].

—— —— **red leaf** = soybean dwarf v., fide Johnstone & McLean 1987, subterranean clover.

—— —— **stunt** Grylls & Butler, *J. Aust.Inst.agric.Sci.* **22**:73, 1956; transm. *Aphis* spp., *Macrosiphum euphorbiae, Myz.pers.*; Australia, severe effects on most cvs. of subterranean clover q.v., Johnstone & McLean 1987 [**36**:248].

subtilissima, *Lachnellula*; **subvinosum**, *Hydnum*, see Hydnaceae.

subviral pathogens, a term used for prions and viroids.

Sudan grass, see sorghum.

sudden death, as for clove sudden wilt; —— **wither**, oil palm, see *Phytomonas*.

sugarbeet, see beet.

sugarcane, *Saccharum*; Edgerton, *Sugarcane and its diseases*, edn. 2, 1958; Martin et al. 1961; Hughes et al. 1964; Hughes, *PANS* **24**:143, 1978;

Agnihotri, *Diseases of sugarcane*, 1983; Sivanesan & Waller 1986.

— **chlorotic streak** Bell 1929; Wilbrink 1929, Abbott & Ingram, *Phytop.* **32**:99, 1942; transm. in setts, *Draeculacephala portola* [**9**:271; **21**:266].

— **Fiji v.** Lyon 1910; Hutchinson & Francki, *Descr.V.* 119, 1973; Phytoreovirus, subgroup 2, isometric *c.* 70 nm diam., transm. *Perkinsiella* spp., persistent, restricted to a few grass genera, infected plants stunted with galls on lower leaf and leaf sheath surfaces; Australasia and Oceania, parts of S.E. Asia, Madagascar; Hays 1974, poss. strs.; Hatta & Francki, *Virology* **76**:797, 1977; *Physiol.Pl.Path.* **19**:337, 1981, particle structure, virus development in host galls; Egan & Hall in Plumb & Thresh 1983, epidemic in Australia (Queensland) [**54**,546; **56**,3692; **61**,3611].

— **grassy shoot** Vasudeva, *F.A.O.Pl.Prot.Bull.* **4**:129, 1956; Rishi et al., *Ann.phytopath.Soc.Japan* **39**:429, 1973; MLO assoc.; David & Alexander 1984, transm. *Matsumuratettix* sp. [**36**:212; **53**,3603; **64**,1732].

— **mosaic v.** Brandes 1919; Pirone, *Descr.V.* 88, 1972; Potyvirus, filament *c.* 750 nm long, transm. sap. aphid spp., many strs., serologically related to maize dwarf mosaic v., unfortunately both virus names have been used for the strs.; maize grown near sugarcane may become infected by SCMV strs. but is seldom severely damaged; Gough & Shukla, *Virology* **111**:455, 1981, coat protein & 4 Australian strs. [**61**,616].

— **sereh** Lyon 1921; transm. in setts, Asia (S.E.) [**1**:184].

— **spike** Sharma & Jha, *Proc.Indian Acad.Sci.*B **45**:16, 1957; transm. in setts, India (Bihar) [**37**:57].

— **streak** (maize streak v.).

— **striate mosaic** Hughes, *Nature Lond.* **190**:366, 1961; *Proc.Qd.Soc.Sug.Cane Technol.* 34:151, 1967; transm. sap by needle prick, Australia (Queensland) [**40**:627; **46**,3187].

— **Trojan mottle** Hughes et al. 1971; Australia (Queensland) [**51**,1853e].

— **white leaf** Ling & Chuang-Yang, *Rep.Taiwan Sug.Exp.Stn.* 1963:69; Lin & Lee, *ibid* 1967-8:17; Chen 1979; MLO assoc., transm. *Matsumuratettix hiroglyphicus* [**44**,825; **47**,3555; **60**,1907].

sugary diseases, see *Claviceps*.

sulcigena, *Lophodermella*.

sulphur dioxide, Ichikawa et al., *Bull.Hokkaido Prefect.agric.Exp.Stns.* 44:90, 1980, crop injury due to SO_2 fumigation.

sulphurem, *Fusarium*; **sulphureus**, *Laetiporus*.

Sumatra wilt, see clove.

summer black stem (*Phoma medicaginis* var. *pinodella*) clover, pea; — **dieback**, see exanthema;

— **dwarf** (*Aphelenchoides besseyi*) strawberry; — **patch** (*Leptosphaeria korrae, Phialophora radicicola* assoc.) Kentucky blue grass.

sunflower, *Helianthus annuus*; Sackston, *PD* **65**:643, 1981; Kolte, *Diseases of annual edible oil seed crops*, vol. 3, 1985.

— **mosaics** Uppal, *Int.Bull.Pl.Prot.* **7**:103, 1933, India (Maharashtra); Sackston, *PDR* **41**:885, 1957, Uruguay; Arnot & Smith, *J.Ultrastruct.Res.* **19**:173, 1967, filament *c.* 480 × 13 nm, USA (Texas) [**12**:611].

— **necrosis** Muntañola, *Revta Invest.agríc.B.Aires* **2**:205, 1948; Traversi, *ibid* **3**:345, 1949; transm. sap, insects, seed; Argentina, wide host range [**29**:366; **30**:165].

— **phyllody** Signoret, Louis & Alliot, *Phytopath.Z.* **86**:186, 1976; MLO assoc., France (S.) [**56**,326].

— **rugose mosaic** Singh, *Sunflower Newsl.* 3:13, 1979; transm. sap, seed; Kenya [**60**,1023].

— **yellowing** Russell, Cook & Bunting, *Pl.Path.* **24**:58, 1975; transm. *Myz.pers.*, persistent, England, beet western yellows v. group [**54**,5030].

sunken bleached rot (*Mycosphaerella caricae*) papaya.

sunnhemp, *Crotalaria juncea*.

— **mosaic v.** Capoor & Varma 1948; Kassanis & Varma, *Descr.V.* 153, 1975; Tobamovirus, rod 300 × 17 nm, transm. sap readily, pollen; infects many legumes, including Bengal bean, cowpea, pigeon pea; Varma in van Regenmortel & Fraenkel-Conrat 1986.

— **phyllody** Bose & Misra, *Indian J.agric.Sci.* **8**:417, 1938; Solomon & Sulochana, *Phytopath.Z.* **78**:62, 1973; transm. sap, India [**18**:181; **53**,2586].

— **rosette v.** Verma & Awasthi, *Curr.Sci.* **45**:642, 1976; *Phytopath.Z.* **92**:83, 1978; rod 400 × 17 nm, transm. sap, India (Uttar Pradesh) [**56**,2519; **58**,251].

sunscald, a superficial form of damage to fruits that results from the action of intense sunlight, especially when the surface is wet, e.g. of apple, grapevine, peach, after FBPP.

superelongation (*Elsinoë brasiliensis*) cassava.

suppressive soils, difficult to define, many types; Hornby, *A.R.Phytop.* **21**:65, 1983, stated: '...strong support for mechanisms involving antagonism has resulted in a marked tendency to consider the subject in terms of pathogen suppression'.

surface canker (*Pezicula corticola*) apple.

suscept, a host affected by, or prone to, disease; the word is displeasing and should be dropped; susceptible, susceptibility, subject to infection, see FBPP.

swamp spot (*Deightoniella torulosa*) banana.

swede, see *Brassica*.

Swedish pine cast (*Lophodermella sulcigena*).
sweet clover, *Melilotus*.
— — **necrotic mosaic v.** Hiruki et al. 1981, 1984;
Descr.V. 321, 1986; Dianthovirus; isometric 34 nm
diam., transm. sap, contact, drainage water,
occurs in lucerne, spreads through soil, Canada
(Alberta); Hiruki, *PD* **70**:1129, 1986, incidence &
distribution in Alberta.
— — **ringspot** Henderson, *Phytop.* **24**:248, 1934;
transm. sap, USA (Virginia) [**13**:519].
— **fern blister rust** (*Cronartium comptoniae*) pine.
— **granadilla**, see *Passiflora*; — **orange**, see
Citrus.
— **pea**, *Lathyrus odoratus*.
— — **mosaic v.** Prasad & Sahambi, *Indian
Phytopath.* **36**:28, 1983; isometric 25 nm diam.,
transm. sap, aphid spp., India (N.) [**63**,2370].
— **potato**, *Ipomoea batatas*; Harter & Weimer,
Tech.Bull.USDA 99, 1929; Steinbauer &
Kushman, *Agric.Handb.USDA* 388, 1971; Arene
& Nwankiti, *PANS* **24**:294, 1978, in Nigeria,
PANS Manual 4, 1978, Centre Overseas Pest
Res.; Sherf & MacNab 1986.
— — **caulimo-like v.** Atkey & Brunt,
J.Phytopath. **118**:370, 1987; isometric 50 nm
diam.; Madeira, New Zealand, Papua New
Guinea, Puerto Rico, Solomon Islands.
— — **chlorotic leaf spot** (sweet potato feathery
mottle v.).
— — — **stunt** Schaeffers & Terry, *Phytop.*
66:642, 1976; transm. *Bemisia tabaci*, Nigeria,
poss. relationship with sweet potato vein clearing
[**56**,508].
— — **curly top**, Australia (New South Wales),
1961 [**41**:429].
— — **dwarf** Summers, *PDR* **35**:266, 1951;
Murayama, *Mem.Fac.Agric.Hokkaido Univ.* 6:81,
1966; Shinkai 1974, abs. *Rev.Pl.Prot.Res.* **1**:65,
1968; Tsai et al., *Ann.phytopath.Soc.Japan* **38**:81,
1972; MLO assoc., transm. *Nesophrosyne
ryukyuensis*, Ryukyu Islands [**30**:625; **46**,724;
49,1865; **52**,1351].
— — **feathery mottle v.** Doolittle & Harter,
Phytop. **35**:695, 1945; Moyer & Kennedy, *ibid*
68:998, 1978; Potyvirus, fide Boswell et al.,
Rev.Pl.Path. **65**:221, 1986, filament *c.* 830–850 nm
long, the particles have been said to differ slightly
in length amongst the strs.; transm. sap, *Aphis*
spp., other aphid spp., non-persistent, widespread,
in *Ipomoea*; Cali & Moyer and Cadena-Hinojosa
& Campbell, *Phytop.* **71**:302, 1086, 1981; *PD*
65:412, 1981, purification, serology, 4 strs.:
chlorotic leaf spot, feathery mottle, internal cork,
russet crack [**25**:93; **58**,2065; **61**, 1519–20, 2606].
— — — **internal cork** Nusbaum, *PDR* **29**:677,
1945; (sweet potato feathery mottle v.) [**25**:97].

— — **leaf curl** Chung et al., *Pl.Prot.Bull.Taiwan*
27:333, 1985; rod assoc., transm. *Bemisia tabaci*
[**65**,2602].
— — **little leaf** van Velsen 1967; Kahn et al.,
Phytop. **62**:903, 1972; MLO assoc., transm.
Halticus tibialis; Pearson et al., *Phytopath.Z.*
109:269, 1984, New Zealand, Papua New Guinea
[**47**,879; **52**,2094; **63**,4689].
— — **mild mottle v.** Hollings, Stone & Bock
1976; *Descr.V.* 162, 1976; Potyvirus, subgroup 4,
filament *c.* 950 nm long, transm. sap, *Bemisia
tabaci*, E. Africa, fairly wide host range, causes
vein mottling in tomato; severe blister, mottle,
distortion and stunt in tobacco.
— — **mosaics A and B** Sheffield, *Phytop.* 47,582,
1957; **48**:1, 1958; transm. *Myz.pers.* and *Bemisia
tabaci*, respectively; A poss. a Potyvirus, E. Africa
[**37**:110, 373].
— — **russet crack** Daines & Martin, *PDR*
48:149, 1964; Campbell et al., *Phytop.* **64**:210,
1974; (sweet potato feathery mottle v.) [**43**,2036;
53,4698].
— — **T** = sweet potato mild mottle v.
— — **vein clearing** Loebenstein & Harpaz,
Phytop. **50**:100, 1960, transm. *Bemisia tabaci*,
Israel, strs.; Clerk, *PDR* **44**:931, 1960; Robertson,
ibid **48**:888, 1964, Nigeria; Schaeffers & Terry, see
sweet potato chlorotic stunt, identified 2 infective
agents in Nigeria from sweet potato. One, SPVC,
was a filament 850 nm long, transm. *Aphis
gossypii*, *Myz.pers.* The other was called SPCS;
together they caused severe symptoms of
puckering leaf strapping, chlorosis, stunting. The
relationships of these forms appears to be obscure
[**39**:496; **40**:383; **44**,1660].
— — — **mosaic v.** Nome, *Phytopath.Z.* **77**:44,
1973; Nome et al., *ibid* **79**:169, 1974; filament
c. 760 × 13 nm, transm. sap, *Myz.pers.*, non-
persistent, Argentina, Convolvulaceae; differs
from sweet potato feathery mottle v. [**53**,1167;
54,640].
— — **witches' broom** Summers 1951; Jackson &
Zettler, *PD* **67**:1141, 1983; MLO assoc., transm.
Nesophrosyne orientalis, *Orosius lotophagorum
ryukyuensis*; China, Korea, Papua New Guinea,
Ryukyu Islands, Solomon Islands, Taiwan, Tonga
[**63**,4213].
— — — — **chlorotic little leaf** Lawson et al.
1970; Dabek & Sagar, *Phytopath.Z.* **92**:1, 1978;
MLO assoc., Guadalcanal; Yang & Chou,
J.agric.Res.China **31**:169, 1982, transm.
Nesophrosyne orientalis, Taiwan [**58**,1491;
62,523].
— — **yellow spot** Liao et al., *J.agric.Res.China*
28:127, 1979; A and N forms, filamentous
particles, transm. sap and A by *Myz.pers.*, non-

persistent; B form, no particles, transm. *Bemisia tabaci*, Taiwan [59,3523].

— **vetch crinkly mosaic v.** Majorana 1967–8; Majorana & Rana, *Annls.Phytopath.* 1 hors sér.:347, 1969; Majorana, *Phytopath.Mediterranea* **10**:86, 1971; poss. Potyvirus, filament 746 nm long, transm. sap, aphid spp., non-persistent, *Hedysarum coronarium* [**48**,1793; **50**,3911].

Swiss needle cast (*Phaeocryptopus gaeumannii*) Douglas fir.

swollen fruit (*Ciboria shiraiania*) mulberry.

sword bean, *Canavalia ensiformis* or *C. gladiata*.

— — **mosaic v.** Mali, *Curr.Sci.* **48**:162, 1979; Mali et al., *Indian Phytopath.* **38**:282, 1985; prob. Potyvirus, filament 790 × 11 nm, transm. sap, aphid spp., non-persistent, India [**59**,558; **66**,1701].

Syagrus calcaratus, transm. okra mosaic v.

sycamore, *Acer pseudoplatanus*; in N. America the planes are called sycamore.

— **leaf scorch**, see elm leaf scorch, Hearon et al.; Sherald et al., *PD* **67**:849, 1983; xylem limited bacterium assoc., Gram negative, see *Xylella* [**63**,899].

Sydowia Bresad. 1895; Dothioraceae; Sivanesan 1984; ascomata uniloculate, asci many spored with 16 or more asco. which are multiseptate.

S. polyspora (Bref. & v. Tavel) E. Müller 1953, fide Cannon et al. 1985; Sutton & Waterston, *Descr.F.* 228, 1970, as anam. *Sclerophoma pythiophila* (Corda) Höhnel 1909; asci 24–32 spored, asco. mostly 3 septate, 9–28 × 3–8.5 μm; con. 4–8 × 2.3 μm; widespread on conifers, pine leaf blight and dieback, prob. a wound pathogen; assoc. with other *Sclerophoma* spp. in Douglas fir dieback, see *Xenomeris abietis*; Butin, *Phytopath.Z.* **48**:298, 1963, causing blue stain [**43**,1457].

symbiosis, this term is neither listed by FBPP nor indexed by Ainsworth 1981. It is only briefly discussed here because it is apparent that some biologists consider that its definition should cover parasitism; if not, by logical extension, pathogenicity. This view appears to be based on the early one of de Bary q.v. This may have been set out in his book of 1866. In the 1887 English translation of his book of 1884, only in part a second edn. of his book of 1866, de Bary refers to the parasitic relationship as that of a common life, a symbiosis. Ainsworth, *Introduction to the history of mycology*, 1976, stated that it was Frank in 1877 who first defined and used the term in its usually accepted biological sense. That is the state of 2 different organisms, attached and living together, and each contributing to the other's support or benefit. Ainsworth also stated that de Bary took up Frank's term in 1879; de Bary therefore may have changed his position. The

term in the restricted sense of meaning a mutual benefit soon came into use. As Marshall Ward, *Diseases in plants*, 1901, put it. '…each organism doing something for the other and each taking something from the other'. Boucher ed., *The biology of mutualism*, 1985.

symphytum stunt Xu et al., *Acta phytopath.sin.* **13**(2):5, 1983; MLO assoc., China, *S. officinale*, comfrey [**63**,182].

sympodial, of a conidiophore or a sporangium; continued growth, after the main axis has formed a terminal spore, and the development of a succession of apices, each of which originates below, and to one side of, the previous apex.

symptom, a visible or otherwise detectable abnormality arising from a disease or a disorder; symptomatology, a displeasing word, is the study of symptoms; FBPP defines symptom types: acute, chronic, local, masked, mild, primary, secondary, systemic. Definitions and descriptions are given in many classical book texts, see especially Harshberger, *A textbook of mycology and plant pathology*, 1917; also: Wheeler 1969 and in Wood & Jellis 1984; Roberts & Boothroyd 1984; Bos 1978 is useful for description, although only concerned with plant viruses.

Synchytriaceae, Chytrideales; thallus holocarpic, endobiotic, forming > 1 sporangium, converted into a prosorus or sorus surrounded by a common membrane, or a resting spore; sporangia inoperculate opening by deliquescence of 1 or more papillae; Sparrow in Ainsworth et al., vol. IV B, 1973; *Synchytrium*.

Synchytrium de Bary & Woronin 1863; Synchytriaceae; thallus large, stimulating multicellular galls on vascular plants, never amoeboid; sorus sessile or a prosorus; sporangial zoospores freed outside host cells; Karling, *Synchytrium*, 1964; *Adv.Frontier Pl.Sci.* **29**:1, 1972, last reference fide *Ainsworth & Bisby's dictionary of the fungi*, edn. 7, 1983.

***S. desmodiae** Munasinghe 1955 or *desmodii*, described in Sri Lanka on *Desmodium ovalifolium* and found causing stem, pod and leaf galls in Colombia where the plant is grown as a pasture legume; Lenné, *PD* **69**:806, 1985 [**65**,777].

S. endobioticum (Schilberszky) Percival 1910; Walker, *Descr.F.* 755, 1983; potato wart is a highly destructive disease with a widespread but limited distribution. Although under control, it remains a potential threat through the spread of strs. differing in pathogenicity. The fungus is subject to the strictest legislative and quarantine control, both nationally and internationally. Large excrescences, at first green or white and later dark brown, form on the tubers; green galls appear on the aerial shoots. Zoospores from thin walled, summer sporangia cause repeated infections of sprouts and tuber initials. The thick walled, resting sporangia can remain dormant in the soil for 30 years and still release infective zoospores.

Spread is mostly through planting infected tubers; actual spread of the pathogen through soil is very slow. Many races ± 17 are present in central Europe; the common race 1 is widespread. Potato wart, characteristically, is found in wet areas, montane regions, smallholdings and geographically isolated sites. Its distribution is: Andean upland areas of S. America, where *Synchytrium endobioticum* is prob. indigenous, Canada (Newfoundland), China, Europe, Falkland Islands, India (N.E.), Mexico, New Zealand (S. Island), South Africa.

Reviews: Karling's monograph, see genus, gives a bibliography; Noble & Glynne, *F.A.O.Pl.Prot.Bull.* **18**:125, 1970; Hampson & Proudfoot, *ibid* **22**:53, 1974; Pratt in Ebbels & King 1979; Hampson, *Can.J.Pl.Path.* **3**:65, 1981; Langerfeld, *Mitt.biol.BundAnst.Ld-u. Forstw.* 219, 1984; Bojňanský, *EPPO Bull.* **14**:141, 1984.

S. lagenariae Mhatre & Mundkur 1945; Walker, *Descr.F.* 756, 1983; on cucurbits, India, the galls on aerial parts are white to light brown.

S. macrosporum Karling 1956; Walker, *Descr.F.* 757, 1983; plurivorous, lavender red galls on leaves and stems, short cycle, only resting spores formed, USA (Texas).

S. phaseoli Weston 1930; Walker, *Descr.F.* 758, 1983; on legumes, short cycle, forms only summer spores; orange pustules, resembling aecidia, on leaves and stems which become distorted; Papua New Guinea, E. Africa, S. America.

S. phaseoli-radiati Sinha & Gupta 1951; Walker, *Descr.F.* 759, 1983; on legumes including pigeon pea, short cycle, forms only resting spores, India.

S. psophocarpi (Racib.) Gäumann 1927; Walker, *Descr.F.* 760, 1983; false rust of winged bean,

prob. confined to this host, short cycle, forms only summer spores; semi-globular, yellow galls on aerial parts, infected leaves are small, curled and thickened. Investigated largely in Papua New Guinea where false rust is important; found in parts of Africa and S.E. Asia; Drinkall & Price, *Pl.Path.* **32**:229, 1983; airborne sporangia showed a diurnal periodicity with a peak at 16.00–18.00 hours; *Trop.Agric.Trin.* **61**:293, 1984, chemical control; *AAB* **109**:87, 1986, infection; and see *Pseudocercospora psophocarpi*, Price et al. [**63**,1516; **64**, 872; **65**,6269].

S. sesamicola Lacy 1950; on sesame, India; long cycle, forms summer sporangia and resting spores, causes stunting and deformation of the crown, leaflets puckered and curled, spindly and curled shoots at the base of the plant; has assumed importance recently in India since the introduction of new cvs.; Variar & Pavgi, *Phytopath.Mediterranea* **18**:201, 1979, resistance of germinating sporangia to desiccation & heat [**62**,4741].

syndrome, the totality of symptoms.

synergism, an assoc. of 2 or more organisms acting at one time and effecting a change which one of them alone does not make, e.g. increasing the severity of a disease, also increased pesticide activity in chemical mixtures; Powell, *A.R.Phytop.* **9**: 253, 1971, nematode & fungus interaction in diseases.

synnema, a conidioma of a ± compacted group of erect, and sometimes fused conidiophores bearing con. at the apex only or at both apex and sides.

syringae, *Phytophthora*, *Pseudomonas*.

Syringa vulgaris, common lilac.

syringomycin, toxin from *Pseudomonas syringae* pv. *syringae*; Sinden et al. and Backman & DeVay, *Physiol.Pl.Path.* **1**:199, 215, 1971; Gross & DeVay, *ibid* **11**:1, 13, 1977; *Phytop.* **67**:475, 1977 [**50**,3984; **56**,5547; **57**,388–9].

Syzygium aromaticum, clove.

T

tabacina, *Peronospora*; **tabacinum**, *Fusarium*;
Microdochium, see *F. tabacinum*.

tabebuia witches' broom Cook, *J.Agric.Univ.P.Rico*
22:441, 1938; Puerto Rico.

tabernaemontana mosaic Lal & Saksena, *Labdev*
J.Sci.Technol. 4:279, 1966; India [46,2046].

— **proliferation** Dollet & Dubern,
Annls.Phytopath. 9:447, 1977; MLO assoc., Ivory
Coast [58,328].

tabtoxin, from *Pseudomonas*, including *P. syringae*
pv. *tabaci*; Stewart, *Nature Lond.* 229:174, 1971;
Mitchell, *A.R.Phytop.* 22:218, 1984; Turner &
Taha, *Physiol.Pl.Path.* 25:55, 1984 [50,1970;
64,767].

tagetica, *Alternaria*.

tagetitoxin, from *Pseudomonas syringae* pv. *tagetis*,
different from phaseolotoxin; Mitchell & Durbin,
Physiol.Pl.Path. 18:157, 1981; Mitchell & Hart
1983, structure; Mitchell, *A.R.Phytop.* 22:234,
1984 [60,6501; 62,3754].

tainanensis, *Stagonospora*; **taiwanensis**,
Leptosphaeria, *Pseudocercospora*.

take all (*Gaeumannomyces graminis*) temperate
cereals and grasses.

Talaromyces C. Benj. 1955; Eurotiales; *T. flavus*
(Klöcker) Stolk & Samson 1972; anam.
Penicillium dangeardii Pitt 1979; antagonistic to
soilborne pathogens; McLaren et al.,
Can.J.Pl.Path. 8:43, 1986, parasitism of
Sclerotinia sclerotiorum; Fravel & Marois, *Phytop.*
76:643, 1986, edaphic parameters; Papavizas et
al., *ibid* 77:131, 1987, in soil & survival in alginate
pellets [65,5672; 66,68,3668].

tamarillo, *Cyphomandra betacea*, also called tree
tomato.

— **mosaic v.** Mossop, *N.Z.J.agric.Res.* 20:535,
1977; poss. Potyvirus, filament av. 745 nm long,
transm. sap, *Myz.pers.*, New Zealand, narrow
host range [57,5603].

tanakae, *Diaporthe*, *Phomopsis*.

tan brown rot (*Pezizella oenotherae*) strawberry.

tangelo, tangerine and tangor, see *Citrus*.

tangle top (*Myriogenospora atramentosa*) grasses;
also used in sugarcane for an abnormal emergence
which leads to the mechanical tangling of the
leaves.

tanier, or tannia, *Xanthosoma*; see taro.

tan spot (*Curtobacterium flaccumfaciens* pv.
flaccumfaciens) soybean.

Taphrina Fr. 1815, fide Cannon et al. 1985;
Taphrinaceae; the spp., many on temperate,

broad leaves trees, e.g. alder, birch, elm,
hornbeam, oak, poplar, do little or no damage.
They cause leaf curls and witches' brooms; Snider
& Kramer, *Mycologia* 66:754, 1974.

T. deformans (Berk.) Tul. 1866; Booth, *Descr.F.*
711, 1981; asco. 3–7 μm diam.; peach leaf curl;
almond, apricot, nectarine; poss. races; Butler &
Jones 1949, Anderson 1956.

T. maculans E. Butler 1911; Sivanesan & Gibson,
Descr.F. 507, 1976; asco. 4–6.5 × 2–3.5 μm; brown
leaf spot of turmeric; on Zingiberaceae but does
not cause a disease in ginger; Upadhyay & Pavgi,
Mycopathologia 69:33, 1979, asco. dispersal
[59,3860].

T. populina Fr. 1832; poplar yellow leaf blister;
Taris 1969 gave a full account of the disease
under the synonym *Taphrina aurea* [*For.Abstr.*
31,6707].

T. pruni Tul. 1866; Booth, *Descr.F.* 713, 1981;
asco. 4–7 × 3.5 μm; plum pockets or bladder, the
fruit shrivels.

T. tosquinetii (Westend.) Magnus 1890; asco.
2.5–5.5 × 2.5–5 μm; alder leaf curl, commonest on
the sucker shoots which may be killed; Bond,
TBMS 39:60, 1956, referred to other *Taphrina*
spp. on alder in Britain [36:114].

T. wiesneri (Rathay) Mix 1954; Booth, *Descr.F.*
712, 1981; asco. 3.5–9 × 3–6 μm; leaf curl and
witches' broom of apricot and cherry; often as
Taphrina cerasi (Fuckel) Sadeb. which is a
synonym.

Taphrinaceae, Taphrinales; mycelium forming a
subcuticular or subepidermal layer of binucleate,
ascogenous cells; asci or spore sacs terminal,
usually on a stalk cell, arranged in a palisade
layer; asco. hyaline, aseptate, subglobose; often
budding before or after discharge to form colonies
like those of yeasts; monogeneric, *Taphrina*.

Taphrinales, Ascomycotina; ascomata absent, asci
or spore sacs arising from mycelium or from
resting spores with thick walls, in a palisade layer;
forms asco. or endospores; causing hyperplasia,
usually on leaves.

tapioca, see cassava.

tap root rot (*Phytophthora drechsleri*) beet.

Taraxacum officinale, dandelion.

tarda, *Ascochyta*.

target leaf spot (*Bipolaris sorghicola*) sorghum,
(*Thanatephorus cucumeris*) rubber, Carpenter,
Tech.Bull.USDA 1028, 1951 [31:294].

— **spot**, a leaf lesion which has concentric rings of

necrotic tissue, e.g. as caused by *Corynespora cassiicola*.

Targioni-Tozzetti, Giovanni, 1712–83; born in Italy, Univ. Pisa, naturalist, chair of botany, Florentine School; director Botanical Garden, Florence. He, independently of Fontana, discovered the nature of wheat stem rust, *Puccinia graminis*, in 1767; *Phytopath.Classics* 9, 1952, English translation by L.R. Tehon; Ainsworth 1981.

tar leaf spots (*Phyllachora* spp., *Rhytisma* spp.).

taro, *Colocasia esculenta*; the edible aroids also include *Alocasia, Cyrtosperma, Xanthosoma*; *PANS Manual* 4, 1978, Centre Overseas Pest Control, London; Jackson, *Diseases and pests of taro*, 1980, South Pacific Commission.

tarocco pit Russo & Klotz, *Calif.Citrogr.* **48**:221, 1963; Italy (Sicily), orange [**42**:546].

taro feathery mosaic Palomar et al. 1983–4; transm. sap, *Tarophagus proserpina*, Philippines [**65**,3114–15].

Tarophagus proserpina, transm. colocasia bobone, taro feathery mosaic.

tarry root rot (*Hypoxylon nummularium*) tea; — spot (*Nectria coccinea*) beech.

tassiana, *Mycosphaerella*.

Taubenhaus, Jacob Joseph, 1885–1937; born in Palestine, now largely Israel; in 1898 went to USA, Univ. Cornell, Pennsylvania; Texas and Delaware Agricultural Experiment Stations 1909–16, 1916–37; pioneer work on Texas root rot of cotton; wrote 3 books on the diseases of onion, with F.W. Mally, sweet potato and truck crops. *Phytop.* **28**:525, 1938; *Pal.J.Bot.* R.Ser. **2**:126, 1938.

taurica, *Leveillula, Oidiopsis*.

tawny blotch (*Stagonospora foliicola*) reed canary grass.

taxis, the movement of a whole organism towards or away from a stimulus; taxes include stages that are motile, e.g. zoospores; tropisms refer to the movement of a part of an organism, e.g. a fungus germ tube, see chemotaxis; Wynn, *A.R.Phytop.* **19**:237, 1981, tropic & taxic responses of pathogens to plants.

taxodii, *Echinodontium*.

taxonomy, the orderly arrangement of replicating entities into units composed of 'likes', after Cowan 1978 who made some very pertinent remarks; and who called classification, nomenclature and identification the trinity of taxonomy. Some taxonomy tends to be subjective; consequently those who practice it may be both dogmatic and quarrelsome. For entities that are universally accepted as organisms the basic unit is the taxonomic species enshrined in the Linnean

binomial system. Virologists may reject species in this sense for plant viruses. They would not accept these viruses as organisms but as, e.g. '– subcellular entities with genomes analagous to such molecules as messenger RNA, plasmids or transposons'; Murant, *Microbiological Sci.* **2**:218, 1985. The taxonomy of plant viruses as a group, and vis à vis animal viruses, has been much discussed: Milne, *ibid* **1**:113, 1984; 4 papers in *Intervirology* **24**:61, 1985; Matthews, *A.Rev.Microbiol.* **39**:451, 1985; Murant 1985 loc.cit. For bacteria: Cowan, *J.gen.Microbiol.* **67**:1, 1971 and 1978; Dye et al. *Rev.Pl.Path.* **59**:153, 1980. For fungi: Hawksworth, *Mycologist's handbook*, 1974; Booth, *TBMS* **71**:1, 1978; *Ainsworth & Bisby's dictionary of the fungi*, edn. 7, 1983.

Taxus, yew.

tea, *Camellia sinensis*; Sarmah, *Mem.Tocklai exp.Stn.* 26, 1960, India (N.E.); Agnihothrudu, *J.Madras Univ.* B **34**:155, 1964, fungi; Shanmuganathan, *Tea Q.* **40**:19, 1969, root diseases in Sri Lanka; Kasai, *Jap.Pest Inf.* **11**:23, 1972; Eden, *Tea*, edn. 3, 1976; T.M. & S.F. Chen, *PD* **66**:961, 1982, in China.

— **phloem necrosis** Bond, *AAB* **31**:40, 300, 1944; **34**:517, 1947; Sri Lanka [**23**:362; **24**:207; **27**:339].

— **rose yellow mosaic**, see camellia yellow mosaic.

— **witches' broom** Uehara & Nonaka, *Bull.Fac.Agric.Kagoshima Univ.* **20**:113, 1970; Japan [**50**,1367].

tears, as for Rio Grande gummosis.

teasel mosaic v. Stoner, *Phytop.* **41**:191, 1951; Gemignani 1965; Hollings 1968; filament 832 × 26 nm, transm. sap difficult, insects including *Myz.pers.*, England, USA (California); *Dipsacus fullonum*, Fuller's teasel [**30**:371; **45**,1318; **48**,1024u].

techniques, see methods.

tecnazene, selective fungicide for *Botryotinia* and *Fusarium*.

tectonae, *Olivea*.

teleomorph, see fungi and holomorph.

telephii, *Phoma*.

telfairia mosaic v. Nwauzo & Brown, *PDR* **59**:430, 1975; Atiri & Varma, *Trop.Agric.Trin.* **60**:95, 1983; Shoyinka et al., *J.Phytopath.* **119**:13, 1987; Potyvirus, filament *c*.800 nm long, transm. *Aphis citricola*, non-persistent, sap to 29 spp. in 6 plant families, Nigeria; *T. occidentalis*, fluted pumpkin [**55**,957; **62**,3683].

— **yellow vein clearing** (pepper veinal mottle v.), Atiri, *J. Pl.Prot.Tropics* **3**:105, 1986, Nigeria; *T. occidentalis*, fluted pumpkin [**66**,4545].

teliospore, or teleutospore; Uredinales, the spore

from which the basidiospores are formed; sometimes used for the spore with the same function in the Ustilaginales, ustilospore is preferred; Mendgen in Bushnell & Roelfs, *The cereal rusts*, vol. 1, 1984; Anikster, *Phytop.* 76:1026, 1986, germination.

temperate viruses, as for cryptic viruses.

temple tree, see frangipani.

temulentum, *Endoconidium*.

tenacity, the property of a deposit or residue to resist removal by weathering or other physical action; hence tenacity index, the ratio of the quantity of residue per unit area at the end of a given amount of weathering to that present at the beginning, FBPP.

Tenericutes, see Procaryotae.

tentoxin, from *Alternaria* q.v., Nishimura & Kohmoto; Saad et al., *Phytop.* 60:415, 1970; Woodhead et al., *ibid* 65:495, 1975; Mitchell, *A.R.Phytop.* 22:231, 1984 [49,2761; 55,622].

tenuazonic acid, toxin from *Alternaria* q.v., Nishimura & Kohmoto; Janardhanan & Husain, *Phytopath.Z.* 111:305, 1984 [64,2672].

tenuissima, *Alternaria*.

teosinte, an annual weed and a close relative of maize, *Euchlaena mexicana* or *Zea mexicana* or *Zea mays* ssp. *mexicana*.

tephrosia symptomless v. Bock et al. 1981; *Descr.V.* 256, 1982; poss. Tombusvirus, isometric *c*.33 nm diam., transm. sap, Kenya (E.), *T. noctiflora, T. villosa.*

tepperianum, *Uromycladium*.

teratology, the study of gross structural abnormalities, e.g. fasciations, galls, witches' brooms, FBPP; hence teratogenic; Harshberger, *A textbook of mycology and plant pathology*, 1917.

Teratosperma H. Sydow & Sydow 1909; Hyphom.; Ellis 1971; conidiomata hyphal, conidiogenesis holoblastic; con. multiseptate, pigmented, solitary, thick appendages arising usually from the proximal cell; Hughes, *N.Z.J.Bot.* 17:177, 1979.

T. oligocladum Uecker, Ayers & Adams, *Mycotaxon* 10:422, 1980; con. mostly 150–260 μm long, including the hyaline prolongation, or appendage, of the multiseptate axis, also 3 arms radiate from the second or third cell from the base; a mycoparasite on sclerotia of *Sclerotinia*; Ayers & Adams, *Can.J.Microbiol.* 27:886, 1981; Parfitt et al., *Pl.Path.* 32:459, 1983 [61,1061; 63,1139].

terebrantis, *Leptographium*; **teres**, *Drechslera*, for *Pyrenophora teres* see *P. trichostoma*.

terminal crook (*Colletotrichum acutatum*) pine.

terminology, the use of special terms; it differs from nomenclature, as part of taxonomy, which is the choice and application of names to particular

biological entities or concepts. All professions need a jargon; but it should never be used in excess for it may then become gibberish. Jargon is always ugly and can be meaningless. The guide for this book has been the FBPP list; and see *TBMS* 33:154, 1950. Naturally not all terms used in plant pathology are given; many can be found in dictionaries of more general subjects, e.g. botany, entomology, fungi, genetics and microbiology. In particular terms proliferate in the field of resistance to pathogens q.v.; see disease names, plant names; Jackson, *A glossary of botanic terms; with their derivation and accent*, edn. 4, 1928; Flood, *Scientific words. Their structure and meaning*, 1960; Usher, *A dictionary of botany*, 1966; Stearn, *Botanical Latin*, edn. 2, 1973; *Ainsworth & Bisby's dictionary of the fungi*, edn. 7, 1983.

terrestris, *Pyrenochaeta*.

test plants, for detection of plant viruses; Gibbs & Harrison 1976 gave a list of 11 of the most useful plant spp., i.e. barley, bean, broad bean, cowpea, cucumber, tobacco, wheat, *Chenopodium amaranticolor, C. quinoa, Nicotiana clevelandii, Petunia hybrida*; see indicator plant.

tetraspora, *Linospora*.

Texas root rot (*Phymatotrichopsis omnivora*) cotton, lucerne, many other herbaceous dicotyledons.

textura, see hyphal tissue.

thaliae, *Puccinia*.

thallic, of conidiogenesis in fungi, one of the 2 basic sorts cf. blastic; characterised by any enlargement, or none, of a recognisable conidial initial after the initial is delimited by a septum. Enterothallic is where only the inner wall of the conidiogenous cell contributes to forming the conidial wall.

Thanatephorus Donk 1956; Ceratobasidiaceae; basidiomata resupinate, parasitic on plant parts in or near soil, but often saprophytic on rotten wood or in soil; sclerotial or sterile mycelial states, *Rhizoctonia*, present; McNabb & Talbot in Ainsworth et al., vol. IV B, 1973.

T. cucumeris (Frank) Donk; sclerotial state *Rhizoctonia solani* Kühn 1858; Mordue, *Descr.F.* 406, 1974; plurivorous; a soil inhabitant which is found in different ecological forms and in different pathogenic, behavioural patterns, i.e. aerial, soil surface and subterranean. The basidial state, a greyish mycelial mat, covers parts of the plant in wet weather; and after some parasitic activity has taken place. The names of diseases caused are given in the order. They involve seed decay and damping off, stem lesion and canker, root and above ground rot, leaf web and thread blights, storage rots.

The fungus is heterothallic, bipolar, with multiple alleles, Adams & Butler, *Mycologia* **74**:793, 1982. It has been divided into anastomosis groups which are genetically distinct and are differentiated as well by other characteristics, Anderson, *A.R. Phytop.* **20**:329, 1982. Each group has forms which infect crops in several families; but group 3 is restricted to Solanaceae. Isolates from 1 crop may be of several groups; some groups anastomose with each other, some do not. Within one group there are different clones; isolates of 1 clone fuse perfectly but fusion between different clones leads to death of the fused cells, i.e. the killer reaction, Ogoshi & Ui, *Ann.phytopath.Soc.Japan* **49**:239, 1983. Many clones may occur in a single field but the distribution of each is limited to a small area.

Parmeter ed., Rhizoctonia solani: *biology and pathology*, 1970; Holliday 1980; Henis et al., *Phytop.* **69**:1164, 1979; Wijetunga & Baker, *ibid* **69**:1287, 1979; Liu & Baker, *ibid* **70**: 404, 1980; Reynolds et al., *ibid* **73**:903, 1983; Homma et al., *Ann.phytopath.Soc.Japan* **49**:184, 1983; Neate & Warcup, *TBMS* **85**:615, 1985, table of anastomosis groups 1–7; Ogoshi, *A.R.Phytop.* **25**:125, 1987 [**59**,4460, 4987; **60**,1849; **62**,568; **63**,46, 88, 292; **65**,1758].

T. sasakii (Shirai) Tu & Kimbrough, *Bot.Gaz.* **139**:457, 1978; rice sheath blight; described by Ou 1985 under *Thanatephorus cucumeris*.

Thaumetopoea pityocampa, see *Elytroderma torresjuanii.*

theae, see *Calonectria pyrochroa, Cercoseptoria ocellata; Elsinoë, Pestalotiopsis, Phomopsis, Sphaceloma, Stigmina.*

Thecaphora Fingerh. 1836; Ustilaginaceae; spore ball a multicellular spore, see *Angiosorus;* Freire, *Fitopat.Brasileira* **11**:543, 1986, described a sp., poss. new, causing galls on *Spilanthes oleracea,* Pará cress, in Brazil.

thelephoroid, flattened, see Aphyllophorales.

Theobroma cacao, cacao, preferred to cocoa.

theobromae, *Botryodiplodia, Oncobasidium, Verticillium.*

thermal death point, the lowest temp. at which heating for a limited period, usually 10 minutes, is sufficient to kill a micro-organism; —
inactivation point, the lowest temp. at which heating for a limited period, usually 10 minutes, is sufficient to cause a virus to lose its infectivity or an enzyme its activity, FBPP.

thermophilic fungi, Cooney & Emerson, *Thermophilic fungi,* 1964; Emerson in Ainsworth & Sussman vol. 3, 1968.

Therrya Sacc. & Penzig 1882, fide Cannon et al. 1985; ? Rhytismatales; asco. fusiform,

multiseptate, lacking a mucilagenous sheath; Reid & Cain, *CJB* **39**:1117, 1961 [**41**:182].

T. piceae Funk, *CJB* **58**:1292, 1980; asco. 3 septate, 30–38 × 10–12 μm; spruce perennial canker, Canada (British Columbia), multiple cankers may cause tree death; Funk, *Can.J.Pl.Path.* **4**:357, 1982 [**62**,2666].

thevetia leaf curl Garga, *Curr.Sci.* **22**:243, 1953; India (N.), *T. neriifolia* [**33**:155].

thiabendazole, similar to benomyl.

thielavioides, *Chalara.*

Thielaviopsis Went 1893; Hyphom.; Ellis 1971; conidiomata hyphal, conidiogenesis holoarthric; con. aseptate, pigmented, catenate, often seceding with difficulty, thick walls.

T. basicola (Berk. & Broome) Ferraris 1912; Subramanian, *Descr.F.* 170, 1968; referred to *Chalara elegans* by Nag Raj & Kendrick, *Chalara* q.v.; and to *Trichocladium basicola* Carmichael, new combination, Carmichael et al. 1980; con. in chains of 4–8, resembling large, multiseptate con., when separated con. 10–17 × 7–12 μm, phialoconidia 7–17 × 2.5–4.5 μm; black root rot of bean, cotton, pea, soybean, tobacco and other crops. The fungus is an unspecialised root inhabitant; it invades the cortex and when infection is severe whole root systems become reduced and blackened; Blume & Harman, *Phytop.* **69**:785, 1979, on pea; Tabachnik et al., *ibid* **69**:974, 1979, soil inoculum & assay, host range & disease; Holliday 1980; Ingold, *TBMS* **76**:517, 1981, phialoconidia; Yarwood, *Mycologia* **73**:524, 1981, soil assay; Anderson, *Can.J.Pl.Path.* **6**:71, 1984, on soybean; Corbaz, *Phytopath.Z.* **113**:289, 1985, variation in pathogenicity [**59**,4060, 4825; **61**,2147; **63**,4172; **65**,884].

thieves, tulip; aberrant plants with withered, small bulbs which can be separated from normal bulbs by floating off in water.

thigmodifferentiation, the response of a fungus germ tube, and appressorium formation, i.e. infection structures, to a contact stimulus, Staples et al., *Phytop.* **73**:380, 1983; also chemodifferentiation for a chemical stimulus.

thigmomorphogenesis, the response of plants to mechanical perturbations, Jaffe 1973; Shawish & Baker, *Phytop.* **72**:63, 1982, found that mechanical shaking of plants induced more severe symptoms when they were infected with *Fusarium oxysporum* ff.sp. [**61**,5220].

thimble berry ringspot Stace-Smith, *CJB* **36**:385, 1958; transm. *Amphorophora* spp., Canada (British Columbia), *Rubus parviflorus* [**37**:728].

thiophanate, protectant and systemic fungicide;

pathogens may develop a tolerance after its extensive use.

thioquinox, contact and protectant fungicide against powdery mildews.

thiram, soil and seed fungicide, used less against foliage; the thiram seed soak has a wide applicability in the control of seedborne fungus pathogens, Maude et al., *AAB* **64**:245, 1969 [49,930].

thistle mottle v. Donson & Hull, *J.gen.Virol.* **64**:2281, 1983; prob. Caulimovirus, England, *Cirsium arvense.*

thompsonii, *Hirsutella.*

thorn apple, or jimson weed, *Datura stramonium.*

thorny stem blight (*Tunstallia aculeata*) tea.

thread blights, see *Marasmius.*

***thrips**, see vector; *Thrips tabaci*, transm. tobacco streak.

thujae, *Kabatina*; **thujina**, *Didymascella.*

thwaitesii, *Xylaria.*

thyme leaf chlorosis Schultz, Harrap & Land, *AAB* **80**:251, 1975; bacilliform particle 219 × 72 nm assoc. but prob. not the causal agent, England [54,5037].

Thymus vulgaris, thyme.

Thyronectria Sacc. 1875; Hypocreaceae; asco. muriform; Seeler, *J. Arnold Arbor.* **21**:429, 1940 [20:81].

T. austro-americana (Speg.) Seeler 1940; anam. *Gyrostroma austro-americanum* Seeler, fide Bedker & Wingfield, *TBMS* **81**:179, 1983; asco. 9.8–15.6 × 5.2–9.1 μm; pycnidia on a stroma, not ostiolate, conidiogenous cells phialidic; con. hyaline, aseptate, 2.4–4.2 × 0.4–1.4 μm; causes cracking and peeling in *Gleditsia triacanthos* var. *inermis*, thornless honey locust, a damaging canker in USA. *Kaskaskia gleditsiae* Born & Crane, *Phytop.* **62**:926, 1972, was described as causing this disease, but later reduced to synonymy by Bedker & Wingfield loc.cit. The name *Gyrostroma* appears to be in some doubt, Sutton, *Mycol.Pap.* 141, 1977; Crowe et al., *PD* **66**:155, 1982, disease patterns in USA (Kansas); Jacobi, *Phytop.* **74**:566, 1984, culture & conidial biology; Riffle & Peterson, *ibid* **76**:313, 1986, effects of temp. & wound age on the disease [52,1287; 61,5263; 63,269, 4283; 65,5185].

T. balsamea (Cooke & Peck) Seeler; asco. 7–30 × 3–5 μm; assoc. with a dieback of balsam fir, *Abies balsamea*, in Canada (Ontario); Raymond & Reid, *CJB* **39**:233, 1961 [40:567].

tiger stripe, a leaf symptom in which clear yellow areas separate marginal and interveinal areas of dark necrotic tissue from persistently green areas along and adjacent to the main veins, giving a striped yellow and black appearance. The term

was applied to a symptom of hop wilt (*Verticillium albo-atrum*), after FBPP.

tigridia latent v. Brunt 1970; rod 790–840 nm long, transm. sap, England, *T. pavonia* [**50**,1555s].

—— **mosaic** (turnip mosaic v.), Brunt, *J.hort.Sci.* **51**:99, 1976; *T. pavonia* [**55**,3619].

tikka (*Mycosphaerella arachidis*, *M. berkeleyi*) groundnut.

Tilia, ornamental and timber lime trees.

tilia cowlforming Blattný, *Ochr.Rost.* **14**:80, 1938; Kláštersky 1951, see rose cowlforming; Czechoslovakia [**17**:568].

—— **mosaic** Smolák, *Ochr.Rost.* **22**:173, 1949; Czechoslovakia [**30**:14].

tiliae, *Elsinoë.*

Tillet, Mathieu, c.1730–91; born in France, farmer, sometime Master of the Mint at Troyes. In 1755 he published the classic account of field experiments on wheat bunt (*Tilletia tritici*), *Phytopath.Classics* 5, 1937, English translation by H.B. Humphrey. This work, and that of Prevost in 1807, laid the foundations of experimental plant pathology. Large 1940; Ainsworth 1981.

Tilletia Tul. & C.Tul. 1847; Tilletiaceae; sori ± dusty at maturity, containing aseptate ustilospores, mostly on Gramineae, usually in ovaries; ustilospores single, usually 16–54 μm diam., often sculptured, forming an aseptate germ tube that bears a crown of primary sporidia which, after fusing in pairs, form sporidia, i.e. basidiospores; destroys the grain; Duran & Fischer, *The genus* Tilletia, 1961; Wiese 1987; Gareth-Jones & Clifford 1983.

T. barclayana (Bref.) Sacc. & Sydow in Sacc. 1899; Ainsworth, *Descr.F.* 75, 1965; Singh et al., *Sydowia* **32**:305, 1979, considered that this smut should be retained in *Neovossia horrida* (Takahashi) Padwick & A. Khan; ustilospores mostly 17–25 μm diam.; rice kernel smut is a minor disease; the airborne basidiospores infect the flowers, germinate and penetrate the ovary; there is no evidence for systemic infection; Holliday 1980.

T. controversa Kühn 1874; Waller & Mordue, *Descr.F.* 746, 1981; ustilospores with polygonal reticulations, 16–25 μm diam., sheathed; larger than the sheathless ustilospores of *Tilletia tritici*; dwarf bunt of winter wheat, can infect barley; present in N. Africa, N. America, N. Asia; Europe, except France, Norway, Portugal, Spain, UK; Argentina, Uruguay. The ustilospores, released from the grain at harvest, contaminate seed and soil; germination is slow, 1–3 months at an optimum temp. of 3–8°. In USA (N.W.) infection is at the seedling stage in December to

April, and is systemic; intact sori survive in soil for up to 10 years; there are *c*.17 races.

Disease outbreaks are localised and the pathogen has very specific requirements. Seed chemical treatment is only effective if systemic fungicides are used. Dwarf bunt is absent from China which has imposed strict quarantine requirements on wheat shipments from USA (W.); these have been questioned. Grey et al., *PD* **70**:122, 1986, found that bunted spikes resulted only from heavily infested seeds ⩾ 1 g ustilospores/kg seeds, equivalent to 20 000 ustilospores/seed; reviews: Purdy et al., *A.Rev.Microbiol.* **17**:199, 1963; Hoffman and Trione, *PD* **66**:979, 1083, 1982 [65,2757].

T. holci (Westend.) Schröter 1877; Waller & Mordue, *Descr.F.* 747, 1983; ustilospores reticulate, 20–30 μm diam., no sheath; covered smut of *Holcus* and *Anthoxanthum odoratum*, little economic importance, found in seed samples.

T. indica Mitra 1931; Waller & Mordue, *Descr.F.* 748, 1983; ustilospores verrucose, 25–43 μm diam.; wheat karnal bunt; India, Iraq, Pakistan, and prob. elsewhere in W. Asia, Mexico; seedborne, occurs in soil, air dispersed basidiospores infect the flowers, usually only some kernels in a spike are attacked, therefore diseased grain is less conspicuous than in other bunts where whole ears are attacked; not systemic, seed chemical treatment effective; Krishna & Singh, *Indian Phytopath.* **35**:544, 1982, support retention in *Neovossia indica* (Mitra) Mundkur; *ibid* **36**:115, 1983, cytology of ustilospore germination & development; Gardener et al., *Mycologia* **75**:333, 1983, sheath structure of ustilospores; Joshi et al., *Bot.Rev.* **49**:309, 1983; Smilanick et al., *Phytop.* **75**:1428, 1985, factors affecting ustilospore germination; Royer & Rytter, *PD* **70**:225, 1986, relationship between kernel & spike infection; Warham, *Trop.Pest Management* **32**:229, 1986, review [62,3828; 63,2266, 5369; 65,2265, 3239].

T. laevis Kühn 1873, fide Mordue & Ainsworth 1984; Mordue & Waller, *Descr.F.* 720, 1981, as *Tilletia foetida* (Wallr.) Liro; ustilospores smooth or with shallow pits, 13–25 μm diam.; wheat common bunt, stinking or covered smut; causes a disease very similar to the one caused by *Tilletia tritici* but is less widespread and apparently absent from UK. The ustilospores are released at harvest and direct infection of the seedling via the basidiospores takes place, infection is systemic; the volatile trimethylamine causes the characteristic fishy smell as the bunt balls are ruptured to release the ustilospores; there are *c*.17 races; control through seed treatment and resistance.

T.lolii Auersw. 1854; Mordue, *Descr.F.* 804, 1984; ustilospores reticulate, 16–23 μm diam.; *Lolium* spp., ryegrass covered smut, prob. seedborne, not economically important; this sp. needs to be distinguished from other members of the genus on temperate cereals.

T. sphaerococca (Rabenh.) A. Fischer v. Waldh. 1867; Mordue, *Descr.F.* 805, 1984; ustilospores reticulate, 20–32 μm diam., *Agrostis* spp., seedborne, prob. of little economic importance.

T. tritici (Bjerk.) R. Wolff 1874, fide Mordue & Ainsworth 1984; Mordue & Waller, *Descr.F.* 719, 1981, as *Tilletia caries* (DC.) Tul.; ustilospores 14–23 μm, reticulate, i.e. rougher than those of *T. laevis* which causes a virtually identical disease; wheat common bunt, stinking or covered smut, also forms trimethylamine; widespread. It was with this smut that Tillet demonstrated in 1755 that a fungus could cause a plant disease. The life cycle is almost the same as for *T. laevis* and seed treatment has almost eliminated the disease. But, since the fungus may also be soilborne, outbreaks of this wheat bunt can take place, e.g. in USA (Pacific N.W.); there are races of a similar number to that in *T. laevis* which can hybridise with both *T. controversa* and *T. tritici*, the 3 fungi forming *c*.40 races.

Tilletiaceae, Ustilaginales; ustilospore germinates by an aseptate promycelium which bears terminal basidiospores; these, after conjugation, form secondary spores or con. The basidiospores, sporidia or primary con. have been interpreted to be highly specialised sterigmata because the secondary spores behave like ballistospores q.v., i.e. typical basidiospores; *Entyloma*, *Tilletia*, *Urocystis*.

Tilletiopsis Derx 1948; Hyphom.; not given by Carmichael et al. 1980 = *Itersonilia* q.v. Hijwegen, *Neth.J.Pl.Path.* **92**:93, 1986, described the control of *Sphaerotheca fuligenea* on cucumber by *T. minor* [65,4607].

timorensis, *Pseudocercospora*.

tinangaja, a disease of coconut similar to the one caused by the coconut cadang cadang viroid; Guam and poss. other islands in the Marianas; Boccardo et al., *Phytop.* **71**:1104, 1981; in Maramorosch & McKelvey 1985 [61,2386].

tinctorium, *Echinodontium*.

tinder rot (*Fomes fomentarius*) beech, birch.

tingens, *Atropellis*; **tingitaninum**, *Oidium*.

tip and twig blight (*Diplodia pinea*) pine, other conifers; —— **blight** (*Myrothecium roridum*) coffee, (*Thanatephorus cucumeris*) pea; —— **over** (*Erwinia chrysanthemi* pv. *paradisiaca*) banana, (*Phomopsis vexans*) eggplant; **tipburn**, disorders: brassicas, Ca deficiency, Maynard et al., *HortScience* **16**:193,

1981; lettuce, etiology doubtful, poss. in part Ca deficiency, Collier & Tibbits, *Hort.Rev.* **4**:49, 1982; Lahoz et al., *Inftore.fitopatol.* **37**(1):47, 1987; potato, excessive moisture loss; strawberry, Ca deficiency, Maas 1984 [**60**,6708].

Tisdale, Wendell H., 1892–1973; born in USA, Alabama Polytechnic Institute, Univ. Wisconsin, USDA cereal office 1919–26, Du Pont Co.; introduced organic fungicides with I. Williams in 1931. *Phytop.* **63**:1323, 1973.

tissue culture, including cell culture and protoplasts; Hollings, *A.R.Phytop.* **3**:367, 1965, virus free stock; Black, *ibid* **7**:73, 1969, insect tissue culture & plant viruses; Zaitlin & Beachy, *Adv.Virus Res.* **19**:1, 1974, protoplasts & separated cells in plant virology; Takebe, *A.R.Phytop.* **13**:105, 1975, protoplasts & plant virology; Brettell & Ingram, *Biol.Rev.* **54**:329, 1979; Henshaw in Ebbels & King 1979; Peberdy, *A.Rev.Microbiol.* **33**:21, 1979, fungal protoplasts; Ingram & Helgeson, *Tissue culture methods for plant pathologists*, 1980; Slack, *PD* **64**:14, 1980, meristem tip culture; Earle & Gracen in Staples & Toenniessen 1981, protoplasts & cell cultures; Helgeson & Deverall ed., *Use of tissue culture and protoplasts in plant pathology*, 1983; Styer & Chin, *Hort.Rev.* **5**:221, 1983, meristem & shoot tip culture; Daub, *A.R.Phytop.* **24**:159, 1986, tissue culture & selection for resistance.

toadskin, tobacco, genetic disorder; Wolf & Sharp, *J.Elisha Mitchell scient.Soc.* **68**:85, 1952 [**32**:456].

tobacco, *Nicotiana tabacum*; Hopkins, *Tobacco diseases with special reference to Africa*, 1956; Wolf, *Tobacco diseases and decays*, edn. 2, 1957; Nakamura, *Bull.Hatano Tob.Exp.Stn.* 59, 1967; Lucas 1975; Gayed, *Publs.Canada Dep.Agric.* 1641, 1978.

—— **ascending necrosis** Fernandez, *Revta.Agric.Cuba* **4**:81, 1971.

—— **broad ringspot** Johnson & Fulton, *Phytop.* **32**:605, 1942; transm. sap, USA (Wisconsin) [**21**:539].

—— **broken ringspot** Smith & Markham, *Phytop.* **34**:324, 1944; Sastry, *Indian Phytopath.* **18**:316, 1965; transm. sap; England, India [**23**:363; **46**,735].

—— **bushy top** Gates, *AAB* **50**:169, 1962, transm. sap, *Myz.pers.* but only in the presence of an assistor which is prob. tobacco vein distorting v., Zimbabwe [**41**,546].

—— **cabbaging** (tobacco leaf curl v.).

—— **chlorotic blotch** Agur et al. 1982; isometric particle 34–45 nm diam. assoc., USSR (Estonia) [**61**,5942].

—— **club root** Valleau & Johnson 1932; Valleau,

Phytop. **37**:580, 1947; Selsky, *ibid* **51**:581, 1961; USA (Kentucky) [**12**:117; **27**:101; **41**:252].

—— **curly leaf** (tobacco leaf curl v.).

—— **distortion mosaic** Azad & Sehgal, *Indian J.agric.Sci.* **28**:373, 1958; transm. sap, India (Simla Hills) [**39**:44].

—— **etch v.** Blakeslee 1921; Purcifull & Hiebert, *Descr.V.* 258, 1982; Potyvirus, filament 730 × 12–13 nm, transm. sap, > 10 aphid spp., non-persistent, infects > 120 spp. in 19 families of dicotyledons, causes diseases in solanaceous crops, including red pepper and tomato.

—— **false broomrape** Valleau, *PDR* **37**:538, 1953; Nielsen, *Phytop.* **68**:1068, 1978; poss. caused by a bacterium; Nicaragua, USA (S.E.) [**33**:325; **58**,1943].

—— **female sterility** Kostoff, *Phytopath.Z.* **5**:593, 1933; USSR [**12**:599].

—— **floral regression** Borges & Martins 1971; MLO assoc. [**52**,4212].

—— **green and yellow ringspots** Valleau, *Bull.Ky.agric.Exp.Stn.* 327:43, 1932 (tobacco ringspot v.) [**12**:471].

—— **leaf curl v.** Peters & Schwartz 1912; Osaki & Inouye, *Descr.V.* 232, 1981; Geminivirus, isometric pairs 25–30 × 15–20 nm, transm. *Bemisia tabaci*, persistent, mainly in the tropics and subtropics, fairly wide host range, also causes diseases in red pepper and tomato.

—— **mild green mosaic v.** McKinney, *J.agric.Res.* **39**:557, 1929; as mild dark green tobacco mosaic v.; Wetter, in van Regenmortel & Fraenkel-Conrat 1986, considered the virus to be a distinct member of the Tobamoviruses [**9**:260].

—— **mosaic v.** Mayer 1886; Zaitlin & Israel, *Descr.V.* 151, 1975; Tobamovirus, type, rod *c.*300 × 18 nm; the first virus to be purified and shown to contain RNA, transm. sap easily, spreads rapidly through plant contact and by man in cultural operations, many strs., Martyn 1968, 1971; very wide host range, causes diseases in many plants, common in field tobacco; Waterson & Wilkinson 1978; Zimmern in Day & Jellis 1987.

—— **mottle v.** Smith 1945; *Parasitology* **37**:131, 1946; Vanderveken, *Parasitica* **19**:65, 1963; **20**:1, 1964; transm. sap, *Myz.pers.*, persistent, but only in the presence of tobacco vein distorting v., one component of tobacco rosette, Africa (Central, E.), see dependent transm. [**24**:208; **26**:318; **43**,828, 3302].

—— **necrosis v.** Smith & Bald 1935; Kassanis, *Descr.V.* 14, 1970; Necrovirus, type, isometric *c.*30 nm diam., contains 1 segment linear ssRNA, transm. sap but not usually infecting plants systemically, also transm. zoospores of *Olpidium brassicae*, soilborne, causes bean stipple streak, tulip necrosis or Augusta, and diseases in citrus, grapevine, pear, potato tubers; Uyemoto in Kurstak 1981; Thomas 1984, epidemiology of cucumber necrosis str. in England; Francki et al., vol. 1, 1985 [**63**,5193].

— **necrotic dwarf v.** Kubo, Tamura & Fukuda 1976; Kubo, *Descr.V.* 234, 1981; Luteovirus, isometric *c*.25 nm diam., transm. aphid, persistent, Japan, narrow host range.

— **rattle v.** Quanjer 1943; Harrison, *Descr.V.* 12, 1970; Tobravirus, rod, longer *c*.190 nm, shorter 45–115 nm, depending on isolate; transm. sap, some isolates readily; and soil inhabiting *Paratrichodorus* and *Trichodorus* spp.; many strs., mostly on light peaty or sandy soils inhabited by the vectors, wide host range, causes: beet yellow blotch, gladiolus notched leaf, potato corky ringspot, potato stem mottle, red pepper ringspot; and in other plants.

— **ring necrosis** Ivancheva-Gabrovska, *Bull.Pl.Prot.Sofia* 3:19, 1955; transm. *Thrips*, Bulgaria, poss. = tomato spotted wilt v. [**37**:420].

— **ringspot v.** Fromme, Wingard & Priode 1927; Stace-Smith, *Descr.V.* 309, 1985; Nepovirus, type, isometric *c*.28 nm diam., transm. sap, *Xiphinema* spp., seed; N. America, wide host range, both herbaceous and woody plants; several diseases are caused and the virus causes problems in this region where the vectors are also present.

— **rosette** (tobacco mottle v. + tobacco vein distorting v.), see both viruses and dependent transm.

— **streak v.** Johnson 1936; Fulton, *Descr.V.* 307, 1985; Ilarvirus, isometric 27–35 nm diam., transm. sap, *Thrips tabaci* and other spp., seed, widespread but not often epidemic, many hosts, some diseases caused: asparagus stunt, bean red node, pea, potato and soybean systemic necroses, rose vein yellowing, strawberry necrotic shock, tomato yellow ringspot and malformation; bean red node v. and black raspberry latent v. are prob. distinct strs.

— **stunt v.** Hidaka 1950; Kuwata & Kubo, *Descr.V.* 313, 1986; rod, commonly 300–340 nm long, 18 nm wide; contains dsRNA, transm. sap, *Olpidium brassicae*, Japan, found naturally only in tobacco but transm. to 35 spp. in 13 plant families through the viruliferous vector; Hiruki, *Pl.Path.* **36**:224, 1987, recovery from air dried, resting spores of *O. brassicae* [**66**,4169].

— **tumour** Nienhaus & Gliem, *Phytopath.Z.* **78**:367, 1973; transm. sap, USA (California) [**53**,3610].

— **veinal necrosis** (potato Y v.); these strs. of PVY have been destructive on potato in Europe and

S. America, fide Klinkowski, *A.R.Phytop.* **8**:41, 1970.

— **vein banding** = potato Y v.

— — — **mosaic v.** Chin, *Ann.Rep.Tob.Res.Inst.Taiwan* 1971; filament 780 nm long, transm. sap, *Myz.pers.*, Taiwan.

— — **distorting v.** Smith 1946, see tobacco mottle v.; Faoro et al., *Micron.* **13**:399, 1982; poss. Luteovirus, isometric *c*.25 nm diam., transm. *Myz.pers.* mainly an assistor virus for tobacco mottle v. and prob. for tobacco bushy top, also a component of tobacco rosette, see dependent transm.

— — **mottling v.** Gooding & Sun 1972; Sun et al., *Phytop.* **64**:1133, 1974; filament 765 × 13 nm, transm. sap, *Myz.pers.*, USA (S.E.), Solanaceae [**54**,1429].

— **wilt v.** Badami & Kassanis, *AAB* **47**:90, 1959; long filament, transm. sap, *Myz.pers.*, India, from *Solanum jasminoides* [**38**:523].

— **yellow dwarf v.** Hill 1937; Thomas & Bowyer, *Descr.V.* 278, 1984; Geminivirus, isometric pairs 35 × 20 nm, transm. *Orosius argentatus*, Australia; causes a lethal necrosis, summer death, in bean, and severe dwarfing in tobacco.

— — **net** Abeygunawardena, Karandawela & Bandaranayake, *Trop. Agriculturist* **123**:37, 1967; Sri Lanka [**48**,914].

— — **vein v.** Adams & Hull, *AAB* **71**:135, 1972; a combination of 2 viruses, transm. aphid spp., persistent; one component TYVV is sap transm. but aphids transm. only when the other or assistor component, which is not sap transm., is present. TYVV is also aphid transm. from plants containing 1 of 2 other assistor viruses, tobacco vein distorting v. and groundnut rosette v., Malawi [**52**,1255].

Tobamoviruses, type tobacco mosaic v., *c*.11 definitive members, rod *c*.300 × 18 nm, 1 segment linear ssRNA, sedimenting at *c*.190 S, thermal inactivation *c*.90°, longevity in sap in years, highly contagious; spread by contact, handling, soil, sometimes in seed; Gibbs, *Descr.V.* 184, 1977; van Regenmortel in Kurstak 1981; Francki et al., vol. 2, 1985; van Regenmortel & Fraenkel-Conrat, vol. 2, 1986.

Tobraviruses, type tobacco rattle v., pea early browning v. is the only other member; rod, a divided genome of 2 segments linear ssRNA, each encapsidated separately, long rod 180–215 × 22 nm which is infective, short rod 45–115 × 22 nm which is non-infective alone but codes for the coat protein, thermal inactivation 70–80°, longevity in sap in months, transm. sap and nematodes in soil, viruliferous nematodes remain infective for weeks; Harrison & Robinson, *Adv.Virus Res.* **23**:25,

1978; in Kurstak 1981; in van Regenmortel & Fraenkel-Conrat, vol. 2, 1986.

toe rot (*Phytophthora verrucosa*) tomato.

togashiana, *Mycosphaerella*.

tolaasii, *Pseudomonas*; **tolerandus**, *Uromyces*, see *U. manihotis*.

tolerant, of a host that is colonised by a pathogen but shows no disease symptoms. Tolerance should not be confused with, and is not a form of, resistance where the colonisation is restricted by some host mechanism. The antonym is sensitive q.v. Clarke, *Adv.Pl.Path.* **5**:161, 1986, discussed the difficulty over defining tolerance; Schafer, *A.R.Phytop.* **9**:235, 1971; Gaunt, *Phytop.* **71**:915, 1981; Mussell in Staples & Toenniessen 1981; Clarke in Wood & Jellis 1984.

tolyofluanid, like dichlofluanid.

Tolyposporidium Thirum. & Neerg., *Friesia* **11**:179, 1977, published 1978; Ustilaginaceae; sori in the inflorescence, usually in ovaries; ustilospores bound together in the spore ball somewhat permanently, not loose or pulverulent. This genus was erected because the type sp. of *Tolyposporium* is a *Sorosporium*.

T. penicillariae (Bref.) Thirum. & Neerg.; Ainsworth, *Descr.F.* 77, 1965, as *Tolyposporium penicillariae* Bref.; sori in ovaries, spore balls 42–325 × 50–175 μm, ustilospores mostly 8–10 μm diam.; bulrush millet smut, infects the inflorescence; Holliday 1980; Rao & Thakur, *TBMS* **81**:597, 1983, morphology & culture; Wells et al., *Phytop.* **77**:293, 1987, effects of inoculation & pollination on smut development. These 2 papers do not notice the transfer made by Thirumalachar & Neergaard [**63**,1249; **66**,3797].

Tolyposporium Woronin ex Schröter 1982; see *Tolyposporidium*.

tomatine, in tomato; its antifungal properties have been much studied in tomato showing resistance to *Fusarium oxysporum* f.sp. *lycopersici*; Pegg & Woodward, *Physiol.Molec.Pl.Path.* **28**:187, 1986, in tomato isolines, resistance to *Verticillium albo-atrum* [**65**,4545].

tomato, *Lycopersicon esculentum*; see desert habitats; *Pest control in tropical tomatoes*, 1983, Centre Overseas Pest Res., London; Martelli & Quacquarelli, viruses, see red pepper; Fletcher 1984; Sherf & MacNab 1986; Pohronezny et al., *PD* **70**:96, 1986, control in USA (Florida).

—— , *Alternaria*.

—— **apical stunt**, as for tomato bunchy top viroid.

—— **aspermy v.** Blencowe & Caldwell 1949; Hollings & Stone, *Descr.V.* 79, 1971; Cucumovirus, isometric *c*.30 nm diam., transm. sap, aphid spp., non-persistent; uncommon in tomato, causing leaf distortion and usually

seedless fruit; common in chrysanthemum, causing severe flower break, dwarfing and distortion; wide experimental host range, many variants, poss. str. is tomato condensed top.

—— **big bud** Cobb 1902; Samuel et al. 1933; Helms, *Aust.J.agric.Res.* **8**:135, 148, 1957; Hill & Helson, *J.Aust.Inst.agric.Sci.* **15**:160, 1949; Bowyer et al., *Aust.J.biol.Sci.* **22**:271, 1969; MLO assoc., transm. *Orosius argentatus*, strs.; see potato purple top wilt, Harding & Teakle [**13**:62; **30**:215; **36**:650; **48**,2242].

—— **black ring v.** Smith 1964; Murant, *Descr.V.* 38, 1970; Boswell & Gibbs 1983; Nepovirus, isometric 26 nm diam., transm. sap readily, pollen, *Longidorus* spp., Europe, wide host range, infects seed of many crops; causes ringspots in: bean, beet, celery, lettuce, raspberry, strawberry and other plants.

—— **Brazilian curly top** Costa, *Phytop.* **42**:396, 1952; poss. str. of beet curly top v. [**32**:220].

—— —— **ringspot and yellow band** (tobacco rattle v.), Silberschmidt, *Phytopath.Z.* **46**:209, 1963 [**42**:634].

—— **bronze leaf** (tomato spotted wilt v.).

—— **bunchy top viroid** McClean, *Bull.S.Africa Dep.Agric.* 100, 1931; Diener 1979; this viroid causes symptoms in potato and tomato that are very like those caused by potato spindle tuber viroid. Both viroids have many common hosts, but there are differences. Diener treated the 2 viroids separately as they may be chemically distinct.

—— **bushy stunt v.** Smith 1935; Martelli et al., *Descr.V.* 69, 1971; Smith 1972; Tombusvirus, type, isometric *c*.30 nm diam., transm. sap readily, by seed from red pepper and tomato, and can be obtained from soil; wide experimental host range, causes diseases in: carnation, chlorotic spot; cherry, fruit pitting, veinal necrosis, stunt; globe artichoke, leaf deformation, mottle, reduced growth, sterility; pelargonium, distortion, yellow stellate spots; petunia, asteroid mosaic; Robinson & Harrison, *Nature Lond.* **297**:563, 1982, structure of expanded state; Fischer & Lockhart, *Phytop.* **67**:1352, 1977, prob. first record of red pepper infection; Tomlinson & Faithfull, *AAB* **104**:485, 1984, from river Thames, England; infection in plants grown from seed of symptomless but infected tomato fruit [**57**,3730; **61**,6273; **63**,4329].

—— **condensed top** Yassin & Nour, *PDR* **49**:599, 1965; Yassin, *Phytopath.Mediterranea* **14**:32, 1975; Sudan, poss. str. of tomato aspermy v. [**44**,3157; **55**,4855].

—— **corky ringspot** (tomato mosaic v.), Mayhew et al., *PD* **68**:623, 1984 [**63**,5550].

—— **curly top** (beet curly top v.); Martin & Thomas, *PD* **70**:136, 1986, levels & dependability of resistance in tomato [**65**,2973].

—— **double streak**, as for tomato mixed streak.

—— **etch** (tobacco etch v.).

—— **fern leaf** (cucumber mosaic v.), Mogendorff, *Phytop.* **20**:25, 1930; also caused by tobacco mosaic v. [**9**:417].

—— **fuzzy vein** (cowpea mild mottle v.), Brunt & Phillips, *Trop.Agric.Trin.* **58**:177, 1981 [**60**,6084].

—— **golden mosaic v.** Costa et al. and Matyis et al. 1975; Buck & Coutts, *Descr.V.* 303, 1985; Geminivirus, isometric pairs *c.*25 × 13 nm, transm. sap, difficult from tomato to tomato, *Bemisia tabaci*, Solanaceae, tropical Brazil, minor economic importance.

—— **leaf curl** (tobacco leaf curl v.).

—— —— **reduction** (potato Y v. + tomato aspermy v.), Schmelzer & Wolf, *Arch.Phytopath.PflSchutz.* **11**:3, 1975 [**54**,4634].

—— **lethal necrosis**, see carna 5 RNA.

—— **mal azul** Borges & David-Ferreira, *Bolm.Soc.broteriana 2* **42**:321, 1968; MLO assoc., like tomato big bud, Portugal [**48**,2566].

—— **malformation v.** El Maataoui & Lockhart, abs. *Phytop.* **72**:989, 1982; bacilliform 360 × 80 nm, Morocco.

—— **marginal flavescence** Singh & Sastry, *Phytopath.Mediterranea* **19**:136, 1980; poss. MLO, transm. *Orosius albicinctus*, India [**63**,248].

—— **mixed streak** (potato X v. + tomato mosaic v.).

—— **mosaic v.** Clinton 1909; Hollings & Huttinga, *Descr.V.* 156, 1976; Tobamovirus, rod *c.*300 × 18 nm, transm. sap readily to many herbaceous plant spp., spread by handling, contaminates seed, can be epidemic on tomato, Broadbent, *A.R.Phytop.* **14**:75, 1976; many strs., symptoms affected by: cv., day length, light intensity plant age, str., temp. Symptoms: light and dark green mosaic leaf mottle, conspicuous yellow mosaic leaf mottle; necrosis of fruit, leaf, stem; in red pepper a severe leaf necrosis and fall, stunt; tobacco mosaic v. strs. seldom occur in tomato and compete poorly with those of TomMV; Martelli & Quacquarelli, *Acta Hortic.* 127:39, 1983, list of diseases caused in tomato; Goto, *Res.Bull.Hokkaido Natn.agric.Exp.Stn.* 140:103, 1984; Brunt in van Regenmortel & Fraenkel-Conrat 1986.

—— **necrosis** (alfalfa mosaic v.), Knorr et al., *Phytop.* **73**:1554, 1983; (pelargonium zonate spot v.), Vovlas et al., *Inftore.fitopatol.* **36**(2):39, 1986 [**63**,1949].

—— **necrotic dwarf v.** Larsen, Duffus & Liu, abs. *Phytop.* **74**:795, 1984; isometric, transm. sap, *Bemisia tabaci*, USA (California).

—— **pale chlorosis** (cowpea mild mottle v.), Antignus & Cohen, *AAB* **110**:563, 1987; Israel.

—— **Peruvian v.** Raymer et al. 1972; Fribourg & Fernandez-Northcote, *Descr.V.* 255, 1982; Potyvirus, filament *c.*750 × 12 nm, transm. sap, *Myz.pers.*, non-persistent; mosaic, epinasty; twisting, crinkling and necrotic spotting on leaves.

—— **Philippine leaf curl** Retuerma, Pableo & Price, *Philipp.Phytopath.* **7**:29, 1971; transm. *Bemisia tabaci* [**52**,4218].

—— **planta macho viroid** Belalcazar & Galindo 1974; Galindo et al., *Phytop.* **72**:49, 1982; distinct from potato spindle tuber viroid which also infects tomato. Pre-inoculation with a mild str. of potato spindle tuber viroid gives little if any protection against tomato planta macho viroid. The latter infects *Gynura aurantiaca*, in which it multiples, but causes no symptoms. But potato spindle tuber and citrus exocortis viroids both cause symptoms in *G. aurantiaca*. This viroid causes severe stunt in tomato, leaves show strong epinasty, become brittle and dry out. The fruit remains small and is unmarketable; Mexico; Orozco Vargas & Galindo Alonso *Revta.Mexicana Fitopat.* **4**:19, 1986, hosts, effect of temp. & agriculture on viroid distribution [**55**,3744; **61**,5215; **66**,313].

—— **pseudo-curly top** Giddings, Bennett & Harrison, *Phytop.* **41**:415, 1951; Simons & Coe, *Virology* **6**:43, 1958; Christie et al., abs. *Phytop.* **74**:800, 1984; poss. Geminivirus, transm. *Micrutalis malleifera* to tomato and *Solanum gracile*, USA (Florida) [**30**:492; **38**:37].

—— **ringspot v.** Price 1936; Stace-Smith, *Descr.V.* 290, 1984; Nepovirus, isometric *c.*28 nm diam., transm. sap readily, seed, *Xiphinema* spp., naturally in N. America but in plant material elsewhere, mostly in perennials, diseases caused: apple union necrosis; peach, and other *Prunus* spp., yellow bud mosaic, stem pitting, decline; grapevine decline, prune brown line, raspberry ringspot and decline; weeds, including seeds, are sources of the virus; Bitterlin & Gonslaves, *PD* **71**:408, 1987, spatial distribution of *X. rivesi* & persistence of virus & vector in soil.

—— **scorch** (potato Y v. + tobacco mosaic v.), Clark et al., *Phytop.* **70**:131, 1980 [**59**,5362].

—— **shatter** Fraser, *Agric.Gaz.N.S.W.* **60**:419, 1949, Australia [**29**:128].

—— **shoestring** Doering, Price & Fenne, *Phytop.* **47**:310, 1957; transm. sap, USA (Virginia); (potato Y v. necrotic str.), Behl et al., *Curr.Sci.* **56**:27, 1987 [**36**:736].

—— **smalling and severe mottle** Vasudeva 1957; India (N.) [**39**:74].

—— **spotted wilt v.** Samuel, Bald & Pittman 1930;

Ie, *Descr.V.* 39, 1970; isometric 70–90 nm diam., enveloped, contains 4 segments linear ssRNA, transm. sap, thrips spp., persistent; characteristic leaf bronzing, and causes tomato tip blight; host range very wide, kromneck of tobacco and tomato, many variants, oak leaf patterns on some plants; Chagas & Vicente in Kranz et al. 1977; Francki & Hatta in Kurstak 1981; Milne & Francki, *Intervirology* 22:72, 1984, poss. member of the Bunyaviridae; Francki et al., vol. 1, 1985; Cho et al., *PD* 71:505, 1987, severe loss in lettuce & other crops in USA (Hawaii).

— streak v. Ladipo & Roberts, *AAB* 87:133, 1977; filament *c.*780 nm long, transm. sap. *Myz.pers.*, Nigeria (Ife-Ife), killed many plants, like pepper veinal mottle v. [57,1842].

— super budding, or giant calyx, Costa, *Biológico* 15:79, 1949; Brazil, like tomato big bud [28:598].

— symptomless Vasudeva & Sam Raj, *Curr.Sci.* 16:348, 1947; transm. sap, *Bemisia tabaci*, India [27:203].

— tisis Adsuar, *J.Agric.Univ.P.Rico* 39:113, 1955; Puerto Rico [35:401].

— top leaf curl Newton & Peiris, *F.A.O.Pl.Prot.Bull.* 2:17, 1953; transm. whitefly, Sri Lanka [33:582].

— — necrosis v. Bancroft, *Phytop.* 58:1360, 1968; poss. Nepovirus, isometric 26 nm diam., transm. sap, USA (Indiana) [48,927].

— vein clearing v. Kano et al., *Ann.phytopath.Soc.Japan* 51:606, 1985; Phytorhabdovirus, bacilliform, *c.*240–250 × 86–88 nm, transm. sap to 7 plant spp. including *Nicotiana glutinosa*, *Petunia hybrida*, tomato; host range different from that of tomato vein yellowing v. [65,4541].

*— — yellowing v. El Maataoui, Lockhart & Lesemann, *Phytop.* 75:109, 1985; Phytorhabdovirus, bacilliform 265 × 86 nm, transm. sap, Morocco, Solanaceae, in the perennial weed *Solanum sodomaeum* [64,2718].

— white leaf Gonslaves, Provvidenti & Edwards, *Phytop.* 72:1533, 1982; an apparent satellite RNA & cucumber mosaic v. [62,2143].

— — necrosis v. Chagas, Vicente & July, *Arq.Inst.biol.S.Paulo* 42:157, 1975; isometric 24 nm diam., transm. sap, Brazil [57,292].

— yellow band (tobacco rattle v.), Salomão et al. 1969 [48,2564].

— — dwarf (tobacco leaf curl v.), Osaki & Inouye, *Ann.phytopath.Soc.Japan* 44:167, 1978 [58,1136].

— — leaf curl v. Cohen & Harpaz 1964; Cohen & Nitzany, *Phytop.* 56:1127, 1966; Russo et al., *J.gen.Virology* 49:209, 1980; Geminivirus, isometric, each half of pair 15–17 nm diam.,

transm. *Bemisia tabaci*, Asia (W.), Makkouk & Laterrot in Plumb & Thresh 1983; Ioannou, *Pl.Path.* 34:428, 1985, epidemic on tomato in Cyprus [44,841; 46,286; 60,1617; 65,370].

— — mosaic v. Debrot, Herold & Dao, *Agronomía trop.* 31:33, 1963; Uzcátequi & Lastra, *Phytop.* 68:985, 1978; Lastra & Gil, *ibid* 71:524, 1981; prob. Geminivirus, isometric, each half of pair 18–20 nm, transm. sap, *Bemisia tabaci*, Venezuela [43,2728; 58,1953; 61,1386].

— — net v. Sylvester, *Phytop.* 44:219, 1954; poss. Luteovirus, transm. *Myz.pers.*, USA (California) [33:766].

— — ringspot and malformation (tobacco streak v.).

— yellows Zitter & Tsai, *PD* 65:787, 1981; Zitter & Everett, *ibid* 66:456, 1982; transm. *Myz.pers.*, poss. a str. of potato leaf roll v. [61,2433, 6577].

— yellow top v. Alstatt & Ivanoff 1945; Costa 1949; Sutton 1955; Braithwaite & Blake, *Aust.J.agric.Res.* 12:1100, 1961; Thomas, *AAB* 104:79, 1984; Hassan et al., *Phytop.* 75:287, 1985; Luteovirus, isometric *c.*24 nm diam., transm. *Macrosiphum euphorbiae*, *Myz.pers.*, persistent; Australia, Brazil, USA; a severe disease, field plants show a bright yellow terminal growth, reduced leaf size, leaf curling and death of flower buds [24:342; 28:598; 35:493; 41:484; 63,2488; 64,5116].

— — vein Ladipo, *PDR* 61:958, 1977; transm. sap, Nigeria (Ife-Ife) [57,3102].

Tombusviruses, type tomato bushy stunt v., poss. 8 other members, isometric 30–34 nm diam., 1 segment linear ssRNA, thermal inactivation 85–90°, longevity in sap weeks to months, transm. sap, seed and sometimes soil, wide host ranges; Martelli in Kurstak 1981; Francki et al., vol. 1, 1985; Gallitelli et al. and Gallitelli & Hull, *J.gen.Virol.* 66:1523, 1533, 1985, relationships between members, characterisation of assoc. satellite RNAs.

tombusvirus neckar, reference as for potexvirus sieg, transm. sap, Germany (river Neckar); Gallitelli & Russo, *J.Phytopath.* 119:106, 1987, properties.

tomentosus, *Inonotus.*

tonic effect, a single spray with Bordeaux mixture, to apparently healthy coffee plants, *c.* doubled the yield in Kenya; but when mistimed the practice aggravated disease situations; Griffiths, *Trop.Sci.* 14:79, 1972 [52,126].

top dying, in *Picea abies*, Norway spruce, in coastal areas of N.W. Europe, prob. due to water stress created by an adverse climate; Diamandis, *Eur.J.For.Path.* 8:337, 345, 357, 1978; 9:78, 175, 183, 1979 [58,4533–5; 59,2378–80].

— **necrosis** (*Gibberella intricans*) cowpea, Ramachandran et al. 1982 [**61**,7259].

topple, tulip, mostly in forced plants and due to Ca deficiency; Moore 1979; Bould et al. 1983.

toppling (*Pratylenchus goodeyi, Radopholus similis*) banana.

top rot (*Pseudomonas rubrilineans*) sugarcane; — **sickness**, tobacco, B deficiency, van Schreven 1934 [**13**:600].

tornatum, *Ganoderma.*

torrendiella, *Catacauma*, see *Phyllachora.*

torres-juanii, *Elytroderma.*

Torula Pers. 1801; Hyphom.; Ellis 1971, 1976; conidiomata hyphal, conidiogenesis holoblastic; con. 0–4 septate, pigmented, catenate, acropetal, frequently verruculose or echinulate; Rao & de Hoog, *Persoonia* **8**:199, 1975.

T. ligniperda (Wilk.) Sacc.; Siggers, *Phytop.* **12**:369, 1922, stain in hardwoods; Campbell & Davidson, *J.For.* **39**:63, 1941, red heart of *Betula papyrifera*, paper birch [**2**:34; **20**:326].

toruloidea, *Hendersonula*; **torulosa**, *Deightoniella*; **tosquinetii**, *Taphrina.*

toxin, any compound formed by a pathogen in its host and which causes a part or all of the disease syndrome, Turner in Wood & Jellis 1984. Scheffer, *Pl.Path.* **31**:193, 1982, discussed terminology and outlined the criteria for establishing the existence of a toxin; and see Turner. Toxins are either host-specific, i.e. toxic only to plants that show disease caused by the pathogens forming them; or non-specific, i.e. toxic also to plants which are not susceptible to the pathogens forming them. Pathotoxin, phytotoxin and vivotoxin are used but are ambiguous and therefore undesirable. Mycotoxin, also ambiguous, usually refers to compounds formed by fungi and which are toxic to animals including man. Aggressins has been proposed as a term for compounds formed by micro-organisms which are not toxic but which have some role in pathogenesis. Some toxins are given in the order. Pringle & Scheffer, *A.R.Phytop.* **2**:133, 1964; Wright, *A.Rev.Microbiol.* **22**:269, 1968; Wood et al. 1972; Strobel, *ibid* **31**:205, 1977; Yoder, *A.R.Phytop.* **18**:103, 1980; Yoder & Pegg in Staples & Toenniessen 1981; Durbin 1981; Strobel, *A.Rev.Biochem.* **51**:309, 1982; Daly & Knoch, *Adv.Pl.Path.* **1**:83, 1982; Daly & Deverall 1983; Mitchell, *A.R.Phytop.* **22**:215, 1984; Scheffer & Livingston, *Science N.Y.* **223**:17, 1984 [**63**,3251].

Toxoptera aurantii, transm. camellia yellow mosaic, coffee blister spot; *T. citricidus*, transm. citrus tristeza, citrus vein enation.

Toya propinqua, transm. cynodon chlorotic streak.

tracheiphila, *Deuterophoma, Erwinia, Phialophora, Phoma, Pseudopeziza.*

tracheomycosis, FBPP prefer the synonym vascular wilt q.v.; another synonym, hadromycosis, is little used; — (*Gibberella xylarioides*) coffee.

Trachysphaera Tabor & Bunting 1923; Pythiaceae; con. echinulate, av. 35 μm diam., oogonia with outgrowths like bosses or fingers; monotypic. *T. fructigena* Tabor & Bunting, Holliday, *Descr.F.* 229, 1970; largely saprophytic, enters hosts through wounds or when tissue is moribund, causes banana finger rot, found in banana ripening rooms; cacao mealy pod, coffee berry rot; central and W. Africa, Madagascar; Holliday 1980.

tradescantia leaf distortion v. Lockhart & Betzold, *Acta Hortic.* 110:55, 1980; Lockhart et al., *Phytop.* **71**:602, 1981; Potyvirus, filament av. 754 nm long, transm. sap, *Myz.pers.*, *Rhopalosiphum padi*, non-persistent, USA, infected *Rhoeo* and *Zebrina* only [**61**,1231].

tragia mosaic Sathiarajan & Gopalan, *Agric.Res.J.Kerala* **14**:88, 1976; transm. whitefly, India, *T. involucrata* [**56**,3924].

tragopogonis, *Albugo.*

transcapsidation, or genome masking; the enclosure of the nucleic acid, the genome, of one virus within a complete protein capsid of a second virus, FBPP.

transfer feeding time, see inoculation feeding time.

transmission, of a virus, see dependent, mechanical and virus.

— **threshold period**, the minimum time necessary for a vector to transmit a virus, beginning with acquisition feeding and finishing at the end of test feeding, after FBPP.

transversalis, *Uromyces.*

Tranzschelia Arthur 1906; Uredinales; Wilson & Henderson 1966; teliospores 1 septate, adhere in groups or fascicles, the lower parts of the pedicels are united to form compound bases; the 2 cells of the teliospores separate readily, are coarsely verrucose, dark brown, one pore in each cell; either heteroecious with telia on Rosaceae and aecia on Ranunculaceae, or autoecious on Ranunculaceae; Bennell & Henderson, *TBMS* **71**:271, 1978, uredospore & teliospore development [**58**,2584].

T. pruni-spinosa (Pers.) Dietel 1922; var. *pruni-spinosa* and var. *discolor* (Fuckel) Dunegan 1938; Laundon & Rainbow, *Descr.F.* 287–8, 1971; var. *pruni-spinosa* has telia on wild *Prunus* spp., aecia on *Anemone* spp., especially *A. ranunculoides* and poss. other genera; teliospores 30–45 × 12–26 μm, both cells strongly verrucose; plum rust. Var. *discolor* has telia on peach and other cultivated stone fruit, aecia on *A. coronaria*; teliospores 28–40 × 17–23 μm, the lower cell smaller, paler, thinner walled, less strongly verrucose; peach and stone fruit rust, rust of florist's anemone. The 2 vars. have not always been distinguished, but since only var. *discolor* attacks economically important crops most work relates to it; mainly a leaf disease which causes premature defoliation. Bolkan et al., *PD* 69:485, 1985, proposed ff.sp. on almond, peach and plum; Kable et al. *ibid* 70:202, 1986, physiologic specialisation & infection efficiency; Michailides & Ogawa, *ibid* page 307, chemical control [64,5441; 65,3405, 5046].

traversiana, *Cercospora*.

tree hoppers, see vector.

trees, see major texts, ornamentals, pine, wetwood, wood decay; Bingham et al., *A.R.Phytop.* 9:433, 1971, resistance; Petersen & Smith, *Forest nursery diseases in the United States*, *Handb.USDA* 470, 1975; Bakshi, *Forest pathology, principles and practice in forestry*, 1976; Lanier et al., *Mycologie et pathologie forestières*, 1976; Hepting & Cowling, *A.R.Phytop.* 15:431, 1977; Bega, *Diseases of Pacific coast conifers*, *Handb.USDA* 521, 1978; Binns et al., *Leaflet For.Comm.UK* 76, 1980, nutrient deficiencies in conifers; Blanchard & Tattar, *Field and laboratory guide to tree pathology*, 1981; Funk, *Parasitic microfungi of western trees*, Canadian For.Sev.Victoria, British Columbia, 1981; Marks et al., *Tree diseases in Victoria*, Handb. For.Comm.Victoria, Australia 1, 1982; Bloomberg, *A.R.Phytop.* 23:83, 1985, nursery diseases.

tree tomato, see tamarillo.

trefoil, *Lotus, Medicago, Melilotus, Trifolium* and other genera.

treleasei, *Atropellis*.

trellis rust (*Gymnosporangium fuscum*) pear.

tremelloides, *Gymnosporangium*; **tremulae**, *Drepanopeziza, Marssonina*.

tretic, of conidiogenesis, an enteroblastic type where each conidium is limited by an extension of the inner wall of the conidiogenous cell; polytretic is the formation of tretoconidia by the extrusion of the inner wall through several channels, monotretic is extrusion through one channel.

triadimefon, systemic fungicide, protectant and curative activity against powdery mildews and rusts.

Trialeurodes vaporariorum, transm. beet pseudo-yellows, lettuce blotchy interveinal yellows, lettuce interveinal yellows, lettuce pseudo-yellows, maple Hungarian mosaic, muskmelon yellows; Bos, *Gewasbescherming* 11:189, 1980; van Dorst et al., *Neth.J.Pl.Path.* 89:171, 1983 [61,1606; 63,2537].

triandrae, *Drepanopeziza*.

trianthema mosaic Singh & Verma, *Pflanzenschutzberichte* 45:1, 1975; transm. sap. India (N.), *T. portulacastrum* [55,3247].

Trichocladium Harz 1871; Hyphom.; Ellis 1971; see *Thielaviopsis basicola*.

Trichoderma Pers. 1801; Hyphom.; Carmichael et al. 1980; conidiomata hyphal, conidiogenesis enteroblastic, phialidic; con. aseptate, hyaline, in mass white or shades of green and yellow; cosmopolitan, especially in soil, not pathogenic to plants, but sometimes antagonistic to, and parasitic on, plant pathogenic fungi, see biological control; Rifai, *Mycol.Pap* 116, 1969, taxonomy; Brasier, *Nature Lond.New Biol.* 231:283, 1971, induction of sexual reproduction in *Phytophthora* by *T. viride*; Haskins & Gardner, *CJB* 56:1651, 1978, effects on sexual reproduction in *Phytophthora* & *Pythium*; Santos & Dhingra, *ibid* 60:472, 1982, pathogenicity on sclerotia of *Sclerotinia sclerotiorum*; Bell et al., *Phytop.* 72:379, 1982, in vitro antagonism to fungus plant pathogens; Seĭketov 1982; Elad et al., *ibid* 73:85, 1983; *Phytopath.Z.* 107:168, 1983, microscopy of parasitism; Papavizas, *A.R.Phytop.* 23:23, 1985, biology & biological control [51,1143; 58,545; 61,5449–50, 6245; 62,1829; 63,438].

T. hamatum (Bonorden) Bainier 1906; Chet & Baker, *Phytop.* 71:286, 1981; Harman et al., *ibid* 70:1167, 1980; 71:569, 1981; Hubbard et al., *ibid* 73:655, 1983 [60,5154; 61,1060, 2509; 62,4516].

T. harzianum Rifai 1969; Abd-El-Moity, Elad, Hadar, Henis, Papavizas, Sivan et al., *Phytop.* 69:64, 1979; 70:119, 1980; 72:121, 396, 1982; 73:1469, 1983; 74:106, 498, 1984; Marshall and Sivan et al., *PD* 66:788, 1982; 71:587, 1987; Abd-El-Moity & Shatla, *Phytopath.Z.* 100:29, 1981; Elad et al., *Crop.Protect.* 1:199, 1982 [59,709, 5476; 60,6199; 61,4695, 6130, 6973; 62,463; 63,1622, 2576, 3813].

T. viride Pers.:Fr. 1829; see *Chondrostereum purpureum*; Bliss, *Phytop.* 41:665, 1951, Garrett 1970; Bolton, *Can.J.Pl.Path.* 2:93, 1980; Orlikowski & Schmidle, *NachrBl.dt.PflSchutzdienst.* 37:78, 1985 [30:608; 60,1480; 64,5009].

Trichodorus cylindricus, *T. minor*, transm. tobacco rattle, European isolates.

T. primitivus (de Man) Micoletzky 1922; Hooper & Siddiqi, *Descr.N.* 15, 1972; transm. pea early browning, English isolates; transm. tobacco rattle, European isolates; assoc. with docking disorder of beet.

T. similis Seinhorst 1963; Wyss, *Descr.N.* 59, 1974; root ectoparasite, can cause stubby root, transm.

tobacco rattle, European isolates; assoc. with docking disorder of beet.

T. viruliferus Hooper 1963; *Descr.N.* 86, 1976; ectoparasite, mostly Europe (W.), transm. viruses as for *Trichodorus primitivus*; assoc. with docking disorder of beet.

Tricholomataceae, Agaricales; basidiospores white, yellowish, pink, buff; never dark; lamellae very rarely free; *Armillaria, Calyptella, Crinipellis, Leucopaxillus, Marasmiellus, Marasmius, Mycena*.

Trichosphaeriaceae, Sphaeriales; ascomata superficial, often setose, asci lacking a distinct apical apparatus, asco. often septate, usually lacking distinct germ pores or slits; *Plectosphaerella*.

trichostoma, *Pyrenophora*; **trichothecioides**, *Fusarium*.

Trichothecium Link 1809; Hyphom.; Rifai & Cooke, *TBMS* **49**:147, 1966; conidiomata hyphal, conidiogenesis holoblastic; con. 1 septate, hyaline, solitary; *T. roseum* (Pers.) Link, common on fruit after harvest, apple pink mould; Deems, *Phytop.* **41**:633, 1951, on tomato fruit [**30**:634].

tricorpus, *Verticillium*.

tricyclazole, systemic fungicide for rice blast.

tridactyla, *Podosphaera*.

tridax mosaic Singh & Verma 1979; Ghana, *T. procumbens* [**62**,2251].

tridemorph, protectant and systemic fungicide, cereal mildews and fruit crops.

trifoliate orange, *Poncirus trifoliata*.

trifolii, *Colletotrichum, Curvularia, Cymadothea, Erysiphe, Leptosphaerulina, Olpidium* see *O. viciae, Polythrincium, Pseudopeziza, Stemphylium, Typhula*; **trifolii-repentis**, *Uromyces*; **trifoliorum**, *Peronospora, Sclerotinia*.

Trifolium, clovers; *T. alexandrinum*, berseem or Egyptian; *T. hybridum*, alsike; *T. incarnatum*, crimson; *T. pratense*, red; *T. repens*, white; *T. subterraneum*, subterranean.

triforine, broad spectrum systemic fungicide.

trillium flower greening Hooper, Case & Myers, *PDR* **55**:1108, 1971; MLO assoc., Canada (Ontario), USA (Michigan, New York State), *T. grandiflorum* [**51**,2306].

trimethylamine, see *Tilletia laevis, T. tritici*.

Trinidad cowpea mosaic (cowpea severe mosaic v.).

Trioza erytreae, transm. citrus greening.

triphenyltin, protectant fungicides with some eradicant action.

Triphragmiopsis Naumov 1914; Uredinales; teliospores ± globose, 3 celled, 2 pores/cell; Monoson, *Mycopath.Mycol.appl.* **52**:115, 1974. Shao et al. and Sun et al. 1983 described a brown rust damaging larch in China and attributed it to *T. laricinum* (Chou)Tai. This rust appears to be

Triphragmium laricinum Chou 1955 on *Larix olgensis, Index Fungi* **3**:392, 1967. *Triphragmium* Link 1825 differs in that the teliospores have one pore/cell. Monoson does not refer to a sp. of either genus on *Larix* [**64**,1304; **65**,1546].

Tripsacum laxum, Guatemala grass.

tritici, *Clavibacter, Septoria, Tilletia, Ustilago*; **triticina**, *Alternaria*; **tritici-repentis**, *Drechslera, Pyrenophora*.

Triticum, wheat.

tritonia mosaic Smith & Brierley, *Phytop.* **34**:593, 1944, USA [**23**:488].

tropaeoli, *Acroconidiella*.

Tropaeolum, nasturtium; Horváth et al., *Acta Hortic.* 110:219, 1980, viruses.

tropaeolum mosaic v. Delhey & Monasterios, *Z.PflKrankh.PflSchutz.* **84**:224, 1977; filament *c.*800 nm long, transm. sap readily, Bolivia, *T. tuberosum* [**56**,5087].

Trophurus imperialis Loof 1956; Siddiqi, *Descr.N.* 22, 1973.

tropism, see taxis.

troyanus, *Marasmiellus*.

true eyespot (*Pseudocercosporella herpotrichoides*), wheat, temperate cereals and grasses.

truncata, *Verticicladiella*; **truncatum**, *Colletotrichum*.

trunk gall and canker (poss. *Cylindrocarpon didymum*) mangrove; — **rot** (*Phytophthora katsurae*) chestnut; — **scab** (*Cryptodiaporthe populea*) poplar.

Tryon's scab (*Sphaceloma fawcettii* var. *scabiosa*) citrus.

Tsai's disease (maize stripe v.).

Tsuga, hemlock.

T toxins, from *Cochliobolus heterostrophus*, Smedegard-Petersen & Nelson, *CJB* **47**:951, 1969; one of the most studied groups of toxic substances formed by a fungus pathogen; Wood et al., 1972; Durbin 1981. A similar toxin is formed by *Phyllostica maydis*, Comstock et al., *Phytop.* **63**:1357, 1973 [**48**,3478; **53**,1798].

tuber black pit (*Alternaria alternata*) potato.

Tubercularia Tode 1790; Hyphom.; Carmichael et al. 1980; conidiomata sporodochial, conidiogenesis enteroblastic phialidic; con. aseptate, hyaline.

T. ulmea J. Carter, *Phytop.* **37**:246, 1947; con. mostly 4.5–6 × 1.5–2.5 μm; canker of *Ulmus pumila*, prob. only attacks weakened trees [**26**:364].

T. vulgaris, teleom. *Nectria cinnabarina*.

tuberculatum, *Phragmidium*.

Tuberculina Tode ex Sacc. 1880; Hyphom.; Carmichael et al. 1980; conidiomata sporodochial, conidiogenesis enteroblastic phialidic; con.

aseptate, hyaline; on rusts; see *Uromyces koae*, Gardner et al.

tuber dry rot (*Fusarium solani* var. *coeruleum*, *Gibberella cyanogena*) potato; —— **rot** (*Botryodiplodia theobromae*, *Rhizopus oryzae*) yam; —— **stem end hard rot** (*Phacidiopycnis tuberivora*) potato.

tuberivora, *Phacidiopycnis*.

tuber rot eelworm, *Ditylenchus destructor*.

Tubeuf, Carl Freiherr von, 1862–1941; one of the founders of plant pathology who worked contemporaneously with Frank, Kühn and Sorauer in German speaking Europe; Univ. Munich; his book *Pflanzenkrankheiten durch kryptogame Parasiten verursacht*, 1895, was translated into English by W.G. Smith, *Diseases of plants induced by cryptogamic parasites*, 1897; Ainsworth 1981.

Tucker, Clarence Mitchell, 1897–1954; born in USA, Univ. Missouri, Puerto Rico and Federal Agricultural Experiment Station, 1920–30; an authority on *Phytophthora*; plant pathologists still await a monograph on this genus to replace Tucker's of 1931. *Phytop.* **45**:351, 1955.

tuckeri, *Oidium*; **tucumanensis**, *Glomerella*.

tulare apple mosaic v., see apple Tulare mosaic v.; TulAMV is apparently the preferred name.

Tulasnellales, Holobasidiomycetidae; distinct external, resupinate basidiomata, basidia subspherical to broad, not forked; with stout, like fingers, or inflated, usually 4, sterigmata; basidiospores repetitive; *Koleroga*.

tulip, *Tulipa*.

tulipae, *Botrytis*; **tuliparum**, *Rhizoctonia*; **tulipifera**, *Mycosphaerella*.

tulip Augusta (tobacco necrosis v.), Kassanis, *AAB* **36**:14, 1949; Mowat, *ibid* **66**:17, 1970, gave a reassessment of the disease and described the factors involved in infection and disease outbreaks [**28**:290; **50**,1276].

—— **breaking v.** Cayley 1928; van Slogteren, *Descr.V.* 71, 1971; Potyvirus, filament *c*.740 nm long, transm. sap, aphid spp., *Lilium, Tulipa*; colour breaking in tepals of tulip cvs. with anthocyanins. These cvs. have been known and prized in Europe since the flower's introduction in the 16th century. The earliest known record of symptoms caused by a plant virus are those due to infection by eupatorium yellow vein; Takahashi et al. 1970, further strs.; Derks et al., *Neth.J.Pl.Path.* **88**:87, 1982, purification & antisera production [**51**,1552; **62**,688].

—— **chlorotic blotch v.** Mowat & Duncan 1982; Mowat, *AAB* **106**:65, 1985; Potyvirus, filament *c*.720 × 12 nm, transm. sap, *Myz.pers.*; unlike tulip breaking v. it is not transm. to *Lilium*

formosanum; but causes symptoms in tulip leaves and flowers identical with those induced by tulip breaking v. in field tulips in Australia [**62**,1058; **65**,1928].

—— —— **stunt** Smookler & Dabush, *PDR* **58**:1142, 1974; MLO assoc., Israel [**54**,2295].

—— **corky fleck** (cucumber mosaic v.), van Slogteren, *Meded.LandbHoogesch.OpzoekStns.Gent* **31**:986, 1966, necrosis of bulb scales [**46**,2369s].

—— **halo necrosis v.** Mowat, *AAB* **69**:147, 1971; prob. not a free nucleic acid, transm. sap to 15 sp. in 5 families of Angiosperms, Scotland [**51**,2592].

—— **mild mosaic**, and tulip grey, Asjes & Segers 1980; rod shaped particles assoc., Netherlands [**60**,2612].

——**mosaic** (tulip breaking v.).

——**mottle and stunt** (tobacco rattle v.), van Slogteren, *Tijdschr.Pl.Ziekt.* **64**:452, 1958 [**38**:479].

—— **necrosis** (tomato bushy stunt v.), Mowat, *Pl.Path.* **21**:171, 1972, symptoms similar to those caused by tobacco necrosis v. in tulip Augusta; (tobacco mosaic v.), Mokrá et al., *Phytopath.Z.* **76**:46, 1973 [**52**,1937, 3726].

—— **top breaking v.** Asjes & Segers 1982; Potyvirus, Netherlands [**62**,3057].

—— **veinal streak** (tobacco ringspot v.), Asjes, *Neth.J.Pl.Path.* **78**:19, 1972; Asjes & Muller, *Phytopath.Z.* **76**:328, 1973 [**51**,2593; **53**,197].

—— **X v.** Mowat, *AAB* **101**:51, 1982; *Descr.V.* 276, 1984; Potexvirus, filament 495 × 13 nm, transm. sap, infection of most herbaceous plant spp. erratic but reliable in *Chenopodium*, causes distinctive elliptical markings on leaves of tulip which is the only known natural host, Scotland but distribution prob. wider [**61**,7057].

tumefaciens, *Agrobacterium, Diplodia, Sphaeropsis*.

tumour, or gall, a localised proliferation of plant tissue forming a swelling or outgrowth, commonly with a characteristic shape and unlike any organ of the normal plant. Galls are usually formed in response to the action of a pathogen or pest, after FBPP; see *Agrobacterium tumefaciens*; Mani, *Ecology of plant galls*, 1964; Braun ed., *Plant tumour research, Progr.exp.tumour Res.* vol. 15, 1972; Beiderbeck, *Pflanzentumoren*, 1977; Kahl & Schell ed., *Molecular biology of plant tumours*, 1982; Dixon in Wood & Jellis 1984.

tundu (*Clavibacter tritici*) wheat.

tung, *Aleurites*; Wiehe, *PDR* suppl. 216:189, 1952, bibliography.

Tunstallia Agnihothrudu, *Phytopath.Z.* **40**:280, 1961; poss. Sphaeriales; ascomata with prominent, conical, raised, ostiolar areas, like thorns; asco. aseptate, ± hyaline. The genus was erected to accommodate *T. aculeata* (Petch)

Agnihothrudu which causes tea thorny stem
blight, asco. 80–110 × 13 μm; a var. has asco.
100–162 × 6–12 μm; a wound pathogen of
weakened plants in India and Sri Lanka; Holliday
1980 [**40**:560].

turcica, *Setosphaeria*; **turcicum**, *Exserohilum*.

turf, see grasses.

turmeric, *Curcuma domestica*.

turnip, see *Brassica*.

—— **crinkle v.** Broadbent & Blencowe 1955;
Hollings & Stone, *Descr.V.* 109, 1972; poss.
Tombusvirus, isometric 28 nm diam., transm. sap;
flea beetles, *Phyllotreta*, *Psylliodes*; fairly wide
host range; Altenbach & Howell, *Virology* **112**:25,
1981, assoc. of a satellite RNA [**61**,446].

—— **mild yellows** (beet western yellows v.).

—— **mosaic v.** Gardener & Kendrick 1921;
Tomlinson, *Descr.V.* 8, 1970; Potyvirus, filament
*c.*750 nm long, transm. sap, many aphid spp.,
non-persistent, causes:mottle, black spots in
Brussels sprout, cabbage, cauliflower; stunt, leaf
distortion, mosaic, necrosis in Chinese cabbage,
horseradish, mustard, radish, rape, swede, turnip,
watercress; mottle, ringspot in rhubarb;
flowerbreak in some ornamentals; Tomlinson &
Ward and Walkey & Webb, *AAB* **89**:61, 435,
1978, infection of swede, internal necrosis of
stored cabbage; Sako & Ogata,
Ann.phytopath.Soc.Japan **47**:68, 1981, helper
factor essential for aphid transm; Provvidenti, *PD*
66:1076, 1982, lethal disease in *Impatiens
balsamina*; Klisiewicz, *ibid* **67**:112, 1983, severe
mosaic in safflower [**57**,4669; **58**,1998; **61**,904;
62,684, 2067].

—— **rape chlorantie** Morvan,
C.r.hebd.Séanc.Acad.Agric.Fr. **44**:505, 1958,
France [**37**:689].

—— **rosette v.** Blencowe & Broadbent 1957;
Hollings & Stone, *Descr.V.* 125, 1973;
Sobemovirus, isometric *c.*28 nm diam., transm.
sap, poss. flea beetles, Scotland, rarely found in
swede and turnip, few hosts outside Cruciferae.

—— **yellow mosaic v.** Markham & Smith 1949;
Matthews, *Descr.V.* 230, 1980; *Intervirology*
15:121, 1981; Tymovirus, isometric 28 nm diam.,
transm. sap; flea beetles, *Phyllotreta*, *Psylliodes*;
few hosts outside Cruciferae; Guy & Gibbs,
Australasian Pl.Path. **10**:12, 1981; *Pl.Path.*
34:532, 1985, str. in *Cardamine*, from the endemic
C. lilacina, transm. *Pedilophorus* sp., Australia;
Benetti & Kaswalder, 1982–3; Hein,
Z.PflKrankh.PflSchutz. **91**:549, 1984, both seed
transm. [**61**,1079; **63**,5595; **64**,2399; **65**,1572].

twig blight (*Phytophthora ilicis*) holly; —— ——**and
wilt** (*P. syringae*) lilac; —— **dieback** (*Macrophoma
theicola*) tea, (*Pseudomonas ficuserectae*) *Ficus*

erecta; —— —— **and stem canker** (*Potebniamyces
coniferarum*) conifers.

twist (*Dilophospora alopecuri*) grasses, temperate
cereals.

twister (*Glomerella cingulata*) onion, Ebenebe, *PD*
64:1030, 1980 [**60**,4795].

twisting rust (*Melampsora pinitorqua*) pine, see
M. populnea.

twitch grass, *Agropyron repens*.

Twort, Frederick William, 1877–1950; born in
England, qualified at St Thomas' Hospital,
London; in 1915 he discovered the lytic
phenomenon which d'Herelle later called
bacteriophagy; FRS 1929. *Brit.med.J.* **1**:788,
1950; *Obit.Not.Fel.R.Soc.Lond.* **7**:505, 1951.

Tylenchorhynchus annulatus (Cassidy) Golden 1971;
Siddiqi, *Descr.N.* 85, 1976; migratory ectoparasite,
warmer regions; rice, sugarcane, other Gramineae.

T. claytonii Steiner 1937; Loof, *Descr.N.* 39, 1974;
cooler regions; cereals, grasses, some trees.

T. cylindricus Cobb 1913; Siddiqi, *Descr.N.* 7,
1972; Egypt, USA (S. and W.); citrus, cotton,
grapevine, *Pinus ponderosa*.

T. dubius (Bütschli) Filipjev 1936; Bridge, *Descr.N.*
51, 1974; browsing ectoparasite, very common,
assoc. with crops in Europe, can cause stunt.

Tylenchulus semipenetrans Cobb 1913; Siddiqi,
Descr.N. 34, 1974; citrus root nematode, causes
citrus slow decline, different races on other hosts.

Tylenchus davainei Bastian 1865; Andrassy,
Descr.N. 97, 1977; cosmopolitan, not parasitic.

tylophora mosaic Sastry & Khan, *Indian J.agric.Sci.*
37:128, 1967; transm. sap, aphid spp., India
(Jammu), *T. indica* [**46**,3532].

Tymoviruses, type turnip yellow mosaic v. *c.*17
definitive members, isometric *c.*30 nm diam.,
1 segment linear ssRNA, 2 particle types
sedimenting at *c.*115 and 54 S, thermal
inactivation 65–95°, longevity in sap a few weeks,
transm. sap and beetle vectors known for several
members; Koenig & Lesemann, *Descr.V.* 214,
1979; in Kurstak 1981; Guy et al., *Pl.Path.*
33:337, 1984, taxonomy of hosts; Francki et al.,
vol. 1, 1985 [**63**,5324].

Tympanis Tode 1790; Helotiaceae; ascomata
erumpent, leathery usually in clusters from bark,
asco. septate or aseptate, but soon disappearing to
be replaced by many secondary, rod shaped
spores which fill the ascus; Groves, *CJB* **30**:571,
1952; Ouellette & Pirozynski, *ibid* **52**:1889, 1974
[**32**:280].

T. confusa Nyl. 1868; asco. > 2 septate,
13–20 × 2–4 μm; canker of red pine, *Pinus resinosa*
in N. America; Hansbrough, *Science N.Y.* **81**:408,
1935; *Bull.Yale Univ.Sch.For.* 43, 1936; on
Korean pine, *P. koraiensis* in China, Cui et al.

1984, Xiang et al. 1985 [**14**:612; **16**:289, **66**,1168–9].

typhina, *Epichloë*, **typhinum**, *Acremonium*.

Typhula (Pers.) Fr. 1818; Clavariaceae; basidiomata slender, filiform, unbranched, with a fertile, subglobose to elongate head, usually arising from sclerotia, up to 2 cm high; mostly saprophytic but can cause snow moulds q.v. The fungi are found on winter cereals, turf and sometimes forage legumes under snow; Bruehl & Cunfer, *Phytop*. **65**:755, 1975; Berthier, *Bull.Soc.linn.Lyon* num.spéc. 45, 1976; Matsumoto & Sato, *Ann.phytopath.Soc.Japan* **49**:293, 1983; Jacobs & Bruehl, *Phytop*. **76**:695, 1986 [**55**,1206; **63**,2796; **66**,155].

T. idahoensis Remsburg 1940; this sp. and *Typhula ishikariensis* Imai 1930 cause speckled snow mould, their sclerotia are black, they cannot be distinguished in vitro. Årsvoll & Smith, *CJB* **56**:348, 1978, considered them to be the same; Christen & Bruehl, *Phytop*. **69**:263, 1979, considered otherwise; Bruehl & Machtmes, *ibid* **70**:867, 1980; Matsumoto et al., *Ann.phytopath.Soc.Japan* **48**:275, 419, 1982 [**57**,3820; **59**,1126; **60**,4301; **62**,588, 1825].

T. incarnata Fr.; sclerotia pink becoming red brown, 0.5–5 mm diam., basidiomata pink; Jacobs & Bruehl, *Phytop*. **76**:278, 1986, compared the behaviour of *Typhula incarnata* with those of *T. idahoensis* and *T. ishikariensis* in winter wheat fields in USA (Idaho, Washington State). The first sp. developed over a wider area and had a competitive advantage in attacking crown and root tissues beneath the soil surface. The 2 other spp. were more frequent on leaves compared with *T. incarnata*; Detiffe & Maraite, *Parasitica* **41**:79, 1985, temp. & germination of sclerotia; Matsumoto & Tajimi, *CJB* **63**:1126, 1985, field survival of sclerotia with *T. ishikariensis*; Wale, *Tests Agrochem.Cvs.* 7 *AAB* **108** suppl.:42, 1986, chemical control in barley [**65**,129, 4853; **66**,1890].

T. phacorrhiza Reichard:Fr, 1821, fide Remsburg, *Mycologia* **32**:52, 1940; Schneider & Seaman, *Can.J.Pl.Path.* **8**:269, 1986, on winter wheat in Canada (Ontario), not previously throught to be pathogenic, seen on necrotic plants [**66**,2314].

T. trifolii Rostrup 1890; sclerotia 1.4–2 mm diam., clover; Noble, *Ann.Bot.* **1**:67, 1937; Ylimäki, *Ann.Agric.Fenn.* **8**:30, 1969 [**16**:388; **48**,3027].

U

udbatta (*Balansia oryzae-sativae*) rice.
udum, *Fusarium*.
ufra (*Ditylenchus angustus*) rice.
ulei, *Aposphaeria*, *Microcyclus*.
Ullstrup, Arnold J. 1907–85; born in USA,
Univ.Wisconsin, Purdue; an authority on maize
diseases. *Phytop.* 76:37, 1986.
ullucus C v. Brunt et al. 1982; Brunt & Jones,
Descr.V. 277, 1984; Comovirus, isometric c.28 nm
diam., M and B particles, transm. sap readily, in
U. *tuberosus* crops in the highlands of Bolivia and
Peru.
—— mosaic v. Brunt et al., *AAB* 101:65, 1982;
Potyvirus, filament c.750 × 12 nm, transm. sap,
Myz.pers., non-persistent, U. *tuberosus*, region as
above [61,7271].
ulmarius, *Rigidoporus*; ulmea, *Stegophora*,
Tubercularia; ulmeum, *Asteroma*; ulmi,
Ceratocystis, *Mycosphaerella*, *Pesotum*,
Phloeospora, *Plectophomella*.
Ulmus, elm.
Ulocladium Preuss 1851; Hyphom.; Ellis 1971,
1976; conidiomata hyphal, conidiogenesis
enteroblastic tretic; con. muriform, pigmented.
Compared with *Alternaria* the con. are mostly
solitary, not rostrate, commonly ellipsoidal or
obovoid, the proximal end is tapered and
narrower than the distal end. U. *atrum* Preuss
1852, con. av. 18.6 × 16 μm; Butler et al. *Pl.Path.*
28:96, 1979, cucumber leaf spot in England. U.
chartarum (Preuss) Simmons 1967, con. av.
24.4 × 14.7 μm; Mimbela-Leyva et al., *Trop.Sci.*
17:61, 1975, assoc. with melon fruit rot;
Simmons, *Mycologia* 59:67, 1967, typification &
distinction, *Alternaria*, *Stemphylium*, *Ulocladium*;
Mycotaxon 14:44, 1982, for U. *cucurbitae* (Let. &
Roum.) Simmons [46,2394; 59,4371; 61,3960].
ultimum, *Pythium*.
ultrastructure, see electron microscopy.
ultraviolet, as a cause of death of fungus spores,
Rotem et al., *Phytop.* 75:510, 1985 [64,4733].
umbrina, *Cryptosporella*; umbrinella, *Discula*.
Uncinula Lév. 1851; Erysiphaceae; mycelium
superficial, ascoma with several asci, appendages
coiled at their tips; conidiophores usually
abundant, con. often borne singly; Spencer 1978.
U. bicornis (Wallr.) Lév.; anam. *Oidium aceris*
Rabenh. 1854; Mukerji, *Descr.V.* 190, 1968,
appendages forked, asco. 15–26 × 7.6–15 μm, con.
24–36 × 15–20 μm; on *Acer*, may cause defoliation
in nurseries.

U. euonymi-japonicae, poss. anam. *Oidium
euonymi-japonicae* q.v.
U. necator (Schwein.) Burrill 1892; anam. *Oidium
tuckeri* Berk. 1855; Kapoor *Descr.F.* 160, 1967;
asco. 15–25 × 10–14 μm; con. 32–39 × 17–21 μm,
usually catenate; grapevine powdery mildew. The
fungus caused damage in Europe soon after the
devastation to potato crops there by *Phytophthora
infestans*. In France the grape harvest was reduced
to one fifth between 1847 and 1854. Sulphur
sprays, already being used against fruit tree
mildews, were found to give effective control of
the grapevine pathogen by the English gardener
Tucker. The use of sulphur by the French growers
enabled them to recover their production by 1858.
Uncinula necator debilitates the vines and causes
the fruit to burst. The teleom. is relatively rare in
Europe, and its biology is less well known
compared with that of the anam. Overwintering is
in the buds as mycelium and sometimes as con.
Most work is now on chemical control which is
essential; sulphur may still be used. Incidence of
the mildew is affected by the site of the vineyard,
pruning pattens and choice of host var.; Large
1940; Dye & Hammett, *N.Z.J.exp.Agric.* 5:63,
1977, yields & chemical control; Bulit & Lafon in
Spencer 1978; Sall, *Phytop.* 70:338, 1980,
epidemiology, a mathematical model; Wicks et al.,
Agric.Rec. 11(16):12, 16, 1984, fungicide
programs & evaluation in southern Australia
[57,721; 60,2706; 65,1428–9].
undulata, *Rhizina*.
Unger, Franz Joseph Andreas Nicolaus, 1800–70,
born in Austria, practised medicine, taught botany
at Univ. Vienna, studied fungi when plant
physiology was tending to supersede the early
dominance of taxonomy; Whetzel 1918.
ungeri, *Chalara*; unicorne, *Seiridium*.
unitunicate, see ascus.
Unkanodes albifascia, U. *sapporana*, transm. rice
black streaked dwarf, rice stripe.
unmottled curly dwarf (potato spindle tuber viroid)
potato.
upland rice soil sickness (*Pyrenochaeta terrestris*
assoc.).
upper stem rot (primary cause *Phellinus noxius*) oil
palm.
urd bean, see *Vigna*.
—— —— leaf crinkles, several of these diseases have
been described and in some cases an isometric
particle 25–30 nm diam.; Kolte & Nene 1972;

Narayanasamy & Jaganathan, *Madras agric.J.*
60:651, 1973; Beniwal & Bharathan, *Indian
Phytopath.* **33**: 600, 1980; Boswell & Gibbs 1983;
Bhardwaj & Dubey, *J. Phytopath.* **115**:83, 1986
[**53**,1630; **54**,1509; **61**,4508; **65**,3083].

—— **mosaics**, several of these uncharacterised,
symptom patterns have been described from
India; sap, aphid transm. [**43**,2144; **50**,396;
54,1061; **59**,1543, 4855].

Urediniomycetes, Uredinales; Basidiomycotina;
macroscopic basidiomata absent, the rust fungi
are a major group of obligate plant parasites and
pathogens; there has been some limited axenic
culture; complex and varied life cycles which are
described with an elaborate and confusing
terminology. There are up to 5 spore stages, 0–IV,
described separately: spermatia from
spermagonia, aeciospores from aecia, uredospores
from uredia, teliospores from telia, and
basidiospores from teliospores. Generally a
macrocyclic rust is one with all 5 stages; it may be
heteroecious, with telial and aecial stages on
taxonomically very different plants; or autoecious
with all stages on one plant or very closely related
plants. A microcyclic rust is usually one where the
aecial stage is absent, in addition the uredial stage
may be absent; even shorter cycles, with only an
aecial stage or a uredial one, exist.

Arthur 1934; Wilson & Henderson 1966;
Laundon in Ainsworth et al. IVB, 1973; Ziller
1974; Cummins 1971, 1978, with Hiratsuka,
Illustrated genera of rust fungi, 1983; Scott &
Chakravorty, *The rust fungi*, 1983; Bushnell &
Roelfs, *The cereal rusts*, 2 vol., 1984–5; Browder,
A.R.Phytop. **23**:201, 1985, specificity in cereal
rusts; Hoch & Staples, *ibid* **25**:231, 1987,
appressorial development.

*Aecidium, Caeoma, Cerotelium, Chrysomyxa,
Gymnoconia, Gymnosporangium, Hemileia,
Kuehneola, Melampsora, Melampsorella,
Melampsoridium, Olivea, Peridermium,
Phakopsora, Phragmidium, Physopella, Puccinia,
Pucciniastrum, Tranzschelia, Triphragmiopsis,
Uredinopsis, Uredo, Uromyces, Uromycladium.*

uredinis, *Kuehneola.*

Uredinopsis Magnus 1893; Ziller 1974;
Uredinales; macrocyclic, heteroecious, spores are
hyaline; uredospores beaked, fusiform; telia
on ferns, aecia on *Abies*, may severely damage
firs.

Uredo Pers. 1801, Uredinales; only uredospores
present.

U. cajani Syd. 1906; Anahosur & Waller, *Descr.F.*
590, 1978; uredospores 21–27 × 18–21 μm; pigeon
pea rust; the fungus has been confused with
Uromyces dolicholi which only forms teliospores

and is prob. restricted to *Rhynchosia*. *Uredo cajani*
can cause defoliation.

U. ficina Juel 1897; Laundon & Rainbow, *Descr.F.*
289, 1971; uredospores 26–35 × 20–25 μm, spines
1.5–3 μm high; *Ficus* rust, distinguished from
Cerotelium fici by the conspicuous paraphyses,
slightly larger uredospores and longer spines.

U. fuchsiae, see *Pucciniastrum epilobii.*

U. musae, see *Uromyces musae.*

uredospore, urediospore; repeating, vegetative spore
of a rust, usually on a dikaryotic mycelium;
typically aseptate, pedicillate, deciduous,
pigmented, echinulate, one or more germ tubes;
Staples & Macko in Bushnell & Roelfs, vol. 1,
1984, see Urediniomycetes.

uredovora, *Erwinia.*

Urnula Fr. 1849; Sarcosomataceae; Dissing,
Mycologia **73**:263, 1981.

U. craterium (Schwein.) Fr. 1851; ascomata stipitate,
asco. 29–35 μm long; Wolf, *Mycologia* **50**:837,
1958, poss. anam. *Conoplea globosa* (Schwein.)
Hughes 1958; Davidson, *ibid* **42**:735, 1950, as
Strumella coryneoidea Sacc. & Winter 1883; con.
7–12 × 4.5–6 μm; oak canker, on hardwoods,
USA; Bidwell & Bramble, *J.For.* **32**:15, 1934;
Sleeth & Lorenz, *Phytop.* **35**:671, 1945; Fergus,
ibid **41**:101, 1951 [**13**:406; **25**:85; **30**:395, 494].

urochloa yellowing Muniyapa, Rao & Govindu,
Curr.Sci. **51**:427, 1982; MLO assoc., India
(Mysore), *U. panicoides* [**61**,7285].

Urocystis Rabenh. 1856; Tilletiaceae; Mordue &
Ainsworth 1984; ustilospores in balls of one or
more spores, ± covered by a cortex of smaller,
lighter coloured, sterile cells; Fischer 1953;
Mundkur & Thirumalachar 1953.

U. agropyri (Preuss) Schröter 1869; Mordue &
Waller, *Descr.F.* 716, 1981; spore balls av. 26.5
μm diam., usually of 1–3 ustilospores av. 15.5 μm
diam.; flag smut; barley, wheat, grasses; linear,
black, erumpent sori on leaves including sheaths;
spore balls survive for 4 years on seed and for 3
years in soil; at least 4 main races on wheat.

U. anemones (Pers.) Winter 1881; Mordue, *Descr.F.*
806, 1984; spore balls each of 1 ustilospore,
ustilospores smooth, globose to ± angular or
elongated 12–26 μm diam., *Anemone, Ranunculus*;
little economic importance.

U. cepulae Frost 1877; Mulder & Holliday,
Descr.F. 298, 1971; spore balls each of 1
ustilospore, ustilospores spherical to ellipsoid,
smooth, av. 13.8 μm diam.; onion smut, infection
of other *Allium* spp. is much less serious, not
likely to be a severe disease in the lowland
tropics; penetration before emergence, each
successive leaf can become infected as it passes
through the susceptible phase, dark spore masses

are formed at the leaf base, spread mostly by setts and transplants, long persistence in soil; Walker 1969; Holliday 1980; Dixon 1981; Sherf & MacNab 1986.
U. gladiolicola Ainsw. 1950; Mordue, *Descr.F.* 807, 1984; spore balls mostly of 1–2 ustilospores 16–34 µm diam., ustilospores globose or slightly angled, av. 13.3 µm diam.; *Gladiolus* spp.; the fungus is present in the corms, on which leaden coloured swellings are formed, and is not always detected; this smut may be severe in commercial production; Moore 1979.
U. occulta Rabenh. 1856; Mordue, *Descr.F.* 808, 1984; sori in leaves, culms and inflorescences, linear blisters; spore balls mostly of 1–2 ustilospores av. 13.6 µm diam.; rye flag, stalk or stripe smut, other temperate Gramineae, seedborne, survives a few months in soil, disease can be severe.
Uroleucon (Uromelan) gobonis, transm. burdock yellows, sonchus Indian yellow vein.
Uromyces (Link) Unger 1833; Uredinales; Wilson & Henderson 1966; Cummins 1971, 1978; differs from *Puccinia* only in that the teliospores are aseptate. Whereas *Puccinia* is particularly important on Gramineae, *Uromyces* is more so on Leguminosae; Holliday 1980; Dixon 1981; Allen 1983; Sherf & MacNab 1986.
U. aloes (Cooke) Magnus 1892; teliospores 31–44 × 28–34 µm, on *Aloe* and *Haworthia*, little damage caused; Nakamura et al., *Trans.mycol.Soc.Japan* 17:342, 1976; Sato et al., *ibid* 21:273, 1980, teliospore germination, infection & hosts [57,1745; 60,6323].
U. appendiculatus (Pers.) Unger 1836; teliospores smooth or sparsely warted or striate, 28–38 × 20–36 µm, macrocyclic, autoecious; *Dolichos*, *Phaseolus*, *Vigna*; the aecial stage is very rare and sometimes the telial stage also; other *Uromyces* spp. on these plants may not be distinct. Bean rust is an important disease but is restricted by high temp. The crop can be lost if infection takes place in the early stages of plant growth. Overcast conditions and a temp. of 20–25° favour bean rust, many races. The biochemical and physiological relationships of host and pathogen have been intensively investigated; Imhoff et al. *Phytop.* 71:577, 1981; 72:72, 441, 1982, epidemiology; Stavely, *ibid* 74:339, 1984; *PD* 68:95, 1984, literature on races; Code et al., *Aust.J.Bot.* 33:147, 1985, comparison of forms on bean & *Macroptilium atropurpureum*; Pohronezny et al., *PD* 71:639, 1987, chemical control [61,2546, 5341, 6069; 63,4155; 64,5193].
U. betae Kickx 1867; Punithalingam, *Descr.F.* 177, 1968; teliospores 26–30 × 18–22 µm, a hyaline

papilla over the germ pore, macrocyclic, autoecious; see *Descr.F.* for other rusts on *Beta*; prob. an unimportant beet pathogen, although it has been locally serious; Emdal & Foldø, *Seed Sci.Technol.* 7:93, 1979, seedborne inoculum [59,980].
U. ciceris-arietini Jacz. in Boyer & Jacz. 1894; Punithalingam, *Descr.F.* 178, 1968; Cummins 1978 has 1893; teliospores 18–30 × 18–24 µm, uredospores 20–28 µm diam., 4–8 germ pores with hyaline papillae, microcyclic; see *Descr.F.* for other rusts on *Cicer*; chickpea rust; can cause defoliation and losses occur in India and Mexico; in the former the disease is more severe in the cooler months.
U. colchici Massee 1892; teliospores 28–40 × 20–28 µm, microcyclic; on *Colchicum* corms, prob. unimportant; Boerema, *Tijdschr. PlZiekt.* 67:1, 1961 [40:539].
U. croci Pass. 1876; teliospores 24–32 × 21–28 µm, no other spore form; on crocus corms and leaves; Boerema & van Kesteren, *Neth.J.Pl.Path.* 71:136, 1965 [45,1396].
U. decoratus Syd. 1907; Punithalingam, *Descr.F.* 179, 1968; teliospores 20–32 × 14–20 µm, uredospores 21–26 µm diam., 4–6 germ pores; see *Descr.F.* for other rusts on *Crotalaria*; sunn hemp rust, losses reported from India.
U. dianthi (Pers.) Niessel 1872; Punithalingam, *Descr.F.* 180, 1968; teliospores 25–29 × 20–23 µm, macrocyclic, heteroecious, telia on *Dianthus* and other Caryophyllaceae, aecia on *Euphorbia*, common, often needs chemical control; Hill, *Pl.Path.* 23:151, 1974, disease development & control; Spencer, *ibid* 28:10, 1979, systemic fungicides; *TBMS* 74:191, 1980, parasitism by *Verticillium lecanii* [54,4518; 59,3797, 4205].
U. dolicholi Arthur 1906; Sivanesan, *Descr.F.* 269, 1970; teliospores 26–32 × 10–15 µm, uredospores 8–12 µm diam., 3–4 germ pores, microcyclic; rust of *Rhynchosia*. This rust has been confused with *Uredo cajani* on pigeon pea, telia unknown; *Uromyces dolicholi* does not occur on pigeon pea.
U. geranii (DC.) Fr. 1849; Sivanesan, *Descr.F.* 270, 1970; teliospores 22–42 × 13–26 µm uredospores 20–30 × 18–26 µm, mostly 1 germ pore; geranium rust.
U. gladioli Henn. 1895; teliospores 20–37 × 18–28 µm, pedicel up to 75 µm long, microcyclic; gladiolus rust. Two *Uromyces* spp. can cause damage to this ornamental; the other is *U. transversalis*; *U. gladioli* has larger teliospores and a longer pedicel. Lindquist et al. 1979 described an epidemic in Argentina; the 2 rusts infect different *Gladiolus* spp. [59,2786].
U. indigoferae Dietel & Holway 1901; teliospores

18–35 × 22–32 μm, microcyclic; can be severe on *Indigofera* spp. in India; Joshi & Reddy, *Indian Phytopath.* **11**:59, 1958; **12**:25, 1959 [**38**:676; **39**:332].

U. **koae** Arthur ex F. Stev. 1925, fide Gardner et al. 1979; teliospores 23–44 × 16–22 μm, macrocyclic, autoecious; on *Acacia koa*, timber tree in USA (Hawaii); causes growths like witches' brooms on shoots, telial pustules on leaves and phyllodes; Gardner, *PDR* **62**:957, 1978; *CJB* **59**:939, 1981; Gardner et al., *Mycologia* **71**:848, 1979 [**58**, 3463; **59**,2348; **60**, 6659].

U. **manihotis** Henn. 1895; prob. the most widespread of the 6 *Uromyces* spp. noted from *Manihot* in tropical America; very little pathology on the cassava rusts has been reported. Laberry et al., *Fitopat. Brasileira* **9**:525, 1984, described them; the other spp. are: *U. carthagenesis* Speg. 1899, *U. jatrophae* Dietel & Holway 1897, *U. manihoticola* Henn. 1895, *U. manihotis-catingae* Henn. 1908, *U. tolerandus* H. Jackson & Holway 1931 [**64**,2878].

U. **mucunae** Rabenh. 1878; Laundon & Rainbow, *Descr.F.* 290, 1971; teliospores 22–26 × 19–25 μm, uredospores 15–21 × 15–19 μm, poss. 4–5 germ pores, prob. microcyclic, autoecious; velvet bean rust, see *Descr.F.* for rusts on *Mucuna*.

U. **musae** Henn. 1907; Mulder & Holliday, *Descr.F.* 295, 1971; teliospores 23–35 × 17–25 μm, uredospores 20–28 × 17–24 μm, microcyclic; banana rust, not to be confused with *Uredo musae* Cummins whose uredia are more crowded, and the uredospores have a thinner wall and more pronounced echinulations; *Uredo musae* is known only from Africa; neither rust is damaging; Firman, *Fiji agric.J.* **38**:85, 1976 [**57**,2598].

U. **pisi-sativi** (Pers.) Liro 1908; Laundon & Waterston, *Descr.F.* 58, 1965; teliospores verrucose, 22–28 × 17–24 μm, uredospores 22–28 × 19–24 μm, 3–6 germ pores; macrocyclic, heteroecious; telia on *Lathyrus*, *Orobus*, *Pisum*, *Vicia*, aecia on *Euphorbia*; rust of broad bean, pea and vetches; usually little damage is caused.

U. **setariae-italicae** Yoshino 1906; teliospores 18–25 × 16–20 μm, uredospores 27–33 × 23–28 μm, 3 germ pores, macrocyclic, heteroecious; telia on grasses, aecia on *Cordia*; foxtail millet rust, can be damaging in India; Ramakrishnan, *Indian Phytopath.* **2**:31, 1949; Ramakrishnan & Sundaram, *Proc.Indian Acad.Sci.* B **41**:241, 1955; Patil & Thirumalachar, *Indian Phytopath.* **22**:110, 1969 [**29**:506; **35**:96; **49**,748].

U. **striatus** Schröter 1870; Laundon & Waterston, *Descr.F.* 59, 1965; teliospores 18–24 × 16–20 μm, uredospores 18–26 × 16–22 μm, 3–4 germ pores,

macrocyclic, heteroecious; telia on *Medicago* and *Trifolium*, aecia on *Euphorbia*. Close to *Uromyces pisi-sativi* but there are fewer germ pores in the uredospores, and the teliospores have striate markings on the walls in contrast to the punctate markings on those of *U. pisi-sativi*. Lucerne rust is considered to be a minor disease in USA but crops elsewhere have been seriously damaged; there are several races; Parmelee, *CJB* **40**:491, 1962, review; Viennot-Bourgin, *Revue Mycologie* **42**:321, 1978, *Medicago* rusts; Goulter, *Australasian Pl.Path.* **13**:58, 1984, on chickpea, a new host [**41**:584; **58**,3352; **64**,4070].

U. **transversalis** (Thüm.) Winter 1884; teliospores 20–34 × 14–21 μm, pedicel up to 45 μm long, microcyclic, gladiolus rust. This rust, like *Uromyces gladioli*, has recently been reported causing damage in Italy and Morocco. It has smaller teliospores; Henderson & Bennel, *Notes R.Bot.Gdn.Edinb.* **37**:475, 1979, stated that it can be distinguished from *U. gladioli* by the paraphysate telia, it also attacks different *Gladiolus* spp.; Viennot-Bourgin 1978; Aloy et al. 1977, 1981; Rolim et al., *Biológico* **51**:29, 1985, chemical control [**58**,3322; **60**,2056; **61**,5794; **65**,4443].

U. **trifolii-repentis** Liro 1908; teliospores 22–26 × 18–22 μm, macrocyclic, autoecious, mostly on *Trifolium hybridum*, *T. repens*; the var. *fallens* (Arthur) Cummins 1977 is mostly on *T. pratense*; *Uromyces trifolii* (Hedw. ex DC.) Fuckel 1870 on *T. repens*, microcyclic; for rusts on clover, and which may reduce yield and quality, see Walker, *Mycotaxon* **7**:423, 1978; Welty et al., *Mycologia* **74**:265, 1982. Rusts which have been described as microcyclic may be macrocyclic as one or more stages may not develop after inoculations [**58**,748; **61**,5815].

U. **viciae-fabae** (Pers.) Schröter 1875; Laundon & Waterston, *Descr.F.* 60, 1965; teliospores 25–40 × 18–26 μm, smooth walls; uredospores 22–28 × 19–22 μm, 3–4 germ pores; macrocyclic, autoecious; on *Lathyrus*, *Lens*, *Orobus*, *Pisum*, *Vicia*. *Uromyces pisi-sativi*, also on these hosts, is heteroecious,, has uredospores with more germ pores and verrucose teliospores; see *Descr.F.* for other spp. on these plants; rust of broad bean, lentil, pea and vetch. de Bary in historic first experiments in 1863 sowed teliospores on broad bean and pea and obtained aecia. The rust is one of cooler temp. but teliospores can remain viable through the Indian hot season on lentil. Some forms seem to be restricted to individual host spp. and races exist on broad bean, lentil and pea. Damage can be severe on the first crop.

Singh & Sokhi, *PD* **64**:671, 1980, 6 races on

lentil; Conner & Bernier, *Phytop.* **72**:687, 1555, 1982, wider host range than had been thought, inheritance of resistance in broad bean; *Can.J.Pl.Path.* **4**:157, 263, 1982, 7 races on broad bean inbreds, using pea 4 more races, slow rusting; Lapwood et al., *Crop Protect.* **3**:193, 1984, damage in UK, chemical control [**60**,2278; **61**,7224, 7231; **62**,1282, 2207; **63**,4628].

Uromycladium McAlp. 1905; McAlpine, *The rusts of Australia. Their structure, nature, and classification*, 1906; Uredinales; teliospores aseptate, 1–2 spores on one pedicel with a hyaline vesicle beneath and attached separately; where there are 3 spores on one pedicel the vesicle is absent; microcyclic, *Acacia, Albizia*; Australasia.
U. tepperianum (Sacc.) McAlp.; teliospores in clusters of 3, vesicle absent, av. 21 × 17 μm diam.; causes galls on stems and reproductive organs, forms witches' brooms on *Acacia saligna*. This tree, native to Australia (W.), has become a major weed in South Africa (Cape Province); Morris, *Pl.Path.* **36**:100, 1987, biology, poss. use in control of *A. saligna* [**66**,35].

Urophlyctis Schröter 1885, fide Sparrow, *TBMS* **60**:339, 1973, types for *Physoderma* & *Urophlyctis*; Sparrow, *Trans.mycol.Soc.Japan* **3**:16, 1962, kept the 2 genera separate. Karling placed *Urophlyctis* in synonymy with *Physoderma* q.v. [**42**:106].

Urtica, *U. dioica*, common stinging nettle.
urtica yellow mottle Schmidt, *Phytopath.Z.* **64**:129, 1969; Germany, *U. dioica* [**48**,3095].

Ustilaginaceae, Ustilaginales; ustilospore germinates by a promycelium which is transversely septate and bears lateral basidiospores, after conjugation they form secondary spores or con.

Ustilaginoidea Bref. 1895; Hyphom.; Ellis 1971; conidiomata hyphal, conidiogenesis holoblastic; con. aseptate, pigmented, solitary, *U. ochracea* Henn. 1899 is found on grasses, especially Panicaceae in the tropics, con. 6–8 μm diam.
U. virens (Cooke) Takahashi 1896; Mulder & Holliday, *Descr.F.* 299, 1971; the conidiomata surround the grain, changing from yellow to orange, olive green to black, con. 4–5 μm diam., verruculose; rice false smut, can be found on maize and other grasses but prob. only locally important on rice; Singh & Gangopadhyay, *TBMS* **77**:660, 1981, flower inoculation; Holliday 1980; Singh 1984 [**61**,3467; **64**,3065].

Ustilaginomycetes, Ustilaginales; Basidiomycotina; see Tilletiaceae and Ustilaginaceae; macroscopic basidiomata absent, the smut fungi are a major group of plant parasites and pathogens; one spore form can be grown in axenic culture; most important on Gramineae. They have simpler life cycles than the rusts, but unlike most rusts cause extensive systemic invasion before sporulation develops. Only related plant spp. are hosts and there are only 2 spore types. The characteristic spore is the ustilospore which is analagous with the teliospore. It is dark, often powdery in mass, aseptate, thick walls which may be characteristically sculptured and important in sp. morphology; ustilospores may arise singly or be united into balls which may contain sterile cells; they are intercalary, arising from a closely septate mycelium; these spores are survival structures, they germinate to form basidiospores or sporidia which can be cultured.

Fischer 1951, 1953; Mundkur & Thirumalachar 1952; Zundel, *Ustilaginales of the world*, 1953; Holton et al., *A.R.Phytop.* **6**:213, 1968, variation; Durán in Ainsworth et al. IV B, 1973; Zambettakis, *Revue Mycol.* **41**:469, 1977; **42**:13, 113, 1978, review & bibliography; Kakashima, *Trans.mycol.Soc.Japan* **21**:423, 1980, surface structure of ustilospores; Mordue & Ainsworth 1984; Vánky, *Illustrated genera of smut fungi*, Cryptogamic studies, vol. 1, 1987.

Angiosorus, Entyloma, Melanopsichium, Sorosporium, Sphacelotheca, Sporisorium, Thecaphora, Tilletia, Tolyposporidium, Tolyposporium, Urocystis, Ustilago.

Ustilago (Pers.) Roussel 1806; Ustilaginaceae; the criteria for the genus were stated by Langdon & Fullerton, see *Sorosporium*. The sori are mostly in the inflorescence; the mycelium is converted entirely into ustilospores after necrosis of host tissue; no columellae or peridia of fungus origin are formed; the exospore is mostly echinulate; ustilospores mostly 4–16 μm diam., not in balls, not mixed with elaters or borne in chains; Huang & Nielsen, *CJB* **62**:603, 1984, hybridisation between forms on barley & oat; Ingold, *TBMS* **81**:573, 1983; **83**:251, 1984, basidium; Kim et al., *CJB* **62**:1431, 1984, chemotaxonomy of spp. on barley, oat, wheat [**63**,1147, 2807; **64**,84, 3784].

U. avenae (Pers.) Rostrup 1890; Punithalingam & Waterston, *Descr.F.* 279, 1970; ovaries destroyed, bare rachis remains, ustilospores av. 6.6 μm diam., minutely echinulate; can be considered as a var. of *Ustilago segetum*; oat loose smut and on *Arrhenatherum*. The spores are mostly released before harvest, infection of the coleoptile leads to systemic invasion and direct infection of the ovaries; races; a chemical seed dressing gives control.

U. bullata Berk. 1855; Mordue & Waller, *Descr. F.* 718, 1981; sori in spikelets are covered by a green to grey membrane which ruptures to expose the spore mass, ustilospores 6–12 μm diam., minutely verrucose or echinulate; head smut of temperate grasses, particularly *Agropyron* and *Bromus*; races; spore contamination of soil and seed, infection of seedlings leads to systemic spread; when older shoots are infected they give rise to diseased tillers; the fungus hybridises with some other *Ustilago* spp.

U. coicis Bref. 1895; sori destroy the ovaries, ustilospores held together by hard, floral glumes, 7–13 μm diam., minutely echinulate; Job's tears smut, can be seriously damaging as reported from India and Thailand; Kuwata et al. 1982, in Japan [**62**,2455–6].

U. crameri Körn. 1873; Ainsworth, *Descr.F.* 78, 1965; sori in ovaries are limited by a thin membrane which ruptures, ustilospores 8–12 μm diam., smooth; head smut of foxtail millet, carried on seed and the seedling is infected; 6 races; Holliday 1980.

U. crus-galli Tracy & Earle 1895; sori, up to 1.2 cm diam., destroy inflorescence, also on stem and young shoot; ustilospores 9–12 μm diam., with blunt echinulations, sometimes verruculose, on barnyard millet, prob. seedborne. Mundkur, *Indian J.agric.Sci.* **13**:631, 1943, described 2 other spp. on this crop: *Ustilago panici-frumentacei* Bref., ustilospores 6–10 μm diam., minutely echinulate; *U. paradoxa* Sydow & E. Butler, ustilospores 8–11 μm diam., smooth [**24**:15].

U. cynodontis (Pass.) Henn. 1891; Mulder & Holliday, *Descr.F.* 297, 1971; sori reduce the inflorescence to a dusty spore mass, ustilospores 5–8 μm diam., smooth; Bermuda grass smut, mainly spread by infected rhizomes; Holliday 1980.

U. esculenta Henn. 1895; sori in hypertrophied tissue of culms, ustilospores 6–9.5 μm diam., finely echinulate; *Zizania* spp., American wild rices. The infected culms become extremely swollen, and in E. Asia this host and parasite structure is cultivated and eaten as a vegetable; Yang & Leu, *Phytop.* **68**:1572, 1978; *Trans.mycol.Soc.Japan* **21**:205, 1980; Chan & Thrower, *New Phytol.* **85**:201, 209, 217, 225, 1980; Terrell & Batra, *Econ.Bot.* **36**:274, 1982 [**58**,5142; **60**,2324–7, 3564; **62**,526].

U. hypodytes (Schlecht.) Fr. 1832; Mordue, *Descr.F.* 809, 1984; sori in stems surrounding internodes and extending to the entire stem, ustilospores 4–7 μm diam., minute verrucose under the electron microscope, otherwise appear smooth; stem smut of temperate Gramineae, prob. of little or no economic importance.

U. maydis (DC.) Corda 1842; Ainsworth, *Descr.F.* 79, 1965; sori in inflorescence, leaf and stem; they form conspicuous galls which can be very large, up to 20–30 cm diam. and attached just above the nodes for 2–5 cm; at first limited by a white to cream or greenish membrane of fungus and host tissue, this soon ruptures to release the typical smut, very dark brown, powdery mass; ustilospores 7–10 μm diam., prominently and bluntly echinulate; maize blister, boil or common smut.

The basidiospores form sporidia in the usual way and these infect young tissue, mostly above soil level. Infection tends to be localised rather than systemic. The fungus has a characteristic saprophytic phase and can persist for long periods in soil, crop debris and refuse; seedborne. The optimum temp. for disease development is moderate so maize blister smut is uncommon in the tropics. Disease is increased by host injury, high N or complete fertiliser. Although *Ustilago maydis* is very variable in pathogenicity, and other characteristics, no races have been delineated; a non-specific type of resistance is used. The disease is somewhat sporadic in its occurrence; avoiding susceptible hybrids is prob. the best method of control; seed treatment, because of the soil phase of the fungus, may not be effective.

Christensen, *Monogr.Am.Phytopath.Soc.* 2, 1963; Walker 1969; Holliday 1980; Shurtleff 1980; Ramberg & McLaughlin, *CJB* **58**:1548, 1980, electron microscopy of promycelial development & basidiospore initiation; Mills & Kotzé, *Phytopath.Z.* **102**:21, 1981, electron microscopy of germination, growth & infection; Jones, *Aust.J.exp.Agric.* **26**:187, 1986, chemical control of ustilospores on seed [**60**,1415; **61**,2252; **66**,2340].

U. scitaminea Sydow 1924; Ainsworth, *Descr.F.* 80, 1965; sori in the inflorescence which becomes a characteristic, reduced structure, like a whip, up to 90 cm long; it is at first covered by a thin, silver grey membrane which flakes away; ustilospores 5–10 μm diam., smooth or punctate. Ainsworth loc.cit., who referred to vars. of the fungus on *Saccharum*, stated that the morphological variation in the ustilospores needs critical re-examination; sugarcane culmicolous smut. A major, and widespread disease of sugarcane, but the pathogen is still absent from Australia and Papua New Guinea. In recent years *Ustilago scitaminea* has spread rapidly in the Caribbean area, Comstock et al., *PD* **67**:452, 1983. It was first reported from Guyana in 1974, and then spread northwards through the island chain and the isthmus to reach USA (Louisiana, Texas) in 1981.

Infection takes place through the stem buds as

they begin development at or below the soil surface. As stem growth proceeds pockets of mycelium form at each primordium which becomes the nodal bud; mycelium is absent from the internodes. Until a bud develops the fungus remains latent. If an infected sett is planted all the shoots from it become smutted and form whips. Gillaspie et al., *PD* **67**:373, 1983, characterised 6 races. Control is through resistance. But when the smut infects a new area, and the cv. being grown is susceptible, it remains to be decided whether to interrupt the planting cycle and replant with resistant cvs., or whether to rely only on replacement without such an interruption, Simmonds, *Int.Sug.J.* **78**:329, 1976, and Whittle, *Trop.Agric.Trin.* **59**:239, 1982, who examined yield loss; Trione, *Phytop.* **70**:513, 1980, described ustilospore formation; reviews: Lee-Lovick, *Rev.Pl.Path.* **57**:181, 1978; Holliday 1980 [**56**,2651; **60**,2175; **61**,6561; **62**,3640].

U. segetum (Bull.) Rousel 1806; infection systemic, sori in spikelets, dark spore masses protected by glumes; covered smuts or exposed, loose smuts, ustilospores 5–10 μm diam. Mordue & Ainsworth 1984 placed these smuts of barley, oat and wheat in 3 vars.; *segetum*, covered smut of barley and oat; *avenae* (Pers.) Brun., loose smut of oat; *tritici* (Pers.) Brun., loose smut of barley and wheat. They are treated here, as plant pathologists may wish, as, respectively: *Ustilago segetum* var. *hordei*, *U. avenae* and *U. tritici*, which are correct under the Code. These smuts hybridise naturally and experimentally; Dickson 1956; Manners in Western 1971; Mathre 1982; Gareth-Jones & Clifford 1983; Wiese 1987.

U.s.var.hordei (Pers.) Rabenh. 1856; Waller & Mordue, *Descr.F.* 749, 1983 as *Ustilago hordei* (Pers.) Lagerh., sori replace ovaries in spikelets within glumes, ustilospores av. 6–7 μm diam., smooth; covered smut, barley and oat; life history as for *U. avenae*; races; a chemical seed dressing gives control.

U. striiformis (Westend.) Niessl 1876; Mordue & Waller, *Descr.F.* 717, 1981; sori on leaves, ustilospores av. 10.8 μm diam., echinulate; stripe of temperate grasses, 6 ff.sp. *agrostidis*, *dactylidis*, *holci*, *hordei*, *phlei*, *poae*.

U. tritici (Pers.) Rostrup 1890; Punithalingam & Waterston, *Descr.F.* 280, 1970, as *Ustilago nuda* (Jensen) Rostrup; sori replace ovaries in spikelets, ustilospores soon dispersed leaving the bare rachis, 6.6 μm diam., minutely echinulate; loose smut, barley and wheat; infection at flowering via

the ovary; since the grain is penetrated before ripening, systemic chemicals, or the older heat treatment, must be used; seed certification procedures are in use; Wray & Pickett, *J.natn.Inst.agric.Bot.* **17**:31, 1985, infection trends of certified barley seed in England & Wales 1976–83 [**65**,4379].

U. vaillantii Tul. & C. Tul. 1847; Mordue, *Descr.F.* 810, 1984; sori in anthers, less frequently in ovaries; ustilospores 6–14 × 5–12 μm, minutely verruculose; anther smut, including *Muscari* and *Scilla*, may cause flower failure but prob. unimportant.

U. violacea (Pers.) Roussel 1806; Waller & Mordue, *Descr.F.* 750, 1983; sori in anthers, ustilospores 4–12 μm diam., reticulate, the spore mass is pinkish to purple; carnation anther smut, on Caryophyllaceae, at least 11 races on different host spp.; 2 vars.: *stellariae* (Sow.) Savile on *Stellaria*, smaller ustilospores; *major* G.P. Clinton on *Silene*, larger ustilospores. Infection of buds, young shoots and rhizomes; it is systemic and plants become stunted with excessive numbers of side shoots. Fletcher 1984 stated that the disease is uncommon in modern carnation cvs., partly because very few of them form anthers. The fungus has often been used for cytological and genetical work.

ustilospore, Ustilaginales, the spore from which the basidiospores are formed; variously and unsatisfactorily also called brand or bunt spores, chlamydospores, teliospores; Mordue, *TBMS* **87**:407, 1986, ornamentation in 26 European genera; Ingold, *ibid* **88**:355, 1987, germination.

Ustulina Tul. & C. Tul. 1863; Xylariaceae; ascomata ± globose, immersed in a massive, sessile stroma which is discoid, irregularly cushion shaped or a spreading crust; asco. aseptate, > 25 μm long, i.e. larger than those of *Hypoxylon*.

U. deusta (Hoffm.) Lind 1913, fide Cannon et al. 1985; Hawksworth, *Descr.F.* 360, 1972; stromata mostly on basal parts of the host, easily removed; asco. 26–40 × 6–13 μm; a conidial state precedes ascomata formation, a greyish, effuse, continuous layer, compact tufts of parallel conidiophores, con. aseptate, hyaline, smooth, 5–7 × 2–3.5 μm; causes charcoal stump, or base, rot; on woody hosts, common on *Fagus sylvatica* in Europe. But prob. more important as a pathogen in the tropics: on tea, particularly in India (N.E.) and Sri Lanka, cacao, oil palm, rubber; Holliday 1980.

uvicola, *Greeneria*.

V

vaccinii, *Diaporthe, Exobasidium, Guignardia, Nocardia, Phomopsis, Phyllosticta, Pucciniastrum.*
vaccinii-corymbosi, *Monilinia.*
Vaccinium, *V. angustifolium*, lowbush blueberry; *V. corymbosum*, highbush blueberry; *V. macrocarpon, V. oxycoccus*, cranberry, often placed in *Oxycoccus.*
vagabunda, *Phlyctema*; vaginae, *Mycovellosiella*; vaillantii, *Ustilago*; valerinellae, *Phoma.*
valeriana ring mosaic v. Richter, H.E. & H.B. Schmidt, *Phytopath.Z.* 58:323, 1967; isometric 24 nm diam., transm. *Myz.pers.*, Germany [46,3162].
vallisumbrosae, *Ramularia.*
vallota mosaic v. Inouye & Hakkaart, *Neth.J.Pl.Path.* 86:265, 1980; Potyvirus, filament c.750 nm long, transm. sap readily, *Myz.pers.*, non-persistent, Netherlands, *V. speciosa* [60,2636].
Valsa Fr. 1825; Valsaceae; ascomata beneath stromatic disc; without blackened, marginal zones; asco. aseptate, narrow, cylindrical, often allantoid, usually not > 30 μm long; anam. *Cytospora*, Cannon et al. 1985 stated that the genus needs critical revision; often on bark and the spp. are prob. not particularly serious pathogens; Défago, *Phytopath.Z.* 14:103, 1944; Wehmeyer 1975; Barr 1978; Spielman, *CJB* 63:1355, 1985, on hardwoods in N. America, morphology & taxonomy.
V. abietis (Fr.) Fr. 1849; anam. *Cytospora* Sacc.; dieback of balsam fir, *Abies balsamea*, Canada; Raymond & Reid, *CJB* 39:233, 1961; Scharpf & Bynum 1975 [40:567; 55, 1972].
V. ceratosperma (Tode: Fr.) Maire 1937, fide Barr 1978; Spielman, see genus; *Valsa mali* Miyabe & Yamada may be a synonym; widespread; apple canker in N. China and Japan; Ye et al., *Acta phytopath.sin.* 11(3):31, 1981; Chen et al., *ibid* 12(1):49, 1982; Tamura & Saito, *Ann.phytopath.Soc.Japan* 48:490, 1982; Tamura, *Rep.Hokkaido Prefect.agric.Exp.Stn.* 49, 1984 [61,3532, 5831; 62, 2037; 63,5010].
V. kunzei (Fr.) Fr. 1849; anam. *Cytospora kunzei* Sacc.; Kamiri & Laemmlen, *Phytop.* 71:941, 1981, described a canker on *Picea pungens* in USA (Michigan); Waterman, *ibid* 45:686, 1955, comparisons with other *Cytospora* spp. & *Valsa* spp. on conifers [35:564; 61,2492].
V. leucostomoides Peck 1884, fide Sproston & Scott, *Phytop.* 44:12, 1954, who described a wood decay of sugar maple, *Acer saccharum* in USA

(Vermont); this sp. is not listed by Spielman, see genus [33:569].
V. mali, see *Valsa ceratosperma.*
V. sordida Nitschke 1870; anam. *Cytospora chrysosperma* Fr. 1923; Spielman, see genus; assoc. with cankers of poplar and other temperate dicotyledonous trees; Christensen, *Phytop.* 30:459, 1940; Müller-Stoll & Hartmann, *Phytopath.Z.* 16:443, 1950; Bloomberg and Bloomberg & Farris, *CJB* 40:1271, 1281, 1962; 41:303, 1963; Walla & Stack, *PD* 64:1092, 1980, control of black stem in poplar cuttings [19:623; 31:523; 42:222, 498; 60,5138].
Valsaceae, Diaporthales; ascomata oblique or horizontal; beaks oblique or lateral, erumpent singly or converging upwards through a stromatic disc; asci soon free, evanescent, asco. simple or multiseptate; *Amphiporthe, Cryptodiaporthe, Cryptosphaeria, Diaporthe, Linospora, Leucostoma, Valsa.*
vanda mosaic Murakishi, *Phytop.* 48:132, 137, 1958; rod av. 117 nm long assoc., USA (Hawaii) [37:484].
— ringspot Jensen, *Calif.Agric.* 6(2):7, 1952 [32:129].
vanderystii, *Cercospora.*
Vanilla.
Vararia P. Karsten 1898; Hymenochaetaceae; basidiomata clavarioid, highly branched, hymenophore smooth to granuliferous; Boidin & Lanquetin, *Mycotaxon* 6:277, 1977; Pascoe et al., *TBMS* 82:723, 1984, reported a sp. causing raspberry white root rot in Australia (Victoria) [63,4002].
variegation, peach, poss. chimera; Helton in Gilmer et al. 1976.
variospermum, *Cladobotryum.*
varnish spot (*Pseudomonas cichorii*) lettuce.
Vasates fockeui, transm. plum latent.
vascular collapse (*Xanthomonas fragariae*) strawberry; — necrosis (*Erwinia carotovora* ssp. *betavascularum*) sugarbeet; — streak dieback (*Oncobasidium theobromae*) cacao.
vascularum, *Sorosphaera.*
vascular wilt, a disease where the pathogen is almost entirely confined to the vascular system; a wilt is the characteristic symptom but it may not always be clearly evident. Where the pathogen is a fungus vascular mycosis or tracheomycosis are sometimes used. *Fusarium oxysporum* ff.sp. frequently cause the typical wilt syndrome; see

Verticillium; Talboys in Horsfall & Cowling, vol. 3, 1978; Mace et al. 1981; Matta in Staples & Toenniessen 1981; Talboys in Wood & Jellis 1984; Pegg, *TBMS* **85**:1, 1985; Beckman, *The nature of wilt diseases of plants*, 1987.

vasinfecta, *Neocosmospora*; **vastatrix,** *Hemileia.*

vector, an organism that carries and transmits a pathogen to a plant, particularly inoculum; and especially a fungus, insect, mite or nematode that transmits a mollicute or a virus. The word has now been extended, very questionably, for a gene carrier in genetic engineering. Vector efficiency is the relative amount of transmission, usually expressed as a percentage, that can be expected from a specific vector and virus and host combination, FBPP. Vector resistance is the resistance of a host plant to a vector; the growth and multiplication of the vector may be inhibited; the vector may avoid the plant, non-preference. Vector specificity is the comparative range of vector spp. that can transmit a given pathogen.

Harrison, *A.R.Phytop.* **15**:331, 1977, soil inhabiting, ecology & control; Harris, *ibid* **19**:391, 1981, arthropods & nematodes; *Adv.Virus Res.* **28**:113, 1983; Thresh, *Phil.Trans.R.Soc.Lond.*B **302**:497, 1983. APHIDS: Swenson, *A.R.Phytop.* **6**:351, 1968, ecology; Rochow, *ibid* **10**:101, 1972, mixed infections; Watson & Plumb, *A.Rev.Entomol.* **17**:425, 1972; Pirone & Harris, *A.R.Phytop.* **15**:55, 1977, nonpersistent transm.; Harris & Maramorosch, *Aphids as virus vectors*, 1977; Simons, *PD* **64**:452, 1980, oils for control. BEETLES: Walters, *Adv.Virus Res.* **15**:339, 1969; Fulton et al., *A.R.Phytop.* **25**:111, 1987; FUNGI: Teakle in Buczacki 1983. LEAFHOPPERS: Bennett, *ibid* **5**:87, 1967; Maramorosch & Harris, *Leafhopper vectors and plant disease agents*, 1979; Harris, *J.econ.Entomol.* **74**:446, 1981. NEMATODES: Harrison et al., *J.Nematol.* **6**:155, 1974, specificity of retention & transm.; *Proc.Am.Phytopath.Soc.* **4**:1, 1977; Lamberti, *PD* **65**:113, 1981; Wyss, *ibid* **66**:639, 1982. WHITEFLIES: Costa, *A.R.Phytop.* **14**:429, 1976; Bird & Maramorosch, *Adv.Virus Res.* **22**:55, 1978; Cohen & Berlinger 1986, abs. *Rev.Pl.Path.* **66**, 1800, 1987, transm. & cultural control; see *Trialeurodes vaporariorum*. For the above vector groups and flies, mealy bugs, membracids, mites, piesmids, psyllids, thrips, see Harris & Maramorosch 1980, 1982.

vegetables, Walker 1952; Ogilvie 1969; Dixon 1981; Conti & Lovisolo, *Acta Hortic.* 127:83, 1982; Garrett and Lund in Rhodes-Roberts & Skinner ed., *Soc.appl.Bact.Sympos.* Ser. 10, 1982; Allen 1983; Fletcher 1984; Sherf & MacNab 1986; Tomlinson, *AAB* **110**:661, 1987, viruses.

vein banding, a change of colour in a narrow zone of leaf tissue alongside the main veins; — **clearing** an increased translucency of the veinal system in a leaf, making the pattern more pronounced, light against dark, by transmitted light, FBPP.

— **spot** (*Gnomonia nerviseda*) pecan; — **streaking,** see cabbage internal necrosis.

Veluticeps (Cooke) Pat. 1894; Corticiaceae or Stereaceae; hymenium with sterile fascicles of agglutinated, brown hyphae, basidiospores 16–24 μm long, not amyloid; pileus sessile, hard, brittle; Gilbertson et al., *Mycologia* **60**:29, 1968, *V. berkeleyi* (Berk. & Curt.) Cooke, causing a brown, cubical, heart rot of pine in N. America; Davidson & Chien, *ibid* **68**:1152, 1976, same sp. causing brown rot of *Chamaecyparis obtusa* var. *formosana* in Taiwan [**47**,2297; **56**,4242].

velvet bean, *Mucuna deeringiana.*

— — **yellow mosaic** Mathew & Balakrishnan, *Madras agric.J.* **69**:119, 1982; transm. *Bemisia tabaci*, India (Kerala) [**62**:2029].

— **grass,** *Holcus lanatus.*

— **spot** (*Phaeoramularia capsicicola*) red pepper.

— **tobacco,** *Nicotiana velutina.*

— — **mottle v.** Randles et el. 1981; Randles & Francki, *Descr.V.* 317, 1986; Sobemovirus, isometric *c.*30 nm diam., naturally occurring isolates contain a circular ssRNA which is a prob. satellite; transm. sap, *Cyrtopeltis nicotianae*, in wild velvet tobacco in south central Australia, serologically related to solanum nodiflorum mottle v. but does not infect *S. nodiflorum.*

veneer blotch (*Deightoniella papuana*) sugarcane.

veneta, *Apiognomonia*, *Elsinoë.*

Venturia Sacc. 1882; Venturiaceae; Sivanesan 1984; ascomata not clustered on a stalked stroma, asco. olive green to shades of brown, position of septum variable, i.e. not strictly septate at the lower or upper end; anam. include spp. in *Cladosporium, Fusicladium, Pollaccia, Spilocaea*; Sivanesan, *Biblthca.mycol.* 59, 1977, monograph; Allitt, *TBMS* **83**:431, 1984, identity & occurrence of spores in air in UK; Morelet, *Eur.J.For.Path.* **17**:85, 1987, spp. on poplars, culture in vitro [**64**,952].

V. carpophila E.E. Fisher 1961; anam. *Cladosporium carpophilum* Thüm. 1877; Sivanesan, *Descr.F.* 402, 1974; asco. 12–16 × 3–5 μm, upper cell wider, con. 12–18 × 4–6 μm; freckle or scab of almond, apricot, peach, plum; on leaves, twigs, fruit; Chandler et al., *PDR* **62**:783, 1978, control of a str. tolerant of benomyl; Lawrence & Zehr, *Phytop.* **72**:773, 1982, conidial biology; Gottwald, *ibid* **73**:1500,

1983, conidial liberation [**58**,3911; **61**,7082; 63,1861].

V. cerasi Aderh. 1900; anam. *Fusicladium cerasi* (Rabenh.) Sacc. 1886; Sivanesan & Holliday, *Descr.F.* 706, 1981; asco. septum below middle, 10–14 × 4–6 μm; con. mostly aseptate, 16–23 × 5–7 μm; cherry scab.

V. inaequalis (Cooke) Winter in Thüm. 1875; anam. *Spilocaea pomi* Fr. 1825; Sivanesan & Waller, *Descr.F.* 401, 1974; Corlett, *Fungi Canadenses* 35, 1974; asco. septum in the upper third, 13–18 × 6–8 μm; con. 0–1 septate, 12–30 × 6–10 μm; principally on apple, scab, and other *Malus* spp. the fungus can cause important losses in fruit quantity and quality. The disease has been one of the most intensively investigated of all diseases on temperate fruit. *Venturia inaequalis* overwinters as ascomata in the leaf litter; the asco. are air dispersed during daylight; con. are splash dispersed. Isolates are highly variable. Warning services for chemical control operations are highly sophisticated, see Mills period; and *Podosphaeria leucotricha*, Jeger & Butt 1984, disease management.

Williams & Kuć, *A.R.Phytop.* **7**:223, 1969, resistance; Boone, *ibid* **9**:297, 1971, genetics of pathogenicity; Hunter, *AAB* **91**:119, 1979, review on biochemical factors in growth in vitro & pathogenicity; Jeger, *ibid* **99**:43, 1981, disease measurement & epidemics; MacHardy & Jeger, *Protect.Ecol.* **5**:103, 1983, integrating control measures; MacHardy & Gadoury, *Phytop.* **75**:381, 1985, forecasting, asco. seasonal maturation; Gadoury & MacHardy and O'Leary & Sutton, *ibid* **76**:112, 199, 1986, forecasting asco. inoculum, effects of moisture & temp. on ascomata formation; O'Leary et al., *PD* **71**:623, 1987, chemical control [**61**,2345; **64**,5003; **65**,3392, 5020, *Hort.Abstr.* **53**,4818].

V. liriodendri Hanlin, *Mycologia* **79**:465, 1987; asco. 13.7 × 5.6 μm, leaf spot of *Liriodendron tulipifera*, tulip tree, USA (Georgia).

V. macularis (Fr.) E. Müller & v. Arx 1950; anam. *Pollacia radiosa* (Lib.) Baldacci & Cif. 1937; Sivanesan, *Descr.F.* 403, 1974; asco. septum just above middle, 8–14 × 4–6 μm, con. 1–2 septate, 15–42 × 6–11 μm; scab of poplar, Leuce section, or leaf blight; damages shoots and leaves; N.A. & R.L. Anderson, *PD* **64**:558, 1980, epidemic on 3-year-old sprouts of *Populus tremuloides* in USA (Minnesota) [**60**,2212].

V. nashicola Tanaka & Yamamoto, *Ann.phytopath.Soc.Japan* **29**:136, 1964; asco. 10–15 × 3.8–6.3 μm, con. 7–23 × 5–7.5 μm; Sivanesan 1977, see genus, and 1984, placed this sp. as a synonym of *Venturia pirina*. Japanese

workers consider it to be distinct on the grounds of the smaller con. and asco.; ascomata that are not so high, i.e. 50–150 μm, compared with 77–200 μm for *V. pirina*; and differing pathogencity to *Pyrus* spp.; Japanese pear scab on *P. serotina* and on Chinese pear, *P. ussurensis*; Ishii & Yanase, *ibid* **49**:153, 1983, teleom. in vitro, tolerance of fungicides & its inheritance; Ishii et al., *Pl.Path.* **34**:363, 1985, tolerance of benzimidazoles, its build up & decline [**43**,3259; **63**,188; **65**,282].

V. pirina Aderh. 1896; anam. *Fusicladium pyrorum* (Lib.) Fuckel 1870; Sivanesan & Waller, *Descr.F.* 404, 1974; asco. septum in lower third, 14–20 × 4–8 μm; con. 0–1 septate 17–30 × 6–10 μm; pear scab, *Pyrus* spp., causes losses in fruit quantity and quality, characteristics similar to those of *Venturia inaequalis*; Latorre et al., *PD* **69**:213, 1985, asco. release in Chile [**64**,3492].

V. populina (Vuill.) Fabric. in Hollrungs 1902; anam. *Pollacia elegans* Servazzi 1939; Sivanesan, *Descr.F.* 483, 1976; asco. septum in or near lower third, 18–30 × 10–16 μm; con. with 2 equidistant septa, a large central cell and 2 smaller end cells, 23–39 × 9–14 μm; poplar blight or dieback; mostly *Populus* spp. in Aigeiros and Tacamahaca sections.

V. saliciperda Nüesch 1960; anam. *Pollaccia saliciperda* (Allescher & v. Tubeuf) v. Arx 1957; Sivanesan, *Descr.F.* 482, 1976; Corlett, *Fungi Canadenses* 251, 1983, anam. only; asco. septum in upper third, 11–14 × 3.5–5 μm; con. 1–2 septate, 16–23 × 6–9 μm; willow scab and shoot dieback; other *Venturia* spp. are found on *Salix* but these do not have a *Pollaccia* state; a severe disease and control may be needed in nurseries, basket willow beds and young trees. The pathogen has been incorrectly referred to *V. chlorospora* (Ces.) Karsten 1873 on willow but with a *Cladosporium* state.

Venturiaceae, Dothidiales; mostly parasitic on green leaves, asco. with thin walls, 1 septate, cells often unequal; *Apiosporina, Arkoola, Phaeocryptopus, Venturia, Xenomeris*.

verbascum mosaic v. Polák & Neubauer, *Biologia Pl.* **9**:360, 1967; rod 616 nm long, transm. sap, Czechoslovakia, *V. thapsiforme* [**46**,3533].

vernal grass, *Anthoxanthum odoratum*.

vernonia phyllody Albrechtsen et al., *Phytopath.Z.* **93**:90, 1978; MLO assoc., transm. *Orosius albicinctus*, India (N.) *V. cinerea* [**58**,3406].

Veronaea Cif. & Montem. 1957; Hyphom. Ellis 1971, 1976; conidiomata hyphal, conidiogenesis holoblastic, conidiophore growth sympodial; con. 1 septate, pigmented. *V. musae* M.B. Ellis 1976, con. 5–10 × 2–3 μm; Stahel, *Trop.Agric.Trin.*

14:42, 1937, as *Chloridium musae* Stahel, banana speckle on leaves [**16**:476].
Verrucalvaceae, *Verrucalvus*.
Verrucalvus Wong & Dick, *Bot.J.Linn.Soc.* **89**:174, 1984; *V. flavofaciens* Wong & Dick. The fungus causes kikuyu yellows of *Pennisetum clandestinum*, a pasture grass, infection is poss. systemic, in Australia, Wong, PDR **59**:800, 1975. It was placed in a new family Verrucalvaceae, with *Sclerophthora*, in the Sclerosporales a new order of the Oomycetes. The vegetative hyphae are uniformly slender, reproductive structures arise from hyphae with enlarged apical domes; sporangia are elongate, variable in size; oogonia *c*.70 μm diam., wall verrucate, plerotic, oospore 29–54 μm diam., antheridia numerous [**55**,1780; **64**,2507].
verrucosa, *Phytophthora*.
verrucose, having small rounded processes or 'warts'; verruculose, delicately verrucose.
verrucosis (*Catacauma torrendiella*, see *Phyllachora*) coconut.
vertical banded blight (*Marasmiellus paspali*) maize.
—— **resistance**, see resistance to pathogens.
Verticicladiella S. Hughes 1953; Hyphom.; conidiomata hyphal, conidiogenesis holoblastic, conidiophore growth sympodial; con. aseptate, ±hyaline, in slimy heads; Kendrick, *CJB* **40**:771, 1962. Wingfield, *TBMS* **85**:81, 1985, reduced the genus to synonymy with *Leptographium* q.v. on the grounds that the 2 genera could not be separated on the basis of conidial development using electron microscopy. The new combinations were made. The teleom. is *Ceratocystis*. The spp. occur in conifers, particularly pines, assoc. with bark beetles and diseased roots, called black stain root. In pathology the most frequently mentioned spp. are: *V. abietina* (Peck) S. Hughes, *V. procera* Kendrick, *V. serpens* (Goid.) Kendrick, *V. truncata* Wingfield & Marasas, *V. wageneri* Kendrick.
Wingfield & Marasas, *TBMS* **75**:21, 1980; **76**:508, 1981; **80**:232, 1983; Wingfield & Knox-Davies, *PD* **64**:569, 1980; Harrington & Cobb, *Phytop.* **73**:596, 1983; Lackner & Alexander, *PD* **66**:211, 1982; **68**:210, 1984; Bertagnole et al., *CJB* **61**:1861, 1983; Witcosky & Hansen, *Phytop.* **75**:399, 1985; Lewis et al., *ibid* **77**:552, 1987 [**42**:59; **60**,2225, 3393; **61**,425, 5997; **62**,4022, 4025; **63**,274, 3607; **64**,4792, 5154].
Verticillium Nees 1817; Hyphom.; Ellis 1971; Carmichael et al. 1980; conidiomata hyphal, conidiogenesis enteroblastic phialidic, branches and phialides commonly in verticils; con. aseptate, hyaline, in slimy masses; root and soil inhabitants. The 2 most important pathogenic

spp. are *V. albo-atrum* and *V. dahliae*; Holliday 1980 compared their characteristics, giving a full bibliography. Both are root inhabitants; both invade the vascular systems of several crops and cause severe diseases, see vascular wilt and these most recent reviews: Beckman, Hastie & Heale, Talboys and Pegg, *Phytopath.Mediterranea* **23**:109, 130, 163, 176, 1984, on, respectively, host & parasite interaction, genetics, chemical control, impact on agriculture; also: Malik & Milton, *TBMS* **75**:496, 1980, survival in monocotyledons; Fitzell et al., *Aust.J.Biol.Sci.* **33**:115, 1980, serological comparisons; Gams & van Zaayen, *Neth.J.Pl.Path.* **88**:57, 1982, fungicolous spp.; Robb et al., *CJB* **60**:825, 1982, cytology of vascular alterations in leaves [**60**,740, 4333; **61**, 5560, 6971].
V. albo-atrum Reinke & Berthold 1879; Hawksworth & Talboys, *Descr.F.* 255, 1970; this sp. has been confused with *Verticillium dahliae* from which it differs in the absence of microsclerotia, no growth at 30° and the dark, resting mycelium. *V. albo-atrum* is a plurivorous pathogen of largely temperate crops; it causes serious diseases at 24° but above this temp. the severity of symptoms falls off. Being non-microsclerotial it tends to decline rapidly in the absence of a host; hence fallow or crop rotation exerts some control. Invasion of the vascular system often leads to the characteristic wilt syndrome but not invariably so, e.g. in hop. Strs. differing in pathogenicity are found within and between crops.
HOP: Pegg & Street, *TBMS* **82**:99, 1984, measurement of pathogen in cvs. of high & low resistance; Sewell & Wilson, *Pl.Path.* **33**:39, 1984, nature & distribution of highly pathogenic strs.; Clarkson & Heale, *ibid* **34**:119, 129, 1985, pathogenicity & colonisation by wild type & auxotrophic isolates; Talboys *Can. J.Pl.Path.* **9**:68, 1987, English hops, review. LUCERNE: Atkinson, *Can.J.Pl.Path.* **3**:266, 1981; Flood & Milton, *Physiol.Pl.Path.* **21**:97, 1982, incompatible & compatible reactions in leaves; Christen et al., *PD* **65**:319, 1981, in USA (N.W.); *Phytop.* **72**:412, 1982, in seed; *ibid* **73**:1051, 1983, virulence of European & N. American isolates; Hawthorne, *N.Z.J.agric.Res.* **26**:405, 1983, variation in pathogenicity; Huang et al., *Can.J.Pl.Path.* **5**:141, 1983, aphid transm.; *Phytop.* **75**:482, 1982, seed contamination; Arny, Heale, Howard, Christie, Peadon et al., *Can.J.Pl.Path.* **7**:187, 191, 199, 206, 211, 1985, spread, genetics & control. TOMATO: Tjamos, *Phytop.* **71**:98, 1981, pathogenicity with *V. dahliae* on tomato in relation to host of origin &

cropping system; Pegg & Young and McGeary & Hastie, *Physiol.Pl.Path.* **21**:389, 437, 1982, chitinases & str. hybridisation; Street & Cooper, *Pl.Path.* **33**:483, 1984, measurement of vascular flow in healthy & infected plants. OTHERS: Parnis & Sackston, *CJB* **57**:597, 1979, in lupin seed; Morehart et al. and Morehart & Melchior, *Phytop.* **70**:756, 1980; *CJB* **60**:201, 1982, wilt of yellow poplar, *Liriodendron tulipifera* [**58**,5419; **60**,3980, 5558; **61**,293, 5261, 5819, 6283; **62**,254, 2535, 2639; **63**,714, 1842, 1902, 2382, 2977; **64**,1756, 3921–2, 4992, 5426–8, 5430–1].

V. biguttatum W. Gams, *Neth.J.Pl.Path.* **88**:65, 1982; a common soil fungus and a hyperparasite of *Thanatephorus cucumeris*, Jager & Velvis, *ibid* **89**:113, 1983; **90**:29, 1984; **91**:49, 1985, poss. control of *T. cucumeris* [**63**,224, 5096; **64**,3943].

V. dahliae Kleb. 1913; Hawksworth & Talboys, *Descr.F.* 256, 1970; this sp. has been confused with *Verticillium albo-atrum* from which it differs by the presence of true microsclerotia and forming colonies which are entirely black in reverse, with growth at 30°. *V. dahliae* is a plurivorous pathogen and causes severe diseases at up to 28°. It is therefore more important than *V. albo-atrum* in warmer environments. Since *V. dahliae* forms sclerotia it survives much longer in the soil than the other sp. It also invades the vascular system but the characteristic wilt syndrome does not always appear. Isolates from one host may or may not be pathogenic towards another. In general monocotyledons are not attacked. Holliday 1980 gave an account of the diseases of cotton, eggplant, sunflower, tobacco and some other crops; other plants in which disease is caused include; apricot, avocado, bean, cacao, castor, cowpea, groundnut, horseradish, mango, mint, olive, pea, pistachio, potato, radish, red pepper, tomato; Evans & Gleeson, *AAB* **95**:177, 1980, evaluation of sampling when estimating soil populations; Green, *Phytop.* **70**:353, 1980, soil factors & sclerotial survival; Baard et al., *TBMS* **77**:251, 1981; **79**:513, 1982, sclerotia in soil, structure, lysis & acidity; Zilberstein et al., *ibid* **81**:613, 1983, effect of microsclerotia source on subsequent pathogenicity; Puhalla & Hummel, *Phytop.* **73**:1305, 1983, vegetative compatibility groups.

APRICOT: Popushoï, *EPPO Bull.* **7**:95, 1977; AVOCADO: Lattorre & Allende, *PD* **67**:445, 1983; COTTON: Friebertshauser & DeVay, *Phytop.* **72**:872, 1982, defoliating & non-defoliating pathotypes; Ashworth et al., *ibid* **73**:1292, 1637, 1983, virulence, internal inoculum & defoliation; str. effects on useful life of cvs.;

DeVay & Pullman, *Phytopath.Mediterranea* **23**:95, 1984, review; El-Zik, *PD* **69**:1025, 1985, control. EGGPLANT: Moorman, *Phytop.* **72**:1412, 1982, effects of plastic mulch on infection & yield. HORSERADISH: Mueller et al., *PD* **66**:410, 1982. MINT: Brandt et al., *Phytop.* **74**:587, 1984. OLIVE: Blanco-López et al., *Phytopath.Mediterranea* **23**:1, 1984. PISTACHIO: Ashworth et al., *Phytop.* **72**:243, 1982; **75**:1091, 1985, effects of plastic mulch & host K nutrition. POTATO: see early dying. RADISH: Kitazawa & Suzui, *Ann.phytopath.Soc.Japan* **46**:271, 1980. RED PEPPER: Regnani & Matta, *Phytopath.Mediterranea* **25**:1, 1986. SUNFLOWER: Sackston and Sedun & Sackston, *Can.J.Pl.Path* **2**:209, 1980; **4**:109, 1982, seed infection; effects of day length, host line & pathogen isolate. TOMATO: Ashworth et al., *Phytop.* **69**:490, 1979, effects of inoculum density & root extension; Grogan et al., *ibid* **69**:1176, 1979, on resistant cvs., virulence of isolates from plant & soil, inoculum density; Bender & Shoemaker, *PD* **68**:305, 1984, races 1 & 2 in USA (N. Carolina); Besri et al., *Phytopath.Z.* **109**:289, 1984, races 1 & 2 in Morocco [**57**,1298; **59**,2926, 5368; **60**,1296, 2434, 4018, 5521; **61**,2722, 5264, 6723, 7036, 7115; **62**,1827, 2247, 3594; **63**,1155, 1181, 1285, 1812, 3558, 4517, 4565; **64**, 1674; **65**,2462].

V. fungicola (Preuss) Hassebrauk 1936; Brady & Gibson, *Descr.F.* 498, 1976; dry bubble of the common cultivated mushroom; called var. *fungicola* fide Gams & van Zaayen, 1982, see genus, who also described vars. *aleophilum* and *flavidum*. In the Netherlands var. *aleophilum* causes brown spots on the cultivated *Agaricus bitorquis*, van Zaayen & Gams, *Neth.J.Pl.Path.* **88**:143, 1982; Fletcher 1984 [**62**,1313].

V. lecanii (Zimm.) Viégas 1939; Brady, *Descr.F.* 610, 1979; parasitises insects, arachnids and rust fungi; on plants and in soil; Spencer & Alkey, *TBMS* **77**:535, 1981, on *Puccinia recondita* & *Uromyces dianthi*; Allen, *ibid* **79**:362, 1982, on *U. appendiculatus*; Hall, *AAB* **101**:1, 1982, control of *Aphis gossypii* & *Trialeurodes vaporariorum* in greenhouse; Jackson et al., *ibid* **106**:39, 1985, on *Macrosiphoniella sanborni*, 18 isolates; Grabski & Mendgen, *Phytopath.Z.* **113**:243, 1985, on *U. appendiculatus* & control; Drummond et al., *AAB* **111**:193, 1987, on *T. vaporariorum* [**61**,3379, **62**,464; **65**,981].

V. nigrescens Pethybr. 1919; Hawksworth, *Descr.F.* 257, 1970; prob. mostly saprophytic but pathogenicity has been described, e.g. Wyllie & DeVay, *Phytop.* **60**:907, 1970, on cotton; Melouk & Horner, *ibid* **64**:1267, 1974, on mint; Vesper et al., *ibid* **73**:1338, 1983, on soybean [**49**,3304; **54**,3436; **63**,1502].

V. nubilum Pethybr.; Hawksworth, *Descr.F.* 258, 1970; chlamydospores 8.5–17 μm diam., those of *Verticillium nigrescens* are 5.5–8 μm diam.; assoc. with potato, see coiled sprout which appears to have several causes that are not due to pathogens. There is late and uneven emergence; shoots below soil level swell and curl; this may continue when stems emerge above ground. *V. nubilum* may attack shoots and cause brown lesions; Hoes, *CJB* **49**, 1863, 1971, chlamydospores [**51**,2246].

V. psalliotae Treschow 1941; Brady & Waller, *Descr.F.* 497, 1976, q.v. for distinction from *Verticillium fungicola*; causes brown spotting of the common cultivated mushroom. Uma, *PD* **65**:915, 1981, described a postharvest rot of red pepper [**61**,3200].

V. theobromae (Turc.) Mason & S. Hughes 1951; Hawksworth & Holliday, *Descr.F.* 259, 1970; one of the fungi involved in cigar end of banana in the field, and in crown rot during shipment of boxed fruit. The rot caused is a dry one compared with the wet rot due to *Trachysphaera fructigena*. The evidence indicates that *Verticillium theobromae* is not a primary pathogen.

V. tricorpus Isaac, *TBMS* **36**:194, 1953; Hawksworth, *Descr.F.* 260, 1970; Isaac reported this sp. to be pathogenic to tomato. He compared it with *Verticillium albo-atrum*, *V. dahliae*, *V. nigrescens* and *V. nubilum* [**33**:185].

vertifolia effect, the loss of race non-specific resistance in the process of breeding for race specific resistance, after the potato cv. Vertifolia, fide Vanderplank, *Disease resistance in plants*, 1968. Crill et al., *PDR* **57**:724, 1973, working with *Fusarium oxysporum* f.sp. *lycopersici*, concluded that the vertifolia effect was not inevitable and that there may be more exceptions to the rule than there are cases to support it [**53**,1556].

vesicarium, *Stemphylium*; **vesicula**, *Phytophthora*.

vetch, *Vicia sativa* and other spp.

vexans, *Exobasidium*, *Phomopsis*.

vicia cryptic v. Kenten, Cockbain & Woods 1979; isometric 30 nm diam., transm. pollen, seed; see cryptic viruses [**59**,592c].

viciae, *Olpidium*, *Peronospora*; **viciae-fabae**, *Uromyces*.

Vicia faba, broad bean q.v.; also: faba, field or horse bean.

vicosae, *Cercospora*.

Victoria blight (*Cochliobolus victoriae*) oat.

victoriae, *Bipolaris*, *Cochliobolus*.

victorin, toxin formed by *Cochliobolus victoriae*; Meehan & Murphy, *Science N.Y.* **106**:270, 1947; Pringle & Braun, *Phytop.* **47**:369, 1957; *Nature*

Lond. **181**:1205, 1958; Wolpert et al., *Experientia* **41**:1524, 1985; Mayama et al., *Physiol.Molec.Pl.Path.* **29**:1, 1986 [**27**:16; **36**:757; **37**:473; **65**,1222; **66**,2335].

victoxinine, a breakdown product of the toxin victorin; Pringle & Braun, *Phytop.* **50**:324, 1960 [**39**:696].

Viennot-Bourgin, Georges, 1906–86; born in France; a prolific worker in mycological taxonomy and pathology, wrote: *Les champignons des plantes cultivées*, 1951; Légion d'Honneur; Gold Medal, Mediterranean Phytopathological Union. *Phytopath. Mediterranea* **25**:169, 1986.

Vigna, *V. angularis*, adzuki bean; *V. mungo*, black gram, urd bean; *V. radiata*, green gram, mung bean; *V. subterranea*, bambarra groundnut; *V. unguiculata*, cowpea; *V. u.* ssp. *sesquipedalis*, asparagus bean. *Dolichos*, *Phaseolus*, *Vigna* are often confused, see Verdcourt, *Kew Bull.* **24**:507, 1970.

vignae, *Cladosporium*, *Entyloma*, *Phytophthora*, *Septoria*.

vigna mosaics Pérez & Bird, *J.Agric.Univ.P.Rico* **55**:468, 1971; **57**:56, 1973; transm. sap, *V. repens*; transm. sap, *Aphis craccivora*, *V. hosei*; Puerto Rico [**51**,4746; **54**,2720].

vignicola, see *Septoria vignae*.

Vinca, periwinkle; there have been several diseases reported and all assoc. with MLO, e.g. **leaf curl** Bisht & Singh 1964; **phyllody** Moll et al. 1977; **virescence** Oldfield et al. 1977; **witches' broom** McCoy & Thomas 1980; **yellows** Maramorosch 1956 [**36**:471; **44**,1593; **57**,1237, 5011; **61**,1764].

vinclozolin, selective fungicide for *Botryotinia*, *Monilinia*, *Sclerotinia*.

vincta, *Poria*.

vinegar rot, a spoilage of onion bulbs assoc. with bacteria; Watson & Hale, *N.Z.J.exp.Agric.* **12**:351, 1984 [**64**,3647].

vinyl film, use of to absorb ultraviolet light and thus inhibit fungus sporulation to exert control of a disease; Honda & Yunoki, *JARQ* **14**:78, 1980; Sasaki et al. and Honda & Nemoto, *PD* **69**:530, 596, 1985 [**60**,3633; **64**,5339; **65**,2818].

violacea, *Ustilago*; **violaceum**, *Phragmidium*.

violae, *Pythium*, see cavity spot.

viola mottle v. Lisa & Dellavelle 1977; Lisa et al., *Descr.V.* 247, 1982; Potexvirus, filament *c*.480 nm long, transm. sap, Italy (N.), *V. odorata*.

violet root rot (*Helicobasidium*).

virens, *Gliocladium*, *Ustilaginoidea*.

virescence, greening of tissue that is normally devoid of chlorophyll; the abnormal development of flowers in which all organs are green and partly or wholly transformed into structures like small leaves. A symptom of disease sometimes induced by MLO.

virginia crab stem grooving = apple stem grooving v.

viride, *Trichoderma*; **viridiflava**, *Pseudomonas*.

viridiol, a toxin, see *Gliocladium*.

virion, the whole virus particle.

viroid, circular, unencapsulated, low molecular weight RNA of one species, resistant to heat, ultraviolet and ionising radiation. Viroids were first characterised, and shown to cause diseases in plants, by Diener in 1971, *Phytop.* **63**:1328, 1973. They have not been found in animals. They were detected, not sought for, in plants with diseases of unknown etiology. Viroids are not detectable as particles in infected plants, but when placed in plants they replicate autonomously in susceptible cells, i.e. no helper is required, and cause the characteristic disease syndrome. Individual viroids are described separately: avocado sunblotch, chrysanthemum chlorotic mottle, chrysanthemum stunt, citrus exocortis, coconut cadang cadang, cucumber pale fruit, hop stunt, potato spindle tuber, tomato apical stunt, tomato bunchy top, the last 2 may be the same, tomato planta macho. This last viroid was found in wild plant spp., which were symptomless, in the area where planta macho was first noticed. Galindo et al., *Physiol.Pl.Path.* **24**:257, 1984, reported a protein assoc. with disease. This antigen accumulated in tomato plants infected with 3 viroids, other than the tomato planta macho viroid, and one satellite RNA. It did so only after symptom appearance; and was not found after tomato was infected by a virus.

Diener 1979; in Kurstak 1981; *A.Rev.Microbiol.* **36**:239, 1982; *Adv.Virus Res.* **28**:241, 1983; *Intervirology* **22**:1, 1984; Semancik, *A.R.Phytop.* **17**:461, 1979; Branch & Robertson, *Science N.Y.* **223**:450, 1984, replication cycle; Sänger in Mahy & Pattison ed., *Sympos.Soc.Gen.Microbiol.* 36, part 1, 1984; Singh, *Can.Pl.Dis.Surv.* **64**:15, 1984, bibliography of reviews; Maramorosch & McKelvy 1985 [**63**,5122].

virology, the study of viruses; for an analysis see Wildy in Mahy & Pattison, as above.

virulence, the observable effects of pathogen on a plant; it is not synonymous with, but describes degrees of, pathogenicity; see FBPP; Michelmore & Hulbert, *A.R.Phytop.* **25**:383, 1987, molecular markers in genetic analysis of virulence.

virulent, highly pathogenic.

viruliferous, of a vector which carries or contains virus; the vector may or may not be infective, after FBPP.

*viruses, transmissible, subcellular entities which replicate only in the living cell and consist of template nucleic acid which is typically surrounded by a protein coat; see taxonomy. The genome has 10^3–10^6 nucleotide pairs. Viruses can cause serious diseases in animals and plants; and they occur in fungi, see mycoviruses. Some need the presence of another for replication, see satellite virus, or for transm. see dependent transm.; many are transm. by animals, see vector. A virus may or may not cause disease symptoms, it may be latent; several viruses may occur in one plant. The host range can be narrow or one virus may cause diseases in many, taxonomically unrelated plants. Some terminology is noted in the order and see FBPP.

Plant virology has no classical binomial system but a taxonomic hierarchy is evolving. Hollings gave a brief historical note in Johnston & Booth 1983. Five levels may be distinguished, in descending order: particle, group, sub-group, virus, str. A virus particle is icosahedral or isometric, a straight rod, a flexuous filament or cylindrical i.e. bacilliform. The groups, Harrison et al., *Virology* **45**:356, 1971, are described separately in the order, they are: ISOMETRIC, ssRNA: Bromovirus, Como-, Cucumo-, Diantho-, Ilar-, Luteo-, Necro-, Nepo-, Sobemo-, Tombus-, Tymo-; dsRNA: Phytoreo-; ssDNA: Gemini-; dsDNA: Caulimo-. FILAMENT or ROD, ssRNA: Carla-, Clostero-, Hordei-, Potex-, Poty-, Tobamo-, Tobra-; BACILLIFORM, enveloped ssRNA: Phytorhabdo-.

An anomaly, compared with any binomial system, is that a virus name conveys no information as to where in the taxonomic order it might be placed. This is recognised. In set 20, 1986, the index of the *Descr.V.*, the group name is given after the virus name. Thus: arabis mosaic nepovirus or carnation etched ring caulimovirus. But these names could also be: nepo arabis mosaic or caulimo carnation etched ring. The latter arrangement has 4 advantages: (1) places the taxonomic label first and therefore, in any list, all names in one group come together; (2) shortens the name; (3) eliminates the unnecessary word virus; (4) in the acronym or sigla the letter v is eliminated, thereby avoiding ugly repetitiveness in a text and confusion with 'viroid', where acronyms are also used. An International Committee on the Taxonomy of Viruses formulates rules and approves new names, Matthews 1982.

See major texts and: Bawden, *Plant viruses and virus diseases*, edn. 4, 1964; Beale, *Bibliography of plant viruses and index to research*, 1976; Klinkowski, *Pflanzliche Virologie*, 4 vol., 1977;

Thresh, *Appl.Biol.* **5**, 1980, origin & epidemiology; *Adv.appl.Biol.* **8**, 1983, progress curves; Walkey, *Acta Hortic.* 88:23, 1980, production of virus free plants; Hamilton et al., *J.gen.Virol.* **54**:223, 1981, identification & characterisation; Bock, *Trop.Pest Management* **28**:399, 1982, identification & partial characterisation in the tropics; Fraenkel-Conrat & Kimball, *Virology*, 1982; Fraser, *Acta Hortic.* 127, 101, 1982, biochemical aspects of resistance; Gould & Symons, *A.R.Phytop.* **21**:179, 1983, molecular biological relationships; Dodds et al., *ibid* **22**:151, 1984, plant virus dsRNA; Boswell et al., *Rev.Pl.Path.* **65**:221, 1986, data bank, keys & viruses in each group; Koenig, *Adv. Virus Res.* **31**:321, 1986, plant viruses in lakes & rivers; Mowat, *Monogr.Br.Crop Prot.Council* 33, 1986, prevention of spread; Ponz & Bruening, *A.R.Phytop.* **24**:355, 1986, resistance mechanisms in plants; Fraser, *Biochemistry of virus-infected plants*, 1987; Matthews, *ibid* **25**:11, 1987.

Viteus vitifoliae, see *Plasmopara viticola*.

viticola, *Cryptosporella*, *Phomopsis*, *Plasmopara*; **vitis**, *Cercospora*.

Vitis, grapevine.

vivotoxin, see toxin.

voandzeia necrotic mosaic v. Fauquet, Monsarrat & Thouvenel 1981, *Descr.V.* 279, 1984; Tymovirus, isometric *c.*28 nm diam., transm. sap; Ivory Coast, Upper Volta; occurs naturally in bambara groundnut, *Vigna subterranea*, experimental hosts in Leguminosae.

Voglino, Piero, 1864–1933; born in Italy, Director, Laboratorio Sperimentale e R. Osservatorio de Fitopatologia, Turin; began some of the first attempts to forecast disease, worked with grapevine downy mildew (*Plasmopara viticola*); wrote: *Patologia vegetale*, 1924. *Riv.Patol.Veg.* **24**:1, 1934.

volatile chemicals, Linderman & Gilbert in Bruehl 1975, effects on soilborne pathogens; for effects on fungi: Hutchinson, *TBMS* **57**:185, 1971; Fries, *ibid* **60**:1, 1973; French, *A.R.Phytop.* **23**:173, 1985.

volumetric spore trap, a device that samples airborne particles, particularly fungus spores. The number of spores in a given quantity of air at any time can be calculated. In most cases the type of spore can be determined; and spore viability can be examined in some traps. Bartlett & Bainbridge gave a brief review in Scott & Bainbridge 1978. The most widely used trap, there are adaptations from it, was described by Hirst, *AAB* **39**:257, 1952; other traps: Pady, *Phytop.* **49**:757, 1959; Schenck, *ibid* **54**: 613, 1964; Kramer & Pady, *ibid* **56**:517, 1966; Stedman, *Mycopathologia* **66**:37, 1978; Morris, *Bull.Br.mycol.Soc.* **16**:151, 1982; Gadoury & MacHardy, *Phytop.* **73**:1526, 1983; Gottwald & Tedders, *ibid* **75**:801, 1985; and see letters by Wili and Gadoury & MacHardy, *ibid* **75**:380, 1985; **76**:127, 1986.

volutum, *Pythium*, see *P. zingiberum*.

vulgaris, *Tubercularia*.

W

wageneri, *Ceratocystis*, *Verticicladiella*.

Waitea Warcup & Talbot, *TBMS* **45**:503, 1962; Corticeaceae; *W. circinata* type sp., see *Rhizoctonia zeae*, Oniki et al.

Wakefield, Elsie M., 1886–1972; born in England, Univ. Oxford, Royal Botanic Gardens, Kew from 1910; took over from George Massee in 1911. She was a mycological taxonomist; and trained many mycologists, as plant pathologists were then called, who went overseas, mostly to the tropics. This was particularly before the establishment of the Imperial College of Tropical Agriculture in Trinidad & Tobago in 1922; OBE 1950. *TBMS* **60**:167, 1973.

Wakker, Jan Hendrick, 1859–1927; born in the Netherlands, Univ. Amsterdam; studied plant diseases in the Dutch East Indies, now Indonesia, particularly in sugarcane. But is mainly notable for making one of the very early studies on a bacterial disease of plants, yellows of hyacinth, now *Xanthomonas campestris* pv. *hyacinthi*, in 1881. *Neth.J.Pl.Path.* **72**:38, 1966.

wakkerii, *Sclerotium*; waksmanii, *Penicillium*.

* waldsterben, a collective term for the serious, widespread and substantial decline in growth and behavioural change in forest ecosystems in central Europe; see acid rain, air pollution. This apparently began in the late 1970s and has developed rapidly since. Three symptom patterns: abnormal growth, decreasing growth and water stress; some of the symptoms have not been described before. These appeared about the same time in Europe, on all tree spp. and some shrubs. The symptoms are still expanding, and are independent of differences in climate, soil type and management practice. The exact cause is unknown. Five possibilities are: acidification and Al toxicity, ozone, general stress, e.g. decrease in net photosynthesis, excess nutrient or excess N, and Mg deficiency. Another, somewhat speculative, cause may be the air transport of organic substances that alter growth. A review, citing major German publications, was given by Schütt & Cowling, *PD* **69**:548, 1985; Blank, *Nature Lond.* **314**:311, 1985, forest decline in Germany [**65**,2028].

wallaby ear, maize, prob. caused by a toxin from an insect, *Cicadulina bimaculata*; Hatta et al., *Physiol.Pl.Path.* **20**: 43, 1982; Ofori & Francki, *AAB* **103**:185, 1983 [**61**,4892; **63**,1753].

Wallace, James Merrill, 1902–79; born in USA,

Univ. Mississipi State, Minnesota, USDA 1929–42, Univ. California Riverside from 1942; an authority on citrus virology. *Phytop.* **69**:781, 1979.

wallflower virescence Le Normand & Gourret, *Annls.Phytopath.* **1**:301, 1969; Gourret et al., *Can.J.Microbiol.* **20**:1617, 1974; MLO assoc., France, *Cheiranthus cheiri* [**49**,502; **54**,2296].

walnut, *Juglans*.

—— black line Schuster & Miller 1933 (cherry leaf roll v.); Mircetich et al., *Phytop.* **70**:962, 1980; de Zoeten et al., *ibid* **72**:1261, 1982; assoc. with declining English walnut scions on *Juglans hindsi* or *J. hindsi* × *J. regia* stocks. The necrosis at the graft union is apparently due to a hypersensitive reaction of stocks to this virus; Rowhani et al., *ibid* **75**:48, 1985, serological detection in English walnut [**60**,3977; **62**,1228; **64**,2741].

—— brooming Hutchins & Wester 1947; Dodge, *J.N.Y.bot.Gdn.* **48**:112, 1947; Seliskar, *Forest Sci.* **22**:144, 1976, walnut bunch, MLO assoc., USA [**26**:362; **56**,1291].

—— line pattern and mosaic Christoff 1937; *Phytopath.Z.* **31**:381, 1958, Bulgaria; Ram & Rathore, *J. Maharashtra agric.Univ.* **12**:113, 1987; India (Himachal Pradesh) [**37**:586].

—— ringspot and yellow mosaic (cherry leaf roll v. assoc.) Savino et al., 1976; *Phytopath.Mediterranea* **16**:96, 1977; Italy (S.) [**56**,1178c; **59**,1421].

Ward, Harry Marshall, 1854–1906; born in England, Science Schools, Kensington, London; Univ. Cambridge, chair of botany 1895. He was an original worker in several fields and a founder of the teaching of plant pathology in England. He will be largely remembered for carrying out prob. the first successful investigation on the etiology of a plant disease in the tropics, coffee rust (*Hemileia vastatrix*). The fungus was seen apparently causing leaf fall in Ceylon, now Sri Lanka, in 1869. But government action was not taken for 10 years. In 1880, fide Large 1940, Ward went out to Ceylon to investigate the disease that was to turn out as a calamity for the island's planters; FRS 1888, Royal Medal 1893. *Gdner's Chron.* **40**:164, 1906; *Kew Bull.* 1906:281; *Ann.Bot.* **21**:ix, 1907; *Proc.Linn.Soc.Lond.* 119th session 1906–7:54; *New Phytol.* **6**:1, 1907; *Proc.R.Soc.Lond.* B **83**:i, 1911.

Wardlaw, Claude Wilson, 1901–85; born in Scotland, Univ. Glasgow, Imperial College of

Tropical Agriculture 1928–40; chairs of botany, Univ. Manchester to 1966; wrote: *Diseases of the banana and of the Manila hemp plant*, 1935; new edn. as *Banana diseases including plantains and abaca*, 1961; reprinted with additional information, 1972; also books on storage of tropical plants, morphogenesis and phylogeny of plants. *The Times* 18 Dec., 1985.

wart (*Synchytrium endobioticum*) potato.

warty (*Botryotinia fuckeliana*) coffee; — **scab** (*Olpidium viciae*) broad bean.

wasabiae, *Phoma*.

washingtonia mosaic Mayhew & Tidwell, *PDR* **62**:803, 1978; filamentous particle assoc., USA (California); *W. robusta*, Mexican fan palm [**58**,3949].

water, fungi and plants, Ayres & Boddy ed., *Sympos.Br.mycol.Soc.* 11, 1986.

— **blister** (*Ceratocystis paradoxa*) pineapple.

— **core**, apple, pear; high temp, water stress, low Ca assoc.; Faust et al., *Bot.Rev.* **35**:168, 1969, carbohydrate metabolic disorders in apple; Marlow & Loescher, *Hortic.Rev.* **6**:189, 1984; in swede, B deficiency Scaife & Turner 1983.

watercress, *Nasturtium officinale*.

— **chlorotic leaf spot** Ward & Tomlinson 1979, 1981–2; Tomlinson & Hunt, *AAB* **110**:75, 1987; transm. sap to *Chenopodium amaranticolor*, *C. quinoa*, *Petunia hybrida*; and by *Spongospora subterranea* f.sp. *nasturtii*, England [**59**,2494; **61**,1535, 6776; **66**,3162].

— **mosaic** Roland, *Parasitica* **8**:1, 1952; transm. sap, *Myz.pers.*, persistent, Belgium [**31**:526].

— **white spot** Loh & Cheo, *Acta phytopath.sin.* **7**:1, 1964; transm. sap, China [**44**,900].

— **yellow leaf band** Fletcher & Hims, *Pl.Path.* **29**:200, 1980; transm. sap, England, plant as *Rorippa nasturtium-aquaticum* [**61**,3191].

— — **spot v.** Spire, *Annls.Épiphyt.* **13**:39, 1962; isometric 27 nm diam., transm. sap, France [**41**:627].

water hyacinth, *Eichhornia crassipes*; Rakvidhyasastra & Visarathanonth, *Kasetsart J.* **9**:170, 1975 [**58**,2175].

water in soil, see anaerobiosis in soil, ecology, soilborne pathogens; Cook & Papendick, *A.R.Phytop.* **10**:349, 1972, water potential of plants & soils & root disease.

watermark (*Erwinia salicis*) willow.

watermelon, *Citrullus lanatus*.

— **curly mottle v.** Brown & Nelson, *Phytop.* **76**:236, 1986; poss. Geminivirus, isometric, paired particles 30 × 18 nm, transm. sap, *Bemisia tabaci*, USA (Arizona), infects several cucurbits, see lettuce infectious yellows [**65**,3539].

— **general mosaic** = watermelon 2 v.

— **mosaic 1**, and watermelon specific mosaic = papaya ringspot v.

*— — **2 v.** Webb & Scott 1965; Purcifull et al., *Descr.V.* 293, 1984; Potyvirus, filament *c.*760 nm long, transm. sap readily, many aphid spp., non-persistent causes diseases in cucurbits and found naturally in other plant families; Yamamoto et al., *Bull.Shikoku Nat.agric.Exp.Stn.* 44:26, 1984, epidemiology in Japan; Romanow et al., *Phytop.* **76**:1276, 1986, acquisition & inoculation as affected by plant resistance to virus & vector [**64**,1816; **66**,4542].

— **silver mottle** (tomato spotted wilt v.), Iwaki et al., *PD* **68**:1006, 1984; Japan [**64**,1337].

watersoaked, plants appear wet, dark and ± translucent; see wetwood in trees.

water spinach, *Ipomoea aquatica*.

watery soft rot (*Sclerotinia sclerotiorum*); — **stipe**, as for La France q.v., Fletcher 1984; — **wound rot** (*Pythium ultimum*) potato.

waxy patch, tomato, see blotchy ripening.

weather, see climate and weather; — **fleck**, tobacco, mainly caused by ozone, Heggestad & Middleton, *Science N.Y.* **129**:208, 1959 [**38**:423].

web blight (*Thanatephorus cucumeris*) bean, cucurbits, Lima bean, soybean, Hepperley et al., *PD* **66**:256, 1982, soybean; Galindo et al., *ibid* **67**:1016, 1983, bean; — **blotch** (*Didymosphaeria arachidicola*) groundnut [**61**,6087; **63**,992].

websteri, see *Mycosphaerella aleuritis*.

weed moulds, in common, cultivated mushroom, Fletcher 1984.

weeds, see culpad viruses, ecology, mycoherbicide, root parasitic weeds.

weeping conk (*Inonotus dryadeus*) temperate hardwoods.

weirei, *Inonotus*; **weirii**, *Chrysomyxa*, *Rhabdocline*, see *R. pseudotsugae*.

Weir's spruce cushion rust (*Chrysomyxa weirii*).

Welsh onion, see *Allium*.

wernsdorffiae, *Coniothyrium*.

Westcott, Cynthia, 1898–1983; born in USA, Wellesley College, Univ. Cornell, well known for advisory work on pathology of ornamentals, books included, *The plant disease handbook*, edn. 3, 1971. *Phytop.* **73**:1601, 1983; *A.R.Phytop.* **22**:21, 1984.

Westerdijk, Johanna, 1883–1961; born in the Netherlands, Univ. Amsterdam, Munich, Zurich; director, phytopathology laboratory 'Willie Commelin Scholten' and Central Bureau voor Schimmelcultures, chairs at Utrecht 1917 and Amsterdam 1930; trained many students and built up the reputation of the Netherlands as a world centre for plant pathology; Otto Appel Medal, Heidelberg 1953. *AAB* **50**:372, 1962;

J.gen.Microbiol. **32**:1, 1963; *A.R.Phytop.* **24**:33, 1986.

Western, John Henry, 1906–81; born in England, Avoncroft Agricultural College, Univ. College Wales, Manchester; Ministry Agriculture UK 1939–51, Univ. Leeds, chair agricultural botany 1959; wrote, with K. Sampson, *Diseases of British grasses and herbage legumes*, 1941; edn. 2, 1954; edited *Diseases of crop plants*, 1971. *AAB* **100**:411, 1982; *Bull.Br.mycol.Soc.* **16**:80, 1982.

western gall rust (*Endocronartium harknessii*) hard pines; —— **yellow rust** (*Phragmidium rubi-idaei*) raspberry.

wet bubble (*Mycogone perniciosa*), common cultivated mushroom.

—— **rot**, tissue rapidly and completely disintegrated, with release of free water from the lysed cells; prob. differs from soft rot only in the rate at which cells lose their capacity to retain water, FBPP.

—— —— (*Choanephora cucurbitarum*) mostly flowers and vegetables, and in storage.

wetter, see spreader.

wet weather blight (*Ascochyta gossypii*) cotton.

wetwood, a watersoaked condition within the wood of living trees; usually in the heartwood but also in the sapwood. Wetwood in the central stem core appears to be of nonpathological origin, although assoc. with a bacterial population. The condition in sapwood may be assoc. with physical injury or pathogen(s). Wetwood is characterised by: a dark colour, excess water, high gas pressures, differing pH, decrease in electrical resistance, presence of bacteria. Exudations of water from wetwood through stem wounds may become slime flux q.v. or wetwood flux. Injection of tree stems with pesticides can lead to this condition.

Carter, *Bull.Ill.Nat.Hist.Surv.* 23:407, 1945, elm; Etheridge & Morin, *CJB* **40**:1335, 1962, *Abies balsamea*; Coutts & Rishbeth, *Eur.J.For.Path.* **7**:13, 1977, *A. grandis*; Schink et al., *J.gen.Microbiol.* **123**:313, 1981; *Appl.Environ.Microbiol.* **42**:526, 1981, microbiology & anaerobic bacteria; Worrall & Parmeter, *Phytop.* **72**:1209, 1982; *Eur.J.For.Path.* **12**:432, 1982, *A. concolor*; Murdoch & Campana, *Phytop.* **73**:1270, 1983; *PD* **68**:890, 1984; elm, assoc. bacteria, wounding & bleeding; Murdoch et al., *ibid* **67**:74, 1983, gas in elm; Scott, *Eur.J.For.Path.* **14**:103, 334, 1984, bacteria in poplar, electrical resistance; Murdoch et al., *Can.J.Pl.Path.* **9**:20, 1987, *Ulmus americana* [**25**:239; **42**:223; **56**,4231; **60**,6095; **61**,3081; **62**,2162, 3292; **63**,1411, 5149; **64**,795, 2181].

wheat, *Triticum*; Manners, *TBMS* **52**:177, 1969, control of rusts; Saari & Wilcoxson, *A.R.Phytop.*

12:49, 1974, dwarf wheats in Africa & Asia; Bingham and Dubin & Rajaram in Jenkyn & Plumb 1981; Dubin & Rajaram, *PD* **66**:967, 1982; breeding for resistance; Bruehl, *ibid* **66**:1090, 1982, resistance to snow moulds; Elekes, *Seed Sci.Technol.* **11**:421, 1983, detection in seed; Joshi et al. ed., *Problems and progress of wheat pathology in South East Asia*, 1986; Wiese 1987.

—— **African streak** = cereal African streak v.

—— **American striate mosaic v.** Slykhuis 1953; Sinha & Behki, *Descr.V.* 99, 1972; Phytorhabdovirus, bacilliform mostly 200–250 × 75 nm, transm. *Elymana virescens, Endria inimica*, high temps. during early summer favour transm., *c*.20 spp. Gramineae are susceptible, N. America.

—— **Australian striate mosaic** (chloris striate mosaic v.).

—— **chlorosis and aspermy** Ploaie 1973; bacterium assoc., fide Hopkins, see phloem limited bacteria.

—— **chlorotic streak**, and wheat chlorotic streak mosaic = barley yellow striate mosaic v.

—— **dwarf v.** Dlabola, fide Vacke, *Biologia Pl.* **3**:228, 1961; Agarkov & Baïdala, *Biol.Nauki* **11**(9):94, 1968; Tomenius & Oxelfelt, *Phytopath.Z.* **101**:163, 1981; Lindsten et al. 1981; Geminivirus, isometric, each half 18–20 nm diam., transm. *Psammotettix alienus*, Europe; Lindsten, *Växtskyddsnotiser* **44**:54, 1980, Sweden & history [**41**:378; **48**,416; **60**,3108; **61**,645; **62**,1431].

—— **eastern striate v.** Nagaich & Sinha, *PDR* **58**:968, 1974; particle 40 nm diam., transm. *Cicadulina mbila*, India (Simla Hills) [**54**,1670].

—— **European striate mosaic** Slykhuis & Watson, *AAB* **46**:542, 1958, transm. *Javesella* sp.; Ammar, *ibid* **79**:195, 203, 1975, vector biology [**38**:315; **54**,5366–7].

—— **Indian chlorotic streak** Nagpal, Upadhyaya & Seth, *Indian J. Mycol.Pl.Path.* **6**:184, 1977; transm. *Cicadulina mbila*, India (Madhya Pradesh) [**57**,1671].

—— **pale green dwarf** (wheat dwarf v.).

—— **rosette stunt** = northern cereal mosaic v., see cereal northern mosaic v.

—— **soilborne mosaic v.** McKinney 1923; Brakke, *Descr.V.* 77, 1971; rod 110–160 and 300 nm long, 18 nm diam., contains ssRNA, transm. sap, *Polymyxa graminis*, narrow host range; Brakke, *Phytop.* **67**:1433, 1977, sedimentation coefficients; Wiese 1987 [**57**,4393].

—— **spindle streak mosaic v.** Slykhuis 1960; *Descr.V.* 167, 1976; Potyvirus, filament 200–2000 nm long, transm. sap, *Polymyxa graminis*. This virus is the same as, or a str. of, the earlier described wheat yellow mosaic v. McKinney, *Science N.Y.* **73**:650, 1931; Usugi & Saito, *Ann.phytopath.Soc.Japan* **45**:397, 1979. Slykhuis &

Barr and Nolt et al., *Phytop.* **68**:639, 1978; **71**:1269, 1981, gave further evidence of the assoc. of *P. graminis* with transm. [**10**:646; **58**,151; **59**,3682; **61**,4031].

— spot chlorosis Nault & Styer 1969; *Phytop.* **60**:1616, 1970; = wheat spot mosaic [**50**,1705].

— — mosaic Slykhuis, *Phytop.* **46**:682, 1956; transm. *Aceria tulipae*, N. America, temperate cereals and annual grasses [**36**:389].

— streak mosaic v. McKinney 1937; Brakke, *Descr.V.* 48, 1971; Potyvirus, subgroup 3, filament *c*.700 × 15 nm, transm. sap, *Aceria tulipae*, many cereals and other Gramineae.

— white spike Caetano, Kitajima & Costa, *Bragantia* **29** Nota 9: XLI, 1970; Brazil (S.) [**51**,1313].

— yellow dwarf Caetano 1972; Diehl et al., *Fitopatología* **9**:100, 1974; Brazil (S.) [**54**,4423].

— yellowing Šutić, *Acta biol.iugosl.B.* **11**:127, 1974; poss. MLO, Yugoslavia [**55**,1176].

— — stripe Fan et al., *Acta microbiol.sin.* **22**:31, 1982; bacterium assoc., China [**61**,5642].

— yellow leaf v. Inouye et al. 1973; Inouye, *Descr.V.* 157, 1976; Closterovirus, filament, *c*.1600–1850 nm long, transm. *Rhopalosiphum maidis*, semi-persistent, Japan, Gramineae.

— — — curl Wu et al. 1981; *Scientia sin.*B **27**:61, 1984; bacterium assoc. China [**61**,2203; **65**,184].

— — stunt Tung & Chang, *Scientia agric.sin.* **3**:74, 1979; transm. aphid spp., China [**60**,216].

— — stunting Radulescu & Munteanu, *Annls.Phytopath.* **2**:403, 1970; transm. *Psammotettix alienus*, Romania [**50**,1704].

Whetzel, Herbert Hice, 1877–1944, born in USA, Wabash College, Indiana; Univ. Cornell where he formed the nucleus of the mycological herbarium, an authority on the Sclerotiniaceae; wrote *An outline of the history of phytopathology*, 1918. *Mycologia* **37**:393, 1945; *Phytop.* **35**:659, 1945; *A.R.Phytop.* **18**:27, 1980.

whetzelii, *Ciborinia*.

whiptail, cauliflower, Mo deficiency; Scaife & Turner 1983.

whiskers (*Rhizopus stolonifer*) fruits and vegetables.

white blister (*Albugo candida*) crucifers; sometimes, inappropriately, called white rust; — blotch (*Bacillus megaterium* pv. *cerealis*) wheat.

— bryony mosaic v. Spire, Bertrandy & Férault, *Annls.Épiphyt.* **17**:Hors. ser. 121, 1966; Hollings et al., 1967; Tomlinson & Carter 1970; filament 620–640 nm long, transm. sap; England, France [**46**,1492m; **47**,390p; **49**,3555j].

— — mottle v. Lockhart & Fischer, *Phytopath.Z.* **96**:244, 1979; Potyvirus, filament 745 nm long, transm. *Myz.pers.*, mostly Cucurbitaceae [**59**,4405].

— clover, see *Trifolium*; Campbell & Moyer, *PD* **68**:1033, 1984, virus infection & yield [**64**,2596].

— — cryptic v. Boccardo et al., *Virology* **147**:29, 1985; particle types, and see cryptic viruses [**65**,768].

— — mosaic v. Pierce 1935; Bercks, *Descr.V.* 41, 1971; Potexvirus, filament 480 × 13 nm, transm. sap, seed; mostly legumes; Beczener & Vassányi, *Acta phytopath.Acad.Sci.hung.* **16**:109, 1981, strs. [**61**,4205].

— — streak mosaic v. Babović, *Zast.Bilja* **26**:199, 1975; *Acta biol. iugosl.* B **15**:175, 1978; Babović & Cekić 1978; isometric 30 nm diam., transm. sap [**55**,4166; **58**,4417; **59**, 3813].

— cob rot (*Fusarium poae*) maize; — ear rot (*Stenocarpella maydis*) maize; — fan blight (*Marasmius palmivorus*) rubber.

— flies, see vector.

— foot rot (*Gibellina cerealis*) wheat; — head (*Fusarium poae*) cereals, grasses; — heads (*Gaeumannomyces graminis*) temperate cereals, grasses; — heart rot (*Inonotus arizonicus*) *Platanus*, (*Phellinus ignarius*) temperate dicotyledonous trees; — heartwood rot (*I. farlowii*) willow; — leaf blotch (*Ascochyta allicepae*) onion; — — spot (*Septoria cannabis*) common hemp, (*S. carthami*) safflower; — — streak (*Mycovellosiella oryzae*) rice; — mottled rot (*Perenniporia fraxinophila*) *Fraxinus pennsylvanica*; — mould (*Erysiphe cichoracearum*) tobacco, (*Ramularia vallisumbrosae*) narcissus, (*Sclerotinia sclerotiorum*) vegetables and other plants.

— mustard, *Sinapsis alba*.

— pine blister rust (*Cronartium ribicola*) 5 needle pines; — pocket rot (*Inonotus tomentosus*) conifers, (*Rigidoporus zonalis*) hardwoods; — root rot (*R. lignosus*) rubber, tropical woody crops, (*Rosellinia necatrix*) temperate fruit woody crops, (*Vararia* sp.) raspberry.

— rot, of wood in trees invaded by fungi that destroy lignin and leave a white cellulose residue, after FBPP; Otjen & Blanchette, *CJB* **64**:905, 1986 [**65**,5172].

— — (*Coniella fragariae*) grapevine, (*Lentinus squarrosulus*) *Shorea robusta*, (*Sclerotiorum cepivorum*) onion, *Allium*; — rust (*Puccinia horiana*) chrysanthemum; — scab (*Elsinoë leucospila*) tea; — spot (*Mycosphaerella fragariae*) strawberry, (*Pseudocercosporella capsellae*) brassicas; — stringy rot (*L. polychrous*) *S. robusta*; — stripe (*Xanthomonas albilineans*) sugarcane; — tip (*Aphelenchoides besseyi*) rice, (*Phytophthora porri*) *Allium*, Cu deficiency in temperate cereals, Bould et al. 1983.

Wiehé, Paul Octave, 1910–75; born in Mauritius,

Imperial College of Science and Technology, London; plant pathologist; first director of the Mauritius Sugar Industry Research Institute which he created and made into an outstanding centre; CBE 1958. Annual Report of the Institute for 1975.

wiesneri, *Taphrina*.

wild cucumber mosaic v. Freitag 1952; van Regenmortel, *Descr.V.* 105, 1972; Tymovirus, isometric *c*.28 nm diam., transm. sap, *Acalymma trivittatum*, USA (California, Oregon); *Marah macrocarpus*, *M. oreganus*.

wildfire (*Pseudomonas syringae* pv. *tabaci*) tobacco.

wildings, potato, genetic disorder; bushy growth; many, often thin, stems; fewer primary leaflets, rudimentary secondary leaflets, flowers are rare; Todd, *Pl.Path.* 3:17, 1954 [**33**:751].

wild plants, see ecology; — **rice**, *Zizania*; Dore, Publ. 1393 Canada Agric. 1969.

willkommii, *Lachnellula*.

willow, *Salix*.

wilpad viruses, see culpad viruses.

wilt, loss of turgidity, usually in leaves; leaf collapse and leaf fall. Typically wilts are caused by pathogens which colonise the vascular system, see vascular wilt; but such colonisation may not lead to loss of turgidity and wilt. Infection of the collar and root cortex, a foot rot, may cause a characteristic wilt syndrome.

Wiltshire, Samuel Paul, 1891–1967; born in England, Univ. Bristol, Cambridge; Imperial Bureau of Mycology, now the CAB International Mycological Institute, 1922–56; director 1940. The fame of the institute was built up by 3 men: E.J. Butler, the first director; E.W. Mason, the first mycologist; and Wiltshire who expanded the services and the publications. He was primarily responsible for the reputation for accuracy and reliability soon achieved by the *Review of Applied Mycology*, later the *Review of Plant Pathology*; and based on meticulous editing. This tradition has still not been completely eroded by the changes brought about by computerisation. *Nature Lond.* **215**:221, 1967; *TBMS* **50**:513, 1967.

wineberry latent v. Jones 1977; *Descr.V.* 304, 1985; Potexvirus, filament *c*.510 × 12 nm, transm. sap, reported only from Scotland in a single, symptomless plant of *Rubus phoenicolasius*.

winged bean, *Psophocarpus tetragonolobus*.

— — **mosaic** (clover yellow vein v.), Fox & Corbett, *PD* **69**:352, 1985 [**64**,4607].

— — **necrotic mosaic v.** Fauquet, Lamy & Thouvenel, *F.A.O.Pl.Prot.Bull.* **27**:81, 1979; filament 620 × 14 nm, transm. sap, Ivory Coast [**60**,586].

— — **ringspot v.**, authors as above; poss. Cucumovirus, isometric 24 nm diam., transm. seed, *Aphis craccivora*, Ivory Coast.

— — **witches' broom** Sumardiyono 1980; MLO assoc., Indonesia [**61**,507].

— — **yellow mosaic** Thompson & Haryono, *HortScience* **14**:532, 1979; Indonesia [**59**,1016].

winter sweet mosaic Foister 1961; Scotland, *Chimonanthus praecox* [**40**:735].

— **wheat Russian mosaic v.** Zazhurilo & Sitnikova 1939, 1941; Razvyazkina & Poljakova, *Proc.6th.Pl.Virol.Conf.*:129, 1969; Phytorhabdovirus, bacilliform 260 × 60 nm, transm. *Psammotettix striatus* [**19**:268; **22**:59].

wire stem (*Thanatephorus cucumeris*) brassicas.

Wisconsin leaf spot (*Pseudomonas syringae* pv. *mellea*) tobacco.

wissadula mosaic Schuster, *PDR* **48**:902, 1964; transm. *Bemisia tabaci*, persistent, USA (Texas), *W. amplissima*, infects cotton but not okra [**44**,1027].

— **proliferation** Dabek, Waters & Lavin, *Phytopath.Z.* **107**:336, 345, 1983; poss. bacterium, transm. *Paracarsidara concolor*, Jamaica, *W. periplocifolia* [**63**,2116–17].

wisteria vein mosaic v. Conti & Lovisolo, *Riv.Patol.veg.* **5**:115, 1969; Bos, *Neth.J.Pl.Path.* **76**:8, 1970; poss. Potyvirus, filament *c*.750 nm long, transm. sap, *Aphis craccivora*, *Myz.pers.*; Europe, USA [**49**,2012–13].

witches' broom, an abnormal proliferation of shoots; (*Caeoma deformans*) *Thujopsis dolobrata*, (*Crinipellis perniciosa*) cacao, (*Elytroderma deformans*) pine, (*Peridermium cedri*) *Cedrus deodara*, (*Taphrina* spp.) temperate hardwoods, (*T. wiesneri*) apricot, cherry.

withertip (*Glomerella cingulata*) citrus.

withertop, flax, Ca deficiency; Millikan, *J.Dep.Agric.Vict.* **42**:79, 1944 [**23**:227].

Wojnowicia Sacc. 1892; Coelom.; Sutton 1980; conidiomata pycnidial, conidiogenesis enteroblastic phialidic, con. multiseptate, pigmented.

W. hirta Sacc.; Punithalingam, *Descr.F.* 773, 1983; con. 7 septate, 30–42 × 4 μm; assoc. with foot or root rot of temperate cereals and grasses, i.e. with *Gaeumannomyces graminis* and *Pseudocercosporella herpotrichoides*.

Wolf, Frederick Adolf, 1885–1975; born in USA, Univ. Nebraska, Texas, Cornell; Alabama and N. Carolina Experiment Stations from 1911, Duke Univ. 1927–54; tobacco diseases, wrote *Tobacco diseases and decays* 1935, edn. 2, 1957; with F.T. Wolf, *The fungi*, 2 vol., 1947. *Mycologia* **68**:229, 1976; *Phytop.* **66**:238, 1976.

Wollenweber, Hans Wilhelm, 1879–1949, born in

Germany, Univ. Berlin, a *Fusarium* world
authority. *Phytop.* **40**:119, 1950.
wood decay, the decay of standing timber is an
important part of forest pathology, see trees.
Scheffer & Cowling, *A.R.Phytop.* **4**:147, 1966,
natural resistance to micro-organisms; Liese, *ibid*
8:231, 1970, fine structure of disintegration;
Merrill, *ibid* **8**:281, 1970, spore germination &
host penetration by heart rot Hymenomycetes;
Wilcox, *Bot.Rev.* **36**:1, 1970, anatomical change in
wood cells attacked by bacteria & fungi; Bakshi &
Singh, *Int.Rev.For.Res.* **3**:197, 1970, tree heart
rots; Shigo & Hillis, *A.R.Phytop.* **11**:197, 1973,
micro-organisms in living trees; Griffin, *ibid*
15:319, 1977, water potential & wood decay
fungi; Rayner & Todd, *Adv.Botanical Res.* **7**:333,
1979, dynamics of fungi in decaying wood;
symposium on wood decay in living trees, *Phytop.*
69:1135, 1979; Gilbertson, *Mycologia* **72**:1, 1980,
wood rotting fungi of N. America; Shigo, *PD*
66:763, 1982; Boddy & Rayner, *New Phytol.*
94:623, 1983, origins of decay in living deciduous
trees, moisture content, concept of decay; Rayner
& Boddy, *Adv.Pl.Path.* **5**:119, 1986, population
structure & the infection biology of wood decay
fungi in living trees.
woodii, *Diaporthe.*
wood pocket, citrus, prob. genetic disorder, like
lime blotch; Calavan, *Calif.Citrogr.* **42**:265, 1957;
Knorr & Childs, *Proc.Fla.Sta.hort.Soc.* **70**:75,
1957 [**37**:166, 659].
— **rot** (*Hypoxylon serpens*) tea.
Wormald, Harry, 1879–1955; born in England, St

John's College, York; Royal College of Science,
London; Wye College, Univ. London 1911–22;
East Malling Research Station 1923–39, later at
Commonwealth Agricultural Bureau of
Horticultural and Plantation Crops; wrote *Brown
rot diseases of fruit trees*, 1935, edn. 2, 1954;
Diseases of fruits and hops, 1939; edn. 2 & 3,
1946, 1955. *Ann.Rep.East Malling Res.Stn.*
1955:15; *Nature Lond.* **177**:649, 1956; *TBMS*
39:289, 1956.
Woronin, Michael Stephanovitch, 1838–1903; Univ.
St Petersburg, worked with de Bary and was an
original investigator. His famous work on
Plasmodiophora brassicae was published in 1878;
the year that he received a gold medal from the
Royal Russian Society of Horticulture.
Phytopath.Classics 4, 1934, English translation by
C. Chupp. *Phytop.* **2**:1, 1912.
woroninii, *Chrysomyxa.*
wound healing, in higher plants; Bloch, *Bot.Rev.*
7:110, 1941; **18**:655, 1952.
— **response**, in tree bark; Biggs, *Phytop.* **75**:1191,
1985.
— **tumour v.** Black 1944; *Descr.V.* 34, 1970;
Prog.expl.Tumour Res. **15**:110, 1972;
Phytoreovirus, isometric *c.*70 nm diam., transm.
sap, few instances; especially by *Agallia constricta*,
Agalliopsis novella to many plants in which it does
not cause disease; the first symptom of systemic
infection is vein enlargement; root tumours are
characteristic; Shikata in Kurstak 1981; Nuss,
Adv.Virus Res. **29**:57, 1984.

X

xanthocephala, *Botryosphaeria.*

Xanthomonas, bacteria; Gram negative rods, motile, a single polar flagellum, no spores, colonies on nutrient media usually yellow; the yellow pigments are characteristic, brominated aryl polyenes, called xanthomonadins; the characteristics of the genus are to be changed; Starr in Starr et al., *Prokaryotae*, 1981; Moffet & Croft in Fahy & Persley 1983; Leyns et al., *Bot.Rev.* **50**:308, 1984, hosts; Bradbury in Krieg 1984, and 1986.

X. albilineans (Ashby) Dowson 1943; Hayward & Waterston, *Descr.B.* 18, 1964; sugarcane leaf scald, a vascular disease with acute and chronic phases. The former shows as a sudden wilt which sometimes affects the whole stool. The latter shows as narrow, pale, leaf stripes, leaves with necrotic tips, proliferation of side shoots, red streaks in the vascular tissue. The bacterium may remain dormant; and the resulting latent infection, without symptoms, presents a problem in the movement of plant material; Sivanesan & Waller 1986.

X. ampelina Panagopoulos 1969; Bradbury, *Descr.B.* 378, 1973; grapevine blight, stunted growth of buds in spring, spur canker; cankers spread to shoots, petioles, flower and fruit stalks; Europe, Mediterranean, South Africa, prob. Argentina; Panagopoulos et al., *EPPO Bull.* **17**:227, 231, 237, 1987.

X. axonopodis Starr & Garces O. 1950; Bradbury, *Descr.B.* 557, 1977; gummosis of *Axonopus affinis*, *A. compressus*, *A. micay*, *A. scoparius*; other grasses are infected after inoculation, including sugarcane; diseased leaves have pale stripes, and stems are elongated and partly bare; the whole tuft may die; Colombia.

X. campestris (Pammel) Dowson 1939; Hayward & Waterston, *Descr.B.* 47, 1965; Moffet & Croft in Fahy & Persley 1983; originally this was the name of the one pathogen which caused black rot of crucifers; but many other names erected in *Xanthomonas* for bacteria causing other diseases have now been rejected since they relate to forms which are not distinct from *X. campestris* except in pathogenicity. These are now called pathovars and Bradbury 1986 lists 136; Liew & Alvarez, *Phytop.* **71**:269, 274, 1981, phage; Daniels, *Microbiological Sci.* **1**:33, 1984, molecular biology; Robinson & Callow, *Pl.Path.* **35**:169, 1986, multiplication & spread of pvs. in hosts &

non-hosts; Daniels in Day & Jellis 1987, molecular genetics of pathogenicity; Lazo et al., *Int.J.syst.Bact.* **37**:214, 1987, distinguishing pvs. by restriction fragment length polymorphism [**61**,1015–16; **64**,440; **65**,5364].

X. c. pv. alfalfae (Riker et al.) Dye in Young et al., *N.Z.J.agric.Res.* **21**:153, 1978; Bradbury, *Descr.B.* 698, 1981; lucerne leaf spot; there are indications that isolates from different geographical areas may differ in pathogenicity characteristics.

X. c. pv. aracearum (Berniac) Dye; leaf spot of *Xanthosoma sagittifolium*, Guadeloupe; Berniac, *Annls.Phytopath.* **6**:197, 1974; Cortés-Monllor, *J.agric.Univ.P.Rico* **70**:225, 1986 [**54**,4298; **66**,815].

X. c. pv. arecae (Rao & Mohan) Dye; areca nut leaf stripe, coconut has been infected by inoculation, India (S.), Kumar, *Trop. Pest Management* **29**:249, 1983, epidemiology [**63**,1355].

X. c. pv. armoraciae (McCulloch) Dye; horseradish leaf spot; McCulloch, *J.agric.Res.* **38**:269, 1929 [**8**:543].

X. c. pv. arrhenatheri Egli & Schmidt 1982; Bradbury, *Descr.B.* 900, 1987; wilt of *Arrhenatherum elatius*, tall oat grass, prob. does not attack other grasses naturally and strongly, Switzerland.

X. c. pv. begoniae (Takimoto) Dye; Taylor et al., *Descr.B.* 699, 1981; begonia leaf spot and wilt; serious in parts of Europe, long periods of latency occur before the sudden appearance of symptoms.

X. c. pv. betlicola (Patel et al.) Dye; spots on leaves, petioles and stems of betel and black pepper, India; Tripathi et al., *Trop.Pest Management* **30**:440, 1984, chemical control [**64**,2662].

X. c. pv. cajani (Kulkarni et al.) Dye; pigeon pea leaf spot and canker; China, India, Sudan.

X. c. pv. campestris (Pammel) Dowson; see *Xanthomonas campestris*; black rot of Cruciferae, a vascular infection, leaf yellowing, blackening of vascular strands, plants dwarfed and misshapen. Williams, *PD* **64**:736, 1980, reviewed black rot with particular reference to USA. It is a severe disease of cultivated brassicas where conditions are wet, warm and humid; it is much less severe in cool, often coastal areas, where seed is mostly produced. The keys to control are the use of seed with minimal infection and hot water treatment of seed. Infection of tropical brassicas is frequent and in these areas seed may have a high level of infection. The bacterium survives on crop debris; Shelton & Hunter, *Can.J.Pl.Path.* **7**:308, 1985, spread & *Phyllotreta cruciferae*; Alvarez et al., *Phytop.* **75**:722, 1985, grouping strs. using monoclonal antibodies; Schultz & Gabrielson, *ibid* **76**:1306, 1986, occurrence & survival in USA (Washington State); Kuan et al., *PD* **70**:409, 1986, aerial dispersal; Sherf & MacNab 1986 [**65**,30, 1583, 5228; **66**,4536].

X. c. pv. carotae (Kendrick) Dye; leaf spot, linear lesions on petioles and stems, and a floral blight, carrot; Kendrick, *J.agric.Res.* **49**:493, 1934; Sherf & MacNab 1986 [**14**:210].

X. c. pv. cassavae (Wiehe & Dowson) Maraite & Weyns 1979; cassava necrosis or leaf spot, not usually systemic unlike *Xanthomonas campestris* pv. *manihotis* on the same crop, E. Africa; Wiehe & Dowson, *Emp.J.exp.Agric.* **21**:141, 1953 [**32**:536].

X. c. pv. cerealis (Hagborg) Dye; water soaked leaf spots becoming necrotic, barley, rye, wheat and other temperate Gramineae; races described by Wallin and tabulated by Bradbury 1986; Argentina, Brazil, Canada, Japan; Hagborg, *Can.J.Res.* C **20**:312, 1942; Wallin, *Phytop.* **36**:446, 1946 [**21**:446; **25**:553].

X. c. pv. citri (Hasse) Dye; Hayward & Waterston, *Descr.B.* 11, 1964, as *Xanthomonas citri* (Hasse) Dowson; citrus canker, a major disease of the crop; young leaves, branches and fruit are attacked, water soaked spots become raised and corky with a chlorotic halo, shoots develop scabbed areas and are killed, fruit lesions develop into extensive cracks, and become heavily scabbed. The bacterium is now the subject of a large scale eradication effort in USA (Florida) where it was found in 1984. Not only has it already been eradicated from this country before, but in this most recent outbreak in Florida a new form of the disease has been detected.

Citrus canker is considered to be of Asian origin. Its seriousness was very early recognised and the bacterium is subject to stringent quarantine. In 1911 an outbreak occurred in the USA Gulf States; the subsequent eradication was successful, but not until c.1933. The disease has been periodically detected in USA but there were no major outbreaks until 1984. Canker has also been eradicated from Australia, New Zealand and South Africa. It is established in Brazil 1957, Paraguay 1967 and Argentina 1972. There are 3 strs. of *X. c.* pv. *citri*, A, B, C. A is highly pathogenic to grapefruit and Mexican lime, less so to lemon and sweet orange. B is highly pathogenic to Mexican lime, less so to lemon. C is

pathogenic, highly so, to Mexican lime only. Thus A is the most virulent and has the widest host range.

The A str. is the Asiatic form; the B str. is present in S. America. But the recent outbreak in USA has led to the detection of a new, called the nursery, form. In this form the leaf lesions are mostly flat, not typically erumpent; although the branch lesions are more typical. The form differs from A, B and C in serology and DNA analyses. Its origin is not known and it has not been detected outside Florida. The Asiatic form is, on the other hand, regularly intercepted at Florida's ports. There is a long history of control in Japan but it is only effective on cvs. which have some resistance. Cu is used in chemical treatment of new, susceptible growth. Control measures, similar to those used in Japan, have been substantiated in Argentina.

Kuhara, *Rev.Pl.Prot.Res.* **11**:132, 1978, status & control of canker in Japan; Stall & Seymour, *PD* **67**:581, 1983, review & threat to citrus in USA Gulf Coast states; Danós et al., *Phytop.* **74**:904, 1984, temporal & spatial spread in Argentina, see for work in S. America; Schoulties et al., *PD* **71**:388, 1987, present outbreak & eradication program in USA (Florida), new nursery form.

X. c. coracanae (Desai et al.) Dye; finger millet leaf spot; Desai et al., *Indian Phytopath.* **18**:384, 1965 [**45**,2100].

X. c. pv. coriandri (Srinivasan et al.) Dye; coriander blight, bacteria enter leaf veins and spread to petioles; Srinivasan et al., *Proc.Indian Acad.Sci.* B **53**:298, 1961 [**41**:164].

X. c. pv. corylina (Miller et al.) Dye; Bradbury, *Descr.B.* 896, 1987; blight of *Corylus* spp., filbert or hazel, leaf spot, bud and twig necrosis, girdling cankers on branches, young trees may be killed; fairly widespread; Gardan & Devau, *EPPO Bull.* **17**:241, 1987.

X. c. pv. cucurbitae (Bryan) Dye; cucurbit leaf spot, stems and petioles can be attacked, lesions like scabs on fruit; Bradbury 1986.

X. c. pv. cyamopsidis (Patel et al.) Dye; leaf spot and blight of *Cyamopsis tetragonolobus*; India, South Africa, USA; Bradbury 1986.

X. c. pv. dieffenbachiae (McCulloch & Pirone) Dye; leaf spot, yellowing and death of leaves in Araceae; Brazil, USA; McCulloch & Pirone, *Phytop.* **29**:956, 1939; Pohronezny et al., *PD* **69**:170, 1985 [**19**:154; **64**,2885].

X. c. pv. eucalyptii (Truman) Dye; eucalyptus dieback, trees can be defoliated; Australia; Truman, *Phytop.* **64**:143, 1974 [**53**,2693].

X. c. pv. glycines (Nakano) Dye; soybean leaf spot

or pustule; seedborne; reports of pustule due to *Xanthomonas campestris* pv. *phaseoli* prob. all refer to *X. c.* pv. *glycines*; Bradbury 1986.

*X. c. pv. graminis** (Egli et al.) Dye; Reay et al., *Descr.B.* 897, 1987; wilt of grasses, wide host range in Gramineae, parts of Europe, New Zealand; Egli et al., *Phytopath.Z.* **82**:111, 1975; Channon & Hissett, *Pl.Path.* **33**:113, 1984, incidence in Scotland (W.) [**54**,4964; **63**,2925].

X. c. pv. guizotiae (Yirgou) Dye; leaf spot of *Guizotia abyssinica*, niger oil seed, Ethiopia, India; Yirgou, *Phytop.* **54**,1490, 1964 [**44**,1637].

X. c. pv. gummisudans (McCulloch) Dye; leaf spot or scorch of *Gladiolus hortulanus*; Australia, Finland, Netherlands, N. America, South Africa; Bradbury 1986.

X. c. pv. holcicola (Elliot) Dye; sorghum leaf spot; Bradbury 1986.

X. c. pv. hyacinthi (Wakker) Dye; hyacinth yellows, a severe disease which causes a complete rot of the bulb with much slime; Moore 1979 stated that only the Dutch hyacinth is susceptible; Australia, Japan, Netherlands, Poland, Yugoslavia; Kamerman 1975, review [**55**,5770e].

X. c. pv. incanae (Kendrick & Baker) Dye; vascular wilt of stock and poss. wallflower, seedborne, Australia, South Africa, USA; Kendrick & Baker 1942; Bradbury 1986 [**22**:25].

X. c. pv. juglandis (Pierce) Dye; Bradbury, *Descr.B.* 130, 1967, as *Xanthomonas juglandis* (Pierce) Dowson; walnut blight, blackened nuts cause severe losses, chemical control.

X. c. pv. malvacearum (Smith) Dye; Hayward & Waterston, *Descr.B.* 12, 1964, as *Xanthomonas malvacearum* (E.F. Smith) Dowson; cotton black arm, also: angular leaf spot, boll rot, bract spot, leaf vein blight. Innes, in a review, *Biol.Rev.* **58**:157, 1983, gave earlier reviews and called this disease potentially one of the most damaging for cotton. It is particularly severe in Africa, and has recently become so in India and Pakistan. The bacterium is seedborne and carried over in plant debris; infection is also spread in wind blown rain and by sprinkler irrigation. Systemic colonisation of the host can occur. At least 18 races have been differentiated and breeding for resistance has a long history; see Brinkerhoff et al., *PD* **68**:168, 1984, on the development of immunity. Control also requires cultural methods and chemical treatments, especially for seed; Watkins 1981; Moffett & Wood, *J.appl.Bact.* **58**:607, 1985, populations on leaves as a source of inoculum [**64**,4959].

X. c. pv. mangiferaeindicae (Patel et al.) Dye; mango black spot; causes fruit fall and postharvest loss in Africa and Asia, also present in Australia and Brazil. Manicom, *AAB* **109**:129, 1986, considered the bacterium to be a wound pathogen of leaves and investigated the factors affecting infection.

X. c. pv. manihotis (Berthet & Bondar) Dye; Bradbury, *Descr.B.* 559, 1977, as *Xanthomonas manihotis* (Berthet & Bondar) Starr; cassava blight has caused severe losses but control measures, that may even lead to eradication of the pathogen, have been successful, see Lozano, *PD* **70**:1089, 1986.

X. c. pv. melhusii (Patel et al.) Dye; teak leaf spot, Patel et al., *Curr.Sci.* **21**:345, 1952, as *Xanthomonas melhusii*, India (Maharashtra) [**32**:368].

X. c. pv. melonis Neto, Sugimori & Oliveira, *Summa Phytopathologica* **10**:217, 1984; soft rot of melon fruit, Brazil (São Paulo) [**64**,4576].

X. c. pv. musacearum (Yirgou & Bradbury) Dye; Yirgou & Bradbury, *Phytop.* **58**:111, 1968, as *Xanthomonas musacearum*; wilt and collapse of *Ensete ventricosum* and banana, Ethiopia [**47**,1389].

X. c. pv. olitorii (Sabet) Dye; Sabet, *AAB* **45**:516, 1957, as *Xanthomonas nakatae* var. *olitorii*; jute leaf spot, Sudan [**37**:169].

X. c. pv. oryzae (Ishiyama) Dye; Bradbury, *Descr.B.* 239, 1970, as *Xanthomonas oryzae* (Uyeda & Ishiyama) Dowson; leaf blight and kresek of rice; the latter name is given to a severe form of the disease in tropical Asia where infection becomes systemic and the plants are killed. Ou 1985 has given a very full account and bibliography of this important disease. Losses in Japan can be 20–30%, and they are higher in India, Indonesia and the Philippines. Considerable variation in virulence is found, and pathotypes may differ in one country when compared with those in another; Mew, *A.R.Phytop.* **25**:359, 1987.

X. c. pv. oryzicola (Fang et al.) Dye; Bradbury, *Descr.B.* 240, 1970, as *Xanthomonas oryziola* Fang et al. 1957, see *Descr.B.* for differences from the pv. *oryzae*; rice leaf streak, distinguishable by the symptoms from rice leaf blight in the early stages of the disease; in tropical Asia; later reviews were given by Devadath in Raychaudhuri & Verma 1984 and Ou 1985 [**64**,1530].

X. c. pv. papavericola (Bryan & McWhorter) Dye; on *Papaver* spp., poppies; leaf spot, blossoms are attacked, and stem lesions may girdle the stem causing death above; Bryan & McWhorter, *J.agric.Res.* **40**:1, 1930; Bradbury 1986 [**9**:456].

*X. c. pv. pelargonii** (Brown) Dye; McPherson et al., *Descr.B.* 560, 1977, as *Xanthomonas pelargonii* (Brown) Starr & Burkholder; *Pelargonium* spp., leaf spot and stem rot, wilt may develop; cuttings may contain the bacterium which then becomes systemic; can be transm. by *Trialeurodes vaporariorum*.

X. c. pv. phaseoli (Smith) Dye; Hayward & Waterston, *Descr.B*. 48, 49, var. *fuscans*, 1965, as *Xanthomonas phaseoli* (E.F. Smith) Dowson; bean common or fuscous blight, different forms or races are found on some other legumes. An important pathogen which is seedborne, remaining viable in seed for years, carried over on crop debris; spread by wind blown rain and soil, and sprinkler irrigation. The disease is most severe at higher temps. *c*.28°. Losses approaching 50 % of the crop have been reported; Allen 1983; Zapata et al., *Phytop*. 75:1032, 1985, resistance; Bradbury 1986; Sherf & MacNab 1986 [65,2542].

X. c. pv. phlei Egli & Schmidt 1982; Bradbury, *Descr.B*. 898, 1987; wilt of timothy grass, *Phleum pratense*; *Phleum* spp. are more susceptible compared with some other grasses, parts of Europe.

X. c. pv. phleipratensis (Wallin & Reddy) Dye; leaf streak of timothy grass, *Phleum pratense*, USA (Iowa, Minnesota, Wisconsin); Wallin & Reddy, *Phytop*. 35:937, 1945 [25:168].

X. c. pv. pisi (Goto & Okabe) Dye; a blight of pea apparently restricted to Japan, fide Bradbury 1986 [39:213].

X. c. pv. poae Egli & Schmidt 1982; Bradbury, *Descr.B*. 899, 1987; wilt of rough stalked meadow grass, *Poa trivialis*; Switzerland and prob. elsewhere in Europe.

X. c. pv. poinsettiicola (Patel et al.) Dye; leaf spot of poinsettia and other *Euphorbia* spp.; Chase, *Pl.Path*. 34:446, 1985, on *Codiaeum variegatum* [65,241].

X. c. pv. pruni (Smith) Dye; Hayward & Waterston, *Descr.B*. 50, 1965, as *Xanthomonas pruni* (E.F. Smith) Dowson; leaf spot and shothole of apricot, cherry, peach, plum; lesions on leaves, branches and fruits; Anderson 1971; Bradbury 1986.

X. c. pv. raphani (White) Dye; leaf spot of radish and turnip, dark lesions on petioles and stems, young plants may be killed; Brazil, USA; may attack tomato and red pepper; White, *Phytop*. 20:653, 1930 [10:72].

X. c. pv. ricini (Yoshi & Takimoto) Dye; Bradbury, *Descr.B*. 379, 1973, as *Xanthomonas ricini* (Yoshi & Takimoto) Dowson; castor leaf spot, causes defoliation.

X. c. pv. sesami (Sabet & Dowson) Dye; Bradbury, *Descr.B*. 700, 1981; sesame leaf spot; India, Sudan, Tanzania, Venezuela.

X. c. pv. tardicrescens (McCulloch) Dye; gladiolus and iris leaf spot; Japan, New Zealand,

N. America, South Africa; Burkholder, *Phytop*. 27:613, 1937; McCulloch, *ibid* 28:642, 1938; Moore 1979 [16:677; 18:31].

X. c. pv. theicola Uehara & Arai 1980; *Xanthomonas theae* Uehara & Arai, cited by Takaya, *JARQ* 12:138, 1978, is prob. the same organism, fide Bradbury 1986; tea canker, Japan [58,5491; 60,1066].

X. c. pv. translucens (Jones et al.) Dye; leaf and sheath streak of barley only, does not infect oat, rye or wheat; Bradbury 1986 described the taxonomic confusion; originally described in *J.agric.Res*. 11:625, 1917.

X. c. pv. undulosa (Smith et al.) Dye; wheat black chaff or leaf streak; a minor, if widespread, disease on which there has been little or no recent work; Bradbury 1986; Wiese 1987.

X. c. pv. vasculorum (Cobb) Dye; Bradbury, *Descr.B*. 380, 1973, as *Xanthomonas vasculorum* (Cobb) Dowson; sugarcane gumming or gummosis, the bacterium infects some palms and maize. The use of resistant sugarcane cvs. has virtually eliminated the pathogen; Sivanesan & Waller 1986.

X. c. pv. vesicatoria (Doidge) Dye; Hayward & Waterston, *Descr.B*. 20, 1964, as *Xanthomonas vesicatoria* (Doidge) Dowson; spot, scab or corky scab of red pepper and tomato; attacks leaves, stems and fruit; occurs on other Solanaceae. A serious disease of both crops. The basis of control is prevention by the use of healthy seed and seedlings, described by Goode & Sasser, *PD* 64:831, 1980. Races differ in pathogenicity to red pepper and tomato; Bashan et al., *Crop Protect*. 4:77, 1985, loss in red pepper; Reifschneider et al., *Fitopat.Brasileira* 10:201, 1985, strs.; Jones et al., *Phytop*. 76:430, 1986, survival; Sherf & MacNab 1986 [64,3650; 65,29, 4804].

X. c. pv. vignicola (Burkholder) Dye; on cowpea, widespread with crop; Allen 1983, who described the disease, referred to 3 syndromes; foliar blight, seedling death and stem canker; the bacterium is seedborne and systemic; Gitaitis et al., *PD* 70:187, 1986, epidemiology & control in USA (Georgia) [65,3605].

X. c. pv. vitians (Brown) Dye; infects lettuce and plants in the Araceae, leaf spot, rot and wilt; Bradbury 1986.

X. c. pv. zinniae (Hopkins & Dowson) Dye; zinnia leaf spot; Bradbury 1986 stated that the bacterium seems to be spreading rapidly because of efficient seed transm.; Hopkins & Dowson, *TBMS* 32:252, 1949 [29:308].

X. fragariae Kennedy & King 1962; Bradbury, *Descr.B*. 558, 1977; angular leaf spot, collapse or vascular decline of strawberry, see Bradbury 1986 for differences from *Xanthomonas campestris*; in Brazil, prob. France, Greece, Italy (Sicily), USA, Venezuela; reported from Australia and New Zealand but eradicated. The bacterium overwinters in leaf litter and may become systemic in planting material; Maas 1984.

Xanthosoma, cocoyam, tanier, tannia, yautia; see taro.

X disease, see peach X.

xenomeria, *Sclerophoma*.

Xenomeris H. Sydow 1924; Venturiaceae; Sivanesan 1984; ascomata small, clustered on a columnar stroma; hypostroma in host tissues.

X. abietis Barr, *CJB* **46**:842, 1968; anam. *Sclerophoma xenomeria* Funk, *Eur.J.For.Path.* **10**:54, 1980; *Hormonema* state in vitro; asco. greenish, rarely pale brown, septate just below the middle, 11–17 × 4–6 μm; con. 8–10 × 5–4 μm; assoc. with a dieback of Douglas fir and western hemlock, *Tsuga heterophylla* in Canada (British Columbia), USA (N.W.); McMinn & Funk, *CJB* **48**:2123, 1970; Funk & Shoemaker, *Mycologia* **63**:567, 1971 [**51**,766; **59**,5415].

Xiphinema, dagger nematodes; Lamberti, *PD* **65**:113, 1981, for transm. of plant viruses by members of the genus, see below.

X. americanum Cobb 1913; Siddiqi, *Descr.N.* 29, 1973; transm. cherry rasp leaf, peach rosette mosaic, tobacco ringspot, tomato ringspot.

X. californicum, transm. tomato ringspot; *X. coxi*, transm. arabis mosaic.

X. diversicaudatum (Micoletzky) Thorne 1939; Pitcher et al., *Descr.N.* 60, 1974; root ectoparasite of woody plants; transm. arabis mosaic, brome mosaic, carnation ringspot, cherry leaf roll, raspberry ringspot cherry str., strawberry latent ringspot.

X. index Thorne & Allen 1950; Siddiqi, *Descr.N.* 45, 1974; most important host is grapevine, trans. grapevine fan leaf, a migratory ectoparasite.

X. italiae Meyl 1953; Cohn, *Descr.N.* 95, 1977; grapevine and other plants, transm. grapevine fan leaf, Mediterranean region, ectoparasite.

X. pachtaicum (Tulaganov) Kirjanova 1951; Lamberti & Siddiqi, *Descr.N.* 94, 1977; in rhizosphere, Mediterranean region and central Asia, reported on grapevine affected by grapevine fan leaf in Spain.

X. rivesi, transm. tomato ringspot; *X. vuittenezi*, transm. cherry leaf roll?

X-ray microanalysis, see methods, Zeyen.

Xylaria Hill ex Schrank 1789; Xylariaceae; ascomata immersed in ±stalked, upright, simple or branched stroma, like a club, thread or strap; becoming black; asci and asco. like those of *Hypoxylon*; virtually all saprophytic. *X.*

polymorpha (Pers.) Grev. 1824, Sivanesan & Holliday, *Descr.F.* 355, 1972, causes a white rot decay of tree stumps and fallen timber; poss. weakly pathogenic through wounds; one of the commonest causes of zone lines in wood. *X. thwaitesii* Cooke has been described as causing black root of rubber, but the taxonomic position of this sp. seems uncertain, Munasinghe, *Q.J.Rubb.Res.Inst.Ceylon* **48**:92, 1971.

Xylariaceae, Sphaeriales; ascomata in a stroma or subiculum, rarely single; asci usually with an apical ring, amyloid; asco. with a germ slit; *Hypoxylon, Kretzschmaria, Nummulariella, Rosellinia, Ustulina, Xylaria*.

xylarioides, *Fusarium, Gibberella*.

Xylella Wells et al., *Int.J.syst.Bact.* **37**:136, 1987; *X. fastidiosa* loc. cit.; fastidious plant bacteria, limited to the xylem, Gram negative, see below. Mostly single, straight rods, 0.9–3.5 × 0.25–0.35 μm, aflagellate, non-motile, strictly aerobic, require special media. The type sp. *X. fastidiosa* was isolated from grapevine with Pierce's grapevine leaf scald. Strs., genotypically and phenotypically similar, were obtained from plants with peach phony, periwinkle wilt and leaf scorches of: almond, elm, mulberry, oak, plum, sycamore.

xylem limited bacteria, and see phloem limited bacteria. These organisms were first found in 1973 in plants showing disease symptoms, see alfalfa dwarf and Pierce's grapevine leaf scald. Both diseases had been studied for a long time before this. The genus *Clavibacter* q.v. was erected in 1984 and some plant bacterial spp. in *Corynebacterium*, and Gram positive, were transferred to it. *Clavibacter xyli*, same year, contains the sspp. *cynodontis* and *xyli* which cause Bermuda grass stunt and sugarcane ratoon stunting, respectively. These bacteria are both limited to the xylem and were not named until the erection of *Clavibacter*. The sugarcane disease was for long thought to be due to a virus. *Xylella*, erected 3 years later, includes other bacteria in this group of organisms, but it is Gram negative. Other diseases, caused by, or assoc. with, *Xylella*, are the leaf scorches, peach phony and plum leaf scald. Another disease caused by a xylem limited bacterium, Gram negative group, is clove Sumatra wilt q.v. and see spruce forest decline.

These bacteria were thought to be close to the animal rickettsias, but the guanine + cystine content is *c*.50.5% mol., quite different from that of rickettsia DNA. The Gram negative forms are spread by leafhoppers which include: *Carneocephala fulgida, Draeculacephala minerva, Helochara communis, Homalodisca, Oncometopia nigricans.* In USA (S.) wild plants are an important source of inoculum; infection usually kills the plant; Hopkins, *A.R.Phytop.* **15**:277, 1977; *Phytop.* **73**:347, 1983; Kamper et al., *Int.J.syst.Bact.* **35**:185, 1985, genetic relationships; Raju & Wells, *PD* **70**:182, 1986 [**64**,4105].

—— **ring discolouration**, potato, caused by a too rapid killing of green plants; Rich, *Am.Pot.J.* **27**:87, 1950 [**29**:380].

xyli, *Clavibacter.*

Y

yam, *Dioscorea*; Noon, *Trop.Sci.* **20**:177, 1978; Noon & Colhoun, *Phytopath.Z.* **94**:289, 1979, postharvest; Nwankiti & Arene, *PANS* **24**:486, 1978; *PANS Manual* 4, 1978; Ono, *TBMS* **79**:423, 1982, rusts [**58**,3606, 6166; **59**,4404; **62**,2250].

yamadae, *Gymnosporangium*.

yam internal brown spot Harrison & Roberts, *Trop.Agric.Trin.* **50**:335, 1973; tubers with brown spot developed foliage with a mosaic. This contained particles that were bacilliform 130 × 29 nm and filamentous *c.*750 × 13 nm. Normal tubers gave uniformly green foliage which contained the filaments only; Barbados, *Dioscorea alata*; Mohamed & Mantell and Mohamed, *ibid* **53**:255, 341, 1976; both particle types, Caribbean, *Dioscorea* spp.; Mantell & Haque, *Expl.Agric.* **14**:167, 1978, incidence during storage [**53**,1168; **55**,6054; **56**,1378; **57**,4745].

— mosaic v. Thouvenel & Fauquet 1977, *Descr.V.* 314, 1986; Potyvirus, filament *c.*785 nm long, transm. sap difficult, to a few plants; *Aphis gossypii, A. craccivora, Rhopalosiphum maidis, Toxoptera citricidus*, non-persistent; differs from dioscorea green banding v. in not infecting *Dioscorea composita* and *D. floribunda*; causes significant losses in parts of W. Africa and the Caribbean region.

— necrotic mosaic v. Fukumoto & Tochihara, *Ann.phytopath.Soc. Japan* **44**:1, 1978; poss. Carlavirus, filament *c.*660 × 12 nm, transm. aphid, non-persistent, *Dioscorea* spp. only; Shirako & Ehara, *ibid* **52**:453, 1986, diagnosis [**57**,5812; **66**,1727].

— Puerto Rican mosaic Adsuar, *J.Agric.Univ.P.Rico* **39**:111, 1955; transm. sap, needle prick, to cucumber, red pepper, tobacco, yam. This paper is not given by Thouvenel & Fauquet 1986, see yam mosaic v., who did not give these plants except yam [**35**:413].

— ring mottle v. Porth & Nienhaus, *Z.PflKrankh.PflSchutz.* **90**:352, 1983; poss. Potyvirus, filament 750 × 15 nm, transm. sap, aphid spp., Togo, *Dioscorea alata*; differs from dioscorea green banding v. Presumably the same as, or similar to, yam mosaic v. but this paper is not given by Thouvenel & Fauquet 1986, see YMV [**63**,1532].

— nematode, *Scutelloma bradys*.

Yarwood, Cecil Edmund, 1907–81; born in USA, Univ. British Columbia, Purdue, Wisconsin, California Berkeley; many papers on the effects of the external environment on host and obligate pathogen, particularly in the Erysiphaceae. *Phytop.* **73**:509, 1983.

yeast spot (*Nematospora coryli*) seeds of bean and soybean.

yellow dwarf (*Heterodera glycines*) soybean; — leaf (*Blumeriella jaapii*) cherry; — — blight (*Mycosphaerella zeae-maydis*) maize; — — blister (*Taphrina populina*) poplar; — — blotch (*Leptotrochila medicaginis*) lucerne, (*Sclerotinia polyblastis*) narcissus; — — mould (*Mycovellosiella puerariae*) Pueraria; — — spot (*Gymnosporangium nelsonii*) Saskatoon berry, (*Pyrenophora tritici-repentis*) temperate cereals, grasses.

— patch (*Ceratobasidium cereale*) turf grasses; — —, tobacco, N toxicity, Lucas 1975; Anderson & Welacky, *Can.Pl.Dis.Surv.* **62**:9, 1982, found that soil fumigation and low greenhouse temp. were assoc. with this disorder [**61**,6573].

— pustule (*Curtobacterium flaccumfaciens* pv. *oorti*) tulip; — ring rot (*Inonotus weirei*) conifers; — rust (*Puccinia striiformis*) barley, wheat.

yellows, a general term for a disease or a disorder where the most conspicuous symptom is a marked and general chlorosis of normally green tissue, particularly leaves. It is commonly used for diseases assoc. with infection by MLO; and, to a lesser extent, for those caused by viruses. Also used for diseases caused by *Fusarium oxysporum*, particularly by *F.o.*f.sp. *conglutinans*. Hyacinth and sunflower yellows are caused by *Xanthomonas campestris* pv. *hyacinthi* and *Phialophora asteris* f.sp. *helianthi*, respectively. Black pepper yellows is assoc. with nutrient factors and attack by *Radopholus similis* but its etiology is not precisely known, de Waard, *J.Plantation Crops* **7**:42, 1979. Tea yellows is caused by S deficiency, Storey & Leach, *AAB* **20**:23, 1933.

yellow slime (*Clavibacter rathayi*) cocksfoot.

— spot (*Mycovellosiella koepkei*) sugarcane; citrus, Mo deficiency, Stewart & Leonard, *Nature Lond.* **170**:714, 1952 [**32**:17].

— strap leaf, chrysanthemum; the pale yellow, narrow leaves commonly appear 3–4 weeks after planting and mostly after pinching. The cause may be a toxin of biological origin; similarities with frenching q.v. of tobacco, see Woltz. Yellow strap leaf is not referred to by Fletcher 1984.

— trunk rot (*Phellinus robustus*) oak, temperate

trees; —— **tuft**, in temperate grasses, assoc. with infection by *Sclerophthora macrospora*, Mueller et al., *PDR* **58**:848, 1974; Dernoeden & Jackson, *J.Sports Turf Res.Inst.* **56**:9, 1980 [**54**,899; **60**, 951].

—— **wilt** (*Sclerophthora macrospora*) rice; —— **witches' broom** (*Chrysomyxa arctostaphyli*) spruce, (*Melampsorella caryophyllacearum*) fir.

yew, *Taxus*.

—— **yellowing** Blattny 1960; Czechoslovakia, fide Cooper 1979.

Yorkshire fog, *Holcus lanatus*.

young tree decline = citrus blight.

yucca yellow spot Milne et al., *Inftore.fitopatol.* **35**(6):43, 1985, bacilliform particles assoc., *Y. elephantipes* from Guatemala [**65**, 3365].

Z

zeae, *Botryosphaeria, Gibberella, Kabatiella, Macrophoma, Physopella, Rhizoctonia*;
zeae-maydis, *Cercospora, Mycosphaerella.*
Zea mays, maize.
zebra leaf spot (*Phytophthora nicotianae* var. *parasitica*) sisal.
zebrina, *Cercospora.*
zeicola, *Phialophora, Physalospora.*
zineb, protectant fungicide sometimes used for seed and soil.
zingiberi, *Pyricularia*; zingiberum, *Pythium.*
Zingiber officinale, ginger
zinniae, *Alternaria.*
zinnia mild mottle v. Padma, Singh & Verma, *Gartenbauwissenschaft* **37**:377, 1972; Padma et al., *Hort.Res.* **14**:55, 1974; Potyvirus, filament 710–750 nm long, transm. sap, *Myz.pers.*, India [52,2967; 54,2298].
— **vein banding** Shreni 1980; Shreni et al., *New Botanist* **10**:7, 1983; transm. aphid, non-persistent, India [64,2588].
— **yellow net** Srivastava et al., *PDR* **61**:550, 1977; transm. *Bemisia tabaci*, persistent, India (Uttar Pradesh) [57,1239].
zinniol, toxin from *Alternaria* spp.; White & Starratt, *CJB* **45**:2087, 1967, from *A. zinniae*; Barash et al., *Physiol.Pl.Path.* **19**:7, 1981, from *A. dauci*; Yu et al., *Ann.phytopath.Soc.Japan* **49**:746, 1983, from *A. porri*; Cotty et al., *Phytop.* **73**:1326, 1983, from *A. tagetica*; Cotty & Misaghi, *ibid* **74**:785, 1984, from 6 *Alternaria* spp.; Sugawara & Strobel 1986, from *Phoma macdonaldii* [61,911; 63,1306, 4667; 64,65; 65,2936].
ziram, a protectant, now largely replaced by other compounds.
Zizania, wild rice; Kohls et al., *PD* **71**:419, 1987, loss in *Z. palustris* caused by *Cochliobolus miyabeanus* in USA (Minnesota).
Zizyphus, *Z. jujuba, Z. mauritiana*; jujube, Chinese date plum.
— **mosaic** Zu et al. 1982; rod assoc. 370–530 × 20 nm, China [62,288].
— **witches' broom** Ciccarone 1957; Kim, *Korean J.Microbiol.* **3**:1, 1965; Pandey et al., *PDR* **60**:301, 1976; La & Chang 1979; Wand et al., *Acta phytopath.sin.* **11**(3):25, 1981, MLO assoc., transm. *Hishimonides chinensis*; Korea, India, China [37:9; 46,2846; 56,1675; 60,1897; 61,3372].
zonalis, *Rigidoporus*; zonata, *Cercospora*; zonatum, *Acremonium, Ganoderma.*

zonate eyespot (*Drechslera gigantea*) Bermuda grass.
— **leaf spot** (*Acremonium zonatum*) plurivorous, (*Aristastoma camarographioides*) soybean, (*Cercospora zonata*) broad bean, (*Gloeocercospora sorghi*) sorghum, (*Grovesinia pyramidalis*) plurivorous, (*Thanatephorus cucumeris*) kenaf, roselle.
zone lines, narrow, dark brown or black lines in decayed wood, especially hardwoods, generally caused by fungi, FBPP.
zoospore, a unicellular propagating or disseminating body with 1 or 2 flagella q.v., and having a limited mobility; characteristic of the Mastigomycotina; Waterhouse, *TBMS* **45**:1, 1962; Hickman & Ho, *A.R.Phytop.* **4**:195, 1966; Fuller, *Mycologia* **69**:1, 1977; Lange & Olson, *Dansk bot.Arkiv.* 33, 1979; Buczaki 1983.
zorniae, *Sphaceloma.*
zucchini squash yellow fleck v. Vovlas, Hiebert & Russo, *Phytopath.Mediterranea* **20**:123, 1981; Potyvirus, filament 700–800 nm long, transm. *Myz.pers.*, infects only Cucurbitaceae; not serologically related to zucchini squash yellow mosaic v.; Mediterranean region [63,961].
— — — **mosaic v.** Lisa et al. 1981; Lisa & Lecoq, *Descr.V.* 282, 1984; Potyvirus, filament *c.*750 nm long, transm. sap readily, *Aphis citricola, A. gossypii, Macrosiphum euphorbiae, Myz.pers.*, causes severe symptoms in courgette, cucumber, melon, watermelon, zucchini squash; not serologically related to zucchini squash yellow fleck v.; Australia, China, England (and Channel Islands), Europe (central), Japan, Mauritius, Mediterranean region, USA; Lecoq & Pitrat, *Phytopath.Z.* **111**:165, 1984, strs. [64, 2211].
zygocactus v. Casper & Brandes, *J.gen.Virol.* **5**:155, 1969; filament *c.*580 nm long, Europe, from a hybrid with *Schlumbergera* showing no symptoms [48,3513].
— **X v.** Giri & Chessin 1972; *Phytopath.Z.* **83**:40, 1975; filament *c.*520 nm long, transm. sap, USA [52,2810; 55, 1281].
Zygomycotina, Zygomycetes, second division of the Eumycota; thallus mycelial, typically aseptate, sexual reproduction results in a resting spore, the zygospore, no motile cells.
Zygophiala Mason, in Martyn, *Mycol.Pap.* **13**:3, 1945; Hyphom.; Ellis 1971; conidiomata hyphal, conidiogenesis holoblastic; conidiophore growth sympodial, with a dark basal part, a less dark

apical cell bearing solitary, 1 septate, hyaline con., *Z. jamaicensis*, teleom. *Schizothyrium pomi*.

zygospore, see Zygomycotina.

Zygotylenchus guevarai (Tobar Jiménez) Braun & Loof 1966; Siddiqi, *Descr.N.* 65, 1975; migratory endoparasite, a serious root parasite of *Cupressus sempervirens* in Spain and may damage other plants.

Zythia Fr. 1825; Coelom.; *Z. fragariae*, teleom. *Gnomonia comari*.

ADDENDA

abutilon mosaic, Schuchalter-Eicke & Jeske, *Phytopath.Z.* **108**:172, 1983; Jeske & Schuchalter-Eicke, *ibid* **109**:353, 1984 [**63**,2219, 4328].

Acremonium strictum, Bandyopadhyay et al., *PD* **71**:647, 1987, systemic infection of sorghum & seed transm., Lebeda et al., *Z.PflKrankh. PflSchutz.* **94**:314, 1987, necrosis and break of carrot flower stems.

air pollution, Roberts et al., *Adv.appl.Biol.* **9**:2, 1983, effects of gaseous air pollutants on agriculture and forestry in UK.

Alternaria carthami, Jackson et al., *Aust.J.exp.Agric.* **27**:149, 1987, inoculum sources [**66**,4409].

A. zinniae, see *Septoria helianthi*, Carson.

Angiosorus solani, Torres & Henfling, *Fitopatología* **19**:1, 1984, chemical control [**66**,2044].

Aphanomyces euteiches, Delwiche et al., *PD* **71**:155, 1987, from lucerne.

Aphis citricola, transm. telfairia mosaic; *A. malvoides*, transm. summer squash mosaic.

Apiognomonia veneta, Milne & Hudson, *TBMS* **88**:399, 1987, infection of London plane with asco. & con.

apple green crinkle, Fridlund & Drake, *PD* **71**:585, 1987, effect on fruit quality of apple cv. Granny Smith.

—— witches' broom Rumbos, *Phytopath.Mediterranea* **25**:54, 1986; MLO assoc., a proliferation disease; Kegler & Meyer, *Archiv.Phytopath.PflSchutz* **23**:199, 1987, testing for resistance.

Armillaria, Jahnke et al., *TBMS* **88**:572, 1987, spp. delimitation by analysis of nuclear & mitochondrial DNA.

artichoke yellow mottle (broad bean wilt v.), Rana et al., *Inftore.Fitopatol.* **37**(4):41, 1987.

Ascochyta lentis, Beauchamp et al., *Can.J.Pl.Path.* **8**:260, 1986, chemical control; Gossen et al., *ibid*, page 154, renamed *A. fabae* f.sp. *lentis* [**66**,2149, 2626].

aster yellows, Jiang & Chen, *Phytop.* **77**:949, 1987, partial purification of assoc. MLO.

Bacillus megaterium pv. cerealis Hosford, *Phytop.* **72**:1453, 1982; wheat white blotch, USA (N. Dakota) [**62**,1451].

banana mosaic (cucumber mosaic v.); epidemiology of the disease, Tsai et al., *Pl.Prot.Bull. Taiwan* **28**:383, 1986 [**66**,2012].

barley yellow dwarf v., Jess & Mowat, *Rec.agric.Res.Dep.Agric.N. Ireland* **34**:57, 1986, transm. *Oscinella frit* [**66**,2283].

beet cryptic v., Kühne et al., *Archiv.Phytopath.PflSchutz.* **23**:95, 1987, purification & serology.

—— necrotic yellow vein v., Abe & Tamada, *Ann.phytopath.Soc. Japan* **52**:235, 1986; assoc. with *Polymyxa* isolates [**66**,1179].

Berkeley, Ainsworth, *The Mycologist* **21**:126, 1987.

biological control, Baker, *A.R.Phytop.* **25**:67, 1987.

Bipolaris australiensis, teleom. *Cochliobolus australiensis*.

bitter gourd witches' broom Singh 1985; MLO assoc., India, *Momordica charantia*, cf. momordica witches' broom [**66**,1667].

black heart, pineapple, Abdulla et al. *MARDI Res. Bull.* **14**:132, 1986, Smith & Glennie, *Trop.Agric.Trin.* **64**:7, 1987.

blast (*Pyricularia angulata*) banana, (*P. zingiberi*) ginger.

blueberry red ringspot, Hepp & Converse, *PD* **71**:536, 1987, detection in crude sap.

Botrytis, Ramsey & Lorbeer, *Phytop.* **76**:599, 604, 612, 1986; epidemic of onion flower blight, 4 spp.

bottom rot, Mahr et al., *PD* **70**:506, 1986, control with iprodione in lettuce, *Thanatephorus cucumeris* [**65**,5820].

brown blot root rot (*Pythium sulcatum*) carrot.

cactus X v., Attathom et al., *Phytop.* **68**:1401, 1978, str. from *Ferrocactus acanthodes*, California barrel cactus [**58**,4404].

carrot chlorotic mosaic (broad bean wilt v.), Fujisawa et al., *Proc.Kansai Pl.Prot.Soc.* **29**:1, 1987; Japan (Hokkaido).

cassava green mottle v. Lennon, Aiton & Harrison, *AAB* **110**:545, 1987; Nepovirus, both RNA species needed for infection, transm. sap to 30 spp. in 12 plant families, seedborne in *Nicotiana clevelandii*; prob. causes an important disease in the Solomon Islands.

cauliflower mosaic v., Hull in Day & Jellis 1987.
Ceratocystis ulmi, Brasier, *Adv.Pl.Path.* **5**:53, 1986, population biology.
C. wageneri, Harrington & Cobb, *Mycologia* **78**:562, 1986, 3 morphological variants that are host specialised; Hessburg & Hansen, *CJB* **65**:962, 1987, histopathology of Douglas fir root black stain [**66**,1627, 4508].
Cercoseptoria, Deighton, *TBMS* **88**:365, 1987, concluded inter alia that this genus cannot be distinguished from *Pseudocercospora* and new combinations into this genus from *Cercoseptoria* were made; these include *C. ocellata*, *C. pini-densiflorae*, *C. theae*, *C. sesami*.
Cercospora beticola, Pundhir & Mukhopadhyay, *Pl.Path.* **36**:185, 1987, epidemiology in India [**66**,4523].
C. carbonacea Miles 1917; Little, *Descr.F.* 913, 1987; con. 3–9 septate, 40–100 × 4.5–5 μm; yam leaf spot, virtually of no economic importance.
C. circumscissa, Little, *Descr.F.* 911, 1987, gave *Cercospora cerasella* Sacc. as a synonym and *Mycosphaerella cerasella* Aderh. as the teleom.
C. citrullina Cooke 1883; Little, *Descr.F.* 917, 1987; con. 2–8 septate, 70–220 × 3–4.5 μm; cucurbit leaf spot.
C. duddiae Welles, *Phytop.* **13**:364, 1923; Little, *Descr.F.* 920, 1987; con. 50–170 × 4.5–6.5 μm; garlic and onion leaf spot [**3**:184].
C. malayensis, Little, *Descr.F.* 916, 1987.
C. zebrina, Barbetti, *TBMS* **88**:280, 1987; *Aust.J.exp.Agric.* **27**:107, 1987, on subterranean clover, disease incidence & chemical control [**66**,3865, 4350].
cereal northern mosaic v., Toryama, *Descr.V.* 322, 1986, as northern cereal mosaic v.; transm. persistent, in China.
Cerotelium fici, McKenzie, *N.Z.J.agric.Res.* **29**:707, 1986; new record for New Zealand, illustrations.
Chaetosiphon, transm. strawberry crinkle.
Chalara australis J. Walker & Kile in *Aust.J.Bot.* **35**:7, 1987; causing wilt of *Nothofagus cunninghamii* in Australia (Tasmania); the taxonomic affinities with *Chalara* spp. and the significance of them in relation to the Fagaceae is described; no teleom. found [**66**,4491].
chemotherapy, Lyr ed., *Modern selective fungicides. Properties, applications, mechanisms of action*, 1987.
chickpea, Bretag & Mebalds, *Aust.J.exp.Agric.* **27**:141, 1987; pathogenicity of fungi in Australia (N.W. Victoria).
Chondrostereum purpureum, Clifford et al., *AAB* **110**:471, 489, 1987, chemical control.
chrysanthemum yellows Conti & Mela, *Difesa Delle*

Piante **10**:171, 1987; MLO assoc., disease long known in Italy.
citrus exocortis viroid, Bitters et al., *PD* **71**:397, 1987; effects on flower & fruit structure & development in *Citrus medica*.
Cladosporium phlei, Shimanuki, *Res.Bull.Hokkaido Natn.agric.Exp.Stn.* 148:1, 1987.
clove sudden wilt, also called clove sudden death; Dabeck & Martin, *J.Phytopath.* **119**:75, 1987, tetracycline therapy.
—— **Sumatra wilt**, Hunt et al., *Pl.Path.* **36**:154, 1987, disease induced after mechanical inoculation with the bacterium [**66**,4408].
cobweb (*Gibberella zeae*) carnation; Linfield, *Pl.Path.* **36**:222, 1987 [**66**,4319].
Cochliobolus australiensis (Tsuda & Ueyama) Alcorn 1983; anam. *Bipolaris australiensis* (M.B. Ellis) Tsuda & Ueyama 1981; Sivanesan, *Descr.F.* 881, 1986; con. usually 3 septate, 14–40 × 6–11 μm, Gramineae and other plants, severe leaf spot of *Cymbopogon winterianus*, *Pennisetum americanum*.
C. bicolor, Sivanesan, *Descr.F.* 882, 1986.
Colletotrichum caudatum (Sacc.) Peck 1909; con. have a characteristic filiform appendage, unbranched, 18.5–25 × 3.5–4 μm excluding the appendage, appressoria with entire edge; Zeiders, *PD* **71**:348, 1987, damaging *Sorghastrum nutans*, Indian grass.
C. graminicola, Ali & Warren, *PD* **71**:402, 1987, 3 races on sorghum.
conifers, Hama, *Bull.Forestry Forest Res.Inst.* 343, 1987, rusts.
Corticium rolfsii, Punja, *Can.J.Pl.Path.* **8**:297, 1986, carrot root rot, disease incidence & inoculum density [**66**:2562].
cotton leaf crumple v., Brown & Nelson, *PD* **71**:522, 1987, infects many plants in Malvaceae and Leguminosae.
Crinipellis perniciosa, Pickering & Hedger, Bastos & Andebrhan and Hedger et al., *TBMS* **88**:404, 406, 533, 1987, basidiocarp formation in vitro, basidiospore formation in vitro, 2 populations & only 1 pathogenic to cacao; Rudgard, *Cocoa Grower's Bull.* 38:28, 1987, epidemiology & control in Brazil.
Cronartium quercuum f.sp. fusiforme, Chappelka & Schmidt and Chappelka et al. 1983–4, inoculum; Kuhlman, *Mycologia* **79**:405, 1987, effects of temp., inoculum & leaf maturation on uredospore formation [**63**,3605; **65**,1551].
Cryphonectria, Micales & Stipes, *Phytop.* **77**:650, 1987, comparison with *Endothia*; Barnard et al., *PD* **71**:358, 1987, eucalyptus basal canker, *C. cubensis* & *C. gyrosa* morphology.
cryptic viruses, these viruses can be divided into 2 particle types; those that are featureless, 29–32 nm

diam.; and those that have prominent subunits, c.38 nm diam.; fide Lisa et al., *Descr.V.* 315, 1986.

Curtobacterium flaccumfaciens pv. flaccumfaciens, Calzolari et al., *EPPO Bull.* **17**:157, 1987, detection in bean seed.

cycas necrotic stunt v., Hanada et al., *Ann.phytopath.Soc.Japan* **52**:422, 1986, properties [**66**,1499].

Cymadothea trifolii, another anam. is cited by Cannon et al. 1985: *Placosphaeria trifolii* (Pers.) Traverso.

dasheen mosaic v., Greber & Shaw, *Australasian Pl.Path.* **15**:29, 1986, in Australia (Queensland); Zettler et al., *PD* **71**:837, 958, 1987, in China & control [**66**,2163].

datura distortion mosaic v., Mali et al., *Indian Phytopath.* **38**: 413, 1985, rod. c.770 nm long, transm. *Dactynotus sonchi*, non-persistent, Solanaceae, India, *D. fastuosa* [**66**, 3607].

Deuterophoma tracheiphila, Somma & Scarito, *Phytopath.Mediterranea* **25**:103, 1983, infections in Italy (Sicily) are more numerous from September to April.

Didymella applanata, Williamson & Pepin, *AAB* **110**:295, 1987, effect of temp. on response of canes of red raspberry cv. Malling Jewel to infection [**66**,3437].

D. lycopersici, Cheah & Soteros, *Tests Agrochem.Cvs.* 8 *AAB* suppl. **110**:78, 1987, chemical control: Fagg & Fletcher, *Pl.Path.* **36**:361, 1987, epidemiology & control.

disease names, *Ann.phytopath.Soc.Japan* **52**:356, 1986, supplements, disease names in Japan [**66**,904].

dodder, Dale & Kim, *Phytop.* **59**:1765, 1969; Siller et al., *J.Phytopath.* **119**:147, 1987, MLO [**49**,1314].

Drechslera nobleae, Sivanesan, *Descr.F.* 890, 1986.

early dying, Hide et al., *AAB* **104**:277, 1984, *Globodera rostochiensis* & *Verticillium dahliae*, effects of soil treatments & cvs.; Rowe, *PD* **71**:482, 1987, several pathogens, management strategies; the disease has been called early maturity wilt [**63**,4049].

echinochloa ragged stunt C.C. & M.J. Chen & Chiu, *Pl.Prot.Bull.Taiwan* **28**:371, 1986, transm. sap, *Sogatella longifurcifera*, Taiwan, *E. crus-galli*, Gramineae [**66**,1959].

ecology, Burdon, *Diseases and population biology*, 1987.

Elisa, Mowat & Dawson, *J. Virological Methods* **15**:233, 1987, virus detection & identification by Elisa using crude sap extracts & unfractionated antisera.

Elsinoë batatas, Nayga & Gapasin 1986, effect on growth & yield in the Philippines [**66**,2653].

Encoelia pruinosa, Juzwik et al., *CJB* **64**:2728, 1986, in USA, comparison of forms from Colorado & Minnesota [**66**,1621].

endophyte, Siegel et al., *A.R.Phytop.* **25**:293, 1987, fungus endophytes of grasses.

Endothia, see *Cryphonectria* in addenda.

Erwinia, Kotoujansky, *A.R.Phytop.* **25**:405, 1987, molecular genetics of tissue maceration.

E. amylovora, Schouten, *Neth.J.Pl.Path.* **93**:49, 55, 1987, disease prediction.

E. carotovora ssp. atroseptica, Elphinstone & Pérombelon, *AAB* **110**:535, 1987, chemical control of potato contamination with airborne bacteria; Pérombelon & Kelman, *PD* **71**:283, 1987, soft rot spp. & potato diseases; Pérombelon et al., *J.appl.Bact.* **63**:73, 1987, soft rots & potatoes in Israel & Scotland.

Erysiphe graminis, Jones et al., *AAB* **110**:591, 1987, f.sp. *avenae*, integration of resistance & chemical control.

eucalyptus little leaf, Dafalla et al., *J.Phytopath.* **117**:83, 1986, Sudan, on *E. microtheca* [**66**,1618].

fungi, Buck in Day & Jellis 1987, viruses in plant pathogenic fungi; Rossman et al., *A literature guide for the identification of plant pathogenic fungi*, 1987; Sigler & Hawksworth, *The Mycologist* **21**:101, 1987, code of practice for systematic mycologists; von Arx, *Beiheft Nova Hedwigia* 87, 1987, taxonomy, morphology & keys in plant pathogenic fungi.

Furoviruses, type wheat soilborne mosaic v.; this group name is given in the indexes to *Descr.V.* 1–324, 1986, but is not yet approved by the International Committee on Taxonomy of Viruses. Other members: beet necrotic yellow vein v., broad bean necrosis v. peanut clump v., potato mop top v.

Fusarium, Mai & Abawi, *A.R.Phytop.* **25**:317, 1987, nematodes & wilt pathogens.

F. oxysporum f.sp. apii, Elmer & Lacy, *Phytop.* **77**:381, 1987, survival & soil populations, race 2 [**66**,4005].

Ganoderma, Petersen, *The Mycologist* **21**:62, 1987, in N. Europe.

G. lucidum, Adaskaveg & Gilbertson, *Mycologia* **78**:694, 1986, culture, genetics of sexuality & taxonomy with *Ganoderma tsugae* [**66**,1371].

garlic mosaic (onion yellow dwarf v.), Graichen & Leistner, *Archiv.Phytopath.PflSchutz.* **23**:165, 1987.

Geminiviruses, Davies 1987, review [**66**,2264].

gene for gene concept, Vanderplank, *Adv.Pl.Path.*
5:199, 1986.
genetics, Kerr, *A.R.Phytop.* **25**:87, 1987, impact of
molecular genetics on plant pathology.
gladiolus, Aly et al., *Phytoparasitica* **14**:205, 1986,
spread & control of cucumber mosaic v., Israel
[**66**, 1947].
— **chloro-necrotic ring** (tobacco streak v.), Vicchi
& Bellardi, *Difesa delle Piante* **10**:153, 1987.
Glomerella cingulata, Ogle et al., *Aust.J.Bot.*
34:537, 1986; on tropical pasture legumes
[**66**,1961].
Gonatophragmium Deighton 1969; Hyphom; Ellis
1971; conidiomata hyphal, conidiogenesis
holoblastic, conidiophore growth sympodial; con.
multiseptate, pigmented; see *Acrospermum*; *G.
mori* teleom. *A. viticola.*
G. mangiferae Mulder, *TBMS* **60**:160, 1973, con.
0–3 septate, mostly 18–33 × 4.5–6.8 μm; mango
zonate leaf spot, Burma.
grapevine Australian yellows, Magarey & Wachtel,
Agric.Record **12**(17):12, 1985; poss. MLO, similar
to grapevine yellows.
grasses, Sivanesan, *Graminicolous species of
Bipolaris, Curvularia, Drechslera, Exserohilum
and their teleomorphs, Mycol.Pap.* 158, in press;
any taxonomic changes given have not been
noted.
Guignardia aesculi, Hudson, *TBMS* **89**:400, 1987,
single asco. isolates formed the anam.
Gymnosporangium juniperi-virginianae, Chen &
Korban, *Pl.Path.* **36**:168, 1987, genetic variability
& inheritance of resistance in apple; Korban
et al., *J.Phytopath.* **119**:272, 1987, interaction
of apple cvs. with rust populations.

hibiscus chlorotic ringspot v., Hurtt, *Phytop.* **77**:845,
1987, electrophorotypes.
— **witches' broom**, Hiruki,
Ann.phytopath.Soc.Japan **53**:1, 1987, MLO assoc.,
Australia (Queensland).
honeysuckle vein yellowing (eggplant mottle dwarf
v.), Martelli & Cherif, *J.Phytopath.* **119**:32, 1987,
Lonicera.
hop chlorosis, Adams et al., *AAB* **111**:365, 1987;
prob. caused by arabis mosaic v. 'But it is not
clear why AMV, which is common in hop, should
induce chlorotic disease only on rare occasions.'
horsegram yellow mosaic v., Muniyappa et al.,
J.Phytopath. **119**:81, 1987, Geminivirus, paired
particles 30 × 15–18 nm.
Hypoxylon serpens, Onsando 1985, on tea
[**66**,2054].

Kentucky blue grass, Smiley, *PD* **71**:774, 1987,
etiology of diseased patches complex.

legumes, Edwardson & Christie,
Agric.Exp.Stns.Univ.Fla.Monogr., vol. 3, 1986,
viruses.
Leptosphaeria maculans, Hammond & Lewis,
Pl.Path. **36**:135, 1987, establishment of systemic
infection in oilseed rape leaves; Mithen et al.,
TBMS **88**:525, 1987, resistance of brassica leaves
[**66**,4530].
lettuce mottle v. Marinho & Kitajima and Marinho
et al., *Fitopat. Brasileira* **11**:923, 937, 1986,
isometric *c.*30 nm diam., transm. sap,
Hyperomyzus lactucae, non-persistent, Brazil.
Leucostoma persoonii, Schulze & Schmidle 1983,
epidemiology, Germany [**62**,4937].
Libertella blepharis, Carter & Bolay, *Australasian
Pl.Path.* **15**:47, 1986, the correct name for the
anam. of *Eutypa armeniacae* or *E. lata* [**66**,1990].
Lima bean mosaic v., Capoor & Sawant, *Indian
Phytopath.* **39**:185, 1986, Cucumovirus, isometric
29–30 nm diam., transm. sap, seed, 4 aphid spp.,
India.
Luteoviruses, D'Arcy, *Microbiological Sci.* **3**:309,
1986.

Macrophoma theicola Petch, tea branch canker and
twig dieback, Petch, *The diseases of the tea bush*,
1923; Hainsworth, *Tea pests and diseases and their
control*, 1952; Chen et al.,
Ann.phytopath.Soc.Japan **53**:198, 1987, in Taiwan.
Marasmiellus paspali (Petch) Singer, *Sydowia*
9:386, 1955; Payak & Sharma, *Curr.Sci.* **55**:1135,
1986, vertical banded blight of maize, India
[**66**, 1899].
melon necrotic spot v., Coudriet et al., *J.econ.Ent.*
72:560, 1979, described transm. by *Diabotrica*
spp. This reference is not given by Hibi & Furuki,
Descr.V. 302, 1985, and they do not give these
cucumber beetles as vectors of the virus
[**59**,5983].
mimosa striped chlorosis Martin & Kim, *Phytop.*
77:935, 1987; rod *c.*95 × 35 nm assoc., transm.
seed, USA (Arkansas), *Albizia julibrissin.*
mycoherbicide, Mortensen, *Can.J.Pl.Path.* **8**:229,
1986; Charudattan 1987 [**66**, 2168].
Mycosphaerella arachidis, Alderman et al., *TBMS*
89:97, 1987, diurnal periodicity of con., release
favoured at 20–24° & a relative humidity of
> 90%, few con. trapped after heavy rain.
M. citri, Ieki, *Ann.phytopath.Soc.Japan* **52**:484,
1986, in Japan [**66**,1483].
M. cryptica, Cheah & Hartill, *Eur.J.For.Path.*
17:129, 1987, asco. release.
M. linicola, Ferguson et al., *Phytop.* **77**:805, 1987,
effects on flower formation & yield; as
Mycosphaerella linorum (Wollenw.) Garcia-Rada
= *M. linicola*, fide Sivanesan 1984.

M. pomi, Sutton et al., *Phytop.* **77**:431, 1987, life history, causing Brook's fruit spot of apple [**66**,3888].

Nectria coccinea, Lonsdale & Wainhouse, *For.Comm.Bull* UK 69, 1987.
N. galligena, Thomas & Hart, *PD* **70**:1117, 1121, 1986, walnut canker, USA (Michigan); Clifford et al., *AAB* **110**:471, 489, 501, 1987, chemical control in apple [**66**, 3046–7].
N. radicicola, Dahm & Strzelczyk, *Eur.J.For.Path.* **17**:141, 1987, effects of pH, temp. & light on pathogenicity to pine seedlings.
nematodes, Stone in Day & Jellis 1987, genetic systems.
Nicotiana, Australian spp. as virus indicator hosts, van Dijk et al., *Neth.J.Pl.Path.* **93**:73, 1987.

Olpidium, transm. pepper yellow vein, see below.
Oncobasidium theobromae, Prior, *Pl.Path.* **36**:355, 1987, chemical control in Papua New Guinea.
ornamentals, Chase, *Compendium of ornamental foliage plant diseases*, 1987.

papaya ringspot v., Wang et al., *PD* **71**:491, 1987, cross protection by mild mutant strs.
Paracercospora fijiensis, and its var. *difformis* considered synonymous, fide Pons, *TBMS* **89**:120, 1987.
parsnip yellow fleck v., Singh & Frost, *Pl.Path.* **36**:415, 1987, poss. causing celery yellow net.
paspali, see *Sorosporium paspali-thunbergii*.
peanut stunt v., Richter et al., *Archiv.Phytopath.PflSchutz.* **23**:127, 179, 1987, serotypes.
—— **witches' broom**, Hobbs et al., *Pl.Path.* **36**:164, 1987, MLO assoc., partial purification, antiserum produced which did not react with datura witches' broom, eggplant little leaf, vinca witches' broom.
pea seedborne mosaic v., Maury & Khetarpal, *Agronomie* **7**:215, 1987, review.
pelargonium zonate spot v., Vovlas & di Franco, *Inftore.Fitopatol.* **37**(3):55, 1987, a virus isolated from water in Italy (Apulia) distantly related to PZSV; outbreaks of the virus in tomato recently.
pepper yellow vein Fletcher, Wallis & Davenport, *Pl.Path.* **36**:180, 1987, transm. *Olpidium* sp., England (S.) [**66**,4604].
Peronospora grisea (Unger) Unger 1847; Francis & Berrie, *Descr.F.* 766, 1983; con. 23–27 × 16–18 μm, oospores 38–49 μm diam.; Whipps & Linfield, *Pl.Path.* **36**:216, 1987, on *Hebe* in England (Sussex), con. broader than given by Francis & Berrie [**66**,4325].
P. sparsa, Hall & Shaw, *N.Z.J.Exp.Agric.* **15**:57, 1987, on cultivated & wild *Rubus* in New Zealand.

Pezizales, Egger & Paden, *CJB* **64**:2368, 1986, pathogenicity of spp. from burnt forest sites [**66**,2087].
Phacidiaceae, Di Cosmo et al., *Mycotaxon* **21**:1, 1984, taxonomy.
Phialophora parasitica, Rumbos, *J.Phytopath.* **117**:283, 1986, cherry dieback in Greece [**66**,3432].
Phoma destructiva, Ciccarese et al., *Inftore.Fitopatol.* **37**(2):53, 1987, epidemics on eggplant in Italy (S.).
P. medicaginis, Morgan-Jones & Burch, *Mycotaxon* **29**:477, 1987.
Phomopsis macrospora Kobayashi & Chiba 1961; *Index Fungi* **3**:122, 1962; poplar canker; Japan, USA; Filer, *Phytop.* **57**:978, 1967; French & Bergdahl 1983 [**41**:66; **47**,332; **62**,4471].
Phytophthora, Ho & Jong, *Mycotaxon* **29**:207, 1987, taxonomic problems.
P. cinnamomi, Weste & Marks, *A.R.Phytop.* **25**:207, 1987.
P. infestans, Shaw in Day & Jellis 1987.
P. mirabilis Galindo & Hohl, *Sydowia* **38**:95, 1985, published 1986; oogonia 21.3 μm diam., group IV, optimum temp. 21°; on wild *Mirabilis jalapa* in Mexico (Chapingo). The fungus was at first thought to be a form of *Phytophthora infestans* which occurs in the same area. But *P. mirabalis* was not pathogenic to potato and *P. infestans* did not infect *M. jalapa*.
Plectophomella concentrica, Grieg & Redfern, *TBMS* **89**:399, 1987, on *Ulmus procera* in England (S.).
poplar witches' broom Sharma & Cousin, *J.Phytopath.* **117**:349, 1986; MLO assoc., France (N.), *Populus alba* var. *nivea* [**66**,3516].
Pseudocercospora, Sutton et al., *Aust.J.Bot.* **35**:227, 1987, described *P. correa*, a new sp., and discussed the criteria in distinguishing between this genus and *Cercoseptoria*.
Pseudomonas cichorii, Janse, *EPPO Bull.* **17**:321, 1987, biology on chrysanthemum.
P. solanacearum, Mayers & Hutton, *AAB* **111**:135, 1987, biovar 3 causing a wilt of custard apple, *Annona*, in Australia (Queensland).
Puccinia sorghi, Pataky, *Phytop.* **77**:1066, 1987, effects on yield of maize.
P. striiformis f.sp. tritici, Wellings et al, *Pl.Path.* **36**:239, 1987, in Australia (E.), poss. means of entry & plant quarantine implications; first reported in Australia in 1979.
Pyrenochaeta lycopersici, Grove & Campbell, *PD* **71**:806, 1987, hosts & survival.
Pyrenopeziza brassicae, Lacey et al., *TBMS* **89**:135, 1987, natural occurrence of teleom. in England.

Pyrenophora tritici-repentis, Hosford et al., *Phytop.* **77**:1021, 1987, effects of wet period & temp. on infection & development in wheats of differing resistance.

raspberry, Duncan et al., *Pl.Path.* **36**:276, 1987, *Phytophthora* spp. & root rot.
red clover necrotic mosaic v., Rao et al., *Phytop.* **77**:995, 1987, new serotype, antigenic relationships among 6 strs.
Rubus, Jones 1986, viruses, particularly in UK [**66**,3434].

Sclerotinia sclerotiorum, Kochman & Langdon, *Aust.J.exp.Agric.* **26**:489, 1986, sunflower seed treatment to inhibit sclerotial germination [**66**,3466].
Spiroplasma, Markham, *Yale J.Biol.Med.* **56**:745, 1983, in leafhoppers; Whitecomb et al., *ibid*, page 453, serology.
Synchytrium desmodii, Price, *TBMS* **89**:333, 1987, liberation & germination of sporangia.

Thrips tabaci, Sdoodee & Teakle, *Pl.Path.* **36**:377, 1987, transm. of tobacco streak v. on pollen. Transm. of TSV by the insect prob. depends on pollen borne virus which presumably infects via wounds made by *T. tabaci*. This is a new method of virus transm.
tomato vein yellowing v., Lockhart, *PD* **71**:731, 1987, in *Hibiscus rosa-sinensis* with vein yellowing; tomato vein yellowing (eggplant mottled dwarf v.), Castellano & Martelli, *Phytopath. Mediterranea* **26**:46, 1987.
Typhula ishikariensis, Honkura et al., *Trans.mycol.Soc.Japan* **27**:207, 1986, can complete life cycle without snow cover [**66**,2365].

Vanilla, viruses, Wisler & Zettler, *PD* **71**:1125, 1987.
viruses, Jones, *AAB* **111**:745, 1987, review, control by vector resistance.

waldsterben, Matzner & Ulrich and Bach, *Experientia* **41**:578, 1095, 1985, reviews [**64**,3552, 5504].
watermelon mosaic v., Chala et al., *PD* **71**:750, 1987, 2 strs. in USA (Texas).

Xanthomonas campestris pv. graminis, Leyns, *J.Phytopath.* **120**:130, 1987, on *Lolium perenne*.
X. c. pv. pelargonii, Kennedy et al., *PD* **71**:821, 1987, on geranium.